QM Library

23 1421596 1

WITHDRAWN
FROM STOCK
QMUL LIBRARY

D1628554

Tropical Asian Streams

Zoobenthos, Ecology and Conservation

WITHDRAWN
FROM STOCK
QMUL LIBRARY

Tropical Asian Streams

Zoobenthos, Ecology and Conservation

David Dudgeon

香港大學出版社

HONG KONG UNIVERSITY PRESS

Hong Kong University Press
14/F Hing Wai Centre
7 Tin Wan Praya Road
Aberdeen, Hong Kong

© Hong Kong University Press 1999

ISBN 962 209 469 4

All rights reserved. No portion of this publication
may be reproduced or transmitted in any form or
by any means, electronic or mechanical, including
photocopy, recording, or any information storage
or retrieval system, without permission in writing
from the publisher.

Printed in Hong Kong by Nordica Printing Co. Ltd.

QM LIBRARY
(MILE END)

Contents

Preface

A personal note. I started this book in October 1995, labouring under the naive conviction that it would take me about a year to complete. The absurdity of this belief soon became evident, and I did not finish writing until Chinese New Year, 1998. By that time, I had long since come to think of the book as my personal folly. Why did I persist? What was the point? This book was written to fill an obvious gap in our ecological knowledge of the rivers and streams of tropical Asia. It must therefore summarise existing information. Despite my best efforts — and the considerable size of this book — it is neither exhaustive nor complete. Some significant papers or monographs are sure to have been omitted, and it is possible that information may been misconstrued or misinterpreted. That said, I think that the main patterns and processes shaping the ecology of tropical Asian streams are adequately covered in these pages. It is time to apply our (albeit limited) knowledge to the conservation and wise management of these endangered ecosystems before they are all irreparably degraded. To stress a point made repeatedly in various guises herein: if we wait until we have complete information about tropical Asian streams and rivers before we try to conserve them, then there will be no trace of the original biota remaining.

I wrote with the intention of providing a source of reference for further study and in this — at least — I believe some measure of success has been achieved. The bibliography provides a comprehensive list of the ecological and (germane) taxonomic literature published up to and including 1996. The interested student will thus be able to begin work with the confidence that much of the relevant material has been referenced, and can be tracked down via the bibliography. Given the scattered and fragmentary nature of the literature pertaining to Asian streams, this is not a trivial matter. The book also provides a series of keys or guidelines for the identification of invertebrates in running waters of the region. This is the first time that such material — much of it unpublished — has been brought together in a single volume. Again, my hope is that the contents of this book will serve as a tool to facilitate further investigation. Whether I have succeeded in this regard, is up to you — the reader — to decide. With luck, subsequent research will soon render this text obsolete or in need of substantial revision.

The scope of this book seems to demand a multi-authored work. I have not followed that path because I wanted to produce a unified overview and synthesis. The resulting text offers the reader both the benefits and shortcomings of 'one man's view'.

Assistance from a number of sources made this book possible. Some of the work reported herein was funded by The Hong Kong University Committee on Research and Conference Grants, and by the Hong Kong Research Grants Council. I am grateful to my colleagues as well as past students for their support, suggestions and questions. In particular, I received a great deal of laboratory assistance from Eddie Chan and Eunice Chung, and it is a pleasure to acknowledge the work of Lily Ng who undertook the painstaking task of digitizing and cleaning up my original drawings for this book. Ignac Sivec gave permission to reproduce his figures of perlid stonefly larvae originally published in *Scopolia*. Help of various kinds was given also by Leo Chan, Tony Chan, Richard Corlett, David Gallacher, Manfred Jäch, Roger Kendrick, Alan Leung, Hans Malicky, Suhkmani Mantel, Brian Morton, Dan Polhemus, John Polhemus, Maria Salas, Yvonne Sadovy, Tong Xiaoli, Manna Wan, Gray Williams, W.D. (Bill) Williams and Peter Zwick. Jon Davies provided a useful review of a draft version of the text.

Finally, this book is dedicated to the one person who has made all the difference — in acknowledgement of her love, support and fortitude.

1

Introduction

We know surprisingly little about the ecology of tropical freshwaters in general, and tropical Asian rivers and streams in particular. This is despite the dependence of humans, livestock and agriculture upon streams and rivers in a region which is poor in natural lakes. The reliance is increased by the monsoonal climate of much of tropical Asia: rainfall and hence stream-flow patterns are strongly seasonal with marked wet and dry seasons. As a result, the region experiences periods of drought and water scarcity interspersed by times of plenitude when damaging floods may occur. This pattern has spurred attempts to regulate or control the flow of streams and rivers, a practice which has been prevalent for many centuries and has reached an apotheosis in recent schemes to dam the Chang Jiang (Yangtze) and Mekong Rivers. It is not surprising, therefore, that few — if any — rivers in tropical Asia remain in their original state. Ever-more ambitious river-engineering projects, driven by demands from burgeoning populations, are being undertaken to contribute to economic development which (it is hoped) will be fuelled by cheap hydroelectric power. These schemes will increase the magnitude and extent of human impacts on streams and rivers, many of which are already degraded by pollution and modification of their drainage basins by forest clearance, overgrazing, urbanization and so on. Habitats have become altered in ways which we do not understand or appreciate fully simply because little was known about them before any changes took place. Loss of biodiversity is a matter for particular concern yet, apart from a few charismatic species (river dolphins, and certain fish such as the mahseer), threats of extinction and the concomitant requirement to frame effective conservation measures have received scant attention. There is thus a need to improve understanding of rivers and streams — whether man-modified or not — in order to provide the information required to predict the impacts of planned developments. We must also formulate environmental management practices which will ameliorate or reverse habitat degradation and loss of biodiversity.

A necessary first step is to organize the scattered and often fragmentary information that is already available, so as to make apparent what is known about tropical Asian rivers and what we have yet to find out. That is one of the four main aims of this book. Its importance stems from the fact that the results of much tropical freshwater research are not disseminated adequately because they are communicated in the form of reports or articles in journals with restricted local or regional circulation (Williams, 1988; 1994). Consequently, a great deal of worthwhile work is neglected. Boon (1995) has illustrated this point by analysing the references ($n = 1,356$) cited in a recent publication on the *Inland Waters of Tropical Asia and Australia: Conservation and Management* (Dudgeon & Lam, 1994). Only 30% were from widely-circulated, international journals; most (56%) were from regional journals (22%), reports and theses (25%), or conference proceedings (9%), with the remainder cited from books (14%).

There are at least two ways of approaching the study of nature. One tactic is to investigate a habitat or any circumscribed local environment with the aim of producing an inventory of the organisms found there, preferably with some indication of their relative abundance or population densities. Such data underpin efforts directed towards the conservation of biodiversity. The second approach concentrates upon what the organisms are doing in the habitat, and might involve measuring attributes of groups of animals which have similar interactions or use the same resources — such as predators or the animals which eat algae — but which are not necessarily closely related. In the first case we are asking descriptive questions: 'Which organisms live in the habitat? How many are there?' We label them using a scientific name, but without additional information this does not provide us with any understanding of what is going on in the habitat. In the second case we ask: 'What are the organisms doing?' — a functional question. The two approaches are complementary, with the descriptive approach often preceding the functional one, but they are nonetheless distinct. The ecological literature exhibits a clear dichotomy between these approaches, and books tend to focus on either one or the other. If the main goal is to introduce the reader to descriptive studies and to assist in their execution, an identification manual (with keys and line drawings) would be the usual publication. If the functional approach is adopted, an introduction to ecological patterns and the processes bringing them about would be the subject of the text. In this book, a combination of both approaches is taken. Firstly, the features of tropical Asian running waters are addressed by adopting a functional approach to review their physical and chemical features, transport processes, the importance and fate of allochthonous and autochthonous organic matter, land-water interactions (particularly on floodplains), and the interactions between major groups of primary producers, consumers and decomposers. It is probably impossible for any individual to do justice to all aspects of stream ecology in a review, and my familiarity with smaller rivers and streams and benthic ecology may have led to an overemphasis on certain topics. Likewise, considerable space has been given to examples from Hong Kong where I have spent most of my research career. That said, I have attempted to review adequately all of the available literature and — given the lacunae in our knowledge of certain important processes (e.g. nutrient spiralling) and elements of the biota in Asian streams (e.g. the biofilm, the hyporheic fauna) — a fully-comprehensive treatment of stream ecology in the region is still some way off.

The functional approach used for the review which constitutes the first part of this book then gives way to a systematic review of the benthic macroinvertebrates, or zoobenthos, of tropical Asian streams and rivers. These are typically the most diverse fauna of running-water habitats, and are used widely as biomonitors of environmental change in Australia, North America and Europe. The intention here is to provide information about — and keys to — the major taxa so as to facilitate their identification and quantification in streams and rivers throughout tropical Asia. The examples illustrated in this section are mostly Hong Kong species, but the keys and supporting biological data were written to be applicable beyond the territory. It is hoped that they will provide a basis for ecological studies or will underpin the use of the zoobenthos as biomonitors of water quality and habitat conditions. Thus the second goal of this book is to serve as a tool to further the descriptive study of zoobenthos in tropical Asian streams and rivers.

The third objective of the book is to review the anthropogenic threats to streams and rivers in the region. Such threats are all-pervasive, and thus such a review cannot be exhaustive. Particular attention has been paid to case studies, including the Mekong River and streams in southern China. This review is followed by an account of sampling techniques and a description of experimental designs and sampling strategies needed for the detection of human impacts and environmental change in streams. This section is included because of the urgent need for well-designed, unambiguous studies of human impacts on streams in Asia. Resources for environmental monitoring and protection efforts are always limited; it therefore behoves us to undertake well-designed studies that test hypotheses about impacts (or, indeed, any ecological process) in a clear, parsimonious, objective and unambiguous fashion. Statistical analysis of the magnitude of putative impacts are also facilitated by good experimental design. The fourth goal of this book is therefore to alert researchers to some sampling strategies and experimental designs which will improve their ability to separate the effects of changes caused by human impacts from those resulting from natural spatial or temporal variations in streams. Application of these designs should improve our ability to detect impacts, and reduce the chances of declaring that a change is due to anthropogenic influences when in fact it has not. Suggestions are also made as to the types of process-orientated ecological studies which are needed to inform the efforts of those charged with managing, restoring or mitigating impacts upon rivers and streams.

The four objectives of this book are complementary and, it is hoped, will provide useful information and some of the methodological tools needed by researchers in the region. Much good science in Asia has been undertaken by individuals and small teams working on limited budgets under less-than-ideal conditions. While funding constraints are unlikely to disappear at any time in the near future, a shortage of books and journals, restricted library budgets and time limitations prevent many scientists from keeping abreast of the literature dealing with the systematics of the organisms they study, the ecology of the habitats where they are found, and the methodological approaches best suited to their investigation. This book was written to provide such information.

In summary, the four goals of this book are:

1. To review the ecology of Asian rivers and streams, adopting a functional approach to describe the interactions between physico-chemical factors, transport processes, the origins and fate of organic matter, and major elements of the stream biota. An additional objective is to provide a compendium or guide to the relevant but widely-scattered ecological literature.

2. To provide a systematic account of the composition of the zoobenthos to facilitate their identification and further study;

3. To give an account of the major human impacts upon tropical Asian streams and rivers.

4. To suggest appropriate research strategies for assessing environmental impacts and undertaking process-orientated investigations of stream ecology.

Throughout this book, the terms 'river' and 'stream' are used interchangeably, with 'stream' being the preferred generic label for running waters. Where one term is used in preference to the other in describing a particular water course, it should be understood that a stream is simply a small river.

2

Scope

In 1984, the Brazilian limnologist J.G. Tundisi wrote of the general perception that our knowledge of tropical freshwaters was far less than was required to understand the mechanisms and processes operating in these ecosystems (Tundisi, 1984). It is certainly less than is desirable and needed for their proper conservation and management (Williams, 1988; 1994). A perennial obstacle to tropical research, Tundisi asserted, was difficulty in obtaining even the basic literature. He exhorted colleagues to disseminate information in the form of reviews and critical syntheses on a regional basis. Such regional syntheses could deal with systematics, biological processes, and ecosystem functioning drawing upon the limited examples available. In that spirit, this book was written.

The freshwater ecology of tropical Asia is not well known (Fernando, 1984a; Dudgeon & Lam, 1994; Boon, 1995; Dudgeon, 1995a), despite Dussart's (1974) review of the pre-1970 literature concerning to inland waters of the region. Lotic habitats (i.e. rivers and streams) have received less attention than standing waters (Crisman & Streever, 1995), notwithstanding the fact that an Asian biologist (Hora, 1923, 1927, 1936) was among the earliest investigators of the stream fauna. For example, Lim (1980) has drawn attention to the paucity of information on Malaysian freshwaters, while Radhakrishna (1984) described the study of the inland waters of the Indian subcontinent as '. . . a virgin field . . .', stating that '. . . the basic knowledge concerning the systematics and ecology of various groups of aquatic animals is still unsatisfactory'. An exception to this generalization is Sri Lanka where the freshwater fauna is better known systematically than that of other tropical countries (Mendis & Fernando, 1962; Fernando, 1984b; 1990). Here too, however, new fish species are still being discovered (Pethiyagoda, 1994) and studies of the ecology of streams and rivers of the island are not well advanced.

Information on the ecology of tropical Asian rivers and streams is fragmentary, and there have been few attempts to gather data about entire communities within particular habitats or to investigate temporal changes over seasonal and annual cycles (but see Bishop, 1973; Furtado & Mori, 1982; Dudgeon, 1992a). Even basic information — such as the longitudinal zonation of fauna — has yet to be gathered for the major rivers of the region. Tropical Asian rivers are made up of a variety of habitats and are host to a diverse array of organisms but, at present, we have no complete studies of any tropical Asian river system of any size. What variety of organisms occur in which habitat type is known in a few cases only, and we have yet to measure how much each habitat contributes to the productivity of the river as a whole. The picture is complicated by the fact that certain fishes and shrimps use various habitats at different times, and some of these habitats are inundated for part of the year only. Further complexity is introduced by year-to-year variations in rainfall which alter the extent and duration of floodplain inundation, and the magnitude of land-water interactions. Furthermore, much of the available data have been generated from studies which were initiated after habitats had become polluted, and were concerned more with documenting the extent of local species loss or environmental degradation than with the composition or functional relationships of the lotic biota. This reflects, in part, a paucity of support for basic ecological research, but is probably due also to the lack of an overall conceptual structure that might serve as a theoretical underpinning for studies of river and stream ecosystems. The importance of this point can be understood by considering the development of lotic ecology.

The first, widely-available, English-language text dealing with lotic ecology was Noel Hynes' encyclopedic *The Ecology of Running Waters* published in 1970 (Hynes, 1970). A wide range of information from an extensive array of sources was synthesized in that volume, providing a well-integrated overview of the field. Hynes' work contributed a firm foundation of knowledge and thereby facilitated the development of innovative research strategies that would further our understanding of lotic habitats. Later investigations gave rise to an appreciation that rivers and streams are the focus of catchment processes, subject to the behaviour of elements within the drainage basin (Hynes, 1975; Vannote *et al.*, 1980; Davies & Walker, 1986). Fish and other animals can range widely throughout the aquatic system exploiting different resources, or may use the same resource at different times or places. It therefore became evident that the appropriate focus for understanding and managing rivers and streams was the drainage basin as a whole rather than the aquatic environment (or segments of it) in isolation. This focus was made explicit in the River Continuum Concept (Vannote *et al.*, 1980; Minshall *et al.*, 1983, 1985) which has stimulated research by stream biologists in North America, Europe, Australia and New Zealand during the last 15 years, and provided a unifying structure against which ideas could be generated and hypotheses tested (e.g. Cushing *et al.*, 1995). Conceptual advances such as this have initiated the transformation of freshwater science from a primarily descriptive discipline to one which has a predictive component that can be brought to bear in formulating strategies for basic and applied research including river conservation and management. Of particular relevance in this book will be the use of the River Continuum Concept (RCC) to provide an underlying structure for a review of our knowledge of streams and rivers in tropical Asia. My main objectives will be to summarize what is known about these lotic habitats, and to show where data are lacking. An additional aim will be to determine whether and how tropical Asian streams and rivers differ from their counterparts elsewhere.

The geographical scope of this book is tropical Asia, a region which lacks discrete physical boundaries but can be defined approximately as those areas experiencing a monsoonal climate and possessing a fauna belonging chiefly to the Oriental Region. The area encompassed extends from west of the Indus River eastward along the Himalaya Mountains to the China Sea at the mouth of the Chang Jiang (Yangtze River), and eastward in the Indies into Wallacea to include Sulawesi (the Celebes) and New Guinea (Fig. 2.1). The incorporation of China south of the Chang Jiang in the Oriental Region follows Banarescu (1972), and concords with demarcation line between the Palaearctic and Oriental mammal fauna of eastern China (Huang, 1985). Coincidentally, this division is consistent with the observation

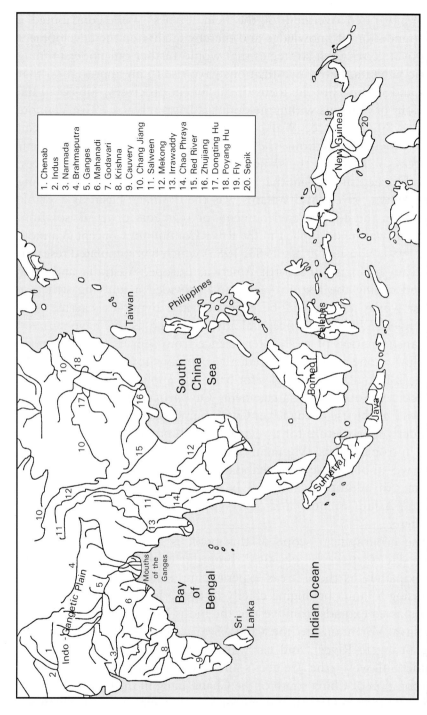

Fig. 2.1 The major rivers of tropical Asia. Poyang Hu and Dongting Hu — large lakes on the Chang Jiang (Yangtze River) floodplain — are shown also.

that two rice crops each year are possible south of the Chang Jiang, but only one north of it (Yang & Morse, 1987). Studies on the biogeography of Chinese fishes (Li, 1981) and Asian caddisflies (Schmid, 1966; Dudgeon, 1987a; Yang & Morse, 1988) confirm that the northern boundary (around 30°N) represents a transition zone for the freshwater fauna. The inclusion of New Guinea is justified on the basis that the great majority of the island's fauna (especially insects) is of Oriental derivation (Gressitt, 1982). For example, the freshwater palaemonid and atyid shrimps of New Guinea include mainly species with a wide geographical distribution, the island forming the eastern — not western — limits of their range (Holthuis, 1982).

3

Ecological Overview

GEOMORPHIC SETTING

The countries of tropical Asia have extensive freshwater resources (summarized by Ali *et al.*, 1987) which are being used increasingly for development purposes. Jalal (1987) records that Bangladesh has over 50 important rivers; India, 400; Indonesia, 200; and Thailand, 10. Six of the longest rivers in the world are found in the region: the Chang Jiang, Mekong, Indus, Brahmaputra, Ganges and Irrawaddy (Table 3.1). There is one estimate that over 80% of sediment transported in the rivers all over the world comes from Asian rivers (including the Palaearctic Huang Ho; Jalal, 1987), although Degens *et al.* (1991) suggest that the proportion may be lower. The following account of the geomorphology and terrain of the lands drained by these rivers is based broadly upon Spencer (1954, 1973) and Stamp (1962).

Southeast Asia

The vast, lofty plateau of Tibet extends over some 1.3×10^6 km^2, dominating southern and central Asia. Three-quarters of the plateau

Table 3.1 Statistics of major tropical Asian rivers; numbers in parentheses indicate rank in the world (data derived from Milliman & Meade [1973] and Van der Leeden [1975]). Note that estimates of river length vary slightly among authorities; Liu (1980), for example, puts the length of the Chang Jiang at 6,300 km.

	Drainage area (km²)	*Mean discharge (m³.s⁻¹)*	*Length (km)*	*Average annual suspended load*	
				t x 10⁶	*t.km⁻²*
Chang Jiang	1,808,500	34,000 (3)	5,980 (4)	499 (4)	257
Brahmaputra	934,990	19,830 (4)	2,900 (31)	726 (3)	1,090
Ganges	1,051,540	18,697 (5)	2,506 (48)	1,451 (2)	1,518
Irrawaddy	429,940	14,079 (12)	2,100 (63)	430 (9)	695
Zhujiang	425,700	12,500 (14)	2,100 (63)	69 (-)	-
Mekong	802,900	11,048 (15)	4,360 (16)	170 (13)	214
Indus	927,220	5,521 (23)	2,900 (31)	435 (5)	449
Red	120,000	3,900 (26)	1,200 (>100)	130 (15)	1,092
Godavari	297,850	3,598 (30)	<1,000 (>100)	96 (-)	-
Krishna	308,210	1,944 (40)	<1,000 (>100)	-	-
Salween	279,720	1,493 (45)	2,400 (52)	-	-

lie above 4,500 m where snows and glaciers give rise to many of the largest rivers of Asia: the Indus, the Brahmaputra, the Salween, the Mekong, and the Chang Jiang, and to major Southeast Asian rivers such as the Irrawaddy and the Chao Phraya (Fig. 2.1). In one part of the Tibetan plateau (at 28° N) the Chang Jiang, Salween and Mekong Rivers flow in deep valleys compressed into a zone less than 80 km wide. The 6,300 km course of the Chang Jiang wanders north and east across China to reach the sea close to Shanghai. The Salween River, by contrast, remains confined for much of its 2,820 km course and maintains a direct southerly flow to the Gulf of Martaban. The Mekong also flows generally south, but follows an easterly trend in the Lao People's Democratic Republic (Lao P.D.R.) due to the Korat Platform. Further south, the Mekong has contributed to the formation of Le Grand Lac — the great lake of Kampuchea (Cambodia) — which is connected to the Mekong by the Tonlé Sap River. The lake is formed in a saucer-shaped basin; drainage from this area was reduced by the Mekong which originally built a delta obstructing outflow from the basin leading to the formation of the lake. Below this obstruction, the banks of the Mekong have been built up by flood deposits, and sand bars and islands are scattered along its course. Further downstream, in Vietnam, the Mekong has formed a new delta, almost 300 km in length, which is a major centre of population and agriculture.

Tonlé Sap and Le Grand Lac have considerable importance in that they are surrounded by rice- and maize-producing areas and yield large fish harvests (Hickling, 1961: pp. 135-140; Ahmed & Tana, 1996). The lake, which receives the waters of the Mekong in the flood season and returns them in the dry season, serves as a huge sump and fluctuates greatly in area and depth, from around 10,000 km^2 in the wet season to 3,000 km^2 in the dry. This process of water backing up in swamps and lakes is a common feature of Asian river floodplains, occurring (for example) in China (Chang Jiang and Poyang Lake), Malaysia (Sungai Pahang and Tasek Bera; Furtado & Mori, 1982), Sulawesi (Lawe Konaweha River and Aopa Swamp; Whitten *et al.*, 1987) and Papua New Guinea (Fly River and Lake Murray; Jaensch, 1994).

The Mekong ranks third in Asia after the Chang Jiang and the Ganges in its minimum flow and drains an area exceeding 795,000 km^2 (Table 3.1). The lower Mekong River basin represents about 77% of the total basin, covering an area of over 600,000 km^2. It drains almost all of the Lao P.D.R. and Kampuchea, one-third of Thailand, and the main agricultural lands of southern Vietnam (approximately 20% of the Socialist Republic of Vietnam). The discharge of the Mekong exceeds 475 x 10^9 m^3.yr^{-1} bestowing great potential for irrigation development and hydroelectric power generation (Chomchai, 1987). The annual fish harvest from the lower Mekong basin has been estimated at about 500,000 t (Pantulu, 1970; 1986a).

Elsewhere in Southeast Asia, the 15,000 km^2 Tonkin Delta lies to the east of the Laotian highlands. It has been formed by three swift rivers — the Clear, the Red and the Black — which rise in Yunnan and join a few kilometres upstream of Hanoi. The Red River (Fig. 2.1) is the major waterway of northern Vietnam and, despite a relatively short course of approximately 1,200 km, carries large quantities of sediment during periods of high water (Table 3.1). The Irrawaddy in Burma also floods for part of the year, submerging a 31,000 km^2 alluvial plain. In this case, local precipitation in the delta region causes inundation which occurs a month before the arrival of floods due to increased discharge from the upper part of the drainage basin (Welcomme, 1979).

The Indian Subcontinent

Hora (1952) distinguished between the wet northeast with its marshes, paddy fields and floods, the drier northwest with greater seasonal variation in temperature and where water conservation tanks are important, the southern tableland with high run-off and rivers drying up for half of the year, and hillstreams with lower temperatures and little aquatic vegetation. Such generalizations reflect the underlying form of the Indian subcontinent which comprises a triangular piece of the ancient Gondwanian land mass. The base of the triangle lies across the north, and between its hard rock and the Tibetan highlands there extends a great alluvium-filled trough. The trough, or North Indian plain, contains the valleys of the Indus on the west, the Ganges in the middle, and the Brahmaputra to the east (Fig. 2.1). They have formed large deltas at either northern corner of the Indian peninsula. These snow- and rain-fed Himalayan rivers rank among the great rivers of the world. They have complicated flow regimes, marked seasonal variability in flow volume and heavy silt loads.

The drainage basin of the Ganges extends over China, Nepal, India and Bangladesh, and that of the Brahmaputra over China, Bhutan, India and Bangladesh. The two rivers have their confluence within Bangladesh and, after joining the Meghna River further downstream, flow into the Bay of Bengal through an expansive estuary. The floodplain is characterized by oxbow lakes and backwater swamps, and '. . . is riddled with old, dead watercourses that the Ganges has forgotten about . . .' (Newby, 1989), a consequence of undercutting, bank erosion and deposition processes that cause the river courses to shift about over the floodplain. Almost all of the rivers and streams of Bangladesh — essentially, a nation created on a gigantic floodplain (Khalil, 1990) — have their headwaters outside the country, and most of the major rivers derive a negligible proportion of their flow from local run-off. As a result, the discharge of many rivers in Bangladesh is affected by impoundment, diversion and regulation of the sections of their courses which flow within India, and negotiations on water sharing are an important aspect of their international relations (see pp. 538–539).

The Indus Basin, of which the Indus River and its tributaries are the principal water courses, drains an area of some 900,000 km^2. Rising in the Tibetan Plateau behind the Himalaya, the Indus flows for almost 3,000 km to empty into the Arabian Sea. Principal tributaries are the Kabul, the Jhelum and the Chenab (the Western Rivers), and the Sutlej, the Beas and the Ravi (the Eastern Rivers). The Indus traverses both India and Pakistan, and agreement has had to be arrived at between these countries to ensure equitable sharing of water resources (Caponera, 1987).

The Indian peninsula is tilted towards the east and most rivers (including the Godavari, Krishna, Mahanadi, and Cauvery Rivers) drain in that direction. The western coast is relatively mountainous and westward-flowing streams have short, steep courses. The west-coast rivers comprise only 3% of total river-basin area in India but 14% of the country's water resources flow through them, reflecting high run-off and torrential flows during the monsoon (Jhingran, 1980). They (like other peninsular rivers) are rain-fed, and flood during the monsoon, when they transport large amounts of sediment, but shrink to straggling streams during the rest of the year (Ramesh & Subramanian, 1993). To conserve monsoonal flood waters, most peninsula rivers have been dammed and the waters thus stored are used for irrigation, water supply, aquaculture and hydroelectric power generation.

China

China has rich inland water resources which include over 50,000 rivers draining catchments of at least 100 km^2. Of these, 1,600 have drainage basins exceeding 1,000 km^2. Approximately 64% of the country's total area is drained by rivers flowing into the sea; the remaining 36% is drained by rivers which flow into lakes from which the water evaporates (Zhao *et al.*, 1990).

The Chang Jiang (Fig. 2.1) is the largest river in China and the third largest in the world by volume of discharge (34,000 m^3.s^{-1}). It is joined by more than 700 sizeable tributaries before it enters the sea (Liu, 1980). The Chang Jiang has an eccentric course of 6,300 km, flowing north and east across China reaching the sea close to Shanghai. As recently as 1985, no one had traversed its entire length. The catchment of 1.8 x 10^6 km^2 is one fifth of China's area, and has a population larger than that of the United States. The Chang Jiang falls over 6,600 m in height from source to estuary, and has great potential for hydroelectric power generation; currently it provides 40% of China's total output.

The upper reaches of the Chang Jiang extend from the headwaters in Tangguala Mountains to Yichang in Hubei Province, and include the 200-km Three Gorges limestone region; the middle reaches extend from Yichang to Hukou in Jiangxi Province; the lower reaches stretch from Hujou to the Chang Jiang estuary, where the Grand Canal to northern China joins the river. The middle and lower course comprise alluvial plains where the river meanders in wide sinuous curves, some of which have caused a section of the Chang Jiang in Hubei and Hunan to be named the 'nine intestine-like bends'. Much of this plain floods during the summer monsoon, and — through processes of erosion and deposition — the river course and lateral lakes are subject to considerable change in shape and size. For example, Dongting Hu (Dongting Lake: 1,728 km²) and Poyang Hu (Poyang Lake: 3,960 km²) — the largest freshwater water bodies in China (Liu, 1984; Wang, 1987; Fang, 1993) — were (respectively) formed no more than 2,000 and 2,500 years ago (Fang, 1993). They act as a natural (although only partially effective) mechanism for flood control in the middle Chang Jiang, and fill as a result of spring run-off (fed by snowmelt) and summer monsoonal rains, inundating marshes and swamps. In autumn and winter, when the river is low, water flows out of the lakes back into the Chang Jiang, the lakes recede, and marshland appears once again. These cycles of rising and falling water levels have important implications for the productivity and ecology of the river floodplain (Van Slyke, 1988; Melville *et al.*, 1992; Melville, 1994). In addition, such conditions are ideal for schistosome-host snails (*Oncomelania hupensis*: Pomatiopsidae; see pp. 138–139) and much of the land around Poyang Hu became habitable only after the elimination of O. *hupensis* (and the associated *Schistosoma japonica* blood flukes) in 1958 (Liu, 1980).

South of the Chang Jiang and its alluvial floodplain, the Chinese landscape is dominated by jumbled hill country, with small patches of lowland set irregularly among the upland masses. Here, the 2,214-km Zhujiang system — which by volume of discharge is China's second largest river — drains a large portion of South China (Guangxi, Guangdong and part of Yunnan Provinces; Fig. 3.1), while numerous smaller rivers traverse the eastern Chinese coastal lands. The Zhujiang comprises three main tributaries: the Xijiang, Beijiang and Dongjiang; the Xijiang (West River) is by far the largest and most important of these. Together they drain an area of 450,000 km² with an total annual run-off averaging 336×10^9 m³. The Xijiang carries most (68%) of this, with the Beijiang and Dongjiang contributing 15% and 8% (respectively) of the total (Xiong *et al.*, 1989). The other major southern Chinese rivers include the Hanjiang in eastern Guangdong, and the Jiulongjiang and Minjiang in Fujian (Fig. 3.1) but, unfortunately, there is relatively little information available on these three rivers (*cf.* Zhang *et al.*, 1987; Xiong *et al.*, 1989).

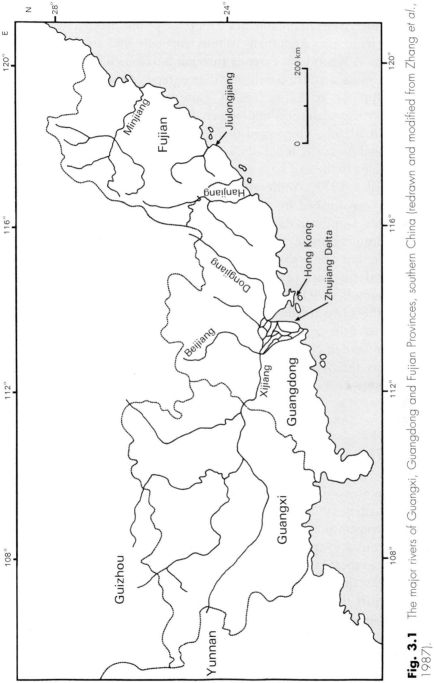

Fig. 3.1 The major rivers of Guangxi, Guangdong and Fujian Provinces, southern China (redrawn and modified from Zhang *et al.*, 1987).

The East Indies

Beyond the Asian mainland lies a pair of island arcs. A western arc extends from the Andaman Islands through Sumatra into eastern Indonesia. A second arc extends through Sulawesi into the Philippines where it joins a north-south arc that extends from Borneo, through the Philippines, to a buttress in Taiwan. These arcs represent the master structural lines along which folding, faulting and volcanism have built islands, submerged them, and rebuilt them, enclosing units of sedimentary rocks lifted above sea level. Consequently, islands like New Guinea to the east have complex histories reflected in their geology (e.g. Loffler 1977; Audley-Charles, 1981), while one view is that Sulawesi was formed by the collision of two land masses which had rifted from Gondwanaland at different times (Audley-Charles, 1987). Rivers draining these islands are smaller than those on the Asian continent, but some do have marshy coastal floodplains which are submerged for part of the year (Welcomme, 1979; Whitten *et al.*, 1984; Jaensch, 1994). Examples among the larger rivers of Sumatra include the eastward-flowing Musi, Hari, Kampar, Solo, and Rokan Rivers; in Borneo the southward-flowing Kahajan, Barito, and Mahakan; and in New Guinea the Fly, Diguil, Purari, and Pulan Rivers, as well as the Sepik River with its extensive lateral floodplain and backswamps.

CLIMATE

The climates of Asian lands between the latitudes of 0° and approximately 30° are dominated by high inputs of solar energy. Mean annual temperatures at low altitudes generally exceed 20°C. As a result, there is rapid evaporation from the warm surfaces of both sea and land. When these large quantities of water vapour condense into rainfall, they release latent heat energy which causes turbulence. The storms which typify this region have high intensities of rainfall, and Southeast Asia has among the heaviest rainfalls of any region on earth, averaging more than 2,000 mm per annum. This climate characterizes the Intertropical Convergence Zone, an irregular and discontinuous belt of low-pressure moist air moving seasonally across the equator, following the sun. The Zone does not form a continuous trough or move in a regular or predictable pattern; it dissipates or reforms to produce a chain of major disturbances which cause highly variable rainfall regimes.

The Oriental tropics is not a single climatological entity, but consists of a wide variety of weather patterns. However, the single major factor which integrates all of the circulations in the area is the summer (southwest) monsoon (Fig. 3.2), although this may be modified by local factors (Koteswaram, 1974). The region is affected also by a winter (northeast) monsoon in its northerly sector comprising dry, cool air moving southwestward from the interior of continental Asia. The air drift extends far enough to influence the island arcs (Fig. 3.2), but its cooling effects disappear south of 15°N. During the summer monsoon (extending, on average, from June to September), warm and moist air moves northeastward out of the Indian and Pacific Oceans. Adiabatic cooling occurs over the landmasses and islands where the hot, humid air releases moisture bringing rain. Rain falls either where these warm air masses meet the cool, dry northeast monsoon travelling southwestward, or where they rise upon encountering high mountain ranges.

The weather pattern of southern China (described by Domrös & Peng [1988], Hulme *et al.* [1992] and Zhang & Lin [1992]) exemplifies consequences of the monsoonal climate on river ecology. The mean annual precipitation of China is approximately 680 mm — most of which falls in the summer wet season. However, the amounts of precipitation vary greatly over different parts of the country, from an annual total of less than 50 mm in the northwest to over 2,500 mm in the south, and the country's rainfall displays a distinct southeast to northwest gradient (Fig. 3.3). Weather patterns in southern China reflect the seasonal alteration of monsoons. In winter, the continental high-pressure region over Siberia and Mongolia results in north or northeasterly winds which bring cool, dry air to the south. In summer, low pressure to the north brings in warmer, moister air from over the tropical oceans. On average, the summer monsoon dominates from early May to the end of September and is replaced by the winter monsoon from November to February. Between the summer and winter monsoons are shorter periods of transitional weather. Summer is the wet season and, in coastal regions (Hong Kong, for example; see Dudgeon, 1992a; Dudgeon & Corlett, 1994), 77% of the total annual rainfall falls between May and September, with 18% in August alone. Summer is also the main typhoon season, and intense typhoons can strongly affect local precipitation patterns. The use of the terms 'summer' and 'winter' for southern China's seasons is perhaps

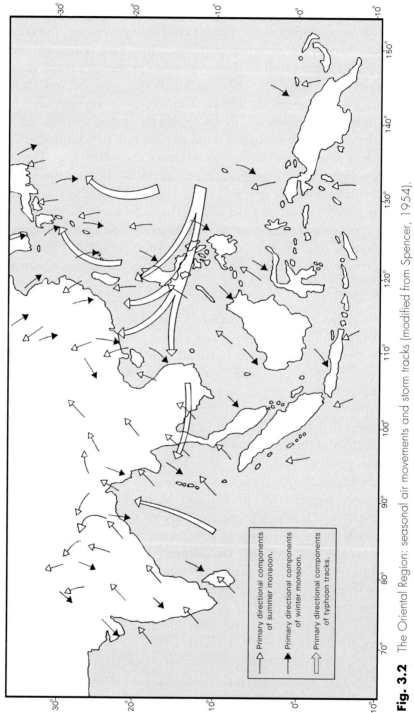

Fig. 3.2 The Oriental Region: seasonal air movements and storm tracks (modified from Spencer, 1954).

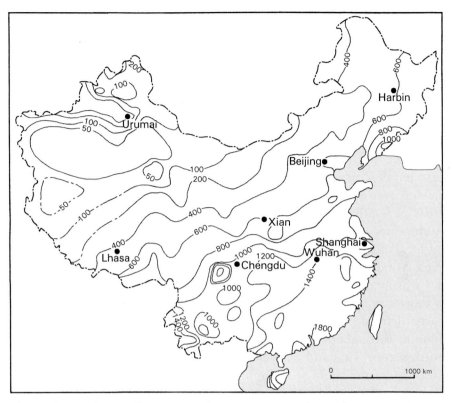

Fig. 3.3 Distribution of mean annual precipitation in China (after Domrös & Peng, 1988).

unfortunate but well established since the region does experience an obvious temperature seasonality. 'Dry season' and 'wet season' would be more appropriate from the stream biologist's perspective because of the seasonality imposed upon discharge which is a direct consequence of the interactions between the northeast and southwest monsoons.

In China and elsewhere in Asia the significance of temperature as a biological limiting factor declines with decreasing latitude and the role of discharge regime increases. The precise climatic sequence and the relative importance of monsoons will vary among localities in response to latitude and the effects of topography, and some places (the more northern and continental) will be more seasonal than others. The timing of peak rainfall will vary also. For example, July is the rainiest month in Kampuchea, Lao P.D.R., Luzon (Philippines), Thailand, and Vietnam; January is the rainiest in Malaysia, the Philippines, and parts of Indonesia, while April and October are the wettest in Sumatra (Koteswaram, 1974). At any one location, the seasonal timing of the onset of summer monsoonal rains is irregular, and the amount of precipitation received may vary greatly — depending especially upon the influence of typhoons (in the western Pacific and South China Sea) and tropical cyclones (in the Bay of Bengal). Floods and droughts are thus inescapable features of tropical Asian latitudes.

The basic pattern outlined above is not entirely applicable to the climates of Malaysia (near the equator), Indonesia (partly in the Southern Hemisphere), or the Philippines. On the west coast of Malaysia, the rainy season occurs between monsoons, while on the east it takes place during the northeast monsoon. A dry season is lacking over much of Indonesia, which is midway between the monsoon areas of Southeast (continental) Asia and Australia. In the Philippines Archipelago, high mountain ranges have a significant modifying effect on local climate. Certain islands (Mindanao, Samar Leyte, Bohol, and West Mindoro) have no dry season, whereas others have either a long dry season and a period of torrential rains, or a long rainy season with a short dry period (Aki & Berthelot, 1974). In addition, there are parts of tropical Asia with relatively low rainfall (500-1,000 mm per annum) throughout the year, including northern Burma, central and northeastern Thailand as well as upper and middle portions of the Mekong basin. These areas lie in rain shadows where the intercepting highland terrain cuts the monsoon short in its path. Here, the native equatorial or monsoon forest is replaced with savannah vegetation. More details are given by Legris (1974) and de Rosayro (1974).

DISCHARGE REGIME

Most tropical Asian streams rivers show a considerable degree of 'peakedness' in discharge (Fig. 3.4), regardless of the timing of peak flows. All large rivers in the more northern part of tropical Asia have a single period of high flood — which can include several peaks — that is caused by rains associated with the summer monsoon; this is followed by a recession which is more or less altered by non-monsoonal precipitation (Aki & Berthelot, 1974). Snow- or ice-fed rivers (such as the Brahmaputra and other Himalayan rivers) have a sustained base flow in their upper reaches but the contribution of meltwater is insignificant in the lower course where is a strong cyclicity in discharge is caused by the seasonal rainfall (Fig. 3.5). Similarly, the Mekong River is fed by snowmelt in its upper reaches and by seasonal monsoonal

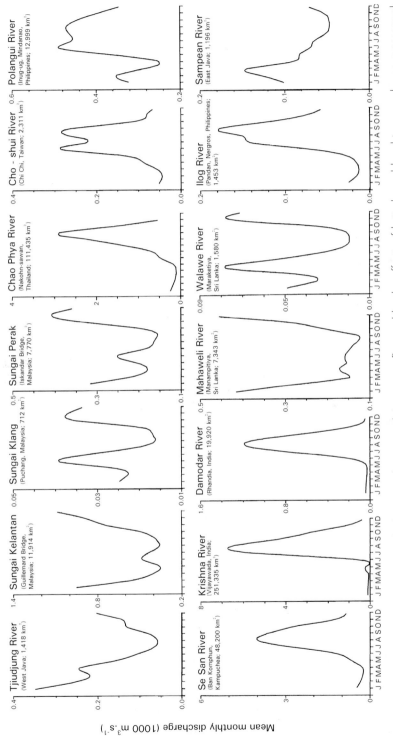

Fig. 3.4 Seasonal changes in river discharge within tropical Asia are influenced by the effects of latitude and local topography on monsoonal rains. Discharge values presented here are means of monthly records for at least six years (raw data derived from van der Leeden, 1975).

Mean monthly discharge (1000 m³·s⁻¹)

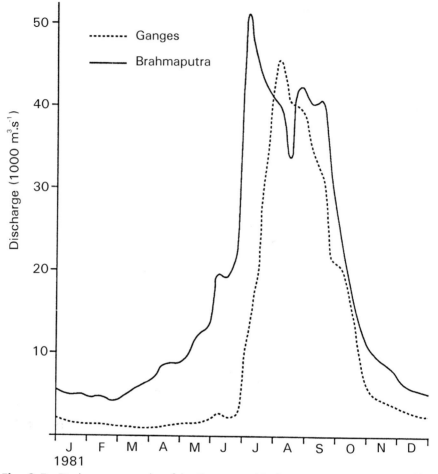

Fig. 3.5 Discharge seasonality of the Ganges and Brahmaputra Rivers, 1981 (modified from Anhwar Khan, 1987).

rains downstream. This results in rather uniform seasonal flow in the upper Mekong, while the lower course has pronounced seasonal variations (Pantulu, 1986a). At Kratie (Kampuchea), for example, the high-water period (September-October) discharge is 33,200 m³.s⁻¹; the minimum discharge is 1,250 m³.s⁻¹ in the dry season (when it never exceeds 1,700 m³.s⁻¹). In 1939, during the highest recorded flood, peak discharge reached 67,000 m³.s⁻¹ (Takenouchi, 1966). In essence, the period of peak flow in these large rivers varies according to whether they are fully rain fed, partly fed by rain and party by snowmelt, or fed predominantly by snowmelt (Bandyopadhyay & Gyawali, 1994), but marked flow seasonality is the rule.

Clearly, the predictability and pattern of river discharge varies from place to place, and reflects latitude, the influence of tropical storms, and local landscape morphometry. In consequence, flow in some rivers will be more seasonal than in others. Where flows are bimodal (e.g. Walawe River, Sri Lanka, and Sungai Klang and Sungai Perak, Malaysia; Fig. 3.4), the relative size of the flood peaks depends upon the amounts of rain associated with the northeast and southwest monsoons. In general, the smaller the stream catchment area, the larger are the deviations from the general pattern of discharge regime.

Floods are regular occurrences along Asian rivers. With rare exceptions, they are not caused by human activities but by typhoons, cyclones, surges in snowmelt due to warm weather, or exceptional atmospheric events (Bandyopadhyay & Gyawali, 1994). Nevertheless, flood severity may be increased significantly by man's alteration of drainage basins (see pp. 549–558). Natural causes include convergence of moist air flows which can concentrate vast amounts of water vapour and lead to its precipitation. A sudden rise in air temperature over snow-clad mountains (such as the Himalaya) can result in a rapid thaw and the release of water at high elevations. Atmospheric disturbances may also generate floods. For example, a deep depression over the semi-desert Indian state of Rajasthan during the 1979 monsoon season delivered more rain (>700 mm) in five days than the annual rainfall (300-500 mm) for the area. This caused the tributaries and mainstream of the Luni River to overflow their banks and reach 5 m above normal levels with the lower course running over 25 km wide (Sharma & Vangani, 1982). Das (1991) and Bandyopadhyay & Gyawali (1994) present similar examples.

Cyclones and typhoons bring a large amount of precipitation concentrated in a short period of intense rainfall. Small catchment areas can experience devastating flash floods, far above average discharge rates, and even large river regimes can be affected noticeably (Takeuchi, 1993). For instance observations at Prek Thont in the Mekong basin (Kampuchea) give a typhoon-related flow rate of 6,000 $m^3.s^{-1}$ — i.e. a specific run-off of about 2 $m^3.s^{-1}.km^{-2}$ for a 3,650 km^2 catchment. Values as high as 10 $m^3.s^{-1}.km^{-2}$ have been recorded from small catchments in Taiwan (Aki & Berthelot, 1974) which compares with a average of specific run-off of approximately 0.02–0.04 $m^3.s^{-1}.km^{-2}$ for monsoonal Asia and 0.015–0.03 $m^3.s^{-1}.km^{-2}$ for equatorial rivers in Indonesia (Meybeck *et al.*, 1989: pp. 29–30). Extreme flood flows in the Solo River can exceed 1,800 $m^3.s^{-1}$ but can fall to 10 $m^3.s^{-1}$ in the dry season (Neame, 1988).

While large rivers show flow seasonality related to the monsoon, some also exhibit short-term fluctuations in discharge which are significant and frequent (Takeuchi, 1993). For example, short-term peaks in discharge during floods along the Narmada and Tapi Rivers — the largest western-flowing rivers in India — range from 10,000 to 60,000 $m^3.s^{-1}$ (Kale *et al.*, 1994) and result from heavy rainfall generated by intense low-pressure systems during the summer monsoon. Less extreme and more predictable fluctuations in flow can occur in other rivers. Constancy of water levels for three days or more in the Rajang River (North Borneo) is exceptional (Lelek, 1985): daily changes ranging between 3 and 5 m are the rule, while the recorded minimum-maximum daily flow rates are 200 — 7,000 $m^3.s^{-1}$. Fluctuations in water level are not synchronized over the entire drainage basin or throughout the river, reflecting the patchy distribution of rainfall (Lelek, 1985), and an analogous pattern is seen within the Mekong basin (Takeuchi, 1993). The catchment topography of the Purari River (New Guinea) produces a similar effect of localized, intense thunderstorms, so that run-off originates from a number of scattered areas rather than from the whole catchment (Pickup & Chewings, 1983).

The example of southern China illustrates the extent of discharge variations seen in tropical Asian rivers. The total annual discharge of Chinese rivers is in the order of 2,700 x 10^9 m^3 (Ministry of Water Resources & Electric Power, 1987). An obvious feature of this discharge — and especially those in the south — is its seasonality which reflects the interactions of the northeast and southwest monsoons. Because the influence of the southwest monsoon wanes with increasing distance from the southern coast, rivers which are more northern and continental show greater flow seasonality. Notwithstanding this intra-regional variation, the discharge of all southern Chinese rivers peaks during the summer wet season. In the Chang Jiang (Fig. 3.6; Table 3.2), for example, over 70% of the annual discharge occurs between June and November (Xiong *et al.*, 1989; Tian *et al.*, 1993) — around 20% in July alone — while January has the lowest flow amounting to < 4% of the total (Wang & Chen, 1987; Zhou, 1987). The degree of 'peakedness' decreases slightly downstream. A similar temporal pattern is seen in the Zhujiang system (Table 3.3) where — if anything — the peak in discharge seasonality is even more pronounced.

In addition to the seasonal variation in climatic means and river discharge in southern China, there are considerable differences between years in the amount and timing of rainfall. To a large degree, these differences reflect variations in the timing and strength of the monsoons.

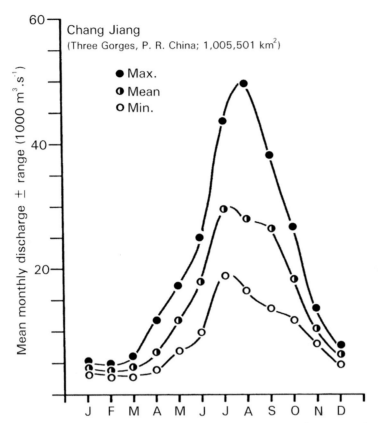

Fig. 3.6 The extent of seasonal variation in maximum, mean and minimum discharge volumes of the Chang Jiang at Yichang close to the site of the Three Gorges Dam (raw data derived from CIDA, 1988).

Table 3.2 Seasonal variations in discharge (% of annual total) of the Chang Jiang (Yangtze River), China. Stations are arranged from upper to lower course; Yichang is located immediately downstream of the Three Gorges Dam. Spring = March-May; summer = June-August; autumn = September-November; winter = December-February. Data derived from Xiong *et al.* (1989).

Station	Spring	Summer	Autumn	Winter
Cun Tan	11.4	46.6	33.7	8.3
Yichang	13.3	45.1	33.3	8.3
Hankou	18.0	39.7	32.7	9.6
Tatung	21.2	38.9	29.5	10.4

At any one location, the seasonal timing of the onset of the monsoon is irregular and the amount of rain received varies greatly, depending especially upon the influence of typhoons which have their strongest

Table 3.3 Seasonal variations in discharge (% of annual total) of the Xijiang (West River) tributary of the Zhujiang (Pearl River), Guangdong Province, southern China. Seasons are defined as in Table 3.1. Data derived from Xiong *et al.* (1989).

Station	Spring	Summer	Autumn	Winter
Nanpan Jiang (upper course)	7.8	45.9	34.3	12.0
Hongshui He (upper course)	12.0	55.5	25.3	7.2
Yujiang (middle course)	10.1	55.0	28.5	6.4

influence along the coast. As a result, the extent of year-to-year variation in maximum and minimum flows can be considerable. For example, Zhang & Lin (1992: p. 330) report a flood flow peak of 110,000 $m^3.s^{-1}$ during 1870 in the Chang Jiang, which is over three times the annual average discharge of 34,000 $m^3.s^{-1}$; more recently, flood flows of 72,000 $m^3.s^{-1}$ were recorded in 1981 after a period of unusually intense rain. A similar pattern is seen in the Xijiang (mean discharge 6,990 $m^3.s^{-1}$) — a major tributary of the Zhujiang in Guangdong Province — where Xiong *et al.* (1989) report that the highest annual average discharge of 11,000 $m^3.s^{-1}$ (in 1915) is over three times greater than the lowest annual average of 3,250 $m^3.s^{-1}$ (in 1963). Like other southern Chinese rivers, the Minjiang (in Fujian Province) shows considerable inter-year variation in discharge with maximum (in 1937) and minimum (in 1971) values of 2,670 and 850 $m^3.s^{-1}$ (Xiong *et al.*, 1989).

In addition to year-to-year fluctuations in discharge, rivers may show longer-term variations in flow. For example, there is recent evidence of a long-term decline (1970-1990) in flow of the Mekong River, resulting from a downward trend in maximum annual flows (Hill, 1995); the 20-year trend for minimum flows has remained relatively constant. The long-term hydrologic record of the Mekong River reflects a number of cycles of both declining and increasing flows and the period of decline from 1970 to 1990 is not unique, and is a natural event (Hill, 1995). Roberts & Baird (1995) believe that the stream flow of the Mekong mainstream is currently more irregular than it has been in the past. This could indicate that the 20-year decline in flow is due to human impact (dam building, deforestation of the drainage basin). Unfortunately, it is unclear whether sufficient data exist to settle this debate.

Discharge regimes in small streams influenced by monsoons are exemplified by data from Hong Kong where the average annual rainfall is 2,225 mm but can exceed 3,000 mm in unusually wet years. Approximately 84% of this rainfall occurs between April and September (the wet season), about 3% in winter (December to February), and approximately 18% — more than twice the monthly average — in August; 25% of the rainfall is associated with tropical cyclones (Dudgeon & Corlett, 1994). Seasonal variation in discharge volume is apparent from comparison of the ratio of run-off during the wet season with that in the dry season (October to March). Data are available for seven Hong Kong catchments, and values of this 'seasonality index' range from 2.4 to 29.2 (Jayawardena & Peart, 1989; Dudgeon, 1992a), indicating both a strong seasonal effect as well as evidence of spatial variation in run-off. If the maximum daily discharge is expressed as a percentage of total annual discharge, an indication of the intensity of spates (or freshets) can be obtained. In one 30-km^2 drainage basin, the maximum daily flow accounted for just under 10% of annual discharge (Jayawardena & Peart, 1989). Values for other catchments were less extreme (mean = 4.2%), but nevertheless show the highly 'pulsed' nature of flow in Hong Kong streams. Rainfall in Peninsular Malaysia (Selangor) is less seasonal than in Hong Kong, with significant quantities falling in all months although there is some concentration during the inter-monsoon period. In consequence, stream discharge in Sungai Gombak is rather aseasonal although there is considerable flow variability both within and between years (Low & Peh, 1987), and the river shows flood peaks which rise and fall rapidly with each intense rainstorm (Bishop, 1973a). Mean discharge (1968-70) was 5.5 m^3.s^{-1}, although mean daily discharge rates ranged from 2.1 — 16.6 m^3.s^{-1}. The highest daily value was only half of the maximum instantaneous discharge (31.0 m^3.s^{-1}) indicating the intensity of flood crests (Bishop, 1973a).

HYDROCHEMISTRY

Composition

The chemical composition (major elements) of rivers with unpolluted drainage basins is highly variable, reflecting watershed lithology which can give rise to water-types that may be dominated by any major ion except potassium (Meybeck, 1984). Other factors such as climate, vegetation, altitude and relief are of secondary importance, except for the regulatory effect of vegetation on plant nutrients which are generally less variable than most major ions (Meybeck, 1984). In summary, the environmental influences on hydrochemistry are as follows (Gibbs, 1970; Meybeck, 1984; 1987; Meybeck *et al.*, 1989):

1. Distance from the ocean (sea spray is rich in sodium, chloride, magnesium, sulphate, etc.);
2. Climate (which influences weathering processes, and concentration of ions through evaporation);
3. Terrestrial vegetation (take up nutrients, influences organic carbon availability and nitrogen speciation, may concentrate ions by evapotranspiration);
4. Lithology (influences the rate of weathering and thus the supply of ions as well as the types available);
5. Aquatic vegetation (take up nutrients, especially nitrogen and phosphate).

Some of the rivers and streams of Southeast Asia (especially those draining primary forest) have many similarities with those of Amazonia, particularly in their apparent '. . . self sufficiency of nutrient supply which (is) based very probably on the quick process of decomposition and mineralization of organic matter originating itself from the extremely rich vegetation surrounding the running waters . . .' (Lelek, 1985). For example, the Rajang River — which drains primary forest in North Borneo — has nutrient-poor characteristics — low conductivity and a slightly acid *p*H — similar to Sioli's (1965) Amazonian 'white waters' (Lelek, 1985). There are parallels also between Asian and Amazonian 'blackwater' streams which have dark, tea-coloured acid waters (see below), although some Asian blackwaters have high concentration of sulphates (contributing to the low *p*H) while comparable Amazonian streams are rich in chloride (Ng *et al.*, 1994).

The waters of the Solo River (the largest river in Java) are boulder controlled (= 'rock dominance'; Gibbs, 1970), being calcium bicarbonate in nature, and close to world mean river waters in terms of equivalent percentages of major ions (Neame, 1988). However, ion concentrations are generally higher and close to those of the River Kwai in Thailand (Tyler, 1984) which, like the Solo, is a turbid river. Fluctuations of ion concentrations in the Solo River reflect large changes in flow volume (Fig. 3.7), with nitrogen peaking during the wet season when other ions are diluted (Neame, 1988). A similar decrease in concentrations of ions (except nitrogen) has been recorded during floods in the Purari

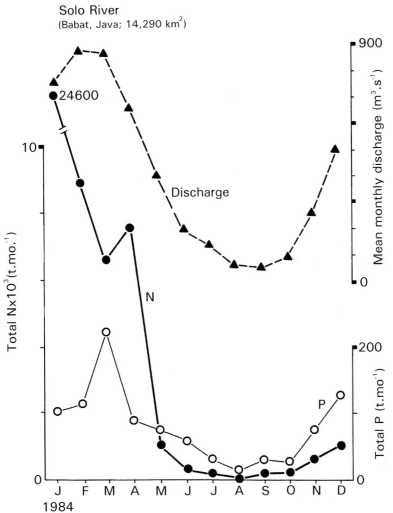

Fig. 3.7 Seasonal changes in discharge volume, and export of total nitrogen (N) and phosphorus (P) from the Solo River catchment, East Java, Indonesia, during 1984 (redrawn from Neame, 1988).

River, New Guinea (Petr, 1983a). However, there is a direct positive correlation between discharge rate and transport of suspended load in the Purari (Pickup, 1983), which must be of significance for transport of nutrients (such as phosphate) bound with solids (Viner, 1982; 1987). The importance of suspended particles in adsorbing nutrients from the surrounding water depends strongly upon particle surface area, and the amounts of phosphate are at least an order of magnitude more on clay relative to equivalent volumes of fine sand or silt (Viner, 1987).

The waters of the River Kwai are dominated by calcium and bicarbonate ions (Tyler, 1984) which reflects the abundance of limestone in the drainage basin, and is characteristic of many Thai rivers (Kobayashi, 1959). In common with the Purari and Solo, River Kwai waters are dilute during the monsoonal rains but ions become more concentrated when flow slackens, and the same pattern is seen in the Cauvery River, India (Somasundaram & Sappanimuthu, 1991). By contrast with the turbid Kwai and Solo, ion concentrations in Purari River waters are relatively low (Petr, 1976). Nevertheless, like most surface waters, they are dominated by calcium and bicarbonate ions which reflects catchment lithology (Petr, 1976, 1983a). Despite relatively low ion concentrations, the Purari (approximate discharge 2,400 $m^3.s^{-1}$) delivers the highest total ionic load of the major New Guinea Rivers (Petr, 1983a; Petr, 1983b), which include the Fly (6,000 $m^3.s^{-1}$) and the Sepik (4,000 $m^3.s^{-1}$). The conductivity of Purari waters correlates negatively with altitude, and concentrations of several major ions increase downstream (Petr, 1983a; Viner, 1987). A similar trend is evident in Sri Lankan streams (Starmühlner, 1984a; 1984b), and a decline in cation concentration with altitude was likewise apparent in three Nepalese stream systems (Rundle *et al.*, 1993) despite interstream differences in conductivity and *p*H. In contrast, there is a downstream decrease in calcium and magnesium along the Mekong which may result from dilution by tributaries with low hardness (Pantulu, 1986a).

The hydrochemistry of major rivers in China is dominated by the weathering of carbonates and evaporites (gypsum, halite) in varying proportions (Hu *et al.*, 1982). The Chang Jiang is significantly more mineralized when compared to the world average (Livingstone, 1963), and ranks second after the Amazon in discharge of dissolved inorganic carbon (Wang & Chen, 1987). There are almost double the world average concentrations of calcium, magnesium and bicarbonates in Chang Jiang waters, reflecting the large extent of exposed limestone in the drainage basin (Hu *et al.*, 1982; Wang & Chen, 1987). In addition, soils are eroded extensively during the flood season when minerals enter the river. Peaks in transported nitrogen, phosphorus, and bicarbonate occur at this time and transport rates are closely related to discharge (Wang & Chen, 1987).

Data on the hydrochemistry of 13 southern Chinese rivers has been summarized by Zhang *et al.* (1987), and highlight the over-riding influence of rock type in the watershed. Zhujiang waters draining Guangxi Province are similar in ionic composition to the Chang Jiang, and contain the weathering products of carbonate rocks. This river is second in China to the Chang Jiang in discharge, and has a similarly high rate of chemical weathering. Rivers in eastern Guangdong and Fujian are characterized by high concentrations of dissolved silica and a low total dissolved solids loads, reflecting weathering of siliceous rocks. Their chemical weathering rates are relatively low (compared to the Zhujiang in Guangxi), but are close to the world average. The Indus River in Pakistan has similar hydrochemistry (Arain, 1987), although annual dissolved phosphate transport and nitrogen loads are high due to anthropogenic modification of the densely-settled Indus floodplain.

The Ganges and Brahmaputra in Bangladesh are less mineralized than the Chang Jiang but do show an influence of the moderate karstification of the Himalaya (Chowdhury *et al.*, 1982; Safiullah *et al.*, 1987). The hydrochemistry of the upper courses of both rivers is dominated by calcium, magnesium, and bicarbonate ions — which is consistent with the regional carbonate lithology and hydrochemistry of many Himalayan rivers (Sarin & Krishnaswami, 1984; Bhatt & Pathak, 1991; Bartarya, 1993) — although some tributaries of the Brahmaputra have been classified as carbonate streams enriched by sodium and chloride (Hu *et al.*, 1982). Further downstream, conditions are similar but sodium, chloride, and sulphate increase considerably during low-flow conditions, and account for nearly a third of the major ions. Sodium and potassium rise more in the lower Ganges reflecting the supply of soil salts in the alluvial plains and effluent seepage from saline ground waters (Sarin & Krishnaswami, 1984). Consequently, waters of the lower Ganges are more than twice as saline as the Brahmaputra. The total dissolved solid content of the Ganges is higher than the global mean river water composition (accounting for 3% of the global supply of dissolved salts to the oceans), but the Brahmaputra is about average (Sarin & Krishnaswami, 1984).

Overall, the chemical composition of river waters in the Himalayan drainage system (comprising mainly the Ganges, Brahmaputra, Indus and Irrawaddy) broadly reflects the world average river-water composition given by Meybeck (1984), although there are some spatial and temporal variations (Chowdhury *et al.*, 1982; Subramanian & Ittekkot, 1991). For instance, flood-season elevations in nitrates occur in the Ganges and Moosi (Hyderbad, India) (Lakshminarayana, 1965a; Venekateswarlu, 1969a), as has been recorded for the Solo River (see above). While this may be a natural phenomenon, anthropogenic factors such as deforestation, increased use of nitrogen fertilizers, and wash-off of sewage and organic wastes may be contributing factors

(Lakshminarayana, 1965a; see also Ittekkot & Zhang, 1989; Gautam, 1990). Note, however, that most of nitrogen transport in large Asian rivers is transported in particulate rather than in dissolved form (Ittekkot & Zhang, 1989). Upland Himalayan rivers and streams have lower concentrations of major ions (i.e. calcium, magnesium, sodium and bicarbonate; e.g. Bhatt & Pathak, 1991; 1992; Bartarya, 1993) than Indian peninsular rivers because the steep slopes, rapid infiltration and quick outflow of rainwater leads to a short residence time and reduces the interaction between water and rocks (Bartarya, 1993). General information on the hydrochemistry of Asian freshwaters can be found in Fernando (1984a), although most of the data pertains to lakes.

Spatial and temporal variation

Some information concerning seasonal changes in the water chemistry of streams and small rivers is given by Bishop (1973a) and Dudgeon (1982a; 1992a). Waters draining an unmodified, forested section of Sungai Gombak (Malaysia) had low conductivity, alkalinity, calcium, and nutrients, circumneutral pH, and some limited buffering capacity provided by abundant silica (Bishop, 1973a). Conditions in Tai Po Kau Forest Stream (Hong Kong) were almost identical (Dudgeon, 1982a; 1987b), and seem to typify many Asian streams (as they were before major human influences) which are usually rather impoverished and poor in nutrients, with low conductivities and, often, acid waters (Bishop, 1973a). There was little seasonal variation in pH in Sungai Gombak or Tai Po Kau Forest Stream, and pH was relatively uniform along the Gombak as has been recorded for Lam Tsuen River, Hong Kong (Dudgeon, 1984a). Progressive downstream enrichment for all ions (except silica) occurred in Sungai Gombak, due in part to anthropogenic activities in the drainage basin. A similar downstream pattern was apparent — for the same reasons — in Lam Tsuen River.

As stated above, seasonality of hydrochemistry in large rivers chiefly reflects greater discharge volumes during the monsoon, although there may be a degree of instability in hydrochemistry (Christensen, 1992) notwithstanding any seasonal pattern. Concentrations of major ions in Himalayan rivers are inversely proportional to flow rate and hence decline during the monsoon season and in storm run-off (Gautam, 1990; Bhatt & Pathak, 1991). Seasonal changes in smaller rivers, such as Sungai Gombak, are less obvious and temporal fluctuations are due largely to spates. Sodium, potassium, silica, chloride, phosphate, and

conductivity decline in Sungai Gombak because of the diluting effects
of large volumes of rain, while aluminium and iron increase during
storms when the soil is waterlogged and colloids are washed out
(Bishop, 1973a). Wet-season elevations in iron concentrations also
occur in larger rivers such as the Moosi and Ganges (Lakshminarayana,
1965a; Venekateswarlu, 1969a). Nitrogenous compounds in Sungai
Gombak increase during floods (Bishop, 1973a), and parallel the wet-
season increases in nitrogen load reported for the Solo, Ganges and
Moosi Rivers (see above). In Himalayan rivers, however, nitrogen
loads peak during periods of moderate discharge (Bhatt & Pathak,
1991). Seasonal changes in nutrient concentrations in Tai Po Kau
Forest Stream are minor but fluctuations were associated with spates,
increasing during the rising phase and falling subsequently because of
dilution (Dudgeon, 1982a; 1987b).

A number of publications provide data on the ionic composition
of river water as background for biological studies (e.g. Chacko &
Ganapati, 1952; Roy, 1955; Lakshminarayana, 1965b; Venekateswarlu,
1969b; Pahwa & Mehrotra, 1966; Ray *et al.*, 1966; Dang, 1970;
Costa & Starmühlner, 1972; Costa, 1974; Starmühlner, 1974; 1984a;
1984b; Weninger, 1972; Jhingran, 1975; Boyden *et al.*, 1978;
Kortmulder *et al.*, 1978; Geisler *et al.*, 1979; Lim, 1987; Soetikno,
1987; Dudgeon, 1988a; Whitton *et al.*, 1988a; Singh & Srivastava,
1989; Shreshtha, 1991; Choudhary *et al.*, 1992; Christensen, 1992;
Gupta & Nichael, 1992; Khan, 1992; Pandey *et al.*, 1992; Dobriyal
et al., 1993; Edds, 1993; Rundle *et al.*, 1993) or as an end in itself
(e.g. Kobayashi, 1959; De Sousa *et al.*, 1981; Petr, 1976; 1983b;
1983c; Agrawal, 1990; Gautam, 1990; Srivastava & Kulshrestha, 1990;
Somasundaram & Sappanimuthu, 1991; Bartarya, 1993 and references
cited above). These studies underscore the conclusion that
hydrochemistry is predominately a consequence of local lithology, and
Hickling (1961: p. 7) notes that higher '. . . fertility of the water in
Java and Sumatra is due to leaching of rich volcanic soil, while in
Malaya . . . volcanic activity ended long ago and both the land and
the water are much less fertile'. Despite the dominant influence of
lithology, high levels of organic matter in some Thai rivers override
the influence of rock type making the waters slightly acidic (Kobayashi,
1959). For the same reason, the waters of the Mekong floodplain are
acidic, with pH down to 4.8 (Takenouchi, 1966) and are generally
soft with a low mineral content (Pantulu, 1986a). It is apparent also
that unpolluted streams which do not experience saline intrusions
from estuarine reaches show rather minor seasonal fluctuations in

ionic composition. Where such fluctuations do occur, dissolved salts reach highest concentrations at times of lowest flow (Kobayashi 1959; De Sousa *et al.*, 1981).

Variations in chemical composition of waters in the tributaries of large river systems are apt to be greater than those shown between main river channels (Welcomme, 1979; Meybeck, 1985). This is because lower-order streams are more sensitive to local geological and vegetation patterns, whereas higher-order channels average conditions from a range of lower-order ones (Vannote *et al.*, 1980). For example, the conductivity of the Musi River (the longest river in Sumatra) is low throughout its course, despite receiving water rich in chloride and phosphate with high conductivity from tributaries flowing from volcanic springs (Ruttner, 1931). Nevertheless, where differences in water chemistry between drainage basins do occur, they may have striking effects on their flora and fauna. For example, in many parts of Malaysia, especially in forested or swampy areas, the soils are highly leached and lacking in calcium. They are associated with acidic 'blackwaters' (Johnson, 1957; 1967a; Mizuno & Mori, 1970) which support a distinctive association of cyprinid fishes (especially *Rasbora* spp.) and shrimps, as well as desmids and the red alga *Batrachospermum* (Johnson, 1967b). Snails are rare or absent (Johnson, 1957). Some fishes are either most abundant in — or confined to — blackwaters with $pH < 5.5$ and several species were present in waters of pH 3.4 — 3.8 (Ng *et al.*, 1993; 1994; see also Ng & Lim, 1992). The blackwater habitats were thought to contain about 10% of species comprising the Malaysian fish fauna (Johnson, 1967b) but a recent, detailed survey of blackwaters uncovered 48 fish species (including 15 stenotopic taxa and five new species) — almost 20% of the total for Peninsular Malaysia (Ng *et al.*, 1994). Similar blackwater habitats are present in Sumatra (Whitten *et al.*, 1984), Sarawak and Thailand, but they have received little or no study.

In contrast to blackwater habitats, limestone outcrops in north and central Malaysia make the soils calcareous; as a result, the stream waters draining them are relatively hard (Johnson, 1967a). Their biota is dominated by macrophytes, blue-green algae, and molluscs; fish are abundant but represented by relatively few species (Johnson, 1957). Elsewhere, Starmühlner (1984a) records that most taxa of Sri Lankan molluscs occur in downstream reaches where calcium, magnesium, hardness and alkalinity values are greatest. Janaki Ram & Radhakrishna (1984) also present data that suggest a relationship between the occurrence of Indian freshwater molluscs and calcium concentrations.

TRANSPORTED INORGANIC LOAD

The copious precipitation and run-off which typifies much of tropical Asia generates high suspended loads in rivers. Loads are elevated by vegetation clearance and poor cultivation practices which are widespread over much of the region, especially in highland areas (e.g. Melkania, 1991; Ahmad, 1993; Chakrapani & Subramanian, 1993). Downstream, sediments deposited in river deltas and floodplains are alternately scoured and silted, producing shifts in the river course and a complex channel morphology. Annual sediment loads of tropical Asian rivers are inadequately documented and estimates of loads transported by some rivers are so recycled that the true data bases are unknown (Milliman & Meade, 1983). An extrapolation of data for world drainage basins suggests that the total suspended sediment delivered by all rivers to the oceans is about 13.5×10^9 t annually; bedload and flood discharges may account for an extra $1 - 2 \times 10^9$ t (Milliman & Meade, 1983). About 70% of this is derived from southern Asia where sediment yields are much greater than for other drainage basins. Because of active tectonism and volcanism, steep slopes, heavy rainfall, and intense human activity, large islands in the region (Taiwan, Philippines, Indonesia, Borneo and New Guinea) contribute prodigious quantities of river sediment to the ocean (Milliman & Meade, 1983; Diemont *et al.*, 1991). Taiwan is the most spectacular of these contributors, with an average annual sediment yield to the ocean of 10,000 $t.km^{-2}$, and a calculated annual sediment load of about 300×10^6 t — only slightly less than that of the United States (Li, 1976).

Suspended loads are influenced strongly by season — such that, in some cases, almost the entire load is carried during the monsoon — but year-to-year variations in maximum and minimum flows also affect the magnitude of transport. For example, one estimate of annual average suspended load in the Ganges-Brahmaputra system is $1,670 \times 10^6$ t, but excessive run-off in 1973 transported $>2,000 \times 10^6$ t (Milliman & Meade, 1983). Such variation may account for the disparity in figures quoted in the literature concerning sediment discharge of tropical Asian rivers (e.g. Holeman, 1968; van der Leeden 1975; Milliman & Meade, 1983; Jalal, 1987; Degens *et al.*, 1990), and the values given in Table 3.1 should perhaps be viewed as an approximation of the relative sediment loads of major rivers. Moreover, these figures do not include bedload transport of sediment. Undoubtedly, monsoon- or flood-related transport of bedload will increase the total output of particulates from a river to the sea and, by modifying the erosive and depositional abilities of the river, also leaves an imprint on channel form which is altered slowly by flows of smaller magnitude during the inter-flood period (Gupta, 1983). The potential importance of bedload transport is apparent from observations of sand waves and dune fields in the Brahmaputra that, during monsoonal rains, can reach heights of 15 m, lengths of 200-900 m, and migrate downstream by as much as 600 m.day^{-1} (Coleman, 1969). Extrapolating this rate across the river channel gives daily bedloads of $10^6 — 10^7$ t during times of peak discharge (Milliman & Meade, 1983). High sediment loads and deposition processes have resulted in the course of the Brahmaputra shifting westwards so that it now meets the Ganges instead of the Meghna as it did 200 years ago (Bandyopadhyay & Gyawali, 1994). Indeed, sediment loads in some Himalayan rivers are so great (with sediment yields often greatly in excess of 10 t.ha^{-1}.yr^{-1}; e.g. Melkania, 1991; Bandyopadhyay & Gyawali, 1994) that stream discharge may best be viewed as a water-sediment combination rather than water flow alone (Bandyopadhyay & Gyawali, 1994).

Compared to other large Asian rivers, the Mekong River carries a relatively small suspended load (Table 3.1): only 100 g.m^{-3} during the dry season and 250 — 300 g.m^{-3} in the wet season; about 20% of this is organic. By contrast, the Red River in Vietnam, which is relatively short and drains a much smaller catchment than the Mekong (Table 3.1), has a maximum silt content (7,000 g.m^{-3}) that is twice the highest ever recorded for the Mekong (Aki & Berthelot, 1974). This reflects high maximum flood discharges in the Red River (up to 35,000 m^3.s^{-1}), which are almost ten times average annual discharge rates. In New Guinea, the Purari River transports 57×10^6 t sediment.yr^{-1}, 5% of which is bedload (Pickup, 1983); this amounts to 88.6×10^6 m^3.yr^{-1}, providing material for a major deltaic complex of global significance. Suspended sediment concentrations in the Purari are, however, lower than some Southeast Asian rivers where a concentrated monsoon coupled with intensive agricultural activity results in severe erosion (Petr, 1983a).

As in most tropical Asian rivers, the months of highest discharge in the Chang Jiang are also the period of greatest sediment transport (Wang & Chen, 1987; Zhou, 1987). Similarly, as much as 90% of the sediment load of Himalayan rivers, which may contribute as much as one-third of the global sediment transport to the world oceans (Subramanian & Ittekkot, 1991), is carried during the three-month monsoon (Degens *et al.*, 1991; Jha *et al.*, 1993). This reflects the steep gradient between the sources of these rivers and their mouths, and as well as deforestation (and subsequently erosion) of their valleys. The

Solo River (Indonesia) is likewise highly turbid after heavy rain, reflecting poor land-use practices in the upper basin and uplands which contribute heavily to the annual average sediment load of over 200 x 10^6 t (Neame, 1988). The Indus River, which drains much of Pakistan, transports an annual load of about 400 x 10^6 t of sediment in its upper reaches, much of it (60%) is sand (Arain, 1987). Large, but indeterminate, amounts of this load settle out on the alluvial plains downstream (below Darband), while an even greater quantity is intercepted by dams at Mangla and Tarbela leading to sedimentation problems in the associated reservoirs (Milliman & Meade, 1983; Khattak, 1983; Pereira, 1989).

Data from small rivers and streams likewise show the influence of flow regime on inorganic suspended loads which, in Tai Po Kau Forest Stream, averaged 1.99 $g.m^{-3}$ (over two years) but ranged from 0.10 to 42.20 $g.m^{-3}$ (Dudgeon, 1982a) with peak loads during spates. Spates carried a large proportion of the total annual sediment load of Sungai Gombak, half of which was exported from a forested part of the catchment in a mere 24 days (Douglas, 1969).

The balance between erosive and depositional tendencies of running waters has profound effects on the bottom sediments. Investigations upon sediment dynamics relevant to ecological studies in the Oriental Region are scarce, but studies undertaken in Malaysia and Hong Kong have come to the same general conclusions (Bishop, 1973a; Dudgeon, 1982b; 1984a; 1992a). Progressive size decrease in riffle substrates occur downstream (in Sungai Gombak and Lam Tsuen River), where depositional areas accumulate silt. Short, discontinuous areas of flow prevail in streams and river headwaters, and sustained current velocities of any kind are rare. The result (in Sungai Gombak and Tai Po Kau Forest Stream) is a 'patch' mosaic of bottom sediments. The stream bed in riffles is characterized by large boulders with interspersed gravels and sands, and by coarse-sand substrates in pool areas. Across-stream sediment gradients occur in Tai Po Kau Forest Stream, with increasing proportions of fine particles close to the banks.

Sustained currents exceeding 1 $m.s^{-1}$ are common during spates in streams and small rivers. Bottom stability is disrupted by increased velocity and pressure, and all sizes of substrate are eroded and transported downstream. Such scouring of the stream bed decreases sediment-patch heterogeneity so that, in Tai Po Kau Forest Stream, bankside and mid-stream substrates become similar following disruption of across-stream gradients during the summer monsoon. Sediment heterogeneity is re-established during the dry inter-monsoonal period when flows are stable or decreasing (Dudgeon, 1982b; 1992a). These changes have important implications for lotic zoobenthos (see pp. 59–63). In Sungai Gombak and Hong Kong streams, substrate stability for all but the largest particles is ephemeral, and spates disrupt the surface sediments in most depositional areas and some riffles.

TRANSPORTED ORGANIC LOAD

The organic matter transported by streams and rivers is made up of contributions from allochthonous and autochthonous sources. The latter arises from the photosynthesis by aquatic plants (especially algae), while allochthonous carbon inputs comprise mainly terrestrial plant litter which is an important source of food and substrate for a variety of consumers. Once in the stream, the transport of this material, whether in dissolved form (particles <45 μm) or as fine (>45 μm — < 1 mm) or coarse (> 1 mm) particulates, reflects the interaction of a variety of factors, including the quality and quantity of autochthonous production and allochthonous imports, channel retentiveness, discharge regime, and in-stream decomposition processes. All of these parameters have been influenced by the anthropogenic alteration of tropical Asian streams and rivers (Milliman *et al.*, 1984; Gan & Kempe, 1987; Safiullah *et al.*, 1987), and the situation is complicated further by the augmentation of 'natural' transported organic loads with waste-water effluents and run-off from agricultural land.

Surprisingly little is known about the transport of organic materials by Asian rivers, and information on nitrogenous organic material is scant (Ittekkot & Zhang, 1989); data from pristine (or near-pristine) systems are particularly scarce. Large rivers, such as the Ganges and Brahmaputra, show flood-season peaks in dissolved organic carbon (flood flows 8.4 — 9.3 g $C.m^{-3}$; Ittekkot *et al.*, 1985) which may arise from the inundation of allochthonous detritus on the floodplain (Safiullah *et al.*, 1987). These rivers contain high concentrations of labile organics such as amino acids (up to 0.6 g $C.m^{-3}$) and sugars relative to other tropical rivers (Subramanian & Ittekkot, 1991), which may reflect (at least in part) human activities on the densely-populated Gangetic plain (Safiullah *et al.*, 1987; Ittekkot & Zhang, 1989). By contrast, the labile fraction of particulate organic carbon in tropical rivers is generally lower than in temperate rivers (20% *versus* 40% respectively; Ittekkot & Laane, 1991).

The limited data from the Indus River suggest that the magnitude of total organic carbon transport (ranging from $0.9 \times 10^6 - 2.03 \times 10^6$ t.yr^{-1}) depends upon year-to-year variations in the wet-season peak discharge (Arain, 1987). This may result from the influence of rainfall on patterns of surface run-off and river-level rises which, to a considerable degree, will determine the magnitude and extent of land-water interactions and allochthonous inputs. Some of the particulate organic material transported by the Brahmaputra during the flood season appears to originate from oxbow lakes on the floodplain (Ittekkot *et al.*, 1985), and high suspended-sediment loads during the monsoon increase river turbidity so reducing autochthonous photosynthesis (Degens *et al.*, 1991) with the consequence that the particulate organic load is dominated by allochthonous material from the inundated land. By contrast, organic matter transported by the Brahmaputra and Ganges at times of low water (when concentrations of particulate organics peak) is of autochthonous rather than allochthonous origin (Ittekkot *et al.*, 1985; Ittekkot & Zhang, 1989), reflecting increased phytoplankton populations during periods of low flow (Safiullah *et al.*, 1987; see also p. 50). Particulate organic transport by the Chang Jiang amounts to 12×10^6 t annually (6 t.km^{-2} drainage basin) although the load is diluted by inorganic particles during the flood season (Milliman *et al.*, 1984). Nevertheless, because of increased discharge volume, transport of organic particles peaks during the wet season. High rates of organic transport by the Chang Jiang probably reflect anthropogenic modification of the drainage basin (Milliman *et al.*, 1984; Gan & Kempe, 1987) and, on the basis of C/N-ratios, Milliman *et al.* (1984) estimated that terrestrial material accounts for at least 25% of the total particulate load. Ittekkot & Zhang (1989) report that most of the world transport of particulate organic nitrogen occurs in the form of relatively low C/N ratio, amino-acid poor material in rivers with high suspended-sediment loads (e.g the Ganges, Brahmaputra, Mekong and Chang Jiang), and attribute this to anthropogenic modification of the drainage basins of large Asian rivers.

The organic load in small rivers and streams consists of a mixture of dissolved organic material and variously comminuted detritus, usually derived from allochthonous leaf litter. Mean total suspended load in Sungai Gombak was approximately 14 g C.m^{-3}; almost 80% of this was in dissolved form (Bishop, 1973), and seasonal variations were not pronounced. Total drift load of allochthonous material (particles >165 µm) from the forested part of the Gombak catchment was rather constant (0.7 m^3.km^{-2}.yr^{-1}; Bishop, 1973) because of the slight seasonal variations in leaf fall in Peninsular Malaysia, although short-term increases occurred during spates when litter was washed from the forest into the river. Concentrations of suspended organic particles (particles >0.5 μm) in Tai Po Kau Forest Stream increased slightly during the dry season when riparian trees shed their leaves (Dudgeon, 1982a, 1982b). Average loads (recalculated from Dudgeon, 1982a) were 1.1 g C.m^{-3} (range 0.3-9.6 g.m^{-3}) and increased during spates. Organic particulates in Lam Tsuen River varied from a mean of 0.4 g C.m^{-3} upstream to 6.0 g C.m^{-3} in the lower course (recalculated from Dudgeon, 1984a): allochthonous detritus dominated the seston upstream, while fine amorphous detritus and particulates from autochthonous sources (algae and macrophytes) were important in the lower course (Dudgeon, 1984a). A similar downstream transition in seston composition was apparent along Sungai Gombak as large fragments of detritus were comminuted (Bishop, 1973; see also Arunachalam & Arun, 1994). Nair *et al.* (1989) recorded a change from coarse to fine organic particles in seston along a first- to fifth-order Indian river, but the actual concentrations of transported organic matter were not given. While only a limited number and size range of streams have been examined, data from the Hong Kong, Indian and Malaysian sites show clearly that allochthonous inputs have a dominant influence on seston loads in low-order streams, and that — in line with predictions of the RCC — the influence wanes downstream.

Data on transported organic loads in tropical Asian streams are set out in Table 3.4. DOM appears to be quantitatively the most important form of organic load and, in streams such as Sungai Gombak, concentrations are relatively stable over time and space (Bishop, 1973a). Note that the concentrations of DOM in large tropical Asian rivers, such as the Ganges and Brahmaputra (see above), are rather high relative to unpolluted European streams (Dudgeon & Bretschko, 1985a) and may reflect the input of material from the inundated (and highly man-modified) floodplain during the monsoon. However, even in small tropical Asian rivers, such as Sungai Gombak, transported DOM loads (approximately 11 mg C.m^{-3}; recalculated from Bishop, 1973a) seem to be higher than in equivalent-sized temperate European streams (Dudgeon & Bretschko, 1985a).

FPOM loads are second in magnitude to loads of DOM but, since sampling is often restricted to the water layers near the surface, FPOM concentrations and transport might be underestimated because FPOM transport is thought to be much greater close to the sediment surface (Schwoerbel, 1974; see also Milliman *et al.*, 1984; Cushing *et al.*,

Table 3.4 Transported organic loads (g C.m^{-3}) in tropical Asian streams. Estimates were made by various methods, and original units have been converted to g C.m^{-3} where necessary. Abbreviations: —, no data; *, mean value;[1].

	FPOM	DOM	*Source*
Ganges River (Bangladesh)			Ittekkot *et al.*,
Dry season	—	1.3	1985
Wet season	—	9.3	
Brahmaputra River (Bangladesh)			Saffiulah *et al.*,
Dry season	—	1.3–2.6	1987
Wet season	—	6.5	
Indus River (Pakistan)			Arain, 1985;
Dry season	0.3	1.2	Ittekot & Arain,
Wet season	16.0	22.0	1986
Chang Jiang (China)			
Dry season	3.0*	—	Milliman *et al.*,
Wet season	17.0*	—	1984
Tai Po Kau Forest Stream	1.1*	—	Dudgeon, 1982a
(Hong Kong)	(0.3–9.6)		
Lam Tsuen River (Hong Kong)			Dudgeon, 1984a
Upstream	0.4*	—	
Downstream (polluted)	6.0*	—	
Sungai Gombak (Malaysia)	3.8*	11.2*	Bishop, 1973

1993). A distinct seasonal pattern in FPOM transport is not discernable in small streams because of the strong correlation between FPOM load and discharge regime which has the consequence that high FPOM export is associated with periodic spates (see above). However, wet season peaks in FPOM transport do occur in the Chang Jiang (Milliman *et al.*, 1984) and may be typical of large rivers with extensively modified floodplains.

Although data are almost entirely lacking, CPOM is quantitatively by far the least important form of organic matter transport in Asian streams (Dudgeon & Bretschko, 1985). This may reflect physical fragmentation or rapid biological comminution of CPOM to FPOM. In north-temperate streams, CPOM loads are highest in autumn following the main period of deciduous leaf fall, but CPOM loads in tropical Asian streams do not show such pronounced seasonal fluctuations as there is no regionally consistent pattern of defoliation and leafing among the forests of the region. As predicated by the RCC, CPOM loads would be expected to become proportionately less important along a river course, while FPOM loads remain relatively constant or increase slightly. Changes in seston composition along

Sungai Gombak and Lam Tsuen River agree broadly with these predictions (see above). In many Asian streams (e.g. Milliman *et al.*, 1984; Safiullah *et al.*, 1987), however, natural patterns of carbon transport (assuming that they exist at all) are more-or-less obscured by human impacts.

BENTHIC ORGANIC MATTER

Inputs and standing stocks

Allochthonous carbon inputs into streams provide food and substrate for a variety of consumers but, in view of the importance of detritus and other allochthonous elements in fish diets (see pp. 82–84), little attention has been paid to quantifying standing stocks of this material in tropical Asia. Quantities of allochthonous organic matter on the stream bed vary in space and time, and CPOM accounts for a variable fraction of the total amount (Table 3.5), according to the density and height of the riparian vegetation, the width of the stream and the slope of the banks. Distribution within the channel is patchy, and depends upon stream discharge regime, the transport and retention capacities of individual reaches of the stream, and on the rate at which particles are consumed or decomposed by the aquatic biota. The retention capacity of the stream channel is especially influential in this regard, and debris dams are particularly effective retention structures contributing to small-scale patchiness in standing stocks of benthic organic matter. Temporal variations in standing stocks are, to some extent, influenced by latitude. In north-temperate streams, up to 80% of annual CPOM imports (mainly leaf litter) enter the stream in a few weeks during autumn but this is not the case in tropical Asia where the timing of litter inputs may be prolonged or differ from the temperate pattern (Dudgeon & Bretschko, 1995a).

In sections of Sungai Gombak draining forest, standing stocks of benthic organic matter (particles > 165 microns) varied over short distances, and were higher in depositional areas (16.3 — 37.0 g $C.m^{-2}$) than at erosional sites (17.4 — 22.8 g $C.m^{-2}$) (Bishop, 1973a). Overall annual mean standing stock in riffles was about 20 g $C.m^{-2}$. Spates influenced the amounts of organic matter in erosional areas by transporting it downstream. Pools accumulated organic matter during small floods, but became barren after severe storms (Bishop, 1973a).

Table 3.5 Benthic organic matter (g C.m^{-2} on and in the upper 5–10 cm of stream bed) in tropical Asian streams. Estimates were made by various methods, and original units have been converted to g C.m^{-2} where necessary. Abbreviations: —, no data; *, annual mean value;[1], particles > 220 μm, excluding woody debris;[2], particles > 165 μm.

	CPOM	FPOM	Total	Source
Tai Po Kau Forest Stream (Hong Kong)[1]				
1977–78	—	—	3.0–12.8	Dudgeon, 1982b
	(5.5)*			
1978–79	—	—	3.7–26.7	Dudgeon, 1982b
	(8.6)*			
Sungai Gombak (Malaysia)[2]				
Erosional reach	—	—	17.4–22.8	Bishop, 1973
Depositional reach	—	—	16.3–37.0	Bishop, 1973
Neyyar River (S. India)	9.0–45.0	—	—	Nair *et al.*, 1989
Achankovil River (S. India)				
Wet season (4th-order)	7.0	35.8	—	Arunachalam &
Dry season (4th-order)	72.3	61.5	—	Arun, 1994
Pachayar River (S. India)				
Wet season (4th-order)	24.2	23.1	—	Arunachalam &
Dry season (4th-order)	19.0	75.7	—	Arun, 1994

Standing stocks of benthic organic matter (particles >220 μm, excluding woody debris) in Tai Po Kau Forest Stream always exceeded 26 g C.m^{-2} (recalculated from Dudgeon, 1982c), and were over 100 times greater than periphyton biomass (Dudgeon, 1982c). Similarly, coarse benthic organic matter (particles >1 mm) ranged from 9.0 — 45.0 g C.m^{-2} from 1st- to 3rd-order sites along an Indian river (Nair *et al.*, 1989).

Spatial and temporal variations in benthic organic matter in Tai Po Kau Forest Stream were slight although a temporary increase in standing stocks at sites close to the stream banks followed abscision by deciduous riparian trees. Direct leaf fall into Tai Po Kau Forest Stream exceeded 600 g C.m^{-2}.yr^{-1} where the canopy was closed, in addition to lateral transport of more than 280 g C.m stream bank^{-1}.yr^{-1} (recalculated from Dudgeon, 1982c). Standing stocks of benthic organic matter declined in the lower Sungai Gombak reflecting the breakdown of particles of forest origin and the partially decomposed state of locally derived organics (Bishop, 1973a). Buried sedimentary organics (but not surface benthic organic matter) increased slightly in downstream reaches of Lam Tsuen River, and seasonal variations were related to the time elapsed since the last major spate (Dudgeon, 1984a); spate-related declines in fine (<220 µm) sedimentary organics occurred also in Tai Po Kau Forest Stream (Dudgeon, 1982b). Similarly, the organic content of sediments in a Javanese mountain stream peaked at the end of the dry season and declined during the period of highest rainfall (Leichtfried & Kristyanto, 1995).

Decomposition and heterotrophy

The decomposition of allochthonous material in tropical Asian rivers and streams has received little attention from researchers, but there is no reason to expect a qualitative difference between decomposition processes in this region and elsewhere. Hyphomycete fungi occur widely on submerged, decaying wood and leaf litter in Asia (Nawawi, 1975; 1976; 1985; Iqbal *et al.*, 1979; 1980; Chang, 1989; Nawawi & Kuthubutheen, 1988; 1989a; 1989b; Sridar & Kaveriappa, 1984; 1989; Khulbe, 1991; Rajashekhar & Kaveriappa, 1993; Firdaus-e-Bareen & Iqbal, 1994a; 1994b); densities and species richness on such substrates increase with duration of submergence (Iqbal *et al.*, 1979). In southwest India, the occurrence of different water-borne fungal species correlates positively with rainfall and leaf deposition, and not with temperature (Sridhar & Kaveriappa, 1984; 1989; Chandrasekar *et al.*, 1990), but different factors may be influential in other Indian streams (e.g. Mer & Sati, 1989; Khulbe, 1991; Rajashekhar & Kaveriappa, 1993) or elsewhere in the region (Firdaus-e-Bareen & Iqbal, 1994a). For example, Khulbe (1991) reports that seasonality varies among species of aquatic fungi depending upon stream temperature because development is inhibited at both low (<15°C) and high (>23°C) temperatures; thus in high-altitude streams fungal diversity will be greatest in summer but, at low altitudes or latitudes, diversity increases during the cooler months.

Bacteria are present on allochthonous leaf litter, and may occur at high densities (Dudgeon, 1982e), but their role in decomposition has not been investigated and their ecology in tropical Asian streams and rivers has received little attention except in relation to pollution (e.g. Ho, 1975; Patralekh, 1991a; Shukla *et al.*, 1992; Ullah *et al.*, 1995). The relative importance of microbes and invertebrates in litter breakdown and mineralization in tropical Asian streams — or indeed in any tropical stream — is unknown, although one view is that the importance of microbial *versus* invertebrate processing may change along a latitudinal gradient with invertebrates being more important at high latitudes (Irons *et al.*, 1994). The limited data from tropical Asia suggest that decomposition rates are rapid — with complete disappearance of leaves within three months — which may reflect high water temperatures (*cf.* Irons *et al.*, 1994) and active feeding by shredder invertebrates on leaf-litter species with rapid rates of breakdown (Dudgeon, 1994a, 1995b). Circumstantial evidence of the importance of invertebrates is apparent from the observation that the snail *Brotia hainanensis* (which is abundant in Hong Kong hillstreams) has a high feeding rate on leaf-litter species which break down quickly *in situ* (Dudgeon, 1982e). The data are not conclusive (and there is a notable lack of information on processing of leaf species with high loads of defensive secondary compounds), but indicate that litter decomposition in tropical Asian streams is similar to this process elsewhere, with breakdown (weight loss) described adequately by a negative-exponential model (e.g. Verghese & Furtado, 1987). In temperate European streams, decay rate coefficients depend upon the quality of the organic matter and environmental parameters such as temperature, and attains highest values when microbial and macroinvertebrate processing is combined (Rosset *et al.*, 1982). It seems reasonable to expect that similar factors will influence breakdown of allochthonous litter in tropical Asian streams.

Data from Tai Po Kau Forest Stream underscore the importance of allochthonous detritus in community metabolism. Together, the large standing stocks of benthic organic matter, rapid leaf-litter processing, and the low biomass of periphyton in this shaded stream (see pp. 51–53) gave rise to a community with a heterotrophic metabolism. Measurements of community respiration (R) and primary production (P) in a shaded riffle reach yielded P/R values of only 0.17 (Dudgeon, 1983a). Even in an unshaded site with abundant periphyton, P/R was only 1.02; apparently community respiration almost balanced production by photosynthesis. Unfortunately, comparable data from elsewhere in tropical Asia are lacking, but the importance of allochthonous materials in the diet of aquatic consumers suggests that a heterotrophic community metabolism will be general feature of shaded streams and small rivers which drain forested catchments.

PRIMARY PRODUCTION

Phytoplankton

Most estimates of primary production in tropical Asian rivers concern phytoplankton, which is confined to large rivers or the floodplain section of smaller systems. Primary production ranged from 1.03 to 1.78 g C m^{-2}.day^{-1}, with mean annual production of 518 g C.m^{-2}, at three sites along the River Mahanadi in India (Patra *et al.*, 1984). Rama Rao *et al.* (1979) report primary production values within the same order of magnitude (0.43-0.47 g C.m^{-2}.6h^{-1}) for three sites along the River Khan (India), while rates in the middle and lower Ganges were similar: 0.91 and 1.02 g C.m^{-2}.day^{-1} respectively (Natarajan, 1989). Ambasht & Srivastava (1994) measured phytoplankton production in the Rihand River (a Ganges tributary) but in this, and each of the studies quoted above, industrial wastes, sewage discharges or turbidity may have contributed to depressed production rates. In the unpolluted Testa River (India), daily production ranged from 0.44 to 2.18 g C.m^{-3} (Venu *et al.*, 1985), but turbidity may have been a limiting factor.

The restricted data support Welcomme's (1979) conclusion that phytoplankton productivity in large tropical rivers is low, and often limited by turbidity (Venu *et al.*, 1985; Shen *et al.*, 1987; Degens *et al.*, 1991; Chopra *et al.*, 1990). Even in a lateral lake on the Chang Jiang floodplain, phytoplankton production (1.15 to 3.61 g C.m^{-2}.day^{-1}; Liang *et al.*, 1988) was scarcely higher than values recorded for main river channels. However, primary production may not be low in all situations: compared to values for total plankton density in the Ganges (1.25 x 10^6 individuals.m^{-3} in 1980; Natarajan, 1989), the abundance of phytoplankton in the Zhujiang (annual mean density 434.7 x 10^6 individuals.m^{-3} and mean biomass 0.8 g.m^{-3}) suggest the potential for relatively high productivity (Liao *et al.*, 1989). Unfortunately, data on production rates in the Zhujiang are lacking.

Seasonal studies of riverine phytoplankton highlight a general trend of greatest abundance and diversity in the dry season when discharge volumes are least; densities decline during the wet season due to the combined effects of dilution, turbidity (causing aphotic conditions) and high current velocity which restricts population densities (Roy, 1955; Chakrabarty *et al.*, 1959; Pahwa & Mehrotra, 1966; Ray *et al.*, 1966; Lakshminarayana, 1965b; Venekateswarlu, 1969b; Mishra & Yadav, 1978; Bhatt *et al.*, 1985; Sharma, 1984a, 1985; Venu *et al.*, 1985; Nautiyal, 1984, 1985; Saffiulah *et al.*, 1987; Chopra *et al.*,

1990; Patralekh, 1991b; Pandey *et al.*, 1992; Dobriyal *et al.*, 1993; Ambasht & Srivastava, 1994). As a result, seasonal changes in phytoplankton densities generally show a unimodal periodicity, although sometimes a bimodal pattern is apparent (Lakshminarayana, 1965b; Pahwa & Mehrotra, 1966; Shrestha, 1991). In addition to such seasonal fluctuations, Natarajan (1989) has recorded a long-term, 400% increase in total plankton abundance between 1960 and 1980 in the middle Ganges which may be a result of organic enrichment of the river increasing the nutrient supply to photosynthesizing algae. These and other papers (e.g. Blache, 1951; Chacko & Ganapati, 1952; Dang, 1970; Dussart, 1974; Liao *et al.*, 1989; Natarajan, 1989, and references therein) present data on the composition of riverine phytoplankton in tropical Asia.

Periphyton

The seasonality of periphyton (attached algae) in tropical Asian rivers is similar to that of the phytoplankton. Periphyton (especially diatom) abundance and diversity in Indian rivers and hillstreams peaks in winter when current velocities are moderate and turbidity is low; reduced standing stocks during the summer monsoon can be attributed to scouring of the substratum by spates (Tiwari, 1990; Nautiyal, 1984; 1986; Ghosh & Gaur, 1991a; Rout & Gaur, 1994). Wet-season declines in periphyton have been recorded in Malaysia and Hong Kong (Bishop, 1973a; Dudgeon, 1982c), with decreases in standing stock from 323.4 to 9.6 mg.m^{-2} in a Tai Po Kau Forest Stream during one month of the wet season. These and other works (e.g. Prowse, 1962; Dussart, 1974; Ratnasabapathy, 1975; Ao *et al.*, 1984; Khan *et al.*, 1987; Phang & Leong, 1987; Wah *et al.*, 1987; Khan, 1990a; 1991; Sharma, 1991; Ormerod *et al.*, 1994 and references therein) present lists of algal taxa and give an indication of periphyton community structure in tropical Asian streams. Whitton *et al.* (1988b) and Das *et al.* (1994) describe the composition of the periphyton associated with aquatic macrophytes on river floodplains.

Diatoms often dominate the periphyton of Indian streams (Nautiyal, 1984; 1986; Rout & Gaur, 1994), which Venkateswarlu *et al.* (1987) consider is related to high silicate concentrations; the same explanation has been proposed to account for the abundance of planktonic diatoms in the Ganges (Lakshminarayana, 1965b). The composition of diatom assemblages in Nepalese streams is strongly influenced by altitude

(Ormerod *et al.*, 1994). Diatom genera characteristic of lower altitudes are mostly motile, epipelic or episammic forms (e.g. *Navicula* and *Nitzchia*) while those characteristic of higher and steeper streams include attached (*Fragilaria*) and prostrate (*Achnanthes*) tolerant of turbulent flow. Longitudinal replacement of algal taxa has been reported along Sungai Gombak, where diatoms, blue-greens, and Rhodophyta predominated in upstream reaches, while chlorophytes and desmids (along with some diatoms) dominated downstream sections (Bishop, 1973a); blue-greens and especially Rhodophyta were virtually restricted to the lower-order, forested reaches. A similar transition — from diatoms and blue-greens to chlorophytes — was seen from shaded, forested reaches to downstream, unshaded sections of an Indian stream (Rout & Gaur, 1994).

There is a paucity of field measurements of periphyton primary production in tropical Asian streams. Data are lacking for large rivers, but periphyton production in streams is likely to be higher (on an area basis) than phytoplankton production in turbid rivers. Values of 81.2 mg $C.m^{-2}.day^{-1}$ and 33.2 mg $C.m^{-2}.day^{-1}$ for gross primary production in an unshaded pool and a shaded riffle (respectively) have been recorded in Tai Po Kau Forest Stream (recalculated from Dudgeon, 1983a). The influence of shade on primary production in this stream is clear, and demonstrated further by significant differences in periphyton standing stock among Hong Kong streams which vary with respect to the development of riparian vegetation (Dudgeon, 1988a). Standing stocks were lowest in shaded streams (< 1 g $C.m^{-2}$), and highest (2.7 g $C.m^{-2}$, dominated by filamentous Chlorophyta) in a stream draining low scrub and grassland. Bishop (1973a) reported that shading by the forest canopy and deficient nutrients restricted periphyton productivity in the upper Sungai Gombak (see also Rout & Gaur, 1994); nutrients (particularly phosphorus) are reported to limit periphyton (under unshaded conditions) in Indian streams also (Ghosh & Gaur, 1990; 1991b; 1994; Rout & Gaur, 1990; 1994). Further downstream, light attenuation and abrasion due to silt from tin-mine effluents restricted production in the lower course of Sungai Gombak (Bishop, 1973a), and similar effects have been reported in other Malaysian streams (Ho, 1976). Net periphyton production was 5, 16, 113, 53, and 16 mg $C.m^{-2}.day^{-1}$ at five sites (stream order 2 — 6) along the river; mean biomass at the same sites was 1.6, 1.3, 3.9, 1.3, and 3.5 g $C.m^{-2}$ (Bishop, 1973a). These figures are within the same order of magnitude as values for periphyton biomass at first- to fifth-order sites along three southern Indian rivers (Nair *et al.*, 1989;

Arunachalam & Arun, 1994), and data from third-order Tai Po Kau Forest Stream (Dudgeon, 1982b) and a fourth order, polluted stream in Malaysia (Ho, 1976). However, as noted above, periphyton standing stocks can vary greatly according to discharge conditions and nutrient availability.

Macrophytes

Tropical Asian rivers and associated wetlands support a considerable diversity of aquatic macrophytes (e.g. Pancho & Soerjani, 1978; Leach & Osborne, 1985; Rothe *et al.*, 1986; Guan, 1987; Gopal, 1988; Pareek & Sharma, 1988; Le, 1994). For example, Bangladesh floodplains have a diverse community of aquatic vascular plants; one site yielded 52 taxa (including five pteridophytes) — as well as 65 algal taxa — in spite of the fact that the land was submerged for only part of the year (Whitton *et al.*, 1988b). Over 130 species of vascular aquatic plants have been identified from the Zhujiang (Liao *et al.*, 1989), although some of the species listed are halophytes and are probably restricted to estuarine sites. Lateral lakes and floodplain sites can be highly species-rich because of the formation of floating islands on which terrestrial plants grow (Soerjani, 1980). The productivity of floodplain macrophytes is demonstrated clearly by the cultivation of deep-water rice (*Oriza sativa indica*[1]), a major crop of India and Southeast Asia. It is grown on river floodplains during the wet season in depths of 0.5 — 3.5 m (Whitton & Rother, 1988: Whitton *et al.*, 1988c). The rice is adapted to deep flooding by elongating as the water level rises; in addition, it can survive short periods of submergence. In Bangladesh alone there are 93,000 km² of floodable land including over 28,000 km² inundated for three to four months each year (Welcomme, 1979). The area covered by deep-water rice is by far the largest for any aquatic macrophyte in the world, with a total of some 11 x 10⁶ ha in Asia. Despite this, little is known of its ecology or its influence on the floodplain habitat (Whitton & Catling, 1986; Whitton & Rother 1988).

There are few studies of macrophytes in lotic sites, but an exception is Hasan's (1988) study of the River Champanala, a side channel of

---

[1] There is no fully accepted classification system for the varieties and subspecies of *Oriza sativa*.

the Ganges. In contrast to Indian lentic systems, where macrophyte productivity peaks in the wet season, most macrophytes were scoured from river during the monsoonal rains so that maximum productivity (>4 g.m^{-2}.day^{-1}) occurred during the winter and spring dry season. The combined production of River Champanala macrophytes (*Eichhornia crassipes, Hydrilla verticillata, Ceratophyllum demersum, Potamogeton crispus* and *Azolla pinnata*) averaged over the year was 3.9 g.m^{-2}.day^{-1}, which is higher than the world mean for cultivated rice fields (Hasan, 1988). By comparison, Liang *et al.* (1988) have recorded lower annual production (103.5 g.m^{-2}; approximately 0.3 g.m^{-2}.day^{-1}) by aquatic macrophytes in a lateral floodplain lake of the Chang Jiang. This community included *Hydrilla verticillata, Ceratophyllum demersum, Potamogeton crispus* which were present also in the River Champanala. Annual net production of *P. crispus* in the Ganges (at Varanasi) is influenced by nutrient levels, ranging from 50 g.m^{-2} in unpolluted reaches to 69 g.m^{-2} at sites enriched with nutrients due to pollution by sewage and cremation ash (Shrivastav *et al.*, 1993).

There are marked spatial and temporal (especially seasonal) fluctuations in autochthonous production in tropical Asian rivers. Although more data are needed, the information at hand (see also Dudgeon & Bretschko, 1995a; 1995b) suggests that autochthonous primary production makes a relatively small contribution to supporting in-stream secondary production — especially during the monsoon season. Allochthonous inputs probably support year-round secondary production in streams shaded by riparian vegetation, and allochthonous carbon sources may be the major energy source in all streams during the monsoon season when — especially in large rivers — autochthonous production falls and the floodplain is inundated. Likewise, in many north-temperate streams, autochthonous primary production is clearly less important as an energy source than allochthonous organic matter (e.g. Nelson & Scott, 1962; Kaushik & Hynes, 1968, 1971; Fisher & Likens, 1973; Cummins, 1974; Ladle, 1981; Iverson *et al.*, 1982), although qualitative aspects (other than energy) of allochthonous and autochthonous energy sources have yet to be addressed fully.

ZOOPLANKTON

Information on the production and seasonality of tropical Asian lotic zooplankton is meagre. A pronounced seasonal pattern in zooplankton abundance is obvious in some Indian rivers (Pahwa & Mehrotra, 1966; Ray *et al.*, 1966; Bhatt *et al.*, 1984; Srivastava *et al.*, 1990; Pandey *et al.*, 1992), and there may also be diurnal fluctuations in population densities of surface zooplankton (e.g. Godavari River: Ahmed & Alireza, 1991). Abundance peaks in the winter dry season and declines during the monsoon which can be correlated with the seasonal pattern in phytoplankton biomass (Pahwa & Mehrotra, 1966; Ray *et al.*, 1966), although Hameed *et al.* (1995) report greatest zooplankton densities during the monsoon in the polluted Kaveri River. In the Zhujiang, the biomass of Protozoa and Rotifera was greater during floods than in the dry season, while the inverse was true for Cladocera and Copepoda; total zooplankton biomass (mean 0.43 g.m^{-3}) was greatest during the dry season (Liao *et al.*, 1989). By contrast, zooplankton at some sites along the Brahmaputra and Ganges Rivers show little seasonality (Saffiulah *et al.*, 1987; but see Pahwa & Mehrotra, 1966; Ray *et al.*, 1966; Sharma & Naik, 1995). Flowing water is a poor habitat for zooplankton (if only because of low phytoplankton productivity), and zooplankton are almost entirely absent from hillstreams and small rivers (e.g. Dobiyal *et al.*, 1993). Populations seem to be sparse in the main channels of large rivers: Sidthimunka (1970; quoted by Welcomme, 1979) recorded 131,000 individuals.m^{-3} in an isolated side arm of the Mekong, but only about 5,000 m^{-3} were present in the main channel. These and other publications (e.g. Chacko & Ganapati, 1952; Dang, 1970; Dussart, 1974; Idris, 1983; Qi & Zhao, 1984; Liang *et al.*, 1988; Sharma & Naik, 1995) list the taxa comprising the zooplankton of tropical Asian rivers and their lateral lakes, while Fernando (1980a; 1980b) describes the biogeographical character and composition of the freshwater zooplankton of the Oriental Region.

ZOOBENTHOS

Composition in large rivers

An Asian scientist — Sunder Lal Hora, one of the earliest contributors to the study of lotic zoobenthos — highlighted the importance of substratum and current as determinants of faunal distribution and morphology in Indian torrential streams (Hora, 1923, 1927, 1936). Likewise, an investigation undertaken in Malaysia was among the first to draw attention to the abundance of lotic animals deep in the bottom sediments, and demonstrate the importance of the hyporheic habitat in some streams (Bishop, 1973b). Despite an early start, however, there has been little research on the zoobenthos of large tropical Asian rivers. In the lower reaches of the Zhujiang in Guangdong Province, the annual mean zoobenthos density was 423 individuals.m^{-2}; further upstream in Guangxi Province (the 'middle reaches') densities increased to >1,000 individuals.m^{-2} (Liao *et al.*, 1989) but detailed information on benthos composition is unavailable. Bandhavong (1990) recorded higher zoobenthos densities (2,620-14,934 individuals.m^{-2}) in the Nam Ngum River (a Laotian tributary of the Mekong) where oligochaetes (followed by gastropods) were the most abundant of 16 taxa. The zoobenthos of the lower Chang Jiang (near Nanjing) was likewise dominated by oligochaetes, although polychaetes, chironomids, gastropods and bivalves were well represented (Wu & Chen, 1986; Liang, 1987); nevertheless, total mean density of the 42 benthic species was only 328 individuals.m^{-2} — considerably less than in the Nam Ngum River. Within this reach of the Chang Jiang, the distribution of benthic molluscs (10 species) was not random. Densities were highest in tributaries emptying into the main channel (67 individuals.m^{-2}), where the bed was muddy and the water rather shallow, and numbers were lower in the marginal zone of the main channel (2.m^{-2}) (Chen & Wu, 1983). Molluscs did not occur in the main channel itself due to the rapid current and shifting, sandy substrate.

Similar patterns in the substrate-relationships of zoobenthos have been recorded in floodplain reaches of the Ganges: benthos abundance fluctuated from 52 — 15,051 individuals.m^{-2} (mean 1,954) at Buxar and were only 29 — 1,329 individuals.m^{-2} (mean 280) at Bihar (Singh & Srivastava, 1989) where the sandy, unstable substrates were unsuitable habitat for most benthic animals. Comparable results were reported by Lim (1987) who found lower species richness on sand substrates compared to rocks in two Malaysian rivers. Pomatiopsid and stenothyrid snails in the Mekong River are substrate limited also, and are restricted to solid objects such as sticks and stones (Davis *et al.*, 1976; Hoagland & Davis, 1979; Kitikoon *et al.*, 1981), although certain species have a preference for stones covered with deposits of fine sediments which would allow for an increased population of attached microbes (Attwood, 1995a). Others are obligate rheobionts that live on rocks in fast current (Brandt, 1974; Davis, 1979).

The general composition of the benthos of large Asian rivers appears similar to that of such habitats the world over, and includes Tubificidae,

Chironomidae (Chironominae), Gastropoda (Prosobranchia), and Bivalvia (Hynes, 1970). However, a distinctive feature of rivers such as the Ganges, Zhujiang and Chang Jiang is the regular occurrence of freshwater polychaetes (Ray *et al.*, 1966; Pahwa, 1979; Qi *et al.*, 1982; Shen & Qi, 1982; Wu & Chen, 1986; Singh *et al.*, 1988; Su & Li, 1988; Ahmad & Singh, 1989; Datta Munshi *et al.*, 1989). To this list can be added freshwater prawns (especially *Macrobrachium* spp.) which, although widespread in the tropics and subtropics, are not collected efficiently by the grabs and devices used to sample large rivers and are therefore under-represented in zoobenthos samples.

Composition and distinctive elements in small rivers

A variety of studies (e.g. Costa, 1974; Boonsom, 1976; Ho & Hsu, 1977; Badola & Singh, 1981; Lin, 1982; Gupta & Michael, 1983; Ao *et al.*, 1984; Chowdhary, 1984; Dobriyal, 1985; Sharma, 1986; Kaul *et al.*, 1987; Kumar, 1987; Julka *et al.*, 1988; Ahmad & Singh, 1989; Mohan *et al.*, 1989; Singh & Nautiyal, 1990; Mohan & Bisht, 1991a; Sehgal, 1983; 1991; Sharma, 1991; Singh *et al.*, 1991; Gupta & Michael, 1992; Kumar & Dobriyal, 1992; Dobriyal *et al.*, 1993; Rundle *et al.*, 1993; Ormerod *et al.*, 1994; Suren, 1994; Tong *et al.*, 1995; Yule, 1995a) give an indication of benthic community composition in tropical Asian rivers and streams. They provide support for Hynes' (1970) assertion that '. . .one of the most striking features of the faunas of stony streams is their remarkable similarity the world over'. To these comments it is possible to add some modifications. The diversity of plecopteran families is reduced in tropical Asian streams where the order is represented largely by Perlidae (especially Neoperlinae) and, to a much lesser extent, Nemouridae, Leuctridae and Peltoperlidae (see p. 318). Plecopteran species richness appears to be low in some streams (Dudgeon, 1992a) — which reflects the restriction of many stonefly taxa to cool-water habitats (Hynes, 1976) — and the order is completely absent from parts of New Guinea (Dudgeon, 1990b, 1994b). However, on-going taxonomic studies (see pp. 323–328) suggest that tropical stoneflies are more diverse than supposed previously. Naucoridae (Hemiptera: Heteroptera) are a conspicuous element of the zoobenthos in some streams where Plecoptera are scarce or lacking (Dudgeon, 1990a, 1994b). Lepidoptera (Pyralidae) appear to make up a greater proportion of the biomass in some areas of tropical Asia (e.g. Sri Lanka, Sulawesi, Papua New Guinea; Reichholf, 1973; Dudgeon,

1990a; 1994b and unpublished observations) than is typical in north-temperate streams. Nevertheless, when Hynes' (1970) statement is applied to morphological-behavioural similarities of stream denizens, the concordance between world-wide stream faunas remains marked.

The lotic mollusc fauna of tropical Asia has some distinctive elements: for instance, the importance of thiarid (and sometimes neritid) gastropods and corbiculid bivalves (e.g. Pace, 1971; Brandt, 1974; Starmühlner, 1974). The Mekong, in particular, contains the richest endemic gastropod fauna known from present-day rivers (Brandt & Temcharoen, 1971; Brandt, 1974; Davis *et al.*, 1976; Davis, 1979; see also pp. 138–139). Two groups dominate this prosobranch assemblage: the Pomatiopsidae (Triculinae, tribe Jullieniini; at least 92 species) and Stenothyridae (at least 19 species). Together these two taxa comprise 92% of the Mekong gastropod fauna. In the Chang Jiang also, pomatiopsids have undergone extensive speciation and adaptive radiation involving the tribes Triculini and Pachydrobiini of the Triculinae (Davis *et al.*, 1986; 1992 and references therein). The genus *Tricula* is particulary speciose in the upper Chang Jiang, while *Neotricula* has its greatest deployment in the middle and lower course. Eleven genera and at least 20 species are known from the river, but new genera of Triculinae are likely to be uncovered as research proceeds (Davis *et al.*, 1992; 1994).

Among meso- and macrocrustaceans, amphipods and isopods are scarce in tropical Asian streams when compared to their abundance in some temperate habitats, although amphipods are important in some highland (>>1,000 m) streams in northern India (e.g. Kumar, 1987; Sehgal, 1983; 1991; Melkania, 1991). Instead, freshwater crabs, shrimps (Atyidae) and prawns (Palaemonidae) occupy a pre-eminent position, and have penetrated fast-flowing upland streams in addition to the lower reaches of rivers. In New Guinea, decapods fill many of the detritivorous roles that are occupied by river fishes elsewhere (Haines, 1983). Lotic palaemonids, atyids and freshwater crabs tend towards suppression of the plankton larval stages which, in extreme cases, is manifested in direct development from egg to miniature adult. The most derived species produce large eggs which hatch into well-developed benthic juveniles, and are generally found far up the course of rivers in hillstreams or swamps. Species in lowland reaches produce many small eggs and some species (in the genera *Macrobrachium* and *Caridina*) migrate to estuaries to spawn. The planktonic larvae may require saline conditions for metamorphosis. Direct development in decapods may be an adaptation to reduce washout of planktonic larvae during spates, and it is significant that many of these animals breed during the summer monsoon (e.g. Dudgeon, 1992a). The matter is discussed in more detail on pp. 174–175 and p. 180 where further examples are given.

Longitudinal zonation and microdistribution

Tropical Asian rivers, like such environments all over the world (Hynes, 1970; Hawkes, 1975; Cushing *et al.*, 1995), exhibit longitudinal changes in zoobenthos communities from headwaters to estuary. Surprisingly, systematic studies of this phenomenon are scarce — especially in unpolluted rivers. Starmühlner (1984a; 1984b) found that total zoobenthos abundance in Sri Lankan rivers declined at low altitudes (< 500 m), and densities fell in slow-moving water. There was a clear succession of mollusc species with altitude, and more taxa occurred at low altitudes where water hardness was greatest. Even within a single snail genus (*Paludomus*: Pleuroceridae), longitudinal replacement of species was apparent (Starmühlner, 1977). Gastropods increased in abundance (relative to insects) in the lower course of the River Ayung, Bali, in parallel with a rise in conductivity (Suryadiputra & Suyasa, 1995). By contrast, most Sri Lankan mayflies lack obvious altitudinal zonation patterns (Hubbard & Peters, 1984), although species richness falls below 300 m. Realon (1980) recorded similar declines in the lower course of a Philippines stream. Mayfly diversity may decrease at high altitudes also: Sivaramakrishnan & Venkataraman (1990) noted that species richness of Indian mayflies declined from 13 species at 400 m elevation to only two species at 2,360 m. However, the relative contribution of each family to the mayfly fauna did not change with altitude, which is similar to the pattern seen in Sri Lanka (Hubbard & Peters, 1984). Furtado (1969) and St Quentin (1973) divided the Odonata of southeast Malaysia and Sri Lanka (respectively) into more-or-less discrete groups based on their association with particular stream biotopes. The number of species varied between biotopes and was lowest at high elevations (1,800 — 2,000 m). Altitudinal zonation and habitat partitioning by elevation has been noted also for neustic *Rhagovelia* (Heteroptera) species in Borneo and Sumatra (J.T. Polhemus & D.A. Polhemus, 1988).

In Nepalese streams, taxonomic or species richness of the zoobenthos (mainly insects) declines with altitude (Rundle *et al.*, 1993; Ormerod *et al.*, 1994; Suren, 1994) but the relative abundance of the major insect orders (Ephemeroptera, Trichoptera, Plecoptera, Diptera and Coleoptera) remained rather constant. Sehgal (1983) likewise reported a lack of clear zonation of zoobenthos in Himalayan streams (altitude 1,000 — 3,000 m), although there were longitudinal variations in the densities of some taxa. By contrast, family composition within zoobenthic orders in Nepalese streams did change with altitude (Suren, 1994). This does not match the pattern reported for mayflies in Sri Lanka or India (Hubbard & Peters, 1984; Sivaramakrishnan & Venkataraman, 1990) but these studies considered only streams at altitudes <2,500 m, whereas the Nepalese studies included streams >3,500 m. In Nepal, at least, Limnephilidae, Rhyacophilidae (Trichoptera) and Taeniopterygidae (Plecoptera) were characteristic of high-altitude streams, whereas Caenidae, Ephemeridae (Ephemeroptera), Hydropsychidae, Lepidostomatidae (Trichoptera), and Elmidae (Coleoptera) typified low-altitude streams (Rundle *et al.*, 1993; Ormerod *et al.*, 1994; Suren, 1994). Baetid mayflies were numerically dominant in alpine streams (>4,000 m) or in high-altitude streams draining scrub and tundra; Simuliidae and Chironomidae (Diptera) also persisted at high altitudes. Nevertheless, most taxa were either confined to lower altitudes or more numerous there (e.g. heptageniid mayflies and hydropsychid caddisflies). A similar trend is seen in the upper (Garhwal Himalaya) sections of the Ganges although the zoobenthos communities up to altitudes of 1,100 m were influenced by organic pollution (Singh & Nautiyal, 1990). In Thailand, longitudinal zonation of caddisflies was manifest in downstream replacement of species, and most families were well-represented at all altitudes between 400 and 2,300 m (Malicky & Chantaramongkol, 1993a). Species richness was highest between 1,200 and 1,700 m, where water temperatures showed the lowest variation. There was no evidence of any 'warm-adapted' families restricted to low altitudes nor 'cool-adapted' families confined to higher elevations (Malicky & Chantaramongkol, 1993a).

In contrast to benthic insects, the longitudinal or altitudinal zonation of decapods — such as *Macrobrachium* spp. (e.g. Costa, 1984) — reflects larval development patterns: only downstream species pass through a prolonged planktonic larval stage. The longitudinal distribution of atyid shrimps is influenced similarly by the need to undergo breeding migrations. However, those species with directly-developing larvae are also affected by water temperatures, and differences in the upper incipient lethal temperature determine the species-specific lower (downstream) limit of distribution among Sri Lankan atyids (de Silva & de Silva, 1988; de Silva, 1989). Unfortunately, there is no information on the temperature tolerances of other Asian stream invertebrates.

The data suggest that, where a sufficiently wide altitudinal range is considered, there are altitudinal changes in the composition of zoobenthos assemblages, and a decline in species richness with increasing elevation. A similar trend has been reported for diatoms, bryophytes and fishes in Nepalese streams (Ormerod *et al.*, 1994). Such transitions arise because of ecological changes that occur simultaneously along a river course with respect to water chemistry, temperature, slope, habitat and substratum diversity, and catchment vegetation. These (and other) factors act upon the zoobenthos synergistically, causing changes in community composition and reductions in diversity with altitude. An important corollary of this pattern is that diversity is concentrated at lower altitudes, where human impacts on the landscape are greatest: this pattern may have important implications for river conservation and management (Ormerod *et al.*, 1994) in view of the fact that local land-use patterns can affect the composition of stream zoobenthos (in Nepalese streams, for example: Rundle *et al.*, 1993; Ormerod *et al.*, 1994; Suren, 1994). Elsewhere in the region, clearance of riparian vegetation resulted in a pronounced loss of species with increasing distance from the forest margin in Sungai Gombak (Bishop, 1973a), while catchment logging reduces the diversity of mayflies in Indian streams (Gupta & Michael, 1992). Dudgeon (1988a) has demonstrated that differences in riparian vegetation (reflecting historical changes in land use) strongly influence zoobenthos community composition in Hong Kong streams.

Our understanding of longitudinal zonation and the zoobenthos communities of large Asian rivers is hampered by the pervasive influence of pollution which confound attempts to distinguish 'natural' communities from those under anthropogenic influences. For instance, there has been faunal change in lower (freshwater) reaches of the Zhujiang during this century, with bivalves such as infaunal *Corbicula* sp. and bysally-attached *Limnoperna fortunei* (Miller & McClure, 1931) being replaced by oligochaetes (Qi, 1987). Oligochaete densities range from $<100 - 700,000$ individuals.m^{-2}, depending upon the pollution load, with *Limnodrilus hoffmeisteri* occurring at both clean and grossly-polluted sites (Qi & Erseus, 1985). Elsewhere, Dutta & Malhotra (1986) attribute intersite differences in zoobenthos in a Kashmiri river to benthic organic matter; highest densities of oligochaetes, molluscs, and chironomids occurred at a station situated near a cremation ground. Pollution effects also determine zoobenthos composition in the middle Ganges (< 200 m altitude), although benthic insect communities at altitudes over 1,000 m (where densities peak at

2000 individuals.m^{-2}) are affected also (Singh & Nautiyal, 1990). At Kanpur (in the mid-Ganges), insects maintain their dominance of the benthic community, and total abundance can reach 3,476 individuals.m^{-2} (insects: 3,405.m^{-2}). Numbers decrease further downstream at Bhaglapur (218 individuals.m^{-2}) where molluscs (mainly gastropods: 124.m^{-2}) predominate (Natarajan, 1989). Below Bhaglapur, further declines (down to only 3 individuals.m^{-2}) have been attributed to the effects of paper pulp and sewage sludge. In general, the abundance and diversity of the benthic community of the middle and lower Ganges has diminished in recent years (Natarajan, 1989). The effect of pollution in Asian rivers has been reviewed elsewhere (Alabaster, 1986; Trivedy, 1988; Dudgeon, 1992b; Gopal & Sah, 1993) and will be considered further on p. 558. In addition to pollution, other human activities — such as river regulation — will influence zoobenthos distribution and abundance (e.g. Dudgeon, 1992b), and river regulation also will have an influence as has been documented extensively in north-temperate regions (e.g. Ward & Stanford, 1979). Boon (1979) has considered this point in tropical Asia, recording that filter-feeding hydropsychid caddisflies (*Amphipsyche* and *Cheumatopsyche*) dominated riffles below a dam on the River Tuntang (Central Java). High caddisfly densities reflected the profusion of suspended food derived from the reservoir upstream, and parallels hydropsychid abundance at temperate-zone lake outlets (e.g. Cushing, 1963; Oswood, 1976).

Bishop (1973a) described marked changes in zoobenthos community composition at second- to sixth-order sites along the lowland Sungai Gombak, although the extent to which these were natural or induced by anthropogenic influences (particularly pollution) was unclear. Among the mayflies, longitudinal distribution of certain families was restricted (e.g. Potamanthidae) while others (e.g. Caenidae and Baetidae) occurred throughout the river. Within the latter group, some species were confined to particular biotopes, and a definite succession of Baetidae along Sungai Gombak was observed. Dudgeon (1990b) has documented the same phenomenon among Baetidae and Caenidae of Lam Tsuen River although, as in the case of Sungai Gombak, sections of the river were polluted. Bishop (1973a) viewed substratum (including both its physical support and accompanying food resources) rather than temperature as the primary factor determining longitudinal distribution of most taxa along Sungai Gombak. Current may influence distribution also, but is an indefinable factor because of its heterogeneity over small areas and the restriction of most benthic animals to the boundary layers of marginal flow. Nevertheless, it must be recognized that substrate characteristics are a product of current velocity and channel characteristics, so that current has an indirect influence on longitudinal zonation. Longitudinal changes in Lam Tsuen River fauna involved downstream declines in species richness from second- to sixth-order sites (Dudgeon, 1984a; 1987a; 1990b). However, the trend alters if invertebrates (93 morphospecies) associated with floating plants (mainly exotic *Eichhornia crassipes*) are included in the analysis. Under these circumstances, the lower course had a richer community than upstream sites, reflecting the greater diversity of microhabitats downstream.

Imposed upon longitudinal changes in zoobenthos communities, there is marked spatial heterogeneity in the distribution of lotic benthos within sites (Benzie, 1984), as is well known in north-temperate streams (e.g. Hynes, 1970). Some of this heterogeneity results from across-stream distribution patterns which are related to the incidence of sediment patches (Dudgeon, 1982b; 1982d). In Tai Po Kau Forest Stream, baetid and leptophlebiid mayflies show species-specific microdistribution patterns, which multiple-regression analysis has demonstrated can be explained largely by sediment grain-size statistics (Dudgeon, 1990b). The microdistribution of Tai Po Kau Forest Stream Odonata and psephenid beetle larvae (water pennies) could also be explained by sediment characteristics (Dudgeon, 1989b, 1989c; Dudgeon, 1995c), although the predictive ability of regression models was less powerful than for mayflies. Elsewhere, Costa (1974) concluded that '. . . substratum seems to be the regulating factor for the distribution of most species. . .' within a small Sri Lankan stream, and a recent, small-scale study in India (Arunachalam *et al.*, 1991) yielded similar findings (see also Negi & Singh, 1990; Yang *et al.*, 1990; Sehgal, 1991; Gupta & Michael, 1994). Indeed, as early as 1936, Hora (1936) had realised the significance of the substratum as a determinant of the distribution of zoobenthos in Asian streams.

Functional organization and the River Continuum Concept

While there have been some studies of the longitudinal changes in zoobenthos composition along tropical Asian rivers, very few investigators have addressed the functional attributes of zoobenthos communities. Their high diversity combined with the difficulties of identifying species of larvae, have encouraged stream ecologists to develop a functional classification of zoobenthos according to how they feed (Cummins, 1973). While the scheme was developed for insects it is applicable, in principle, to other aquatic animals such as crabs, shrimps, snails and fishes. Functional classification of animals has the advantage of reducing the difficulty of dealing with taxonomic groups which are poorly known — a definite benefit for researchers in tropical Asia. Simplification of community structural data is an additional advantage as it facilitates pattern recognition in ecosystems. The main functional feeding groups are:

1. Grazers and scrapers — herbivores feeding on periphyton (i.e. attached algae and associated material), including those which pierce plant tissues or cells and suck out fluids;
2. Shredders — detritivores feeding on coarse particles (especially decomposing terrestrial leaf litter), with a significant reliance on the associated microbes;
3. Collectors — feeding on fine (<1 mm) particles (and associated microbes) suspended in the water (= filtering-collectors or filter-feeders) or deposited on the substratum (= collector-gatherers);
4. Predators — including those which swallow or engulf prey, and those which pierce their animal victims extracting cell and tissue fluids.
5. Piercers and suckers (of plant cells) — specialized microherbivores which pierce plant (usually algal) cells with their mouthparts and ingest the contents.

Some animals will change feeding mode during their development, and so cannot be referred to a single functional group but a given age class tends to fall predominately into one functional group (Cummins, 1973). Nevertheless, a few species can perhaps best be classified as generalists. In running waters receiving organic enrichment (pollution) an additional functional category — deposit-feeders — can be added to include those species ingesting fine bottom sediments and the organic material that they contain. Alternatively, these animals may be considered as specialized collector-gatherers.

As an example of the application of functional approach, a general classification of the trophic relations of zoobenthic macroinvertebrates in Hong Kong streams is given in Table 3.6. The categories used are broad, and generalizations included therein are likely to suffer from some exceptions due to the relatively small number of species whose trophic ecology has been well studied. Additional complications are added when feeding habits change according to local circumstances or during growth: for example, *Brotia hainanensis* — a thiarid snail — can feed as a shredder or a grazer and is therefore assigned to a

Table 3.6 Functional-feeding group classification of stream invertebrates, based upon major taxa commonly encountered in Hong Kong streams.

ANNELIDA		LEPIDOPTERA	
All Oligochaeta	D	Pyralidae	Sc
All Hirudinea	P	TRICHOPTERA	
MOLLUSCA		Rhyacophilidae	P
All Bivalvia	F	Glossosomatidae	Sc
Gastropoda		Hydroptilidae	Sc
Thiaridae	ShSc	Philopotamidae	F
Viviparidae	CoSc	Psychomyiidae	Co
Lymnaeidae	G	Xiphocentronidae	Co
Physidae	G/Sc	Polycentropodidae	P
Planorbidae	G	Dipseudopsidae	Co?
Ancylidae	CoSc	Ecnomidae	P
CRUSTACEA		Hydropsychidae	F
Potamidae	Sh	Brachycentridae	CoSc
Parathelphusidae	P	Lepidostomatidae	Sh
Atyidae	Co	Odontoceridae	Sc
Palaemonidae	P	Helicopsychidae	Sc
EPHEMEROPTERA		Calamoceratidae	ShSc
Baetidae	CoSc	Leptoceridae	CoSc
Oligoneuriidae	F	COLEOPTERA	
Heptageniidae	CoSc	Gyrinidae	P
Ephemerellidae	CoSc	Haliplidae	Sc
Caenidae	CoSc	Dytiscidae	P
Leptophlebiidae	CoSc	Hydrophilidae	CoSc
Ephemeridae	Co	Hydraenidae	Co?
Prosopistomatidae	Sc	Scirtidae	Co
ODONATA		Psephenidae	Sc
All Zygoptera	P	Dryopidae	CoSc
All Anisoptera	P	Elmidae	CoSc
PLECOPTERA		Eulichadidae	CoSh
Nemouridae	CoSh	DIPTERA	
Leuctridae	Sh	Nymphomyiidae	CoSc
Perlidae	P	Tipulidae	Sh/P
HETEROPTERA		Culicidae	Co/P
All Gerromorpha	P	Psychodidae	Co
Nepomorpha		Ceratopogonidae	P
Nepidae	P	Chironomidae	
Belostomatidae	P	Tanypodinae	P
Naucoridae	P	Orthocladiinae	CoSc
Notonectidae	P	Chironominae	Co
Pleidae	P	Simuliidae	F
Helotrephidae	P	Dixidae	Co
Corixidae	Co/P	Stratiomyidae	Co?
MEGALOPTERA		Empididae	P
Corydalidae	P	Syrphidae	Co

KEY: Co = collectors; Sh = shredders; F = filter-feeders; P = predators; Sc = scrapers; ShSc, CoSh, CoSc = mixed feeding modes (shredder-scraper, collector-shredder, etc); Sh/P, Co/P = alternative feeding modes (shredder or predator, collector or predator, etc); G = generalists; D = deposit-feeders.

combined functional group. Other taxa, such as hydropsychid caddisflies, are primarily filter-feeders which sieve particles from the current with the aid of a silken capture net; nevertheless, they behave as predators when an animal is caught on the mesh.

The RCC makes predictions (albeit rather broad or approximate) about downstream changes in functional feeding group representation in response to changes in the food base (especially allochthonous inputs) and thus serves as an appropriate template for comparison of Asian rivers and streams with those elsewhere. Put simply (Fig. 3.8), the RCC groups stream communities according to channel size into headwaters, medium-sized streams and large rivers. Many headwater streams are influenced strongly by riparian vegetation, which reduces aquatic primary production by shading and contributes large amounts of leaf litter. Shredders are predicted to be co-dominant with collectors in such streams, reflecting the importance of the riparian zone and the detritus derived from it. As stream width increases and shading decreases, the reduced importance of litter inputs coincides with a greater significance of aquatic primary production and the import of fine organic particles from upstream. In medium-sized streams grazer-scraper biomass is maximized, but collectors are numerous also. The transition from headwaters, dependent on terrestrial inputs, to medium-sized rivers, relying on algal or aquatic macrophyte production, is associated with a change in the ratio of gross primary productivity (P) to community respiration (R). The position at which the stream shifts from heterotrophic ($P/R < 1$) to autotrophic ($P/R > 1$) is dependent upon the degree of shading and thus the form and extent of the riparian vegetation.

Large rivers receive fine particulate organic matter from upstream, and there is a general reduction in detrital particle size as stream width increases. As a result, collector-gatherers and filter-feeders dominate macroinvertebrate assemblages downstream. Although the shading effect of riparian vegetation is insignificant, primary production may be limited by water depth and turbidity; here the river once again becomes heterotrophic ($P/R < 1$). Further details of the RCC and predictions arising from it are given by Vannote *et al.* (1980) and Minshall *et al.* (1983; 1985).

Rather general agreement with the predictions of the RCC have been recorded in Lam Tsuen River (Dudgeon 1984b), but changes in the relative abundance of functional groups did not match expectations closely and there were departures caused by organic pollution of the lower course which were particularly notable during the dry season.

The River Continuum Concept

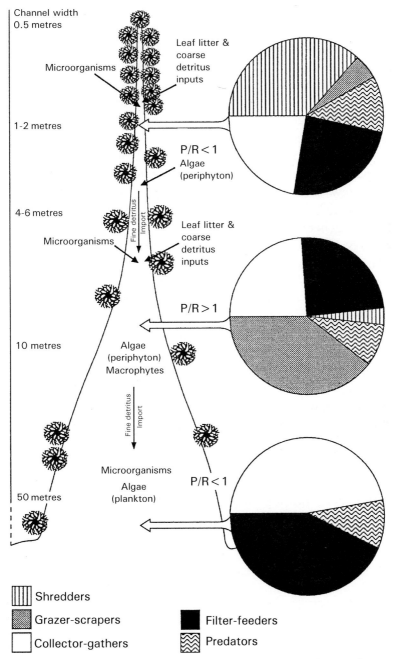

Fig. 3.8 A simple representation of the River Continuum Concept, showing the expected relative abundance of invertebrate functional groups along an unpolluted river. (Further explanation of the Concept and its predictions are given in the text.)

More significant was the paucity of shredders (*sensu* Cummins, 1973) — which comminute coarse detritus — in upper Lam Tsuen River when compared to a prediction of the RCC that shredders and collectors should co-dominate in upstream reaches. In this and other Hong Kong streams shredders comprise between 0.1 and 8.8% of the benthic community (Dudgeon, 1989a) and make up only about 1% of the benthos in partly-shaded streams (Table 3.7). Even in primary rainforest streams in New Guinea, shredders do not exceed 2% of benthic populations (Dudgeon, 1994b). Likewise, shredders were no more abundant in forested than unshaded streams in Nepal (Ormerod *et al.*, 1994), and data from a Taiwanese stream (Yang *et al.*, 1990) reveals dominance of the benthic community by collectors and predators with relatively few shredders. It seems that the under-representation of shredders which is typical of Hong Kong streams (Dudgeon, 1989a; 1992a) is a general feature of tropical Asian rivers (Dudgeon, 1995a; Dudgeon & Bretschko, 1995a; 1995b). A possible explanation for this phenomenon is trophic flexibility and hence functional feeding-group misclassification — i.e. the same taxon acting as a shredder or a collector of fine organic material under different circumstances. Alternatively (or in addition) a lack of shredders could reflect limited stream retentiveness for leaf litter, an increased importance of mycoflora in litter breakdown in tropical streams (Padgett, 1976; Irons *et al.*, 1994), or higher investment in phytochemical defense by tropical leaves making them unpalatable to shredders (Stout, 1989). Significantly, Stout (1980) remarked upon the lower biomass of invertebrates on leaf packs in a tropical Costa Rican stream compared with mid-latitude North American sites. A deficiency of shredders has been recorded

Table 3.7 Relative abundance (%) of functional feeding groups comprising the benthos of four Hong Kong streams across a gradient of shading by riparian vegetation. ***, significant differences among sites ($P < 0.001$) as revealed by ANOVA (data from Dudgeon, 1989).

	Shaded stream	*Partly -shaded stream*	*Almost unshaded stream*	*Unshaded stream*	*ANOVA*
Predators	10.3	8.6	7.2	5.2	-
Scrapers	18.1	14.1	18.4	21.1	-
Collector-gatherers	32.4	55.1	42.0	52.6	***
Filter-feeders	30.4	21.1	31.2	19.4	-
Shredders	8.8	0.9	1.2	0.1	***
Piercers	0	0	0.1	1.6	***

also in New Zealand and Australian streams (e.g. Winterbourn *et al.*, 1981; Marchant *et al.*, 1985; Bunn, 1986). Nair *et al.* (1989) state that the abundance of functional feeding groups along the first- to fifth-order Neyyar River in south Indian matched the predictions of the RCC, despite the presence of a dam in their third-order study reach. However, they present no data or figures in support of their assertion.

Although studies of longitudinal community changes in Hong Kong streams do not match the predictions of the RCC closely, the functional organization of benthic communities in shaded streams — with high standing stocks of allochthonous detritus — differ (in a rather general way) from communities in unshaded streams (Dudgeon, 1989a; 1992a; see also Yule, 1995b). The zoobenthos of New Guinea streams, by contrast, shows rather minor changes in functional organization among sites with different riparian condition (Dudgeon, 1994b). Likewise, studies of changes in community organization in European streams along a dense- to open-canopy gradient do not match the RCC predictions of a decrease in shredders and an increase in grazers (Statzner & Resh, 1993). The fact that a mismatch frequently arises between the RCC and field data is a matter for some concern, and may reflect a variety of factors including historical constraints upon the fauna or over-simplifications in the RCC itself (Statzner & Higler, 1985, 1986; Statzner & Resh, 1993; Bretschko, 1994). While community organization in Asian and European streams do not agree very closely with the predictions of the RCC (Dudgeon & Bretschko, 1995a; 1995b), it is not immediately obvious what kind of field data or sampling scale would be required to test the RCC unequivocally. Moreover, although the predictions seem reasonable (in that they make 'biological sense'), a dissatisfaction with the poor performance of the RCC as a predictive model remains.

While the relatively low proportion of shredders in tropical Asian streams suggests that consumption of allochthonous organic matter is unlikely to make a major contribution to the secondary production of the zoobenthos, other data imply that this supposition is incorrect. Food webs of zoobenthos in streams and small rivers (e.g. Costa & Fernando, 1967; Bishop, 1973a; Costa, 1974; Ho & Hsu, 1977) generally involve a widespread use of allochthonous foods by primary consumers, and a relatively high percentage of nominally predatory forms. In this regard, tropical Asian streams are remarkably similar to those of Africa and Central Amazonia. Nevertheless, the importance of autochthonous algal food should not be underestimated, and a small-scale manipulative study in a Hong Kong stream showed that reductions in periphytic algae caused by shading led to declines in zoobenthos (especially mayfly) abundance (Dudgeon & Chan, 1993). In other feeding studies, Rathore & Rama (1979) have listed the algae present in the gut of Indian lotic *Chironomus* larvae, Yule (1995b) has constructed a food web for the zoobenthos of a stream on Bougainville Island (Papua New Guinea), Datta Munshi *et al.* (1990) have drawn attention to the importance of detritus in the diet of the mayfly *Ephemerella*, and Jayachandron & Joseph (1989) and Krishna Murthy & Rajagopal (1990) have described ontogenetic dietary changes (detritivory/herbivory to carnivory) in *Macrobrachium idella* and *M. equidens* respectively.

Edington *et al.* (1984) and Dudgeon (1992a) have investigated the relationship between capture net mesh-size and diets of filter-feeding hydropsychid caddisflies in Malaysian and Hong Kong streams. The species spinning the finest nets had a diet dominated by fine detritus and algae. Species with wider net dimensions had partly carnivorous diets and occupied microhabitats in fast current. Mesh-size differentiation *may* facilitate species coexistence wherever food is limiting (e.g. Edington *et al.*, 1984), but there are other explanations for this phenomenon (Alstad, 1987 and references therein). By contrast with the net-spinning caddisflies, there is no evidence of interspecific dietary separation among lotic Odonata in Hong Kong (Dudgeon, 1989c). The odonates are unselective predators, consuming all prey species that share their microhabitat, and the size of prey ingested increases with predator size (Dudgeon, 1989b, 1989c; Dudgeon & Wat, 1985). Accordingly, dietary spectrum reflects microhabitat specialization (or lack of it).

Seasonality

Seasonal variations in zoobenthos density in the large rivers such as the Ganges reflect monsoonal influences and are broadly similar to the patterns described for phytoplankton and zooplankton. Zoobenthos populations peak in early summer before the floods are lowest during the summer monsoon (Ray *et al.*, 1966; Singh & Srivastava, 1989; Singh & Nautiyal, 1990). Sharma (1984a) recorded the same trend in the Bhagirathi River (Garhwal Himalaya) where densities fell from a peak of 1,006 to 33 individuals.m^{-2}. The tendency of zoobenthos abundance to peak in the winter-summer interphase (i.e. the end of the dry season) and decline during the summer monsoon (or wet season) appears to be a general characteristic of perennial streams and rivers in tropical Asian (Pahwa, 1979; Realon, 1980; Dudgeon, 1983b; 1992a; Bhatt *et al.*, 1984; Dutta & Malhotra, 1986; Nautiyal, 1986; Sunder & Subla, 1986; Anwar & Siddiqui, 1988; Arunachalam *et al.*, 1991; Mohan & Bisht, 1991a; Gupta & Michael, 1992; Dobriyal *et al.*, 1992). Species richness changes in the same fashion (Soetikno & Atmowidjojo, 1985; Dobriyal *et al.*, 1992). The inhabitants of the hyporheic zone (Bishop, 1973b) could be an important source of colonists for surface substrates defaunated during monsoonal floods, and seem to be at least as important as drift in contributing to animals arriving on introduced substrates (Benzie, 1984). However, drift from upstream is an important recolonization mechanism for those downstream reaches where surface flow ceases or becomes intermittent during the dry season (Dudgeon, 1992c).

The importance of sediment characteristics as a determinant of zoobenthos microdistribution (see pp. 62–63) implies that there will be strong seasonal effects on community organization during the wet-season spates when sediment-patch heterogeneity is disrupted (Dudgeon, 1982b; 1992a), and this is in accordance with the general observation that zoobenthos diversity and abundance decline during the wet season. Bishop (1973a) states that the abundance of a particular benthic taxon at any one time in Sungai Gombak was a reflection of changes in habitat stability in the period immediately prior to sampling. Zoobenthos populations in Sungai Gombak built up to high densities as long as recruitment continued and there was no intervening catastrophic washout of sediments (Bishop, 1973a). After substrate disruption, recolonization was probably rapid (see Benzie, 1984), but population recovery of animals with specialized diets (e.g. periphytic diatoms) might be delayed while food supplies built up (Bishop, 1973a). The incidence of spates, or disturbance, can influence zoobenthos densities indirectly by way of an effect on the intensity of biotic interactions. For example, Dudgeon (1993) reported that spate-induced disturbance during the wet season reduced the impacts of predatory fish on benthic invertebrates in a small (second-order) Hong Kong stream. Impacts were greater during the dry season, when zoobenthos densities and species richness *declined*. Whether this decline was due to increased predation, or reduced discharge producing unfavourable conditions for rheophilic taxa is not known. However, this study shows that (for a variety of reasons) there may be inter-stream variation with respect to seasonal fluctuations in zoobenthos densities such that either the wet or the dry season may be the period of greater abundance.

Life-history patterns and production

The seasonality of tropical Asian zoobenthos reflects, in part, washout and scouring during spates caused by monsoonal rains combined with variations in the intensity of predation, but the timing of life-cycle events will also influence population densities and community composition. Information on life-history patterns and productivity for most groups is scarce. Taxonomic difficulties, especially concerning insects, are a major obstacle to those undertaking such research. For example, detailed investigation of Ephemeroptera systematics in tropical Asia have been undertaken for only one large family — the Leptophlebiidae; these studies are at the generic level only (Peters & Edmunds, 1970) and new genera continue to be described (e.g. Peters & Peters, 1990; Grant & Peters, 1993). The available data on life histories of lotic zoobenthos show a diversity of patterns which have been reviewed recently by Dudgeon (1992a). In short, where reproductive seasonality occurs (and frequently it does not), it seems to be related to the pattern of monsoonal rains. A strategy of emergence in late spring or early summer is common to some Hong Kong Odonata, Trichoptera and Ephemeroptera (Dudgeon, 1988b), and it is noticeable that such seasonality involves relatively large species. There is evidence that small larvae are less liable than larger conspecifics to be crushed and mutilated by the movement of stones during spates (Harker, 1953). Concordance of life histories among large species from three insect orders lends support to the suggestion that emergence may be timed to avoid spate-induced mortality of mature larvae, a degree of synchrony arising from physiological responses to temperature and/or increasing day length. Some gastropod life-histories can be interpreted in a similar way. Coincidence of reproduction with the monsoon — as seen in decapods (Raman, 1967; Rajyslakshmi, 1980; Pillai & Subramoniam, 1984; Dudgeon, 1985a) — could reflect also increased habitat availability (e.g inundated floodplain) for hatchlings in rivers swollen by monsoonal rains. However, in some pomatiopsid snails (e.g. *Neotricula*) in the Mekong, at least, copulation is coincident with the period of high water but oviposition (recruitment) occurs after water levels have fallen (Attwood, 1995b). Rheophilic pomatiopsids (e.g. *Paraprososthenia*) living in the same river are thought to reproduce before the flood season, and development of the eggs is delayed so that they hatch after the restoration of low-flow conditions (Attwood, 1995b).

By contrast with larger invertebrates, the strategy of small, multivoltine insects, certain atyid shrimps, and some molluscs, involves flexible, poorly-synchronized life histories of the type that are thought to represent an adaptive response to streams with variable and unpredictable discharges (e.g. Winterbourn *et al.*, 1981; Lake, 1982; Yule, 1995b). Periods of extended recruitment and multiple overlapping cohorts seem to be typical of many invertebrates in tropical and subtropical streams (e.g. Bishop, 1973a; Edmunds & Edmunds, 1980; Statzner, 1976; Gupta, 1993; Yule, 1995b), and may be the norm where water temperatures remain above 15°C for a major portion of the year (Benke & Jacobi, 1986). Indeed, generation times of approximately one month (less in some cases) have been recorded for mayflies (especially Baetidae) and chironomid (Diptera) larvae growing at temperatures > 20°C (J.D. Hynes, 1975; Marchant, 1982; Sweeney & Vannote, 1984; Benke & Jacobi, 1986; Maheshwari, 1989; Stites & Benke, 1989).

Information on the secondary production of Asian zoobenthos is meagre. Assuming two or three generations per year, Bishop (1973a) calculated production of riffle zoobenthos in Sungai Gombak to be in the range of 6-12 g dry wt.m^{-2}.yr^{-1}. In view of rapid life-cycle completion by some warm-water insects (e.g. Benke & Jacobi, 1986; Maheshwari, 1989; Stites & Benke, 1989), this estimate may be conservative and hence production will be underestimated (Benke, 1993). Dudgeon (1989b, 1989d; 1995c, 1995d, 1996a, 1996b; 1996c; 1997) has made direct measurements of the production of Odonata, Psephenidae and Helodidae (Co2leoptera), Ephemeridae and Heptageniidae (Ephemeroptera), and Stenopsychidae and Hydropsychidae (Trichoptera) in Tai Po Kau Forest Stream (Table 3.8). Production in some species was quite variable from year-to-year, and production: biomass (*P:B*) ratios were within the range reported for related species in other regions. Interestingly, inter-year differences in the total production of net-spinning caddisflies (all species combined) was minor, but there was considerable variation in the production of individual species within that functional group. The Hong Kong figures must be viewed with caution, as they are by no means representative of the zoobenthos as a whole. The data are derived mainly from larger species which have clear life-history patterns (usually univoltine or bivoltine), and production estimates for small, fast-growing species are lacking. Multivoltinism was suspected for certain Psephenidae and Heptageniidae

Table 3.8 Summary of production statistics - in terms of mg DW.m⁻² or individuals.m⁻² - calculated by the size-frequency method for macroinvertebrates in Tai Po Kau Forest Stream, Hong Kong, during the years 1977-78 and 1978-79; N = mean density; B = mean biomass; P = annual production; *, production estimates are available for a single year only; !, P and $P:B$ estimates unreliable because of small sample size.

Species	1977-78				1978-79			
	N	B	P	$P:B$	N	B	P	$P:B$
Euphaea decorata	28.7	33.5	173.6	5.1	20.7	33.9	167.9	5.0
Ophiogomphus sinicus	5.7	97.9	724.4	7.4*	-	-	-	-
Heliogomphus scorpio	3.4	50.6	364.2	7.2*	-	-	-	-
Zygonyx iris	2.6	83.1	606.8	7.3!	-	-	-	-
Ephemera spilosa	11.1	10.2	37.3	3.7*	-	-	-	-
Electrogena sp.	119.8	26.8	163.6	6.0	81.8	21.8	127.8	5.9
Cinygmina sp.	59.0	46.0	218.2	4.8	7.2	5.0	28.2	5.7
Epeorus sp.	27.4	10.8	75.6	7.0	35.4	15.2	113.8	7.5
Iron sp.	5.6	4.6	51.6	11.5	11.6	16.2	145.0	9.0
Paegniodes cupulatus	4.2	2.4	14.4	6.0	11.0	12.4	67.2	5.4
Stenopsyche angustata	82.2	278.4	2457.4	8.8	77.2	260.2	2265.4	8.7
Macrostemum fastosum	27.4	101.8	452.0	4.5	31.4	79.8	383.6	4.8
Polymorphanisus astictus	9.0	46.2	175.2	3.8	3.8	7.8	37.8	4.9
Cheumatopsyche criseyde	242.8	24.0	197.4	8.3	319.8	31.6	265.2	8.4
Cheumatopsyche ventricosa	148.2	31.4	327.0	10.4	75.6	15.4	150.0	9.8
Herbertorossia quadrata	1.4	2.4	60.8	25.4	12.2	26.2	387.0	14.8
Hydatopsyche melli	2.6	14.8	49.4	3.3	13.2	40.2	147.4	3.7
Hydropsyche chekiangana	0.8	3.6	20.8	5.8	7.6	8.6	38.0	4.4
Hydrocyphon sp.	189.2	9.9	129.5	13.1	121.0	7.5	88.2	11.8
Eubrianax sp.	52.6	23.5	311.8	13.3	51.8	22.8	353.2	15.5
Sinopsephenus chinensis	88.6	139.8	2695.1	19.3	34.2	70.2	858.0	12.2
Mataeopsephus sp.	1.8	3.1	55.5	17.9!	0.8	1.3	16.5	12.6!
Psephenoides sp.	22.2	1.0	7.6	7.5	2.0	0.2	0.9	5.5

but, because of the presence of multiple overlapping cohorts and continuous recruitment, could not be confirmed in the absence of *in situ* growth measurements (which are entirely lacking for Asian stream invertebrates). Consequently, production estimates were based on conservative assumptions about voltinism and, as is the case for Bishop's (1973a) estimates, are almost cetain to be significantly lower than the actual values for some species.

Overall, Asian zoobenthos communities seem to show many of the ecological characteristics of their temperate counterparts (e.g. Hynes, 1970). Studies of the behaviour of zoobenthos also reveal more similarities than differences. For example, investigations of diel and seasonal fluctuations in drift (Bishop, 1973a; Dudgeon, 1983c; 1990c; Brewin & Ormerod, 1994) have uncovered day-night rhythms and drift densities which are broadly similar to those in temperate streams. While the factors influencing timing of reproduction may be different, and the proportions of various functional feeding groups and the magnitude of production may show inter-regional variations, it is apparent that zoobenthos ecology in tropical Asian streams and rivers does not differ from that of equivalent temperate habitats in most qualitative aspects.

FISH

Composition

Tropical Asia has a rich freshwater fish fauna (Table 3.9) which includes the world's largest and smallest lotic fishes (Smith 1945; Roberts, 1988); cyprinids are especially diverse (Rainboth, 1991a). Within the region, the Indochinese Peninsula has 930 species of native fishes in 87 families (Kottelat, 1989; see also Table 3.9) but, despite such richness, existing inventories of fish biodiversity are far from complete (Kottelat & Whitton, 1995; Zakaria-Ismail, 1994). The individual states have diverse faunas: Thailand, for example, has over 500 species belonging to 49 families (Smith, 1945); 206 of these species are Cyprinidae. The Mekong alone sustains 244 species (Kottelat, 1989), with over 500 expected from the drainage basin (Zakaria-Ismail, 1994); new species of large fish (and many smaller ones) are still being described from the river (e.g. Compagno & Roberts, 1982; Roberts & Karnasuta, 1987; Monkolprasit & Roberts, 1990; Rainboth, 1991b; Roberts &

Table 3.9 Native freshwater fishes in tropical Asian fresh waters, based on data complied from Indochina by Kottelat (1989). The geographic coverage includes Peninsular Malaysia, the Mekong and the Salween Basins (outside China), the Chao Phraya and Mae Klong Basins in Thailand, and coastal streams in the intervening areas. This is not a complete inventory of the Oriental Region, but gives an overall impression of fish diversity and composition. Families which live in both fresh and salt water (i.e. 'peripheral' freshwater fishes) are indicated with an*, but certain species within some of these families (e.g. the Gobiidae) are entirely confined to stream habitats.

	Genera/Species		*Genera/Species*
Carcharhinidae*	3/5	Scorpaenidae*	1/1
Pristidae*	1/1	Platycephalidae*	1/1
Dasyatidae*	3/7	Centropomidae*	1/1
Osteoglossidae	1/1	Ambassidae*	4/13
Notopteridae	1/3	Serranidae*	2/2
Megalopidae*	1/1	Teraponidae*	2/4
Anguillidae*	1/3	Apogonidae*	1/2
Ophichthidae*	2/3	Sillaginidae*	1/1
Congridae*	1/1	Leiognathidae*	2/4
Muraenidae*	1/1	Lutjanidae*	1/5
Chirocentridae*	1/1	Lobotidae*	2/3
Clupeidae*	11/18	Gerreidae*	1/1
Pristigasteridae*	2/3	Haemulidae*	1/1
Engraulididae*	5/21	Sciaenidae*	9/18
Chanidae*	1/1	Monodactylidae*	1/1
Cyprinidae	65/280	Toxotidae*	1/4
Balitoridae	16/78	Scatophagidae*	1/1
Cobitidae	9/30	Nandidae	2/3
Gyrinocheilidae	1/1	Badidae	1/1
Bagridae	6/31	Pomacentridae*	1/1
Siluridae	8/24	Mugilidae*	3/4
Schilbidae	1/1	Polynemidae*	2/8
Pangasiidae	6/25	Blenniidae*	3/7
Amblycipitidae	1/1	Callionymidae*	1/1
Akysidae	3/10	Eleotridae*	7/12
Sisoridae	10/25	Gobiidae*	30/65
Clariidae	2/8	Gobioididae*	3/8
Heteropneustidae	1/1	Trypauchenidae*	3/3
Chacidae	1/1	Microdesmidae*	1/1
Olyridae	1/1	Siganidae*	1/2
Ariidae*	5/27	Scombridae*	1/1
Plotosidae*	1/2	Anabantidae	1/1
Sundasalangidae	1/1	Belontiidae	6/23
Bregmacerotidae*	1/1	Helostomatidae	1/1
Batrachoididae*	2/2	Osphronemidae	1/1
Hemirhamphidae*	5/15	Luciocephalidae	1/1
Belonidae*	2/4	Channidae	1/9
Oryziidae	1/3	Mastacembelidae	2/15
Aplocheilidae	1/1	Chaudhuriidae	1/1
Atherinidae*	2/2	Soleidae*	4/9
Phallostethidae*	4/6	Cynoglossidae*	1/5
Indostomidae	1/1	Triacanthidae*	1/1
Syngnathidae*	5/12	Tetraodontidae*	7/15
Synbranchidae*	3/4	**87 families; 316 genera; 930 species**	

Vidthayanon, 1991; Roberts, 1992; 1994). The Ganges River system has 141 species (Natarajan, 1989), the Salween some 150 (Zakaria-Ismail, 1994) and the Kapuas River basin (Borneo) supports at least 290 species (Roberts, 1989). There are over 250 species in Peninsular Malaysia (Johnson, 1967b; Kottelat, 1989; Ng *et al.*, 1994), 182 species in Nepal (Edds, 1983), and 132, 272 and 394 species in Java, Sumatra and Borneo respectively (Kottelat & Whitton, 1995). Li (1981) records 717 primary freshwater fish species in 33 families from China's rivers; a further 66 species spend part of their lives in rivers. Data collated from Li (1981) show that 586 species of the former group have been recorded from the Chang Jiang River or further south. The Zhujiang supports 381 species, including 262 primary freshwater fishes and 119 brackish-water and diadromous species (Liao *et al.*, 1989). Kottelat (1989) gives additional information on the composition of fish faunas in some large Asian rivers. Despite these studies, the general picture that emerges is one of a rather poorly-known fauna, and it is clear that the diversity of Southeast Asian fishes has been underestimated (Siebold, 1991; Zakaria-Ismail, 1994).

Many freshwater fish families in tropical Asia are found in Africa: for example, the cyprinids, certain siluriform catfishes, notopterids, anabantids, channids and mastacembelids. There are some fish in common at the generic level: *Mastacembelus, Barilius, Labeo, Garra, Clarius* and *Anabas* (Lowe-McConnell, 1970). Characoid fishes are absent from Asia and cichlids are represented by the single genus *Etroplus* in Sri Lanka, although introduced African *Oreochromis* and *Tilapia* are now widespread. The Balitoridae (previously Homalopteridae: Kottelat, 1988) are an exclusively Oriental group of benthic fishes that are widespread in fast-flowing waters throughout the region. They rank second only to the Cyprinidae in terms of species and generic richness (Kottelat, 1989). Aspects of their bionomics have received some attention from Hora (1923, 1932), Alfred (1969) and Dudgeon (1987d) but, considering their diversity and potential importance in Asian streams, very little is known of their ecology.

Wallace's and Weber's Lines serve as geographical/ecological boundaries separating the fish of the Asian (continental) realm from those further east. While Borneo, Java and Sumatra have fish of freshwater origin (Kottelat & Whitton, 1995), most Sulawesi and New Guinea species are derived from marine ancestors (Roberts, 1978; Berra *et al.*, 1985; Coates, 1985; 1987; 1993; Lowe-McConnell, 1987; Allen & Coates, 1990; Allen, 1991; Parenti & Allen, 1991; Kottelat & Whitton, 1995), some of which must undertake breeding migrations into estuaries or the sea. These secondary freshwater fishes include representatives of the Ariidae, Plotosidae, Melanotaeniidae, Eleotridae and Gobiidae, and New Guinea is the richest area in the world for freshwater eels (McDowell, 1981). The island is divided into two regions by the Central Highlands; some of the species inhabiting southward draining rivers are shared with Northern Australia, but those north of the mountains are not (Whitley, 1938; Allen & Coates, 1990; Allen, 1991). Like the New Guinea species, most lotic fishes in Sulawesi migrate between fresh- and saltwater or are of marine origin (Whitten *et al.*, 1987).

Cyprinids are preeminent over much of tropical Asia and are by far the most important food fishes, particularly the major carp in India (*Catala catala, Cirrhinus mrigala, Labeo rohita, L. calbasu*), *Osteochilus* and *Puntius* in Indonesia and Malaysia, and the Chinese carp (*Ctenopharyngodon idella, Aristichthys nobilis* and *Hypophthalmichthys molitrix*) which have been introduced throughout the region. Gouramies (*Osphronemus* and *Trichogaster*) are significant food fishes over a wide area, as are *Pangasius* catfish in Thailand and Vietnam (Lowe-McConnell, 1970). Together with cyprinids, these fishes comprise the major component of valuable river and flood fisheries over much of tropical Asia (Hickling, 1961; Welcomme, 1979).

There is an extensive literature on freshwater fish in Asia, most of it concerned with capture fisheries or aquaculture (see reviews by Jhingran, 1975; Welcomme, 1979; Fernando, 1984a; Datta Munshi & Srivastava, 1988). Taxonomic studies are well-developed in some countries, and still proceeding in others (e.g. Kottelat, 1985; Yen, 1985; Yen & van Trong, 1988), while much revisional work on widespread groups is required (e.g. Kottelat, 1984; Roberts, 1986; Roberts & Vidthayanon, 1991). Basic biological research has been neglected by comparison (Dussart, 1974; Rainboth, 1991a), and investigations of Asian rivers are restricted in comparison to those at African or Neotropical sites (Welcomme, 1979). Although information on artisanal fisheries in some rivers is available (e.g. Vaas *et al.*, 1953; Hickling, 1961: pp. 122-124; Watson, 1982; Lelek, 1987; Roberts, 1993a; Christensen, 1993a; 1993b; 1993c; Hoggarth & Utomo, 1994; Roberts & Warren, 1994; Baird & Roberts, 1995; Hortle, 1995; Ahmed & Tana, 1996), resource assessments are almost completely lacking over much of Southeast Asia (Welcomme, 1987), and data on the ecology of many species of Southeast Asian fishes are very limited (Kottelat, 1989; Roberts, 1993a; Hill, 1995). Nevertheless, relative to other consumers (such as zoobenthos), a considerable amount of information has been published. The results of investigations undertaken in the Mekong river are illustrative of fish ecology — especially breeding behaviour — in large tropical Asian rivers.

Breeding

Fishes in the Mekong River take part in upstream breeding migrations during the rainy season when water levels begin to rise; downstream migrations occur when water levels fall in the dry season (Shiraishi, 1970). The pattern arises because upstream migrants spawn in inundated areas during the wet season, and gather in the river channel or lateral lakes (such as Le Grand Lac) during the dry season. The importance of inundated areas as spawning and rearing grounds is shown by the fact that the Mekong delta accounted for 70% of all fish production in what was formerly South Vietnam (Dang, 1970; see also Hickling, 1961: pp. 135-146). The delta includes a large zone flooded by the Mekong waters during the rainy season (May-October); over 60 fish species have been recorded from this area (Shiraishi, 1970). Fishes follow the rising water and enter flooded forests, rice fields, oxbows and swamps which serve as feeding grounds, shelters and spawning sites, and thus account for a significant proportion of fish production (Dang 1970; Mizuno & Mori, 1970; Pantulu 1970; Rainboth & Kottelat, 1987; Roberts, 1993a). However, the oxygen content of the water varies across the floodplain and waters far from the main channel may become stagnant and anaerobic, causing death of fry and fingerlings (Pantulu, 1970; 1986b). The timing of migrations differs slightly according to species and among different parts of the Mekong basin: for example, *Probarbus jullieni* migrates and spawns from November through February (Roberts & Warren, 1994), and other cyprinids undertake non-reproductive migrations in January and February (Roberts & Warren, 1994). Migrations in the Mekong at southern Laos can be divided into three periods, late January to February, May to July, and November to December, with major differences in size and species composition, reproductive condition, and direction (Roberts, 1993a).

Pelagic fishes that are generally planktivorous or piscivorous ('whitefishes': e.g. *Ambassis wolfii*, *Puntius* spp., *Cyclocheilichthys enoplos*) are thought to cover a greater distance when migrating than the bottom-dwelling species ('blackfishes': e.g. *Mystus nemurus*, *Kryptopterus apogon*, *Channa micropeltis*) which feed on benthic organisms and favour floodplain habitats, making lateral movements between the main channel and fringes (Shiraishi 1970; Welcomme, 1979; Pantulu, 1986b; Hill, 1995). The two groups differ further with respect to breeding habits: blackfishes show some degree of parental care (e.g. mouthbrooding, bubble-nest building), while whitefishes are egg scatterers (Welcomme, 1979; Regier *et al.*, 1989). Some blackfishes can tolerate floodplain deoxygenation, whereas whitefishes avoid severe conditions by long-distance migration (Welcomme, 1979). The downstream post-breeding migration of Mekong blackfishes takes place at night, and follows a lunar rhythm being concentrated (in Kampuchea, at least) in a short period before each full moon in the months of November-February (Shiraishi, 1970; Taki, 1978). Whitefishes migrate downstream during the day (Shiraishi, 1970). Regier *et al.* (1989) have expanded the whitefish-blackfish classification to give an intermediate assemblage — the 'greyfishes' — including *Labeo* spp. in the Mekong. Greyfishes inhabit backwaters, the edges of larger floodplain lakes and the stagnant main channel during the dry season. Their behaviour is facultative, and they have both migratory and static/territorial components enabling them to respond readily to changing hydrological conditions in the river.

Despite the generalizations made above, it should be emphasized that little is known about the ecology of Mekong River fishes (Roberts, 1993a; Roberts & Baird, 1995), and the relationship between migrations, feeding and breeding, although information is beginning to accumulate (Hill, 1995; Roberts & Baird, 1995). For example, Roberts & Baird (1995) have shown recently that *Cirrhinus lobatus* is the lead species of almost all of the major migrations moving up the Mekong, and is exceptional among migratory cyprinids (which are all dioecious with separate sexes) in that it seems to be a protogynous hermaphrodite (females change into males).

The wider ecological implications of these large-scale fish movements have yet to receive much attention. For example, although Mekong fishes are known to feed on a range of fruits in flooded forests (e.g. various species of *Allophyllus, Ardisia, Artabotrys, Eugenia, Ficus, Hydnocarpus, Morindopsis, Olax, Physalis* and *Quassia*) it is unclear whether dispersal and germination of seeds is facilitated by passage through fish guts (Roberts, 1993a). Of interest in this context is *Leptobarbus hoevenii* — known as the 'mad fish' — which waits under *Hydnocarpus* trees for fermented fruit to fall. 'Eating this fruit soon renders the fish drunk and schools can be seen floating helplessly in the water, safe only because their flesh has become inedible' (Banarescu & Coad, 1991).

Engineering works planned or in progress along the Mekong involve the construction of a series of dams for power generation (Chomchai, 1987; Petersen & Sköglund, 1990; see p. 537). Changes in flow, temperature regime and inundation patterns (and hence food availability) may remove important directive factors which stimulate migratory and breeding behaviour (Roberts, 1992; 1993a; 1993b; Robert & Baird, 1995). The dams will obstruct the passage of several long- and medium-range whitefish migrants. By contrast, greyfishes may be less affected by planned modifications of the Mekong River because of their behavioural flexibility (Regier *et al.*, 1989). Elsewhere in tropical Asia, river control projects and declining water quality (e.g. the Ganges; Natarajan, 1989) have led to diminished fisheries with distinct changes in community structure. For example, mahseer (*Tor* spp.) make long upstream migrations from the Ganges Plain to spawn over gravel substrates (Hora, 1952; Natarajan, 1989; Dobriyal & Kumar, 1988). Habitat changes (e.g. barrage construction), pollution and overfishing have caused a decline in stocks of mahseer and snow 'trout' (*Schizothorax* spp.) during recent years (Sehgal, 1983; Natarajan, 1989; Joshi, 1991; Shrestha, 1991; Singh *et al.*, 1991) to such an extent that '. . . Himalayan mahseers are now considered endangered species' (Singh *et al.*, 1991) and '. . . many schizothoracid species . . . have become rare and may become extinct' (Sehgal, 1983). With other river control projects planned (e.g. the Three Gorges Project on the Chang Jiang), human impacts on fish populations will be heightened. The matter will be considered further on pp. 521–549 and p. 563 *et seq.*

In common with Mekong River species, most lotic fishes in tropical Asia synchronize breeding activity with the monsoon or flood season (e.g. Khan, 1924; Qasim & Qayyum, 1961; Welcomme, 1979; Tsai *et al.*, 1981; Sharma, 1984b, 1984c, 1984d; Baloni & Tilak, 1985; de Silva *et al.*, 1985; Silva & Davies, 1986; Baloni & Grover, 1987; Lowe-McConnell, 1987; Dobriyal & Singh, 1987; 1989; 1990; Soetikno, 1987; Coates, 1990a; de Silva, 1991a; Smith, 1991; Rainboth, 1991a; Edds, 1993; but see de Silva, 1991b), and migrations, especially onto inundated floodplains, are important (Joseph & Job, 1983; Ali & Kathergany, 1987; Coates, 1990a; Smith, 1991; Rainboth, 1991a; Christensen, 1992). Where there are two monsoonal floods each year (e.g. in south India, Sumatra and Sri Lanka) fish may reproduce twice (Alikunhi, 1953; Soetikno, 1987; de Silva, 1991a). Thus widely-distributed species (such as the snakehead *Channa marulius*) vary in reproductive timing, breeding once (Parameswaran & Murugesan, 1976) or twice (Alikunhi, 1953) each year according to the local monsoon frequency and pattern of land-water interactions. Exceptions to the general pattern of spawning during the wet season are three Sri lankan *Barbus* spp. and Indian *Tor putitora* (Cyprinidae) which breed throughout the year (de Silva *et al.*, 1985; Kortmulder, 1987; Qasim & Qayyum, 1961), and *Barbus melanampyx* in southern India which spawns during the dry, inter-monsoon period (Harikumar *et al.*, 1994). *Gnathopogon argentatus* (Cyprinidae) in the Zhujiang River also spawns during the period of low water (Liang *et al.*, 1986), while spawning by the Taiwanese minnow *Zacco pachycephalus* takes place before the onset of summer spates which reduce recruitment success (Wang *et al.*, 1995).

Feeding

Many migratory fish show highly seasonal feeding activity, with intense feeding during the flood and fasting at times of low water. In some rivers, fishes which do not move onto floodplains (or inhabit rivers without floodplains) suffer food shortages during the spawning period (Dobriyal & Singh, 1990), as food is not easily obtained in turbid, flood-swollen waters (Bhatnagar & Karamachandani, 1970; Sharma, 1984b). For certain Indian fishes, however, feeding is interrupted during breeding regardless of whether they move onto the floodplain or not (Desai, 1970; Dobriyal & Singh, 1990). In the Sepik River, ariid catfish — which are secondary freshwater species that do not exploit the floodplain — show no marked seasonality in relation to the flood cycle (Coates, 1991), and a lack of pronounced seasonality also characterizes the reproduction of the freshwater cardinalfish (*Glossamia gjellerupi*) in the Sepik (Van Zwieten, 1995).

River fishes in tropical Asia exploit a wide range of foods, both of animal and vegetable origin (e.g. Inger & Chin, 1962; Costa & Fernando, 1967; Bishop, 1973a; Chatterji *et al.*, 1977; de Silva & Kortmulder, 1977; Roberts, 1978; Geisler *et al.*, 1979; Haines, 1983; Kumar & John, 1984; Moyle & Senanayake, 1984; Sehgal *et al.*, 1984; Sharma, 1984b; 1984c; 1990; Langer, 1986; Baloni & Grover, 1987; Dudgeon, 1987b; Butt & Khan, 1988; Rachmatika & Soetikno, 1988; Majhi & Dasgupta, 1989; Wikramanayake & Moyle, 1989; Afser, 1990; Agarwal *et al.*, 1990; Datta Munshi *et al.*, 1990; Bhuiyan & Islam, 1991; Coates, 1991; Rainboth, 1991a; Anwar & Siddiqui, 1992; Shrivastava *et al.*, 1992; Van Zwieten, 1995). Allochthonous sources such as terrestrial insects and fruits (e.g. Inger & Chin, 1962;

Costa & Fernando, 1967; Mizuno & Mori, 1970; Bishop, 1973a; Tan, 1980; Haines, 1983; Kumar & John, 1984; Coates, 1990a; 1990b; Rainboth, 1991a; Coates & Van Zwieten, 1992; Roberts, 1993a) are important dietary items and may constitute the base of fish production in rainforest streams (Watson & Balon, 1984a; Lowe-McConnell, 1987), although endogenous benthos is eaten also (Bishop, 1973a; Van Zwieten, 1995). Welcomme (1979) has reviewed dietary studies in the Asian fisheries literature which show a heavy bias towards allochthonous food in the forest streams and floodplains (see also Hickling, 1961: pp. 55-71). Indeed, the exploitation of allochthonous food by Asian river fishes seems to be much greater than of lotic fishes in temperate latitudes. The availability of allochthonous foods may stimulate fish movements onto inundated floodplains in the tropics, and Rainboth (1991a) stresses that these are not simply breeding migrations but rather are migrations to new or superior living space. It is significant that in northwest European rivers, at least, such migrations do not occur (Dudgeon & Bretschko, 1995a; 1995b). This may be due, in part, to the alteration of inundation patterns by channelization and regulation of European rivers, but — as far as can be ascertained — seasonal migrations on to flooded land is not essential for the feeding or reproduction of European fishes.

One feature of tropical Asian streams which has attracted attention is the extent to which resource partitioning occurs among the diverse fish communities. Information is available from streams rather than large rivers, but indicate that niche segregation can occur on the basis of seasonality, diet or habitat use (Alfred, 1969; de Silva & Kortmulder, 1977; Moyle & Senanayake, 1984; Schut *et al.*, 1984; Hartoto, 1986; Kortmulder, 1987; Wikramanayake & Moyle, 1989; Kortmulder *et al.*, 1990). Many species show a high degree of morphological specialization, especially in structures related to feeding (Moyle & Senanayake, 1984; Kortmulder *et al.*, 1990; Wikramanayake, 1990). In general, fishes in headwater streams are euryphagous, taking whatever food drops into the water and feeding at different levels in the water column and at the surface (Lowe-McConnell, 1987); specialized feeders comprise a minor part of the community (Bishop, 1973a). Further downstream, benthic omnivores become important. Longitudinal changes in community composition involves species replacement and an increase in fish diversity with greater stream order and size (e.g. Bishop, 1973a; Roberts, 1978; Sehgal, 1983; 1991; Lowe-McConnell, 1987; Soetikno, 1987; Allen & Coates, 1990; Allen, 1991; Edds, 1993; Ormerod *et al.*, 1994), reflecting the greater complexity of downstream reaches (Bishop, 1973a; Lowe-McConnell, 1987; Soetikno, 1987) or severe conditions (steep gradients, high rainfall, and cool temperatures) in some highland streams (e.g. Sehgal, 1983; 1991; Allen, 1987; 1991; Allen & Coates, 1990; Edds, 1993; Ormerod *et al.*, 1994). An interesting feature of the downstream increase in fish diversity in Southeast Asia is dominance of the species composition of lower-order streams or upper tributaries of large rivers by cyprinids (Rainboth, 1991a; Zakaria-Ismail, 1994); the percentage of the ichthyofauna made up by cyprinids declines in lower reaches, and may also fall (relative to anabantoids) in soft, acidic streams (Rainboth, 1991a). Downstream species replacement has led to the development of classification schemes for streams at different altitudes in the Himalaya according to the characteristic ichthyofauna (Sehgal, 1983), but the demarcation and definition of these fish zones are of doubtful ecological value (Melkania, 1991).

Specialization in feeding habits is especially pronounced among the fishes of forested Sri Lankan hillstreams (which include many endemic species), and may reflect their reliance on autochthonous foods (Moyle & Senanayake, 1984; Kortmulder *et al.*, 1990). The circumstantial evidence suggests that Sri Lankan fish assemblages are at or near equilibrium and structured by deterministic, biotic processes (Moyle & Senanayake, 1984; Wikramanayake & Moyle, 1989), with ecomorphological relationships that reflect evolutionary adjustments to facilitate resource partitioning and thereby reduce competition (Kortmulder *et al.*, 1990; Wikramanayake, 1990). Because Sri Lanka has a relatively depauperate fish fauna compared to India (Wikramanayake, 1990), it is reasonable to assume that similar niche partitioning occurs in more diverse continental assemblages although some workers consider that Sri Lankan hillstreams offer fish an unusually wide array of microhabitats (Kortmulder *et al.*, 1990). However, Dudgeon (1987b) found little evidence of niche segregation in food or habitat among four benthic fishes in a Hong Kong stream. By contrast, vertical position in the water column was an important axis of specialization in a Sarawak rainforest river (Watson & Balon, 1984b) where increases in species richness and niche packing were associated with a reduction in body size and length of life, a shift to more sedentary habits, and the development of complex life history and reproductive styles (Watson & Balon, 1984b). These specializations make tropical Asian fish particularly vulnerable to environmental perturbations or over-exploitation (Watson & Balon, 1984a).

It is important to note that circumstantial evidence for the role of competition in structuring Asian tropical fish assemblages is derived from studies of small rivers and streams. There are no comparable data on the significance of competition in large rivers, where fishes undergo longitudinal or lateral migrations and where there is an extensive inundated floodplain during the monsoon. The incidence of migrations and seasonal fluctuations in habitable area make it less likely that assemblages of fish in large rivers are structured by deterministic processes (such as competition) when compared to assemblages of relatively sedentary species in smaller rivers and streams. In this regard it is significant that Coates (1991) has reported a degree of dietary segregation among five species of ariid catfish in the Sepik River. These fishes inhabit higher-order tributaries and do not use the river floodplain to any great extent, and this restriction to main channels could have lead to competition for limited food resources and concomitant selection for dietary specialization.

There is a surprising paucity of data on fish production and biomass in tropical Asian rivers. Productivity of fish in Sri Lankan rivers and streams seems to be low (Indrasena, 1970), and the average annual production of a rainforest river in Sarawak (76.7 kg.ha^{-1}) was less than that of rivers in Canada or Poland, which might reflect food limitation at the tropical site (Watson & Balon, 1984a). Biomass values given by Bishop (1973a) and Geisler *et al.*, (1979) for Malaysian and Thai streams (81 — 186 kg.ha^{-1}) are similar to those reported from tributaries of the Sepik-Ramu River Basin in New Guinea (52.8 kg.ha^{-1}; van Zwieten, 1995) but seem low in comparison to European waters with similar hydrochemistry (Geisler *et al.*, 1979; see also de Silva, 1991a, 1991b). Fish biomass in Malaysian blackwaters (5.6 — 40 kg.ha^{-1}; Ng *et al.*, 1994) is even lower and generally less than in larger, 'clear water' streams; a probable consequence of low primary production due to poor light penetration and high acidity (Johnson, 1967b; Ng *et al.*, 1994). Standing stocks in Nepalese Himalayan streams seem low also (2.3 — 25 and 8.6 — 28 kg.ha^{-1} wet mass in two river systems; Ormerod *et al.*, 1994). By contrast, Lelek (1987) considers fish yields in the Rajang River (North Borneo) may reach 100 kg.ha^{-1}.yr^{-1}. Fish biomass on river floodplains may be higher, but are subject to great temporal variation and are extremely sensitive to human activities within the drainage basin (Welcomme, 1979; 1987). Annual production of river fishes is likely to fluctuate greatly from year to year. Species which depend upon the floodplain for food and breeding sites will be subject to vagaries in the intensity of the monsoon. Thus in some years there will be extensive flooding and relatively high fish production; in years when a smaller area is inundated, production will be lower.

FLOODPLAIN ECOLOGY

Biota and productivity

Monsoonal rains increase the volume of water carried by tropical Asian rivers, leading to lateral overspill which inundates the floodplain and low-lying ground. During the dry season, much of this water drains away or evaporates, and only the main river, relics of previous channels, oxbow lakes and isolated depressions remain filled. Their role in the carbon balance of the floodplain remains largely unknown (Safiullah *et al.*, 1987). Geochemical evidence demonstrates that lateral lakes (backswamps) on the Fly River (Papua New Guinea) receive most allochthonous material from the river and not the surrounding land (Osborne *et al.*, 1988), and similar results have been reported for the Chang Jiang where riverine inflows and decomposition of macrophyte mats were the main sources of organic carbon in a floodplain lake (Yuan, 1992a; 1992b). Research at a floodplain site in Bangladesh has shown that the water column is an important source of nutrients (nitrogen and phosphorus) for the floodplain sediments and the ecosystem as a whole (Whitton *et al.*, 1988c). However, the relative importance of recycled versus imported nutrients in floodplains, and their source (river channel or land), probably varies between systems (Junk *et al.*, 1989) and requires further study. For example, Gan & Kempe (1987) report that the lower Chang Jiang River receives organic material from the floodplain and inundated rice paddies during the wet season which may reflect the predominance of manure fertilization in China (Wang & Chen, 1987). Similarly, large amounts of dissolved organic carbon drain from the heavily-cultivated floodplain into the main channels of the Ganges and Brahmaputra during flooding (Chowdhury *et al.*, 1982).

   Human influences are a general phenomenon of tropical Asian floodplains which are more highly modified than those of Africa and South America (Welcomme, 1979). Works for irrigation, drainage and flood protection have caused the disappearance of many of the original features (Natarajan, 1989). Using the Negara River, a tributary of the Barito River in Kalimantan (southeast Borneo), as a case study, Klepper

(1992) has highlighted the flooding danger that could arise from draining and impoldering wetlands; acidification of potential acid sulphate soils — which has deleterious effects on fishes (Klepper *et al.*, 1992) — would also take place. Riverine wetlands tend to mitigate flood severity (Ogawa & Male, 1990), and anthropogenic modification of the floodplain has important effects on the timing and duration of inundation. Although it can be difficult to distinguish natural seasonal changes from those brought about by human activities, investigations of Lake Chenhu — a managed lateral lake of the Chang Jiang — give some indication of the annual patterns of changes on that river floodplain (Liang *et al.*, 1988) and, by inference, the pattern of events on the floodplains of other monsoonal Chinese rivers. Flooding occurs from May until September when the river contributes organic matter to the lake (Yuan, 1992a; 1992b) and the inundated area increases to about six times the dry season extent. Annual net primary production of phytoplankton (for the lake as a whole) during the flooded period is low but approximately ten times that of aquatic macrophytes (Liang *et al.*, 1988); however, decomposing macrophytes comprise an important source of detritus for aquatic consumers during the dry season (Yuan, 1992a). The production of wetland plants (mainly Cyperaceae, Graminae and Cruciferae) growing on the exposed mud after water levels fall exceeds 800 $g.m^{-2}.yr^{-1}$ (or 15,400 $t.yr^{-1}$), more than twice phytoplankton production. The wetland vegetation is an important food for secondary consumers during the next flood season (Liang *et al.*, 1988), especially Cyprinidae which eat terrestrial hydrophytes readily (Dudgeon 1983d). Freshwater shrimps and prawns as well as fish benefit from the organic matter made available by rising water levels. Singh & Srivastava (1989) report that *Macrobrachium birmanicum* migrate onto vast shallow inundated areas in the middle Ganges, where the availability of food promotes rapid growth rates (see also Prakash & Agarwal, 1985). Inundation of terrestrial vegetation on floodplains may also be of importance to ecosystem functioning in cases where the leachate influences aquatic macrophyte productivity (e.g. Kushari & Sinhababu, 1987).

The zoobenthos of Chang Jiang floodplain lakes is composed mainly of gastropods (Chen, 1988; Liang *et al.*, 1988), which are also numerous on floodplains in Bangladesh and along the Mekong River (Mizuno & Mori, 1970; Whitton *et al.*, 1988b; 1988c). Estimated (shell-free) production in Lake Chenhu was 16.8 $g.m^{-2}.yr^{-1}$ (Liang *et al.*, 1988), and biomass was 10.5 $g.m^{-2}$. Forty-one species of gastropods have been recorded from Poyang Hu (Chen, 1988) and molluscs (45 species) dominated the benthos of Lake Hwama, another lateral lake of the middle Chang Jiang, where they attained a total density of 87 individuals.m^{-2} (Chen, 1979). Gastropods (34 species) in this lake appeared to be favoured by dense growths of submerged aquatic macrophytes. A benthic biomass of 1.75 g m^{-2}, 20% of which comprised molluscs, has been recorded in a Mekong River backswamp at a time when benthic biomass in the river was only 0.12 g m^{-2} (Sithimunka, 1970; quoted by Welcomme, 1979). Additional comparative data are lacking, but available information suggests that abundance and productivity of invertebrates (especially molluscs) on the floodplain exceeds that in the river channel.

Freshwater swamp forests, dominated by *Homalium brevidens* and *Hydnocarpus anthelminthica* (Flacourtiaceae), occur in Kampuchea along the shores of Le Grand Lac and along some banks of the Mekong in its floodplain (Pantulu, 1986a). In the rainy season, the 11,000 km^2 of inundated forest serves as an important fish spawning ground and a source of allochthonous food (Mizuno & Mori, 1970; Pantulu, 1986b; Roberts, 1993a). Because of the considerable annual variation in water levels, all (or almost all) fishes in the area must migrate moving between spawning grounds in the inundated forest and the lake, and from the lake to Tonlé Sap River and thence to the Mekong, and *vice versa* (Pantulu, 1986b; Hill, 1995).

Annual flooding dominates fish production in the Mekong, inundating soil and vegetation and introducing millions of tonnes of suspended and dissolved solids to the waters of the floodplain from June through October. The production of fishes which breed at this time is sustained by allochthonous organic matter inundated by the flood (Pantulu, 1986b). Detrital decomposition also stimulates macrophyte production, and forms a food base for certain Cyprinidae. As well as species feeding on invertebrates, a variety of carnivores crop the system: among them snakeheads (Channidae) and the freshwater 'shark' — actually, a silurid catfish — *Wallago attu*. The timing of the floods and their duration is critical, because rapid flood recession will strand large numbers of eggs and hatchlings. Prolonged flooding may cause large-scale fish mortality due to deoxygenation of the floodplain waters and drastic pH changes which are a result of the decomposition and leaching of allochthonous detritus (Pantulu, 1986b). This mortality may have been a selective pressure for the construction of floating bubble-nests by some floodplain fishes (Anabantidae, Channidae). Indonesian floodplain lakes such as those of the Rivers Ogan and Komering (Sumatra), Upper Kapuas (Borneo) and the Lakes

Tempe and Sidenreng system (Sulawesi) are productive sites also, forming the basis of important fisheries (Vaas *et al.*, 1953; Hickling, 1961; Giesen, 1994 and references therein) with yields, on a per hectare basis, comparable to artificial fishponds (Giesen, 1994). In addition to the use of floodplains by fishes, freshwater dolphins (*Platanista gangetica*) enter flooded fields on the Ganges floodplain (Pelletier & Pelletier, 1980), while newly-flooded swampland is an important habitat for juvenile crocodiles (*Crocodylus novaeguineae*) on the Strickland River, New Guinea (Montague, 1983).

One important floodplain site in tropical Asia which has received a significant amount of study is Tasek Bera in Malaysia. This is a peat swamp-forest complex with an arborescent system of shallow watercourses draining into a southern tributary of Sungai Pahang, the longest river on the east coast of Malaysia. Tasek Bera is a predominately heterotrophic, blackwater system with food chains based mainly on allochthonous inputs. Invertebrates are confined largely to submerged plants (*Utricularia flexuosa*), rather than the bottom mud, while periphyton is associated with stem bases in reed (*Lepironia articulata*) swamps. Tasek Bera was designated Malaysia's first Ramsar site in March 1995 (Chew, 1995). Research work on this ecosystem has been reviewed extensively by Furtado & Mori (1982), and will not be considered further here. Tasek Bera represents a relict of a once extensively-distributed riparian/innundation forest ecosystem in Southeast Asia, now largely removed by humans (see Corner, 1978; Claridge, 1994; Ng *et al.*, 1994), dominated by *Eugenia* spp. (Myrtaceae) and *Pandanus helicopus* (Pandanaceae) trees which line the river and trap detritus and silt. Rivers draining these environments are blackwaters, with a highly adapted and specialized flora and fauna. In 1997, Wetlands International (Indonesia) estimated that — based on 1988 figures — only 10% of the remaining peat swamps in Indonesia were undisturbed, while at least 35% had been totally converted to other uses. Massive forest fires which raged in Borneo (Kalimantan) and Sumatra during 1997 burned huge areas of peat swamp forest, and those areas which had been drained (or partially drained) were extremely vulnerable. The environmental consequences of these fires for the aquatic biota of peat swamps has yet to be assessed, but preservation of the limited areas that remain undisturbed should be a conservation priority.

River-floodplain interactions

Consideration of floodplain ecology suggests that, for many rivers, the conventional longitudinal continuum of change in geomorphic and biological features from headwaters to mouth (as embodied in the RCC), must be augmented by a second axis running perpendicular to the first across the floodplain and encompassing the lateral extent of aquatic, semiaquatic and terrestrial habitats. This river-floodplain ecosystem would include the river channel(s), backwaters, oxbows and relict channels, floodplain pools, and seasonally-inundated areas of the terrestrial environment. The importance of this lateral axis is implicit in the view that large floodplain-river systems cannot be addressed adequately using the RCC alone (Sedell *et al.*, 1989). In rivers which have no geomorphic or engineering constraints on the extent of lateral exchanges, floodplain- or riparian forest-river interactions are likely to be critical for carbon processing and fish habitat. This applies regardless of the size of the drainage basin (Sedell *et al.*, 1989). The significance of such interactions will vary according to the extent and duration of inundation, which can be predicted from a knowledge of hydrology and floodplain morphometry. Where the floodplain is well-developed, transport of carbon from the upstream catchment will be of little importance to downstream productivity. Instead, primary and secondary production on floodplains will be essential to fauna in the main channel. The seasonal pulsing of flood flows onto the floodplain provides the driving force for the river-floodplain interaction (Welcomme, 1979; Junk *et al.*, 1989) because it determines the magnitude and extent of land-water interactions and allochthonous inputs. As water inundates the floodplain it produces a 'moving littoral' unit which prevents permanent stagnation and allows rapid recycling of nutrients and organic matter. This results in higher productivity than in systems which are either permanently wet or dry (Junk *et al.*, 1989).

The extent to which primary production by plankton or macrophytes in floodplain water bodies contributes to secondary production in the river channel has yet to be quantified. The rich periphyton associated with these macrophytes has also received little attention (but see Whitton *et al.*, 1988b; Das *et al.*, 1994). However, an energy budget constructed for a floodplain section of the Amazon (Bayley, 1989) suggests that 6.9% of total carbon supply is contributed by phytoplankton and periphyton; <1% is transported from upstream and the remaining >90% was derived from the inundated floodplain and flooded forest. Similarly, about 80% of the organic carbon supply to a Georgia blackwater river originated in riparian swamps (Meyer & Edwards, 1990). Given the relatively limited area of the main channel as well as the high silt load and turbidity of many tropical Asian rivers, primary production can be only a fraction of that on the floodplain. Moreover, river phytoplankton biomass and productivity declines further during the flood season (see p. 50) when floodplain productivity is increasing. In addition, a considerable proportion of fisheries yields from large rivers is derived from floodplain habitats (Welcomme, 1979; Junk *et al.*, 1989), and it is significant that the development of flood-control measures (canals, levees and dykes) on the Ganges floodplain have deprived major carp of their extensive breeding sites causing a marked decline in the fishery (Natarajan, 1989).

An important task for stream biologists is to integrate the RCC with the floodplain-forest interaction of entire basins (Sedell *et al.*, 1989; Dudgeon, 1995a). If the river is constrained from interaction with the surrounding landscape along most of its reach (as is the case if river-valley form prevents inundation of extensive areas during floods), then the RCC may be an adequate descriptor of river ecology. However, if certain reaches have an extensively-inundated floodplain, and others do not, then each reach will have to be considered more-or-less separately from those up- and downstream of it, and in localities where river-floodplain interactions are important the RCC will have limited predictive power. Extra complications are added where there has been anthropogenic alteration of natural floodplain inundation patterns. Despite the over-riding influence of lateral interactions with the floodplain in certain reaches, the unidirectional flow of water ensures that the lower course of a river is influenced by its headwaters and tributaries. Accordingly, the floodplain cannot be viewed in isolation from the river basin as a whole.

4

The Zoobenthos: A Systematic Review

KEY TO MACROINVERTEBRATE GROUPS

This book is concerned particularly with the zoobenthic invertebrate communities of tropical Asian streams. Microscopic forms are not considered, and the focus is on macroinvertebrates defined broadly as animals with body length ≥ 0.5 mm which would be retained by a net of 200 µm mesh size. They can be divided informally into two large groups: the lower and the higher invertebrates. The category 'lower invertebrates' includes all invertebrates except molluscs and arthropods, and is less speciose than the 'higher' grouping. For convenience, my treatment of the zoobenthos will begin with a key to the major groups of lotic macroinvertebrates.

I took a pragmatic approach while developing this key and all of the other keys in the book. Many keys use 'natural' characters in the couplets that are designed to convey phylogenetic information about the degree of relatedness among groups and the sequence of evolutionary branching. While this approach can enhance the usefulness of a key, it may hinder identification if the relevant features are obscure or difficult to see. In the following keys, 'artificial' characters have been employed where they will aid identification; in other words, any feature or characteristic which 'works' has been included. As an additional aid I have labelled a taxon as 'rare' in cases where it is infrequently encountered or has few lotic species (e.g. the phyla Cnidaria and Polyzoa), or occupies a specialized microhabitat (branchiobdellidan worms, for example, live upon the bodies of decapod crustaceans), or where most members of the group (e.g. nematode worms) are too small to be collected by the usual methods for sampling macroinvertebrates.

While constructing the various keys reproduced herein, I consulted all keys to aquatic invertebrates (devised mainly for the Palaearctic Regions and Australia) that I could find, but the final form of the keys which have been included differs substantially from those published elsewhere. They are designed to permit a quick and accurate means of identifying most families of benthic macroinvertebrates from streams across tropical Asia. Invertebrates are first keyed to major groups (phylum, class or order) and then keys are given for most of the major groups allowing identification to family. In some cases, taxa within each family are keyed out separately. General information together with an introduction to the specialized literature has been given in the text to assist in the many cases where it is not yet possible to provide a reliable guide to the identity of genera or species within a family. All of the keys in this book are dichotomous and progress through a series of steps at each of which a decision is made between two more-or-less contrasting characters or groups of characters. Although dichotomous keys are, in principle, simple to use, taxa not included in the keys may not be recognized as alien elements and it is difficult to ensure accuracy of identification as each path in the key leads to a name. In an attempt to overcome these problems, a range of relevant characters for each option has been given at many of the choice points rather than selecting only one or two principal features. While this approach will not obviate misidentification of taxa, it may reduce the incidence of mistakes. Where there is uncertainty about generic limits or status, the contrasting characters of known taxa have been listed.

1. Encrusting forms with an irregular, asymmetric shape, lacking discrete organs or tentacles; spongy to the touch; rare ....
    .... freshwater sponges: phylum **Porifera**

    Without the above combination of characters; bilaterally symmetrical with organs and organ-systems 2

2. *Either* free-floating and transparent (Fig. 4.1A), like an umbrella-

shaped disc (< 30 mm diameter); *or* sessile and rather small (generally < 5 mm, but individuals may be solitary or colonial) with a short stem and several tentacles arising from it (Fig. 4.1B); rare .... coelenterates: phylum **Cnidaria**

Without the above combination of characters 3

3. Sessile, colonial forms made up of numerous small individuals each bearing several retractile tentacles on a horse-shoe shaped structure (the lophophore) around the mouth (Figs 4.3A & B); the rest of the body is enclosed in a gelatinous structure; the colony may be encrusting, compact, or branching and twig-like; rare ... . moss animals: phylum **Polyzoa** (= **Ectoprocta**)

Form not as above; free-living and not colonial...................... 4

4. Body segmented, or with jointed legs, or bearing a shell, or any combination of these features 8

Body unsegmented, lacking jointed legs or a shell 5

5. Elongate flattened body (< 30 mm long) pressed to the substratum (Fig. 4.2); moves with a gliding motion; often with a pair of anterior eyespots .... flatworms (planarians): phylum **Platyhelminthes**; class **Turbellaria**

Without the above combination of characters 6

6. Body thin, flattened and 'bootlace-like'; with a long, eversible proboscis (inconspicuous when retracted) and three pairs of eyes; rare .... ribbon or proboscis worms: phylum **Nemertea**

Form not as above; if body is long and thin it is approximately circular in cross-section 7

7. Body long (may be >20 cm long) and threadlike; anterior and posterior ends blunt (not tapering); usually dark brown; rare .... horsehair worms: phylum **Nematomorpha**

Worm-like cylindrical body tapering at both ends and lacking external segmentation; move in a 'whip-like' fashion; usually <1 cm long; rare.. phylum **Nematoda**

8. Body completely or largely enclosed by an unsegmented calcareous shell which may be coiled, spherical or bi-valved; body is soft and unsegmented with a ventral muscular foot .... snails, clams and mussels: phylum **Mollusca** 9

Body segmented and not enclosed in a calcareous shell **11**

9. Body completely enclosed by a bi-valved shell ....
   ....clams and mussels (bivalves): class **Bivalvia**

   Shell not bi-valved, body may not be entirely enclosed by the shell
   .... snails and limpets: class **Gastropoda** **10**

10. Shell aperture can be closed by a horny operculum attached to the
    foot; shell shape usually turbinate or turriculate with dextral coils
    (i.e. the aperture is on the right when the shell is held with the
    opening towards the observer and the spire uppermost); relatively
    large (often > 15 mm long) .... order **Prosobranchia**

    Operculum lacking; shell is rather thin and delicate and may be
    globose, flat or planispiral and with sinistral (i.e. the aperture is on
    the left) or dextral coils; relatively small (usually ≤ 15 mm long)
    .... order **Pulmonata**

11. Body more-or-less elongated or worm-like with obvious
    segmentation and generally >30 similar segments; may have anterior
    and posterior suckers; if suckers are lacking then the segments
    bear paired, fleshy, lateral outgrowths or bundles of fine bristles
    (chaetae) .... true worms and leeches: phylum **Annelida** **12**

    Generally with jointed legs and a segmented body (which is often
    hardened) and *usually* divided into two or more discrete regions
    (e.g. head, thorax and abdomen); if legs are lacking then there are
    < 15 body segments and the head bears paired mandibles ....
    .... phylum **Arthropoda**[1] **15**

12. Body with anterior and posterior suckers and usually dorsoventrally
    flattened (Figs 4.5A-G); 34 segments (which may be externally
    subdivided) all lacking chaetae (setae)...
    .... leeches: class **Hirudinea** (= Hirudinoidea)

    Either lacking suckers, and the body is usually worm-like with
    many segments; *or* with a sucker on the posterior end, and only
    11 body segments) **13**

---

[1] Some workers treat the arthropods as an artificial group of unrelated phyla whose
members happen to have segmented bodies and jointed legs. In that case, the
component subphyla (Crustacea, Uniramia and Chelicerata) are raised to phylum
rank.

13. Small (<1 cm long); body rod- or flask-shaped with a posterior sucker (Figs 4.7A-C), and may have short tentacles on the head (Figs 4.7D & E); lacking chaetae or parapodia; living on the body surface of decapod crustaceans; rare .... class **Branchiobdellida**

    Body worm-like with no suckers **14**

14. With paired, fleshy lateral outgrowths (= parapodia) bearing bristles on most segments; may have palps or other protuberances on the head .... class **Polychaeta**

    Lacking parapodia but segments have fine chaetae ....
    .... class **Oligochaeta**

15. Four pairs of jointed legs; antennae lacking; body either somewhat globose (made up of a single unit: Figs 4.24A & B and 4.25A & B) or divided into two regions (tagmata: Figs 4.23A & B) with indistinct segmentation (some minute forms [$\leq$ 2 mm long] with globose bodies may have six legs only) ....
    water mites and spiders: subphylum **Chelicerata**: class **Arachnida**
    .... **21**

    Body not as above; antennae present (but may be inconspicuous)
    .... **16**

16. **Either** with three pairs of jointed legs **or** legless; one pair of antennae
    .... insects: subphylum **Uniramia**; class **Insecta**[2]

    More than three pairs of jointed legs; two pairs of antennae.....
    .... crustaceans: subphylum **Crustacea**

17. Five pairs of legs, of which at least the first (and often the second and third also) is chelate; eyes are on stalks and the carapace encloses a branchial chamber; may be quite large (> 40 mm body length) ....
    crabs, shrimps and crayfishes: order **Decapoda** (class **Malacostraca**)
    .... **18**

    Five or seven pairs of legs which are not chelate; eyes not on stalks and carapace absent; relatively small (< 40 mm body length)
    .... **20**

---

[2] As used here, the insects includes the order Collembola which are presumed to be primitively wingless in contrast to other aquatic insects which (if wingless) have lost their wings secondarily. For this reason, the Collembola are generally regarded as belonging to a separate class: the Apterygota (or Entognatha).

18. Well-developed abdomen and tail fan (Figs 4.17A & B and 4.18A & E); the rostrum (which projects from the front of the cephalothorax between the eyes) is conspicuous (Figs 4.15 & 4.20); body more-or-less laterally compressed **19**

Abdomen reduced and folded beneath the cephalothorax; body more or less flattened or rounded — 'crab-like' (Figs 4.21A — D); rostrum reduced or absent and tail fan lacking.........................
.... crabs: **Decapoda: Brachyura**

19. First three pairs of legs chelate — first pair of chelae much larger than the succeeding two; abdomen more-or-less dorsoventrally flattened; second segment of abdomen overlapped by the first laterally; rare ....
.... crayfishes: **Decapoda: Parastacidae and Cambaridae**

First two pairs of legs chelate (Figs 4.17A & B, 4.18A & E and 4.19C & D); cephalothorax and abdomen laterally compressed; second abdominal segment overlaps the first and third laterally
.... shrimps: **Decapoda: Caridea**

20. Five pairs of legs; body laterally compressed ....
.... order **Amphipoda** (class **Malacostraca**)

Seven pairs of legs with the posterior ones generally longer than the anterior ones; body dorsoventrally flattened.........................
.... order **Isopoda** (class **Malacostraca**)

21. Body comprises two distinct divisions or tagmata (Figs 4.23A & B) .... spiders: order **Araneae**

Body comprises a single unit or tagma (Figs 4.24A & B and 4.25A & B) .... water mites: suborder **Hydrachnida**[3] (order **Acarina**)

---

[3] Termed suborder here for convenience, although acarologists usually consider water mites to be a taxon of intermediate rank between superfamily and suborder (Smith & Cook, 1991).

LOWER INVERTEBRATES

Porifera

While most sponges are marine, around 100 species are confined to freshwater. The vast majority of them, and all of the tropical Asian genera bar one (*Pachydictyum*), are in the family Spongillidae (Penney & Racek, 1968). Most are confined to lentic habitats and their importance as stream inhabitants is typically minor. Where present, they encrust upon the upper or lower surfaces of rocks, submerged wood or the leaves of aquatic macrophytes. Penney & Racek (1968) list the known species and their global distribution, but there are certain to be additional, undescribed species. Frost (1991) has recently reviewed the ecology and systematics of North American freshwater sponges (see also Pennak, 1989: pp. 91–109; Hutchinson, 1993: pp. 62–71). Larvae of spongillaflies (Neuroptera: Sisyridae; see p. 512) feed on sponges, and are entirely dependent upon them.

Sponge species in the genera *Corvospongilla, Dosilia, Ephydatia, Eunapius, Metania, Pectispongilla, Spongilla, Stratospongilla, Trochospongilla* and *Umborotula* occur in tropical Asia (Annandale, 1911; Penney & Racek, 1968), but several genera of spongillids have a wide, almost cosmopolitan, distribution so others might be expected. Sponges are difficult to identify because their size and colour can vary according to season, maturity of the specimen and the ecological conditions. In flowing water, for instance, the growth form tends to be more encrusting. Some sponges are green because of the presence of algae (zoochlorellae) among the cells, but a particular species may not serve as an algal host under all circumstances. Identification depends mainly upon microscopic examination of the structure of the skeletal elements (siliceous spicules); these include megascleres which support the body tissues and are found in all species, and microscleres or dermal spicules. Sponges produce special spicules — gemmoscleres — during asexual reproduction: the gemmoscleres form a resistant wall around a resistant resting stage or gemmule which contains a mass of undifferentiated sponge tissue. Gemmosclere structure can be an important aid to sponge identification. Sponge life histories in Asia have yet to be studied in detail, but Annandale (1911: pp. 3–5) reports that some Indian species produce gemmoscleres during the warm months and flourish in winter, others grow most actively during the summer monsoon, while at least one species flourishes throughout the year. The genus *Spongilla* contains species which have adopted each of these three strategies.

Cnidaria

In marine environments, the Cnidaria (or coelenterates) are a remarkably successful phylum, including the jellyfish, corals, anemones and other polyps. The phylum is represented in freshwater by a few species of the class Hydrozoa, and apparently only three genera in tropical Asia: *Hydra* (order Hydroida: family Hydridae) — a solitary polyp — and *Craspedacusta* (Fig. 4.1A) and *Limnocnida* (order Limnomedusae) — 'jellyfish' with minute polypod larvae. The brackish-water hydrozoan genus *Cordylophora* may occur in the tidal reaches of Asian rivers; it has the same basic plan as *Hydra* but the individuals are joined in a continuous colonial structure (Fig. 4.1B). *Cordylophora* is known from Australia, Europe and America (Williams, 1980: pp. 41–42; Slobodkin & Bossert, 1991)

Because they are small and have soft bodies, freshwater cnidarians are often overlooked; in addition, they are not preserved well by the methods employed during routine sampling. All are predators on microinvertebrates and (possibly) fish larvae, but their ecological significance in streams has yet to been assessed. Species of *Hydra* are widespread in freshwater bodies (e.g. Annadale, 1911), and are often associated with submerged macrophytes. Some have algal cells (*Chlorella*) occupying vacuoles in their endodermal cells, and these animals are assigned to the genus *Chlorohydra*. While species are difficult to distinguish (body size, relative tentacle length, and reproductive details are diagnostic), the two genera *Hydra* and *Chlorohydra* are easily separated by the bright green colour of the latter. Hutchinson (1993: p. 76) remarks that a third hydrid genus — *Pelmatohydra* — contains a cosmopolitan species (*P. oligactis*) which presumably occurs in tropical Asia.

Craspedacusta (Olindiidae), like most cnidarians (except *Hydra*), has a life cycle including two stages: a polyp and a medusa. The medusa or jellyfish form of the life cycle (Fig. 4.1A) is generally confined to lakes, ponds and floodplain pools, while the polyp stage (which in *Craspedacusta* lacks tentacles) is found in streams also (Slobodkin & Bossert, 1991; Hutchinson, 1993: p. 71). *Craspedacusta sowerbii* has been introduced widely to freshwaters in Europe, North and South America and Australia. This species is thought to be native to the Chang Jiang drainage basin, with *C. sowerbii* and *C. sinensis* in the upper basin and *C. sowerbii* in downstream areas. The genus includes two additional species confined to Japan (Dumont, 1994). The reason for placing the origin of *C. sowerbii* in the Chang Jiang relates to

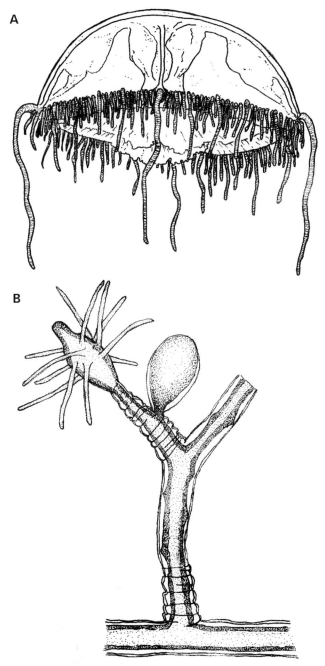

Fig. 4.1 Freshwater cnidarians (redrawn from Pennak, 1989): A, the freshwater 'jellyfish' *Craspedacusta sowerbii* (approximate diameter 12 mm): B, one hydranth (a feeding individual approximately 2.5 mm long) and an immature gonopore (or ovoid reproductive polyp) of the colonial hydrozoan *Cordylophora*.

sexuality: in the Chang Jiang populations generally comprise both males and females; elsewhere they are single-sexed and either entirely male or female (Dumont, 1994 and references therein). A second genus of Limnomedusae — *Limnocnida* (Limnocnididae; referred to as *Limnocodium* by Slobodkin & Bossert, 1991) — occurs in India and Africa, and is found at low densities in rivers although numbers can build up in backwaters on floodplains. The validity of two further genera of Limnomedusae from India (*Keralica* and *Mansariella*) requires confirmation (Dumont, 1994). The ecology of and global distribution of Limnomedusae is reviewed by Dumont (1994), and accounts of the biology and classification of freshwater Cnidaria in North American are given by Pennak (1989: pp. 110–123) and Slobodkin & Bossert (1991); Hutchinson (1993: pp. 71–85) gives an account of their natural history in (mainly) standing waters.

Turbellaria

The Turbellaria or flatworms are the only non-parasitic class in the phylum Platyhelminthes (which also includes the flukes and tapeworms: Trematoda and Cestoidea). They are found in a range of freshwater habitats but most commonly in streams where, in the Palaearctic at least, different species show characteristic patterns of longitudinal zonation and replacement (e.g. Kawakatsu, 1964; Kolasa, 1991). Because they are soft-bodied and fragile, flatworms require special attention if they are to be preserved intact, and these animals are often overlooked during examination of benthic samples. It is therefore likely that they are more diverse and abundant in tropical Asian streams than has been supposed.

To date, most turbellarian records from the Oriental region (e.g. Kawakatsu & Iwaka, 1968; Ball, 1970; de Beauchamp, 1973; Kawakatsu & Basil, 1971; Kawakatsu, 1973; Kawakatsu & Ogawara, 1974; Kawakatsu *et al.*, 1979; 1989; Kawakatsu & Mitchell, 1989a; 1989b; Lue, 1989; Vyas *et al.*, 1989) are species of the genus *Dugesia* (family Dugesiidae: Fig. 4.2) in the order Tricladida which are found most frequently on the underside of stones in upland streams. At least seven epigean *Dugesia* species are known from the region (*D. andamanensis*, *D. annandalei*, *D. bactriana*, *D. burmaensis*, *D. indica*, *D. lindbergi* and *D. nannophallus*: Ball, 1970), and a further two from Japan, one of which — *D. japonica* — seems to extend south into Taiwan (Kawakatsu & Iwaki, 1968). *Dugesia batuensis* — which lacks

Fig. 4.2 Gross anatomy of the turbellarian genus *Dugesia* (Tricladida: Dugesiidae) showing the three divisions of the gut and the pharynx lying within the pharyngeal chamber in the middle of the body (redrawn from Pennak, 1989).

pigment — is hypogean (Ball, 1970; Kawakatsu *et al.*, 1989), as are other members of the genus described more recently (e.g. Kawakatsu & Mitchell, 1989a; 1989b). Other triclads include the family Planariidae (e.g. *Fonticola*, *Phagocata* and *Polycelis*) which is restricted to the northern part of Asia and does not seem to penetrate south of Japan (e.g. Ichikawa & Kawakatsu, 1963) into tropical latitudes, although there are records of a species of *Planaria* (*P. arborensis*) from the Indian subcontinent (Fernando, 1969 and references therein; Gupta & Ghosh, 1979: p. 253). The Dendrocoelidae (*Bdellocephala*, *Dendrocoelopsis* and *Sphalloplana*) likewise occur in the north of eastern Asia beyond the tropics. The low generic diversity of tropical triclads is not surprising since — even in temperate latitudes — the majority of species are cold-water stenotherms. Significantly, sexual reproduction by Sri Lankan and Indian *Dugesia* occurs only during those months when temperatures fall below 24°C, although asexual reproduction by binary fission is possible throughout the year (de Silva & de Silva, 1980; Aditya & Mahapatra, 1991).

Triclads are characterized by the possession of a three-branched intestinal cavity (Fig. 4.2) and a muscular pharynx which is extruded posteriorly from the mouth during feeding. They consume a variety of invertebrates and carrion, but a degree of prey selection is indicated by the report that *Dugesia nannophallus* in Sri Lankan streams specializes on gastropod prey (de Silva & de Silva, 1980). Triclad classification and biology, and that of other orders of North American turbellarians, has been reviewed by Pennak (1989: pp. 124–151) and Kolasa (1991). Living *Dugesia* can be separated from *Planaria* by the appearance of the head: the former has lateral outgrowths or auricles giving the head a triangular shape; it also possesses relatively large eyes. Identification of triclad species is difficult for the non-specialist: for example, Ball (1970) has constructed a key to the Asian species of *Dugesia* which depends on the availability of serial histological sections of the reproductive organs of sexually-mature animals.

Apart from the Tricladida, there are reports of other orders of non-parasitic Turbellaria from Sri Lanka and elsewhere in Asia. The majority of species are most common in standing water where some are planktonic. Records include *Catenula* sp., *Convoluta anostica*, *Mesostoma rostratum*, *M. erhenbergi*, *Stenostomum unicolor* and *Macrostomum tuba* (Mendis & Fernando, 1962; Fernando, 1969; 1974; Kawakatsu & Basil, 1971). The latter three species are widely distributed in the tropics. In earlier classifications these genera were placed in the order Rhabdocoela, but recent texts treat them as belonging to three separate orders: the Catenulida (*Catenula, Stenostomum, Convolutum*), Neorhabdocoela (*Mesostomum*) and Macrostomida (*Macrostomum*) (e.g. Pennak, 1989: pp. 124–151; Kolasa, 1991). They differ from the triclads in their smaller size (< 6 mm long) — hence they are sometimes referred to as the microturbellarians — and the fact that their bodies are covered with cilia; in addition, the intestinal cavity comprises a simple sac. They feed on protozoans, zooplankton and algae.

Of peripheral interest (although not belonging strictly to the category of 'benthic macroinvertebrates') is the order Temnocephalida (sometimes considered a class: the Temnocephalidea) comprising tiny (< 3 mm long) transparent flatworms with short tentacles anteriorly and a muscular adhesive organ posteriorly. They live inside the branchial chambers of atyid shrimps (*Caridina* is host to species of *Caridinicola*, *Monodiscus* and *Paracaridinicola*: Annandale, 1912; Mendis & Fernando, 1962) and freshwater crabs (which harbour *Temnocephala* spp.), as well as on the antennae, antennules, leg bases and major chelae of parastacid crayfish (*Cherax* spp. which host *Diceratocephala boschmai*: Baer, 1953; Jones & Lester, 1992). Temnocephalids are probably not parasitic, but use their hosts as a substratum for attachment and to gain access to food. Further information is given by Williams (1980: pp. 46–49) and Jones & Lester (1992) who provide an account of the biology of Australian Temnocephalida.

Nemertea

Little attention has been paid to the phylum Nemertea, or ribbon worms, by stream ecologists. Like turbellarians, they are almost unrecognizable after preservation of field samples in formalin or alcohol because of excessive contraction, but their relative neglect is likely to be a reflection of low global diversity. Most nemerteans are marine, but at least 12 species in two families and six genera live in freshwater (Moore & Gibson, 1985). Of these, the genus *Prostoma* (Tetrastemmatidae) contains most of the freshwater species, including some which live in slow-flowing rivers and streams in North and South America, Japan, Europe, Africa and Australia (Young & Gibson, 1975; Gibson & Moore, 1976; Williams, 1980: pp. 53–55; Moore & Gibson, 1985; 1988; Kolasa, 1991); *Prostoma* is likely to occur in tropical Asia also. An additional, monotypic genus *Planolineus* (Valenciniidae) is known from a single locality in Java (Gibson & Moore, 1976), and *Amniclineus zhujiangensis* (Lineidae) was described recently from the Zhujiang, China (Gibson & Qi, 1991). Most freshwater nemerteans are hermaphroditic and reproduce sexually (Gibson & Moore, 1976), but virtually nothing is known of their life histories or ecology (*cf.* Young & Gibson, 1975). They feed on other invertebrates using a long, eversible proboscis armed with a terminal stylet. The functional importance of nemerteans in stream ecosystems is unknown (Kolasa, 1991) and probably minor. Gibson & Moore (1976) give a key to the freshwater nemerteans of the world that were known in 1975.

Nematoda

Nematodes (or roundworms) are among the most numerous and diverse group of organisms on earth. Many are parasitic and of great economic importance, and members of the phylum inhabit almost all types of environment. Most free-living freshwater species are small (< 10 mm long) and, because they are thin, tend to pass through the mesh of all but the finest nets used for benthic sampling. Consequently, nothing can be stated with confidence about their distribution, abundance and importance in tropical Asian rivers and streams. Many freshwater genera are cosmopolitan and, although a number of families have freshwater representatives, three families attain their greatest diversity in fresh waters: Dorylaimidae, Plectridae and Trilobidae. Of additional interest is the family Mermithidae (e.g. *Hydromermis* in Asia) which parasitise aquatic insects (including Trichoptera, Ephemeroptera and Diptera — especially Chironomidae); the larval worm is parasitic, and the adult (which may reach 5 cm length) kills the host upon exiting to take up a free-living aquatic existence.

Nematode identification depends on microscopical examination (often under oil immersion) of the stoma (mouth), pharynx, cuticle and genital structures of worms which have been anaesthetized and relaxed before death. The taxonomy of the phylum is intricate and in a state of flux, and reliable generic determinations of Asian species are likely to prove problematic. Fernando (1964) records several free-living genera from Sri Lankan freshwaters (*Plectus, Monohystera, Monochromadora, Tobrilus, Actinolaimus, Bathyonchus, Dorylaimus* and *Mononchus*; see also Kreis, 1936; Schneider, 1937) but gives no information on their ecology; however, all but *Bathyonchus* are recorded from North America (Poinar, 1991) and are probably more-or-less cosmopolitan. Keys for Nearctic genera are given by Pennak (1989: pp. 226–245) and Poinar (1991) who have reviewed the biology of freshwater nematodes. A useful key to American genera is provided by Tarjan *et al.* (1977) also. Poinar (1991) laments the lack of attention nematodes have received from freshwater biologists in general (and stream ecologists in particular), citing difficulties associated with sampling, extraction and identification.

Nematomorpha

The phylum Nematomorpha (= hairworms) contains two orders, one of which — the Gordioidea — is confined to fresh water and has more than 100 species in five families. Little has been published on their biology or ecology (for recent summaries see Pennak, 1989: pp. 246–253; Poinar, 1991), but life cycles of all species involve a parasitic stage. Nematomorph eggs may be ingested directly if the host is an aquatic invertebrate. Alternatively, larvae encyst on vegetation at or near the water's edge soon after hatching, and are eaten accidentally by a herbivorous terrestrial insect. The cyst disintegrates, releasing the larva which bores through the gut wall into the haemocoel where it lives as an endoparasite. Upon reaching maturity (when it may attain lengths of 10 cm to one metre), the worm breaks out of the host's body and returns to the water. It is probable that the parasite's chances of getting back to the aquatic environment are increased by the manipulation of host behaviour in some way, compelling it to seek water. Poinar (1991) gives further details of the life cycles and hosts of nematomorphs (see also Pippet & Fernando, 1961).

The cosmopolitan genus *Gordius* occurs in tropical Asian streams (e.g. in Malaysia: Yang *et al.*, 1990), in addition to *Baetogordius*, *Chordodes* and *Paragordius* (Mendis & Fernando, 1962 and references therein; Bishop, 1973: p. 198; Gupta & Ghosh, 1979: p. 318), but other nematomorphs are certain to be present; for example, the genus *Gordionus* has been reported recently from central Vietnam (Spiridonov, 1993). Nematomorph identification depends on details of the cuticular ornamentation, shape of the anterior and posterior ends of the body, and structure of the male genital region.

Ectoprocta (Polyzoa)

The phylum Ectoprocta or Polyzoa comprises mainly marine forms, but about 50 species occur in fresh water. They are most diverse in equatorial regions (Bushnell, 1973), and widely-distributed species (e.g. *Plumatella javanica*) tend to show more luxuriant growth in tropical than temperate regions. All ectoprocts are colonial and sessile comprising tiny zooids which build up colonies by budding. These may be encrusting or branching structures, or united into a compact gelatinous form. The zooids feed on suspended particles by means of an oval or horseshoe-shaped lophophore which bears numerous ciliated tentacles (Figs 4.3A & B). Ectoprocts are common in standing waters and stream backwaters attached to submerged wood, rocks or aquatic macrophytes, but certain species (e.g. *Plumatella emarginata*: Plumatellidae) thrive in both lentic and lotic habitats (Wood, 1991) — even in turbulent riffles (Karlson, 1991). Certain species can tolerate organic pollution and flourish under eutrophic conditions (Bushnell, 1974; Shrivastava *et al.*, 1985); they sometimes foul water pipes and the screens and intake grates of hydroelectric plants. Diversity in tropical Asian streams can be quite high: Rao (1973) records 10 species from the Narmada River basin in central India (although some of them may have been collected in ponds).

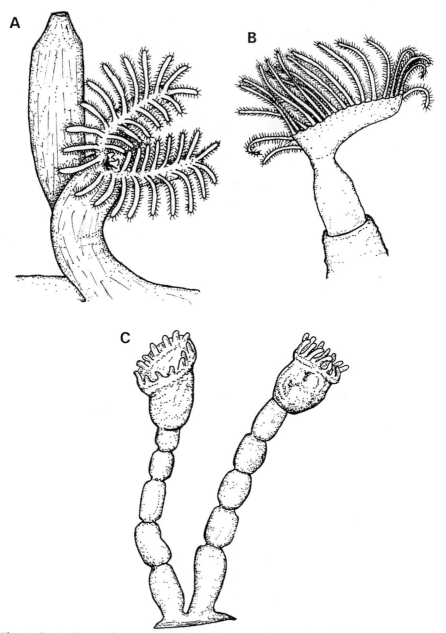

Fig. 4.3 Freshwater bryozoans (redrawn from Pennak [1989] and Wood [1991]): A, two zooids of a tubular ectoproctan colony, one retracted, the other extended in a feeding position (zooid length 3–4 mm); B, a *Plumatella* (Ectoprocta) zooid, and details of the lophophore (one side shown only); C, two zooids of the entoproct genus *Urnatella* (zooid length approximately 4 mm).

Studies of the ecology or basic biology of ectoprocts in tropical Asia are sparse (*cf.* Rao *et al.*, 1978; Shrivastava & Rao, 1985), as is true of the group elsewhere (Wood, 1991; but see Hutchinson, 1993: pp. 85–99; Mukai *et al.*, 1990; Oda, 1990), and there are rather few records of their occurrence in Asia (summarized in Table 4.1). Species such as *Victoriella pavida* and *Paludicella articulata* might be expected throughout the region as they have an almost cosmopolitan distribution; *V. pavida* lives in both fresh- and brackish water (hence it is not included in Table 4.1) and these two genera are in the same class (the Gymnolaemata) as the marine Ectoprocta. Annandale (1911) records a third genus of this class — *Hislopia* — from Indian fresh waters; it occurs throughout southern Asia (see also Rao, 1973 and references therein). Most strictly freshwater ectoprocts belong to the Phylactolaemata including the widely-distributed *Lophopodella carteri* (Lophopodidae) which is known from India and Sri Lanka (Annandale, 1911; Rao, 1973; Fernando, 1980). The Indian ectoproct fauna consists of at least 18 species (Annandale, 1911: pp. 7–10; Rao, 1973; Wiebach, 1974). Genera such as *Plumatella*, *Afrindella*, *Stolella*, *Stephanella*, *Gellatinella*, *Hyalinella* (Plumatellidae), *Pectinatella*, *Lophopus* (Lophopodidae) and *Fredericella* (Fredericellidae) have a broad geographic distribution (Annadale, 1911; Wood, 1991; Hutchinson, 1993: p. 86) and may be quite prevalent in tropical Asia. Of the total of 31 species recorded from the region (Table 4.1), at least ten (including three *Hislopia* species and the genus *Varunella* which was established by Wiebach [1974]) are endemic. *Plumatella* (Fig. 4.3B) seems to be particularly speciose and widespread.

Ectoproct identification can be difficult due to the extent of variation in colony form (in *Plumatella repens* or *P. emarginata*, for example: Mukai *et al.*, 1990), and examination of the resting bodies or statoblasts (sclerotized bi-valved structures containing a mass of undifferentiated cells which are produced by asexual reproduction) is usually required (Williams, 1980: pp. 72–78); these are present in the Phylactolaemata only. Keys to the Nearctic species (including some expected to occur in tropical Asia) are given by Pennak (1989: pp. 269–289) and Wood (1991). Life histories of tropical Asian ectoprocts have received little attention, although Shrivastava & Rao (1985) provide some information on the occurrence and growth of *Plumatella emarginata* in central India (Madhya Pradesh). Annadale (1911: pp. 3–5) reports that, like the Indian freshwater sponges, some ectoproct colonies flourish in the cooler months while others grow best during the summer monsoon; still others grow throughout the year. In species where statoblast

Table 4.1   Species of freshwater Ectoprocta recorded from tropical Asia (data derived from Bushnell, 1973; Rao, 1973; Weibach, 1974; Rao *et al.*, 1985). Of the total of 31 species, those with a cosmopolitan distribution and those endemic to the Oriental Region are indicated; all species treated as non-endemics are shared with at least one other biogeographic region.

Species	Cosmopolitan	Endemic
Gymnolaemata		
Paludicellidae		
Paludicella articulata	-	-
P. pentagonalis	-	-
Hislopiidae		
Hislopia cambodgiensis	-	+
H. lacustris	-	-
H. malayensis	-	+
H. moniliformis	-	+
Phylactolaemata		
Fredericellidae		
Fredericella sultana	+	-
Plumatellidae		
Plumatella casmiana	-	-
P. emarginata	+	-
P. ganapato	-	+
P. javanica	-	-
P. longigemmis	-	-
P. repens	+	-
P. vorstmani	-	-
Hyalinella diwaniensis	-	+
H. indica	-	-
H. minuta	-	-
H. punctata	+	-
Afrindella bombayensis	-	+
A. philippinensis	-	-
A. tanganyikae	-	-
A. testudinicola	-	+
Varunella coronifera	-	+
V. indorana	-	+
Stollela indica	+	-
Gelatinella toanensis	-	-
Lophopodidae		
Lophopus sp.	-	-
Lophopodella carteri	-	-
L. pectinatelliformis	-	+
L. stuhlmanni	-	-
Pectinatella gelatinosa	-	-

production is seasonal, it tends to occur at the approach of the period of poorest growth. Note, however, that these remarks apply to ectoprocts in lentic habitats in India; apparently Annadale's monograph was based upon specimens obtained in '. . . stagnant water' (Annadale, 1911: p. 23).

The phylum Entoprocta is a small group of animals, similar to — but distinct from — the Ectoprocta. Despite a macroscopic resemblance, two phyla differ in that ectoprocts have the anus located outside the lophophore, but the endoproct anus is located on an erect, stalk-like segment within the lophophore. This seemingly small difference has considerable phylogenetic importance. Most entoprocts are marine and the only freshwater genus is *Urnatella* (Fig. 4.3C): *U. gracilis* has an almost cosmopolitan distribution, and a second species is confined to India (Pennak, 1989: p. 269). *Urnatella* can be distinguished easily from any of the freshwater ectoprocts because the zooids have an externally-segmented stalk. Moreover, they do not produce statoblasts. The ectoprocts and entoprocts were formerly treated as classes within a single phylum — the Bryozoa — and thus are often referred to collectively as bryozoans.

Polychaeta

Most polychaetes are marine animals but some species live in freshwater sections of rivers and streams; typically they are found at low altitudes or close to the coast. Wesenburg-Lund (1958) lists 17 species of freshwater polychaetes in six families, plus a further eight species in both fresh- and brackish water, but the total number of freshwater species is probably nearer 50. Their biology has been reviewed most recently by Davies (1991) who notes the tendency for freshwater polychaetes to undergo direct development with suppression of the normal (marine) planktonic trochophore larvae. This is analogous to the suppression of planktonic larval stages in lotic crabs and shrimps and some bivalves (see p. 149, pp. 174–175 and p. 180).

Most freshwater polychaetes (or those which can live in both fresh and brackish waters) belong to the family Nereidae (*Nereis, Chinonereis, Lycastis, Lycastopsis, Namalycastis, Tylorrhynchus*), but the Phyllodocidae (*Eteone*), Nephtyidae (*Nephtys*), Spionidae (*Pseudopolydora*), Ampharetidae, Sabellidae (*Caobangia, Laonome*) and Serpulidae contain freshwater species also. The latter three families are tube dwellers while the former two are active crawlers and swimmers (errant polychaetes). As a generalization, only the errant polychaetes are found in running waters, and they appear to be quite common in the lower course of large rivers such as the Ganges, Chang Jiang and Zhujiang among others (Ray *et al.*, 1966; Pahwa, 1979; Qi *et al.*, 1982; Shen & Qi, 1982; Wu & Chen, 1986; Singh *et al.*, 1988; Su & Li, 1988; Ahmad & Singh, 1989; Datta Munshi *et al.*, 1989). Some genera (*Eteone, Laonome, Namalycastis* and *Nephtys*) are quite tolerant of organic pollution, while *Pseudopolydora* thrives at highly-enriched sites (Shen & Qi, 1982). Nevertheless, polychaetes do not occur throughout the course of Asian rivers, and they seem to be absent from upstream sections of all streams. Because the freshwater polychaetes are so poorly known (Davies, 1991), there may be additional families and certainly further genera in tropical Asian freshwaters, and thus the brief list given here should not be considered as definitive in any way. Furthermore, some marine polychaetes may penetrate brackish or fresh water although they are unable to breed there.

The polychaetes are distinguished from other classes in the phylum Annelida by the presence of parapodia: muscular, lateral projections from the body wall borne in pairs on most of the body segments. The parapodia have chaetae also. Additional distinguishing features are tentacles, jaws or other appendages on the anterior end. Their ecology has received little attention, but some genera (e.g. *Namalycastis* and *Nephtys*) thrive in polluted reaches of the Ganges (Datta Munshi *et al.*, 1989) where they co-occur with oligochaetes such as *Tubifex* (see pp. 112–118). The Nereidae and Nephtyidae are predators or omnivores; most other species are filter-feeders or deposit-feeders (Davies, 1991). The sabellid, *Caobangia billeti*, bores into the shell apex of prosobranch snails such as *Brotia* and *Paracrostoma* in Thai rivers (Brandt, 1974).

Oligochaeta

Oligochaetes are segmented worms with bundles of chaetae (or setae) on every segment except the first. Unlike the polychaetes, the anterior portion of the body has no other projections or appendages, and most oligochaetes (apart from the family Naididae) lack eyes. Compared to the familiar earthworm, most freshwater oligochaetes are rather small (< 30 mm long) and delicate with thin, almost transparent, body walls. Gaseous exchange occurs through the general body surface in all species, but this is supplemented by external gills in some species. All oligochaetes are hermaphrodites but cross-fertilization is usual; there are no larval stages. The biology of the group has been reviewed recently by Brinkhurst & Gelder (1991; see also Pennak, 1989: pp. 290–307).

A catalogue of global aquatic oligochaete species is given in Brinkhurst & Jamieson (1971; supplemented by Brinkhurst & Wetzel, 1984), while the oligochaetes of the lower courses of the Chang Jiang and Zhujiang are listed by Erseus & Qi (1985), Qi & Erseus (1985), Liang (1987) and Qi (1987) (Table 4.2). More recently, Brinkhurst *et al.* (1990) have catalogued the freshwater oligochaetes of China, noting a general similarity with the Indian fauna (e.g. Naidu & Srivastava, 1980; Naidu & Naidu, 1981a; 1981b). Many Chinese and Asian genera (and species) are cosmopolitan (e.g. *Aeolosoma, Aulodrilus, Bothrioneurum, Branchiura, Chaetogaster, Dero* — which is especially speciose in Asia — *Eiseniella, Haplotaxis, Limnodrilus, Nais, Pristina, Slavina* and *Tubifex*) or distributed widely in the northern hemisphere (e.g. *Allonais, Haemonais, Lumbriculus, Rhyacodrilus, Stylaria* and *Uncinais*) (Timm, 1980). The total Chinese fauna amounts to 73 species in 31 genera and 4 families (Naididae, Tubificidae, Lumbriculidae and Haplotaxidae). Most (29 species in 13 genera) are Naididae, and Brinkhurst *et al.* (1990) believe that further collecting in the southern provinces would yield additional tropical representatives of the family. They note also that there is debate over the specific limits and identities of a number of Chinese species — especially Naididae — and that these problems apply to the Oriental oligochaete fauna in general.

Tropical Asia contains about 170 species of freshwater oligochaetes (24% of the world fauna; Timm, 1980) including mainly Tubificidae and numerous naidids. The Glossoscolecidae (*Glyphidrilus*) are found in the region also although the majority of the family are terrestrial and confined to damp mud at water margins or may be accidental in freshwater habitats. Most of the 700 or so species of oligochaetes known from fresh water are confined to temperate latitudes because they must undergo hibernation in cold conditions before successful sexual reproduction can take place (Timm, 1980). The prevalence of Naididae in the tropics can thus be explained, at least in part, by their ability to undergo asexual reproduction (by budding to form chains of individuals). Asexual reproduction by fragmentation occurs also in the Tubificidae which contains a number of successful tropical genera. Significantly, among the thermophobe Lumbriculidae, which are Holarctic in distribution, the only species to have colonized tropical

Table 4.2   Aquatic oligochaetes from the Chang Jiang (Sichuan and Hubei Provinces) and Zhujiang (near Guangzhou, Guangdong Province), People's Republic of China. Data from Erseus & Qi (1985), Liang (1987) and Qi (1987).

Species	Chang Jiang	Zhujiang
Naididae		
Paranais frici	+	-
Nais inflata	+	-
Nais papardalis	+	-
Nais sp.	+	-
Arcteonais lomondi	+	-
Aulophorus furcatus	+	+
Pristina accuminata	-	+
Pristina spp.	-	+
Dero digitata	-	+
Dero nivea	-	+
Dero dorsalis	-	+
Brachiodrilus hortensis	-	+
Tubificidae		
Tubifex tubifex	+	-
Tubifex sp.	+	-
Limnodrilus claparedeianus	+	+
Limnodrilus hoffmeisteri	+	+
Limnodrilus grandisetosus	+	+
Limnodrilus silvani	+	-
Limnodrilus udekemianus	-	+
Aulodrilus pluriseta	+	-
Aulodrilus prothecatus	-	+
Telmatodrilus sp.	+	-
Rhyacodrilus riabuschinskii	+	-
Rhyacodrilus sodalis	-	+
Monopylephorus limosus	+	-
Branchiura sowerbyi	+	+
Teneridrilus mastix	-	+
Tectidrilus achaetus	-	+
Enchytraeidae		
Enchytraeus (?) sp.	+	-
Total number of species	18	16

latitudes (*Lumbriculus vareigatus*) can reproduce asexually (Timm, 1980).

Much data on oligochaete ecology concerns feeding behaviour, and is based on studies of tubificids and the sediments which they ingest (and therefore defaecate) continuously (Brinkhurst & Gelder, 1991); the organic component and bacteria are digested during passage through the gut. There is much less information on other families (with the exception of the deposit-feeding Holarctic lumbriculid *Stylodrilus*), but *Chaetogaster* (Naididae) is unusual in that it is a predator on microinvertebrates; some other naidids feed mainly on algae. Unfortunately, most ecological studies of oligochaetes have been concerned mainly with lentic species. In running waters, oligochaetes are most numerous in slow-flowing reaches with muddy beds, especially in the floodplain sections of larger Asian rivers (e.g. Ray *et al.*, 1966; Pahwa, 1979; Qi & Erseus, 1985; Qi, 1987; Wu & Chen, 1986; Ahmad & Singh, 1989; Bandhavong, 1990). Average densities of oligochaetes along the Chang Jiang in its floodplain are 231 — 343 individuals.m^{-2} for tubificids, and 21 — 4,982 individuals.m^{-2} for naidids (Liang, 1987), but densities of Naididae — chiefly *Nais inflata* — can reach 90,504 individuals.m^{-2}. Certain species (such as the tubificids *Branchiura*, *Limnodrilus* and *Tubifex*) can be extremely abundant (e.g. up to 700,000 individuals m^{-2} in the Zhujiang near Guangzhou; Qi & Erseus, 1985) at organically-polluted sites (e.g. Qi *et al.*, 1982; Qi & Lin, 1985; Qi, 1987; Su & Li, 1988).

Oligochaetes are often overlooked in stony streams, because their fragile bodies are fragmented during most benthic sampling procedures. Investigations are hampered further by the fact that identification can depend on details of the reproductive system which may involve preparation of serial histological sections of the relevant body segments. Nevertheless, identification to families, at least, is practical, and a short key to the major families of freshwater oligochaetes (which follows Brinkhurst & Gelder, 1991) is given below. Note that the key does not include families (e.g. Lumbricidae, Megascolecidae and Glossoscolecidae) which are primarily terrestrial. However, the southern hemisphere Phreodrilidae (which has a single Sri Lankan representative of *Phreodrilus*) and the Enchytraeidae have been included. The latter family is mainly terrestrial but they may be part of the Indian and Chinese freshwater oligochaete fauna (Naidu & Srivastava, 1980; Brinkhurst *et al.*, 1990). The Haplotaxidae, which seem to occur mainly in groundwater, are included also because they may be encountered occasionally in streams.

1. Ventral chaetae large, sickle-shaped, arranged singly with two in each segment; dorsal chaetae are small and straight, often missing from some or all segments; prostomium (the anterior-most segment) long with a transverse furrow; mouth large and muscular; body long (up to 15 cm) and slender (like a member of the Nematomorpha); uncommon .... **Haplotaxidae** (*Haplotaxis*)

All chaetae paired or in bundles; prostomium short, often conical 2

2. Chaetae usually paired; if not, simple in structure and pointed 3

   Chaetae more than two per bundle; usually bifid (Fig. 4.4A) 6

3 Long, hair-like chaetae included in the dorsal bundles of chaeta; ventral chaetae simple or bifid, may be one of each in every bundle; rare (Sri Lanka only) .... **Phreodrilidae**

   Lacking hair chaetae in the dorsal bundles 4

4 Chaetae paired and bifid (although the teeth are asymmetric: Fig. 4.4A) .... **Lumbriculidae** (in part)

   Chaetae simple and pointed 5

5 Relatively large worms with nodulate chaetae .... **Lumbriculidae** (in part)

   Relatively small worms with straight chaetae (may be missing from some segments) .... **Enchytraeidae**

6 Usually < 1 cm long; with or without eyes; long, hair-like chaetae included in the dorsal bundles (Fig. 4.4B) which are usually made up of simple or bifid chaetae (rarely oar-shaped or palmate); however, dorsal chaetae may be limited to only two per bundle or absent altogether (in *Chaetogaster*); dorsal chaetal bundles may start on segment II, V or VI or further posteriorly; ventral chaetae are bifid, and posterior ventral chaetae (> segment V) may differ in structure from those anteriorly; perianal gills may surround the anus (in *Dero*: Figs 4.4E & F); asexual reproduction involves budding to form chains of individuals (zooids) .... **Naididae**

   Usually > 1 cm (width 0.5–1 mm); without eyes; dorsal chaetae begin at segment II; dorsal bundles may or may not include hair chaetae, but other dorsal chaetae (Fig. 4.4C) are pectinate, bifid, palmate or oar-shaped; ventral chaetae usually bifid; may be dorsal and ventral gill filaments on some posterior segments (in *Branchiura*: Fig. 4.4D) .... **Tubificidae**

Many — but not all — genera of Asian freshwater oligochaetes can be identified using the family keys given by Brinkhurst & Gelder

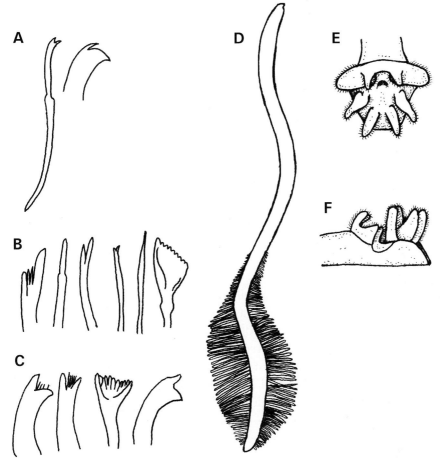

Fig. 4.4 Diagnostic features of aquatic oligochaetes (redrawn from Pennak [1989] and Brinkhurst & Gelder [1991]): A, bifid chaetae typical of Lumbricidae; B, dorsal chaetae typical of Naididae (the ventral chaetae are usually bifid); C, dorsal chaetae typical of Tubificidae; D, a diagram of *Branchiura sowerbyi* showing the posterior gills; E & F, dorsal and lateral views (respectively) of the posterior end and perianal gills of *Dero digitata*.

(1991); in particular *Allonais, Amphichaeta, Arcteonais, Chaetogaster, Dero, Haemonais, Nais, Paranais, Pristina, Pristinella, Ripistes, Slavina, Specaria, Stephensoniana, Stylaria, Uncinais* and *Vejdovskyella* (Naididae), and *Aulodrilus, Branchiura, Ilyodrilus, Limnodrilus, Rhyacodrilus, Spirosperma, Teneridrilus* and *Tubifex* (Tubificidae).

The Aphanoneura is a small group of worms of uncertain taxonomic affinity but probably related to the Oligochaeta, consisting of the family Aeolosomatidae and including the cosmopolitan genus

Aeolosoma (Timm, 1980) which is found in tropical Asia (e.g. Naidu, 1963; Costa, 1967; Naidu & Srivastava, 1980). Although treated as oligochaetes by some workers, aeolosomatids differ from them in that they lack a clitellum and move in a gliding motion using cilia. They are generally only 1–2 mm in length, but may occur in chains of individuals (produced asexually by budding) up to 10 mm long. Hair chaetae are present in both the dorsal and ventral chaetal bundles, in contrast to the oligochaetes which have hair chaetae in the dorsal bundles only. A conspicuous distinguishing character is the presence of coloured (or refractive) epidermal glands in the body wall. Aeolosomatids feed on suspended organic particles that are swept towards the mouth by ciliary currents.

Hirudinea

Leeches are distinguished easily from other annelid classes by the presence of an anterior and a posterior sucker (Figs 4.5A-G). The mouth is situated in the middle or towards the front of the anterior sucker. Most leeches are dorsoventrally flattened to some extent, and there is a fixed number of 34 body segments although they are subdivided superficially and hence are not easy to count (see below). Furthermore, there is considerable fusion of the primary segments of the body at the extremities — especially the posterior end. Locomotion is by swimming (in some species) or, more commonly, by 'looping' using a combination of the body muscles and the two suckers. Reproduction is sexual and involves cross-fertilization of hermaphrodites: sequential protandrous hermaphroditism (males changing to females) is common. Eggs are enclosed by a cocoon and fixed to the substratum or, in the Glossiphoniidae, the eggs (inside a membraneous capsule) and young are brooded on the ventral surface of the body. Hatchlings may feed upon mucous secretions produced by the parent. Some aspects of the life history and growth of *Glossiphonia weberi* in India are described by Raut & Saha (1986a; 1986b). Many leeches breed once and then die (i.e. they are semelparous), but iteroparity (repeated breeding) is known in some larger sanguivorous species. General information on leech biology is given by Davies (1991) and Pennak (1989: pp. 314–335), while the group has been reviewed at length by Sawyer (1986).

The class Hirudinea consists of two orders. The Rhynchobdellida have a small pore-like mouth and an extrusible proboscis. The

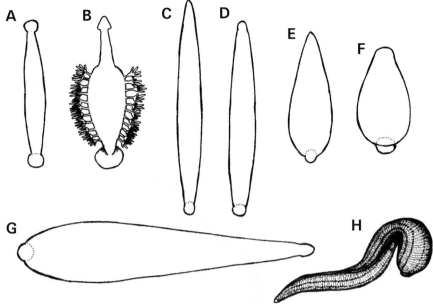

Fig. 4.5 The body forms of freshwater leech (Hirudinea) families and some distinctive Asian genera: A, Piscicolidae; B, Piscicolidae (*Ozobranchus*); C & D, Erpobdellidae (*Erpobdella* and *Barbronia* respectively); E, Glossiphoniidae (*Glossiphonia*); F, Glossiphoniidae; G, Hirudinidae; H, Hirudinidae or Haemopidae (*Whitmania*).

Arhynchobdellida (sometimes divided into the Gnathobdellida — including the Hirudinidae — and the Pharyngobdellida — comprising the Erpobdellidae) have a relatively large mouth which usually bears toothed jaws; they lack a proboscis. Arhynchobdellidan saliva contains a polypeptide — hirudin — which prevents clotting of the host/prey blood. Most leeches occur in fresh water, but some are marine or — in the tropics — terrestrial (e.g. the Haemadipsidae). The majority of freshwater species occur in still or slow-flowing waters, but others are confined to rapid streams. Most are found on (or under) stones or aquatic macrophytes, but some (e.g. *Whitmania*) live in mud. Because they sometimes attach to humans, it is well known that certain leeches are blood-feeding ectoparasites of mammals, but sanguivorous species may specialize also upon fish, amphibians, terrapins or waterfowl. The duration of feeding on the host is typically rather short, and the majority of ectoparasitic leeches spent much of their time as free-living members of the benthos. Predation upon other invertebrates (especially oligochaetes, gastropods and chironomid larvae) — and not sanguivory — is a frequent mode of feeding in the class (especially in the Glossiphoniidae and Erpobdellidae).

Leech body segments are divided externally into rings or annuli which do not correspond to internal segmentation. The basic (or primary) number of annuli per mid-body segment is three, but there may be further subdivision to produce four-, five- or six-annulate segments. In a given specimen the number of annuli per segment is abbreviated progressively towards either end of the body; this pattern is species specific. A complete segment is one which has the full complement of annuli characteristic of mid-body segments of the species in question. The mid-body segments of molluscivorous *Whitmania laevis* (Fig. 4.5H), for example, are subdivided into five annuli, and this species has 17 such complete segments. Interestingly, the tri-annulate condition is found in leeches that are relatively sedentary (e.g. the Glossiphoniidae) whereas the five-annulate state occurs in relatively active leeches (such as the Hirudinidae) which may be good swimmers.

Some leeches are rather brightly coloured, with spots, mottling or bands; the ventral and dorsal surfaces often differ, and the underside is always paler. Colour can be an aid to leech identification but will fade when the animal is placed in preservative. Species identification may require dissection after the specimen has been relaxed (by placing it in soda water, for example) and killed, but recognition of families is possible from external features. The following key will permit separation of families of aquatic leeches in tropical Asia; identification of some genera in the Glossiphoniidae, Hirudinidae and Erpobdellidae is possible also. A conservative leech classification scheme has been followed in this key, but some indication of possible alternative placement of genera within families (and family divisions) is given below. One narrowly-distributed freshwater leech family — the Richardsonianidae (Arhynchobdellida) — has not been included in the key because it is confined to southern New Guinea (and Australia) where it is represented by a single genus (*Goddardobdella*); Sawyer (1986) treats this group as a subfamily within the Hirudinidae. The generic-level portion of the key has been modified from Sawyer's (1986) keys to leeches of the Sino-Japanese Region (*op. cit.*: pp. 738–748), the Bornean Subregion (pp. 748–751), and the Indian Region (pp. 763–769). Soos (1965; 1966; 1967; 1969a; 1969b) has developed useful keys to the genera within each leech family also. Where identification requires dissection and examination of the reproductive organs the interested reader is referred to Sawyer (*op. cit.*) for further guidance. Production of a reliable key to all leech genera in tropical Asia — based solely on external features — is not possible at present.

1. Mouth is a small pore on the anterior sucker, or situated towards the forward rim of the sucker; with a protrusile muscular proboscis but no teeth or jaws; body often strongly flattened; usually < 50 mm long .... order **Rhynchobdellida** 2

   Mouth large, within a deep cavity comprising much of the anterior sucker; no proboscis; jaws or teeth *may* be present; body often not strongly flattened; with red 'blood'; various sizes but may reach > 80 mm body length .... order **Arhynchobdellida** 3

2. Large anterior sucker noticeably distinct from remainder of the body which tends to be cylindrical and usually long and narrow (Figs 4.5A & B); **may** lack eyes (or one or two pairs are present) .... **Piscicolidae**

   Anterior sucker not noticeably distinct from the rest of the body (Figs 4.5D & E) which is usually strongly flattened; eyes (one, two, three or four pairs) always present (Figs 4.6D-H); eggs and young are brooded on the ventral surface of the body .... **Glossiphoniidae** 4

3. Five pairs of eyes arranged in an arch on annuli (= apparent segments) 1-8 (Fig. 4.6A); jaws may be present ....**Hirudinidae** 20

   Three or four pairs of eyes arranged in transverse rows (Figs 4.6B & C), or eyes absent; body elongate; lacking jaws ....**Erpobdellidae** 29

4. Four pairs of eyes (Fig. 4.6D); sometimes found in the nostrils of waterfowl .... *Theromyzon*

   Three or fewer pairs of eyes (Figs 4.6E-H) 5

5. Mid-body segments biannulate dorsally (may be triannulate ventrally) .... *Torix*

   Mid-body segments distinctly triannulate 6

6. Brown, horny plate (= nuchal scute or nuchal gland) in mid-dorsal 'neck' region (Fig. 4.6I); one pair of eyes (Fig. 4.6H); confined to northern part of tropical Asia .... *Helobdella*

   No such plate; often with more than one pair of eyes 7

7. Three pairs of eyes on annuli 3, 4 and 7 .... *Paraclepsis*

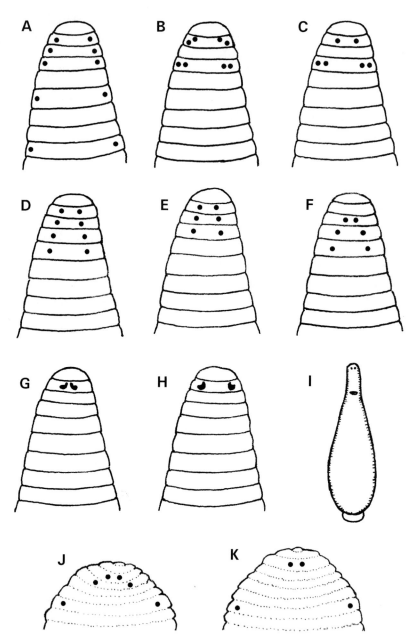

Fig. 4.6 Arrangement of eyes in freshwater leech (Hirudinea) families and in some distinctive Asian genera (J & K redrawn from Nesemann, 1995): A, Hirudinidae; B, *Dina* (Erpobdellidae); C, *Erpobdella* (Erpobdellidae); D, *Theromyzon* (Glossiphoniidae); E, *Glossiphonia* (Glossiphoniidae); F, *Alboglossiphonia* (Glossiphoniidae); G, *Batracobdella* (Glossiphoniidae); H & I, *Helobdella* (Glossiphoniidae) showing the nuchal plate (I); J, *Dina japonica* (Erpobdellidae); K, *Barbronia weberi* (Erpobdellidae, but some workers view this genus as part of a separate family — the Salifidae).

Eyes not as above (if there are three pairs the last two pairs are not separated by two annuli) 8

8. Crop with 7 or fewer pairs of caeca 9

   Crop with 9–11 pairs of caeca; two or three pairs of eyes ....*Hemiclepsis*

9. Crop with 7 pairs of caeca which *may* be branched or lobate 10

   Crop with 6 or fewer pairs of simple, unbranched caeca 18

10. Gonopores (on ventral surface) separated by 1 or $1\frac{1}{2}$ annuli ....*Parabdella* (in part)

    Gonopores separated by at least 2 annuli............................... 11

11. Gonopores on annuli, not in furrows; one pair of eyes ....*Placobdelloides* (in part)

    Gonopores in furrows; may be more than one pair of eyes ... 12

12. Mouth terminal or subterminal (i.e. situated near the tip of the oral sucker) 13

    Mouth towards the centre of the oral sucker, not terminal.... 16

13. Two pairs of eyes; seven pairs of gut caeca 14

    One pair of eyes 15

14. Anterior pair of eyes relatively small; the two pairs *may* be coalesced and appear as one pair ....*Parabdella* (in part)

    Eyes on annuli 2 and 3 ....*Oosthuizobdella* (in part)

15. Caudal sucker large (at least one-half body width); dorsum *may* be strongly papillated .... *Placobdelloides* (in part)

    Caudal sucker not especially large or well-developed; eyes (one pair) on annulus 4 or 5 ....*Batracobdella* (in part)

16. One pair of eyes (Fig. 4.6G) ....*Batracobdella* (in part)

    Two or three pairs of eyes 17

17. Two pairs of eyes on annuli 4 and 5; sometimes appearing as a single pair of large, well-separated eyes ....*Oosthuizobdella* (in part)

Two or three pairs of eyes, on annuli 2, 3 or 4 (Fig. 4.6E) ......
.... *Glossiphonia* (in part)

18. Two pairs of eyes, but anterior pair relatively small (*may* be coalesced and appear as one pair) .... *Parabdella* (in part)

    Three pairs of eyes **19**

19. A significantly smaller space between the first pair of eyes than between the second and third so that the first pair appear noticeably closer together than the others (Fig. 4.6F) .... *Alboglossiphonia*

    No significant difference in spacing between the eyes (Fig. 4.6E) .... *Glossiphonia* (in part)

20. Well-developed jaws; numerous (> 30) small teeth arranged in a single series **23**

    Jaws small and weak; teeth (if present) mostly in two irregular rows **21**

21. Crop with well-developed caeca; caudal sucker large (almost equal to or greater than body width) **22**

    Crop lacking caeca or caeca minimally developed in posterior portion; 17 complete segments; head attenuated; jaws *may* have 15–20 thin, chitinoid plates which resemble teeth; gonopores on annuli and not in furrows .... *Whitmania*

22. With 16 complete segments; mid-body segments perfectly five-annulate; teeth lacking .... *Dinobdella*

    With 12–14 complete segments, sometimes 16; mid-body segments are imperfectly five-annulate, and the annular furrows are of unequal depth; 3–6 teeth set in two irregular rows in each jaw .... *Myxobdella*

23. Gonopores (on the ventral surface) separated by seven or nine annuli .... *Hirudinaria* (in part)

    Gonopores separated by five annuli **24**

24. Unpaired female organs with caecum, usually in the form of a large 'pouch' (= vaginal caecum); usually with large salivary papillae on the jaws; median longitudinal furrow on ventral surface of upper lip; > 80 teeth; body length may be > 80 mm **25**

    Unpaired female organs lacking a large 'caecal pouch'; jaws usually

— but not always (e.g. *Limnatis*) — lack salivary papillae and a median longitudinal furrow on ventral surface of upper lip; 30–60 teeth; body length 40–50 mm 27

25. Well-developed vaginal stalk, equalling or exceeding the length of the vaginal sac (including the vaginal caecum) 26

    Vaginal stalk lacking; ducts of vaginal caecum and common oviduct opening directly into bursa .... *Hirudinaria* (in part)

26. Large vaginal caecum (longer than vaginal duct) which also receives the short common oviduct .... *Poecilobdella*

    Vaginal duct with a slight caecum (poorly-developed 'pouch') ... . *Illebdella*

27. Salivary papillae on jaws numerous; 30–47 teeth; median longitudinal furrow on ventral surface of upper lip .... *Limnatis*

    Jaws lacking salivary papillae (rarely a few small ones); no median longitudinal furrow on ventral surface of upper lip 28

28. Around 55 teeth (43–63) in each jaw; found in the Indian subcontinent .... *Asiaticobdella*

    Around 65 teeth (60–100) in each jaw; confined to the northern portion of tropical Asia (China, Taiwan) .... *Hirudo*

29. Ejaculatory duct with a pre-atrial loop extending anteriorly; gonopores (on the ventral surface) separated by $2\frac{1}{2}$–4 annuli; lacking accessory (= postcephalic) eyes and pharyngeal stylets; eyes usually arranged as in Fig. 4.5C; confined to the northern part of tropical Asia (easily confused with *Dina* which has a southern distribution, and eyes usually arranged as in Fig. 4.6B) .... *Erpobdella*

    Ejaculatory duct lacks a pre-atrial loop extending anteriorly; gonopores separated by $2\frac{1}{2}$–5 annuli; usually have retractile stylets associated with the anterior end of the pharynx; may have accessory eyes and/or a dorsal gastropore 30

30. Mid-body segments basically five-annulate 31

    Mid-body segments six- or seven-annulate 32

31. Gonopores (on the ventral surface) separated by $4\frac{1}{2}$–5 annuli; one or two retractile stylets on anterior end of each pharyngeal ridge;

accessory copulatory pores (pits) anterior and posterior to the gonopores; lacking accessory eyes .... *Barbronia*

Gonopores separated by $2\frac{1}{2}$–3 annuli; no copulatory pores; four or five pairs of accessory eyes .... *Salifa* (in part)

32. Mid-body segments six-annulate; gonopores separated by five annuli; accessory eyes present but dorsal gastropore lacking ......
.... *Salifa* (in part)

Mid-body segments seven-annulate; gonopores separated by seven annuli; dorsal gastropore present but no accessory eyes; may be amphibious .... *Salifa* (in part)

The Glossiphoniidae (Figs 4.5D & E and 4.6E-I) is the largest family of Rhynchobdellida and the largest family of freshwater leeches; a few species (e.g. *Glossiphonia heteroclita* and *Helobdella stagnalis*) have world-wide distributions. Asian genera include *Alboglossiphonia, Batracobdella, Batracobdelloides, Glossiphonia, Helobdella, Hemiclepsis, Oosthuizobdella, Parabdella, Paraclepsis, Placobdelloides, Theromyzon* and *Torix* (e.g. Harding, 1927; Moore, 1924; 1930a; 1930b; 1935; 1938; Mendis & Fernando, 1962; Song & Yang, 1978; Wu, 1979; 1981; Yadav & Mishra, 1982; Sawyer, 1986; Hechtel & Sawyer, 1991). Some genera are predators of invertebrates such as aquatic insects; others take blood meals from a variety of hosts including snails (*Glossiphonia, Helobdella*), fish (*Hemiclepsis*), frogs (*Parabdella, Torix*) and birds (*Theromyzon*). *Paraclepsis* enters the branchial chambers of freshwater crabs to feed, but is also reported from terrapins (Harding, 1927; Mendis & Fernando, 1962). *Batracobdella kasmiana* can be found inside the mantle cavity of unionid bivalves (e.g. *Anodonta*), while *B. gracilis* occurs in crab (e.g. *Parathelphusa*) branchial chambers. The Piscicolidae (Figs 4.5A & B) is a primarily-marine family with some freshwater representatives (e.g. *Cystobranchus, Ozobranchus, Piscicola* and *Zelanicobdella*); most feed on fish or crustaceans but *Ozobranchus* is an ectoparasite of terrapins. Note that some workers (e.g. Sawyer, 1986) place *Ozobranchus* within a separate family — the Ozobranchidae — represented in Asia by a single genus. *Ozobranchus* is distinguished by the presence of 11 pairs of finger-like, branched branchiae ('gills') along the lateral margins of the body.

The Hirudinidae (Figs 4.5F & G and 4.6A) is the most diverse family of Arhynchobdellida (Sawyer, 1986), including *Asiaticobdella, Dinobdella, Hirudo, Hirudinaria, Illebdella, Limnatis, Myxobdella,*

Poecilobdella and *Whitmania*. They are either predators upon invertebrates (as in the case of molluscivorous *Whitmania*, or *Myxobdella* which sucks the body fluids of aquatic insects) or sanguivorous ectoparasites of frogs and mammals (e.g. *Hirudo* and *Hirudinaria*). *Limnatis* and *Dinobdella* crawl into the nostrils and throat of mammals to feed on blood, and then return to a free-living mode of life (Moore, 1927). Some workers place certain hirudinid genera (e.g. *Limnatis* — the Indian cattle leech -and *Whitmania* in Asia: Fig. 4.5G) into a separate family — the Haemopidae — whose members possess a well-developed penis, have diffuse salivary glands, and rather poorly-developed jaws. Under this scheme the Hirudinidae consist of blood-feeding leeches with well-developed (often toothed) jaws, large salivary glands and salivary papillae on the lips. There are differences also in the detailed anatomy of the reproductive system. The Erpobdellidae (Figs 4.5C & D and Figs 4.6B, C, J & K) — which includes *Barbronia, Dina, Erpobdella, Odontobdella, Salifa, Scaptobdella* and *Sinobdella* (Sawyer, 1986; Nesemann, 1995) — is made up of leeches which lack jaws and swallow small invertebrates whole, or suck the fluids from larger prey. Sawyer (1986) and Nesemann, 1995) treat some of these genera (*Barbronia, Odontobdella, Salifa, Scaptobdella, Sinobdella* and *Trocheta*) as members of a separate family: the Salifidae. They are thought to occupy a more southerly latitudes than the erpobdellids, but occur in analogous microhabitats and hence have evolved similar morphologies and annular structures (Nesemann, 1995). Family separation is based on details of the internal morphology (testicular structure, etc.). It *may* be possible to distinguish the two families on the basis of external features: the first pair of eyes in all known salifids is somewhat larger and closer together than succeeding pairs (Fig 4.6K). By contrast, many erpobdellids have eight eyes of similar size (Fig. 4.6B), although the number is reduced in some species (Fig. 4.6 C) and the anterior pair may be closer together (Fig. 4.6J).

Branchiobdellida

These are small (< 10 mm long), leech-like worms with 17 body segments (some of which are fused) lacking chaetae (Figs 4.7A-E). The head usually bears a circle of finger-like projections (Figs 4.7D & E). There are two strong-toothed jaws at the pharyngeal opening, and a posterior sucker. All branchiobdellidans are confined to freshwater and are epizoic on decapod crustaceans. Most live as commensals (or ectosymbionts) feeding on accumulated detritus and microorganisms (especially diatoms), but some species appear to be ectoparasites on the host gills or feed on their eggs. The taxonomic position of the Branchiobdellida within the phylum Annelida has been problematic. Some workers (e.g. Sawyer, 1986) consider that they should be included within the Hirudinea, but this view is disputed (Holt, 1989) and the group is now considered to constitute a separate order, class or subclass within the phylum (Holt, 1986; Gelder & Brinkhurst, 1990). It consists of a single family — the Branchiobdellidae. Branchiobdellidan biology has been reviewed recently by Brinkhurst & Gelder (1991), and most

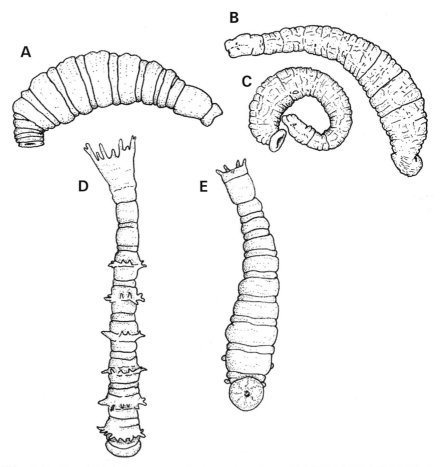

Fig. 4.7 Branchiobdellidan worms (redrawn from Liang [1963] and Holt [1986]): A, *Branchiobdellida* sp.; B & C, *Caridinophila unidens*; D & E, *Cirrodrilus* spp.; body length in all cases approximately 2 mm.

of what little is known of their life history, ecology or physiology (Holt, 1986) has been summarised by Sawyer (1986: pp. 424–434). They are hermaphroditic, and eggs are deposited in cocoons attached to the host. It is therefore unlikely that branchiobdellidans are ever free-living members of the stream benthos except for a short period should the host become injured or die. Nevertheless, they are likely to be encountered by anyone collecting decapod crustaceans from Asian streams.

Asian branchiobdellidan genera include *Caridinophila* (Figs 4.7B & C), *Cirrodrilus* (= *Stephanodrilus*: Figs 4.7D & E) and *Branchiobdella* (Fig. 4.7A) (Liang, 1963; Liu, 1964; Liu & Zhang, 1983; Liu, 1984; Holt, 1986) — although the latter is probably confined to temperate latitudes — and the group is particularly diverse in northeastern Asia where cambarid crayfish (*Cambaroides*) hosts — which do not penetrate the tropics — are found. In tropical areas, branchiobdellidans are associated with freshwater shrimps (*Caridina* and *Neocaridina*) and crabs, although the full range of potential hosts has yet to be investigated in detail, and much less than half of the total world branchiobdellidan fauna is known (Holt, 1986). A key to known families and genera — which depends upon examination of the internal structures of these worms — is given by Holt (1986).

MOLLUSCA

Prosobranchia

Prosobranch snails, which can be recognized externally by the operculum attached to the dorsal surface of the foot, respire by means of an internal gill (ctenidium) inside a mantle cavity within the shell. Ampullariids supplement the gill with a vascularized 'lung' formed from part of the mantle cavity. A siphon developed by the nuchal lobe (an outgrowth of body wall behind the head) is used to transport air from the water surface to the 'lung' in some ampullariids. The majority of prosobranchs have separate sexes (i.e. they are dioecious or gonochoristic), although some families include parthenogenetic species (Thiaridae, Hydrobiidae) whose populations may lack males and a few (the Valvatidae, some Pomatiposidae) are hermaphroditic (Heller, 1993). Hermaphroditism usually involves sex reversal and protandry: i.e. small males change sex and become female as they grow (Heller,

1993). Most freshwater prosobranchs lay eggs in capsules of various forms, but some release hatchlings (i.e. they are 'live-bearing' or ovoviviparous) after a period of egg development within a brood pouch in the parental body. All species of Viviparidae and Thiaridae — which are widespread and abundant in tropical Asian fresh waters — are ovoviviparous. The typical prosobranch life history involves iteroparity: that is breeding several times, each breeding episode being separated by a resting periods which may last several months.

As a broad generalization, tropical regions are richer in freshwater prosobranch snails than in pulmonates (Hutchinson, 1993: p. 135). In Java, for example (van Benthem Jütting, 1956), 89% of species are prosobranchs, and only 11% are pulmonates; equivalent figures for Canada are 16% and 84% respectively. In Taiwan also, prosobranchs make up 74% of the total of 27 species of freshwater snail recorded from the island (Pace, 1973). Running waters tend also to be richer in prosobranchs, while pulmonates occur at least as often in lentic waters. The trend towards greater diversity of prosobranchs in Asian streams is apparent also from monographs cataloguing the freshwater mollusc fauna of Java (van Benthem Jütting, 1956), Thailand (Brandt, 1974) and Sri Lanka (Starmühlner, 1974; 1984a), and is especially marked in stream habitats. However, pulmonates can become extremely abundant in streams that are organically polluted where elevated nutrient levels contribute to dense algal growths and increased food availability. Prosobranchs have been the subject of considerable attention because of the role of some species (e.g. *Melanoides, Oncomelania, Semisulcospira, Tricula*) as intermediate hosts of trematode parasites which infect birds and mammals (including man) as definitive hosts. While such parasitological significance is beyond the scope of this book, it has had the consequence of directing research towards the classification and ecology of freshwater snails.

Given that snails have calcareous shells, we would expect some correlation between the occurrence of gastropods in streams and rivers and the calcium content and alkalinity (or pH) of the water. Unfortunately, as Hutchinson (1993: pp. 146–154) points out, the chemical ecology of freshwater snails is by no means straightforward (e.g. Yipp, 1991), and unfavourable conditions can be reflected in increased mortality, slower growth, reduced reproductive effort, longer hatching period (e.g. Liu, 1993), or combinations thereof. The correlation of calcium with other factors (such as salinity), the possible importance of bicarbonate and other ions, and the varying availability of calcium in different food sources will tend to confound simple correlative (and causal) relationships between snail occurrence and water chemistry. Additional factors such as predation, disturbance (e.g. spates, intermittent stream flow) and (perhaps) competition will influence snail distribution and community composition also (Lodge *et al.*, 1987). Predation may have a strong influence on community structure because heavy-shelled species are less vulnerable to predation by freshwater crabs (e.g. Dudgeon & Cheung, 1990) which may eliminate populations of thin-shelled, poorly-defended snails. Further discussion of prosobranch assemblages in tropical Asian streams can be found on p. 58.

Feeding by freshwater snails involves the use of a radula (a file-like structure to which muscular protractors and retractors are attached allowing rasping feeding movements) to scrape up algae growing on the surface of rocks or macrophytes; detritus, fresh vegetation, carrion, bacteria and fine organic material are eaten also. Some prosobranchs (e.g. Bithyniidae and Viviparidae) supplement radula feeding by trapping food particles from their respiratory current in mucus, which is then ingested. The composition of the diet depends to some extent on the form and arrangement of the radular teeth. Brown (1991) gives a general account of the diets of freshwater snails, while Dudgeon & Yipp (1985) have documented the food eaten by a variety of prosobranchs and pulmonates from Hong Kong streams. There is considerable dietary overlap among species, but those with relatively specialized diets tend to exhibit more directed (or less random) foraging behaviour than generalist feeders (Dudgeon & Lam, 1985). Note that specialized algivores may have important effects on the composition and productivity of the periphyton communities that they graze, tending to favour dominance by adherent diatoms or toxic blue-green algae (see Brown, 1991 and references therein). Other aspects of freshwater gastropod biology and life history are reviewed by Brown (1991) and Hutchinson (1993: pp. 127–275).

The shells of freshwater snails can vary greatly in form and shape within a species, and this is especially noticeable in the ornamentation and degree of elongation of the shells of parthenogenetic species of the family Thiaridae. Care should therefore be taken in the identification of putative species, and many specific names present in the scientific literature have been (and will continue to be) reduced to synonyms of other species as the extent of intraspecific variation in shell form becomes better appreciated. Species determination and even generic separation of a number of freshwater prosobranchs (pomatiopsids, for example) is not possible on the basis of shell morphology alone, and may require dissection and detailed examination of the body parts and especially the structure of the genitalia, but this is not necessary for identification to the level of family or (usually) genus. Extra difficulties face the non-specialist because the higher taxonomy of prosobranchs is rather unstable, and the relationship among families is by no means clear (e.g. Hutchinson, 1993: p. 129). The following key to the major families of freshwater prosobranchs is modified from Pace (1973) and Brown (1991). Note that the shell characteristics mentioned in the key apply to adult snails or well-grown individuals. Since it is frequently impossible to determine the maturity of a single snail specimen, caution should be exercised in using the key when only one or a very few individuals can be examined.

1. Shell typically cap-shaped; the body whorl is large and the spire is small or almost non-existent (in which case the shell is limpet-like); shell may be brightly coloured or with spines; operculum more-or-less semicircular (i.e. shaped like a half-circle), thick and calcareous with one or two projections on the inner surface; shell aperture usually semicircular also, with small teeth or projections on the inner (straight) margin .... **Neritidae**

    Shell of various shapes and not brightly coloured or spiny; operculum corneous[4] without projections on the inner surface; aperture is not semicircular and lacks teeth or projections **2**

2. Growth lines on the operculum are arranged in a spiral **3**

    Growth lines on the outer portion of the operculum are concentric (i.e. the inner ones are enclosed completely by the outer ones) **7**

3. Shell (of adult) usually < 10 mm long; males have a penis (or verge) **4**

    Shell usually > 12 mm long; males lack a penis; shells *usually* elongated (i.e. narrowly conical or 'carrot-shaped' : Figs 4.8B-D) with an aperture that may be quite small relative to the total shell length **6**

4. Body whorl narrowing to aperture (Fig. 4.9L); may have a papilla on dorsal surface of the foot near posterior margin; the inner

---

[4] *Pila* (Ampullariidae) and the Bithyniidae are exceptional in that they have a calcareous operculum, but the other characters match this couplet.

surface of the operculum is ridged .... **Stenothyridae**

Body whorl increasing (or, at least, not decreasing) to aperture (Figs 4.9A-K, M & N); posterior dorsal surface of foot lacking papillae; the inner surface of the operculum is smooth 5

5. Rostrum short (i.e. an extended portion of the head bearing the mouth); tentacles are reduced to stubby eyestalks; foot with lateral groove from mantle cavity to the base of the rostrum; ctenidium degenerate .... **Assimineidae**

   Rostrum long; tentacles elongate or filamentous with eyes at the outer base; foot lacks any lateral grooves; well-developed ctenidium .... **Pomatiopsidae** (many species) & **Hydrobiidae** (a few species)

6. Mantle edges smooth; may be oviparous or ovoviviparous; males present; brood pouch (if present) is uterine and associated with the roof of the mantle cavity .... **Pleuroceridae** (in part)

   Mantle edges papillate; ovoviviparous; males often absent; brood pouch is not associated with the oviduct and is located behind the head in the floor of the mantle cavity .... **Thiaridae**

7. Shell generally > 20 mm long; operculum *usually* corneous 8

   Shell generally < 15 mm long; shell rather globose with a rather large aperture (i.e. the length of the aperture and spire are almost equal: Figs 4.9G & H); operculum is calcareous[5] ..... **Bithyniidae**

8. Tentacles *either* long and filiform (in which case the male has a well-developed penis) *or* tentacles relatively short, the right tentacle of the male is thickened and modified for use as a penis; shell globose or subglobose (Fig. 4.8A); aperture *may* be large; common and widespread 9

   Shell somewhat conical or subglobose with a large aperture; tentacles not especially long and not modified for copulation; penis lacking; not especially common or widespread ....
   .... **Pleuroceridae** (in part; some Paludominae[6])

---

[5]   Any snail keying to this couplet which has a *corneous* operculum and a ridged or sculpted shell is likely to be a species of Iravadiidae; they are usually found in brackish water or streams with a tidal influence.

[6]   Species of *Paludomus* in Thailand have a concentric operculum (Brandt, 1974), but this does not appear to be typical of members of the genus elsewhere in Asia (e.g. Burch, 1980) or Pleuroceridae in general. Accordingly, most snails that key to this couplet are likely to be ampullariids or viviparids and not pleurocerids.

9. Shell large, may exceed 100 mm; strongly globose (apple-shaped) with the shell spire depressed; with well-developed penis (on the right side of the body) and long, filiform tentacles; oviparous, eggs laid in masses above the water line .... **Ampullariidae**

Shell subglobose (Fig. 4.8A) and usually < 40 mm long; spire somewhat elongate (making up more than half of the total shell length); tentacles relatively short, the right tentacle of the male is thickened and modified for use as a penis; ovoviviparous .... **Viviparidae**

The Neritaceae is represented in Asian streams by several freshwater genera of the primarily-marine Neritidae (*Clithon, Neritina, Neritilia, Septaria,* and *Smaragdia,* plus *Neritodryas* from Malaysia which is apparently amphibious). Most species are euryhaline and coastal (especially *Septaria* which has a cap-shaped, almost limpet-like bilaterally-symmetrical shell) but they occur also in running (not standing) fresh water. For example, *Clithon retropictus,* which is widely distributed from Japan through Taiwan to southern China (Pace, 1973), spends its life in fresh water although it may migrate downstream into brackish reaches to oviposit (Nishiaki, 1991a; 1991b). *Clithon corona* is likewise confined to streams in Java (van Benthem Jütting, 1956). Other members of the genus (especially the more euryhaline species) show colourful shell polymorphism, as does *Smaragdia* (Burch, 1980), or have rows of projecting spines on the shell surface. Apart from shell shape, neritids differ from other freshwater prosobranchs in possessing a hemispherical, calcareous operculum with a pair of tooth-like projections on the inner surface which seem to lock it in position when the foot is retracted.

The Viviparaceae live principally in freshwater and supplement radula feeding by trapping and ingesting suspended particles with the use of ciliary tracts. The relative importance of the ciliary collecting mechanism as a feeding mode has not been established in the vast majority of species. There are three freshwater families in the Viviparaceae: the Vivparidae, Ampullariidae and Bithyniidae. All are gonochoristic (i.e. with separate sexes). The Viviparidae is particularly well represented in tropical Asia by the subfamily Bellamyinae, and viviparid genera in the region include *Angulyagra, Anulotaia, Bellamya, Cipangopaludina, Eyriesia, Filopaludina, Idiopoma, Margarya, Mekongia, Rivularia, Robinsonia, Siamopaludina, Sinotaia* (Fig. 4.8A), *Taia, Tototaia* (endemic to the Philippines and Sulawesi) and *Trochotaia* (e.g. Pace, 1973; Brandt, 1974; Starmühlner, 1974; Chen, Q., 1979, Burch, 1980; Chen & Wu, 1983; Chen, Y., 1988). Viviparids occur most frequently in standing waters such as floodplain lakes where they may be partially buried in the soft sediments. Nevertheless, some genera (e.g. *Sinotaia, Bellamya*) can be abundant in slow-flowing streams — especially close to the banks where they may be found in mud or on rocky substrata. Individual species seem to be stenotopic and prefer particular habitat types (Brandt, 1974), although the basis for such selection is not known. Mature viviparids usually exceed 40 mm in shell height. Females are ovoviviparous with a uterine brood pouch, and give birth to well-developed young which resemble miniature adults. The right tentacle of male viviparids is modified into a copulatory organ which serves as a penis.

The Ampullariidae (sometimes termed Pilidae or 'apple snails') includes *Pila* which is the most common and widely-distributed genus of the family in tropical Asia; it has a calcareous (and nacreous)

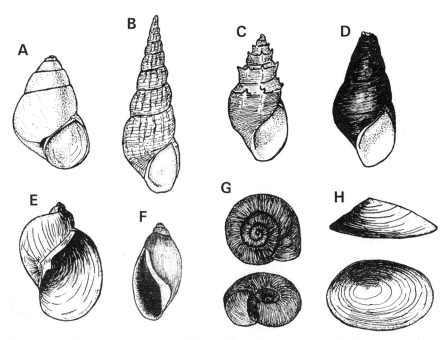

Fig. 4.8 Native and exotic snails from Hong Kong streams: A, *Sinotaia quadrata* (Viviparidae); B, *Melanoides tuberculata* (Thiaridae); C, *Thiara scabra* (Thiaridae); D, *Brotia hainanensis* (Thiaridae); E, *Radix plicatulus*; F, *Physa* (= *Physella*) *acuta*; G, *Biomphalaria straminea* (Planorbidae — an exotic, Brazilian species); H, *Ferrissia baconi* (Ancylidae). Scale lines = 5 mm (A, B, E & F), 4 mm (G), 3 mm (C), 2 mm (H) or 10 mm (D).

operculum. These large snails (which can reach > 10 cm in height or diameter) have a complete siphonal tube, and inhabit slow-flowing streams and ponds. *Pila* may leave the water at night to forage for plants; on such occasions these snails breath with the 'lung' (see above) but do not extent the siphon. Oviparous females of some ampullariids deposit their eggs (which may have a calcareous shell) in masses above the water line on rocks or plants; such behaviour might minimize cannibalism upon the eggs. The Indian hill-stream genus *Turbinicola* lives on rocks at the water surface and lacks a siphon, while a third ampullariid genus, *Forbesopomus*, is endemic to Lake Lanao in the Philippines. A South American species, *Pomacea canaliculata*, has been introduced into Hong Kong and southern China, where it abounds in polluted streams and irrigation ditches (Liang, 1984; Yipp *et al.*, 1992). This snail occupies similar types of habitat in Singapore where it has become established also (Ng *et al.*, 1993).

The Bithyniidae are the smallest snails in the Viviparaceae, and rarely exceed 20 mm in shell height. Females lay eggs in small masses with each egg enclosed by a capsule. The family is represented in tropical Asia by *Alocinma* (Fig. 4.9G), *Bithynia* (which is widespread with several subgenera including *Bithynia*, *Digoniostoma* and *Gabbia*; Figs 4.9H & I), *Emmericiopsis*, *Hydrobioides*, *Mysorella*, *Parabithynia*, *Parafossarulus* (Figs 4.9J & K), *Pseudovivipara*, *Sataria* and *Wattebledia*. One genus, *Petroglyphus*, is apparently confined to Lake Mainit, Mindanao (the Philippines). Most bithyniids are found in standing water or slow-flowing streams, particularly those with muddy beds, but some (e.g. *Mysorella*) tolerate brackish water. Chitramvong (1992) provides a key to the Bithyniidae of Thailand, and draws attention to the wide range of shell colour between populations of the same species. Some workers (e.g. Starmühlner, 1974) use the generic name *Bulimus* in preference to *Bithynia*, and term the family the Bulimidae.

The superfamily Valvataceae — comprising the single family Valvatidae containing hermaphroditic species — occurs in fresh water in Palaearctic Asia, but does not penetrate tropical latitudes. The Littorinacea is represented by a single genus of Littorinidae in tropical Asian streams, although the family is diverse in mangroves and brackish waters. *Cremnoconchus*, found only in India (the Western Ghats) and northern Vietnam, lives on wet (water-splashed) rocks, and breathes aerially by means of a ctenidium modified into a series of stiff filaments through which gaseous exchange occurs (Hutchinson, 1993: p. 156).

The Rissoacea (= Trancatelloidea) includes freshwater species in the Hydrobiidae, Pomatiopsidae, Stenothyridae, Assimineidae, Tornidae

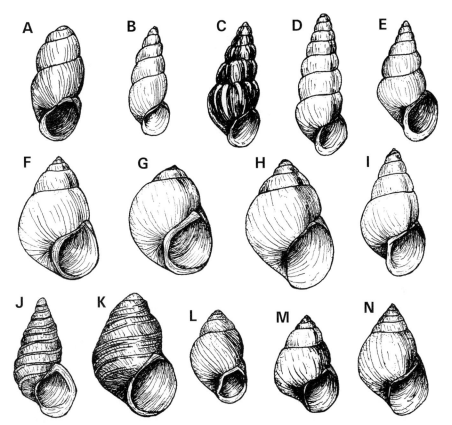

Fig. 4.9 Genera, species and subspecies of freshwater microgastropods (Prosobranchia) from southern China: A, *Pseudobythinella* (shell height 2 mm); B, *Tricula* (4.5 mm); C, *Oncomelania hupensis hupensis* (10 mm); D, *Oncomelania hupensis fausti* (10 mm); E, *Oncomelania hupensis tangi* (6.5 mm); F, *Oncomelania hupensis chiui* (5 mm); G, *Alocinma* (8.5 mm); H, *Bithynia fuchsiana* (10 mm); I, *Bithynia misella* (7 mm); J, *Parafossarulus sinensis* (10 mm); K, *Parafossarulus eximius* (17 mm); L, *Stenothyra* (3 mm); M, *Assiminea lutea* (6 mm); N, *Assiminea latericea* (12 mm). A - F are Pomatiopsidae; G - K are Bithyniidae; L is a member of the Stenothyridae, and M & N are Assimineidae.

and Iravadiidae. (Some workers place the Bithyniidae in this group also, and the precise composition of the superfamily has been the subject of some controversy.) The Iravadiidae (*Iravadia* and *Fairbankia*) occurs mainly in mangroves and estuaries, and the family lacks any exclusively freshwater species. The Tornidae, likewise, is a predominately marine family, although the euryhaline genus *Chamlongia* may enter fresh water occasionally. A second genus — *Phaneta* in Kalimantan (Borneo) — may, however, be confined to rivers. A notable characteristic of tornids is the distinctly swollen tips of their tentacles.

The Asian Hydrobiidae are little known, perhaps because these tiny snails (usually < 6 mm shell height) are easily overlooked. Hydrobiids are gonochoristic or (rarely) parthenogenetic, and most are oviparous although a few are ovoviviparous. They feed on periphyton and fine organic matter; many live in brackish habitats. Tropical Asian genera include *Clenchiella* and *Rehderiella,* which are widely distributed in fresh and brackish water, as well as *Fluviopupa* and *Tatea* which are confined to southern New Guinea (and Australia). Many genera treated as Hydrobiidae by Brandt (1974) and Chinese workers (e.g. Liu & Zhang, 1979; Liu *et al.,* 1980a; 1982a; 1983; 1993; Kang, 1983; 1986a; 1986a) have been reassigned to the Pomatiopsidae — a family which has been the subject of considerable taxonomic study (Davis, 1979; Davis *et al.,* 1992; and see below). Brandt (1974) gives a key to the pomatiopsids and hydrobiids known from Thailand and neighbouring regions, but many recently-described species and new genera are not included.

The Pomatiopsidae is the largest family of prosobranchs in the region, having undergone extensive speciation in south and east Asia, but is much less diverse over the rest of its extensive geographic range (Africa, North and South America). One of the two pomatiopsid subfamilies, the Triculinae, is confined to Asia. *Tricula* (Fig. 4.9B) extends from India through China and the Philippines and through Burma to Malaysia, and the Triculinae consists of many rather poorly-known genera (e.g. *Delavaya, Fenouilia, Gamatricula, Guoia, Halewisia, Hydrorissoia, Jinhongia, Jullienia, Lithoglyphopsis, Lacunopsis, Neotricula, Pachydrobia, Paraprososthenia, Robertsiella* and *Wuconchona*) which are endemic to the region. The Pomatiopsinae includes *Akiyoshia, Oncomelania* (Figs 4.9C-F) and *Pseudobythinella* (Fig. 4.9A). Significant variation in shell form within this group can occur in the absence of major changes in the soft parts, as the divergence in the shell morphology of various subspecies of *Oncomelania hupensis* from China demonstrate (Figs 4.9C-F).

Pomatiopsids are small — usually less than 10 mm shell height — but the shell form is rather variable. The tentacles are elongate (relative to those the Assimineidae, for example) but generally shorter than the filiform tentacles of the Hydrobiidae. Pomatiopsids are gonochoristic (or hermaphrodites with protandrous sex change: Attwood, 1995b) and oviparous; the eggs are laid singly and coated with sand or mud. They appear to be confined to fresh water, especially riverine habitats. There have been major pomatiopsid radiations in the Chang Jiang and Mekong basins (Davis, 1979; Davis *et al.,* 1976; 1986; 1992 and references therein). The great majority of snail species in the Mekong assemblage comprise Pomatiopsidae (Triculinae, tribe Jullieniini; at least 92 species), but the Stenothyridae (at least 19 species) are diverse also. The Triculinae have likewise undergone extensive speciation and adaptive radiation of the tribes Triculini and Pachydrobiini in the Chang Jiang. Eleven genera and at least 20 species were known from the river by 1992, but it is certain that new genera of Triculinae await discovery (Davis *et al.*, 1994). Attwood (1995b) has reviewed the existing information on pomatiopsid life histories in the Mekong, and has investigated the substrate preferences and diet of *Neotricula aperta* (Attwood, 1995a).

Most species of Stenothyridae inhabit brackish water, but both Asian genera — *Gangetica* and *Stenothyra* — have freshwater representatives. *Stenothyra* (Fig. 4.9L) has many more species than *Gangetica* (which is found in India, Burma and Sri Lanka) and has undergone active speciation in the Mekong Basin (Brandt & Temcharoen, 1971; Brandt, 1974; Hoagland & Davis, 1979). The family is characterized by an operculum with ridges or keel-like projections on the inner surface, and (in *Stenothyra*) a filament extending from the posterior end of the foot; *Stenothyra* shells have a constricted aperture and, like the Pomatiopsidae, are small. Stenothyrids are gonochoristic and oviparous; they are often found on sandy substrates but may be associated also with rocks, submerged wood and even muddy sediments where they feed on decomposing organic matter. Most freshwater *Stenothyra* are confined to lotic habitats, and individual species have much more restricted geographic ranges than those species of *Stenothyra* which live in brackish water (Brandt, 1974).

The Assimineidae includes both marine and terrestrial species, but the circumtropical genus *Assiminea* (Figs 4.9M & N; some workers recognize several subgenera within this genus) contains species which penetrate fresh waters although their preferred habitat seems to be brackish (e.g. Pace, 1973; Brandt, 1974). Most are amphibious, and may be found just below, or just above the water surface. Assimineids rarely exceed 10 mm shell height, and are rather globose or ovate. They are gonochoristic and oviparous, with planktonic larvae. Brandt (1974) reports that the brackish-water genera *Paludinella* and *Cyclotropis* are found in fresh water in Thailand, but some tidal influence seems to be necessary for their survival. *Acmella* and *Ekadanta* may be exclusively fresh water, and the latter is apparently confined to the Salween River, Burma.

The Superfamily Cerithioidea contains two freshwater families. Both have elongate, turriculate shells (which may exceed 40 mm in height), and the apex of the spire is usually eroded and truncate. Most are gonochoristic but some species are parthenogenetic. The Pleuroceridae (termed Pachychilidae by some workers) consists of two subfamilies in Asia: the oviparous Pleurocerinae and the Lavigeriinae (including *Semisulcospira*) which are ovoviviparous (e.g. Takami, 1991 and references therein). *Semisulcospira* is widely distributed in northeast Asia and Japan (Pace, 1973) but extends south into northern Vietnam and Guangdong Province in China (Liu *et al.*, 1993), while *Chlorostracia* and *Paludomus* (with three subgenera) inhabit streams in Burma, Thailand, the Philippines, India and Sri Lanka (Brandt, 1974; Starmühlner, 1974; 1977; Burch, 1980). *Paludomus* was previously placed in the 'Paludomidae' (e.g. Burch, 1980), although Starmülner (1974) treats them as a subfamily of the Thiaridae (which is also in the Cerithioidea). They are considered to be a separate subfamily — the Paludominae — of the Pleuroceridae by others (Brandt, 1974), and this placement is followed here. *Paludomus* shells are rather ovate or more broadly conical and less elongated than those of other pleurocerids; the aperture is relatively large (nearly half of the total shell length) and ovate. Most species of Pleuroceridae and the related Thiaridae are inadequately known, but show considerable intraspecific variation in shell morphology. The anatomy, ecology (especially reproductive biology) and taxonomic relationships of the majority of species have yet to be investigated in any detail.

The Thiaridae is a taxonomically-unstable group of snails. This reflects the marked degree of intraspecific variation in shell form among species in certain genera (*Melanoides*, *Thiara*) which is confounded by the extreme similarity in anatomy of different species. The family contains a number of genera (e.g. *Adamietta*, *Antimelania*, *Brotia* [with the subgenera *Brotia* and *Senckenbergia*], *Neoradina*, *Paracrostoma*, *Sermyla*, *Stenomelania* and *Sulcospira*), some of which are very widely distributed (*Melanoides*, *Tarebia* and *Thiara*) while others have confined ranges (e.g. *Tylomelania*, which is endemic to Sulawesi). They occupy a variety of freshwater habitats, although *Sermyla* and *Stenomelania* seem to prefer slightly brackish water, whereas *Brotia* (Fig. 4.8D) is usually restricted to flowing water. *Melanoides* (Fig. 4.8B), *Tarebia* and *Thiara* (Fig. 4.8C) have been introduced widely to regions of the tropics beyond their natural ranges.

A number of thiarid species are parthenogenetic (e.g. *Melanoides tuberculata* and *Thiara scabra*; Abbot, 1952; Morrison, 1954; Jacob,

1957; Berry & Kadri, 1974; Dudgeon, 1986), but some populations of nominally parthenogenetic species include males (Livshits & Fishelson, 1983) although they do not necessarily produce viable sperm (Chaniotis *et al.*, 1980). For instance, Jacob (1958) found that 3% of *M. tuberculata* in an Indian population had male organs, but all were sterile and unable to produce ripe spermatogonia. Nevertheless, some populations of this species are apparently gonochoristic (Heller & Farsey, 1990; Hodgson & Heller, 1990). The reproductive biology of most species has not been studied in detail, but some genera (e.g. *Brotia*) are not parthenogenetic (Davis, 1971; Dudgeon, 1989e). Polyploidy occurs in *Melanoides tuberculata* (also in *Tarebia* and *Thiara*; Jacob, 1957; 1958), and is manifested by a reduction in the extent of sculpturing of the shell whorls in pentaploid individuals — thereby increasing the extent of intraspecific variation in form. Significantly, thiarid populations which have established themselves successfully following introduction outside their natural ranges are composed of parthenogenetic individuals.

Most thiarids are ovoviviparous and have a cephalic brood pouch in which the hatchlings (which emerge resembling tiny replicas of the adult) are incubated. In this respect they differ from ovoviviparous pleurocerids which have a uterine brood pouch. *Stenomelania*, which inhabits brackish and fresh water in the lower course of streams, releases planktonic veliger larvae rather than tiny snails following incubation of eggs in the brood pouch. There is evidence that the timing of recruitment in certain thiarids (e.g. *Melanoides tuberculata*) varies according to prevailing temperatures: where these remain high throughout the year reproduction is continuous (e.g. Malaysia: Berry & Kadri, 1974), but hatchling release is periodic in seasonal tropical climates such as Hong Kong (Dudgeon, 1986). Interestingly, hatchling release by other Hong Kong thiarids (*Brotia hainanensis* and *Thiara scabra*) is periodic also, but does not take place at the same time as in *M. tuberculata* (Dudgeon, 1982f; 1989e).

The Melanopsinae (formerly Melanopsidae) is a distinct subfamily of oviparous snails within the Thiaridae (Houbrick, 1991). *Faunus ater* (which Brandt [1974] places in the Potamididae) is the only Asian representative. It is widely distributed (in coastal Southeast Asia, the Philippines, Indonesia, Papua New Guinea as well as southern India and Sri Lanka) and found in the lower course of streams and rivers in fresh and slightly-brackish water. Males are aphallate and produce spermatophores; females lay large eggs using a grooved 'ovipositor' on the right side of the foot (Starmülner, 1974; Houbrick, 1991).

Apart from the major prosobranch groups mentioned above, a few other mainly-marine families have freshwater species in tropical Asia. The Buccinidae is represented by *Clea* in rivers in Southeast Asia which — unusually for a freshwater snail — is a predator (van Benthem Jütting, 1956). This buccind genus is found China also (Liu *et al.*, 1980b). A species of *Clea* from the Philippines was, until recently, placed in the Planaxidae (as *Quadrasia*: Houbrick, 1986). A single marginellid genus — *Rivomarginella* — is known from Thailand (Brandt, 1974). Marine species of the family are predators, and it is conceivable that, like *Clea*, *Rivomarginella* feeds on small invertebrates. In addition to these two families, a single, monotypic genus (*Potamacmaea*) of Lottiidae — a limpet family typical of shallow marine habitats — is known from tidal reaches of the Irrawaddy and Ganges drainages (Lindberg, 1990), while the nassariid *Pygmaenassa* occurs in fresh- and brackish waters of India.

Pulmonata

Pulmonate snails respire by means of a 'lung-like' structure consisting of the vascularized lining of the mantle (within the shell) which encloses an air-filled space. Communication with the exterior is by way of a small pore — the pneumostome — through which the air supply can be renewed at the water surface (although this does not seem to be necessary for all pulmonates). In the Ancylidae (= Ferrissiidae: the freshwater limpets) and most Planorbidae the mantle cavity is vestigial and respiration is effected by a single vascularised pseudobranch projecting from the foot on the left side of the body. In those planorbids which lack a pseudobranch (e.g. *Camptoceras*) it is substituted by a flap-like process on the left side of the animal. Physids have finger-like projections arising from the mantle edges which may enhance gaseous exchange. Pulmonate respiratory physiology has been reviewed by Hutchinson (1993: pp. 157–172).

All freshwater pulmonates are oviparous hermaphrodites (i.e. they are monoecious; Heller, 1993) with cross-fertilization (although most *can* self). Many are short-lived and semelparous (i.e. they breed once and then die) although iteroparity is known also (Calow, 1978; Hutchinson, 1993: pp. 206–216). Pulmonate diets tend to be more generalized than those of prosobranchs, reflecting the rather uniform and relatively simple structure of the radula within the group. Algivory, detritivory and bacterial feeding are the main modes of nutrition (reviewed by Hutchinson, 1993: pp. 174–186) and any one species may show a mixture of all three. Growth and reproductive performance varies, however, according to the precise composition of the diet. As is the case for prosobranchs (see above), pulmonate grazing can have marked effects on the composition and productivity of periphytic algal communities. Other aspects of pulmonate biology have been summarized recently by Brown (1991).

There are four families of freshwater pulmonates in Asian streams: Lymnaeidae, Physidae, Planorbidae and Ancylidae. They are less widespread and diverse than prosobranchs, and tend to be most abundant in slow-flowing streams, especially those supporting growths of submerged or emergent macrophytes, or at sites where organic pollution or nutrient enrichment has enhanced food availability (e.g. Dudgeon, 1983b). Lymnaeids and planorbids are of great parasitological importance because several species are intermediate hosts of trematodes which infect man and his domestic animals. The Succineidae — which have extremely thin and fragile shells — are associated with streams also in that they are found at the margins of freshwater habitats, but are never submerged in the water and hence are not considered here. Some workers believe that these animals should be included among the Opisthobranchia rather than the Pulmonata (Hutchinson, 1993: p. 128). The following key to pulmonate families is modified from Pace (1973) and Brown (1991), and is based upon the characteristics of adult snails. Identification to species can be difficult because, like prosobranchs, pulmonates (especially lymnaeids) show marked intraspecific variation in shell form, and populations from different habitats may differ considerably in shape (Hutchinson, 1993: pp. 217–232).

1. Shell limpet-shaped (Fig. 4.8H) and small (< 6 mm long) ....
    .... **Ancylidae**

    Shell obviously asymmetrically coiled 2

2. Shell dextrally coiled (Fig. 4.8E) and globose or subglobose[7]; tentacles flat and triangular .... **Lymnaeidae**

    Shell sinistrally coiled (Figs 4.8F & G); shape varies considerably; tentacles filamentous 3

---

[7] One exceptional planorbid, *Indoplanorbis exustus* (Fig. 4.10F), is dextrally coiled (all other planorbids are sinistral), but can be separated from lymnaeids by the flat, planispiral shape of the shell.

3. Shell somewhat elongate and globose with a raised spire (i.e. not coiled in one plane) although this may be secondarily flattened to some extent (e.g. *Amerianna*) **4**

   Shell planispiral (like a ram's horn: Figs 4.8G and 4.10B-F) and coiled in one plane .... **Planorbidae** (most species)

4. Shell with raised spire (Fig. 4.8F); mantle margin with finger-like projections (on one or both sides); lacks a pseudobranch (false gill) ....**Physidae**

   Shell spire may be flattened (Figs 4.8G and 4.10B-F) or raised (Fig. 4.10A); mantle margin lacks projections; a single pseudobranch is present .... **Planorbidae** (a few species)

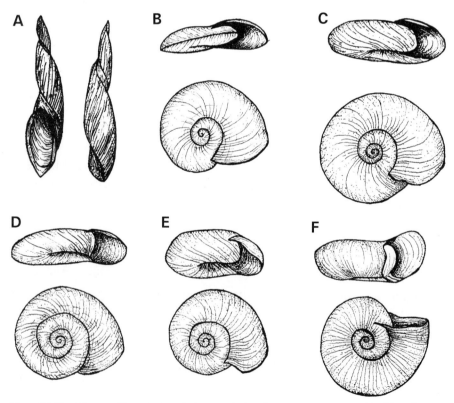

Fig. 4.10 Chinese Planorbidae (Pulmonata): A, *Camptocerus austeni* (shell height 3.5 mm); B, *Hippeutis cantonensis* (8 mm diameter); C, *Hippeutis umbilicalis* (8 mm); D, *Gyraulus convexiusculus* (8 mm); E, *Segmentina (Polypylis) succinea* (7 mm); F, *Indoplanorbis exustus* (15 mm).

The Lymnaeidae have globose to elongate, thin shells up to 45 mm in height (but usually much less), with a large aperture (Fig. 4.8E) containing the broad, oval foot. The tentacles — which are distinctive — are short, flattened and triangular with wide bases where the eyes are situated. Lymnaeid eggs are deposited in large, gelatinous, elongate masses. They inhabit both running and standing waters but are usually more abundant in the latter. There is great intraspecific variability in shell shape and thickness but a general conservatism in anatomy which has led to the description of numerous nominal species. The majority of them are synonyms and, in a revision of the family, Hubendick (1951) was able to reduce the number of species by over 90%; he believed the majority of valid species should be incorporated into the genus *Lymnaea*. There is a lack of agreement about generic limits in the Lymnaeidae, and the names *Amisus*, *Austropeplea*, *Bullastra*, *Cerasina*, *Galba*, *Lymnaea* and *Radix* (Fig. 4.8E) have been applied to various tropical Asian members of the family, in some cases (e.g. *Radix*, *Austropeplea*) both as genera and as subgenera. Brandt (1974), for example, treats *Radix* as a subgenus of *Lymnaea*, while Pace (1973) and Burch (1980) give it full generic status. To complicate matters, Zheng *et al.* (1983) view *Radix* as a subgenus of *Amisus* — a genus which does not seem to occur outside China (and which is not mentioned in a recent monograph on Chinese snails by Liu *et al.*, 1993). Pace (1973) separates the Taiwanese lymnaeids into two genera (*Austropeplea* and *Radix*) on the basis of chromosome number, and this may provide a reasonable basis for grouping the known species within genera when a regional revision of the family is undertaken.

The Physidae (or Physellidae) is a primarily Palaearctic family of pulmonates, and only one genus is found in tropical Asia. *Physa* (= *Physella*) *acuta* (Fig. 4.8F) is almost cosmopolitan and, because it has been introduced widely to parts of the world beyond its natural range, the original geographic distribution cannot be ascertained with certainty. In general form the smooth globose physid shell resembles a small (< 25 mm shell height), elongate lymnaeid, but the coiling is sinistral instead of dextral. The tentacles are long and thin, and the eyes are situated at their bases on the median surface. The body is uniform black, and the foot is long and narrow, tapering to a point posteriorly. The eggs are laid in rather irregular, crescent-shaped masses which are smaller than those of the Lymnaeidae.

The Planorbacea comprises the Planorbidae and Ancylidae in tropical Asia. The Planorbidae is the more diverse of the two families, and the majority of members are characterized by their discoidal or planispiral, sinistral shells (Figs 4.8G and 4.10B-F) with rounded whorls which rarely exceed 25 mm in diameter (and are generally no more than 10 mm). The aperture is small and sometimes has internal lamellae (e.g. in *Segmentina*), and the typical shell form resembles a ram's horn. In some cases (e.g. *Campoceras, Glyptophysa* and *Physastra*), however, the shell is ovate or somewhat elongate (Fig. 4.10A) or globose like a physid. In *Amerianna* this form is modified further in that the spire and whorls are depressed into a flat dorsal surface. Planorbids have long, filiform tentacles, and the eyes are situated on small swellings at the inner base of the tentacles. Their bodies are generally grey or sand-coloured and pigmented with dark spots, but some species (e.g. *Indoplanorbis exustus*) possess haemoglobin as a respiratory pigment and thus their bodies may appear reddish. The eggs are laid spirally in a gelatinous mass which usually has a membraneous cuticle. Species identification and placement within genera depends to a large extent on reproductive anatomy, especially that of the male (Hubendick, 1955).

Planorbid genera (and subgenera) that are widespread in tropical Asia include *Amerianna, Camptoceras (Culmenella)* (Fig. 4.10A), *Gyraulus* (Fig. 4.10D), *Hippeutis (Helicorbis)* (Figs 4.10B & C) and *Segmentina (Polypylis)* (Fig. 4.10E). The Asian (and world) species of *Gyraulus* have been revised by Meier-Brook (1979; 1983); although the genus is made up of several subgenera, only *Gyraulus sensu stricto* occurs in tropical Asia. *Glyptophysa* and *Physastra* are east Indonesian-New Guinea genera, and *Miratesta* — which has an unusually heavy, wrinkled shell — is endemic to Sulawesi. *Patelloplanorbis* is known so far from a single lake in New Guinea only. It has a conical, limpet-like shell with a reduced spire, and some authorities (e.g. Hubendick, 1957) place it in a separate family — the Patelloplanorbidae — within the Planorbacea, although this position is questionable. The native planorbid fauna of tropical Asia has been augmented by exotic species, notably *Biomphalaria straminea* (Fig. 4.8G) from South America which is an intermediate host of the blood fluke *Schistosoma mansoni* (Platyhelminthes: Trematoda) in its home range. This snail was introduced into Hong Kong during the early 1970s (Meier-Brook, 1974), probably by way of the aquarium trade (Dudgeon & Yipp, 1983). It has since established itself and spread into southern China (Liu *et al.*, 1982b; 1993; Dudgeon, 1983b; Yipp, 1990).

The monotypic genus *Indoplanorbis* (containing only *I. exustus*: Fig. 4.10F), which is found naturally in Indochina but not the

Philippines or New Guinea (Starmühlner, 1974)[8], is the only Asian member of the subfamily Bulininae (treated by some authors — e.g. Brandt [1974] — as a separate family: the Bulinidae). The external features of this animal are those of a typical planorbid. Apart from details of the anatomy, only its relatively large size (shell diameter > 10 mm), and the fact that the egg mass is not enclosed within a capsule, set *Indoplanorbis* apart from the other Asian Planorbidae.

The Ancylidae, or freshwater limpets, are tiny (< 6 mm long), pale snails with a limpet-like (patelliform) shell. The apex is located medially or inclined a little to the right in the middle of the shell (or slightly anterior). Ancylids have short flat tentacles with eyes located at their inner bases. Eggs masses are small (that is, there are few eggs per mass) and enclosed by a thin membranous cuticle. Tropical Asian genera include *Ferrissia* (Fig. 4.8H) and *Pettancylus* (Burch, 1980); the latter is generally treated as a subgenus of the former (e.g. Pace, 1973; Brandt, 1974). An additional genus — *Gundlachia* — which was previously thought to be restricted to the Americas (Hubendick, 1964) has been reported from Thailand (Brandt, 1974). Ancylids are found in rather slow-flowing streams crawling on stone surfaces, leaf-litter or aquatic macrophytes. Their anatomical organization shows a little or no sign of the coiling of the visceral mass typical of other snails. This is one reason why the familial placement of freshwater limpets of the genus *Protancylus* (containing two species from Sulawesi and Malaysia) is unclear, because their internal organs are sinistrally coiled. In addition, the shell is relatively strong and has a posterior apex They may justify isolation in a separate planorbacean family — the Protancylidae — although the ranking and placement are moot.

Bivalvia

Bivalves are characterized by possession of a shell that is divided into two valves which are hinged dorsally by an elastic ligament. Adjacent to the ligament is a raised area — the umbo or beak — which represents the oldest part of the shell from which growth has proceeded outwards. The valves usually have teeth on their inner margins at the umbo (adjacent to the hinge), although in some species (especially members of the family Unionidae) these teeth are reduced or degenerate. The

---

8   *Indoplanorbis* has been introduced to Java, Sulawesi and the Philippines (Brandt, 1974; Burch, 1980).

inner surface of the valves may be nacreous (i.e. lined with 'mother of pearl') and bear features of taxonomic importance such as the scars of the pallial line (see below) and various muscles (Fig. 4.11). The adductor muscles — which close the shell valves — produce the most conspicuous scars, and almost all freshwater bivalves have two of them (the dimyarian condition) located anterior and posterior to the visceral mass. In addition, there are smaller scars left by the pedal retractor muscles.

The shell valves enclose the visceral mass which is surrounded by the folds of the mantle. These folds may be fused along their margins to a greater or lesser extent, but complete fusion never occurs. The pallial-line scar on the shell marks the fastening of the mantle to the inner surface of the valves. The mantle folds are modified posteriorly to form a pair of apertures which may be drawn out into siphonal tubes. Water enters the mantle cavity through the ventral inhalant siphon (or incurrent aperture) and leaves via a dorsal exhalant siphon (or exhalant aperture). Large ctenidia (gills) lie on either side of the

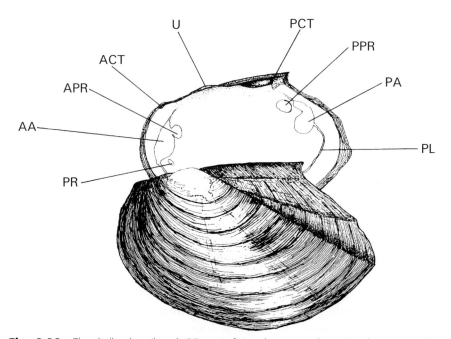

Fig. 4.11 The shell valves (length 80 mm) of *Anodonta woodiana* (Bivalvia: Unionidae) showing the main features of the inner surface: AA, anterior adductor muscle scar; ACT, anterior subcardinal tooth; APR, anterior pedal retractor muscle scar; PA, posterior adductor muscle scar; PCT, posterior subcardinal tooth; PL, scar of pallial line; PPR, posterior pedal retractor muscle scar; PR, protractor muscle scar; U, umbo.

visceral mass. Each ctenidium is composed of two lamellae or demibranchs joined at the lower but not at the upper margins, and thus the ctenidium is W-shaped in cross-section. They are well ciliated and fulfil a respiratory function, but play a vital role also in filtering food particles from water entering the mantle cavity. Particles are trapped in mucus on the ctenidial surfaces and travel to the mouth along ciliary tracts. Some sorting of this material occurs on the surface of paired palps which surround the mouth. The foot is located ventrally on the anterior part of the visceral mass; it can be protruded or extended anteroventrally between the folds of the mantle beyond the shell valves allowing the animal to burrow in sand or mud.

Most freshwater bivalves are gonochoristic, but the Sphaeriidae and some Corbiculidae are hermaphrodites and are capable of selfing. The majority are ovoviviparous also: fertilized eggs develop within a brood chamber or marsupium (modified from part of the ctenidia) prior to their release. This is evidently an adaptation to reduce the mortality resulting from the production of planktonic veliger larvae which would be swept downstream into potentially unfavourable habitats. In the Sphaeriidae, at least, the brooded larvae receive nutrition from the parent during incubation; upon release, these juveniles resemble miniature replicas of the adults and each one may be up to 5% of the mean adult weight (McMahon, 1991). Males of the Unionidae, Hyriidae and Margaritiferidae (in the superfamily Unionacea) release sperm into the surrounding water, which is taken into the mantle cavity of the female where the eggs are fertilized. The resultant zygotes are brooded within a marsupium where they develop into shelled glochidia larvae (Fig. 4.12) which, upon release, are ectoparasites of the gills or fins of fish. Glochidia may or may not have hooks or denticles that aid attachment to the host but, following attachment, the larvae encyst and commence tissue reorganization. After some time (usually days to weeks), the reorganized larva leaves its host and falls to the bottom sediments where it adopts the filter-feeding mode of life and anatomy of the adult bivalve. Brooding of young occurs also in freshwater species of Corbiculidae (e.g *Corbicula fluminea*: Figs 4.14F & G) which release shelled, benthic juveniles, but freshwater Mytilidae (e.g. *Limnoperna*: Fig. 4.14E) have planktonic larval stages (Morton, 1977b). McMahon (1991) provides a recent, comprehensive review of the biology of freshwater bivalves, although the emphasis is on North American species (see also Pennak, 1989: pp. 569–603).

Freshwater bivalves are most diverse in large rivers and floodplain lakes, but some taxa (Corbiculidae and Sphaeriidae) occur in small

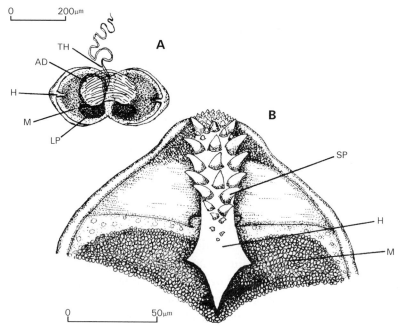

Fig. 4.12 The glochidium larvae of *Anodonta woodiana* (Bivalvia: Unionidae) showing details of one of the shell hooks: AD, adductor muscle; LP, lateral pit cells (which give rise to the ctenidia); H, hook; M, mantle cells; SP, spine; TH, thread (of possible importance in host detection).

streams also. A few specialized forms (the mytilids *Limnoperna* and *Sinomytilus*, and the etheriid *Pseudomulleria*) are attached to rocks, but most bivalves burrow in stable gravel, sand or mud substrates. Certain unionaceans which require a particular type of fish host for glochidial development may be limited in their distribution by the availability of suitable host species, and some unionids are highly stenotopic and habitat specific (McMahon, 1991). Aside from regional monographs on mollusc faunas (e.g. van Benthem Jütting, 1953; McMichael, 1956; Brandt, 1974; Wu, 1982) and the compilation of species lists for various water bodies which sometimes include provisional investigation of sediment-species abundance relationships (e.g. Ahktar, 1978; Chen, 1979; Chen & Wu, 1983; Zheng *et al.*, 1983; Djajasasmita & Budiman, 1984; Janaki Ram & Radhakrishna, 1984; Begum & Rizvi, 1987; Qi & Lin, 1987; Marwoto, 1987; Huang & Zhang, 1990; Nazneen & Begum, 1994), the ecology of freshwater bivalves has received little attention in tropical Asia.

Data on the population dynamics of members of four bivalve families from Hong Kong and southern China reveal a lack of correspondence in seasonality among species. Three species of small sphaeriid bivalves inhabit lowland, slow-flowing streams: *Musculium lacustre*, *Pisidium clarkeanum* and *P. annandalei*. The predominately holarctic species, *M. lacustre*, exhibits two periods of recruitment per year — in 'spring' (before the summer monsoon) and 'autumn' (after the summer monsoon) — which represents life-cycle completion by two overlapping generations. The spring recruits give birth to the autumn recruits which in turn give birth to the succeeding spring recruits (Morton, 1985). *Musculium lacustre* is essentially temperate-zoned, but has a distribution extending southwards into the tropics. Temperatures in Hong Kong do not seem to limit growth or reproduction in this species, and the timing of recruitment may be seen as a strategy to reduce mortality of newly-released larvae: peaks of recruitment precede and immediately follow the wettest months so minimizing the chance of larvae being swept away during spates. The Hong Kong *Pisidium* spp. reproduce throughout much of the year, and three breeding periods (during the summer monsoon, the dry season, and the 'spring') have been identified (Morton, 1986). Each represents reproduction by a separate generation, with most individuals dying after breeding, although some *P. clarkeanum* may survive long enough to produce a second brood. Typically, the peak in recruitment during the summer monsoon is relatively small because of spate-induced larval mortality (Morton, 1986). *Pisidium* seasonality (and, indeed, sphaeriid seasonality in Hong Kong) appears to result from a short life span rather than climatic constraints on growth and reproduction.

Settlement of larvae of the freshwater mytilid, *Limnoperna fortunei* — which occurs in some Hong Kong streams and reservoirs and in the Zhujiang River in Guangdong Province — takes place twice each year, during the warmest (June — August) and the coolest (November — December) months (Morton, 1977b). Like *L. fortunei*, the infaunal clam, *Corbicula fluminea* (Corbiculidae), inhabits lentic and lotic waters. It also breeds twice a year, producing larvae before and after the summer monsoon (Morton, 1977c). The expression of sexuality of *C. fluminea* differs between habitats: lotic populations comprise equal numbers of hermaphrodites and females, but reservoir populations includes males and females in addition to hermaphrodites (Morton, 1983). The relative importance of the hermaphroditic strategy in streams may reflect the frequency of catastrophic mortality caused by spates, and the success of hermaphrodites in recolonizing depopulated areas. Significantly, *Corbicula* spp. (Figs 4.14F-H) occur throughout tropical Asia and the widespread *C. fluminea* can reach very high densities

(7,500 individuals.m^{-2}: Qi & Lin, 1987; see also Chen & Wu, 1983) in lotic habitats. This clam has been introduced to the United States and Mexico where it has '. . . become the dominant benthic species in many habitats' (McMahon, 1991); it is established in some European rivers also (e.g. Bij de Vaate & Greijdanus-Klaas, 1990).

Studies on the population dynamics of bivalves in Hong Kong streams yield no clear indication of a uniform response to climate among species, although some evidence of life-cycle strategies which ameliorate the effects of spate-induced mortality can be adduced. Additional data on bivalves from lentic or large lotic habitats (the Zhujiang River) likewise do not indicate any general pattern of seasonality: *Corbicula fluminalis*, for example, breeds during the coldest months of the dry season (Morton, 1982), while *Anodonta woodiana* (Unionidae: Fig. 4.12) females release glochidia larvae early in the summer monsoon; males have mature gonads throughout the year (Dudgeon & Morton, 1983). Certain Indian unionids (*Indonaia caerulea*) have mature gonads throughout the year also, and spawning is continuous although there is a period of peak activity (Arrawal, 1980). Because the glochidia larvae of *A. woodiana* — and all unionids — are ectoparasites of certain fish (Dudgeon & Morton, 1984), the timing of spawning might reflect host availability rather than a climatic constraint on recruitment.

The following key to families (and selected genera) of tropical Asian bivalves deals with widespread taxa only (e.g. the distinctive but localised etheriid *Pseudomulleria* is not included), and depends upon the availability of mature or well-grown specimens with the adult shell form. It is not possible to provide a key to all genera or species in the region because the taxonomy of the largest family (the Unionidae) is unstable, and intraspecific variation in shell form is common. For example, individuals from small streams are less swollen than conspecifics from large rivers; the former group tend to have smooth shells whereas shells from inhabitants of large rivers may be more strongly sculptured, marked with tubercles or ridges. In the Corbiculidae also, many putative species are merely local variants or ecotypes of widely-distributed but variable species (e.g. *Corbicula fluminea*; see below). A list of freshwater bivalve taxa in tropical Asia (excluding those which are confined to lakes or lentic habitats or are of dubious taxonomic status) is given after the key.

1. Relatively large (shell length > 30 mm) and somewhat elongate; the internal shell surface is nacreous (mother-of-pearl) 5

Relatively small (shell length < 35 mm) and not particularly elongate; internal shell surface is not nacreous 2

2. Shell length > 12 mm, coloured yellowish brown to dark olive-green or black; the inner surface *may* be purple or violet; pallial line sinuate; shell thick and massive with strong concentric sculpture on the valves .... **Corbiculidae:** *Corbicula*[9] (Figs 4.14F-H)

   Shell length < 12 mm; shell cream or translucent with weak sculpture; pallial line simple .... **Sphaeriidae** 3

3. Umbo (beak) anterior or slightly to the anterior end of centre (note: the siphons or apertures are located posteriorly); inhalant and exhalant siphons present and partially fused 4

   Umbo posterior or slightly to posterior end of centre; inhalant (ventral) siphon lacking or represented by a slit .... *Pisidium*

4. Beaks prominent or swollen .... *Musculium*

   Beaks neither particularly prominent nor swollen ....
   .... *Sphaerium* (Fig. 4.14I)

5. Markedly elongate with umbo (beak) situated at the anterior end of the shell (Fig. 414E); ligament internal; anterior adductor muscle greatly reduced; attached to hard surfaces by means of a thread-like byssus secreted by a gland in the foot; generally < 45 mm long .... **Mytilidae:** *Limnoperna* and *Sinomytilus*

   Not markedly elongate and umbo not situated at the anterior end (although it may be anterior of centre); ligament external; anterior adductor muscle not greatly reduced; free-living (not attached to the substratum) and lacking a byssus; generally > 45 mm long
   5

5. Posterior mantle margins are not fused, and there is no thickening or other structures associated with the development of siphons; shell laterally compressed and elongated; usually < 180 mm long; glochidia are brooded both the outer and inner pairs of ctenidial demibranchs ....**Margaritiferidae**

---

[9] Large, greenish shells lacking concentric ribs on the surface are likely to be *Polymesoda (Geloina)*, while black ones may be *Villorita*. Brown shells over 40 mm long with growth lines only (i.e. no concentric ribs) are probably *Batissa (Cyrenobatissa)*. These three corbiculids are found in brackish water or rivers with some tidal influence.

Posterior mantle margins are fused (at least above the exhalant siphon), and there are distinct inhalant and exhalant siphons; shell form varied and may be rather swollen or globose; often > 180 mm long (but may be smaller); glochidia are *usually* brooded in *either* the outer *or* inner of the two pairs of demibranchs only 6

6. Usually < 150 mm long; glochidia brooded in the inner demibranchs only; extensive fusion of the mantle margins, especially around the siphons, which may extend ventrally; in New Guinea only ....**Hyriidae**

   Usually > 150 mm long (sometimes considerably larger); glochidia brooded in outer demibranchs (or, less often, in both inner and outer demibranchs); fusion of posterior mantle margins not particularly extensive (only above the exhalant siphon); widespread in tropical Asia (with one genus in New Guinea) .... **Unionidae**

The Mytilacea are represented by two genera of exclusively freshwater Mytilidae: *Limnoperna* (Fig. 4.14E) and *Sinomytilus* (although the latter may be a junior synonym of the former). Both occur in flowing and standing water (especially in rivers) where they attaches by the byssus to almost any hard substrate, although a Thai species of *Limnoperna* is apparently confined to the decollated spires of *Brotia* (Thiaridae) snails and one *Sinomytilus* species attaches only to the shells of the unionid *Modellnaia* (Brandt, 1974). Neither genus is found in brackish water. *Limnoperna* is highly invasive and fouls water-supply lines (Morton, 1975; Tan *et al.*, 1987); *L. fortunei* has been introduced into Argentina (Pastorino *et al.*, 1993). According to Brandt (1974), *Sinomytilus* produces free-swimming veliger larvae while *Limnoperna* is ovoviviparous and incubates shelled larvae; however, such brooding does not occur in *Limnoperna fortunei* from southern China nor does the gill morphology support such a proposition (Morton, 1977b). Some workers (e.g. Brandt, 1974) put *Sinomytilus* in the Dreissenidae, but the filamentous structure of the gills (*versus* lobes in dreissenids) and the reduction of the anterior adductor muscle indicate that Mytilidae is the correct placement. A third mytilid genus, *Brachidontes*, is sometimes encountered in the lower reaches of Asian rivers, but seems to be confined to tidal reaches with some saline influence. *Modiolus*, which also occurs in the region (Morton, 1977a), is likewise restricted to brackish water.

The Unioncea consists of the Unionidae, Amblemidae,

Margaritiferidae and Hyriidae. The Unionidae is the most diverse family of freshwater bivalves, consisting of several hundred species worldwide. There is much disagreement among authorities concerning grouping and delimitation in presumed phyletic lineages. The subfamilial divisions are controversial, and the validity and limits of a number of genera are the subject of debate. The shell is equivalve but variable in shape from nearly circular to elongate, and from compressed to globular. Adult size can vary from 25 to 300 mm in length, but most species are in the range 100 — 200 mm. The internal surface is nacreous mother-of-pearl, and the shell is usually greenish or brown externally with an obvious ligament. Concentric growth lines may be present but sculpture is generally weak or lacking. The muscle scars of the anterior and posterior adductor muscles are subequal; anterior and posterior pedal retractor muscles are present also, the former usually producing a distinct scar. The pallial line is entire and there is limited fusion of the mantle margins; siphonal tubes are rarely present. Glochidia larvae are brooded in a marsupium which occupies all four demibranchs or (more commonly) the outer pair only. Some Southeast Asian genera with hookless glochidia that have been included within the Unionidae (e.g. *Discomya*, *Lamprotula*, *Potomida*, *Pseudobaphia*, *Rhombuniopsis* and *Schepmania*; see below) may be more appropriately placed in the closely-related Amblemidae — a family which is mainly confined to the Nearctic Region.

Some unionid genera and species have restricted distributions which may reflect adaptation (and restriction) to particular local conditions. In North America, this specialization has meant than many unionids are endangered or near extinction in habitats modified by human activities (McMahon, 1991). Nevertheless, certain Asian unionids have been spread beyond their original geographic ranges as a result of accidental introductions — possibly as glochidia encysted on fish. *Pseudodon vondembuschianus*, for example, has been introduced into Singapore and, in the absence of native freshwater bivalves, has become well established (Ng *et al.*, 1993), while *Anodonta woodiana* has been introduced from China to Indonesia and Europe (Djajasasmita, 1982; Girardi & Ledoux, 1989). The following list of genera (and subgenera) of tropical Asian unionids is based largely upon the work of Haas (1969) and Brandt (1974). It is not exhaustive and undoubtedly includes some taxa which will be subject to revision or synonymy in the future: *Acuticosta* (Fig. 4.13C), *Anodonta* (Fig. 4.11), *Arcidopsis*, *Arconaia* (Fig. 4.13H), *Balwantia*, *Caudiculatus*, *Chamberlainia*, *Contradens* (*Contradens* and *Sprickia*), *Cristaria* (*Cristaria* and *Pletholophus*:

Fig. 4.14B), *Ctenodesma, Cuneopsis* (Fig. 4.13D), *Diaurora, Discomya, Elongaria* (*Elongaria* and *Nannonaia*), *Ensidens* (*Ensidens* and *Uniandra*), *Haasodonta* (which is endemic to New Guinea and is the only unionid present on that island), *Harmandia, Heudeana, Hyriopsis* (*Hyriopsis* and *Limnoscapha*: Fig. 4.13F), *Indonaia, Lamellidens, Lamprotula* (*Lamprotula* and *Parunio*: Fig. 4.14A), *Lanceolaria* (Fig. 4.13E), *Lepidodesma* (Fig. 4.14C), *Modellnaia, Oxynaia, Parreysia* (*Parreysia* and *Radiatula*), *Physunio (Physunio, Lens* and *Velunio), Pilsbryoconcha, Potomida (Potomida), Pressidens, Prohyriopsis, Protunio, Pseudobaphia, Pseudodon (Pseudodon, Bineurus, Chrysopseudodon, Cosmopseudodon, Diplopseudodon, Indopseudodon, Monodontina, Nasus* and *Trigonodon), Ptychorhynchus, Rectidens, Rhombuniopsis, Scabies, Schepmania, Schistodesmus* (Fig. 4.13G), *Simpsonella, Sinanodonta, Solenaia* (Fig. 4.14D), *Sulcatula, Trapezoideus, Unio* (Fig. 4.13B), *Unionea* and *Unionetta*.

Although the ecological importance of unionids in tropical Asian rivers has not been investigated in detail, they do represent a considerable biomass of secondary consumers. Typically, they are not vulnerable to aquatic predators when in the adult stage, and thus 'block' the flow of energy to higher trophic levels. However, they do participate in ecological interactions: unionids are egg repositories for certain small cyprinids — the Rhodeinae (also known as Acheilognathinae) — which have been studied intensively in Japan (e.g. Kondo *et al.*, 1984; 1987), but have received little attention elsewhere in Asia except in terms of their potential as aquarium fish (e.g. Grondejs, 1978; Wezeman, 1978; but see Dudgeon, 1985b). Other unionids (e.g. *Hyriopsis* and *Pilsbryoconcha*; Vidrine, 1984) are hosts to parasitic water mites (Hydrachnida: Unionicolidae: *Unionicola*; see p. 190), glossiphoniid leeches, and commensal baetid mayflies (*Symbiocloeon* in *Hyriopsis*; Müller-Liebenau & Heard, 1979).

Margaritiferids resemble unionids superficially, although there are differences in details of the anatomy and especially the ctenidial structure which is relatively primitive. In addition, they lack siphons. The shell is generally elongate, compressed and rather robust (Fig. 4.13A). It is typically less than 150 mm long, blackish externally with a nacreous inner surface. The posterior hinge teeth tend to be reduced (feature which distinguishes them from most — but not all — unionids). The glochidia are incubated in both inner and outer demibranchs of the ctenidia, and the larvae are hookless but have small denticles on the ventral rim of the valves. The Margaritiferidae is a relatively small, prevalently Holarctic family, but some genera occur in tropical Asia:

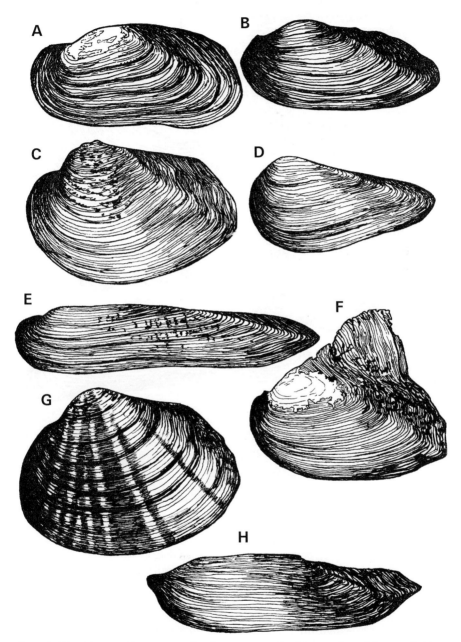

Fig. 4.13 Chinese Unionacean bivalve genera: A, *Margaritifera* (Margaritiferidae, shell length 115 mm); B, *Unio* (55 mm); C, *Acuticosta* (285 mm); D, *Cuneopsis* (70 mm); E, *Lanceolaria* (100 mm); F, *Hyriopsis* (140 mm); G, *Schistodesmus* (35 mm); H, *Arconaia* (90 mm); B - H are Unionidae.

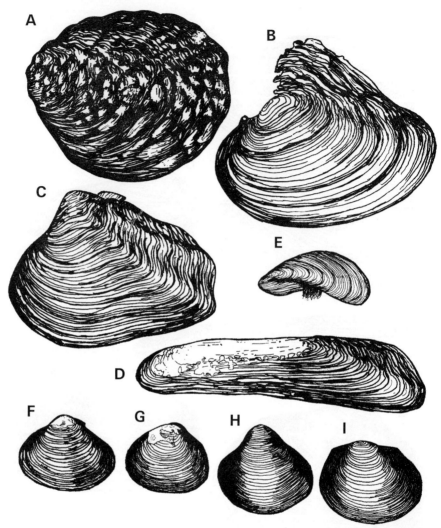

Fig. 4.14 Chinese freshwater bivalve genera and species: A, *Lamprotula* (Unionidae, shell length 90 mm); B, *Cristaria* (Unionidae, 150 mm); C, *Lepidodesma* (Unionidae, 125 mm); D, *Solenaia* (Unionidae, 80 mm); E, *Limnoperna* (Mytilidae, 25 mm); F & G, *Corbicula fluminea* (Corbiculidae, 30 mm); H, *Corbicula fluminalis* (Corbiculidae, 40 mm); I, *Sphaerium* (Sphaeriidae, 11 mm).

Heudeana, *Margaritana* and *Ptychorhynchus* (in central China), and *Margaritanopsis* (in Thailand and Burma). The validity or status of these margaritiferid genera is questionable: Haas (1969) records only *Margaritanopsis* — as a subgenus of *Margaritifera* (Fig. 4.13A) — from tropical Asia, and treats *Margaritana* as a junior synonym of *Margaritifera*.

The Hyriidae is confined mostly to Australia and South American but a few genera occur in New Guinea. The shell is usually < 150 mm long and of variable shape but resembling that of the Unionidae (rather than the Margaritiferidae). The greater extent of fusion of the mantle margins (especially around the inhalant and exhalant apertures), and details of ctenidial structure distinguish these bivalves from other Unionacea. In addition, larvae are incubated within the inner demibranch of the ctenidia. Genera include *Alathyria, Hyridella* — which has the umbonal region sculpted with v-shaped ridges — *Velesunio* and *Westralunio* in New Guinea and Australia (Williams, 1980: p. 83); and *Leiovirgus, Microdontia* and *Virgus* which occur in New Guinea but not Australia (McMichael, 1956).

A single family of Mutelacea, the Etheriidae (= Mulleriidae), is found in Asia, and is represented by only one genus, the monotypic *Pseudomulleria* from Burda River, southern India. (This genus is sometimes treated as a subgenus of *Acostaea*.) Etheriids are unmistakeable: the shell is oyster-like and irregularly shaped with one valve cemented to the substrate. The foot, which is no longer required for locomotion, has degenerated. *Pseudomulleria* lacks an anterior adductor muscle and hence is monomyarian, in contrast to the dimyarian condition of all other freshwater bivalves. The larval form and reproductive biology of *Pseudomulleria* has not been investigated, but they are presumed to be ovoviviparous with incubation of fertilized eggs within a marsupium in the inner demibranch.

Together with the Unionacea, the Corbiculacea — consisting of the Corbiculidae and Sphaeriidae (or Pisidiidae) — dominates the freshwater bivalve fauna of tropical Asian streams in terms of both diversity and abundance. Corbiculids and sphaeriids resemble each other in terms of shell form, hinge teeth and gross anatomy; in addition, both families contain some hermaphroditic species. However, some authorities regard corbiculids as relatively advanced, considering that the Sphaeriidae warrant placement in a separate superfamily. Corbiculids are larger than sphaeriids (mature individuals are generally > 12 mm long and can be considerably bigger), with yellowish, brown or olive-green shells and a strong ligament. The hinge teeth are distinctive, with three cardinal teeth and well developed anterior and posterior lateral teeth. The family includes exclusively freshwater species (e.g. *Corbicula fluminea*: Figs 4.14F & G), some which tolerate brackish water (*Corbicula fluminalis*: Fig. 4.14H), and others which are associated with tidal rivers (e.g. *Batissa, Cyrenobatissa, Cyrenodonax* and *Villorita*) or mangroves (*Polymesoda [Geloina]*) (Morton, 1984).

Villorita cyprinoides is of some economic importance in India, and aspects of its growth, salinity tolerance and burrowing behaviour have been studied (Sivankutty-Nair, 1975; Sivankutty-Nair & Synamma, 1975; Ansari *et al.*, 1981; Chatterji *et al.*, 1984). Morton (1992) has investigated the salinity tolerance of *Cyrenobatissa subsulcata* from southern China — a genus which is sometimes treated as a subgenus of *Batissa* (Dudgeon, 1980) or even *Corbicula* (Prashad, 1929; Chen *et al.*, 1992); *Cyrenodonax* is likewise placed in *Corbicula* by some workers (Chen *et al.*, 1992). Djajasasmita & Budiman (1984) have documented the substrate relationships of *Batissa violacea* in the Pisang River (Sumatra), a species which occurs in tidal rivers throughout most of Indonesia, Papua New Guinea and some Pacific Islands (e.g. Fiji; Raj & Fergusson, 1980). *Batissa violacea* seems to prefer muddy sediments, and avoids substrates overlain with organic detritus. The ionic composition of the haemolymph and tissues is different from that of other freshwater bivalves (Raj & Fergusson, 1980), and suggests that *B. violacea* is unlikely to be able to persist in rivers which lack a tidal influence; moreover, the larvae are apparently planktonic. Only the wholly freshwater corbiculids (*Corbicula* spp.) are ovoviviparous, and larvae are brooded within the inner demibranch.

There is considerable variation in shell form within species of the genus *Corbicula*, and a huge number of nominal species have been described from tropical Asia. Many of them are likely to be junior synonyms of *C. fluminea* (= *C. leana*, etc.), which is found in a wide variety of lotic habitats, and *C. fluminalis* (= *C. japonica*, *C. manilensis*), which inhabits the lower course (and tidal reaches) of rivers. Indeed, Prashad (1929) synonymised 42 of 50 Chinese species as *C. fluminea* (see also Morton, 1979).

The Sphaeriidae — or fingernail clams — are the smallest freshwater bivalves and rarely greater than 10 mm long; adults of several species are do not exceed 6 mm. There are around 100 species of sphaeriids worldwide. Most are rather eurytopic, being generally more widespread and cosmopolitan than the Unionidae. The shell is porcelaineous (usually pale yellow or white), and often rather thin and fragile. The ligament and hinge teeth are small, and the outer demibranch of the ctenidium is relatively poorly developed or even absent (in *Pisidium*). Larvae are brooded in marsupia within the inner demibranchs. The mantle is modified posteriorly to form siphons in all sphaeriids. Some (e.g. *Sphaerium*) have well-developed siphonal tubes, but in most genera, the siphons are short; only the (dorsal) exhalant siphon is present in *Pisidium*. The tropical Asian Sphaeriidae are poorly known, and are often overlooked because of their small size. Genera include *Pisidium* and *Musculium* in Hong Kong streams (see above) and *Sphaerium* in the north of the region in the Chang Jiang (Chen, 1979). *Afropisidium* and *Odhneripisidium* — usually treated as subgenera of *Pisidium* (e.g. Brandt, 1974) — are present in southern Asia, and *Pisidium* is the most widespread sphaeriid genus in the region (Brandt, 1974; Chang & Lin, 1978; Kuiper, 1982). *Eupera*, which is distributed across Eurasia, may penetrate the northern fringes of the Oriental Region.

In addition to the families listed above, the Glauconomidae (as *Glauconomya*) and Mactridae (*Tanysiphon* in India) have been reported from the lower reaches of some tropical Asian rivers but are probably confined to habitats with some tidal influence. However, the Arcidae (Arcacea) is represented in tropical Asia by four species of *Scaphula* which occur in the lower, non-tidal reaches of rivers in India, Burma and Thailand (Janaki Ram & Radhakrishna, 1984; Habe, 1985), where the animals are attached by the byssus to rocks or wood. Although quite widespread, they are known from few localities (Brandt, 1974). Other predominantly-marine bivalve families sometimes occur in tropical Asian rivers but they do not seem to include any true freshwater species: among them are the Cuspidariidae, Pholadidae, Psammobiidae, Scrobiculariidae, Solenidae, Ungulinidae and Veneridae.

CRUSTACEA

The most conspicuous crustaceans in tropical freshwater streams are members of the order Decapoda: the shrimps, crabs and crayfishes. The rest are microcrustaceans from various orders, as well as a few amphipods and isopods. Microcrustaceans comprise mainly zooplankton (see p. 54) and will not be considered here. Readers are referred to Fernando (1980a; 1980b) for an account of the composition and biogeography of the freshwater zooplankton of the Oriental Region. A few other microcrustaceans are inhabitants of the hyporheic zone of stony streams. They are considerably smaller than the macroinvertebrates which comprise the bulk of the lotic zoobenthos, and have received virtually no attention from biologists. Because of their diversity and undoubted ecological significance, the following account of freshwater crustaceans is focused mainly upon the decapods.

Most of the larger crustacean species belong to the class Malacostraca, and the Decapoda — with approximately 10,000 species, of which about 10% live in freshwater — is the most diverse malacostracan order. Although the body form can vary greatly, all decapods have five pairs of legs (peraeopods), of which the first (in crabs) and often the second (in shrimps) and third (in crayfishes) are chelate. Anterior to these limbs are three pairs of maxillipeds which aid in feeding. The gills are contained in a branchial chamber enclosed by the carapace. The anatomy and physiology of freshwater crustaceans has been reviewed succinctly by Covich & Thorp (1991), and will not be rehearsed here. Additional information is given also by Hobbs (1991) and Pennak (1989: pp. 489–513) who have summarized knowledge on the ecology of freshwater decapods in North America. Those accounts are, however, concerned largely with crayfishes which have a very limited distribution in tropical Asia.

There are two crayfish families in the Asian region. The Cambaridae (subfamily Cambaroidinae, including four species of *Cambaroides*) occur in the northeast (Korea, Japan and China) and are not found in tropical latitudes. *Procambrus clarkii* (Cambarinae) has been introduced into China and Taiwan from the United States by way of Japan (Dai, 1983; Welcomme, 1988: pp. 229–231; Hobbs *et al.*, 1989); it is cultured widely for food and some feral populations have become established in China (Dai, 1983; Shu Xinya, Hubei Fisheries Science Research Unit, pers. comm.). The possible ecological impacts of this invader have yet to be investigated, but are a matter for concern given the damaging effects of introductions of exotic crayfishes — including *P. clarkii* — elsewhere (e.g. Hobbs *et al.*, 1989 and references therein). Parastacid crayfish (13 species of *Cherax* in the subgenera *Astaconephrops* and *Cherax*) are confined to rivers south of the Central Highlands on the island of New Guinea, and have not spread northwards beyond this biogeographic barrier (Fig. 1 in Holthuis, 1982; see also Baer, 1953; Jones & Lester, 1992). The family is well represented in Australia (Williams, 1980: pp. 167–172).

Decapoda: Caridea

The Infraorder Caridea includes the Atyidae, Alpheidae and Palaemonidae, and these two families contain all of the truly freshwater shrimps. The higher classification of the Caridea have been revised recently by Chace (1992) who gives a key to the superfamilies, families and subfamilies of the group. Carideans can be distinguished from prawns and other shrimps (e.g. Penaeidae, Sergestidae, Stenopodidae) because the second abdominal pleura (the lateral part of the segment or somite) overlap those of the first and the third segments. Any shrimps lacking this feature that are collected from the lower course of rivers are probably marine or brackish-water species and not permanently resident in fresh water. Note that penetration of rivers and streams by brackish-water shrimps (such as planktonic *Acetes* spp. [Sergestidae] which live in estuaries: e.g. Ravindranath, 1980; Zafar & Mahmood, 1989) is facilitated by high alkalinity, and is much more restricted in soft-water streams (Johnson, 1961b). Atyids inhabit fresh or low-salinity brackish water (Johnson, 1961a; 1961b) and none of them are marine *sensu stricto*. There are few atyid species with adults that can survive or live in salinities approaching those of sea water, although the larvae may do so (Smith & Williams, 1981). Some palaemonids are entirely marine, while others are confined to brackish or fresh water; still others move between fresh and brackish waters. The following key includes only those epigean genera of Caridea which are likely to be found in fresh water; identification depends upon examination of well-grown individuals and follows the classification used by Chace (1983; 1992) and Chace & Bruce (1993). Terminology and morphological features follow Fig. 4.15.

1. First pair of chelate limbs extremely well developed (although unequal in size); one of the pair is massive and approximately the same length as the carapace (Fig. 4.16); second pair of chelate limbs relatively very small and delicate; body somewhat dorsoventrally flattened; rostrum short and inconspicuous; rare ....**Alpheidae** (*Alpheus cyanoteles*)

    First pair of chelae not as above; either the two pairs of chelate limbs are the same size, or the second pair is longer; body typically more-or-less laterally compressed; rostrum usually conspicuous; very commo... 2

2. Second pair of chelate limbs usually longer than the first (4.17A & B); without terminal hair tufts or bristles; lacking exopodites on the peraeopods ....**Palaemonidae** 3

    First and second chelae approximately the same size, with terminal hair tufts (Figs 4.18A & E and 4.19C & D); exopodites *may* be present on some of the peraeopods ....**Atyidae** 9

3. Carapace with a branchiostegal spine on each side; mandibular palp *may* be lacking... 4

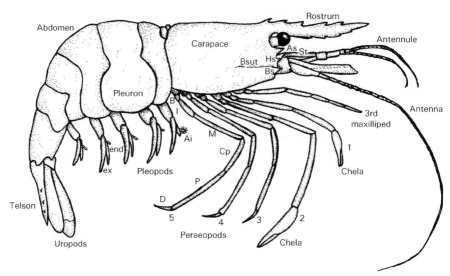

Fig. 4.15 Lateral view of a generalized caridean shrimp (after Hobbs, 1991): Ai, appendix interna; As, antennal spine; B, basis; Bsut, branchiostegal suture; Bs, branchiostegal spine; Cp, carpus; D, dactyl; end, endopodite; ex, exopodite; Hs, hepatic spine; I, ischium; M, merus; P, propodus; St, stylocerite. Peraeopods ('walking legs') are numbered sequentially 1 - 5 from the anterior end.

Carapace lacking branchiostegal spines; mandibular palp present
.... 8

4. Elevated, dentate (toothed) crest at the base of the rostrum 5
   Lacking an elevated crest at the base of the rostrum 6

5. Carapace with branchiostegal suture extending posteriorly from anterior margin at a point dorsal to the branchiostegal spine ....
   ....*Exopalaemon*
   Carapace without a branchiostegal suture .... *Nematopalaemon*

6. Mandible with a palp ....*Palaemon*

   Mandible without a palp 7

7. Telson with three pairs of spines on the posterior margin; endites of maxillae and first maxilliped markedly modified; chelae of second peraeopods are weakly developed and only slightly larger than those of the first (Fig. 4.17B) .... *Coutierella*

   Telson with two pairs of spines on the posterior margin; endites of maxillae and first maxilliped of 'normal' form; chelae of second peraeopods distinctly larger than the first .... *Palaemonetes*

8. Carapace without a hepatic spine; rostrum with elevated basal crest; second pair of chelate limbs (i.e. the second peraeopods) longer than the first but not exceptionally so; not especially common or widespread; relatively small (carapace length ≤ 10 mm) .... *Leptocarpus*

    Carapace with a hepatic spine; rostrum rarely with elevated basal crest; second pair of chelate limbs much longer than the first; common and widespread (Fig. 4.17A); *may* be quite large (carapace length > 30 mm) .... *Macrobrachium*

9. Carapace armed with supraorbital spine; peraeopods bearing exopods; rare .... *Paratya*

    Carapace not armed with supraorbital spine; peraeopods without exopods 10

10. Carpus (i.e. the segment adjacent to the chela) of second peraeopod is deeply excavated (or cleft) and shorter (or little longer) than broad; carpus (and chelae) of first and second peraeopods similar in size and shape 11

    Carpus of second peraeopod is not deeply excavated (or cleft) and distinctly longer than broad (Fig. 4.19B); carpus of first peraeopod differs somewhat in size and shape from that of the second peraeopod (Fig. 4.19B); chela of second peraeopod *may be* less setose than the first 12

11. Third maxilliped terminates in a single apical spine (sometimes partially concealed by setae: Fig. 4.18C); endopod of first pleopod of male tapering from base to tip (Fig. 4.18A); third peraeopod of large males lacking a spur on the merus (i.e. the longest segment: Fig. 4.18D); well-developed epipodites on the third and fourth peraeopods; posterolateral margins of telson not drawn out into points (Fig. 4.18B); females often larger than males; chelae may be heteromorphic (i.e. not all individuals have chelae with the same morphology) and modified for filter-feeding or for collecting and scraping food from the substrate .... *Atyoida*

    Third maxilliped terminates in numerous stout setae rather than in a single spine (Fig. 4.18F); endopod of first pleopod of male is expanded into a lamina which is somewhat oval in shape (Fig. 4.18E) with a submarginal row of spines; third peraeopod of large males bears a massive spur on the merus (Fig. 4.18H);

epipodites on the third and fourth peraeopods are vestigial; posterolateral margins of the telson drawn out into short points (Fig. 4.18G); females and males of approximately the same size; chelae are monomorphic and tipped with brushes of long setae for filter-feeding .... *Atyopsis*

12. Stylocerite of the antennular peduncle may be very long and slender (reaching the middle or the end of the second segment of the peduncle); endopodite of the first pleopod in males modified markedly (and unlike the endopodite of the second pleopod) consisting of a hemispherical lamina with marginal spines (Fig. 4.19D); appendix interna (a thumb-like projection on the endopodite) situated more-or-less basally .... *Neocaridina*

Antennular stylocerite short (not reaching the second segment of the antennular peduncle); endopodite of the first pleopod not markedly modified and rather similar in shape to the exopodite (and the endopodite of the second pleopod); appendix interna (if present) situated distally .... *Caridina*[10]

There is only one freshwater species of Alpheidae (the snapping or pistol shrimps): *Alpheus cyanoteles* (Fig. 4.16) reported from Peninsular Malaysia (Yeo & Ng, 1996). The massive first chelae are so conspicuous that this animal cannot be confused with other Asian freshwater shrimps; their burrowing habit is distinctive also. The Alpheidae contains many marine and mangrove representatives, but only *A. cyanoteles* has the large eggs and highly-abbreviated non-pelagic larval development typical of freshwater atyids and palaemonids. This species has been described and well-illustrated by Yeo & Ng (1996), who give a key to *Alpheus* species found in Southeast Asian freshwaters and mangroves. The alpheid genus *Potamalpheops*, which is known from West African freshwaters (Powell, 1979), may occur in Asian streams also (Yeo & Ng, 1996) although this has yet to be confirmed.

The Atyidae comprises 35 genera world-wide, including *Atyopsis*, *Atyoida*, *Caridina*, *Edoneus*, *Neocaridina*, *Paratya* and *Typhlocaridina* in tropical Asia. Larval habitats of atyids vary, but adults are almost exclusively confined to freshwater. *Antecaridina* has a wide and disjunctive distribution on islands in the Pacific (the Solomons, Fiji, Kuro-shima Island, the Ryukyus and Maui) and the Red Sea (Suzuki,

---

[10] Chace (1997) provides a key to the 12 species of *Caridina* known from the Philippines.

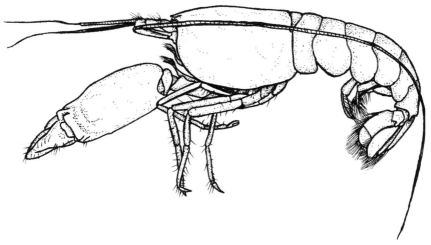

Fig. 4.16 *Alpheus cyanoteles* (Alpheidae), a freshwater snapping shrimp (body length 31 mm; redrawn from Yeo & Ng, 1996).

1980; Smith & Williams, 1981; Choy, 1991), but there are as yet no records from tropical Asia. *Edoneus*, a monotypic genus is confined to the Philippines, is hypogean (Holthuis, 1978a; Balete & Holthuis, 1992; Chace) as is *Typhlocaridina* from southern China (Liang & Yan, 1981; Liang & Zhou, 1993 — but see Guo *et al.*, 1996). There are hypogean species of *Caridina* also (Holthuis, 1978a; Guo *et al.*, 1992; 1996) — which are characterized by their partially or completely unpigmented eyes — but most species in this large genus are epigean. *Edoneus* resembles *Caridina* (and will fall out at the couplet for *Caridina* in the key given above), but has degenerate, unpigmented eyes (Chace, 1997). *Atya* (*sensu stricto*) is not found in Asia, although some *Atya*-like shrimps from the region (*Atyopsis* and *Atyoida*) have been confused with this genus (examples are given in Chace [1983] who provides a key to the species in the two genera); *Pseudatya* (mentioned by Johnson, 1961b) is a junior synonym of *Atyoida*. *Atyoida* (Fig. 4.18A) occurs in mountain streams in the Philippines and Indonesia but is absent from the Asian mainland (Chace, 1983). *Atyopsis* (Fig. 4.18E) has a similar distribution but is present on Sri Lanka, Taiwan and the Asian mainland from India to Thailand and the Peninsular Malaysia. The genus occupies a range of lotic habitats from torrential streams to larger rivers in their plains reaches (Chace, 1983). *Paratya* occurs in mainly northern Asia (Japan), Australia, New Zealand and New Caledonia (Carpenter, 1978; Williams, 1980: pp. 164–166; Nishino, 1981; Walsh, 1993), but there are a couple of records (Vietnam and

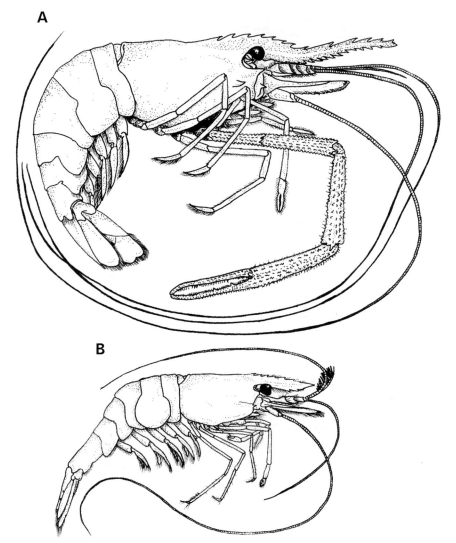

Fig. 4.17 Palaemonid freshwater shrimps (redrawn from Mendis & Fernando [1962] and Bruce [1989]): A, *Macrobrachium rosenbergii* (body length 30 cm); B, *Coutierella tonkinensis* (body length 27 mm).

Flores Island, Indonesia) from within the Oriental Region (Chace, 1997).

Unfortunately, the Atyidae — and especially the largest genus, *Caridina* — is '. . . in urgent need of detailed revision . . .' (Ng, 1990a; see also Johnson, 1961b). Of *Caridina*, Chace (1997) writes: '. . . few caridean groups offer greater difficulty than do the approximately 160 species and subspecies that are currently recognized in this genus

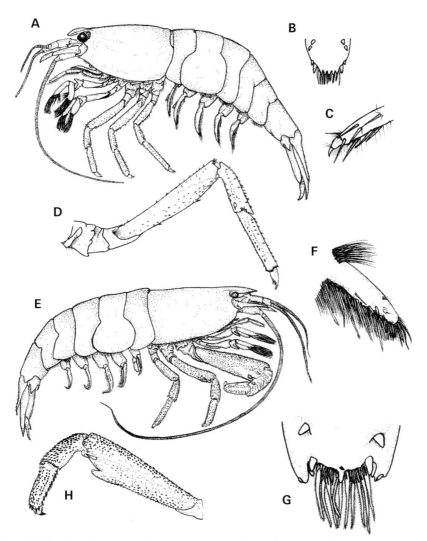

Fig. 4.18 Atyid shrimps of the genera *Atyoida* and *Atyopsis* from Asian streams (redrawn from Chace, 1983): A, *Atyoida* (carapace length 25 mm); B, posterior end of the telson ; C, distal portion of the third maxilliped; D, third peraeopod of a mature male; E, *Atyopsis* (carapace length 22 mm); F, distal portion of the third maxilliped; G, posterior end of the telson; H, third peraeopod of a mature male.

. . .'. The situation is complicated further because new species of *Caridina* continue to be discovered from the region (e.g. Liang & Yan, 1983; Liang *et al.*, 1987; 1990; Liang & Zhou, 1993; Choy & Ng, 1991; Choy, 1992; Guo *et al.*, 1992; 1996; Hung *et al.*, 1993; Ng, 1995). A particular issue is the status of *Neocaridina* (Fig. 4.19), which has been treated variously as a subgenus (or synonym) of

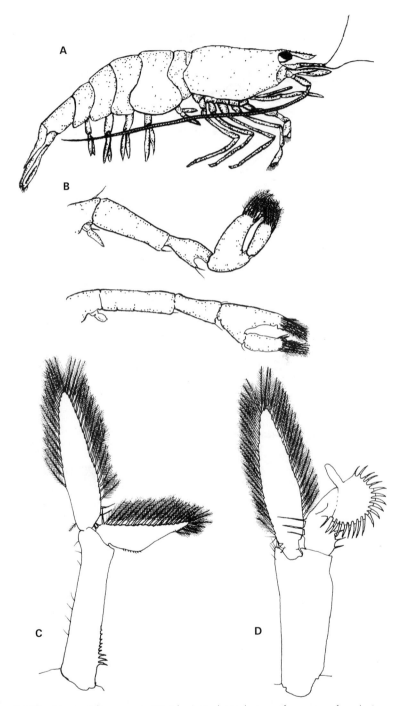

Fig. 4.19 *Neocaridina serrata* (Atyidae): A, lateral view of a mature female (carapace length 10 mm); B, distal portion of the chelate first and second peraeopods; C, first pleopod of a mature female; D, first pleopod of a mature male.

Caridina (e.g. Johnson, 1961b; Liang & Yan, 1983; Shokita, 1976), or as a separate genus (e.g. Kubo, 1938; Mizue & Iwamato, 1961; Zhang & Sun, 1979; Dudgeon, 1987c; Hung *et al.*, 1993). Moreover, even in cases where *Neocaridina* has been accepted as a separate genus, the placement of certain *Caridina* species (e.g. '*Caridina' serrata* of Hung *et al.*, 1993) is contentious. The two genera resemble each other closely, and cannot be distinguished without examination under a stereomicroscope. Nevertheless, following Zhang & Sun (1979) and Holthuis (L.B. Holthuis, Rijmuseum van Natuurlijke Historie, pers. comm.), *Neocaridina* is treated as a valid genus in the key given above.

The Palaemonidae consists of two subfamilies: Palaemoninae and Pontoninae; only the former has freshwater species (Chace & Bruce, 1993). *Macrobrachium* (especially) and *Palaemonetes* include both wholly freshwater and some brackish-water species although it appears that only one freshwater species of *Palaemonetes* (*P. sinensis*) is present in Asia, and is confined to the north of the region (Bruce, 1989). *Palaemonetes* is usually restricted to coastal streams (although they may be entirely fresh water), but certain *Macrobrachium* (Fig. 4.17A) can live in steep, rocky hillstreams (e.g. *M. hainanense* in Hong Kong). Most species of *Palaemon* and *Leptocarpus* are euryhaline (e.g. Johnson, 1961a), although a few inhabit streams (Holthuis, 1978b; Chace & Bruce, 1993). Some freshwater *Palaemon* species in tropical Asia have been reclassified as *Macrobrachium* (Ng, 1990a; Chace & Bruce, 1993) and *Cryphiops* is apparently a junior synonym of that genus (Ng, 1990a). *Coutierella* (Fig. 4.17B) is known only from Hong Kong and Vietnam (the Red River) where it is found in brackish water at river mouths (Bruce, 1989); whether it can live permanently in fresh water is not known. Likewise, *Nematopalaemon* is reported from tidal reaches of rivers in Bangladesh, but is probably not truly freshwater. *Leandrites* has been recorded also from brackish (but not fresh) waters within the region (Bruce, 1989; Chace & Bruce, 1993). *Troglindicus* is hypogean and known from a single locality in India (Chace & Bruce, 1993), and there are some cavernicolous species of *Macrobrachium* also (Holthuis, 1978a; 1984). Chace & Bruce (1993) provide a key to the Indo-West Pacific Palaemoninae, which includes many (but not all) of the species found in tropical Asian streams.

Macrobrachium is the most diverse genus of Palaemoninae (> 175 valid species world-wide) and contains the largest species of freshwater shrimps. For this reason they are usually referred to as 'prawns'. *Macrobrachium rosenbergii* (Figs 4.17A and 4.20K) — which is propagated widely for food and may be an important component of arisanal fisheries (e.g. Hickling, 1961; Kurian & Sebastian, 1982; Christensen, 1993a) — can exceed 100 mm in carapace length. Most *Macrobrachium* species inhabit tropical freshwaters although some live in tidal rivers and a few are marine as juveniles. Holthuis (1950: pp. 105–111) gives a key to all species recognized prior to 1950, while Chace and Bruce (1993: pp. 20–23) provide one for fully-grown males of Philippine and Indonesian *Macrobrachium*. As Fig. 4.20 shows, species identification is based upon (among other things) the form of the rostrum, and each species has a characteristic rostral-formula describing number and arrangement of the teeth along the dorsal and ventral margins. For many species, however, reliable identification is possible for mature males only and determination of preserved females or juvenile males can be problematic.

Considering the importance of *Macrobrachium* as a source of human food, and the conspicuousness of atyids and palaemonid shrimps in Asian rivers and streams, we know surprisingly little about their ecology — particularly their role in energy flow and organic-matter transfer. *Macrobrachium* are omnivorous, but some species (at least) become largely predatory as they mature (e.g. Krishna Murthy & Rajagopal, 1990). Their impact on populations of benthic prey has yet to be investigated, but may be substantial in view of the large size and biomass of *Macrobrachium* species relative to most other stream invertebrates. Atyids feed on fine organic material which they obtain by filtering, gathering deposited material, or scraping the surface of submerged leaf litter with the setae of the first and second chelate peraeopods. Interestingly, *Atyoida serrata* (which is found in the Seychelles, although the genus occurs in tropical Asia) shows interspecific variation in the chelae (i.e. they are heteromorphic), and individuals may have chelae that are modified for picking and scraping, or sweeping and filter-feeding, or filter-feeding only (Chace, 1983). Manipulation of atyid densities in Neotropical montane streams has been shown to cause significant reductions in sedimentation rates and enhancement of algal standing stocks, thereby influencing the abundance of other macroinvertebrates indirectly (Pringle *et al.*, 1993; Pringle & Blake, 1994). It is tempting to suggest that atyids play a key role in structuring benthic communities in Asian streams also, but supporting evidence is lacking.

Individual *Macrobrachium* species tend to be rather stenotopic, and are often confined to particular types of stream habitat (e.g. Johnson, 1964; Ng, 1990a). However, *M. lanchesteri* and *M. nipponense* have been introduced into Singapore (Ng *et al.*, 1993)

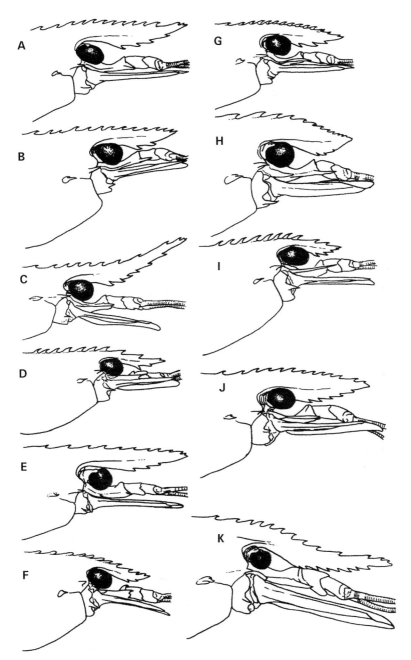

Fig. 4.20 Interspecific variation in the rostrum of Asian *Macrobrachium* (based on male specimens from the Philippines and Borneo redrawn from Fenner & Chace, 1993): A, *Macrobrachium australe* (carapace length 20 mm); B, *M. bariennse* (13 mm); C, *M. equidens* (21 mm); D, *M. gracilirostris* (15 mm); E, *M. idae* (17 mm); F, *M. jaroense* (16 mm); G, *M. latidactylus* (20 mm); H, *M. latimanus* (32 mm); I, *M. lepidactyloides* (16 mm); J. *M. mammillodactylus* (25 mm); K, *M. rosenbergii* (66 mm).

where they have established themselves in disturbed environments. They do not co-occur with the native *Macrobrachium* species which prefer softer, acidic waters draining relatively undisturbed catchments (Ng, 1990a). Some *Caridina* (Atyidae) species show strong habitat preferences also (e.g. Johnson, 1964; de Silva, 1982; 1994; Ng, 1990a; Choy, 1992) — based in part upon their differential responses to temperature, *p*H, salinity and temperature (e.g. de Silva, 1988; 1989; de Silva & de Silva, 1988) — and some of them have potential as bioindicators of environmental disturbance and water quality (Choy, 1992). Both *Caridina* and *Macrobrachium* show longitudinal (and altitudinal) zonation along rivers, with species replacement — in *Macrobrachium*, for example (Costa, 1984) — that reflects larval development patterns and/or the need to undertake breeding migrations. In essence, only downstream species pass through a prolonged planktonic larval stage, and such a development pattern often requires brackish or saline conditions and may involve breeding migrations. The longitudinal distribution of some atyid shrimps is influenced similarly by the need to undergo breeding migrations, but water temperatures determine the downstream distribution limit of certain species which lack planktonic larval stages (de Silva & de Silva, 1988; de Silva, 1989). Chace (1997) considers that '. . . . although tropical freshwater shrimps have rather finely drawn habitat preferences there is some indication that nearly all of the species occurring in a broad geographic region may be found in a single stream, if that stream offers the required habitats . . .'.

Although ovigerous females of many *Macrobrachium* species undertake a seaward migration for mating, this is not always a necessity and life-history patterns exhibit intraspecific — as well as interspecific — variation. For instance, the widely-distributed, *Macrobrachium nipponense* shows an increase in egg volume and larval survival, a shortening of larval stages, and a decrease in clutch size and larval duration for freshwater compared to brackish-water populations (Mashiko, 1983a; 1990; Wong, 1987). In addition, the growth rate of females in freshwater populations is noticeably depressed (Mashiko, 1983b). Both *M. nipponense* and *M. hainanense* occur in Hong Kong and ovigerous females are found in May and June (Dudgeon, 1985a). Reproduction seems to be stimulated by rising water temperatures, but has the consequence that recruitment takes place at the start of the summer monsoon. Unlike *M. nipponense*, larval development in *M. hainanense* is abbreviated (Wong, 1989) and lacks the planktonic stages seen in the former species. *Macrobrachium hainanense* is confined to hillstreams and does not undertake breeding migrations, whereas *M. nipponense* occurs in the lower course and tidal reaches of Hong Kong streams. The suppression of larval stages in *M. hainanense* may be a selective advantage for species which produce young close to the onset of the wet season when streams experience spates. The difference in patterns of larval development between species of *Macrobrachium* inhabiting upstream and downstream sections of Hong Kong streams is reflected across the genus as a whole, and egg size decreases along the transition from riverine to brackish water habitats and estuaries (Mashiko, 1990). This egg-size variation is associated with intraspecific genetic differentiation among *Macrobrachium* populations (Mashiko & Numachi, 1993). Jalihal *et al.* (1993) divide the 40 *Macrobrachium* species for which information on development patterns is available into three groups, reflecting the greater trend towards occupation of entirely freshwater riverine habitats: the prolonged or 'normal' type (with eight to 20 larval stages); the partially-abbreviated type (with two or three stages); and the completely abbreviated type (with only one stage). This trend towards 'freshwaterization' is accompanied by a decrease in average body size from estuarine (range = 50 — 320 mm body length) to hillstream (25 — 70 mm) habitats.

Two atyid species — *Caridina lanceifrons* and *Neocaridina serrata* — are found in Hong Kong hillstreams where *Macrobrachium hainanense* occurs. Unlike some other atyids (e.g. certain *Caridina* spp.: Benzie, 1982), both shrimps breed in fresh water (Mizue & Iwamoto, 1961; Shokita, 1976; Dudgeon, 1987c) and have abbreviated larval development, producing hatchlings (from large eggs) in an advanced stage of growth (Mizue & Iwamoto, 1961; Dudgeon, 1987c). As postulated for *M. hainanense*, this may be an adaptation to reduce washout of larvae during wet-season spates, and ovigerous females of *N. serrata* occur from April to October (Dudgeon, 1985a) — a period including the summer monsoon. An alternative or supplementary explanation is that freshwater shrimps produce relatively large eggs (from which well-developed juveniles hatch) in response to the paucity of planktonic larval food in streams (Wong, 1989; see also p. 50 and p. 54). As is the case for *Macrobrachium*, atyids (including *Caridina*) show a trend towards reduction of planktonic larval stages in those species confined to upstream sections of lotic habitats (e.g. Benzie, 1982; Choy, 1991; Hung *et al.*, 1993). Even within a single species (for example, in the genus *Paratya*), riverine populations have larger eggs and early-stage larvae, and smaller brood sizes, than conspecifics in estuarine environments (Nishino, 1981; Walsh, 1993).

Hong Kong *Neocaridina serrata*, like other atyids, is rather short-lived with a life span of approximately 12 months. Consequently, each individual participates in only one summer breeding season, although females may spawn more than one batch of eggs during that time. The abundance of *N. serrata* and *Macrobrachium hainanense* in Hong Kong is correlated with water temperature (Dudgeon, 1985a), and a similar relationship has been noted for *Caridina singhalensis* in Sri Lankan mountain streams (de Silva 1982). Warmth stimulates reproduction in *N. serrata*: ovigerous females are present only when water temperatures exceed 20°C and the percentage incidence of brooding females increases as temperatures rise (Dudgeon, 1985a). The hypothesis that breeding is stimulated by rising temperatures gains support from the observation that *Caridina* and *Macrobrachium* in subtropical latitudes reproduce in summer (e.g. Kamita 1956; 1957; 1959; Truesdale & Mermilliod, 1979) while, in tropical regions with equable temperatures, recruitment of freshwater shrimps occurs throughout the year (Babu, 1963; Hart, 1981) although 'hot season' peaks in the abundance of ovigerous females have been recorded (Hart, 1980). This tendency towards year-round reproduction is modified in regions with a marked wet season, where breeding is synchronized with the onset of the rains (Raman, 1967; Abele & Blum, 1977; Rajayalakshmi, 1980) which increases habitat availability for juveniles (Rajayalakshmi, 1980). In some species (e.g. *Macrobrachium birmanicum*) the wet season is associated also with migrations onto inundated river floodplains where food availability is high (Singh & Srivastava, 1989; see also Prakash & Agarwal, 1985).

Aside from the timing of breeding and the pattern of larval development, freshwater shrimps vary with respect to mating behaviour and sexuality. *Macrobrachium* species are sexually dimorphic: males have larger second chelipeds than females and (in *M. nipponense*) attain sexual maturity later and so are bigger (Mashiko, 1981; 1983b). The female is guarded between the long chelipeds of the male during pair formation which may involve males in the defense of a territory. Interference with pairing individuals is reduced by the male nipping potential rivals with his large chelipeds (Mashiko, 1981). Moreover, because the female moults immediately before spawning, the large chelipeds of the male are important for fending off cannibalistic attacks on his soft-shelled mate (Mashiko, 1981). It is of interest that those species of *Macrobrachium* with abbreviated development which reproduce in streams tend to exhibit a reduction in the degree of sexual dimorphism (i.e. there is a decline in the relative size of the body and the second cheliped of males) compared to species which breed in estuaries (Jalihal *et al.*, 1993). The ecological significance of this difference is not known.

Atyids seem to lack complex mating rituals and, in contrast to dioecious *Macrobrachium*, the family includes some hermaphroditic species. For example, males of *Atyoida pallipes* are smaller than females suggesting protandrous hermaphroditism (Chace, 1983), which has been confirmed for *Paratya curvirostris* in New Zealand (Carpenter, 1978), but appears to be ruled out in *Atyopsis* by the relatively large size of males (Chace, 1983). *Neocaridina serrata* females in Hong Kong grow to a larger size than males, but protandry has not been reported for atyids (other than *Atyoida pallipes*) in tropical Asia. Whether this reflects an absence of hermaphroditism — or indicates that the phenomenon has been widely overlooked — is not clear.

Decapoda: Brachyura

Freshwater crabs are a distinctive element of stream faunas in tropical latitudes. True freshwater crabs spend their entire lives in streams and rivers, and never return to the ancestral marine habitat. Nevertheless, a variety of estuarine or mangrove crabs in the family Grapsidae (and especially the Sesarminae which are mainly denizens of mangroves) may move into entirely fresh water. They include *Varuna litterata* and *Varuna yui* (Ng, 1988a; Hwang & Takeda, 1986), species of *Utica*, *Ptychognathus*, *Pseudograpsus* (Varuninae) and *Holometopus* (Sesarminae), as well as some poorly-known sesarmines (*Labaunium*, *Neosarmatium* and *Sesarmoides*) which have been reported some distance from the sea; *Sesarmoides* has even been found in cavernicolous waters (Ng, 1988a). *Pseudograpsus crassus* is known from fresh waters in New Guinea and eastern Indonesia (including Sulawesi and the Moluccas), while *Ptychognathus riedelii* is more widely distributed from New Guinea through the greater part of Indonesia to Sumatra (Holthuis, 1982); *Ptychognathus demani* is confined to New Guinea. The breeding habits of these 'freshwater' grapsids have not been investigated.

The majority of estuarine grapsids are occasional visitors to — or temporary residents in — rivers and streams, but there are (at least) two genera which are exceptional in this regard. Species of *Eriocheir* (the hairy-clawed or mitten crabs: *E. formosa*, *E. japonicus* [Fig. 4.21A; including *E. recta*], *E. leptognathus* and *E. sinensis* [Chan *et al.*, 1995]) inhabit the rivers of eastern Asia. There is some disagreement over the

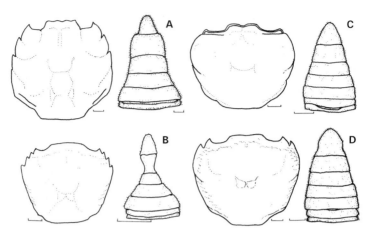

Fig. 4.21 Carapace and abdomens of male specimens of Hong Kong freshwater crabs (scale line = 4 mm): A, *Eriocheir japonicus* (Grapsidae); B, *Somanniathelphusa zanklon* (Parathelphusidae); C, *Nanhaipotamon hongkongense* (Isolapotamidae); D, *Cryptopotamon anacoluthon* (Isolapotamidae).

classification of the genus *Eriocheir*, and one extreme view is that all nominal species represent races of a single, highly-variable species: *E. sinensis* (Li *et al.*, 1993; but see Chan *et al.*, 1995). Recent research suggests that the genus *Eriocheir* consists of three species only: *E. japonica*, *E. sinensis* and *E. hepuensis*, and that *E. formosa* and *E. leptognathus* should be placed in other genera (Guo *et al.*, 1997). These anadromous crabs migrate to the sea in order to release their eggs which hatch into planktonic zoea larvae (e.g. Terada, 1981; Lai *et al.*, 1986; Shy & Yu, 1992). After larval development is complete, the young crabs move upstream. *Eriocheir sinensis* has invaded and become well established in European rivers, where it has received a considerable amount of study (e.g. Paepke, 1984; Ingle, 1986; Rasmussen, 1987; Anger, 1991). It has been accidentally introduced to Lake Erie in North America also (Nepszy & Leach, 1973), but does not seem to have become established or widespread. *Eriocheir sinensis* is cultivated in ponds for food in China (Wang & Zhou, 1989a; 1989b), and has been introduced into the southern provinces where *E. japonicus* is the native species. The cultivation and consumption of *E. sinensis* takes place despite the fact that it serves as an intermediate host for lung flukes (*Paragonimus* spp.; Platyhelminthes: Trematoda), especially *Paragonimus westermani* which causes paragonimiasis. Indeed, most tropical Asian freshwater crabs seem to be potential hosts for these parasites (e.g. Fernando, 1960; 1974; Mendis & Fernando, 1962; Dai *et al.*, 1975; 1979; Ng, 1988a).

A second grapsid genus — *Geosesarma* (subfamily Sesarminae) — has become more fully adapted to non-marine environments than *Eriocheir*, and does not require a seaward migration for reproduction (Ng, 1988a). Instead, these crabs produce a few (around 50) large (1–1.5 mm diameter), yolky eggs and most of the zoea stages are suppressed. The larvae do not feed, and development is completed rapidly in entirely fresh water. While information on most species in the genus is lacking, it appears that larval development takes place in the bottom of water-filled burrows (Soh, 1969; Ng, 1988a) where the free-swimming zoea avoid the risk of being swept downstream by the current. The genus *Geosesarma* consists of around 40 species which occur throughout Southeast Asia and the Indopacific. All are fresh water and at least partly semiterrestrial; many make burrows along stream banks. Individual species tend to be confined to particular altitudes and hence show longitudinal zonation along rivers (Ng, 1988a). The semi-terrestrial species scavenge for food on land (Collins, 1980). One of them (*G. malayanum*) has become emancipated from streams entirely, and visits pitcher plants (*Nepenthes* spp.: Nepenthaceae) to replenish the water supply inside its branchial chambers (Ng & Lim, 1987). *Geosesarma notophorum* is terrestrial also and unique among the Grapsidae in that it exhibits completely abbreviated development: i.e. the eggs (there are usually 8–12 of them) hatch into small crabs without any larval stages (Ng & Tan, 1995). The female carries her young on the *dorsal* surface of her carapace for two to three days after they hatch, a behaviour that has not been reported in any other crab (Ng & Tan, 1995).

The Ocypodidae — a family of semiterrestrial crabs confined mainly to depositional shores in marine environments — has a single freshwater representative (*Potamocypoda pugila*) in Peninsular Malaysia. It is apparently confined to rivers with some tidal influence (Ng, 1988a) and, since other species of *Potamocypoda* live in the sea (e.g. Tai & Manning, 1984), it is probable that *P. pugila* has planktonic, marine larvae. The Hymenosomatidae (false spider crabs) is likewise a mainly-marine family (Lucas, 1980; Chuang & Ng, 1994), but has at least five freshwater species: *Amarinus wolterecki* (in the Philippines), *A. angelicus* (New Guinea: Holthuis, 1982), *Cancrocaeca xenomorpha* (cavernicolous in Sulawesi: Ng, 1991a), *Elamenopsis introverta* (in China) and *Limnopilos naiyanetri* (in Thailand: Chuang & Ng, 1991). *Elamenopsis exigua* is known from tidal reaches of rivers in Thailand and India, but may not survive in completely fresh water (Chuang & Ng, 1994). The breeding habits of freshwater hymenosomatids are unknown (although some members of the family produce zoea larvae: Muraoka, 1977; Terada, 1977; Kakati, 1988), but *A. wolterecki, C. xenomorpha* and *L. naiyanetri* have large eggs so that direct development — with suppressed larval stages — is suspected (Chuang & Ng, 1994).

Six families of crabs — arranged in two superfamilies (the Potamoidea and Gecarcinucoidea) — live permanently in the freshwaters of tropical Asia: the Gecarcinucidae, Isolapotamidae, Parathelphusidae, Potamidae (not Potamonidae: see Ng, 1988a), Sinopotamidae and Sundathelphusidae (Bott, 1970). They are quite diverse: for example, at least 160 species are known from China alone (Ng & Dudgeon, 1992). The recent discovery of 25 new potamid species in Taiwan (Shy *et al.*, 1994) — where the freshwater crabs were supposed to be quite well known (Hwang & Mizue, 1985) — suggests that the biodiversity of the group has been underestimated greatly. Ng & Yang (1986) and Ng (1987; 1988a) consider that the Isolapotamidae and probably the Sinopotamidae should be included within the Potamidae, their separation being based largely on structural details of the male gonopods (the first and second pleopods which are used for copulation). Thus Ng (1988a) recognizes only a single family — the Potamidae — within the Potamoidea of tropical Asia, although this opinion is not accepted by all workers (e.g. Dai & Chen, 1979; 1987; Dai *et al.*, 1984; Hwang & Mizue, 1985).

True freshwater crabs have large (> 1 mm in diameter) yolky eggs, and the planktonic larval stages typical of marine crabs are suppressed so that hatchlings are miniature — but fully-formed — replicas of the adult. The eggs are held in place under the abdomen by setaceous pleopods, where they are incubated for several weeks. Brooding females are reclusive and retreat into burrows or cavities beneath large stones. Juveniles are released near the onset of the wet season in Hong Kong streams (Dudgeon, 1992a), and production of young during the monsoon seems to be typical of freshwater crabs in tropical Asia (Fernando, 1960; Mendis & Fernando, 1962; Pillai & Subramonian, 1984; Ng, 1988a; Dayakar & Ramana Rao, 1992; see also Gangotri *et al.*, 1978; Joshi & Khanna, 1982a; 1982b). Regrettably, detailed information on the population dynamics of these animals is lacking, and much remains to be discovered about their natural history (Ng, 1988a).

Freshwater crabs vary in the extent to which they are confined to the aquatic environment: in Taiwan, sinopotamids (and grapsids) are entirely aquatic and feed in water, potamids and parathelphusids are amphibious and feed in water and on land, while isolapotamids are largely terrestrial and feed on land much of the time (Hwang & Mizue, 1985). Some gecarcinucids (*Thelphusula*) and sundathelphusids (*Perbrinckia* and *Terrathelphusa*) are more-or-less terrestrial also (Collins, 1980), as are some potamids (*sensu* Ng [1988a]: e.g. *Nemoron* and *Pudaengon*). There is some preliminary evidence of habitat segregation between potamid, parathelphusid and isolapotamid crabs in Hong Kong (Dudgeon, 1992a). Potamids and isolapotamids, such as *Cryptopotamon anacoluthon* (Fig. 4.21D) and *Nanhaipotamon hongkongense* (Fig. 4.21C), are most numerous in upland habitats with clear, fast-flowing water and accumulations of leaf litter. However, *N. hongkongense* is markedly more terrestrial than *C. anacoluthon* (which very rarely leaves the water): it makes burrows in damp soil and forages on land during heavy rain. The parathelphusid *Somanniathelphusa zanklon* (Fig. 4.21B) inhabits riverine habitats and slow-flowing low-gradient streams where it burrows in mud and clay banks. *Cryptopotamon anacoluthon* feeds largely upon allochthonous leaf litter, while *N. hongkongense* and *S. zanklon* are omnivorous with strongly carnivorous tendencies. The latter shows marked selectivity when a choice of gastropod prey is available, and preferentially consumes pulmonates with relatively light, fragile shells (Dudgeon & Cheung, 1990).

Despite different habitat occupancy, all freshwater or amphibious crabs exhibit similar life cycles with complete suppression of planktonic larval stages. This obvious adaptation to life in running waters is shared with some atyids and palaemonids (see pp. 174–175). Apart from this feature, and the ability to regulate their salt and water balance in a hypo-osmotic environment, freshwater crabs do not exhibit conspicuous adaptations for life in inland waters (Ng, 1988a), although the branchial chambers of amphibious species are modified and enlarged to facilitate air breathing (e.g. Diaz & Rodriguez, 1977; Farrelly & Greenaway, 1993). In most external aspects, however, these crabs resemble their marine relatives rather closely. The following key provides an aid to the identification of the higher taxa of crabs that may be encountered in tropical Asian rivers and streams, but depends largely on the availability of specimens of mature males. Following Ng (1988a), it includes the Isolapotamidae and the Sinopotamidae within the Potamidae. A key to genera has not been given because freshwater crabs are quite alike in external morphology, this convergence arising because of similar habits and ecology (Ng, 1988a). In addition, the classification of freshwater crabs is unstable. New genera and species continue to be described from the region, particularly from China and Taiwan (Dai *et al.*, 1975; 1979; 1980; 1984; 1985; 1995; Dai & Chen, 1979; 1985; 1990; Chen & Chang, 1982; Dai & Song, 1982; Huang *et al.*, 1986; Dai & Yuan, 1988; Ng & Dudgeon, 1992; Dai & Naiyanetr, 1994; Shy *et al.*, 1994), Southeast Asia (Naiyanetr & Ng, 1995; Ng, 1986a; 1986b; 1986c; 1986d; 1988b; 1989a; 1989b; 1989c; 1990b; 1990c; 1991b; 1992a; 1992b; 1993; 1995a; 1995b; 1996; Ng & Lim, 1986; Ng & Yang, 1986; Ng & Goh, 1987; Ng & Stuebing, 1989; 1990; Ng & Wowor, 1990; Ng & Tan, 1991; Naiyanetr, 1989; 1992; Dang & Tran, 1992; Ng & Takeda, 1992a; Ng & Naiyanetr, 1993; 1995), and the Philippines (Ng, 1991c; Ng & Takeda, 1992b; 1992c; 1993a; 1993b). A list of known genera based on these and other sources (Bott, 1970; 1974; Chuensri, 1974; Holthuis, 1982; Hwang & Mizue, 1985; Dai *et al.*, 1986; Ng, 1988a, 1991d; Dai, 1990) is provided at the end of the key.

1. Spider-like crabs with carapace as long or longer than broad (wide) — the front part is rounded and a small rostrum is present; long legs; body flat, thin and not·well calcified; with six abdominal segments .... false spider crabs: **Hymenosomatidae**

    Crabs of typical form with carapace usually somewhat broader than long; legs not especially long; rostrum lacking; body not especially flat and often rather inflated; always well calcified; with seven abdominal segments 2

2. The 'palms' of the chelipeds are bearded with soft tufts of 'hair'; lateral margin of the carapace with four teeth behind the orbital angle (the fourth is minute and inconspicuous) .... mitten crabs: *Eriocheir* spp.: **Grapsidae**

    Chelipeds and lateral margins of the carapace not as above 3

3. Carapace tends to be rather square (box-like) and not much broader than long, with straight antero- and posterolateral margins; eyes are large and the orbits of the eye 'sockets' reach the edge of the carapace; the inner margins of the third maxillipeds (which are the outer — external — pair of maxillipeds, covering the mouth and buccal cavity) are indented (Figs 4.22A & C) forming a distinct gap between them when closed .... **Grapsidae** (in part) 4

    Carapace usually distinctly broader than long, with the posterolateral margins converging more-or-less distinctly; orbits of the eye 'sockets' do not reach the edge of the carapace; the inner margins of the third maxillipeds are straight (Fig. 4.22B) and do not have a distinct gap between them when closed 5

4. Carapace rough and covered with granules; frontal area of carapace below eyes and on either side of the maxillipeds has a reticulate pattern of short hairs; merus of third maxillipeds (Fig. 4.22C) with a distinct oblique ridge; walking legs are not flattened noticeably nor fringed with long hair; with large eggs and abbreviated larval development .... *Geosesarma* spp.: Sesarminae

   Carapace very smooth and lacking hairs; merus of third maxillipeds lacking an oblique ridge; walking legs flattened and fringed with long, soft hairs; with small eggs and planktonic larvae ....
   .... *Varuna* spp.: Varuninae

5. Mandibular palp has two lobes (Fig. 4.22M); frontal margin of the carapace (Figs 4.22D-F) often dilated to form a median triangle (although this may be indistinct) the margins of which may be thickened to form a rim; the abdomen of the male (Figs. 4.21B and 4.22H) is 'T'-shaped (rather than triangular); the terminal and subterminal segments of the first gonopore (first pleopod) of the male are usually not clearly demarcated and may be fused ....
   .... **Gecarcinucoidea** 6

   Mandibular palp with a single lobe (Fig. 4.22L); frontal margin of the carapace is entire without any trace of a median triangle (Fig. 4.21C); the abdomen of the male is triangular (Figs. 4.21C & D and 4.22G); terminal and subterminal segments of the first gonopore (first pleopod) of the male are usually clearly demarcated..........
   .... **Potamoidea: Potamidae**

6. Anterolateral margin of the carapace with one tooth; frontal median triangle rather small and indistinct (Fig. 4.22D); exopod of third maxilliped with flagellum vestigial or lacking (first gonopod of male has four segments, the ultimate and penultimate are sometimes fused) .... **Gecarcinucidae**

   Anterolateral margin of the carapace with one or more teeth; frontal median triangle relatively large and prominent (Figs 4.22E & F), and *may* be well demarcated with raised margins; exopod of third maxilliped with long flagellum 7

7. Frontal medial triangle present (Fig. 4.22F), but without raised

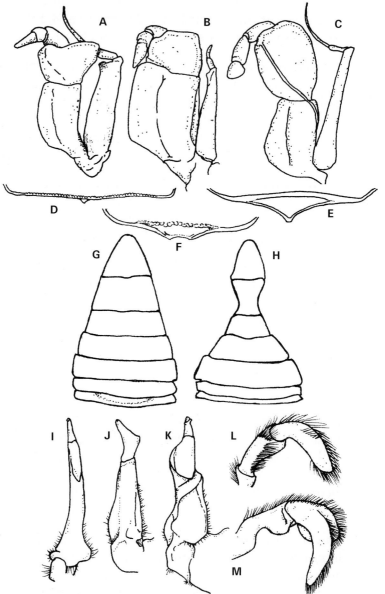

Fig. 4.22 Diagnostic features of Asian freshwater crab taxa (redrawn from Hwang & Mizue [1985] and Ng [1988a]): A, B & C, the third maxillipeds of (respectively) *Varuna litterata* (Grapsidae: Varuninae), *Nanhaipotamon* (Isolapotamidae) and *Geosesarma* (Grapsidae: Sesarminae); D, E & F, the frontal median triangle of Gecarcinucidae, Parathelphusidae and Sundathelphusidae (respectively); G & H, abdomen of adult male specimens of Potamoidea and Gecarcinucoidea (respectively); I, J & K, dorsal views of the first male gonopods of (respectively) Potamidae (*Geothelphusa*), Isolapotamidae (*Nanhaipotamon*) and Sinopotamidae (*Candidiopotamon*); L & M, mandibular palps of Potamoidea and Gecarcinucoidea (respectively).

margins (or raised margins incomplete); the terminal segment of the first gonopod of the male is short and not fused with the penultimate one .... **Sundathelphusidae**

Frontal medial triangle distinct and complete with raised margins (Fig. 4.22E) the segments of the male first gonopod are fused, and it is usually armed with terminal spines .... **Parathelphusidae**

The superfamily Potamoidea (sometimes written, incorrectly, as Potamoniodea) is made up of the Isolapotamidae (comprising *Cryptopotamon, Isolapotamon, Malayopotamon, Nanhaipotamon, Sinolapotamon, Tenuilapotamon* and *Trichopotamon*), the Sinopotamidae (*Candidiopotamon, Parapotamon* and *Sinopotamon*) and the Potamidae. The Potamidae consists of *Acanthopotamon, Allopotamon* (from the Tambelan Islands, Indonesia; Ng, 1988b), *Aparapotamon, Demanietta, Dromothelphusa, Geothelphusa, Huananpotamon, Ibanum, Insulamon, Johora, Kanpotamon, Larnaudia, Lobothelphusa, Mindoron, Nemoron, Ovitamon, Potamiscus* (which includes *Ranguna*: Naiyanetr, 1992) *Potamon, Pudaengon, Stoliczia, Terrapotamon, Thaiphusa, Thaipotamon, Tiwaripotamon* and the hypogean genera *Cerberussa* (which is confined to Kalimantan) and *Phaibulamon. Insulamon, Mindoron* and *Ovitamon* appear to be restricted to the Philippines (Ng & Takeda, 1992c; 1993b). Separation of the families is on the basis of the first gonopod of the male (e.g. Bott, 1970; Hwang & Mizue, 1985): in the Potamidae the ultimate (terminal) segment of the male gonopod is tapered and acute; it is shorter than the penultimate segment (Fig. 4.22I). The gonopod of isolapotamids has a distinct terminal portion (which *may* be relatively long) and a basal section; the distal end of the gonopod is truncate and oblique (not conical), and it is often expanded basally (Fig. 4.22J). Sinopotamids have a gonopod with a short and blunt terminal portion which is twisted inwards or bent medially (Fig. 4.22K). As mentioned above, a more recent opinion about potamoidean classification is that all species in the Isolapotamidae and Sinopotamidae should be treated as members of the Potamidae.

The superfamily Gecarcinucoidea (= Parathelphusoidea) comprises three families including the Gecarcinucidae which is represented in tropical Asia by *Adeleana, Arachnothelphusa, Barytelphusa, Cylindrothelphusa, Gecarcinucus, Gubernatoriana, Inglethelphusa, Lepidothelphusa, Liothelphusa, Phricotelphusa, Sartoriana, Stygothelphusa* (which is cavernicolous: Ng, 1989b), *Thaksinthelphusa,*

Thelphusula (which includes cavernicolous species also [Ng & Goh, 1987; Ng, 1989c] some of which have recently been transferred to *Arachnothelphusa* [Ng, 1991b]) and *Travancoriana*. The Parathelphusidae consists of *Balssiathelphusa, Ceylonthelphusa* (in Sri Lanka only), *Chulathelphusa, Esamthelphusa, Geithusa, Heterothelphusa, Mekhongthelphusa, Irmengardia, Nautilothelphusa* (endemic to Sulawesi), *Oziotelphusa, Parathelphusa* (which includes *Palawanthelphusa* [Ng & Goh, 1987], and is often misspelt '*Paratelphusa*' [e.g. Fernando, 1960]), *Salangathelphusa, Sayamia, Siamthelphusa, Somanniathelphusa* and *Spiralothelphusa*. The Sundathelphusidae is well-represented on the eastern Indonesian islands and in New Guinea. It includes the speciose genus *Sundathelphusa* as well as *Archipelothelphusa, Bakousa, Currothelphusa* (which is cavernicolous: Ng, 1990b) *Mainitia, Perbrinckia, Perithelphusa* and *Terrathelphusa*; *Gleevinkia, Holthuisiana* (including a cavernicolous species) and *Rouxana* are confined to New Guinea (or, in the case of *Holthuisiana*, New Guinea and Australia: Holthius, 1974; 1982), while *Sendleria* is known only from New Britain (the Bismark Archipelago) and the Solomons. Ng (1988a) is doubtful about the separation of the Parathelphusidae and Sundathelphusidae into separate families because the diagnostic features of each merge gradually into one another.

Other Malacostraca

Aside from the Decapoda, the orders Isopoda and Amphipoda are the only malacostracans of ecological significance in epigean freshwaters. However, neither order is well-represented in tropical Asian streams. As a group, the amphipods are cold stenotherms, photonegative and strongly thigmotactic. Most species are marine, and many of the freshwater representatives are hypogean and restricted to subterranean waters and springs. The family Gammaridae is widespread in the Holarctic region, but these amphipods are not an important component of the zoobenthos of tropical streams although some gammarids penetrate as far south as northern India and Yunnan Province, China (Karaman, 1984). For example, Sehgal (1991) records '*Gammarus pulex*' as associated with mosses and macrophytes in two of 11 tributaries of the Indus and the Jhelum in northwest Himalaya (north of 32°N). These animals are apparently confined to rather high altitudes: so-called 'trout streams' above 1470 m (Sehgal, 1983; Melkania, 1991). Karaman (1984) considers that most reports of '*Gammarus pulex*' in

Asia are misidentifications and almost certainly refer to other species of that genus. *Gammarus* (= *Rivulogammarus*) is the only genus in the family in Asian freshwaters; Karaman (1984) gives a key to the northeast Asian species which are essentially extra-tropical. Other amphipods with a similar north-Asian distribution include the Crangonyctidae (*Crangonyx* occurs in the far north of China), and some freshwater representatives of Anisogammaridae; neither family is likely to occur in tropical streams.

Among the rest of the amphipods, the Neoniphargidae includes a southern-Indian genus *Indoniphargus* which is hypogean, but no other genera (hypogean or epigean) are known from tropical Asia. Likewise, the Bogidiellidae is represented by *Bogidiella* from subterranean waters in Thailand, and two blind, unpigmented, hypogean species of Melitidae (*Psammogammarus fluviatilis* and *P.* [= *Niphargus*] *philippensis*) have been reported from the Philippines (Stock, 1991). The Pontogeneidae (= Eusiriidae) includes one marine genus, *Paramoera*, that contains species which may enter rivers in northeast Asia; however, they are essentially temperate in distribution. The primarily-marine and circumtropical Hyalidae (*Hyale*, *Parhyale* and probably *Parhyalella*) penetrates freshwaters also, and members may be encountered in the tidal reaches of rivers. Readers are referred to Barnard & Barnard (1983) for a review of the freshwater amphipods of the world.

Isopods are notably absent from tropical Asian streams (Fernando, 1974). The Asellidae — which is the most speciose family of freshwater isopods — is confined to Palaearctic Asia and do not occur in the tropics. The protojanirid genera *Enckella* and *Aneckella* have been reported from Sri Lanka (Enckell, 1970; Sket, 1982), but all known species in the family are eyeless and inhabit subterranean freshwaters (Sket, 1982). Likewise there are a few hypogean species of Stenasellidae (*Stenasellus*) in Southeast Asia (Thailand and Borneo), and a subterranean species of Nichollsiidae (*Nichollsia menoni*) is known from India (Gupta, 1981; 1985; Knott, 1986).

The isopod family Anthuridae has marine and brackish-water representatives in Asia (e.g. Müller, 1992), but *Cyathura* occurs in subterranean waters in New Guinea and Borneo (Andreev, 1982a; 1982b). At least one genus of brackish-water Sphaeromatidae — *Cymodetta* — ascends upstream beyond the tidal reaches of rivers in the region but cannot breed in freshwater. Overall, therefore, it is clear that isopods and amphipods are of very minor ecological significance in tropical Asian streams where the niches that they occupy in temperate running waters (see Pennak, 1989: pp. 462–488; Covich & Thorp, 1991) have been filled by decapods.

CHELICERATA

One class of chelicerates — the Arachnida (or Arachnoidea) — is found in stream habitats, where it is represented by two orders: the Araneae (spiders) and the Hydrachnida (water mites). The latter is an extremely speciose but poorly-known group and is treated below. By contrast, only one spider genus — *Argyroneta* (containing a single Asian species, *Argyroneta aquatica*: Agelenidae) — can be regarded as truly aquatic, and it is mainly confined to standing water. *Argyroneta* spins an underwater web from which it fashions a 'diving bell' filled with air brought (as a series of bubbles) from the surface; the spider spends much of its time within the air-filled bell (Balasuriya, 1979; Shunmugavelu & Palanichamy, 1992). A few members of one web-spinning family of terrestrial spiders — the Tetragnathidae (e.g. *Tetragnatha mandibulata* and *Eucta javana*) — are found along streams. Their webs are spun horizontally over the water surface between boulders and partly-submerged woody debris; consequently, they are able to trap aquatic insects emerging from the stream or flying in the vicinity (e.g. Hutchinson, 1992).

Several spider genera associated with stream habitats dwell along the water margins where they lie in wait for prey (e.g. Clark, 1986; Hutchinson, 1992). While not benthic species, these predators are of considerable interest. The most conspicuous of them are the raft or fishing spiders, *Thalassius* and *Dolomedes* (Fig. 4.23A); both are members of the Pisauridae, although some workers (e.g. Song & Chen, 1991) place *Dolomedes* in a separate family: the Dolomedidae). Pisaurids detect their prey by sight and through reception of vibrations on the water surface which, in effect, acts like a giant web (Bleckmann & Barth, 1984; Bleckmann & Lotz, 1987). In addition, there are reports that *Dolomedes* dabbles the front legs in the water so that they act as lures to attract small fish; the spider plunges into the water to grab the prey and drag it ashore to feed (Le Roy, 1978; Noonan, 1980; Bleckmann & Lotz, 1987; see also Williams, 1979; Sierwald, 1988). Tadpoles, aquatic insects and especially neustic forms such as water skaters (Heteroptera: Gerridae) and other spiders are eaten also (Zimmermann & Spence, 1989; Hutchinson, 1992). Certain wolf

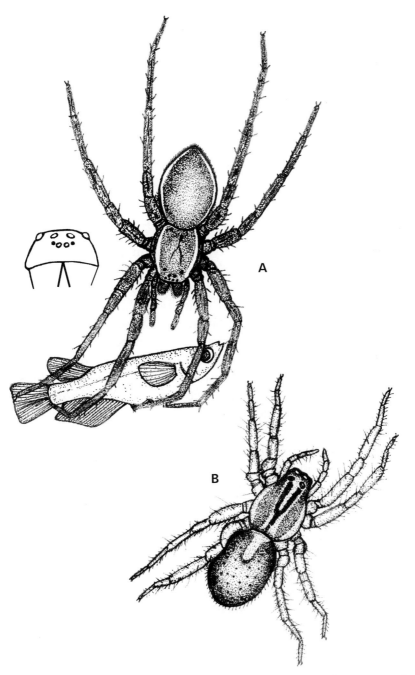

A

B

Fig. 4.23 Chinese semiaquatic spiders (Araneae): A, *Dolomedes pallitarsis* (Pisauridae) capturing a ricefish (*Oryzias curvinotus*); B, *Pirata subpiraticus* (Lycosidae).

spiders (Lycosidae) — mainly *Pirata* (Fig. 4.23B) but also some *Lycosa* — likewise occur at the edges of freshwater habitats. They feed mainly on aquatic insects (e.g. Mohan & Bisht, 1991b), although tadpoles and crustaceans can be included in the diet (Gettmann, 1978). Pisaurids and lycosids are supported by the surface film (Shultz, 1987) during short, rapid forays across the water to grab emerging aquatic insects; their ability to dive beneath the surface is quite limited. Other aspects of the behaviour of north-temperate species of semiaquatic spiders is described by Roland & Rovner (1983), Bleckmann & Bender (1988), Zimmermann & Spence (1992) and Arqvist (1993).

Hydrachnida

The water mites are an extremely diverse group. Approximately 5,000 species in 300 genera and over 100 families and subfamilies have been described worldwide (Viets, 1987), but the total global fauna is certainly much greater than this. The vast majority of them are in the suborder Hydrachnida (= Hydracharina, Hydrachnellae or Hydrachnidia) of the order Acarina (mites and ticks). Water mites display a passing resemblance to minute spiders. All lack antennae, possess piercing, retractile chelicerae (= 'jaws') instead of mandibles, and adults have four pairs of jointed legs (larvae have three pairs) plus a pair of short, five-segmented palps anteriorly. The segmentation of the body (the idiosoma) is indistinct (Figs 4.24A & B and 4.25A & B). Movement is by swimming — in which case the legs may have rows of long setae — or creeping and crawling along the stream bed; the legs of these forms bear spines and short setae. All Hydrachnida are predators and ectoparasites of other invertebrates, particularly aquatic insects and most especially dipterous Chironomidae (Smith & Oliver, 1986), although members of one family (the Unionicolidae: *Unionicola*) are parasitic in unionid bivalves. Water mites occur in a wide range of aquatic habitats, but individual species can be highly stenotopic. Stream-dwelling taxa tend to be flattened and well sclerotized, but species which occupy hyporheic (interstitial) habitats within the stream bed are more varied in body form, and some may have reduced eyes.

All water mites are gonochoristic and oviparous, but their growth and development is complex. In general, the life cycle is as follows (Fig. 4.26), although there are deviations from this pattern (e.g. in *Unionicola* spp.). Eggs hatch into a hexapodous larvae which seeks out and becomes ectoparasitic upon an aquatic insect. It remains

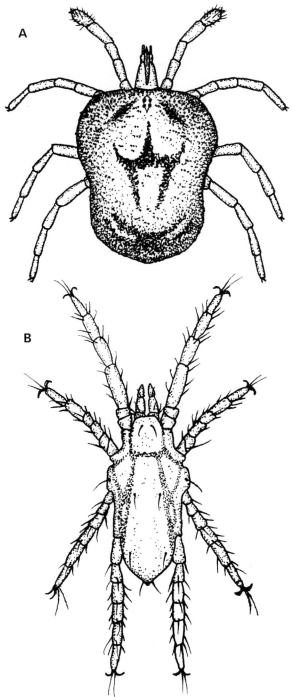

Fig. 4.24 Adult water mites (redrawn from McCafferty, 1981): A, Eylaoidea; B, Halacaridae (body lengths approximately 0.5 mm).

A

B

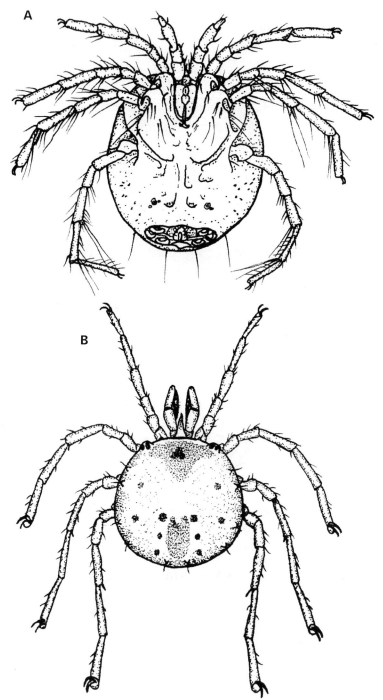

Fig. 4.25 Hygrobatoid water mites (redrawn from Uchida & Imamura [1951] and McCafferty [1981]); A, ventral view of *Axonopsis* (body length 350 μm); B, dorsal view of an adult hygrobatoid.

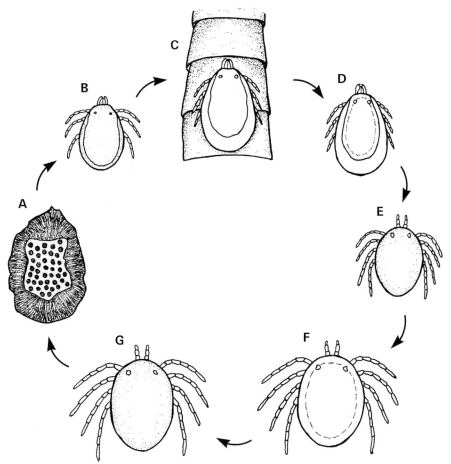

Fig. 4.26 Diagrammatic representation of a generalised water mite life cycle (redrawn from Smith & Cook, 1991): A, egg mass; B, larva; C, parasitic larva; D, nymphochrysalis; E, deutonymph; F, imagochrysalis; G, adult.

attached to the host for some time when it feeds and grows. The duration of attachment varies from around two weeks to considerably longer, and the larva may transfer to the terrestrial adult insect upon its emergence from the stream. Engorged larvae then enter a quiescent stage — the nymphochrysalis — when radical tissue reorganization occurs, giving rise to the deutonymph. Like the adult, this stage has eight legs and is a free-living predator. The deutonymph feeds and grows before entering another quiescent stage — the imagochrysalis — when it metamorphoses into a sexually-mature adult. Smith & Oliver (1986), Smith (1988) and Smith & Cook (1991) give further details of the life histories and host preferences of water mites, and

Smith & Cook (1991) provide a comprehensive review of the biology and systematics of the North American species (see also Pennak, 1989: pp. 514–540).

Although mites are often collected in samples taken by stream biologists, they are almost always under-represented in terms of both numbers and diversity (Smith & Cook, 1991). This is because collecting devices are usually designed to capture animals — such as insects, molluscs, and crustaceans — which are larger than water mites, less cryptic in their habits, more sedentary, and not so inclined to cling to the substrate. The situation is exacerbated because, for most purposes, ethanol is an unsuitable fixative for water mites. Instead they should be killed and preserved in modified Koenike's solution comprising glycerine (5 parts), acetic acid (1 part) and water (four parts). This allows them to be cleared (in 10% potassium hydroxide), dissected and slide-mounted for identification and study. Identification depends on microscopical examination of the features of adult specimens, particularly the mouthparts (which may require dissection), the dorsal and ventral plates (if present) on the body, and the setae on the legs.

The extra effort required to process samples of water mites has contributed to their neglect by stream biologists, who tend to view them as being poorly-known and difficult taxonomically. However, their diversity — and the fact that all species are carnivorous — suggests that water mites may play an important role in structuring zoobenthic communities (Smith & Cook, 1991). The fact that a high proportion of individuals in a wide array of aquatic insect taxa are parasitised by larval water mites (Smith & Oliver, 1986; Smith, 1988) implies that reductions in the growth rates, fecundity and vitality of host populations may be substantial. A recent behavioural study of the effects of adult water mites on their blackfly (Diptera: Simuliidae) prey indicates that their impacts on populations are considerable (Mwango *et al.*, 1995). Because they are highly diverse, water mites have potential also as bioindicators of water quality (Imamura & Kikuchi, 1985; Kumar & Dobriyal, 1992).

The water mites comprise eight superfamilies world-wide: the Stygothromdidoidea, Hydrovolzioidea, Hydrachnoidea, Eylaoidea (Fig. 4.24A), Hydraphantoidea, Lebertioidea, Hygrobatoidea (Figs 4.25A & B) and Arrenuroidea. There are in addition, a few soil mites (Oribatida) and some members of the predominately-marine family Halacaridae (Fig. 4.24B) that have independently invaded freshwater and become adapted for living there (Smith & Cook, 1991). As will be evident from the comments made above, the water mites of tropical Asian streams are incompletely known, and the fragmentary data do not provide a reliable indication of the composition of the lotic fauna. They are certainly insufficient to permit construction of a useful key. Information from Japan is somewhat more comprehensive, and may give some intimation of the taxa likely to be encountered further south. Although Japan's stream fauna is Palaearctic rather than tropical, there is a degree of similarity in the generic composition of aquatic insects (see, for example, Kawai, 1985a). I have therefore drawn upon Japanese studies of water mites (Imamura, 1986; Imamura & Kikuchi, 1985 and references therein) — as well as some data from northern China (Uchida & Imamura, 1951), Sri Lanka (Fernando, 1974), India (Cook, 1967), Papua New Guinea (Imamura, 1983), Indonesia and the Philippines (Walter, 1930; Viets, 1987 and references therein), and Peninsula Malaysia (Wiles, 1991: Table 4.3) — to compile a list of the genera (and subgenera) which might be encountered in tropical Asian streams. It is by no means definitive or exhaustive, and is intended as an aid to further study of these rather obscure animals. Genera of water mites which appear to be confined to lakes and ponds (e.g. *Atax, Eylais, Hydrachna* and *Oxus*) or brackish waters have been omitted, but a few genera (e.g. *Arrenurus, Limnesia* and *Piona*) contain species which are lotic as well as some found in lentic habitats. Many other genera (e.g. *Albia, Diplodontus, Ecpolopsis, Encentridophorus, Nilotania* and *Vietsatax*) are recorded from tropical Asia, but whether they live in streams is uncertain. *Unionicola* is confined to habitats where suitable molluscan hosts are found.

Strongly rheophilic water mites from Asian streams include species in the genera (and subgenera) *Arrenurus (Megaluracarus, Micruracarus* and *Rhinophoracarus), Atractides (Atractides), Aturus, Axonopsis (Axonopsis:* Fig. 4.25A), *Bandakia, Barbaxonopsalbia, Bharatalbia (Japonalbia), Calonyx, Feltria (Feltriella), Hydrodroma, Hydrovolzia (Hydrovolzia), Hygrobates (Hygrobates* and *Rivobates), Japonothyas, Javalbia, Javathyas, Koenikea, Kongsbergia (Kongsbergia), Lebertia (Ptilobertia* and *Septlebertia), Limnesia, Lundbladia, Mamersella, Megapus, Mundamella, Neumania, Protzia (Protzia), Spercheon (Palpispercheron), Spercheronopsis, Sumatralbia* and *Torrenticola (Allotorrenticola, Heteratractides, Monatractides* and *Torrenticola).* Among the water mite genera that may be encountered living interstitially in stream-bed sediments are species of *Atractides (Atractides), Axonopsis (Hexaxonopis), Bharatalbia (Japonalbia), Bharatohydracarus, Chappuisides (Neochappuisides), Feltria (Feltria), Guineaxonopsis, Kantacarus, Kawamuracarus (Kawamuracarus), Lethaxona (Lethaxona), Morimotacarus, Neumania (Neumania),*

Table 4.3 The 30 rheophilic water-mite (Hydrachnida) species known from Peninsular Malaysia. Data from Wiles (1991). Some species cannot be placed in a subgenus until a male has been collected and described. Arrangement within families and superfamilies follows Smith & Cook (1991).

HYDRYPHANTOIDEA
   Hydrodromidae
      Hydrodroma monticola
LEBERTIOIDEA
   Anisitsiellidae
      Anisitsiellinae
         Badankia wendyae
   Torrenticolidae
      Torrenticolinae
         Torrenticola (Allotorrenticola) abnormipalpis
         Torrenticola (Allotorrenticola) bahtilli
         Torrenticola (Allotorrenticola) malayensis
         Torrenticola (Heteratractides) orientalis
         Torrenticola (Monatractides) circuloides
         Torrenticola (Monatractides) minor
         Torrenticola (Monatractides) neoapratima
         Torrenticola (Torrenticola) dentifera
         Torrenticola (Torrenticola) semisuta
HYGROBATOIDEA
   Limnesiidae
      Limnesiinae
         Limnesia volzi
   Unionicolidae
      Pionatacinae
         Neumania nodosa
         Neumania supina
   Hygrobatidae
      Hygrobatinae
         Atractides (Atractides) conjunctus
         Atractides (Atractides) putihi
         Atractides (Atractides) spatiosus
         Hygrobates (Hygrobates) limi
   Aturidae
      Aturinae
         Aturus viduus
         Mundamella cataphracta
      Axonopsinae
         Axonopsis baumi
         Axonopsis hornseyi
         Bharatalbia (Japonalbia) darbyi
         Barbaxonopsalbia pilosa
         Javalbia sunyi
ARRENUROIDEA
   Arrenuridae
      Arrenurinae
         Arrenurus madaraszi
         Arrenurus (Megaluracarus) bicornicodulus
         Arrenurus (Megaluracarus) rostratus
         Arrenurus (Micruracarus) gibberifrons
         Arrenurus (Rhinophoracarus) luxatus

Nipponacarus (Hexanipponacarus), Nudomideopsis, Stygolimnesia, Stygomomonia (Stygomomonia), Torrenticola (Torrenticola), Tsushimacarus, Uchidastygacarus (Uchidastygacarus) and *Wandesia (Allowandesia)*. Some genera appear in both lists because they contain hyporheic as well as benthic species. The interested reader is referred to Viets (1987) for lists of the species in each of these genera, and their placement within families.

INTRODUCTION TO THE AQUATIC INSECTS

Among the arthropods and, indeed, among all living organisms, the insects are preeminent. Estimates of the number of insect species range from one to three million, higher totals of over 30 million have also been suggested. The success of the group is not in question: insects can be found in almost every kind of habitat, subsisting upon all potential food sources. They have been particularly successful in colonizing freshwater environments — for example, over 10,000 species of aquatic insects have been recorded in North America alone (Merritt & Cummins, 1978). The distinction between aquatic or semiaquatic and terrestrial insects is somewhat arbitrary, but here the former category will be taken to comprise species which spend one or more developmental stage (egg, larva, pupa or adult) in freshwater. Forms dwelling on the water surface (= neuston) will also be included, but they will be considered in less detail than taxa which live on or among the bottom sediments. The majority of evidence indicates that the insects evolved on land from a centipede-like ancestor, and subsequently colonized freshwater (Rolfe, 1985). Paradoxically, the success of many groups through geological time is a result of their exploitation of aquatic habitats. For example, the most ancient groups of flying insects, the mayflies (Ephemeroptera) and dragonflies (Odonata) depend upon freshwater habitats. The majority of aquatic insects have aquatic larvae and terrestrial adults. While many others have both adults and immature stages which live in water, relatively few species (in the beetle families Hydraenidae and Dryopidae) have aquatic adults and terrestrial juveniles (Hutchinson, 1981). The aquatic bugs (Hemiptera) and beetles include almost all of those insects which have aquatic adults and larvae but, curiously, beetle larvae usually pupate on land before returning to the water as adults.

The high incidence of terrestrial adults among aquatic insects reflects the fact that, ultimately, nearly all freshwater bodies are transient.

Consequently, some opportunity for dispersal is necessary for the continued existence of a species. The nearer such dispersal is to the breeding period, the more effective it will be. Since, to a flying insect, the period of flight is also the period of maximum dispersal, aeroterrestrial life would be expected at the time of reproduction. This may also explain the rarity of species with terrestrial larvae and aquatic adults.

The 13 insect orders which contain aquatic or semiaquatic species are set out in Table 4.4. Of these, the Ephemeroptera, Odonata, Plecoptera, Trichoptera, Coleoptera and Diptera are the most diverse. The suborder Heteroptera of the order Hemiptera (but see comment on p. 334) includes numerous surface-dwelling (neustic) forms and it is moot as to whether they should be regarded as aquatic or semiaquatic. Note that there are no aquatic species in the Homoptera — the other hemipteran suborder.

All insects, including aquatic forms, fall into one of two divisions according to their life cycle: they are either exopterygote or endopterygote. Insects in both divisions grow by periodic shedding of the rigid exoskeleton, a process called moulting. The act of shedding

Table 4.4 Insect orders which include species that are aquatic or semiaquatic. Orders marked by * contain semiaquatic (rather than aquatic) species, and most species within those orders are terrestrial. Many Heteroptera associated with streams are neustic, dwelling on the water surface.

Phylum Uniramia
   Class Insecta
      Subclass Apterygota
            Order Collembola (Springtails) *
      Subclass Pterygota
         Infraclass Palaeoptera
            Division Exopterygota
               Order Ephemeroptera (Mayflies)
               Order Odonata (Dragonflies and damselflies)
         Infraclass Neoptera
            Division Exopterygota
               Order Plecoptera (Stoneflies)
               Order Hemiptera: suborder Heteroptera (Bugs)
               Order Orthoptera (Grouse locusts, mole crickets, etc.) *
            Division Endopterygota
               Order Neuroptera (Spongillaflies)
               Order Megaloptera (Alderflies and fishflies)
               Order Trichoptera (Caddisflies)
               Order Lepidoptera (Moths)
               Order Coleoptera (Beetles)
               Order Diptera (Flies)
               Order Hymenoptera (Wasps) *

the old exoskeleton is called ecdysis, and the form of an insect between successive ecdyses is an instar. The process of maturation may or may not occur in parallel with growth. All members of the exopterygote orders have a hemimetabolous life cycle, and the wings develop externally as wing pads (or buds) in the immatures. Both larvae (sometimes termed nymphs) and adults have compound eyes, and the immatures resemble the adults in general form. In these insects with a hemimetabolous life cycle, the larvae undergo growth and maturation simultaneously, and the degree of morphological and ecological difference between juvenile and winged adult varies considerably among the orders. The hemimetabolous life cycle is regarded as primitive, and it is notable that — unlike most other flying insects — adult Ephemeroptera and Odonata (comprising the infraclass Palaeoptera) cannot fold their wings over the back when at rest. The Ephemeroptera are unique among the insects in having two winged instars. The mature larva moults to a terrestrial subimago which, within hours, moults again to the reproductive adult or imago.

Endopterygote insects have a holometabolous life cycle in which larvae grow but do not mature; i.e. successive larval instars resemble each other in all respects except for size. Maturation is delayed until the final-instar larva has moulted to yield the inactive pupal stage. This type of life cycle involves a complete metamorphosis, rather than the gradual change seen in hemimetabolous insects, and differences between larvae and adults in body form and ecology are usually extreme.

Respiration is a major problem associated with living in water. In most insects, respiration is accomplished by way of a network of air-filled tubes or trachea through which gaseous oxygen is distributed to various parts of the body. The trachea bifurcate and narrow at the tips to form tracheoles which penetrate the organs and tissues, so supplying oxygen directly. The tracheal system of terrestrial insects is connected to the surrounding air by a series of openings (spiracles) along the thorax and abdomen. To permit survival under water, this external respiratory system has been variously modified. Minor adjustment is required in certain beetles and bugs which come to the surface at intervals to renew the supply of air in the trachea. An extension of this pattern involving the use of air tubes to obtain atmospheric oxygen, has evolved independently in the Hemiptera and Diptera but tends to restrict insect activity to near the water surface. Bubble-like air stores under the wings or beneath hydrophobe hairs on the bodies of beetles and bugs keeps the spiracles in contact with air, oxygen diffusing into the bubble as it is consumed by the insect for so long as the nitrogen in the bubble has not dissipated. Unfortunately, carbon dioxide produced by insect respiration is highly soluble and quickly passes into the water so that the bubble constituting the air store contracts. The partial pressure of nitrogen in the bubble therefore increases, so nitrogen diffuses out. Ideally, the process can continue until all the nitrogen is lost but, in fact, the bubble is renewed at the surface once it dwindles below a critical size (Thorpe, 1950). Nevertheless, provided that dissolved oxygen is freely available, these physical gills or (where hydrofuge — 'unwettable' — hairs are involved) plastrons permit the insect to remain below the surface for hours or even days. Outgrowths of the thoracic spiracles, the spiracular gills of certain Diptera, function in the same way and oxygen is obtained from the water via a spiracle-associated plastron.

Aeropneustic insects, such as those described above, have at least one pair of functional spiracles with which they breathe atmospheric oxygen (although this may be from a bubble). Hydropneustic insects, including most aquatic insect larvae, use dissolved oxygen obtained by way of cutaneous respiration through the body surface. They do not rely on functional (i.e. open) spiracles and are independent of aerial respiration. Most species carrying out cutaneous respiration require well-aerated water, although certain Diptera can survive periods of oxygen depletion because their haemolymph ('blood') contains haemoglobin which increases the animal's affinity for oxygen. Many hydropneustic insects have thin-walled, often lamellate projections from the body containing tracheae. Others possess filamentous bunches or tufts which appear to be outgrowths of the tracheal system. They may be combined with tracheal tufts. The outgrowths are variously termed gills (e.g. in Ephemeroptera, Plecoptera, Trichoptera and Lepidoptera), filaments (in Megaloptera) or lamellae (in damselflies). The ability to ventilate the gills varies in different groups: some mayflies beat the gill lamellae rhythmically thereby causing water to flow over the body surface; other mayflies have immovable gill lamellae. Many caddisflies undulate the abdomen drawing water over the gills, while Plecoptera jerk the body up and down (as if doing 'push ups') when suffering respiratory stress. Note that the inability of many hydropneustic insects to ventilate the gills effectively confines them to running water where the unidirectional current ensures a continuous supply of oxygen. Not all thin-walled projections from the bodies of aquatic insects are respiratory structures, even though they may be referred to as gills (Hynes, 1984); some are concerned with salt uptake and ionic regulation. The interested reader should consult Hynes (1970, chapter 9), Richards & Davies (1977a, chapter 13) or the recent review by Eriksen & Lamberti (1996) for a detailed account of respiratory mechanisms.

Identification, and thus the study, of aquatic insects is hampered by the fact that the adult and larval stages may have different habits and habitats. It is often difficult to match a particular type of larva with the adult of a known species. Such difficulties arise, in part, from a tendency among entomologists to describe new species of adult or (less frequently) larval insects without reference to other life stages. Species separation among adult insects is relatively straightforward because variation in the form of the genitalia between (for example) two male insects will ensure that they are unable to mate with the same type of female. Since biologists believe that males and females of the same species can potentially interbreed, a structural difference in the genitalia can usually be interpreted as a species-rank difference. The significance of morphological variations among larvae is, however, much less clear. This problem is exacerbated for larvae of holometabolous species which do not resemble the adult and may be grub-like, lacking characters that would facilitate species identification. Often, however, examination of the morphological features of many aquatic insects does permit identification to family or genus level. In this book I have emphasized features of the larvae because it is mostly the larval stages which inhabit the aquatic milieu.

How good is our taxonomic knowledge of aquatic insects? The study of aquatic insects began in western Europe over a century ago, and much of the fauna of that region is fairly well known so that larval stages can be identified to genus or even species. By contrast, tropical Asia has received only rather cursory attention (Hynes, 1984), largely from visiting specialists in certain groups who have done confined their attention to adult insects. Most publications describe new species and extend known distributions; in the majority of instances, immature stages receive little or no attention. For this reason, identification of aquatic insects to family or subfamily level is frequently the best that can be achieved by the stream biologist or ecologist. This is not always an obstacle to understanding because, in many cases, family boundaries define the limits of ecological niches. However, there are numerous exceptions. The mayfly family Baetidae, for example (see p. 216 *et seq.*), includes a majority of species that feed on algae, but a few are predators on other insects, and one genus lives as a commensal within the mantle cavity of bivalve molluscs. Clearly we must not become complacent about taxonomy, because ecological work is severely limited without an ability to identify species correctly. Such identification depends upon (among other things) a basic understanding of morphology.

The insect body is segmented, usually comprising three functional divisions or tagmata (Fig. 4.27): the head (six or seven fused segments), the thorax (three segments) and the abdomen (eight to 11 segments). Where the thorax and abdomen cannot be distinguished (as in some Diptera larvae), they are jointly termed the trunk. The segments may be though of as separate boxes, each with a top (the tergum), a bottom (the sternum) and sides (the pleura).

The head bears important sensory structures (Fig. 4.27): a pair of antennae, and a pair of multifaceted compound eyes and/or some ocelli (simple eyes). Several ocelli may be grouped together on each side of the head forming stemmata (common in endopterygote larvae), while insects with compound eyes may posses two or three ocelli anteriorly (mayfly and stonefly larvae) or dorsally (certain adult caddisflies). The mouthparts are variable, reflecting feeding modes, but the basic structure — well illustrated by mayfly larvae — is derived from several components (Figs 4.28 & 4.29). The plate-like labrum (which is analogous to the upper lip) is attached to the front margin of the head, while the paired mandibles and maxillae and attached on either side of the mouth. The mandibles are variously adapted for piercing, chewing, tearing, scraping or crushing food, and have a toothed incisor area and a grinding molar area (sometimes with a spine or brush-like prostheca between). The form of the maxillae reflects their role in manipulating food particles and, may be armed with spines and/or stout setae (Fig. 4.28D) or fringed with long setae and complex arrangements of pectinate bristles (Fig. 4.29D). The bottom of the mouth is formed by a single structure, the labium, resulting from the midline fusion of the second maxillae. Both the maxillae and the labium bear sensory palps; the maxillary palps may also be used to collect small food particles (e.g. in some mayflies). The main body of the maxilla bears teeth, spines and/or setae, especially on the apical part which is termed the galea-lacinia. In addition to these four major structures (i.e. the labrum, mandibles, maxillae and labium), the hypopharynx — roughly equivalent to a tongue — is located just anterior to the labium. Like all mouthpart components, the hypopharynx is subject to considerable modification — especially in larvae of the Hemiptera and Diptera.

The thorax often bears three pairs of jointed legs (if these are

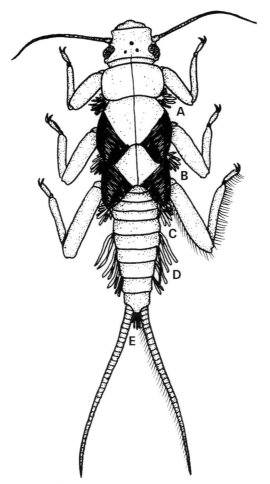

Fig. 4.27 Dorsal view of an imaginary aquatic insect showing the general body features Tracheal gills are situated on the prothorax (A), mesothorax (B), metathorax (C) and abdomen (D), and there is a tuft of gills arising from the anus (E).

present at all there are always three pairs) — one pair each on the pro-, meso- and metathoracic segments (Fig. 4.27). The wing pads of exopterygote larvae are present on the terga of the two posterior (meso- and metathoracic) segments, but the metathoracic pair are secondarily lost in some mayflies (e.g. certain Baetidae and all Caenidae). The jointed legs comprise a basal segment, the coxa, followed by the trochanter, femur, tibia and tarsus. The latter are be three to five segmented and terminates in one or two tarsal claws which may bear small teeth or denticles. The legs may be armed with spines or fringed with hairs which, in some cases, are used for swimming. The forelegs

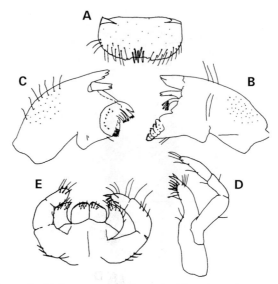

Fig. 4.28 *Caenodes* (Ephemeroptera: Caenidae) larval mouthparts: A, labrum: B & C, right and left mandibles; D, maxilla; E, labium.

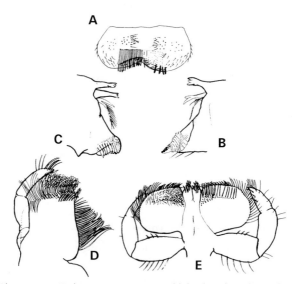

Fig. 4.29 *Choroterpes* (Ephemeroptera: Leptophlebiidae) larval mouthparts: A, labrum: B & C, right and left mandibles; D, maxilla; E, labium.

may be variously modified for burrowing or prey capture. Certain species bear gills at the base of the thoracic legs. Where jointed thoracic legs are not present, some insect larvae have short, fleshy, unsegmented prolegs which may bear rings of hooklets at the tip. Many Diptera have a single pair of prolegs on the prothoracic segment, while others have several pairs along the trunk; still others lack limbs of any sort. Lepidoptera larvae have abdominal prolegs in addition to jointed thoracic legs.

Primitively, the insect abdomen has 11 segments but fusion or subdivision confuse the picture. Abdominal structures include gills, filaments and prolegs. The identity and number of abdominal segments bearing such processes can be an important clue to insect identification. The gills of Ephemeroptera and Megaloptera arise from the pleural regions and are extensions of the tracheal system that are usually in the form of lamellae of various shapes and sizes, sometimes combined with (or replaced by) tracheal tufts. Terminal abdominal structures can include filamentous paired cerci or caudal filaments (often termed 'tails'), gills, hooks, prolegs (= anal prolegs) and breathing tubes or siphons. Genitalia are generally only apparent in the adult insect, but can sometimes be seen in the final instar of exopterygote larvae. Additional information on features of individual orders will be given in the relevant sections below. The identification of aquatic insects to order (or suborder) can be made by the following key. Pupae (which can best be described as resembling a pharonic mummy and are often enclosed within a silk cocoon or a case) are not included.

1. Thorax bears three pairs of segmented legs; head fully formed 2

    Thorax lacks segmented legs (but unsegmented prolegs may be present); head may be fully formed but is often inconspicuous and incompletely formed or retracted into the thorax .... larval **Diptera**

2. Minute insects (generally < 5 mm long), with an obvious midventral tube-like appendage on the abdomen and a small clasp-like structure anterior to it; generally very rare in stream habitats .... **Collembola**

    Body generally considerably longer than 5 mm; ventral abdominal appendages lacking (but a ventral gill series *may* be present)... 3

3. Hindlegs long with expanded femora and modified for jumping ....**Orthoptera**

    Hindlegs not modified for jumping; femora not expanded 4

4. Wings or wing pads present 6

   External wings or wing pads completely absent 12

5. Two pairs of short (approximately as long as the body) membranous wings with few veins .... **Hymenoptera**: family **Agriotypidae**[11]

   Wings or wing pads present but, if wings are present, the first pair are leathery or hardened and overlie the second pair 6

6. Abdomen ends in three long filiform 'tails' (two cerci and a median terminal filament) .... **Ephemeroptera**

   Abdomen ends in one, two or no such 'tails' 7

7. Mouth in the form of an elongate beak, or tube or cone-like structure .... **Hemiptera**: suborder **Heteroptera**

   Mouth not as above 8

8. Labium (lower 'lip') is modified into a large mask-like structure which covers the other mouthparts when viewed from below ....**Odonata** ...... 9[12]

   Labium not modified into a large mask-like structure 10

9. Abdomen ending in caudal lamellae (leaf-like gills)..................... ....suborder **Zygoptera**

   Abdomen without caudal lamellae .... suborder **Anisoptera**

10. Forewings modified into hard protective coverings, concealing most of the dorsal surface .... adult **Coleoptera**

    Forewings not thus modified; abdomen ends in a pair of segmented terminal filaments or 'tails' 11

11. Most abdominal segments possess lateral gills ..... **Ephemeroptera**

    Lateral abdominal gills are absent (thoracic gills in the form of tracheal tufts usually present) .... **Plecoptera**

---

[11] If the wings are completely membranous with numerous veins the specimen concerned is a terrestrial insect or the ovipositing adult of a species with aquatic larvae.

[12] This couplet will not allow separation of the odonate suborder Anisozygoptera, which is rare and has a restricted distribution. A more detailed key, which includes these animals, is given on p. 292.

12. Mouth parts adapted for sucking: may be in the form of a pair of long, needle-like stylets, a tube, a beak or cone-like structure.... .... **13**

   Mouthparts not as above, usually mandibulate (= chewing)... **14**

13. Mouth parts in the form of a tube, a beak or cone-like structure .... **Hemiptera**: suborder **Heteroptera**

   Mouthparts are a pair of long, needle-like stylets; rare, associated with freshwater sponges.................. **Neuroptera**: family **Sisyridae**

14. Abdomen bears four pairs of short, fleshy prolegs, each bearing a ring of tiny hooks, on the underside of segments 3–6 .... **Lepidoptera**: family **Pyralidae**

   Abdomen lacks *ventral* prolegs (but may have lateral projections, or prolegs at the *end* of the abdomen) **15**

15. Abdomen ends in a pair of short or long, fleshy prolegs (sometimes fused together) that end in a single hook .... **Trichoptera**

   Abdomen ends variously, but never in a pair of fleshy prolegs each ending in a single hook **16**

16. Abdomen is fleshy with well-developed lateral filaments **17**

   Abdomen hardened and lacks well developed lateral filaments .... larval **Coleoptera**

17. Abdomen ends in a pair of prolegs, each with a pair of hooks .... **Megaloptera**

   Abdomen ends variously but never in a pair of prolegs each having a pair of hooks (although there may be two pairs of prolegs which each have hooks) .... larval **Coleoptera**

EPHEMEROPTERA

Mayflies have a world-wide distribution and occur in a variety of freshwater habitats; they are most diverse in streams. All species have aquatic larvae and short-lived terrestrial adults (imagoes) and subimagoes. The primitive characteristics of the order are manifest in the inability of the palaeopterous adult to fold the net-veined wings flat when at rest. Instead, the they are held straight above the body. The adults are characterized also by relatively long forelegs (especially in the male), and fragile bodies with an elongate abdomen terminating in two or three filaments. The eyes are usually sexually dimorphic, those of the male being larger and (in the Baetidae and Leptophlebiidae) turbinate (i.e. the upper part is raised on a stalk-like portion). The mouthparts are vestigial, and the short-lived adults do not feed.

Adult mayflies are generally weak fliers but, under appropriate conditions, certain species aggregate in mating swarms. Groups of males congregate in a clearing or close to a specific landmark, where they fly rhythmically up and down to attract mates (Allan & Flecker, 1989). Females attracted to the vertical nuptial dance are seized when they enter the swarm, and mating takes place in flight (Allan & Flecker, 1989). Males collected from mating pairs have significantly greater body lengths than males collected randomly from mating swarms, which implies that large males have an advantage in pursuing females (Flecker *et al.*, 1988). After mating, the eggs are shed underwater which can entail full or partial submergence of the ovipositing female; death follows. Males may mate more than once and, in long-lived species, can join swarms on successive days (Allan & Flecker, 1989).

Swarming behaviour is more common in temperate than in tropical regions. While the details vary among species, many tropical mayflies emerge as subimagoes in the first two hours of darkness, transform to imagoes before dawn, and mate and oviposit by mid-morning or earlier. By so doing, they may minimise losses to diurnal predators. Predation by dragonflies and birds is intensive in the lowland tropics (Bishop, 1973a; Edmunds & Edmunds, 1979) — spiders also seem to be of considerable importance (Soldán, 1983; Mohan & Bisht, 1991b). Seasonal coordinated mass emergence has evolved in a few tropical species, as well as some temperate ones, possibly as a mechanism for satiating predators (Sweeney & Vannote, 1982). All tropical mass-emergent species swarm over rivers (Edmunds & Edmunds, 1979), while species lacking such synchronization carry out remote nuptial flights some distance from the larval habitat. Such behaviour may function as a means of avoiding concentrations of predators near stream or river margins.

Although mayfly adults are rather uniform in structure, larvae exhibit a variety of morphologies: the body of active swimmers is streamlined, while taxa sprawling on or clinging to rocks are dorsoventrally flattened; many burrowing mayflies have cylindrical bodies with mandibular tusks and broad forelimbs for digging.

Respiration in all species is hydropneustic (see p. 199–201) and involve gills consisting of outgrowths of the tracheal system (tracheated lamellae, tufts of tracheae, or a combination thereof) borne on the abdomen. Exchange of dissolved gases with the water column may be passive, or (especially in burrowing or standing water taxa) the gills may beat to produce a current of oxygenated water over the body. Most mayfly larvae are omnivorous primary consumers, ingesting algae and fine detritus, and belong to the scraper or collector-gatherer functional-feeding group (see Table 3.6). For this reason, and because they are generally abundant, mayflies play an important role in freshwater food chains. A few genera (e.g. *Echinobaetis, Raptobaetopus, Anepeorus* and probably *Protobehningia*) are predators upon small invertebrates (Müller-Liebenau, 1978; Mol, 1989), and *Symbiocloeon* (Baetidae) lives as a commensal inside unionid bivalves (Müller-Liebenau & Heard, 1979).

There has been considerable disagreement about mayfly classification above the family level (McCafferty & Edmunds, 1979; McCafferty, 1991a, and references therein), and some dispute over family classification (e.g. Sivaramakrishnan & Venkataraman, 1987; McCafferty, 1991a; Hubbard, 1994). While the details of such arguments need not be rehearsed here, they have had important implications for the composition of families within the superfamily Ephemeroidea (see p. 281). Edmunds *et al.* (1976) have provided a list of valid names for North American mayflies, and Hubbard & Peters (1976) have compiled a list of families of recent mayflies of the world, together with an indication of the numbers of genera and species included in each. Hubbard (1990) has expanded this list into a catalogue of family and genus-group taxa of the world. Unfortunately, the limits of many Asian mayfly genera are poorly defined, and much revisionary work is needed (which will include the establishment of new genera). It is therefore no surprise that reliable keys to the mayfly fauna of the region are lacking. Nevertheless, Gose (1985) provides a key to the larvae of Japanese genera, and Ulmer (1940) has described larvae of some of the Indonesian (Sunda Islands) fauna. Ulmer (1932a) has also produced a key to the Chinese genera of adult mayflies — albeit based on classification scheme now superseded (see also Ulmer, 1920). Together with Hsu (1931; 1935; 1936a; 1936b; 1936c; 1936d; 1937a; 1937b; 1937c), Ulmer (1912; 1919; 1925; 1932a; 1935) made a major contribution to the study of Chinese Ephemeroptera during the first half of this century, and his monograph on Indonesian mayflies (Ulmer, 1940) is essential reading for those interested in the Oriental fauna.

Ulmer (1919) also gives descriptions of some adult Ephemeroptera from Asia (and elsewhere).

The following key can be used to separate mayfly families in tropical Asia and, when combined with the material given under the sections concerning individual families (pp. 219–290), may allow preliminary identification of genera in the region. However, given our incomplete knowledge of the fauna and uncertainty over larval taxonomy, identifications should be regarded as — at best — provisional. Misidentification remains a real possibility while the larvae of many mayfly species remain undescribed. Wherever possible, identifications made with these (or any) keys should be checked by an expert in the relevant group. The key is intended for use with mature larvae (i.e. with well-developed wing pads), and the reader is cautioned that the developing turbinate eyes of the adult male may be visible in baetid and leptophlebiid larvae. Obviously, these males are the same species as female larvae without the large eyes. Note also that the gills are numbered according to the abdominal segment on which they occur: thus gill II is borne on the second abdominal segment, but may be the first gill in the series if segment I lacks gills. If specimens are damaged during collection, some of the gills may be missing. In order to increase the chances of encountering individuals with an intact gill array, identifications should be based on a number of larvae from the same locality.

1. Body smooth and hemispherical (like a beetle); all of the gills (six pairs) and much of the abdomen covered by a thoracic shield .... **Prosopistomatidae**

    Body form not as above; abdominal gills partially or completely exposed 2

2. Mandibles bearing tusk-like projections (which may be rather slender or broad and flat); gills II-VII doubled and uniform in structure with fringed margins, gill I variable .... superfamily **Ephemeroidea** 3

    Mandibular tusks lacking; gill form otherwise 6

3. Legs slender for running or crawling, tibiae cylindrical; gills project laterally 4

    Legs (especially tibiae) robust or flattened for digging; gills angled backwards dorsally over the body 5

4. Mandibular tusks long and sickle-shaped, bearing many long setae;

maxillary palp more than twice as long as the galea-lacinia (the apical part of the maxilla); rare ....

.... **Polymitarcyidae: Euthyplociinae**[13]

Mandibular tusks otherwise, bearing short bristles; maxillary palp as long or slightly longer than the galea-lacinia ... **Potamanthidae**

5. Tusks curved outwards, inner edges convex .... **Ephemeridae**[14]

Tusks curved inwards, inner edges concave .... **Polymitarcyidae**

6. Gills on abdominal segment II large and plate-like (= operculate), touching or overlapping along the dorsal midline and covering all or some of the gills arising posteriorly, gills III-VI with fringed margins **7**

Gills on abdominal segment II not greatly enlarged and, if larger, not much flatter or noticeably different from the rest of the series, never touching along the dorsal midline even if overlapping some of the more posterior gills **8**

7. Gills on abdominal segment II fused along the midline and concealing most of the succeeding (III-VI) gills; terminal filament densely clothed with setae on both margins, cerci (= lateral filaments) with setae on the inner margins only; small hindwing pads are present on the metathorax beneath the forewing pads of mature larvae; rare .... **Neoephemeridae**

Gills on abdominal segment II not fused but *overlapping* along the midline and covering all of the succeeding (III-VI) gills; cerci and terminal filament bearing rather short and sparse setae on the inner and outer margins; hindwing pads lacking; common and widespread .... **Caenidae**

8. Forelegs with conspicuous rows of long setae along the inner margins of femora and tibiae .... **Oligoneuriidae**

Forelegs without rows of long setae along inner margins **9**

---

[13] Until recently (McCafferty, 1991a) the Euthyplociidae was treated as a distinct family, but incorporation within the Polymitarcyidae is a better reflection of phylogenetic relationships.

[14] The mandibular tusks of *Afromera* are atrophied and are therefore inconspicuous (McCafferty & Edmunds, 1973). Following McCafferty (1991a), the Ephemeridae as used herein includes the Palingeniinae (formerly Palingeniidae).

9. Gills on abdominal segments II-VII are elongate (lanceolate) and paired with fringed margins; head with two conspicuous crowns of bristles; legs are modified for burrowing (this *may* involve fusion of the tarsi and tibiae of the forelegs) and lack tarsal claws; rare .... **Behningiidae**

    Gills otherwise (but never elongate, paired and fringed in combination); head lacking crowns of bristles; legs not modified for burrowing **10**

10. Flat plate-like head, with dorsally situated eyes, concealing the mouthparts when viewed from above; labium robust and strongly developed; body dorsoventrally compressed (flattened); gills plate-like and never doubled although they may bear a dorsal tuft of tracheae (= gill tufts) at the base of the lamellae .... **Heptageniidae**

    Mouthparts (labrum and at least the outer margins of the mandibles) clearly visible from above; labium neither robust nor expanded (although the labial palps may be well-developed); head not plate-like; gill form various but if plate-like then never bearing a dorsal tracheal tuft at the base of the lamellae **11**

11. Labium fused into a single semicircular structure, with well-developed palps bearing long setae (easily observed when examining the underside of the head); gills on abdominal segments II-V or II-VII, gill II *may* overlay and partially conceal the rest of the series; a terminal filament is present; rather rare .... **Tricorythidae**

    Mouthparts and gills not as above; terminal filament *sometimes* reduced or lacking; extremely common **12**

12. Gills borne on abdominal segments I-VII (rarely I-VI or II-VII), form diverse but usually projecting laterally beyond the abdomen; forewing pads (of mature larva) free for at least half their length **13**

    Lamellate gills borne on abdominal segments III-VII or IV-VII or, more rarely, II-V, II-VI or I-V (gill on segment I rudimentary if present), usually consisting of a dorsal lamella and a ventral pair of tufts or lamellules, first lamellate gill may conceal some or all of the posterior gills; forewing pads fused to the thoracic notum for more than half their length .... **Ephemerellidae**

13. Gills lamellate or plate-like, rarely doubled, all members of the

series rather similar in general form; terminal filament *sometimes* reduced or lacking **14**

Either the gills are similar in form along the series in which case they are long, slender and bifurcate (forked), *or* the first pair is rudimentary (thread-like) with the remainder plate-like (usually with apical prolongations or fringes) and doubled; terminal filament is well-developed and not significantly shorter than the cerci (= lateral filaments) .... **Leptophlebiidae**

14. Median terminal filament well developed; antennae short (less than twice as long as the width of the head); hind corners of the last few abdominal segments are drawn out into spines; rather uncommon (apex of galea-lacinia of maxilla bears a row of pectinate bristles) ....**Siphlonuridae**

Median terminal filament often much reduced and always shorter than the cerci (lateral filaments); antennae long (over twice as long as the width of the head); hind corners of the last few abdominal segments are not drawn out into spines; very common and abundant (apex of galea-lacinia of maxilla narrowed, and lacking pectinate bristles) .... **Baetidae**

There are great differences between the mayfly families with respect to species richness and abundance in streams. The Baetidae (with well over 500 described species globally), Heptageniidae and Leptophlebiidae (both at least 400 species) are the most numerous and diverse families in tropical Asian streams, while the Ephemerellidae (around 200 species), Caenidae and Ephemeridae (both with approximately 100 species) are also well represented. Oligoneuriids, potamanthids, polymitarcyids and prosopistomatids are widely distributed but less diverse. By contrast, some families or subfamilies *appear* to be lacking from large areas of Asia. For example, the polymitarcyid subfamily Euthyplociinae, as well as the Behningiidae and Tricorythidae, have yet to be recorded within China. The Behningiidae is one of the smallest mayfly families comprising just seven described species worldwide (Hubbard, 1994). Only the genus *Protobehningia* is found in tropical Asia, represented by *Protobehningia merga* in Thailand (Peters & Gillies, 1991). Hubbard (1994) gives a key to larval Behningiidae. Based on the habits of Palaearctic behningiids (e.g. Fink *et al.*, 1991), *P. merga* larvae probably burrow in sand and are likely to be predaceous.

The Tricorythidae are mainly African (including Madagascar) in distribution (McCafferty & Wang, 1995a), but include two Asian genera: *Neurocaenis* (incorrectly referred to and figured as *Tricorythus* by Ulmer, 1939) and *Teloganella*. The family is known from the Peninsular Malaysia, the Philippines, Sabah, Borneo, Java, Sumatra, Sulawesi, Sri Lanka and southern India (Hubbard & Pescador, 1978; Hubbard & Peters, 1978; Wang *et al.*, 1995). Tricorythids bear a superficial resemblance to certain Ephemerellidae, but can be distinguished by couplets 11 and 12 in the key given above. In particular, the labial structure separates these families easily. *Neurocaenis* is distinctive because mandibles are large and extend laterally to give the head a rather broad appearance, and the abdominal gills on segments II-VII overlap but are not semioperculate. Gills II-VI each consist of a lamella (with entire — not fringed — margins) plus paired tracheal tufts, while gill VII is single and threadlike. There has been particular confusion over the familial placement of *Teloganella*: this genus was put in the Tricorythidae (Edmunds & Polhemus, 1990), then moved to the ephemerellid subfamily Teloganodinae (Peters & Peters, 1993), but has recently been reinstated in the Tricorythidae (McCafferty & Wang, 1995a; Wang *et al.*, 1995). *Teloganella* larvae are uncommon but distinctive. The body is rather stout and the legs are short but have unusually broad and swollen femora — most notably, the forefemora. There are gills on abdominal segments II-V, plus a minute, vestigial gill on segment I. Gill II is semioperculate and oblong, covering all but the tips of the succeeding gills. The abdominal segments of *Teloganella* have well-developed spinous lateral projections and dorsal tubercles, and there is a terminal filament. A full diagnosis of the genus is given by Wang *et al.* (1995).

The Neoephemeridae is a rather poorly-known family with only a few described species in three genera; two of them inhabit the Asian region (Edmunds & Polhemus, 1990; Tiunova & Levanidova, 1989). *Neoephemeropsis* occurs in Sumatra, Bali, Borneo, the Philippines, Vietnam, Malaysia and India (Edmunds & Polhemus, 1990), and the larvae have been described recently (Tiunova & Levanidova, 1989; see also illustrations in Ulmer, 1939; Dang, 1967). The larvae of *Potamanthellus* (= *Rhoenanthodes*) have received little attention (but see Tiunova, 1991) and their habits are obscure, but the genus is reported from Korea, China, Burma, Thailand, Vietnam, Malaysia and Borneo (Hsu, 1936a; Hubbard & Peters, 1978; Edmunds & Polhemus, 1990; Bae *et al.*, 1994).

The Prosopistomatidae is the smallest family of mayflies, both in terms of generic diversity and body size (typically less than 8 mm long). Their beetle-like larval morphology is so divergent (Fig. 4.30) that these animals are unmistakeable. The sole genus, *Prosopistoma*, is widespread and known from India, Sri Lanka, Malaysia, Vietnam, Hong Kong, the Philippines, Sulawesi and New Guinea (Peters, 1967; Hubbard & Pescador, 1978; Hubbard & Peters, 1978; Soldán & Braasch, 1984; Edmunds & Polhmeus, 1989; Dudgeon, 1990a; 1990b). Soldán & Braasch (1984) provide a key to the eight species of *Prosopistoma* larvae known (at that time) from the Oriental region,

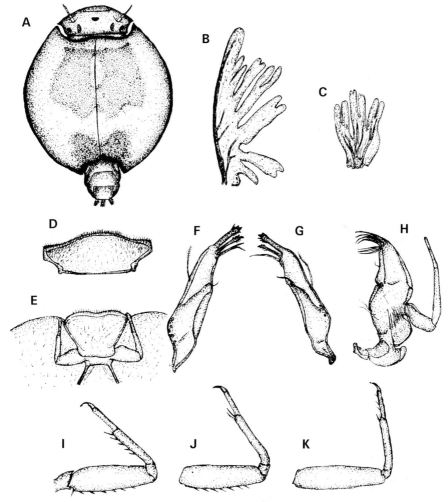

Fig. 4.30 *Prosopistoma* (Ephemeroptera: Prosopistomatidae): A, dorsal view of larva; B & C, gills; D, labrum; E, labium, F & G, mandibles; H, maxilla; I, J & K, fore-, mid- and hindlegs (respectively).

and they (as well as Peters, 1967) give detailed illustrations of the larval morphology. Little is known of the ecology of *Prosopistoma*, but they are generally rather scarce in the streams they inhabit. A substratum of gravel and cobbles, and moderate- to fast-flowing shallow water seems to be favoured. Preliminary data indicate a varied diet (Soldán & Braasch, 1984), but some species are predators in the later larval instars (Yule, 1995b).

The Siphlonuridae, which occur in northern China, are essentially Palaearctic Asian in distribution and thus will not be treated in detail. At the southern-most extent of their range, the family is represented by three species of *Ameletus* known from mountain streams (> 1,000 m altitude) in Taiwan (Kang & Yang, 1994a); two siphlonurid species have been recorded from the Himalaya also (Traver, 1939). Adults of *Parameletus* have been collected from Tibet (= Xizang; You, 1987), but the genus seemingly does not penetrate the Oriental Region. Adults of one other genus and species (*Siphluriscus chinensis*) have been described by Ulmer (1919) from Guangdong Province, southern China (see also Ulmer, 1925) and a second species that may belong to this genus (*S.? davidi*) occurs in Sichuan Province (Ulmer, 1935). The larval stages of these two species are not known. Recent work (e.g. McCafferty, 1991; Studemann *et al.*, 1994; Kluge *et al.*, 1995) suggests that *Ameletus* should be placed in a distinct family — the Ameletidae — and siphlonurid classification is rather unstable (McCafferty & Wang, 1995b). Because these animals are outside the geographic scope of this book, the matter will not be considered further here.

A list of genera of Oriental mayflies is given in Table 4.5. The generic limits of some of taxa included are poorly defined and determination of correct placements must await extensive collection from the region followed by years of systematic study. There is no doubt that many more genera will be established as the fauna of the region is investigated more fully. Nevertheless, the data are included in the hope that they may serve as an aid to further research.

Baetidae

Larval baetids inhabit a wide range of freshwater habitats. Typically, they are the most abundant mayfly family in Holarctic streams (Clifford, 1980) and are numerous in Asian running waters where they dominate drift samples (Bishop, 1973a; Dudgeon, 1990a; 1990c). At least one genus (*Cloeon*) thrives in unpolluted lentic habitats. Baetid systematics

Table 4.5 A checklist of tropical Asian mayfly genera. Mayflies present in New Guinea are treated as part of the Oriental fauna (see Gressitt, 1982). Palaearctic genera which may penetrate the northern parts of the region, or which are present as a few isolated records, are indicated by *, and genera that seem to be entirely confined to Palaearctic Asia are indicated by **. Classification and synonymies largely follow Hubbard (1990), but include some revisions suggested by McCafferty (1991a) and Bae & McCafferty (1991). Placement of *Teloganella* within a subfamily of the Tricorythidae is problematic, and must await reclassification of the family (Wang *et al.*, 1995).

BAETIOIDEA
  Ametropididae
    Ametropus**
  Baetidae
    Acentrella
    Afroptilum?
    Alainites
    Baetiella
    Baetis
    Baetopus**
    Centroptilum*
    Centroptella
    Chopralla
    Cloeodes
    Cloeon
    Echinobaetis
    Gratia
    Indobaetis
    Indocloeon
    Jubabaetis
    Labiobaetis
    Liebebiella
    Nigrobaetis
    Platybaetis
    Procloeon
    Pseudocentroptiloides
      s.g. *Psammonella*
    Pseudocloeon
    Raptobaetopus
    Symbiocloeon
  Siphlonuridae
    Acanthametropodinae
      Acanthametropus**
      Siphluriscus*
    Metretopidinae
      Metretopus**
    Siphlonurinae
      Ameletus*
      Dipteromimus**
      Parameletus*
      Siphlonisca**
      Siphlonurus**
HEXAGENIOIDEA
  Oligoneuriidae

    Chromarcyinae
      Chromarcys
    Isonychiinae
      Isonychia
        s.g. *Isonychia*
    Oligoneuriinae
      Oligoneuriella*
  Heptageniidae
    Anepeorinae
      Anepeorus*
    Heptageniinae
      Afghanurus*
      Afronurus
      Asionurus
      Atopopus
      Belovius*
      Belptus*
      Cinygma*
      Cinygmina
      Cinygmoides**
      Cinygmula*
      Compsoneuria
      Ecdyonurus*
      Electrogena
      Epeiron*
      Epeorella
      Epeorus
      Heptagenia
      Iron
      Ironodes*
      Nixe
      Notacanthurus*
      Ororotsia*
      Paegniodes
      Rhithrogena*
      Rhithrogeniella
      Thalerosphyrus
      Trichogenia
LEPTOPHLEBIODEA
  Leptophlebiidae
    Atalophlebiinae
      Barba
      Chiusanophlebia**
      Choroterpes

Table 4.5 *(cont.)*

s.g. *Choroterpes*
s.g. *EuthraulusChoroterpides*
Cryptopenella
Edmundsula
Indialis
Isca
   s.g. *Isca*
   s.g. *Minyphlebia*
   s.g. *Tanycola*
Magnilobus
Megaglena
Nathanella
Nonnullidens
Notoplebia
Petersula
Simothraulus
Sulawesia
Sulu
Thraulus
  Leptophlebiinae
  Dipterophlebiodes
  Gilliesia
  Habrophlebiodes
  *Leptophlebia**
  *Paraleptophlebia**
BEHNINGIODEA
Behningiidae
  Protobehningia
EPHEMEROIDEA
Polymitarcyidae
  Euthyplociinae
  Mesoplocia
  Polyplocia
  Asthenopodinae
  Povilla
    s.g. *Languidipes*
    s.g. *Povilla*
  Polymitarcyinae
  *Ephoron**
Potamanthidae
  Rhoenanthus
    s.g. *Potamanthidus*
    s.g. *Rhoenanthus*
  Potamanthus
    s.g. *Potamanthodes*
    s.g. *Potamanthus**
    s.g. *Stygifloris*
Ephemeridae
  Ephemerinae
  Afromera
  Ephemera

s.g. *Aethephemera*
s.g. *Ephemera*
  Hexageniinae
  Eatonigenia
  Palingeniinae
  Anagenesia
  *Palingenia***
  Plethogenesia
CAENOIDEA
Ephemerellidae
  Ephemerellinae
  Cincticostella
    s.g. *Cincticostella*
    s.g. *Rhionella*
  Crinitella
  Drunella
    s.g. *Drunella*
    s.g. *Tribrochella*
  Ephacerella
  Ephemerella
  Eurylophella
  Hyrtanella
  Serratella
  Torleya
  Teloganopsis
  Uracanthella
  Vietnamella
  Teloganodinae
  Ephemerellina
    s.g. *Ephemerellina*
  Macafertiella
  Teloganodes
Tricorythidae
  Tricorythinae
  Neurocaenis
  Teloganella
Caenidae
  Brachycerus
  Caenis
  Caenodes
  Caenoculis
  Caenomedia
  Cerobrachys
  Clypeocaenis
  Tasmanocaenis
Prosopistomatidae
  Prosopistoma
NEOEPHEMEROIDEA
Neoephemeridae
  Neoephemeropsis
  Potamanthellus

are in a state of flux (e.g. Waltz & McCafferty, 1987a; 1987b; 1987c; 1987d; 1987e; 1989; McCafferty & Waltz, 1990; Gillies, 1990; 1991; Kluge & Novikova, 1992; Waltz *et al.*, 1994), and recent developments involve reassignment of species among genera and subfamilies (although there is a lack of consensus here), as well as erection of new genera. These changes have arisen, in part, from the contradiction that species-level systematics of Baetidae is now primarily based on larval morphology (e.g. Morihara & McCafferty, 1979; Müller-Liebenau, 1981; Waltz & McCafferty, 1985a; Müller-Liebenau & Hubbard, 1986; Kang *et al.*, 1994) but, to a large degree, generic limits (and inclusion within subfamilies) have remained adult-orientated (Waltz & McCafferty, 1987a). It is clear that a greater number of baetid species and morphological characters must be examined before phylogenetic analysis can be attempted successfully (Lowen & Flannagan, 1992).

Among recent changes is a proposal — based initially on the study of larval characteristics of Nearctic species — to split the cosmopolitan genus *Baetis* into several genera (Waltz & McCafferty, 1987b; Waltz *et al.*, 1994). The recognition that *Baetis* is not a natural entity will have implications for the systematics of Oriental Baetidae, as a number of larval *Baetis* have been described recently from the region (Müller-Liebenau, 1981; 1982a; 1984a; 1984b; 1985; Müller-Liebenau & Hubbard, 1986; Kang *et al.*, 1994). Similar difficulties arise when considering the genus *Pseudocloeon*, which has been viewed as widespread in the Holarctic and Oriental region. Waltz & McCafferty (1987a) suggest a restriction of the use of *Pseudocloeon* to the type species, and have transferred many of the Oriental species previously placed in *Pseudocloeon* (e.g. Müller-Liebenau, 1981; 1982a; 1982b; 1984b; 1985; Braasch, 1983a) to the new genus *Liebebiella*. Other *Pseudocloeon* have now been included within the genera *Acentrella* and *Baetiella*. Extra complications arise because, according to Hubbard (1989), the genus *Pseudocloeon* was established three times by Klapálek (in 1905), Bengtsson (in 1914) and Matsumura (in 1931). Klapálek's use of the name *Pseudocloeon* has precedence, and Hubbard (1989) has therefore proposed the substitute name *Matsumuracloeon* for the Japanese species described by Matsumura. Further confusion was caused by the (initially incorrect) designation of a new type (= lectotype) specimen for *Pseudocloeon* because the original type specimen (of *Pseudocloeon kraepelini*) had been lost; however, that matter has now been resolved (Waltz & McCafferty, 1985a). Nevertheless, the case of *Pseudocloeon* demonstrates that there is some way to go before the systematic affinities and generic placement of Asian Baetidae can be clarified. Certainly, '. . . the proper combination of (a) species with any of several Asian genera, including *Baetis* and *Pseudocloeon*, cannot ultimately be resolved with out the study of its larval stage . . .' (Waltz & McCafferty, 1984).

Complications have arisen also with the Oriental genus *Centroptella*, erected on the basis of baetid larvae collected from southern China (Braasch & Soldán, 1980) and currently including seven species from Sri Lanka (Müller-Liebenau, 1983), Malaysia and Vietnam (Soldán *et al.*, 1987). Recent research suggests that *Centroptella* is not a natural grouping, and Waltz & McCafferty (1987c) have transferred two species (including the type from southern China) to the genus *Cloeodes*. They place the remaining '*Centroptella*' in the new genus *Chopralla* (Waltz & McCafferty, 1987d), although a recent revision of Taiwanese Baetidae (Kang *et al.*, 1994) retains the 'old' baetid classification.

These examples indicate both the scale of difficulties facing those wishing to understand the ecology of Oriental Baetidae, and the many research opportunities that await the student of aquatic entomology. Because of such problems, the following account of baetid genera must be regarded as provisional, requiring modification in the light of on-going and future research.

Baetis larvae have two-segmented maxillary palps and tarsal claws bearing a single row of denticles. The femur, tibia and tarsus lack long bristles, and the abdominal terga are often patterned. The median terminal filament is shorter than the cerci, and bears lateral hairs. Single, lamellate gills with minutely serrate margins are borne on segments I-VII or II-VII. Hindwing pads may be present or absent, and it is worth stressing that some workers believe that this character has been over-emphasized in studies of baetids (Jacob & Glazaczow, 1986). Larvae with 'typical' *Baetis* features are widespread (Fig. 4.31), and Müller-Liebenau & Hubbard (1986) have summarized the known species-groups of larvae of those Oriental species for which adequate descriptions exist. Pending further investigation as to whether Oriental *Baetis* are a natural group (see Waltz & McCafferty, 1987a), or should be divided into subgenera (as proposed by Kang *et al.*, 1994; but see Kang & Yang, 1996a), larvae referred to this genus should be viewed as *sensu lato*.

Identification beyond the level of genus may be necessary in ecological studies of baetids because individual *Baetis* species show different patterns of longitudinal zonation along rivers (Dudgeon, 1984a; 1990b); some species are quite tolerant of organic pollution. Details of mouthpart morphology may also be used as an aid to

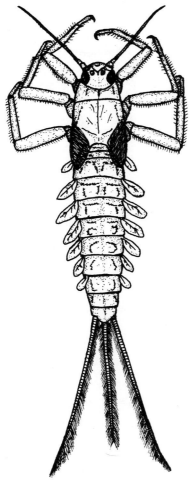

Fig. 4.31 A typical *Baetis* (Ephemeroptera: Baetidae) larvae.

species or generic separation separation, and can provide additional characters to support identification on the basis of more conspicuous features or general *habitus*. However, differences in mouthpart morphology may be subtle as Figs. 4.32 & 4.33 show. Some baetids may be distinguished by the abdominal markings (Fig. 4.34), although the patterns fade during long-term storage of specimens and may be less distinct in small larvae or individuals which have moulted recently.

The larvae of *Baetiella (= Neobaetiella)* are distinctive in that the terminal filament is very short or lacking, and the femora and tibiae bear a single row of long bristles (Fig. 4.35). The gills are simple and

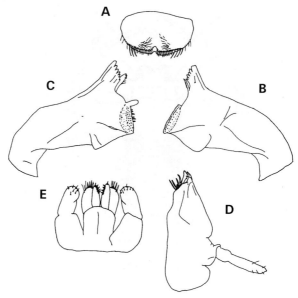

Fig. 4.32 *Baetis* (Ephemeroptera: Baetidae) larval mouthparts: A, labrum: B & C, mandibles; D, maxilla; E, labium.

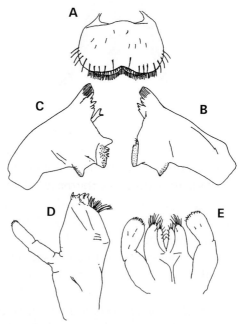

Fig. 4.33 *Liebebiella* (Ephemeroptera: Leptophlebiidae) larval mouthparts: A, labrum: B & C, mandibles; D, maxilla; E, labium.

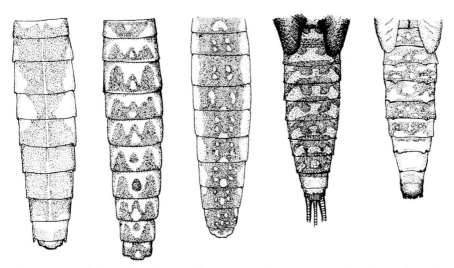

Fig. 4.34 Abdominal markings of five species of *Baetis* (*sensu lato*) larvae from the Lam Tseun River, Hong Kong.

without serrate margins. The abdomen is cylindrical in cross section (not dorsoventrally flattened) and bears single or paired dorsal tubercles (Fig. 4.35N) which are a convenient aid to recognition of the genus. The terminal segment of the labial palp segment is somewhat conical which also distinguishes *Baetiella* from other baetid genera (Waltz & McCafferty, 1987a; see also Braasch 1983a; Braasch & Soldán, 1983). Larvae favour rocky, fast-flowing streams.

Liebebiella larvae have a short terminal filament, and the femora bear a single row of bristles. There are two rows of bristles on the tibiae (contrasting with *Baetiella*). The tarsi of all legs bear single subapical bristles. The abdomen is not flattened and does not bear tubercles; however, a median row of long bristles is present dorsally and some dark-coloured markings may be present. The gills are simple and lack serrate margins, and the third segment of the labial palp is rounded. Most Oriental species that were originally assigned to *Pseudocloeon* belong to this genus (Waltz & McCafferty, 1987a). Larvae inhabit a variety of running waters but are generally associated with the upper surfaces of stones in faster current. They have some tolerance for organic pollution.

Like *Liebebiella*, *Acentrella* larvae have a short terminal filament, and the femora bear a single row of bristles. There is a single row of bristles on the tibia and the tarsus lacks a subapical bristle (contrasting with *Liebebiella*). The abdomen may be dorsoventrally flattened and

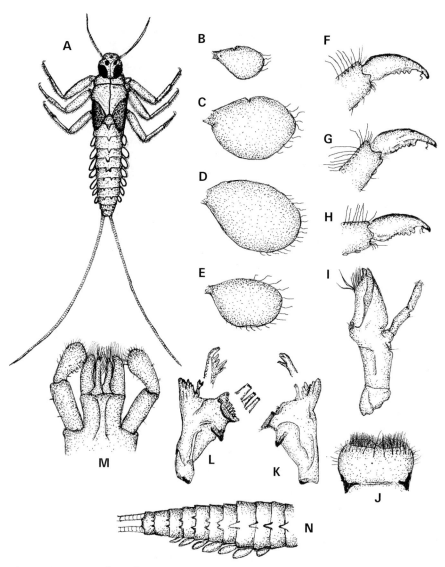

Fig. 4.35 *Baetiella* (Ephemeroptera: Baetidae): A, dorsal view of larva; B, C, D & E, gills I, III, V & VII (respectively); F, G & H, tarsal claws of fore-, mid- and hind legs (respectively); I, maxilla; J, labrum; K & L, mandibles; M, labium; N, dorsal spination of the abdomen.

may have a dorsal median row of bristles. The gills are simple and lack serrate margins. The labium is rather compact (compared to *Baetiella*) and the labial palps are apically rounded (Waltz & McCafferty, 1987a). Published records indicate that *Acentrella* is largely Holarctic in distribution, although at least one Oriental *Pseudocloeon* species and *Baetis feminalis* from Sri Lanka have been transferred to this genus (Waltz & McCafferty, 1987a; Waltz, 1996).

The terminal filament of *Gratia* larvae is very reduced and appears to lack segmentation. The tergites of the metathorax and the abdomen (I-IX) bear a singe median spiniform tubercle which distinguishes these larvae from other Asian Baetidae (Thomas, 1992). The tubercles are covered by minute scales. Only *Jubabaetis* (see below) has single spiniform tubercles on these abdominal segments, but it has a distinctive, almost circular and flattened head with the frontal area enlarged considerably (Müller-Liebenau, 1980a). The hind margins of each femur (but not the tarsus or tibia) in *Gratia* are fringed with pectinate bristles or setae, and the tarsal claws are robust, bearing a number of well-developed teeth on the inner margin and a pair of subapical setae. The detailed structure of the mouthparts is distinctive (illustrated by Thomas, 1992); in particular, the labial palps are large, strongly developed and broad at the apex terminating in a small tubercle. There are seven pairs of robust, rounded gills which are fringed by marginal scales along the leading edge. To date, *Gratia* has been reported only from cascades in a shaded hillstream in Thailand; Thomas (1992) has interpreted the form of the larvae as an adaptation to maintaining position in strong, turbulent flows.

Platybaetis larvae have a large head with a comparatively small, ventrally-directed labrum. The thorax (especially the pronotum) is somewhat rounded laterally and the body is rather flattened. Thus from general appearance (Fig. 4.36), *Platybaetis* can be distinguished easily from other Baetidae. The larval *habitus* is somewhat reminiscent of a heptageniid. The legs are stout and the femora have a dense row of fine ciliated bristles along the outer margins; there is a pair of long setae near the apex of each tarsus. Hindwing pads may be absent or much reduced. Larvae have seven pairs of clearly tracheated, elongate gills, which lack marginal denticles. There is a distinct pattern on the dorsum of the abdomen, and tubercles are not *usually* present. The median terminal filament is greatly reduced while the cerci are approximately the same length as the body. Müller-Liebenau (1980a; 1980b; 1984b) and Braasch (1981a) record *Platybaetis* spp. from the Philippines, Nepal, Sabah and Peninsular Malaysia, and provide detailed descriptions of the larval characteristics of the genus. Larvae occur most commonly on the upper surface of large flat rocks in moderate to swift current where the stream is not heavily shaded by riparian vegetation. They can be numerous in small rivers experiencing mild organic pollution.

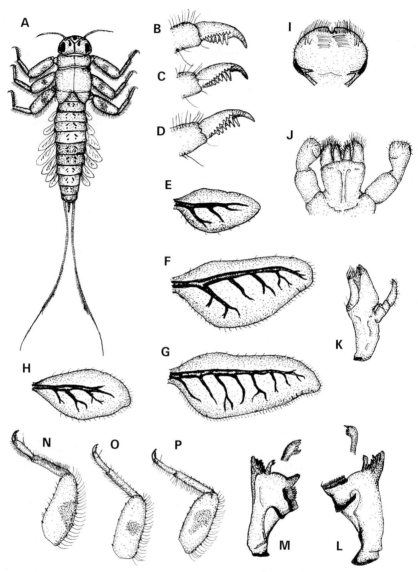

Fig. 4.36 *Platybaetis* (Ephemeroptera: Baetidae): A, dorsal view of larva; B, C & D, tarsal claws of fore-, mid- and hind legs (respectively); E, F, G & H, gills I, III, V & VII (respectively); I, labrum; J, labium; K, maxilla; L & M, mandibles; N, O & P, fore-, mid- and hindlegs (respectively).

Echinobaetis larvae stand out among the Baetidae because they have a flattened head that is almost circular in dorsal view. The anterior margin is convex and bears downward-curving, bristle-like setae (Mol, 1989). The mandibles and maxillae are adapted for a predaceous habit with long, narrow apical teeth. There are no hindwing pads, and the posterior margin of the femur is fringed with a double row of long bristles; the tarsal claws have a single, long apical seta. A notable feature is the presence of three long spines on abdominal terga I-VIII: one spine in the centre of each tergum and two smaller spines near the middle of the posterior margin. The median terminal filament is well-developed. Mol (1989) gives further structural details of this monotypic genus — represented only by *Echinobaetis phagas* from Sulawesi — and reports that the gut contents contained mainly other mayflies and blackfly (Simuliidae) larvae.

The genus *Jubabaetis* was erected by Müller-Liebenau (1980a) on the basis of larvae collected from the Philippines (Luzon). Like *Echinobaetis*, *Jubabaetis* has a large head, with the frontal area expanded and covering the mouthparts entirely; it bears a dense row of setae along the anterior margin. However, the head is quadrangular (rather than circular as in *Echinobaetis*) and the mouthparts are of the 'typical' baetid type (i.e. not adapted for predation). *Jubabaetis* has small hindwing pads, and the femur bears a posterior fringe of long setae with a row of stout setae parallel to it. The tarsal claws lack apical setae, and (as in *Platybaetis*) there are a pair of long ventral setae near the apex of the tarsus. The abdominal terga (I-IX) bear a single long spine at the posterior margin (*cf.* the single median spiniform tubercle of *Gratia*) and the terminal filament is reduced to a single segment.

The terminal filament of *Chopralla* larvae is subequal to (i.e. only slightly shorter than) the cerci unlike, for example, *Baetiella*, *Liebebiella*, *Acentrella*, *Gratia* and *Playtbaetis*, where the median filament is greatly reduced (or lost). Gills I-VII are asymmetric and rounded apically, lacking marginal spination. The parallel-sided femora have long dorsal bristles; the tibiae bear two arcs of long, fine setae. The foreleg has a single subtending bristle, and the setal arcs may be more weekly developed than on the succeeding legs. The claws are distinctive and bear three to five paired denticles. The labrum is emarginate and (unlike most baetids) the left mandible lacks a setal tuft between the incisor and molar areas. The maxillary palps have broadly rounded apices. Species in the genus *Chopralla* include mayflies originally assigned to *Centroptella* from Malaysia and Sri Lanka (Müller Liebenau, 1983; 1984b; Waltz & McCafferty, 1987d) as well as (at least) one in Hong Kong. They occur in shaded and unshaded hillstreams and can be collected in areas of slow current close to the banks and in the lee of obstructions to flow.

Cloeon larvae (Fig. 4.37) are easily recognized because the lamellae of gills I-VI are doubled while gill VII is single. The gill margins are smooth, lacking denticles. There is a setal tuft between the incisors and molars of the left and right mandibles; the maxillary palps are three-segmented. Additional distinguishing features include rather long gently-curving (somewhat sickle-shaped) tarsal claws, and spines on the lateral margins of the abdominal segments (segments VI-IX of the

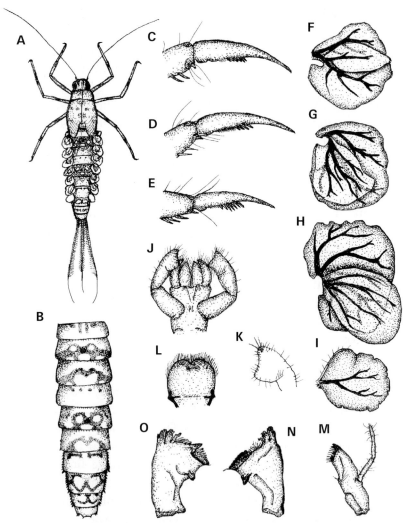

Fig. 4.37 *Cloeon* (Ephemeroptera: Baetidae): A, dorsal view of larva; B, dorsal view of the abdomen; C, D & E, tarsal claws of fore-, mid- and hind legs (respectively); F, G, H & I, gills I, III, V & VII (respectively); J, labium; K, labial palp (detail); L, labrum; M, maxilla; N & O, mandibles.

Hong Kong species: Fig. 4.37B). Kluge & Novikova (1992) give further information on characteristics of the genus including detailed descriptions of the larvae of some Palaearctic species. *Cloeon* larvae inhabit clean, well-oxygenated standing waters or slower flowing areas of rivers. They may also be taken from roots of the blades of terrestrial grasses trailing in the water along stream and river banks. The genus is widespread in tropical Asia and the Palaearctic.

Procloeon larvae (Fig. 4.38) have rather large, broad, smooth-margined gill lamellae which are distinctly asymmetrical and pointed apically; they may be unilamellate (single) or bear a small dorsal flap so that they resemble the bilamellate gills of *Cloeon*. Like *Cloeon*, the abdominal segments have spines on the lateral margins, and the abdominal terga are strongly patterned although the markings differ among species. The larvae resemble certain Holarctic *Centroptilum* (McCafferty & Waltz, 1990) but larvae of that genus either lack lateral abdominal spines or bear spines which are relatively small (Lowen & Flannagan, 1992). The terminal segment of the maxillary palp in *Procloeon* is no more than half the length of the preceding one, but in *Centroptilum* the two segments are subequal (see, for example, Braasch & Soldán, 1983). In addition, *Procloeon* larvae are characterized by setal tufts on the mandibles, partly-fused canines on the left mandible (at least), and long, sickle-shaped tarsal claws. Unlike *Cloeon*, the terminal filaments are robust and approximately equal length. They bear well-developed bristles which aid the larva in swimming. *Procloeon* larvae have been described from Sri Lanka (Müller-Liebenau & Hubbard, 1985) and Taiwan (Waltz & McCafferty, 1985b). Larvae occupy a range of habitats in Hong Kong, from rocky-bottomed pools in hill streams to trailing grasses along the banks of riffles. Kluge & Novikova (1992) consider that *Procloeon* is a subgenus of *Cloeon*, and include some Holarctic *Centroptilum* species within it. However, this view is not shared by all workers.

A recent revision of the genus *Cloeodes* (Waltz & McCafferty, 1987d) has led to the transfer of some *Centroptella* species (see above) into what had been viewed as a New World genus. *Cloeodes* is pantropical (Waltz & McCafferty, 1994), and larvae can be recognized by a combination of characters: ventral tufts of setae (i.e. with contiguous bases) on abdominal segments II-VI (although these setae are not always conspicuous); gills I-VII that are rather elongate (often twice the length of the associated tergum), asymmetric and broadly pointed, lacking marginal spination; an emarginate labrum; a left mandible which lacks a tuft of setae between the incisor and molar

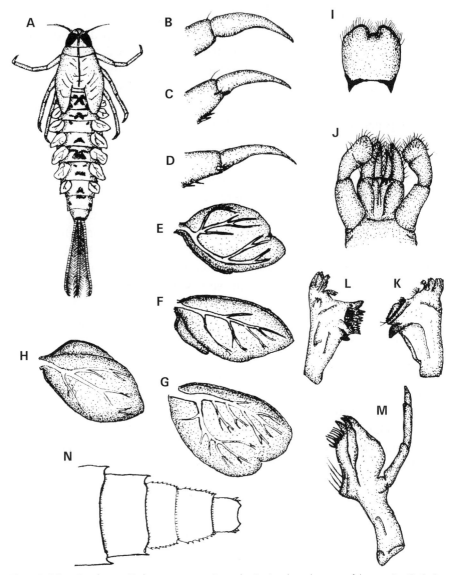

Fig. 4.38 *Procloeon* (Ephemeroptera: Baetidae): A, dorsal view of larva; B, C & D, tarsal claws of fore-, mid- and hind legs (respectively); E, F, G & H, gills I, III, V & VII (respectively); I, labrum; J, labium; K & L, mandibles; M, maxilla; N, dorsal view of the tip of the abdomen.

areas (similar to *Chopralla* but contrasting with *Cloeon*); and, labial palps with rounded or oblique apices. In addition, the femora have short, dorsal bristles (these bristles are long in *Chopralla*); the tibiae bear a subproximal arc of long, fine setae with a subtending bristle on the foreleg only; and, the tarsal claws lack denticles but may possess microspines. The terminal filament is somewhat shorter than the cerci. *Cloeodes* are similar in general appearance to *Chopralla* larvae but can be separated from them by the presence of ventral tufts of setae on the abdominal segments II-VI, the broadly-pointed, asymmetrical gills (versus apically rounded in *Chopralla*), the edentate tarsal claws, and the presence of a single arc of setae on the tibia (*versus* arcs two found in *Chopralla*). *Cloeodes* larvae inhabit slow-flowing, sandy-bottomed streams in Hong Kong, but elsewhere they seem to occupy a range of lotic habitats (Waltz & McCafferty, 1994).

Larvae of *Indobaetis* (Fig. 4.39) are distinctive in that the head is bowed vertically and the head and thorax are laterally compressed. This combination of features gives the larvae a streamlined appearance. The mandibular incisors are fused together and the right mandibular prostheca is slender and pointed (Figs. 4.39D & E). The apical teeth on the galea-lacinia of the maxilla are long and sharp, and the second segment of the maxillary palp has a slight indentation on the inner margin near the apex. The labial palps are rather slender, and the third segment is not broadly rounded as in *Baetis* (Müller-Liebenau & Morihara, 1982). The tarsal claws have a single long seta between the apical denticles (Waltz *et al.*, 1994). Hong Kong *Indobaetis* are uniformly pale brown in colour, and the terminal filament is subequal to the cerci in length. They have seven pairs of gills (with minutely serrated margins) although the first pair are reduced. Sri Lankan *Indobaetis* have lost the first pair of gills (Müller-Liebenau & Morihara, 1982). Note that the presence or loss of the first pair of gills (and hindwing pads) also varies among species of the genus *Baetis* (*sensu lato*; see above). *Indobaetis* larvae lack hind wing pads (Waltz *et al.*, 1994). Larvae inhabit clean hillstreams in swift currents and cascades.

Some Asian species previously assigned to *Baetis* have recently (Waltz *et al.*, 1994) been placed in two other genera related to *Indobaetis*: *Alainites*, known from China, and *Nigrobaetis* from Peninsular Malaysia. Like *Indobaetis*, the body of *Alainites* is laterally compressed but the mouthparts differ in that the right mandibular prostheca is reduced and the mandibular canines are not fused. In addition, the tarsal claws lack setae, and a hind wing pad *may* be present (Waltz *et al.*, 1994). *Nigrobaetis* has the cylindrical body of a 'typical' *Baetis* (i.e. it is not laterally compressed), and the genus is characterized by a combination of features of the mouthparts and tarsal claws as well as the structure of the surface and details of marginal

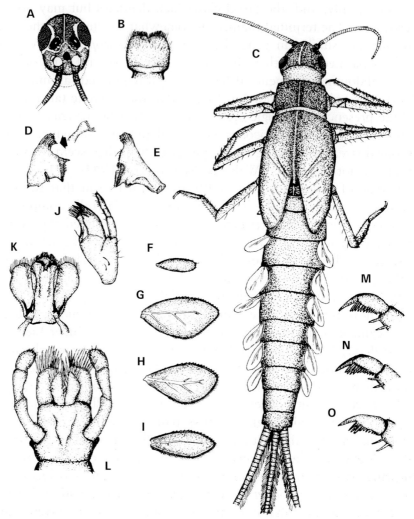

Fig. 4.39 *Indobaetis* (Ephemeroptera: Baetidae): A, anterior view of larval head; B, labrum; C, dorsal view of larva; D & E, mandibles; F, G, H & I, gills I, III, V & VII (respectively); J, maxilla; K, hypopharynx; L, labium; M, N & O, tarsal claws of fore-, mid- and hind legs (respectively).

spination of the abdominal terga; gill I and hindwing pads may be present or absent (Waltz *et al.*, 1994; Novikova & Kluge, 1995; see also illustrations of '*Baetis*' *minutus* in Müller-Liebenau, 1984a).

As exploration of Asian streams continues, new genera of Baetidae have been collected and described. Among these have been *Indocloeon* and *Pseudocentroptiloides* (subgenus *Psammonella*) from Sri Lanka (Müller-Liebenau, 1982c; Jacob & Glazaczow, 1986; Waltz &

McCafferty, 1989). The latter is closely related to *Procloeon*, and treated as a subgenus of *Cloeon* by Kluge & Novikova (1992). *Indocloeon* is most akin to *Procloeon* and *Centroptilum* but is easily distinguished from these and all other baetid genera by the detailed structure of the surface and posterior margin of the abdominal terga (see Müller-Liebenau, 1982). Other tropical Asian baetids include *Symbiocloeon* from Thailand which lives inside the mantle cavity of unionid bivalves (Müller-Liebenau & Heard, 1979), and the predatory *Raptobaetopus* from Malaysia (Müller-Liebenau, 1978) which is considered to be a subgenus of *Baetopus* by Kluge & Novikova (1992). *Baetopus* (*sensu stricto*) does not appear to occur south of Mongolia. The African genus *Afroptilum* — species of which were previously placed in *Centroptilum* — may also occur within the Oriental Region (Gillies, 1990). For example, 'Genus no. 1' of Müller-Liebenau (1984a) from Malaysia might be *Afroptilum* (Gillies, 1990). There are certain to be other Asia baetid genera which await description: one distinctive larvae from Hong Kong streams which does not appear to fit into any genus yet described is shown in Fig. 4.40. Given the changeable state of baetid systematics, it is not yet possible to devise a reliable or comprehensive key to the genera. This instability is reflected in the recent transfer of several Oriental *Baetis* to the new genus (formerly a subgenus) *Labiobaetis*, including species from Peninsular Malaysia, the Philippines, India, Sri Lanka and (possibly) Indonesia (McCafferty & Waltz, 1995).

Oligoneuriidae

The genera *Chromarcys*, *Oligoneuriella* and *Isonychia* are known from Asia, but *Oligoneuriella* is probably restricted to the Middle East (including Pakistan: Hubbard & Peters, 1978) and the Palaearctic; Soldán & Landa (1977) have figured the larvae. *Chromarcys* and *Isonychia* occur in the Oriental region. *Chromarcys* larvae have a broad, flat head (which conceals the labrum when viewed from above), and an expanded labium with well-developed palps bearing long setae. The femora of the mid- and hindlegs are broad, while the forelegs are fringed with long setae along the inner margins of the femur and tibia. The gills are rather small. The terminal filament is reduced and is much shorter than the cerci. These mayflies inhabit larger rivers and are reputed to be rather rare. *Chromarcys* is known from China (Yunnan Province), north Thailand, Sri Lanka and Sumatra (Hsu, 1936a; Edmunds & Polhemus, 1990); Ulmer (1939) gives a detailed

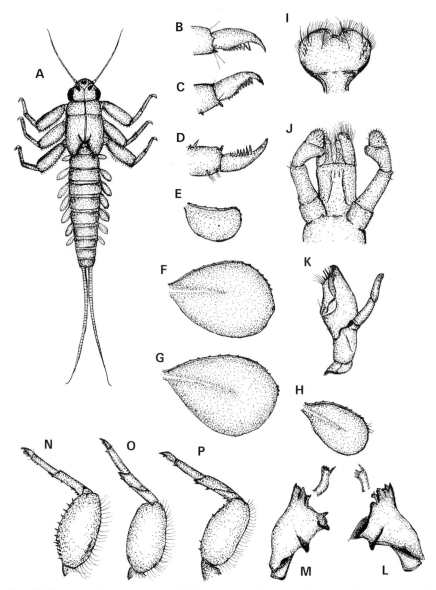

Fig. 4.40 Baetidae genus indet. (Ephemeroptera): A, dorsal view of larva; B, C & D, tarsal claws of fore-, mid- and hind legs (respectively); E, F, G & H, gills I, III, V & VII (respectively); I, labrum; J, labium; K, maxillae; L & M, mandibles; N, O & P, fore-, mid- and hindlegs (respectively).

description of the larvae of *Chromarcys* (as *Pseudoligoneuria*) *feurborni* from Sumatra.

Isonychia was previously included among the Siphlonuridae as *Chirotonetes* but is now treated by some authors (e.g. McCafferty,

1991a) as belonging to its own family: the Isonychiidae. It is widespread within Palaearctic Asia and the Oriental Region but does not reach as far east as Sulawesi (Edmunds & Polhemus, 1990). Four species are known from China (Hsu, 1935; She & You, 1988) including one — *Isonychia kiangsinensis* (Fig. 4.41) — in Hong Kong: all four have been tentatively placed in the subgenus *Isonychia* (Kondratieff & Voshell, 1983; see also McCafferty, 1989). You & Su (1987a) give a brief description of *Isonychia kiangsinensis* larvae from China, and Kang & Yang (1994b) have figured the larva of *I. formosana* from Taiwan. The streamlined body, possession of coxal gill tufts, form of the abdominal gills, and the long, double row of bristles on the inner margin of the forelegs (tibiae and femora) make *Isonychia* larvae distinctive (Fig. 4.41). Mature larvae of *I. kiangsinensis* are unmistakable: they are chocolate brown with a cream streak along the dorsal midline.

Isonychia larvae are filter-feeders. The animals orientate facing the current and use the well-developed fringe of bristles on their forelegs to collect suspended particles from the stream flow. These bristles, in turn, bear two kinds of microtrichia, that act as an efficient coupling mechanism latching the long filtering setae to those adjacent. This coupling mechanism results in an effective pore size of 0.1 — 0.7 microns for part of the filtration device (Wallace & O'Hop, 1979). When feeding, the forelegs of *Isonychia* are periodically brought within reach of the labial and maxillary palps. The latter bear numerous setae which undoubtedly aid in the removal of particles filtered onto the foreleg bristles. Gut contents of North American *Isonychia* comprise over 90% fine detritus (Wallace & O'Hop, 1979). *Isonychia* larvae are apparently confined to clean, unpolluted streams where they occur on stones and boulders areas of swift (but not torrential) flow. They are excellent swimmers and on occasions can be quite abundant; at other times, the same habitat yields few specimens.

Heptageniidae

The Heptageniidae (= Ecdyonuridae of some authors) is a conspicuous component of benthic communities in Palaearctic and Oriental streams. In Asia, the family is rather poorly known. Many genera are recorded from the region (including *Afghanurus, Afronurus, Asionurus, Atopopus, Belovius, Bleptus, Cinygma, Cinygmoides, Cinygmina, Cinygmula, Compsoneuria, Compsoneuriella, Ecdyonuroides,*

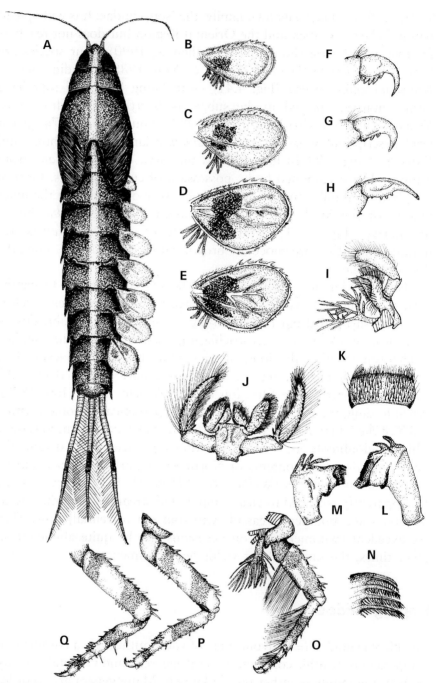

Fig. 4.41 *Isonychia kiangsinensis* (Ephemeroptera: Oligoneuriidae): A, dorsal view of larva; B, C, D & E, gills I, III, V & VII (respectively); F, G & H, tarsal claws of fore-, mid- and hind legs (respectively); I, maxilla; J, labium; K, labrum; L & M, mandibles; N, molar area of left mandible (detail); O, P & Q, fore-, mid- and hindlegs (respectively).

Ecdyonurus, Electrogena, Epeiron, Epeorella, Epeorus, Heptagenia, Iron, Ironodes, Notacanthurus, Ororotsia, Paegniodes, Rhithrogena, Rhithrogeniella, Thalerosphyrus and *Trichogenia*; see, for example, Tomka & Zurwerra, 1985), but a some of these (e.g. *Asionurus, Compsoneuriella, Ecdyonuroides, Epeiron, Rhithrogeniella* and *Trichogenia*) were established on the basis of a few specimens from a small range of localities. As a result, their status, generic limits and geographic distribution are unclear. Phylogenetic relationships within the Heptageniidae are not well understood (Mol, 1987; see also Tshernova, 1976) and the family is in need of taxonomic revision. For example, Tshernova (1976) drew attention to a lack of clarity in discrimination among the genera *Afronurus, Ecdyonurus* and *Thalerosphyrus*, and highlighted the need to resolve the interrelationships of other species- or genus-groups. Kluge (1988) synonymised many heptageniid genera on the basis of adult characters: in particular, *Afghanurus, Afronurus, Asionurus, Cinygmina, Compsoneuriella, Ecdyonuroides, Electrogena, Nixe, Notacanthurus* and *Thalerosphyrus* were included with *Ecdyonurus*; the synonymy of *Afghanurus* and *Ecdyonurus* had been suggested earlier by Kluge (1982). The assimilation of these genera under *Ecdyonurus* is rejected by most other workers (e.g. Belfiore, 1994; Kang & Yang, 1994c; Braasch, 1986c; 1990; Braasch & Soldán, 1986a; 1986b; Soldán & Braasch, 1986) because it does not take full account of larval differences and egg morphology. Moreover, biochemical (electrophoretic) evidence strongly suggests that some of these genera (at least) are distinct from *Ecdyonurus* (Zurwerra *et al.*, 1987; Hefti & Tomka, 1989). Although there is some divergence of opinion, it is evident that *Ecdyonuroides* (described from Vietnam; Dang, 1967) is a synonym of *Thalerosphyrus* (Braasch & Soldán, 1984b; Hubbard, 1990; but this is disputed by Tonka & Zurwerra, 1985). Likewise, monotypic *Compsoneuriella* from Java (Ulmer, 1939) is a junior synonym of *Compsoneuria* (see Braasch & Soldán, 1986b; Hubbard, 1990).

Heptageniid larvae are flattened and sprawl on hard surfaces; they are active crawlers but, unlike some baetids, they cannot swim. Although some research on these animals has been undertaken in Palaearctic Japan (e.g. Imanishi, 1938; 1941; Tanida, 1989a), data on the ecology of heptageniids in tropical Asian streams are meagre comprising, in the main, observations made in the wider context of studies of biogeography or the adaptations and community structure of lotic benthos (e.g. Hora, 1930; Bishop, 1973; Realon, 1980; Dudgeon, 1988a; Gupta & Michael, 1990; Hubbard & Peters, 1984;

Sivaramakrishnan & Venkataraman, 1990). Dudgeon (1996c) gives some preliminary data on life histories of five species of Hong Kong heptageniids, and estimates of their secondary production. Because these larvae showed asynchronous growth, cohorts cannot be distinguished and production estimates were based on the conservative assumption that larval development took one year to complete. However, it is probable that some heptageniids in Hong Kong have multivoltine life histories.

A definitive key to the Heptageniidae is some way off, and the one presented here is provisional. It is adapted, with modifications, from keys developed by Tshernova (1976), Tomka & Zurwerra (1985) and Hefti & Tomka (1989) and must be used with caution. Genera included are those reported from the Oriental Region, in addition to some with a mainly Palaearctic distribution which *may* be encountered along the boundary between these two regions. A few Palaearctic Asian genera (e.g. *Afghanurus, Belovius, Cinygmoides, Epeiron, Ororotsia*) have been excluded entirely, and there are some tropical Asian genera (e.g. '*Heptagenia*-ally' reported by Edmunds & Polhemus [1990] from Sulawesi) which have yet to be classified adequately[15]. Readers may wish to refer to published descriptions of larvae (especially among the Ecdyonurini) once the key has been used to arrive at a preliminary identification. For this reason, I have made reference to publications that include figures of the larvae in the following account of the Heptageniidae. In addition, the distribution and characteristics of each Asian genus are indicated. All distributional data are provisional and should be regarded as incomplete.

1. Two filaments at the end of the abdomen (i.e. the lateral cerci are present only) 2

    Three filaments at the end of the abdomen (i.e. the lateral cerci plus a terminal filament are present) 5

2. Tracheal tufts of gills weakly developed; lamella of gill I expanded; apex of galea-lacinia of maxilla with three, large articulated teeth 3

    Tracheal tufts of gills well developed; lamella of gill I small; apex

---

[15] I have seen heptageniid larvae from Sulawesi which match '*Heptagenia*-ally' of Edmunds & Polhemus (1990). They have well developed tracheal tufts, but extremely reduced lamellae on gill I and lanceolate lamellae on gills II and III. A dense row of setae fringes the anterior margin of the head capsule.

of galea-lacinia of maxilla without large articulated teeth but may bear spines **4**

3. Gill-pairs I and VII are modified to form a 'sucking disc': the first pair are strongly broadened anteriorly and the opposite members of the pair contact each other under the body; gill VII curves inward and often has a longitudinal fold; usually paired spines or a row of hairs along the median dorsal line of the abdomen ....
.... *Iron*

Gill-pairs I and VII are not modified to form a 'sucking disc', although they may be well developed .... *Epeorus*

4. Maxilla with a few spines at the apex; anterior margin of the head with a dense row of setae; gill lamellae rather small, gill I the smallest; paired tubercles or spines present on abdominal terga I-IV; rather rare (Himalaya only) .... *Ironodes*

Maxilla with a few setae at the apex; anterior margin of the head lacks setae; gill lamellae rather small, and lamella of gill I very much smaller than the tracheal tuft; tubercles or spines on abdominal terga lacking or single (not paired); probably confined to Japan and Korea .... *Bleptus*

5. Gills overlap ventrally to form a 'sucking disc'; mainly Palaearctic
.... *Rhithrogena*

Gills not as above **6**

6. Gills II-VI (and *usually* gill VII), but not necessarily gill I, are well developed and more-or-less leaf-like; they *usually*[16] have a tracheal tuft **7**

Gills I and VII have lanceolate lamellae, but the rest consist of tracheal tufts with small, rather inconspicuous lamellae (gill VII lacks a tracheal tuft entirely, while the tuft of gill I is small); rare
.... *Trichogenia*

7. Lamella of gill I more-or-less leaf-like (although it may be slightly shorter than the tracheal tuft, if this is present) **10**

Gill I otherwise (either tiny and scale-like or slender and lanceolate)
.... **8**

---

[16] Except some species of *Cinygmula*.

8. Gill I reduced to a tiny scale-like platelet and a short tracheal tuft; all gills joined laterally to the body 9

Gill I is slender and lanceolate; abdominal gills II and III inserted ventrally; *may* have tubercles on the head, thorax and abdomen; mainly Palaearctic (mouthparts adapted for predation) ....
.... *Anepeorus*

9. Anterior margin of head capsule thickened ventrally; a distinct suture separating the pronotum and mesonotum, the posterior margin of the pronotum being distinctly concave; gill VII lacks a tracheal tuft; Sabah and the Philippines only .... *Atopopus*

Anterior margin of head capsule not thickened ventrally; pronotum broadest at its anterior margin where the angles are rounded, thereafter gradually narrowing posteriorly and joining directly to the mesonotum so that the posterolateral margins of the pronotum are indistinct; Asian mainland only .... *Paegniodes*

10. Abdomen with spiniform lateral (paranotal) processes pointing backwards 11

Abdomen lacks spiniform lateral processes 13

11. Spiniform lateral processes well developed: those of segment VII *may* extend slightly beyond the posterior boundary of segment VIII .... *Thalerosphyrus*

Lateral abdominal processes less well developed and overlap no more than half the length of the next segment (often considerably less) 12

12. Tibia lack spines; maxillary palp with two segments; gill VII with rounded tip and no more than twice as long as broad; cerci bear spines as well as lateral bristles and segments of the cerci with stout spines alternate with those lacking such spines ....
.... *Rhithrogeniella*

Spines borne on the apex of the tibia; a three-segmented maxillary palp; gill VII lanceolate (three times longer than broad) with pointed tip; cerci not as above .... *Asionurus*

13. A ridge runs mid-dorsally along the abdominal tergites (II-IX) terminating in a small spinule on each segment; the head is noticeably ellipsoid; mainly Palaearctic .... *Notacanthurus*

Abdomen lacks a mid-dorsal ridge **14**

14. Maxilla with a dense row of long setae on the apical margin of the galea-lacinia; lamella of gill I rather small, tracheal tufts of all gills well developed; Palaearctic .... *Cinygma*

    Maxilla with pectinate spinules on the apex of the galea-lacinia **15**

15. Anterior margin of head capsule concave; head rather small so that the maxillary palps *may* protrude from the side; labrum rather small and weakly expanded laterally; gill I well-developed and lacks a tracheal tuft, tufts on remaining gills poorly-developed or absent .... *Cinygmula*

    Anterior margin of head capsule entire and expanded so that the maxillary palps are concealed when viewed from above; most gills with well developed tracheal tufts **16**

16. Galea-lacinia of maxilla with a row of long setae on the ventral side (paraglossae of labium weakly expanded laterally) **17**

    Galea-lacinia of maxilla with scattered hairs (short setae) on the ventral surface (paraglossae of labium strongly expanded laterally) **18**

17. Maxillary palp with three segments, and the apical segment is small, slender and acutely pointed; cerci with sparse setation at the articulation of each segment (only) but spines are absent; gill VII may lack a tracheal tuft; tibia of the forelegs *may be* longer than the tarsus; acute, supra-coxal spurs present above and behind the coxae of the mesothoracic legs; Oriental .... *Compsoneuria*[17]

    Maxillary palp with two segments; cerci with spines and long setae; gill VII *usually* includes a tracheal tuft; tibia of the foreleg not unusually long; supracoxal spurs lacking; mainly Palaearctic .... *Heptagenia*

18. Lamellae of gill V (at least, possibly some or all of II-VI) with acutely pointed apical prolongation; cerci with rings of short, black bristles, and alternating bands of light and dark pigment on the

---

[17] *Atopopus* may key out here; gill VII lacks the tracheal tuft which is usually present in *Compsoneuria*. In addition, there is a gradual increase in the size of the lamellae of gills II-VI of *Atopopus*, and the lamella of gill I is rather reduced.

segments; mainly Oriental .... *Cinygmina*

Gill lamellae without apical prolongations; cerci not as above
.... **19**

19. Lateral borders of the pronotum are dilated and extend posteriorly onto the sides of the mesonotum; Palaearctic .... *Ecdyonurus*[18]

    Lateral borders of the pronotum not elongated posteriorly **20**

20. Fine setae on the cerci, but no spines; cerci rather short (less than the length of the body); lamella of gill I distinctly rounded ....
    .... *Nixe*

    Cerci without fine setae but may have spines; cerci long (up to twice the length of the body); lamella of gill I lanceolate or rather blunt **21**

21. Lamella of gill I lanceolate, that of gill VII *usually* broad and rounded; anterior border of labrum with a median notch ....
    .... *Afronurus*

    Lamella of gill I oblique oval or blunt, that of gill VII *usually* rather pointed; anterior border of labrum straight or concave but lacking a median notch .... *Electrogena*

The Heptageniidae comprises two subfamilies in Asia: the Heptageniinae, which is relatively diverse, and the Anepeorinae represented by only one genus in the region (of two in the world). *Anepeorus* (= *Spinadis*; Hubbard, 1990) is reported from the mountainous regions of northern Sichuan and southern Gansu Provinces in China (Ulmer, 1935). It is essentially a Holarctic genus which will not be considered further. The taxonomy of the Heptageniinae is confused but, on the basis of larval characteristics, Soldán & Braasch (1986) recognize three distinct lineages in the Heptageniinae (only Asian genera are included here):
i)   *Cinygma — Rhithrogena — Cinygmula*;
ii) *Bleptus — Epeorus — Iron*;
iii) *Afronurus, Cinygmina, Ecdyonurus, Heptagenia, Rhithrogeniella, Thalerosphyrus* and probably *Compsoneuria* and *Nixe*.

---

[18] *Afghanurus*, viewed by some workers as a junior synonym of *Ecdyonurus*, may key out here.

More recently, Braasch (1990) erected the tribe Ecdyonurini to include a number of related genera: *viz. Afghanurus, Afronurus, Asionurus, Atopopus, Cinygmina, Compsoneuria, Nixe, Electrogena, Rhithrogeniella, Notacanthurus* and *Thalerosphyrus*. On the basis of comments by Tshernova (1976), *Paegniodes* (see below) should be added to group-i.

Rhithrogena occurs mainly in Palaearctic Asia (e.g. Japan and Korea: Imanishi, 1936; Yoon & Yeon, 1984), but it is found in Pakistan (Hubbard & Peters, 1978), Nepal (Braasch, 1981b; 1984a), Kashmir (Braasch, 1981b) and China (You, 1990; You & Gui, 1995). Surprisingly, the genus has been reported from Sumatra by Ulmer (1939), and this distribution is confirmed by Braasch & Soldán (1986d) although Edmunds & Polhemus (1990) consider that *Rhithrogena parva* of Ulmer (1939) may be *Rhithrogeniella*. *Rhithrogena* larvae are figured by Braasch (1984a), Yoon & Yeon (1984) and Sartori & Sowa (1992). The shape of the head, presence of a terminal filament, and the expansion of the first pair of gills which overlap to form a ventral 'sucking disc' characterize the genus *Rhithrogena*. In addition, the apex of the galea-lacinia of the maxilla bears pectinate spines and combs (Sartori & Sowa, 1992). The presence of a terminal filament distinguishes *Rhithrogena* and other group-i genera (see above) from *Epeorus* and the group-ii genera. Furthermore, the galea-lacinia of *Epeorus* is covered with fine hairs and movable teeth (Sartori & Sowa, 1992).

Although Kluge (1988; 1993) and others (e.g. Imanishi, 1934; Tshernova, 1976; Yoon & Yeon, 1984; Gui & Zhang, 1992) consider that *Iron* is a subgenus of *Epeorus*, Ulmer (1925; 1935), Hsu (1935), Kapur & Kripalani (1961), Braasch (e.g. 1979; 1980c; 1981c; 1981d; 1981e; 1984c), Braasch & Soldán (1984c), Zurwerra & Tomka (1985), Hubbard (1990) and Bae *et al.* (1994) treat *Iron* (Fig. 4.42) as distinct from *Epeorus* (Fig. 4.43). Electrophoretic studies of phylogenetic relationships support this view (Zurwerra *et al.*, 1986; 1987). Larvae of both genera lack a terminal filament, and the frontal region of the head is expanded to a greater or lesser degree. They can be distinguished from each other by the gills (see couplet 3 of the heptageniid key given above) which are modified to form a 'sucking disc' in *Iron* and are therefore larger than the equivalent gills in *Epeorus* (see Fig. 4.42H *versus* 4.43D). In addition, the front margin of the head of some *Iron* species is more markedly expanded than in *Epeorus* (see Fig. 4.42A *versus* 4.43A). Imanishi (1934) reports that there is considerable intraspecific variation in the form of gill I in one Japanese species of

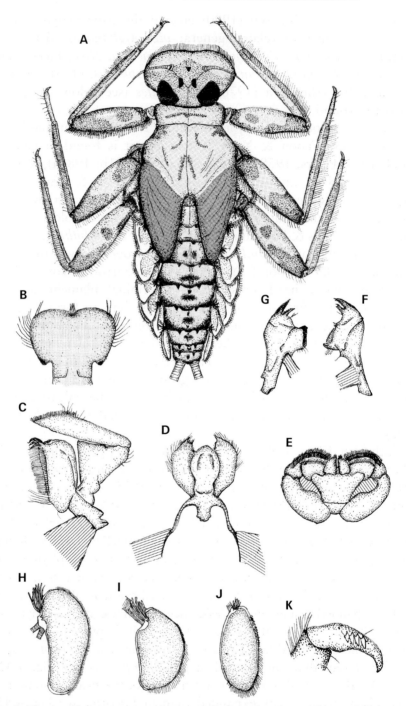

Fig. 4.42 *Iron* (Ephemeroptera: Heptageniidae): A, dorsal view of larva; B, labrum; C, maxilla; D, hypopharynx; E, labium; F & G, mandibles; H, I & J, gills I, III & VII (respectively); K, tarsal claw of foreleg.

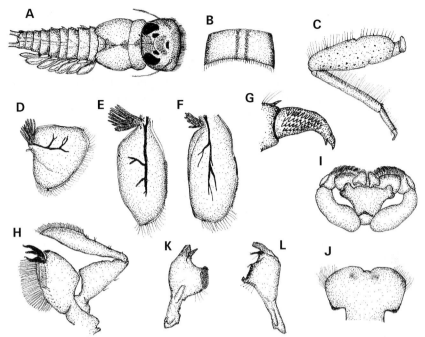

Fig. 4.43 *Epeorus* (Ephemeroptera: Heptageniidae): A, dorsal view of larval body; B, tergum of the abdominal segment VIII; C, foreleg; D, E & F, gills I, III & VII (respectively); G, tarsal claw of foreleg; H, maxilla; I, labium; J, labrum; K & L, mandibles.

Epeorus, with some individuals resembling *Iron*; however, such variation has not been recorded for other members of the genus. Descriptions of *Epeorus* larvae are given by Braasch (1979; 1980c; 1981c; 1990), Braasch & Soldán (1984), Yoon & Yeon (1984) and Kang & Yang (1994c). The genus is widely distributed in Tibet (You, 1987), Nepal (Braasch, 1980c; 1981e), India (Hubbard & Peters, 1978; Braasch, 1981c; Braasch & Soldán, 1987), Vietnam (Braasch, 1979; Braasch & Soldán, 1984), Thailand (Braasch, 1990), Malaysia (Braasch & Soldán, 1986b), Hong Kong (Dudgeon, 1990b; 1996c), Taiwan (Braasch, 1981d; Kang & Yang, 1994c) and into the Palaearctic. Tomka & Zurwerra (1985) recognize the related genus *Ironodes*, which occurs in the Himalaya, although it is viewed by most workers as either a synonym or a subgenus of *Epeorus* (e.g. Tshernova, 1976; Kluge, 1988; 1993). Hubbard (1990), however, treats *Ironodes* as valid, and the small lamellae of gill I seem to set these larvae apart from *Epeorus* and *Iron*.

Iron is widespread, occurring in China (Ulmer, 1925; 1935; Hsu, 1935; Gui & Zhang, 1992); Hong Kong (Dudgeon, 1990b; 1996c);

Vietnam (Braasch & Soldán, 1984), Nepal (Braasch, 1980c; 1981e; 1983b), India (Kapur & Kripalani, 1961; Braasch, 1981b; 1983b) and further north (e.g. Bae *et al.*, 1994). Kapur & Kripalani (1961) record the Nearctic genus *Ironopsis* from India, but Kluge (1988) considers that *Ironopsis* as a junior synonym of *Iron* since one feature that was cited as distinguishing *Ironopsis* (a row of long hairs along the midline of the abdominal tergites) is, in fact, shared by larvae of *Iron* (Tshernova, 1976). It is therefore clear that this record from India is actually *Iron* (Braasch, 1983b). Braasch (1979; 1980c; 1981e; 1983b) provides illustrations of *Iron* larvae, as do Kapur & Kripalani (1961: as *Ironopsis*), Braasch & Soldán (1984) and Yoon & Yeon (1984).

Among the Ecdyonurini (group-iii heptageniids: see above), *Compsoneuria* (including *Compsoneuriella*) is known from China, Java, Sumatra, Thailand and Malaysia (Ulmer, 1932a; 1939; Braasch & Soldán, 1986b; 1986c; Braasch, 1990). Some details of the larval structure are given by Ulmer (1939), but a more complete account is provided by Braasch & Soldán (1986b). The legs are rather distinctive, in that the femora are broad and strongly stippled while the tibia of the foreleg is much longer than the tarsus. In addition, acute, supra-coxal spurs are present above and behind the coxae of the mesothoracic legs. The cerci bear rather sparse setae at the articulation of the segments but spines are absent (*cf.* many other Ecdyonurini: e.g. *Afronurus*, *Ecdyonurus*, *Rhithrogena*, *Thalerosphyrus*). Rather distinct markings on the head and abdomen also seem typical of *Compsoneuria*, and the maxillary palp has three segments setting apart these animals apart from the majority of the Heptageniinae (except *Asionurus*). The tracheal tuft and lamella of gill I is usually well developed and gill VII includes a tuft as well as a lamella in some species of *Compsoneuria*. *Afronurus* is widely distributed from China, Tibet and the Indian Himalaya to the Philippines, Java and Sulawesi (Ulmer, 1939; Hubbard & Peters, 1978; You *et al.*, 1982a; Flowers & Pescador, 1984; Mol, 1987; You, 1987; Edmunds & Polhemus, 1990); it has been reported from Europe also (Hefti & Tomka, 1989). Flowers & Pescador (1984), Hefti & Tomka (1989) and Kang & Yang (1994c) provide larval descriptions of *Afronurus*, and Realon (1979) gives an account of some aspects of the ecology of these animals in a Philippines stream. Diagnostic features include a labrum which strongly expanded laterally and somewhat pronounced posteriorly, with a median notch on the anterior border. Gill I is lanceolate, Gill VII usually lacks a tracheal tuft and is rather broad and rounded, and the cerci lack fine setae (Hefti & Tomka, 1989; Kang & Yang, 1994c). *Thalerosphyrus* occurs from China through Southeast Asia (Vietnam, Malaysia, the Philippines and Sulawesi) to India (Hubbard & Pescador, 1978; Braasch & Soldán, 1984b; 1986b; Venkataraman & Sivaramakrishnan, 1987; Edmunds & Polhemus, 1990), and larval descriptions are given by Braasch & Soldán (1986b) and Venkataraman & Sivaramakrishnan (1987); Braasch & Soldán (1986b) provide a larval diagnosis for the genus. Of particular note is the presence of long lateral projections on the sternites of abdominal segments III-VIII. There are spines on the cerci but few setae.

Nixe was thought to be confined to the Nearctic Region until its discovery in Europe (Zurwerra *et al.*, 1987) and recent records from Taiwan (Kang & Yang, 1994c). Larvae (see Kang & Yang, 1994c) are distinguished by their rather short cerci (less than the length of the body) which bear long lateral fringes of setae (Hefti & Tomka, 1989; Kang & Yang, 1994c). In addition to the presence of cercal setae, the lamella of gill I is distinctly rounded which may help to differentiate *Nixe* from superficially-similar genera such as *Afronurus* or *Electrogena* (although in the latter case the separation by gill shape is more problematic). Gill VII of *Nixe* seems to lack a tracheal tuft (Kang & Yang, 1994c). Until recently, *Electrogena* was known only from Europe (Belfiore, 1994; Zurwerra & Tomka, 1985). The presence of *Electrogena* in Taiwan (Kang & Yang, 1994c) increases the known range of this genus, which has since been reported from Hong Kong (Dudgeon, 1997). Kang & Yang (1994c) figure the larval stage. Diagnostic features of *Electrogena* larvae (Fig. 4.44) include a labrum with a border that is straight or concave (without a median notch; *cf. Afronurus*). Gill I may be blunt; gill VII usually lacks a tracheal tuft and may be rather pointed; and, the cerci are twice as long as the body and lack lateral setae (*cf. Nixe*: Hefti & Tomka, 1989; Kang & Yang, 1984c).

Cinygmina (Fig. 4.45) is recorded from India (Hubbard & Peters, 1978; Braasch & Soldán, 1987b), Nepal (Braasch, 1981b; 1984b), Vietnam (Braasch & Soldán, 1984a; 1987a), Thailand Braasch, 1990) and China (You *et al.*, 1981; Wu & You, 1986a; Zhang & Cai, 1991). Braasch (1984b, 1990), Braasch & Soldán (1984a; 1987a; 1987b) and Wu *et al.* (1985) have figured the larval stages, and Braasch & Soldán (1984a) give a generic diagnosis. Notably, gills I-VI comprise a lamella and a tracheal tuft, but gill VII lacks the tuft and is more lanceolate than the preceding gills. In addition, there is a small filament arising from the apex of (at least) the fifth gill lamella, and the shape of the lamella is subovate or almost 'square-tipped'. The hind margin

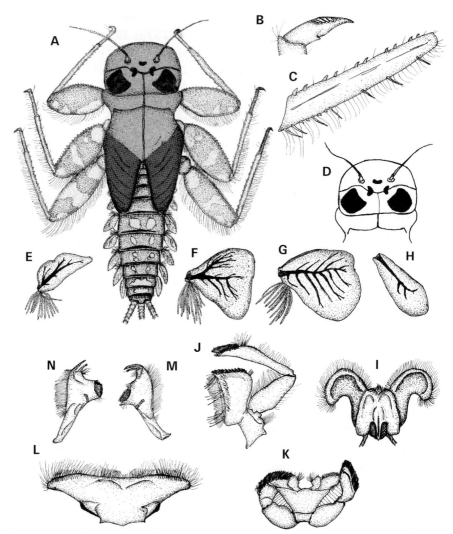

Fig. 4.44 *Electrogena* (Ephemeroptera: Heptageniidae): A, dorsal view of larva; B, tarsal claw of foreleg; C; spination of hind tibia; D, diagrammatic view of head and prothorax; E, F, G & H, gills I, III, V & VII (respectively); I, hypopharynx; J, maxilla; K, labium; L, labrum; M & N, mandibles.

of the head is concave (which differs from most other Ecdyonurini), but the mouthparts are similar to those of *Ecdyonurus* and *Nixe*. The cerci bear rings of short, black bristles, and have alternating bands of light and dark pigment on the segments (Braasch & Soldán, 1984a).

In addition to the widely-distributed genera discussed above, a number of small and/or poorly-known heptageniid genera are confined

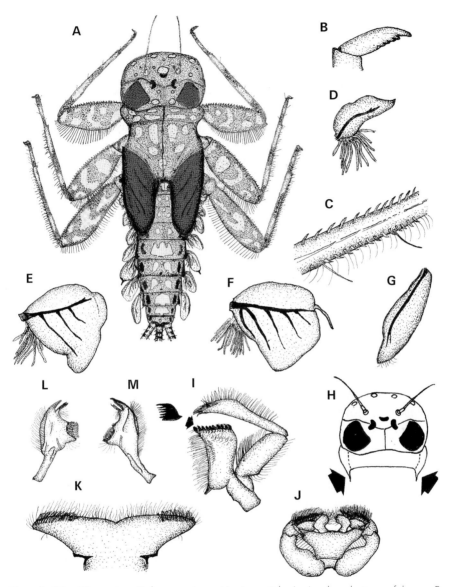

Fig. 4.45 *Cinygmina* (Ephemeroptera: Heptageniidae): A, dorsal view of larva; B, tarsal claw of foreleg; C; spination of hind tibia; D, E, F & G, gills I, III, V & VII (respectively); H, diagrammatic view of head and prothorax; I, maxilla; J, labium; K, labrum; L & M, mandibles.

to the Oriental Region. The larva of *Epeorella* — reported only from Indonesia (Ulmer, 1939) — is unknown. *Atopopus* occurs in the Philippines (Ulmer, 1919), Borneo (Sabah) and Vietnam (Edmunds & Polhemus, 1990) but published descriptions of the larva were lacking until a recent diagnosis of the genus (Wang & McCafferty, 1995a). *Atopopus* larvae may be characterized by gill I with a reduced lamella, a gradual increase in the size of the lamellae from gill II through VI, and the lack of a tracheal tuft on gill VII; the cerci bear whorls of spine-like setae but no hairs (Prof. G.F. Edmunds, pers. comm.; Wang & McCafferty, 1995a). *Asionurus* is recorded from Sumatra (as *Thalerosphyrus sinuosus* in Ulmer, 1939), Malaysia and Vietnam (Braasch & Soldán, 1986a; 1986b); Braasch & Soldán (1986a) give a detailed description of the larval stage of one of the two known species of *Asionurus*. The genus differs from most other Ecdyonurini in the detailed structure of the gills and mouthparts (e.g. a three-segmented maxillary palp, although this is a feature shared by of *Compsoneuria*), and the presence of spines on the apex of the tibia. Gill VII lacks a tracheal tuft but has an unusually long lamella, and there are paranotal spines on the abdomen similar to — but less well developed than — those of *Thalerosphyrus*. *Trichogenia* is a monotypic genus known only from the larval stage which has been figured by Braasch & Soldán (1988). To date, these animals appear confined to Vietnam. The anterior and lateral margins of the labrum bear long bristles, and gills I and VII of *Trichogenia* appear as lanceolate lamellae while the others consist of tracheal tufts with small, rather inconspicuous lamellae. Gill VII lacks tracheal tufts entirely, while the tuft of gill I is small (Braasch & Soldán, 1988).

Rhithrogeniella is recorded from Java, Sumatra, Thailand, Vietnam and China (Soldán & Braasch 1986; Braasch, 1990; see also Edmunds & Polhemus, 1990; You & Gui, 1995): Soldán & Braasch (1986) give a description and diagnosis of the larva. *Rhithrogeniella* is distinguished from other Ecdyonurini by several characters: the long, well-developed lamella of gill I; the tendency for gills II-VII — which are ovate or triangular — to increase in size posteriorly (although gill VII lacks tracheal tufts); tarsal claws which have a rounded subapical tooth; the presence of lateral bristles on the cerci; and, the alternation of segments of the cerci bearing stout spines with those lacking such spines (Soldán & Braasch, 1986). *Paegniodes cupulatus* (Fig. 4.46) occurs in Hong Kong (Hsu, 1936b; Uéno, 1969; Dudgeon, 1990b; 1996c) southern China and Tibet (Hsu, 1931; 1936b; Tshernova, 1976). The larva was first described by Tshernova (1976). Kluge (1988; 1993) treats *Paegniodes* as a subgenus of *Rhithrogena*, but this seems unlikely given their disparate larval morphologies. Although both genera have a terminal filament, the lamella of the first pair of abdominal gills in *Rhithrogena* are expanded and overlap to form a ventral disc, whereas

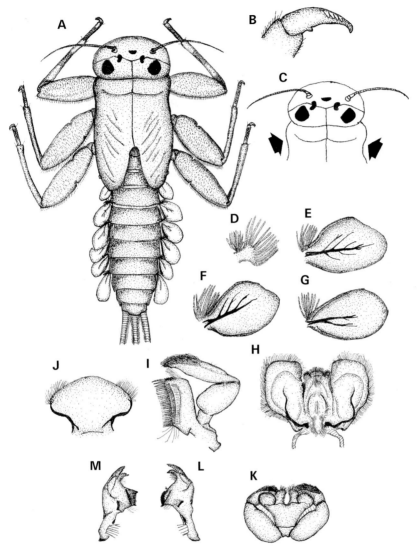

Fig. 4.46 *Paegniodes cupulatus* (Ephemeroptera: Heptageniidae): A, dorsal view of larva; B, tarsal claw of foreleg; C; diagrammatic view of head and prothorax; D, E, F & G, gills I, III, V & VII (respectively); H, hypopharynx; I, maxilla; J, labrum; K, labium; L & M, mandibles.

the same gills in *Paegniodes* are represented by small tracheal tufts and lack lamellae entirely. Gills II-VII are lamellate with tracheal tufts, and gill morphology clearly distinguishes these larvae of this monotypic genus from other heptageniids. Only in *Bleptus* (see below) — a group-ii Heptageniinae which lacks a terminal filament — does gill I resemble that of *Paegniodes* somewhat, but *Paegniodes* has a terminal filament and appears to be a group-i Heptageniinae. Significantly, Tshernova (1976) likens *Paegniodes* larvae to *Cinygmula*, although the adults are more similar to *Rhithrogena*. Curiously, You & Gui (1995) assign *Paegniodes* to the Palingenidae (*sic*).

Several Asian heptageniid genera have a primarily or almost exclusively Palaearctic distribution. *Bleptus* is confined to Japan and Korea (Yoon & Yeon, 1984; Kluge, 1988; Bae *et al.*, 1994); Yoon & Yeon (1984) give illustrations of the larvae which is characterized by the lack of a terminal filament and the strongly-developed tracheal tufts (and small lamellae) of the first pair of gills. *Ororotsia* occurs in the Indian Himalaya (Hubbard & Peters, 1978) but has not been reported elsewhere in the Oriental Region. *Afghanurus*, which has been recorded in Kashmir (Braasch, 1981b) has a similar distribution, as does *Notacanthurus* — a predominately Central Asian genus that penetrates south to the Indian Himalaya. Braasch (1986c) provides a diagnosis of this genus and figures the larvae of two species of *Notacanthurus* from Kashmir. The head is notably ellipsoid, and there is a ridge running mid-dorsally along the abdominal tergites. The cerci and terminal filaments are longer than the body, bearing small spines but very few setae. Gills have a well-developed lamella and a tracheal tuft; gill VII is elongate and lacks the tuft. *Epeiron* has also been found as far south as Kashmir (Braasch, 1981b; Braasch & Soldán, 1982) but is not known in the larval stage. Kluge (1988; 1993) and Sartori & Sowa (1992) consider that *Epeiron* is a junior synonym of *Rhithrogena*, while Hubbard (1990) lists it as synonymous with *Epeorus*. *Heptagenia* is primarily Palaearctic (Imanishi, 1936; 1938; Yoon & Yeon, 1984). Braasch (1986b) states that there are some erroneous reports of *Heptagenia* in India (e.g. Hubbard & Peters, 1978), but provides one definite record from Kashmir and gives a larval description (see also Yoon & Yeon, 1984). Like other Ecdyonurini, *Heptagenia* larvae have a terminal filament with spines and resemble *Ecdyonurus* from which — according to Imanishi (1936) and Yoon & Yeon (1984) — they can be distinguished by details of mouthpart structure. Some *Heptagenia* larvae have a tracheal tuft on gill VII and lack denticles on the tarsal claws. *Afronurus*, *Electrogena* and *Nixe* apparently lack this tuft (e.g. Kang & Yang, 1994c), and *Nixe* has denticles on the tarsal claws (although some *Nixe* do not have denticles on the foreleg claws despite their presence on the metathoracic legs; Kang & Yang, 1994c). Like *Heptagenia*, *Ecdyonurus* seems to be mainly Palaearctic, occurring in Japan (Imanishi, 1936), Korea (Yoon & Yeon, 1984), Mongolia (Braasch, 1986a), although it has been reported from Nepal (Braasch, 1980a; 1981b; 1984b), the Indian Himalaya (Kapur & Kripalani, 1961; Hubbard & Peters, 1978) and the Philippines (Hubbard & Pescador, 1978). Braasch (1980a; 1984b) and Yoon & Yeon (1984) have figured the larvae of *Ecdyonurus*.

Cinygma does not appear to extend beyond Palaearctic Asia (Kluge, 1988); moreover, most species of *Cinygma* which have been described are attributable to *Cinygmula* or *Epeorus* (Belov, 1982; 1983). Thus the genus may not be valid, although You (1987) has described a new *Cinygma* from Tibet. *Cinygmula* occurs in Tibet (You, 1987), Afghanistan, Mongolia, India (Braasch, 1979; 1980b; 1986a; Braasch & Soldán, 1987b) and Korea (Yoon & Yeon, 1984). Illustrations of the larvae (Braasch, 1979; 1980b; 1986a; Yoon & Yeon, 1984; Braasch & Soldán, 1987b) reveal that the head of *Cinygmula* is distinctly more ovate and less expanded than other heptageniids (with the result that maxillary palps may protrude at the sides of the head), so these mayflies are unlikely to be confused with other Asian genera; the labrum is smaller and much less expanded laterally, while the gill form is rather variable and the tracheal tufts are poorly developed (or even absent). Although Kluge (1988; 1993) treats *Cinygmula* as a subgenus of *Rhithrogena*, this seems unlikely because the larval morphology (head, gill structure) of these two genera differs markedly (see, for example, Yoon & Yeon, 1984).

Leptophlebiidae

Like all mayfly families in Asia, the Leptophlebiidae is incompletely known. However, the situation with this group is less confusing than that for baetids or heptageniids. This reflects the early publication of an authoritative monograph on leptophlebiid larvae (Peters & Edmunds, 1970) describing many of the genera and setting out the larval characters that can be used to discriminate them. Thus research on this family has been based on a firm foundation, and generic limits, as well as the arrangement of species within genera, has been quite stable and relatively uncontroversial. Despite the advances which have been made in leptophlebiid classification, knowledge of the ecology of the group in tropical Asia is meagre, concerned mainly with longitudinal zonation and microdistribution (e.g. Bishop, 1973; Hubbard & Peters, 1984; Dudgeon, 1990b; Sivaramakrishnan & Venkataraman, 1990). Life-history data are lacking, but most species are likely to be multivoltine.

The diet includes algae and fine organic matter which is collected from the stream bed (i.e. they are collector-gatherers), but *Choroterpides* and (probably) *Notophlebia* filter-feed by means of their greatly elongated and setaceous labial and maxillary palps (Sivaramakrishnan & Peters, 1984).

The bodies of leptophlebiid larvae are more-or-less depressed. The tendency is not as marked as in heptageniids although some members of the two families are superficially similar. They can be distinguished easily by the fact that the mandibles of heptageniids are hidden beneath the head capsule when viewed from above; those of leptophlebiids are not concealed by the head (although *Nathanella* approaches the heptageniid condition). Leptophlebiid larvae are active crawlers, but most are rather weak swimmers. They occupy a variety of habitats, but most species are confined to streams where they are associated with bottom sediments, or found on submerged wood and among trailing roots at the banks. The family as a whole shows species replacement along streams but, even within a particular reach, there is specialization in habitat occupancy. In Hong Kong, for example, *Isca* is confined to sites with swift current, whereas *Thraulus* occurs in slow-flowing microhabitats close to the banks; *Choroterpes* and *Habrophlebiodes* occur in places where the flow is moderate and these mayflies seem to be relatively eurytopic (Dudgeon, 1990b). There is a degree of correlation between habitat occupancy and gill morphology in leptophlebiids. *Thraulus* has large, oval gills with elaborately fringed margins which increase the surface area, while the gills of *Isca* are small and lanceolate. *Choroterpes* gills are leaf-like with apical prolongations.

The following key to leptophlebiid larvae draws heavily upon Peters & Edmunds (1970) monograph which gives many illustrations of the details of the mouthparts and distinctive features of most genera. The original key has modified somewhat in the light of recent research (see below) on leptophlebiid classification. Where it will facilitate identification, the geographic range of genera has been included in the key. It is unlikely that, for example, a genus confined to India will turn up in samples collected from China, although the recent description of a species of *Indialis* based on adults from Hainan Island (You & Gui, 1995) may be just such an instance. Once again, the reader is reminded that our knowledge of the Asian mayfly fauna is far from complete.

1. Abdominal tergites extending round to the ventral surface of the

abdomen on segments III-VII, so that the slender gills arise ventrally (Fig. 4.47N); widespread from India to China through Indonesia to Sulawesi ....*Isca*[19]

Abdominal tergites extend to the lateral margins of the body; gills arise dorsally or laterally (Figs 4.48A & 4.49.A) 2

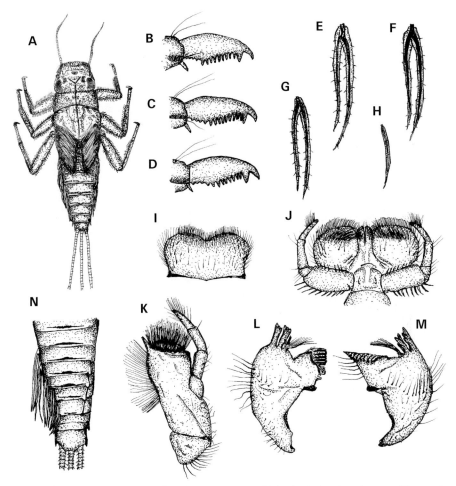

Fig. 4.47 *Isca purpurea* (Ephemeroptera: Leptophlebiidae): A, dorsal view of larva; B, C & D, tarsal claws of fore-, mid- and hindlegs (respectively); E, F, G, & H, gills I, III, V & VII (respectively); I, labrum; J, labium, K, maxilla; L & M, mandibles; N, ventral view of abdomen showing gill insertion.

---

[19] Three subgenera (*Isca*, *Minyphlebia* and *Tanycola*) are recognized on the basis of characteristics of the adults.

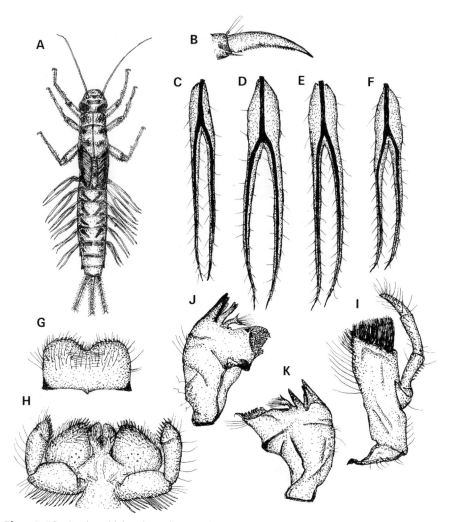

Fig. 4.48 *Habrophlebiodes gilliesi* (Ephemeroptera: Leptophlebiidae): A, dorsal view of larva; B, tarsal claw of foreleg; C, D, E & F, gills I, III, V & VII (respectively); G, labrum; H, labium; I, maxilla; J & K, mandibles.

2. Maxillary and labial palps elongated and highly setaceous, extending well beyond the side of the head so that they are conspicuous when viewed from above 3

   Palps neither greatly elongated nor extending any distance beyond the sides of the head 4

3. Gills present on abdominal segments I-VI; the apical denticle on the tarsal claws is greatly enlarged; India only ....

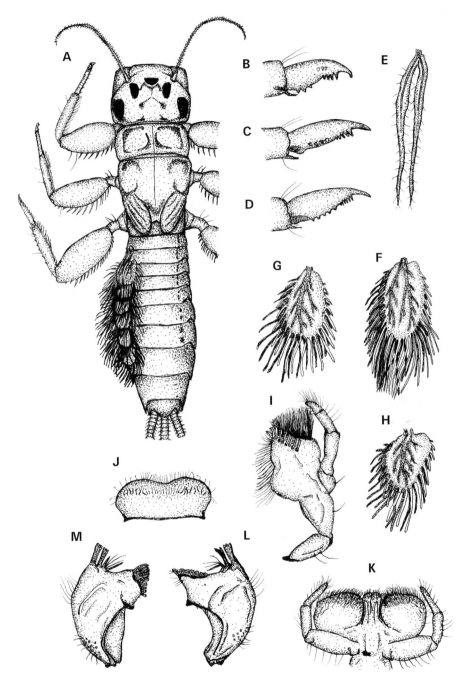

Fig. 4.49 *Thraulus* (Ephemeroptera: Leptophlebiidae): A, dorsal view of larva; B, C & D, tarsal claws of fore-, mid- and hindlegs (respectively); E, F, G, & H, gills I, III, V & VII (respectively); I, maxilla; J, labrum; K, labium; L & M, mandibles.

.... *Notophlebia*[20]

Gills present on abdominal segments II-VII; denticles on the tarsal claws are approximately equal in length; widespread.................
.... *Choroterpides*

4. Middle abdominal gills (III-V) flat and rather plate-like and fringed with filamentous processes.. 5

   Middle abdominal gills without a fringed margin, may be plate- or leaf-like, or bifurcate and long and slender 8

5. Middle abdominal gills plate-like and broadly oval and fringed with thin filamentous processes (e.g. Figs 4.49F-H); posterolateral spines on abdominal segments VII-IX or VIII-IX; widespread 6

   Middle abdominal gills plate-like but irregular in shape, unevenly fringed with rather broad filamentous processes; posterolateral spines on abdominal segments III-IX, which are especially long and well-developed on the last two segments; India only ....
   .... *Petersula*[21]

6. Ventral lamella of abdominal gills II-VII greatly reduced; labrum appears triangular in shape due to strong convexity of the posterior margin; New Guinea only .... *Nonnullidens*[22]

   Ventral lamella of abdominal gills II-VII subequal to or slightly larger than dorsal lamella; posterior margin of labrum not strongly convex 7

7. Apical margins of abdominal gills II-VII fimbriate (= fringed); New Guinea only.. *Barba*[22]

   Entire margins of abdominal gills II-VII fimbriate (Fig. 4.49F-H; in exceptional cases fimbriae are lacking from the outer margin at the gill base); widespread .... *Thraulus*

---

[20] Larvae of this genus, placed in *Nathanella* by Peters & Edmunds (1970), were reassigned by Sivaramakrishnan & Peters (1984). Thus Fig. 6 of Peters & Edmunds is *Notophlebia* and not *Nathanella* as labelled. The larvae of *Nathanella* has been described by Sivaramakrishnan *et al.* (1996).

[21] Sivaramakrishnan (1984) provides a detailed description of the larva of *Petersula*.

[22] Grant & Peters (1993) give detailed descriptions of the larvae of *Nonnullidens* and *Barba*.

8. Gill I similar to the others in the series 9

   Gill I differing from the others.. 16

9. Middle abdominal gills plate-like and broad 10

   Middle abdominal gills long, slender and bifurcate (Fig. 4.48D) or forked 11

10. Margins of the head capsule are broadly expanded (giving it a rather rounded appearance); apex of gills I-VII with three filamentous processes, the median process is longer than the laterals; India only .... *Nathanella*

   Margins of the head capsule not noticeably expanded; apex of middle portion of gills I-VII is cleft with a median filamentous process arising from the cleft; Sri Lanka only .... *Kimminsula*

11. Gills forked and the two portions of the lamellae overlap basally; trachea do not branch in each gill portion; a moveable denticle at the base of the tarsal claw gives the appearance of a second claw; Sulawesi only ....*Sulawesia*[23]

   Gills *and* tarsal claws not as above 12

12. Head prognathous; larvae from India and Sri Lanka only...... 13

   Head hypognathous; widespread 15

13. Third segment of the labial palp has a row of short stout spines on the inner dorsal margin; outer margin of the mandibles with a median tuft of around 10 long setae; posterolateral spines on abdominal segments IV-IX; India only .... *Edmundsula*[24]

   Third segment of the labial palp without a row of spines on the inner dorsal margin; rather sparse or scattered setae on the outer margin of the mandibles; posterolateral spines on abdominal segments V-IX 14

14. Denticles on tarsal claws increase in size apically; tip of tarsal claw strongly hooked; trachea of gills branched; larvae known from India only .... *Indialis*

---

[23] Peters & Edmunds (1990) provide a detailed description of the larva of *Sulawesia*.
[24] Sivaramakrishnan (1985) provides a detailed description of the larvae of *Edmundsula*.

Denticles on tarsal claws increase in size apically but the last one in the series is very much larger than the others; tip of tarsal claw slightly hooked; trachea of gills unbranched; Sri Lanka .... *Megaglena*

15. Anterior margin of labrum with a shallow emargination; gills fringed with setae; mainly Palaearctic Asia .... *Paraleptophlebia*

    Anterior margin of labrum deeply emarginate (Fig. 4.48G); gills lack a setal fringe (only scattered setae are present); widespread .... *Habrophlebiodes*

16. Middle gills plate-like, terminating in a single filamentous process; head hypognathous; probably only Palaearctic Asia .... *Leptophlebia*

    Middle gills leaf- or plate-like, terminating in more than one process; head prognathous (or not obviously hypognathous) 17

17. Posterolateral spines on abdominal segments III-IX (Fig. 4.51A), those on VIII and IX with a curved (convex) inner edge; a large tooth-like process on the anterior apex of the maxilla; China (Hong Kong), Taiwan and Thailand .... *Cryptopenella*

    Posterolateral spines on abdominal segments IV-IX or V-IX, those on VIII and IX do not have a curved inner edge; large tooth-like process on apex of the maxilla lacking; widespread .... *Choroterpes* 18

18. Gills III-V terminate in three processes, the middle one being longer than the laterals (Fig. 4.50D-F) .... subgenus *Choroterpes*

    Gills III-V terminate in three slender processes of approximately equal length (Fig. 4.52F-H) .... subgenus *Euthraulus*

In addition to the genera included in the key, *Gilliesia* is present in India (Peters & Edmunds, 1970), *Dipterophlebiodes* has been recorded in Borneo and Peninsular Malaysia (Peters & Edmunds, 1970; Edmunds & Polhemus, 1990), *Simothraulus* is found in Sabah (Peters & Edmunds, 1993), and *Chiusanophlebia* is apparently confined to Japan (Grant & Peters, 1993). The larvae of these four genera are unknown. *Paraleptophlebia* is mainly Palaearctic (e.g. Bae *et al.*, 1994) but has been found in northern India, Nepal (Peters & Edmunds, 1970) and Taiwan (Kang & Yang, 1994d). Kang & Yang (1994d) have figured the larvae of two species *Paraleptophlebia* and suggest

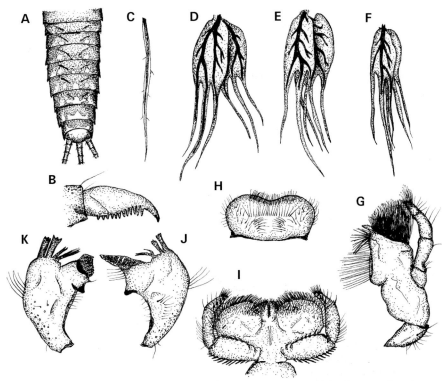

Fig. 4.50 *Choroterpes (Choroterpes)* (Ephemeroptera: Leptophlebiidae): A, dorsal view of larval abdomen; B, tarsal claw of foreleg; C, D, E & F, gills I, III, V & VII (respectively); G, maxilla; H, labrum; I, labium; J & K, mandibles.

that the genus is in need of revision. *Leptophlebia* is known from China (Hsu, 1935; Ulmer, 1936) but has not been reported south of Jiangxi Province.

Knowledge of Asian Leptophlebiidae has been enhanced by recent work which includes the establishment of new genera that were not dealt with by Peters & Edmunds (1970). These include *Sulawesia* from Sulawesi (Peters & Edmunds, 1990), *Petersula* and *Edmundsula* from India (Sivaramakrishnan, 1984; 1985), and four new genera in the *Thraulus* group: *Barba, Nonnullidens* (both confined to New Guinea), *Magnilobus* and *Sulu* (Grant & Peters, 1993). *Sulu* (including one species formerly placed in the Neotropical genus *Hagenulus*) has been reported from Borneo and the Philippines, but *Magnilobus* has, to date, been found only on the island of Manua (New Guinea); the larvae of both genera are unknown. Grant & Peters (1993) have figured the larvae of *Barba* and *Nonnullidens*, while Sivaramakrishnan *et al.* (1996) have described *Nathanella* larvae for the first time.

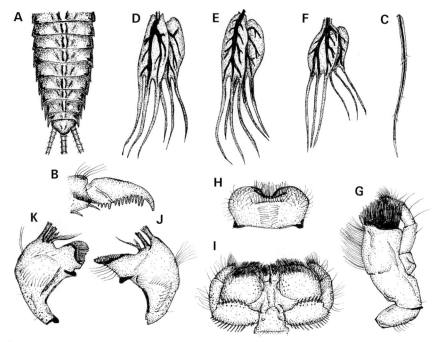

Fig. 4.51 *Cryptopenella* (Ephemeroptera: Leptophlebiidae): A, dorsal view of larval abdomen; B, tarsal claw of forelegs; C, D, E & F, gills I, III, V & VII (respectively); G, maxilla; H, labrum; I, labium; J & K, mandibles.

Illustrations of various leptophlebiid larvae which supplement those given by Peters & Edmunds (1970) include *Choroterpides* (Ulmer, 1939; Dang, 1967; Kang & Yang, 1994d) which is known from Java, Sumatra, Thailand, Nepal and Taiwan; *Choroterpes* (Wu *et al.*, 1987; Yu & Su, 1987; Kang & Yang, 1994d) from China (You *et al.*, 1980a; You & Su, 1987a; Wu & You, 1989; 1992; You & Gui, 1995), Taiwan, India, Malaysia, Borneo, Sumatra and Java; *Habrophlebiodes* (Kang & Yang, 1994d; see also Tsui & Peters, 1970) from China, Taiwan, Sumatra, Java, Sulawesi and the Philippines (Fig. 4.48); and, *Thraulus* (Peters & Tsui, 1972; Grant & Sivaramakrishnan, 1985; Arumuga Soman, 1991; Kang & Yang, 1994d) from India, China (Hong Kong), Taiwan, Peninsular Malaysia, Borneo, Java, the Philippines and Papua New Guinea (Fig. 4.49). Note that some New Guinea *Thraulus* have since been transferred to *Barba* and *Nonnullidens* (Grant & Peters, 1993).

Some uncertainty surrounds the status of *Cryptopenella* recorded in Thailand and Hong Kong (Gillies, 1951; Peters & Edmunds, 1970). The single described species, *Cryptopenella fascialis*, is reported from

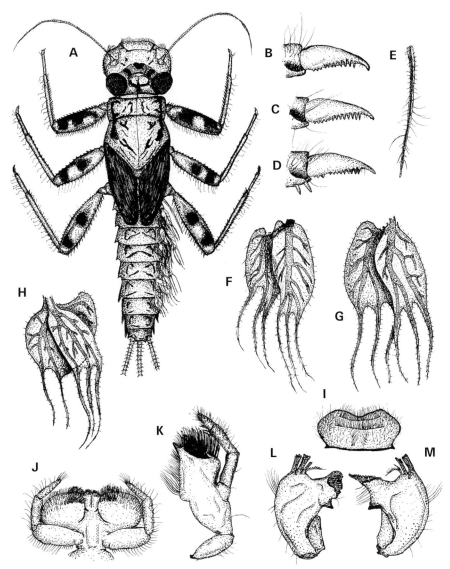

Fig. 4.52 *Choroterpes (Euthraulus)* (Ephemeroptera: Leptophlebiidae): A, dorsal view of larva; B, C & D, tarsal claws of fore-, mid- and hindlegs (respectively); E, F, G, & H, gills I, III, V & VII (respectively); I, labrum; J, labium; K, maxilla; L & M, mandibles.

Taiwan by Kang & Yang (1994d) as a junior synonym of *Choroterpes trifurcatus*. This follows Kluge (1985) who stated that *Cryptopenella* might be a synonym of the subgenus *Euthraulus* (= *Thraululus*), and that *Cryptopenella fascialis* was, in fact, *Choroterpes (Euthraulus) trifurcatus*. In Kluge's view, the extent of development of the abdominal posterolateral spines was insufficient to justify generic status for *Cryptopenella*, leaving only the large projection on the apex of the larval maxilla as diagnostic (see Fig. 336 in Peters & Edmunds, 1970; Fig. 4.51H). This seems to be a weak character, given the presence of similar projections on the maxillae of some *Choroterpes (Euthraulus)* (see Fig. 186 in Peters & Edmunds, 1970; Fig. 4.52K) and *Choroterpes (Choroterpes)* (see Fig. 2D in Kang & Yang, 1994d; Fig. 4.51G). The gills of *Cryptopenella* match those of *Choroterpes (Euthraulus)* also (see Fig. 342 in Peters & Edmunds, 1970; Figs 4.51D-F & 4.52F-H). Since the matter cannot be resolved here, I have treated *Cryptopenella* as a valid genus in the key given above.

Ephemerellidae

Like several other mayfly families, the classification of the Ephemerellidae has been in a rather muddled state. As is the case in the Baetidae, larval characteristics have been of great importance in developing a robust classification because the evolution of larval stages proceeds rapidly while the adults remain essentially conservative (Edmunds, 1959). Much confusion was removed by Allen (1980; 1984) who reclassified the group, raising many subgenera proposed by Edmunds (1959) and Allen (1971) to generic rank. The classification of some Holarctic genera is now relatively stable (McCafferty & Wang, 1994) although the situation in Asia is still unsettled. Yoon & Bae (1988) believe that Allen's (1980; 1984) generic classifications are rather unsatisfactory because they were based mainly on phenetic similarities of the larvae and geographic evidence, without any phylogenetic data.

The Ephemerellidae as it is now known comprises the subfamilies Ephemerellinae and Teloganodinae in Asia. According to Allen (1984) and Hubbard (1990), the genera present in the region are as follows (with subgenera [or synonyms] in parentheses): *Acerella, Cincticostella* (*Cinctocostella* [= *Asiatella*] and *Rhionella*), *Crinitella, Drunella* (*Drunella* and *Tribrochella*), *Ephemerella, Hyrtanella, Serratella, Torleya, Teloganopsis* and *Uracanthella* (both of which have been treated as a subgenera of *Ephemerella*), and *Vietnamella* — all Ephemerellinae — as well as *Ephemerellina* (*Ephemerellina*) and *Teloganodes* in the Teloganodinae, and *Eurylophella* [= *Melanameletus*] in the Timpanogae (a subdivision of the Ephemerellinae). Of these genera, *Acerella* has recently been renamed *Ephacerella* (Paclt, 1994), while *Ephemerella* is primarily Palaearctic. Genera such as *Cincticostella* and *Drunella* seem to comprise mainly Palaearctic species, but some representatives occur far south. Wang & McCafferty (1995b) are of the view that Chinese species previously assigned to *Ephemerellina* (*E. ornate* and *E. sinensis*: Hsu, 1935; Tshernova, 1972) belong to the genus *Vietnamella*, with the consequence that *Ephemerellina* now appears to be confined to temperate southern Africa. However, You & Gui (1995) recognize an additional species of *Ephemerellina* (*E. xiaosimaensis*) from Xizang.

Ephemerellid larvae are small, rather stocky insects. There are spines or tubercles on the legs and body, especially on the lateral margins of the abdomen which may also bear paired tubercles or spines on the tergites. The degree of spination varies greatly among genera. They are poor swimmers, and the cerci and terminal filament bear spines rather than long swimming hairs (as seen in, for example, many Baetidae). The terminal filament is usually well developed, but is lost in a few cases (e.g. *Teloganodes*). Lamellate gills are arranged dorsally on the abdomen, usually on segments III-VII or IV-VII. There may be a rudimentary gill on the first segment, but gills are rarely found on segment II (*Ephemerellina*, *Teloganodes* and *Vietnamella* are exceptional: Allen & Edmunds, 1963; Tshernova, 1972) and are sometimes missing from abdominal segment III (e.g. in *Eurylophella* and the Holarctic '*Timpanoga* complex': McCafferty & Wang, 1994) There is a tendency for the gills earlier in the series to overlap the succeeding ones. In extreme cases, gill III (in *Hyrtanella*; Allen & Edmunds, 1976) or gill IV (in *Eurylophella*) is operculate but this tendency is not as strongly developed as in the Tricorythidae (and far from that in the Caenidae and Neoephemeridae). Virtually nothing has been written of ephemerellid ecology in tropical Asia, but these animals occupy a range of lotic habitats from rapid torrential streams to more slow-flowing reaches where larvae may occur among filamentous algae or leaf litter. Some ephemerellids (e.g. *Uracanthella* in Hong Kong) appear to be specialized consumers of allochthonous detritus (i.e. they function as shredders) but this is not the sole feeding mode of all genera.

In developing the following key, I relied upon published descriptions to separate the genera of ephemerellid larvae. Identifications made with the key should be corroborated by consulting the relevant literature set below out. The key should be regarded as provisional, and will require modification as our knowledge of Asian Ephemerellidae increases.

1. Larvae lacking a terminal filament (and maxillary palps; lamellate gills on abdominal segments II-V; femora greatly expanded; see Fig. 4.53) .... *Teloganodes*[25]

   Larvae with a well-developed terminal filament 2

2. Lamellate gills on abdominal segments II-V and a rudimentary gill on segment I *or* lamellate gills on segments II-VI; maxillary palps lacking .... *Ephemerellina*

   Gill arrangement not as above; maxillary palps *usually* present 3

3. Lamellate gills on abdominal segments II-VII; conspicuous cephalic horns present .... *Vietnamella*

   Lamellate gills do not occur on abdominal segment II, but are present on segments III-VII, IV-VII or III-VI; a rudimentary gill may occur on segment I; cephalic horns short or lacking 4

4. Larvae with lamellate gills on abdominal segments III-VI or III-VII 5

   Larvae with lamellate gills on abdominal segments IV-VII, gill IV more-or-less operculate (covering more than half of gill V); maxillary palps lacking .... *Eurylophella*[26]

5. Larvae with lamellate gills on abdominal segments III-VI; gill III is operculate and completely covers the rest of the gill series; the head bears tubercles .... *Hyrtanella*

   Larvae with lamellate gills on abdominal segments III-VII; gill III *may* be semioperculate but does not cover the rest of the series completely; the head may or may not bear tubercles or spines 6

6. Femora of mid- and hindlegs greatly expanded and much broader than those of the forelegs; anterolateral corner of the pronotum projecting forward, and the anterior portion of the mesonotum is expanded .... *Cincticostella*

   Femora of forelegs either at least as broad or broader than those of the mid- and hindlegs; pronotum and mesonotum not projecting or expanded 7

[25] *Macafertiella*, a new genus of Teloganodinae from Sri Lanka, will key out here but, unlike *Teloganodes*, has gills on abdominal segment VI (Wang & McCafferty, 1996).

[26] Funk & Sweeney (1994) provide descriptions and a key to the known larvae of North American *Eurylophella* which indicate the features that are likely to be useful to separate Asian species of this genus.

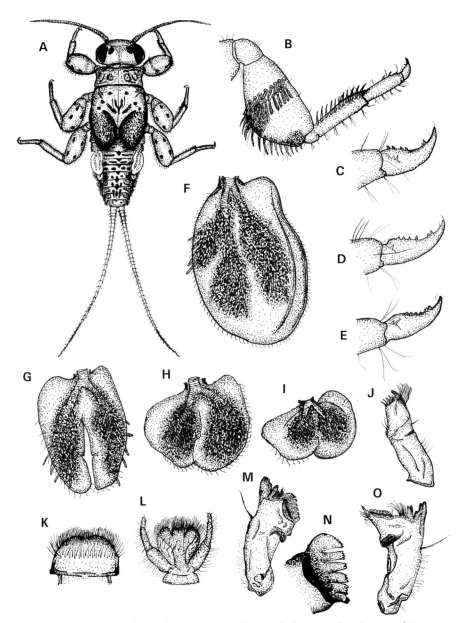

Fig. 4.53 *Teloganodes* (Ephemeroptera: Ephemerellidae): A, dorsal view of larva; B, foreleg; C, D & E, tarsal claws of fore-, mid- and hindlegs (respectively); F, G, H & I, gills II, III, IV & V (respectively); J, maxilla; K, labrum; L, labium; M & O, mandibles; N, molar area of left mandible (detail).

7. Body and appendages covered with long hair; body lacks tubercles; mandibles are asymmetrical .... *Crinitella*

   Body not covered with long hair; with tubercles or spines on the head, thorax or (most typically) the abdomen 8

8. Larvae with spines or tubercles on the head; forefemora usually expanded bearing spines along the anterior margin; body robust .... *Drunella*

   Head lacks spines or tubercles; forefemora not markedly expanded and without spines along the anterior margin 9

9. Larvae with an anterolateral spine or projection on the mesonotum .... *Ephacerella*

   Mesonotum lacks lateral spines or projections 10

10. Gill III is enlarged and semioperculate, covering most of the rest of the gill series (Fig. 4.54) .... *Torleya*

    Gill III is not enlarged or semioperculate 11

11. Cerci with whorls of setae at the end of each segment, and with lateral intersegmental setae; maxillary palps present *Ephemerella* Cerci with whorls of setae at the end of each segment, but lateral intersegmental setae are lacking; maxillary palps may be lost 12

12. Abdominal terga lack paired tubercles; dorsal surface of the body strongly marked with paired longitudinal stripes (Fig. 4.55A) .... *Uracanthella*

    Abdominal terga with paired tubercles; body *may* lack stripes .... *Serratella*

Cincticostella (= *Asiatella*) has been reported from Palaearctic Asia, including Japan and Korea (Allen, 1971; Tshernova, 1972; Bae *et al.*, 1994), as well as from Thailand (Allen, 1971). Tshernova (1972) has illustrated a larvae of *Cincticostella* (as *Asiatella*) which has a large prothorax with projecting anterior angles; there are lateral protuberances at the anterior margins of the expanded mesothorax also. The legs are unusual in that the femora of the second and third pairs are greatly expanded with spinous posterior margins, while those of the forelegs are more-or-less 'normal' (or considerably less expanded; see Allen, 1971). Paired spines are present on the abdominal tergites,

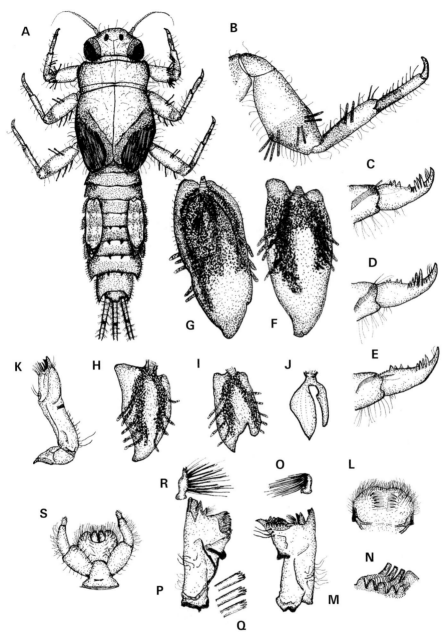

Fig. 4.54 *Torleya* (Ephemeroptera: Ephemerellidae): A, dorsal view of larva; B, foreleg; C, D & E, tarsal claws of fore-, mid- and hindlegs (respectively); F, G, H, I & J, gills III, IV, V, VI & VII (respectively); K, maxilla; L, labrum; M, right mandible; N, molar area of right mandible (detail); O, prostheca of right mandible (detail); P, left mandible; Q, molar area of left mandible (detail); R, prostheca of left mandible (detail); S, labium.

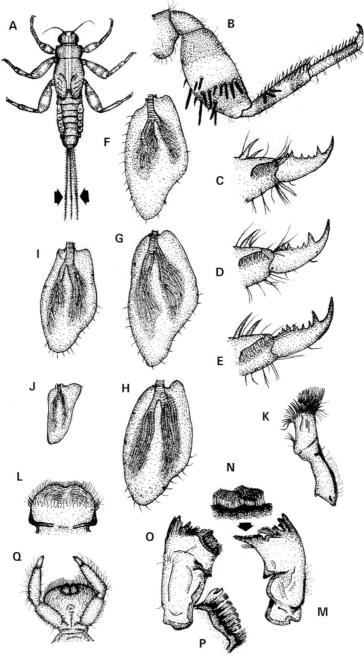

Fig. 4.55 *Uracanthella* (Ephemeroptera: Ephemerellidae): A, dorsal view of larva; B, foreleg; C, D & E, tarsal claws of fore-, mid- and hindlegs (respectively); F, G, H, I & J, gills III, IV, V, VI & VII (respectively); K, maxilla; L, labrum; M, right mandible; N, molar area of right mandible (detail); O, left mandible; P, molar area of left mandible (detail); Q, labium.

but there are no spines or tubercles on the head and this seems typical of most *Cincticostella* (Yoon & Bae, 1988). Nevertheless, Allen (1971) notes that some *Cincticostella* may have tubercles on the head, and that a maxillary palp may be present or lost. Diagnostic features of this genus seem to be the projecting anterolateral corners of the prothorax and the expanded anterior part of the mesothorax (figured by Allen, 1971; Yoon & Bae, 1988); the form of the legs seems to identify some, but not all, members of the genus (see Yoon & Bae, 1988). *Crinitella* occurs in Nepal, India and Peninsular Malaysia (Allen & Edmunds, 1963; Hubbard & Peters, 1978; Allen, 1980; Braasch, 1981f). Allen & Edmunds (1963) have illustrated the larvae. The body and appendages are covered with long hair, and there are no tubercles on the body although the abdominal terga are expanded laterally. The mandibles of *Crinitella* are asymmetric (the left has a reduced molar surface), and the tarsal claws have a palisade of short denticles.

Drunella (including the subgenera *Drunella* and *Tribrochella*) is widespread from Korea and Palaearctic Asia (Tshernova, 1972; Bae *et al.*, 1994) through Vietnam to Nepal and the Indian Himalaya (Allen & Edmunds, 1963; Allen, 1971; 1986; Edmunds & Peters, 1978; Braasch, 1981f). The body form is robust. Tshernova (1972) states that *Drunella* has paired median tubercles on all abdominal tergites but, based on drawings given by Allen (1971) and Braasch (1981f), the usual situation is for tubercles to be present on segments III-VII (although they may occur on segments I-IX). Allen (1980) characterizes larvae of the subgenus *Tribrochella* as having submedian ridges rather than tubercles on the abdomen. The forefemora of *Drunella* larvae are usually somewhat enlarged with a spiny anterior margin (Allen & Edmunds, 1963; Allen, 1971; Braasch, 1981f; Yoon & Bae, 1988). The tarsal claws of *Drunella* have rather few denticles (between 1 and 7, and usually less than 5), or may lack them entirely; if present, they are arranged in a single row (Allen & Edmunds, 1963; Allen, 1971; Braasch, 1981f). The head bears tubercles which can be sharp or blunt; their positioning, development and arrangement varies (e.g. Allen, 1971; Yoon & Bae, 1988). The thoracic nota *may* have tubercles. In Vietnamese *Drunella (Drunella)*, for example, paired tubercles are present on the head, pronotum and mesonotum, as well as on the abdomen (Allen, 1986). Variation within the genus is shown also by *Drunella gilliesi*, *D. uenoi* and *D. corpulenta* which lack the distinctively spined forefemora (Allen & Edmunds, 1963; Braasch, 1981f) that are typical of *Drunella*. Nevertheless, the robust body form and tubercles on the head and body clearly define these larvae as *Drunella* (Allen &

Edmunds, 1963). Based on the arrangement of the gills, the enlarged and spiny forefemora, the tarsal claws and paired abdominal tubercles, the larvae of *'Ephemerella' nasiri* described by Ali (1971) are probably *Drunella*. If so, the distribution of this genus extends into Pakistan.

Ephacerella (= Acerella) occurs in Japan and Korea (Allen, 1971; Bae *et al.*, 1994), China (You & Gui, 1995), Vietnam (Allen, 1986), Thailand (Allen, 1971; 1980) and Nepal (Hubbard & Peters, 1978). An anterolateral mesonotal spine or projection (which is not attached to the coxae as in certain *Drunella* species) characterizes larvae of this genus (Allen, 1971; 1986; Yoon & Bae, 1988). Paired dorsal tubercles occur on the abdomen, the head is with or without tubercles, and the spination of the forefemora is variable. Allen (1971) and Yoon & Bae (1988) provide illustrations of *Ephacerella* larvae.

Ephemerella (sensu stricto) is known from Korea, Japan (Allen, 1971; Bae *et al.*, 1994) and China (Ulmer, 1935; Su & You, 1989). It is essentially Palaearctic. Su & You (1989) give some information on the larvae which have paired submedian tubercles on abdominal terga III-VIII and distinct posterolateral projections on segments III-IX. According to the descriptions given by Allen (1971), *Ephemerella* larvae lack tubercles or spines on the head, and they are often (but not always) absent from the thoracic nota. In addition, the forefemora have a subapical band of stiff bristles (thin spines), and the tarsal claws have 5–10 denticles. The number of abdominal segments with paired tubercles or protuberances varies somewhat among species, but they may be present on segments I-IX and are usually seen on (at least) segments III-VII (Allen, 1971). *Serratella* occurs in Korea (Bae *et al.*, 1994), Thailand (Allen, 1980) and China (Xu *et al.*, 1980; 1984) — including Hong Kong (Dudgeon, 1990b; see also Fig. 4.56). The larvae are similar to *Ephemerella*. According to Yoon & Bae (1988), who give descriptions of both genera, *Serratella* larvae have paired tubercles on the abdominal terga, and whorls of setae at the end of each segment of the cerci. However, *Ephemerella* has lateral intersegmental setae while *Serratella* does not. In addition, *Serratella* may lack maxillary palps, but these are always present in *Ephemerella*. *Teloganopsis*, which seems to be confined to Sumatra and Java (Ulmer, 1939; Edmunds & Polhemus, 1990), was previously treated as a subgenus of *Ephemerella* (Allen, 1971), although this is no longer the case (Allen, 1984; Hubbard, 1990). The larvae are figured by Ulmer (1939), but I have been unable to find a satisfactory point of inclusion for this genus in the ephemerellid key given above.

Torleya is primarily Palaearctic but occurs at least as far south as

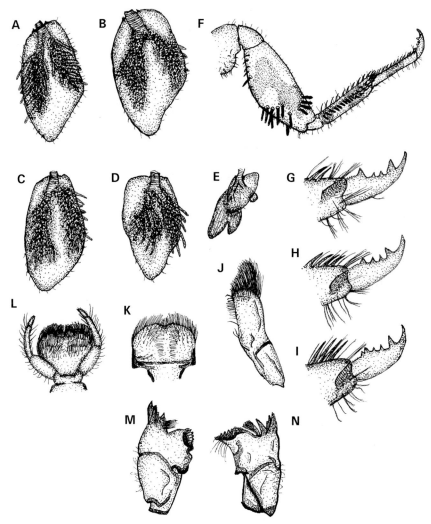

Fig. 4.56 *Serratella* (Ephemeroptera: Ephemerellidae): A, B, C, D & E, gills III, IV, V, VI & VII (respectively); F, foreleg; G, H & I, tarsal claws of fore-, mid- and hindlegs (respectively); J, maxilla; K, labrum; L, labium; M & N, mandibles.

Nepal (Allen, 1980; Braasch, 1981f; Tiunova, 1995). Allen & Edmunds (1963) describe the larvae as having semioperculate abdominal gills and long, hair-like spines on the cerci; paired occipital tubercles *may* be present but short, paired spines on the posterior margins of the abdominal terga (at least on segments IV-VII) appear typical. Illustrations of larvae given by Allen & Edmunds (1963) and Tiunova (1995) indicate that the maxillary palp may be absent or present (but reduced to two segments) and occipital tubercles likewise may be developed or lacking, and ephemerellid larvae from Hong Kong have characteristics typical of *Torleya* (Fig. 4.54). Ali (1971) describes the larva of *'Ephemerella' wahensis* from Pakistan which resembles *Torleya* and lacks maxillary palps.

Uracanthella is known from Palaearctic Asia, through Peninsular Malaysia to Borneo (Edmunds & Polhemus, 1990; Bae *et al.*, 1994). *Uracanthella* was raised to the level of genus by Allan (1984), and although this ranking is not recognized by Hubbard (1990) it has been adopted by others (Edmunds & Polhemus, 1990; Bae *et al.*, 1994). Based on a description of *Uracanthella rufa* given by Yoon & Bae (1988), larvae of this genus resemble *Serratella* and *Ephemerella* but lack abdominal tubercles and (at least in the case of *U. rufa*) their bodies are distinctively marked with paired longitudinal stripes (see Fig. 4.55A). Note that the placement of *Uracanthella rufa* in *Serratella* by Yoon & Bae (1988) was corrected by Bae *et al.* (1994). Hong Kong ephemerellids assigned to *Serratella* by Dudgeon (1990b; 1992a) may be *Uracanthella* also (Figs 4.55 & 4.56).

Vietnamella occurs in Vietnam and China (Tshernova, 1972; You & Su, 1987; Wang & McCafferty, 1995b; You & Gui, 1995). Following, Allen (1980), You & Gui (1995) treat *Vietnamella* as a subgenus of *Cincticostella*, but this classification was abandoned by Allen (1984) who reinstated the genus *Vietnamella* and placed it within its own subtribe (Vietnamellae). The cephalic horns of larval *Vietnamella* are diagnostic (Wang & McCafferty, 1995b) and there may be as many as three pairs; the pair on either side of the antennae is especially well developed (Tshernova, 1972; You & Su, 1987). Six pairs of gills are present on segments II-VII — an unusual combination among ephemerellid larvae. The forefemora are broad with a large proximal tooth, and the anterior angeles of the pronotum are elongated into spinous cusps. Tshernova (1972) gives a larval diagnosis and Tshernova (1972) and You & Su (1987) figure *Vietnamella*. *Hyrtanella* is known from Borneo (Sabah), Peninsular Malaysia, Thailand and southern India (Allen, 1980; Edmunds & Polhemus, 1990); a description and larval diagnosis are given by Allen & Edmunds (1976). The head with frontal and occipital tubercles and the operculate gill on segment III seem sufficient to separate *Hyrtanella* from other Asian ephemerellids.

Teloganodes is distinctive among Asian ephemerellids because it lacks a terminal filament, and maxillary palps, but has expanded femora (Fig. 4.53). A newly discovered Sri Lankan genus — *Macafertiella* — has similar characteristics but has an additional pair of gills on abdominal segment VI (Wang & McCafferty, 1996). *Teloganodes* is widespread, occurring in Sri Lanka, India, Nepal, China (Hong Kong), Peninsular Malaysia, the Philippines and through the East Indies to Sulawesi (Ulmer, 1940; Hubbard & Pescador, 1978; Hubbard & Peters, 1978; Braasch, 1981f; Edmunds & Polhemus, 1990; Dudgeon, 1990b). *Teloganodes* is the only ephemerellid to have reached Sulawesi. Ulmer (1940) has figured the larva. Like *Teloganodes*, *Ephemerellina* larvae have lamellate gills on segments II-V (gill I is rudimentary) or on segments II-VI (in which case the rudimentary gill I is absent) and lack maxillary palps. In addition, Allen & Edmunds (1963) characterize *Ephemerellina* larvae (which they figure) as having a head and thorax lacking tubercles, and tarsal claws with a double row of denticles (each 6–10 denticles). There are paired abdominal tubercles on segments I-VII, which are longest on segments IV and V, but the tubercles may be single in some species. There are short spines at the apex of each segment of the cerci.

Caenidae

The Caenidae is widespread in tropical and Palaearctic Asia. *Caenis* is almost cosmopolitan (excluding Australia), *Brachycerus* is mainly Palaearctic (e.g. Korea; but with a single species in Sri Lanka), *Caenodes* occurs throughout Southeast Asia and most of Indonesia to Bali, *Caenomedea* is known from Sulawesi and New Guinea, and *Tasmanocoenis* (= *Pseudocaenis*) occurs in New Guinea and Australia (Ulmer, 1940; Thew, 1960; Hubbard & Pescador, 1978; Soldán, 1978a; Edmunds & Polhemus, 1990; Soldán & Landa, 1991; Bae *et al.*, 1994). *Clypeocaenis* has been reported from India, Sri Lanka and Vietnam (Soldán, 1978a; 1983; Soldán & Landa, 1991), *Cercobrachys* from Thailand, and *Caenoculis* from Peninsular Malaysia and Vietnam (Soldán, 1986). As in the Baetidae and Ephemerellidae, the adults have few distinguishing characteristics, so studies of caenid classification must be based mainly on larval morphology (Thew, 1960; Soldán, 1986).

Caenids are small mayflies, typically less than 10 mm body length. The adults lack hind wings, and parthenogenetic reproduction is quite common. Where they are present, males are significantly smaller than females, and mating flights of *Caenis* and *Clypeocaenis* take place just before dawn almost immediately after emergence (Soldán, 1983). The larvae occur in a wide variety of habitats (Soldán, 1978a; 1986), including standing water, and are quite tolerant of organic pollution (Dudgeon, 1990b). Most species are probably collector-gatherers, but *Clypeocaenis* larvae filter-feed by means of setae on the legs and mouthparts (Soldán, 1983; Provonsha & McCafferty, 1995). The specialization of the legs and mouthparts of *Clypeocaenis* to strain food from the current is nearly identical to, and convergent with, that seen in the Oligoneuriidae (e.g. *Isonychia*). Many caenids inhabit silty substrata, and their distribution may reflect an adaptation of the morphology of the abdominal gills and the way that they beat. The beating of gills draws a current of water over the body of some mayflies, bringing it in from the front and sometimes the side also. Caenids have a reduced first gill (Fig. 4.57D), and the second forms an robust operculum (Fig. 4.57E) overlying the abdomen and succeeding gills which are thin and delicate (Figs 4.57F-H). When they beat, these gills cause a current from side-to-side across the body. If the animal rests on the substratum with one side slightly raised, it can draw water in from that side without disturbing the silt (Hynes, 1970). *Cercobrachys* appears to be particularly specialized for life in silty habitats: the body bears many long bristles, and the lamellate gills are protected and enclosed beneath the gill covers within a 'basket' of recurved, lateral abdominal spines (Soldán, 1986).

The Caenidae has received little attention in Asia (Kang & Yang, 1994e). For example, only a single species (*Caenis nigropunctata*[27]) is known from mainland China (Hsu, 1936a; You & Gui, 1995), but the family is well represented in Hong Kong (Dudgeon, 1990b) and Kang & Yang (1994e; 1996b) have described and illustrated eight species of *Caenis* larvae from Taiwan. Two undescribed species of *Caenodes* and one *Caenis* from Hong Kong are shown in Figs 4.57–4.59. Elsewhere in Asia, Soldán (1983) gives a key to the known species of *Clypeocaenis* and figures the larvae (see also Soldán, 1978a; Soldán & Landa, 1991), and Soldán (1986) provides illustrations of the larvae of *Cercobrachys* and *Caenoculis*. Soldán & Landa (1991) have figured the larvae of the only Oriental species of *Brachycerus* which was recorded from Sri Lanka. Malzacher (1984; 1992) and Alba-Tercedor & Zamora-Munoz (1993) give keys to the European and Ethiopian species of *Caenis* and indicate the morphological characters which might be used to distinguish among Asian species. A key to Taiwanese *Caenis* larvae is provided by Kang & Yang (1994e). Malzacher (1991; 1993) believes that the generic classifications of Thew (1960) and Soldán (1978a) need modification, and has proposed the synonymy of *Caenodes* and *Caenomedea* with

---

[27] Thew (1960) is of the opinion that this species should be placed in the genus *Caenodes*.

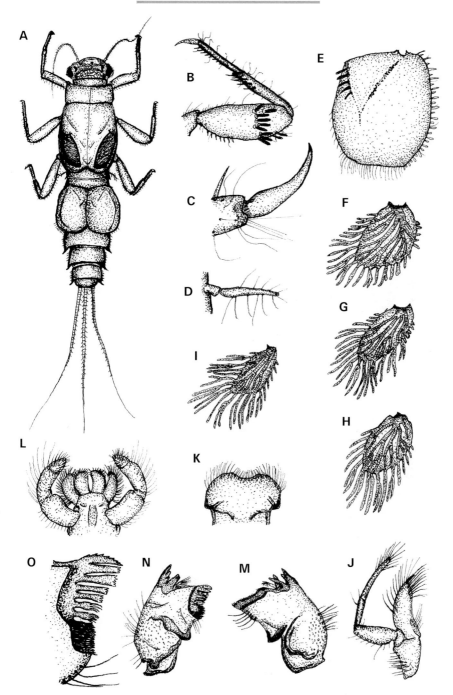

Fig. 4.57 *Caenodes* sp. 1 (Ephemeroptera: Caenidae): A, dorsal view of larva; B, foreleg; C, tarsal claw of foreleg; D, E, F, G, H & I, gills I, II, III, IV, V & VI (respectively); J, maxilla; K, labrum; L, labium; M & N, mandibles; O, molar area of left mandible (detail).

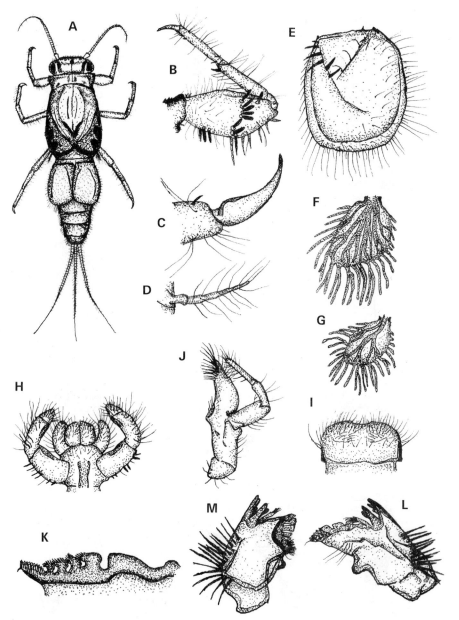

Fig. 4.58 *Caenodes* sp. 2 (Ephemeroptera: Caenidae): A, dorsal view of larva; B, foreleg; C, tarsal claw of foreleg; D, E, F & G, gills I, II, III & VI (respectively); H, labium; I, labrum; J, maxilla; K, molar area of right mandible (detail); L & M, mandibles.

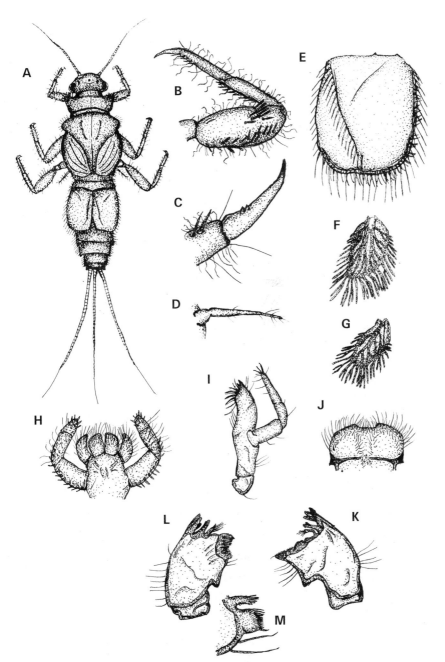

Fig. 4.59 *Caenis* (Ephemeroptera: Caenidae): A, dorsal view of larva; B, foreleg; C, tarsal claw of foreleg; D, E, F & G, gills I, II, III & VI (respectively); H, labium; I, maxilla; J, labrum; K & L, mandibles; M, lower edge of molar area on left mandible (detail).

Caenis. Unfortunately, the application of correct names within the family remains problematic (e.g. Edmunds & Polhemus, 1990).

The following key is derived mainly from Soldán (1978a; 1986), and his view of caenid classification (which agrees with the catalogue given by Hubbard, 1990) has been adopted. Because caenids are small, slide mounting of mouthparts and gill covers will facilitate identification.

1. Head with ocelli on raised tubercles 2

   Head lacks ocular tubercles (although ocelli are present) 4

2. Femora wide (three times wider than the tibiae); fore coxae nearly contiguous; maxillary and labial palps with three segments; posterolateral spines on abdominal segments IV-VII .. *Caenoculis*

   Femora narrow (no more than twice as wide as the tibiae); fore coxae widely separated (and the prosternum is much wider than long); maxillary and labial palps with two segments; posterolateral spines on abdominal segments II-VII, III-VI or III-VIII 3

3. Posterolateral spines on abdominal segments II-VII or III-VI, those on segment VI are strongly bent medially (i.e. towards the midline); the anterior margin of the mesosternum bears numerous long bristles; legs with long bristles; second antennal segment (the pedicel) approximately the same length as the basal one (the scape) ....
   .... *Cercobrachys*

   Posterolateral spines on abdominal segments III-VIII, those on segment VI not bent medially although spines of segments III-V overlap slightly; the anterior margin of the mesosternum lacks bristles; bristles on legs not unusually long; pedicel of antenna approximately four times longer than the scape; Sri Lanka only
   .... *Brachycerus*

4. Foretibiae with two transverse rows of filtering setae; long hairs protruding from the front of the head (the clypeus) between and below the antennae; gill covers with a simple ridge; maxillary palps with two segments .... *Clypeocaenis*

   Forelegs with scattered setae and hairs; head lacks long hairs; gill covers *may* have a triangular or Y-shaped ridge; maxillary palps with three segments 5

5. Mandibles lack lateral marginal setae (but maxillae labium and labrum are hirsute); third segment of labial palp short (about quarter

the length of the second segment); may be confined to New Guinea and Sulawesi 6

Mandibles with lateral marginal setae; third segment of labial palp at least one third the length of the second; common and widespread 7

6. Gill covers with a triangular ridge and marginal fringe of setae; Sulawesi and New Guinea only .... *Caenomedea*

Gill covers lack a triangular ridge and a marginal fringe of setae (only a few scattered setae along the margin); New Guinea (and Australia) .... *Tasmanocoenis*

7. Gill cover with stout spines on the mesal fork of the triangular ridge (i.e. the fork closest to the midline: Figs 4.57E & 4.58E); submarginal spines lacking but a marginal fringe of hairs is present .... *Caenodes*

Gill cover lacks stout spines (but not setae: Fig. 4.59E) on the upper surface, but a row of submarginal spines is present .... *Caenis*

Ephemeridae

The mayflies with mandibular tusks (Fig. 4.60) constitute the superfamily Ephemeroidea which consists of the families Ephemeridae, Polymitarcyidae and Potamanthidae. The composition of these families has recently been changed to reflect current views of phylogenetic relationships within the superfamily (McCafferty, 1991a). Of relevance here is the reduction of the Palingeniidae to subfamily rank — the Palingeniinae — within the Ephemeridae. As it is now recognized, the Ephemeridae has an almost cosmopolitan distribution and, by comparison with other geographical regions, the number of Asiatic species is high (McCafferty, 1973). They are arranged in three subfamilies: the Ephemerinae (consisting of *Ephemera* and *Afromera*), the Hexageniinae (represented by *Eatonigenia*), and the Palingeniinae (*Anagenesia* and *Plethogenesia*). All have mandibular tusks and frontal processes for burrowing, but the development of these two features varies considerably (Fig. 4.60).

Ephemera (Fig. 4.61) is the most speciose and widely distributed ephemerid genus, ranging from Korea and Japan (Yoon & Bae, 1985;

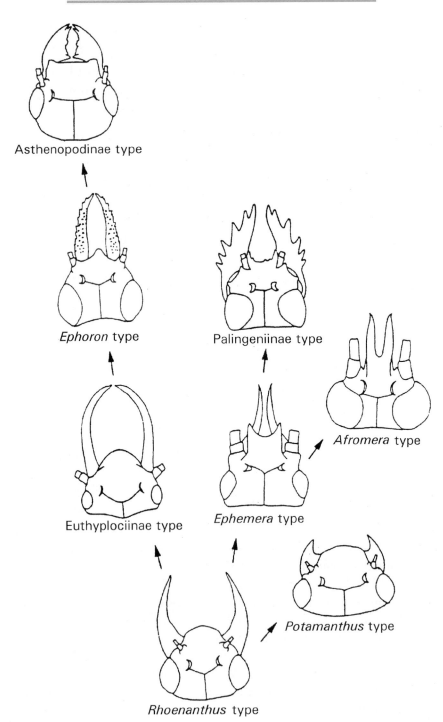

Asthenopodinae type

Ephoron type

Palingeniinae type

Afromera type

Euthyplociinae type

Ephemera type

Potamanthus type

Rhoenanthus type

Fig. 4.60 General trends in the evolution of tusk form in Asian Ephemeroidea (after Bae & McCafferty, 1991).

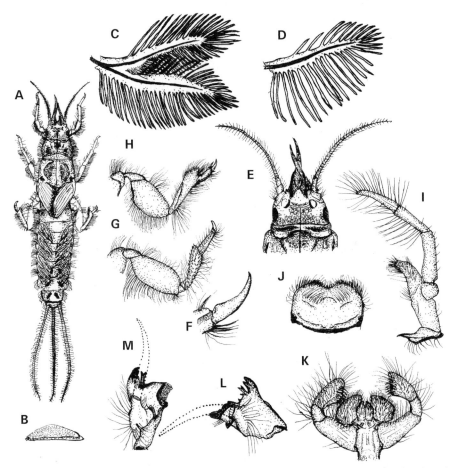

Fig. 4.61 *Ephemera (Ephemera) spilosa* (Ephemeroptera: Ephemeridae): A, dorsal view of larva; B, C & D, gills I, II & VII (respectively); E, dorsal view of the larval head showing projecting mandibular tusks; F, tarsal claw of hindleg; G, foreleg; H, hindleg; I, maxilla; J, labrum; K, labium, L & M, mandibles (with tusks omitted for clarity).

Bae *et al.*, 1994) through China (Hsu, 1936a, 1936c, 1936d; 1937a), Hong Kong (Hsu, 1936b; Dudgeon, 1990b; 1996a), Taiwan (Kang & Yang, 1994f), the Southeast Asian mainland, the Philippines, Java, Sumatra and India (Ulmer, 1940; McCafferty, 1973; McCafferty & Edmunds, 1973; Edmunds & Polhemus, 1990). The genus is in need of complete revision because '. . . there has always been an obvious lack of consistency in species descriptions of Ephemeridae in general, making species comparisons difficult . . .' (Hubbard, 1983). Two subgenera of *Ephemera* are recognized: *Ephemera* and *Aethephemera*. They can be distinguished easily by virtue of the strongly asymmetric mandibular tusks of *Aethephemera* and the generally larger size of these larvae (for a description, see McCafferty & Edmunds, 1973). The tusks of *Ephemera (Ephemera)* are subequal (Fig. 4.61E). Although the subgenus *Aethephemera* was established initially on the basis of larval characteristics only (McCafferty & Edmunds, 1973; Hubbard, 1983), the features of the adults are now known (Balasubramanian *et al.*, 1991). Illustrations of the larvae of *Ephemera (Ephemera)* from Asia are given by McCafferty & Edmunds (1973), Yoon & Bae (1985), Kang & Yang (1994f) and Bae (1995).

The other Asia genus of Ephemerinae — *Afromera* — is known from Thailand and India. They were originally placed in a subgenus of *Ephemera* — *Dicrephemera* — by McCafferty & Edmunds (1973), but this subgenus has since been synonymised with the predominately African genus *Afromera* (Hubbard, 1990). The larvae are characterized by a well-developed, bifurcate frontal process on the head (larger than those of other ephemerids) combined with small, somewhat atrophied, mandibular tusks. In addition, abdominal gill I, which consists of two equal lobes (or forks) in *Ephemera* and *Aethephemera* is markedly asymmetrical in *Afromera* with a large outer lobe. Other details and illustrations are given by McCafferty & Edmunds (1973). The Hexageniinae consists of *Eatonigenia* which is confined to tropical Asia (McCafferty, 1991b). It contains seven species, some of which had been placed incorrectly within the genera *Heterogenesia* and *Hexagenia* (Zhang, 1988; McCafferty, 1991b). *Eatonigenia* occurs on Hainan Island (Zhang, 1988), the Philippines (as *Hexagenia philippina*: Hubbard & Pescador, 1978), Java, Borneo, Thailand, Vietnam and India (McCafferty, 1973; 1991b). The Vietnamese record is that of Dang (1967) who described what he believed was a new palingeniid genus — *Heterogenesia* — based on the anterior wing venation and structure of the larval mandibles. However, McCafferty (1991b) has synonymised *Heterogenesia* with *Eatonigenia*. *Eatonigenia* larvae are described and illustrated by McCafferty (1973): they have reduced tarsal claws on the prolegs; mandibular tusks which are triangular (not round) in cross-section and almost 'knife-like'; a highly convex labrum; and, a markedly truncate frontal process on the head. All records of the Holarctic genus *Hexagenia* from tropical Asia appear to be misidentifications of *Eatonigenia* (McCafferty, 1973; 1991b), and it is unlikely that *Hexagenia* occurs in the Oriental region.

Asian members of the Palingeniinae are represented by *Anagenesia* — which ranges from China over much of tropical Asia through Indonesia to Sulawesi (Hubbard & Peters, 1978) — and *Plethogenesia* which is known from New Guinea (Hsu, 1936a; Edmunds & Polhemus, 1990; Dudgeon, 1990b). The larva of *Plethogenesia papuana* has been described in detail by Ulmer (1940). The Palaearctic and Middle Eastern (Iranian) genus *Palingenia* has been reviewed by Soldán (1978b) who gives descriptions of the larvae where they are known. Although it may occur in the north of the region, *Palingenia* has yet to be reported from tropical Asia, and it is not included in the following key to ephemerid larvae.

1. Tusks smooth and slender (or rather reduced and atrophied); gill I doubled (although the two forks or lobes may be of markedly different sizes) 3

   Broad, flat tusks with outer edge notched ... **Palingeniinae** 2

2. Tusks with distal prongs and deep indentations along the outer margins; New Guinea only .... *Plethogenesia*

   Tusks with shallow indentations along outer margins; widespread .... *Anagenesia*

3. Frontal process on the head reduced, rather inconspicuous and not bifid; mandibular tusks are triangular in cross-section; tarsal claws on forelegs reduced relative to those of the fore- and hindlegs .... **Hexageniinae** .... *Eatonigenia*

   Frontal process on the head well developed and bifid; mandibular tusks are circular in cross-section (or are atrophied); tarsal claws on forelegs not reduced .... **Ephemerinae** 4

4. Mandibular tusks reduced or atrophied; abdominal gill I asymmetrical with a relatively large outer lobe .... *Afromera*

   Mandibular tusks well-developed (but may be asymmetric); abdominal gill I with two lobes of roughly equal sizes .... *Ephemera* 5

5. Mandibular tusks of approximately the same length.................. .... subgenus *Ephemera*

   Mandibular tusks markedly asymmetrical (the right is shorter) .... subgenus *Aethephemera*

All ephemerids burrow, using the tusks and frontal processes (which are especially important in *Afromera*: Bae & McCafferty, 1995) together with the powerful legs to dig and maintain the burrow. Within the genus *Ephemera*, there is variation in behaviour, with some species simply dwelling interstitially and others forming a well-defined burrow (Edmunds & McCafferty, 1996). Most ephemerids ingest fine detritus using the labium (and its setae) to collect food that is carried into the burrow on water currents generated by beating the abdominal gills (Ladle & Radke, 1990). In general, they burrow into sand (Ephemerinae) or silt and mud (Hexageniinae and Palingeniinae), but one species of *Palingenia* excavates wood (Edmunds & McCafferty, 1996) — a behaviour which is otherwise confined to the polymitarcyid subfamily Asthenopodinae (see below).

Investigations of ephemerid ecology in Asia have been restricted mainly to the genus *Ephemera* in Japan and Korea (e.g. Gose, 1970a; Kuroda *et al.*, 1984; Watanabe, 1985; Watanabe & Kuroda, 1985; Ban & Kawai, 1986; Ban, 1988; Lee *et al.*, 1995). Papers by Balasubramanian *et al.* (1992) and Dudgeon (1996a) reporting upon work undertaken in India and Hong Kong are the only exceptions. Species of *Ephemera (Ephemera)* in Palaearctic Asia and Hong Kong are univoltine, although the emergence and larval recruitment periods may be prolonged. The rate of growth can vary with stream temperature, being more rapid downstream, and this may be the basis for the observed longitudinal zonation of different *Ephemera (Ephemera)* species in Japanese and Korean rivers (Kuroda *et al.*, 1984; Watanabe, 1985; Watanabe & Kuroda, 1985; Ban & Kawai, 1986; Bae, 1995; Lee *et al.*, 1995). The same pattern of longitudinal replacement occurs between the two Taiwanese *Ephemera (Ephemera)* species (Kang & Yang, 1994f). No unequivocal evidence of bivoltinism has been found in these mayflies (Dudgeon, 1996a). *Ephemera (Aethephemera) pictipennis* in Hong Kong is univoltine with highly synchronised emergence, but *Ephemera (Aethephemera) nadinae* in southern India lacks the clear seasonal cycle of development seen in Hong Kong or in Japanese ephemerids, and larval growth is asynchronous with adults emerging throughout the year (Balasubramanian *et al.*, 1992). This ephemerid may be multivoltine, but the available size-frequency data (Balasubramanian *et al.*, 1992: p. 74) merely display asynchronous growth and give no hint as to voltinism. In other words, Indian *Aethephemera* could be univoltine (as in Hong Kong) with asynchronous larval growth and development.

Polymitarcyidae

A recent reclassification of the Polymitarcyidae (McCafferty, 1991a) has led to the inclusion of the Euthyplociidae (now Euthyplociinae) as a subfamily. It is represented by in Asia *Polyplocia* from India and Borneo (Edmunds & Polhemus, 1990) and *Mesoplocia* from Vietnam, the larvae of which are described by Soldán & Braasch (1986) as sprawlers on sandy sediments. The habits of both of these genera are obscure, but Edmunds & McCafferty (1996) believe that some Euthyplociinae may burrow and feed within the stream substrate using filtering setae on the mandibular tusks. Other polymitarcyids are burrowers: they include the Polymitarcyinae — represented by *Ephoron* (*Polymitarcys* and *Eopolymitarcys* are junior synonyms) — and the Asthenopodinae. You & Gui (1995) include two species of *Ephoron* among the Chinese mayfly fauna (recorded as *Polymitarcys* by Hsu, 1936a). *Ephoron* occurs in India, Sri Lanka, Burma, Vietnam, Java, Sumatra and Borneo (Hubbard & Peters, 1978), although there are rather few records from the Oriental Region, and it is known also from Korea (Yoon & Bae, 1985; Bae *et al.*, 1994). Yoon & Bae (1985) have illustrated a Korean *Ephoron* larva which is characterized by mandibular tusks with tubercles and setae (Fig. 4.60), and a single gill on abdominal segment I. The gills on segments II-VII are forked. Larval behaviour can be flexible and, according to circumstances, a single species may be an interstitial dweller or construct U-shaped burrows; some *Ephoron* have also been reported as burrowing along clay-rock interfaces — in such cases the burrow form is variable (Bae & McCafferty, 1995; Edmunds & McCafferty, 1996).

The pantropical genus *Povilla* is the only member of the Asthenopodinae in Asia. The larvae burrow into wood, a behaviour that — among mayflies — is almost entirely confined this group (Edmunds & McCafferty, 1996). *Povilla* can be separated from *Ephoron* by the form of the tusks. In both genera, the tusks are curved inwards, with rather smooth inner edges (Fig. 4.60). In *Ephoron*, the tusks are slender with setae and small tubercles. *Povilla* has broad and flat tusks which are strongly toothed close to the apex. *Povilla* occurs in India, Sri Lanka, Thailand, Cambodia, Borneo, Sumatra and Java (Ulmer, 1940; Gillies, 1951; Hubbard, 1984; Edmunds & Polhemus, 1990). Ulmer (1940) gives a detailed description of the larva. More recently, Hubbard (1984) has revised the genus, establishing two subgenera (*Povilla* and *Languidipes*) and providing an illustrated key to the larvae. Both subgenera of *Povilla* are widespread. The mandibular tusks of the subgenus *Languidipes* are toothed close to the apices of the inner margin only, and the first abdominal gill is uniramous. The tusks of *Povilla (Povilla)* larvae have at least one tooth on the outer margin, and the first abdominal gill is biramous. Hubbard (1984) has noted an interesting sexual dimorphism in *Povilla* larvae. The terminal filament of males is shorter than the cerci and bears many small spines but, in females, the terminal filament is longer than the cerci and all three caudal filaments are heavily spinose.

Potamanthidae

The Potamanthidae comprises three genera — *Rhoenanthus*, *Anthopotamus* and *Potamanthus* — of which *Rhoenanthus* and *Potamanthus* occur in tropical Asia (Bae & McCafferty, 1991). *Rhoenanthus* includes the subgenera *Rhoenanthus* and *Potamanthindus*, while *Potamanthus* has three subgenera: *Potamanthus*, *Stygifloris* and *Potamanthodes* (Bae *et al.*, 1990; Bae & McCafferty, 1991). Many of these subgenera have been treated as valid genera in the literature (e.g. Hsu, 1937b; You *et al.*, 1980b; 1982b; You, 1984; You & Su, 1987; You & Gui, 1995), and other genera of doubtful validity (e.g. *Neopotamanthus, Neopotamanthodes* and *Rhoenanthopsis*: Hsu, 1937c; Wu & You, 1986b) are still accepted by some workers (You & Gui, 1995). In addition, You & Gui (1995) continue to treat *Potamanthellus* — which is actually a neoephemerid (Hubbard, 1990; Bae & McCafferty, 1991) — as a member of the Potamanthidae. In consequence, much confusion has surrounded the correct naming of potamanthids. A recent revision of the family by Bae & McCafferty (1991) has clarified the situation greatly, and these authors present a detailed account of the potamanthid phylogeny and geographic distribution. *Rhoenanthus (Rhoenanthus)* occurs in India, Vietnam, Thailand, Malaysia and Indonesia (but does not reach Sulawesi), while *Rhoenanthus (Potamanthindus)* ranges from Korea through China to Vietnam and Thailand. *Potamanthus (Potamanthus)* is predominately Palaearctic but penetrates at least as far south as Anhui Province in China. *Potamanthus (Stygifloris)* is confined to Borneo (Sabah), and *Potamanthus (Potamanthodes)* extends from the eastern Palaearctic through southern China and Taiwan to Vietnam, Laos, Cambodia, Thailand, Burma and the Peninsular Malaysia (Bae & McCafferty, 1991).

Although they have flattened bodies, larvae of all potamanthid genera burrow in interstitial sediments (Bae & McCafferty, 1991; 1994; 1995). They are relatively crude burrowers, using their tusks to excavate the substrate and their gills to generate an interstitial current (Bae & McCafferty, 1994). This mode of burrowing occurs despite the varied development of tusks in the group, which includes a degree of reduction in the genus *Potamanthus* (Bae & McCafferty, 1995; see also Fig. 4.60). Some species — including certain *Potamanthus (Potamanthodes)* and *Rhoenanthus (Rhoenanthus)* in Asia — burrow in silt at the interface of rocks, probably because they lack the adaptations to move around within the substrate unless they can obtain a firm purchase on a rock surface (Edmunds & McCafferty, 1996). Potamanthid burrows are never U-shaped, in contrast to larval Ephemeridae which can make more 'advanced' U-shaped burrows that are usually independent of rocks (Bae & McCafferty, 1995; Edmunds & McCafferty, 1996). It seems that ephemeroid mayflies with flattened bodies (e.g. the Euthyplociinae, Polymitarcyinae and Potamanthidae) are relatively 'primitive', unspecialized burrowers (Bae & McCafferty, 1995), although wood-excavating *Povilla* (Polymitarcyinae) seems to be an exception to this generalization.

All potamanthids are filter-feeders (McCafferty & Bae, 1992). They are able to utilise both active deposit filter-feeding and passive seston filter-feeding in their interstitial habitat (McCafferty & Bae, 1992) although the relative importance of each feeding mode may vary among species (Elpers & Tomka, 1994). Most feeding takes place at night (Meier, 1980). The filtering organs are long, pectinate setae on the forelegs and palps, and detritus dominates the diet (Bae & McCafferty, 1992). Little is known of potamanthid life histories, but Watanabe (1988; 1989) has reported bivotinism in Japanese *Potamanthus (Potamanthodes)*, with adults emerging throughout the night. Bae & McCafferty (1994) speculate that tropical Asian potamanthids may be multivoltine with year-round emergence.

The larvae of all potamanthid genera and subgenera have been illustrated by Bae & McCafferty (1991), who also provide a key for the identification of known larvae. The larval stages of *Rhoenanthus (Anthopotamus) coreanus* (= *rohdendorfi*) from Palaearctic Asia have also been described by Yoon & Bae (1985) and Tiunova (1992), while *Potamanthus (Potamanthodes) yooni* is figured by Yoon & Bae (1985; misidentified as *Potamanthus kamonis*) and Bae & McCafferty, 1991). *Rhoenanthus (Rhoenanthus) speciosus* larvae from Sumatra were first illustrated by Ulmer (1940), and Bae *et al.* (1991) give the original diagnosis and illustrations of *Stygifloris* (now a subgenus of

Potamanthus) from Sabah. Kang & Yang (1994f) have figured the larvae of a *Potamanthus (Potamanthodes)* species from Taiwan. Illustrations of the larval stages of potamanthids from the Chinese mainland are provided by Yu & Su (1987) for *Potamanthus (Potamanthodes) kwangsiensis* (referred to in their paper by the junior synonym of *Potamanthodes fujianensis*); by Wu & You (1986b) for *Rhoenanthus (Potamanthindus) youi* (albeit incorrectly treated as a new genus, *Neopotamanthus*); and, by Wu (1987) for *Potamanthus (Potamanthus) huoshanensis*. The following key to genera and subgenera is derived from Bae & McCafferty (1991) to which the reader is referred for further information.

1. Mandibular tusks roughly equal to — or longer than — the length of the head .... *Rhoenanthus* 2

   Mandibular tusks shorter than one half of the length of the head .... *Potamanthus* 3

2. Mandibular tusks with a large lateral apical spine (so that the tusks appear forked); maxillary palp slender with weakly-developed setae on the terminal segment .... subgenus *Rhoenanthus*

   Mandibular tusks lack a large lateral apical spine (not appearing apically forked), but sometimes have a small lateral subapical spine; maxillary palp thick, with strongly-developed setae on the terminal segment .... subgenus *Potamanthindus*

3. A distinct dorsal row of well-developed stout setae running transversely across the middle of the forefemora; foretibiae with a ventral, subapical tuft of hair-like setae; Oriental and eastern Palaearctic 4

   Forefemora lacking a dorsal transverse row of setae (a small cluster of setae may be present); foretibiae without a subapical tuft of setae; mainly Palaearctic .... subgenus *Potamanthus*

4 Terminal segment of labial palp symmetrically pointed; forewing pads of older larvae lack obvious markings; widespread in Asia .... subgenus *Potamanthodes*

   Terminal segment of labial palp falcate (somewhat hooked or sickle-shaped with a distinct apicomedial margin); forewing pads of older (but not final-instar) larvae each have a distinct median spot; Borneo only ... subgenus *Stygifloris*: *Potamanthus (Stygifloris) sabahensis*

ODONATA

Members of the Odonata — the dragonflies and damselflies (collectively termed dragonflies) — are conspicuous insects that are often large and colourful. As a result they have been collected extensively, and adults of the group are quite well known although larvae are less familiar. The adult stages have been well served by regional monographs and identification guides, among them Needham (1930: China), Fraser (1933; 1934; 1936; India), Askew (1988: Europe), Zhao (1990: China; Gomphidae only), Watson *et al.* (1991; Australia), Hirose & Itoh (1993: Japan), Wilson (1995a: Hong Kong). A great deal of work on the Asian fauna has also been undertaken by Syoziro Asahina (e.g. Asahina, 1993; see also Inoue & Eda, 1984) and M.A. Lieftinck (see van Tol, 1992). Although new species continue to be described from the region (e.g. Zhu, 1991; Zhu & Han, 1992), the taxonomic composition and geographic distribution of the Odonata are well known (see, for example, Prinratana *et al.*, 1988; Tsuda, 1991), especially when compared to other aquatic insect groups. Nevertheless, some taxonomic problems remain and considerable confusion surrounds the higher classification of the group. For example, the dragonfly families Corduliidae and Libellulidae intergrade, and there appears to be no way that the larvae can be reliably categorised at family level. In addition, some workers have treated the Macromiidae as a separate family (e.g. Nel & Paicheler, 1994) — an opinion which I have adopted (Table 4.6) — but it is often viewed as a subfamily of the Corduliidae. There is also some disagreement over whether the Chlorogomphinae — a subfamily of the Cordulegasteridae — warrants family status. These issues will not be addressed here.

The order Odonata is diverse in tropical and warm-temperate latitudes. There are three suborders (Table 4.6), but one — the Anisozygoptera — contains only a single genus (*Epiophlebia*) with few species. These insects are rare, occurring to high-altitude (> 2,000 m) streams in Japan and the Himalaya (Tani & Miyatake, 1979; Kumar & Khanna, 1983). The putative larval stage was first described by Tillyard (1921), and a Japanese *Epiophlebia* larva is figured by Ishida & Ishida (1985). According to Hutchinson (1993: p. 406), Japanese *Epiophlebia* larvae are terrestrial in the last instar but aquatic in early stages. The Zygoptera — damselflies — are slender insects, with fore- and hindwings that are similar in general appearance and held above the body when perching. Their larvae have three (occasionally two) external gills, situated at the tip of the elongated abdomen, and take

Table 4.6 Genera of Odonata in tropical Asia. Data derived mainly from Lieftinck *et al.* (1984), Pinratana *et al.* (1988), Zhao (1990), Tsuda (1991) and Asahina (1993). Subfamilial placing largely follows Carle (1986; for Gomphidae) and Tsuda (1991). Genera marked with an * are confined to the Philippines, Indonesia, and/or Papua New Guinea.

ZYGOPTERA
  Calopterygoidea
    Amphipterygidae
      Amphiteryginae
        Devadatta
      Philoganginae
        Philoganga
    Calopterygidae
      Caliphaeinae
        Caliphaea
        Noguchiphaea
      Calopteryginae
        Archineura
        Calopteryx
        Echo
        Matrona
        Mnias
        Neurobasis
        Psolodemus
        Vestalis
    Chlorocyphidae (= Libellaginidae)
      Chlorocyphinae
        Calocypha
        Cyrano *
        Indocypha
        Libellago
        Melanocypha *
        Pachycypha *
        Rhinocypha
        Sclerocypha *
        Sundacypha
      Disparocyphinae
        Disparocypha *
    Euphaeidae (= Epallagidae)
      Anisopleura
      Bayadera
      Cyclophaea *
      Dysphaea
      Euphaea
      Heterophaea *
      Schmidtiphaea
  Lestoidea (= Lestinoidea)
    Lestidae
      Lestinae
        Lestes
        Orolestes
        Sinhalestes
      Sympecmatinae
        Austrolestes *
        Indolestes

        Sympecma
    Megapodagrionidae
      Argiolestinae
        Agriomorpha
        Argiolestes *
        Burmargiolestes
        Calilestes
        Celebagriolestes *
        Podolestes
        Podopteryx *
        Rhinagrion
      Megapodagrioninae
        Mesopodagrion
        Rhipidolestes
    Synlestidae (= Chlorolestidae)
      Megalestinae
        Megalestes
      Synlestinae
        Sinolestes
  Coenagrionoidea
    Coenagrionidae
      Agriocnemidinae
        Agriocnemis
        Mortonagrion
      Amphicnemidinae
        Amphicnemis *
        Papuagrion *
        Pericnemis
        Teinobasis
      Argiinae
        Onychargia
        Moroagrion *
        Palaiargia *
      Coenagrioninae
        Cercion
        Coenagrion
        Himalagrion
        Rhodischnura
      Ischnurinae
        Aciagrion
        Austroallagma *
        Enallagma
        Ischnura
        Xiphiagrion *
      Pseudagrioninae
        Archibasis
        Ceriagrion
        Pseudagrion
        Pyrrhosoma
        Stenagrion *

Table 4.6   *(cont.)*

Xanthagrion *
Isostictidae
Cnemisticta *
Selysioneura *
Tanymecosticta *
Titanosticta *
Platycnemididae
Calcinemiinae
Asthenocnemis *
Calicnemia
Coeliccia
Idiocnemis *
Indocnemis
Risiocnemis *
Platycnemidinae
Copera
Platycnemis
Platystictidae
Drepanosticta
Platysticta
Protosticta
Protoneuridae
Caconeurinae
Caconeura
Esme
Phyllonera
Disparonerinae
Disparoneura
Elattoneura
Nososticta (= Notoneura) *
Prodasineura
Protoneurinae
Melanoneura
ANISOZYGOPTERA
Epiophlebiidae
Epiophlebia
ANISOPTERA
Aeshnoidea
Aeshnidae
Aeshninae
Aeshna
Amphiaeschna *
Indaeschna
Anactinae
Anaciaeschna
Anax
Hemianax
Brachytroninae
Aeschnophlebia
Cephalaeschna
Gynacanthaeschna
Periaeschna
Petaliaeschna
Planaeschna

Gomphaeschninae
Oligoaeschna
Gynacanthinae
Agyrthacantha *
Gynacantha
Heliaeschna
Plattycantha *
Tetracanthagyna
Polycanthaginae
Polycanthagyna
Gomphoidea
Gomphidae
Epigomphinae
Heliogomphus
Leptogomphus
Macrogomphus
Microgomphus
Gomphinae
Anisogomphus
Anormogomphus
Asiagomphus
Burmagomphus
Cyclogomphus
Eogomphus
Gastrogomphus
Gomphus
Labrogomphus
Merogomphus
Platygomphus
Shaogomphus
Sinogomphus
Stylogomphus
Stylurus
Hageniinae
Sieboldius (= Hagenoides)
Lindeniinae
Gomphidia
Gomphidictinus
Ictinogomphus
Sinictogomphus
Octogomphinae
Davidius
Dubitogomphus
Fukienogomphus
Trigomphus
Onychogomphinae
Acrogomphus
Amphigomphus
Davidioides
Lamelligomphus
(= Lamellogomphus)
Megalogomphus
Melligomphus
Nepogomphus

Table 4.6   *(cont.)*

Nihonogomphus
Nychogomphus
Onychogomphus
Ophiogomphus
Orientogomphus
Paragomphus
(= Mesogomphus)
Perissogomphus
Phaenandrogomphus
Scalmogomphus
Cordulegasteroidea
Cordulegasteridae
Chlorogomphinae
Chlorogomphus
Cordulegasterinae
Anotogaster
Cordulegaster
Neallogaster (= Allogaster)
Libelluloidea
Corduliidae
Hemicordulia
Heteronaias *
Metaphya *
Procordulia
Somatochlora
Synthemistidae
Synthemis *
Macromiidae
Idionychinae
Idionyx
Idiophya (= Idionyx?)
Macromiinae
Epophthalmia
Macromia
Macromidia
Libellulidae
Brachydiplacinae
Brachydiplax
Brachygonia
Chalybeothemis
Nannophya
Nannophyopsis
Raphismia
Tyriobapta
Libellulinae
Agrionoptera
Amphithemis
Cratilla
Diplacina
Epithemis
Lathrecista
Libellula
Lyriothemis

Nesoxenia
Orchithemis
Orthetrum
Pornothemis
Potamarcha
Protorthemis *
Palpopleurinae
Palpopleura
Sympetrinae
Acisoma
Brachythemis
Bradinopyga
Crocothemis
Deielia
Diplacodes
Indothemis
Nannodiplax *
Neurothemis
Sympetrum
Tetrathemistinae
Celebophlebia *
Hylaeothemis
Microtrigonia *
Nannophlebia *
Palaeothemis
Risiophlebia *
Tetrathemis
Trameinae
Aethriamanta
Camacinia
Hydrobasileus
Macrodiplax
Pantala
Pseudotramea
Rhodothemis
Rhyothemis
Selysiothemis
Tholymis
Tramaea
Urothemis
Zyxomma
Trithemistinae
Celebothemis *
(= Parathemis)
Huonia *
Onychothemis
Phyllothemis
Pseudagrionoptera *
Pseudothemis
Trithemis
Zygonychinae
Zygonyx

the form of expanded sacs, flat blades, or lamellae which may be divided into a proximal (or basal) and a distal 'segment'. The Anisoptera — or dragonflies — are relatively stout (and usually larger) strong-flying insects, with fore- and hindwings that differ somewhat in shape and which are held flat when the insect perches. Their larvae have internal gills within the rectum, and are relatively squat-bodied with a short, broad abdomen. Rectal gills occur in the Anisozygoptera also. Water projected from the anus by sudden contraction of the rectal muscles can cause dragonfly larvae to shoot through the water as though they were jet propelled. An account of odonate respiratory physiology is given by Hutchinson (1993: pp. 469–480).

Lotic specialists among the Odonata include amphipterygid, calopterygid, chlorocyphid, euphaeid, platystictid, protoneurid and synlestid damselflies, and cordulegasterid (especially *Chlorogomphus*), gomphid and macromiid dragonflies. Certain genera of Aeshnidae (*Tetracanthagyna*) and Libellulidae (*Onychothemis* and *Zygonyx*) are also confined to streams, but these two families are more diverse in ponds and marshes. Preliminary information on the habitat preferences of Malaysian and Indian odonates is given by Furtado (1969) and Kumar & Kanna (1983; see also Kumar & Prasad, 1981). The Gomphidae, which are most diverse in streams, are extremely well represented in Asia (Table 4.6) and over 160 species are known from China alone (Chao, 1984; Zhao,1990). The Euphaeidae are almost entirely confined to the Oriental Region, and the majority of the world fauna of Chlorocyphidae and Calopterygidae are found in tropical Asia. Other families, such as the Lestidae, are poorly represented, while the Petaluridae are confined to the Palaearctic portion of East Asia. Overall, it is clear that tropical Asian streams have a richer odonate fauna than comparable north-temperate habitats and, although figures on relative biomass are not readily available, it seems likely that these insects play a more important ecological role in tropical than temperate latitudes.

All odonates undergo a similar life cycle (Corbet, 1980; see also Hutchinson, 1993: pp. 407–435). The vast majority have aquatic larvae although a few damselflies (certain Megapodagrionidae and a few Platycnemididae) are terrestrial or semiterrestrial as larvae. Eggs are laid in water (or sometimes nearby), and may be attached to rocks or wood, inserted into the tissues of aquatic macrophytes, or simply dropped into the water. The egg hatches into a legless 'prolarva' which almost immediately moults to give the larva proper. There are around 12 larval instars, and development may take between several weeks and a year or two. Bivoltine life histories are known for some damselflies (e.g. the coenagrionid damselflies *Ceriagrion* and *Ischnura*; Kumar, 1973a), and rapid development is typical of species associated with standing waters or temporary pools. In libellulids, for example, *Neurothemis tullia* develops from egg to adult in 82 days (Begum *et al.*, 1990a), whereas *Pantala flavescens* requires only 72 days (Begum *et al.*, 1990b); *Orthetrum pruinosum* takes slightly longer: 155 days (Kumar, 1970). *Diplacodes*, which breeds in temporary pools and marshes, can complete three generations within a year, with development time varying from 52 to 169 days according to temperature (Kumar, 1984). Univoltine life histories are, however, common — especially in lotic species.

Adult emergence involves the mature larva crawling up some kind of support (a twig, rock or sloping stream bank) where the final moult to the imago takes place. The adult phase begins with a pre-reproductive period of sexual maturation that may last from several days to about three weeks. Adults are rather pale during this stage, and spend their time feeding on small insects while the adult pigmentation develops. Colours often vary between the sexes, and the males are usually brighter. Certain species may show seasonal colour polymorphism (as in the libellulid *Neurothemis tullia*: Asahina, 1981); polychromatism of the wings is known also; for example, in the calopterygid damselfly genus *Mnias* (e.g. Suzuki, 1984). Teneral (immature) adults may fly far from the emergence site during the prereproductive phase, and this is common in species which occupy riverine wetlands or floodplains. Some dragonflies (e.g. *Pantala flavescens*: Libellulidae) concentrate in conspicuous swarms. Long dispersal flights are uncommon in most lotic odonates, especially among those damselflies that are confined to stony streams. The reproductive phase involves a return to larval habitats, and the manifestation of often rather complex courtship or mating behaviour. Males commonly establish territories, which they defend from rivals, and where they attempt to intercept and mate with females that arrive after their eggs have matured (see Hutchinson, 1993: 507–513, and references therein).

Investigations of Hong Kong Odonata — the damselfly *Euphaea decorata* (Euphaeidae), and the dragonflies *Zygonyx iris* (Libellulidae), *Heliogomphus scorpio* and *Ophiogomphus sinicus* (Gomphidae) — indicate that they complete one generation per year (Dudgeon, 1989b; 1989c; 1989d; Dudgeon & Wat, 1986). The flight period of *Euphaea decorata* adults extends over five months, and larvae grow throughout the year. This pattern appears to be rather consistent from year-to-year (Dudgeon, 1989b). Data on other damselflies, although less detailed, indicate a similar univoltine pattern: *Rhinocypha perforata* (Chlorocyphidae) adults are present for around six months each year, adult *Mnias mneme* and *Neurobasis chinensis* (Calopterygidae) emerge prior to the summer monsoon and, while the flight period of *Mnias mneme* is no more than a month, *Neurobasis* adults are present for eight or nine months. The flight period of *Zygonyx iris* appears to extend over five to six months but the seasonality of gomphid adults — which do not remain close to emergence sites (Kumar & Prasad, 1981) — is difficult to determine with certainty. Gomphidae from Indian hillstreams spend three or four months as adults (Kumar, 1976), and available data on Hong Kong species (e.g. Wilson, 1995) are in accordance with this. Although there is good evidence that *Ophiogomphus sinicus* and *Heliogomphus scorpio* are univoltine, first-stage gomphid larvae are present in Hong Kong streams throughout much of the year (Dudgeon, 1989e). A similar pattern is seen in other Odonata with extended flight periods. This could be indicative of undetected adult emergence and subsequent oviposition, but staggered hatching of eggs laid in summer would give rise to a similar pattern. At least one other Hong Kong gomphid — *Paragomphus* sp. (probably *Paragomphus capricornis*) — emerges prior to the monsoon; oviposition begins soon after. While most of the new *Paragomphus* generation do not mature until the following year, a small proportion of the population of larvae grow quickly enough to emerge six months later. After mating, they contribute to a second generation of dragonflies within a single year (Dudgeon & Corlett, 1994).

The life histories of Hong Kong Odonata seem to be typical of dragonflies and damselflies inhabiting perennial streams in the tropics. Adult emergence begins before the onset of the summer monsoon which may reduce mortality or wash-out of large larvae during summer spates. Timing of emergence could, however, reflect an influence of rising water temperatures on larval maturation (Dudgeon, 1992a). Significantly, many dragonflies in Indian hillstreams are univoltine and adult emergence in summer is timed to precede the monsoon, thereby reducing spate-induced mortality of larvae (Kumar, 1976; see also Dudgeon, 1995a). Available information does, in fact, indicate that odonate reproduction in the tropics coincides with seasonal rains (Corbet, 1980; 1981).

While the seasonality of some Oriental Odonata is relatively straightforward, this is not true of at least one species — the migratory *Pantala flavescens* (Libellulidae). These insects migrate across the Old

World tropics and, in Hong Kong and southern China, huge swarms arrive in Hong Kong around the beginning of the summer monsoon. Some of these individuals mate and oviposit in temporary pools and ponds but the swarms soon move on. The dragonflies return towards the end of the wet season, when swarms are joined by emerging larvae that have developed during the summer months. Some weeks later the insects depart and few of them are seen again until the following year. The phenomenon of migration by *Pantala flavescens* has been well documented (Corbet, 1980), but the distance flown by individuals, their flight routes, and the length of adult life are not known.

Odonate larvae are predators (see Hutchinson, 1993: pp. 449–469), and prey are captured with the aid of a highly modified and elongate, prehensile labium or 'lower lip'. It can be projected forward rapidly to grasp prey with a pair of toothed or spinose palps mounted on the anterior corners (Fig. 4.62). The labium is then retracted bringing the hapless victim within reach of the mandibles whereupon it is torn apart and devoured. When not in use, the retracted labium is folded flat beneath the head, with the expanded palps covering the lower part of the larval 'face' in some dragonfly families (especially the Corduliidae and Libellulidae). Studies of the diets of co-occurring Odonata in Hong Kong streams (Dudgeon, 1989b; 1989c; Dudgeon & Wat, 1986) indicate that they are generalist predators exhibiting little dietary selection, although large larvae tend to have a more varied diet than smaller ones. Microhabitat occupancy influences dietary composition also. The damselfly *Euphaea decorata* (Euphaeidae), for example, has rather unspecialized microhabitat requirements. It encounters consumes a variety of prey types with the consequence that the diet is broad. By contrast, the libellulid dragonfly *Zygonyx iris* occupies a more specialized microhabitat which results in a relatively narrow diet made up of typically rheophilic taxa (simuliid larvae and mayfly genera such as *Baetiella* and *Epeorus*) which have the same environmental requirements. Significantly, interspecific dietary overlap among Hong Kong odonates with similar labium morphology is high even if these species live in different microhabitats. The Hong Kong data also show that dietary breadth of odonates increases as the predator grows, which reflects an ability to handle a wider range of prey sizes and the failure to drop small prey items from the diet. In other words, dragonfly larvae eat anything they can handle, and do not ignore smaller, less profitable items. Unfortunately there is no information on the effects of odonate predation on benthic communities in Asia, and manipulative studies on the impacts of these predators are lacking.

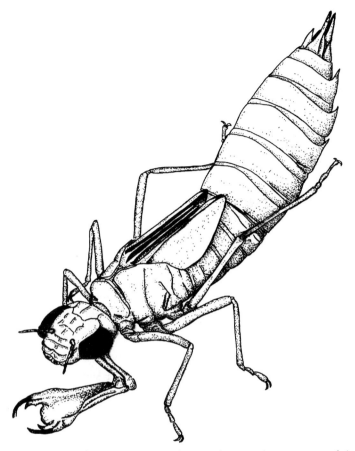

Fig. 4.62 Larval aeshnid (Anisoptera: Odonata) showing the extension of the labium during prey capture.

The following key to odonate larvae will allow identification to family level, and is most effective when used with well-grown or mature larvae. I have omitted the Isostictidae which is a small family of rather obscure damselflies; all four genera from the region are confined to Indonesia and/or New Guinea, and two of them were described only recently (Donnelly, 1993). For the same reason, the Synthemistidae — which are confined to the Southwest Pacific — have been excluded. (Watson *et al.* [1991] give a larval illustration of *Synthemis*.) Identification of genera and species of larval Odonata can be difficult. More adults than larvae have been described and named, and a reliable determination certainly cannot be achieved without consulting the primary literature (see references given below). Careful examination of the labium is essential, as differences in the location and number of setae on the central lobe (or mentum) — as well as the disposition of teeth and hooks on the lateral palpal lobes — are important diagnostic characters. It is often necessary to rear larvae to the adult stage in order to confirm an identification, but this can be achieved for many Odonata as they are relatively easy to maintain in captivity. Note that there is no single character which will reliably separate all larvae of the Corduliidae from the Libellulidae, and for this reason some workers recognize only one family — the Libellulidae — with two subfamilies — the Libellulinae and Corduliinae.

1. Abdomen rather short and stout, lacking caudal gills (although internal rectal gills are present), and terminating in five, short, spine-like processes, the three largest of which form an 'anal pyramid' at the tip of the abdomen with the shorter pair representing the cerci 2

    Abdomen long and slender, and terminating in three (more rarely, two) leaf- or sac-like lamellae (= caudal gills) .... suborder **Zygoptera** 3

2. Larvae somewhat slender and elongate, with a slight petiolation at the base of the wing pad; antennae with five segments; body covered with tubercles, but lacking bristles; extremely rare (Japan and the Himalaya only) .... suborder **Anisozygoptera** ...... Epiophlebiidae

    Larvae *usually* rather short and squat (although Aeshnidae have elongate bodies), without any petiolation at the base of the wing pad; antennae with four, six or seven segments; common .... suborder **Anisoptera** 13

3. Two forceps-like caudal gills (the median gill is minute) which are triangular in cross-section .... **Chlorocyphidae**

    Three caudal gills that are sac-, leaf- or blade-like 4

4. Filamentous gills (Fig. 4.63B) on the underside of abdominal segments II-VIII (the caudal gills are saccoid) .... **Euphaeidae**

    Without filamentous gills on abdominal segments II-VIII 5

5. First antennal segment (the scape) longer than the combined length of the subsequent segments; labium lacks long setae on the inner face of the mentum, and the anterior margin has a well-developed median cleft; body often slender and elongate; caudal gills are usually blade-like with a distinct dorsal ridge (or carina) ....

....Calopterygidae

First antennal segment much shorter than the combined length of the subsequent segments; without the above combination of characters 6

6. Labium distinctly spoon-shaped and strongly tapered posteriorly, with palpal lobes that are divided into large, sharp teeth...........
.... **Lestidae**

Labium quadrate or more-or-less triangular in shape, with palpal lobes bearing moveable hooks or spines at the tip.................... 7

7. Labium lacks setae on the mentum or palpal lobes; mainly found in stony streams 8

Labium with setae on the mentum and/or palpal lobes; gills lamellate and may have conspicuous tracheal branching; common, but most often found in standing water or slow-flowing streams **11**

8. Gills more-or-less saccoid; median cleft in the anterior margin of the labial mentum is rather shallow (Figs 4.64B and 4.65B) 9

Gills leaf-like with rounded apices; anterior margin of the labial mentum is distinctly cleft; a distinct 'breaking joint' or area of weakness occurs at the base of each caudal gill .... **Synlestidae**

9. Sturdy larvae with cylindrical bodies and short antennae; legs thick and rather stout bearing few setae; often semiaquatic or terrestrial; rare .... **Megapodagrionidae**

Delicate aquatic larvae with flattened bodies and relatively long antennae; long, slender legs fringed with setae 10

10. Larvae *may* be large and deeply pigmented; head with a well-developed postocular lobe, and a row of a dozen spines *may* occur along the ventral margin of the eye; palpal lobes of the labium with three spines and one moveable hook, and lateral margins of the mentum *may* bear a row of small teeth; a pair of retractable gill tufts *may* be present between the anus and the caudal 'gills'
.... **Amphipterygidae**

Pale, somewhat spindly larvae with large, bulbous eyes (lacking a post-ocular lobe or associated spines); palpal lobes of the labium with a single spine and one moveable hook, and lateral margins of the mentum lack hooks ....**Platystictidae**

11. Gills clearly divided into a thickened dark proximal portion and a thin, paler distal part; anterolateral margins of the labial mentum fringed with tiny teeth, one seta is situated on either side of the midline of the mentum, and there are three setae on the palpal lobes .... **Protoneuridae**

Gills not usually clearly divided into proximal and distal portions; anterolateral margins of the labial mentum are not toothed; usually more than one seta on either side of the midline of the mentum (if only one seta is present there are six setae on the palpal lobes) **12**

12. Caudal gills long (approximately the same length as the abdomen), with apices somewhat pointed or attenuated, and inconspicuous tracheal branching; third segment of the antenna longer than the second; usually in slow-flowing streams .... **Platycnemididae**

Caudal gills shorter than the abdomen, with rounded apices and conspicuous tracheal branching; proximal margins of the gills may be fringed with tiny teeth, and the gills may show signs of being divided into proximal and distal portions; third segment of the antennae shorter than the second; mainly in standing water .... **Coenagrionidae**

13. Labial mentum and palpal lobes more-or-less flat; without setae on the mentum or (usually) the palpal lobes **14**

Labial mentum and palpal lobes are mask- or bowl-shaped; setae usually occur on the mentum and are always present on the palpal lobes **15**

14. Antennae four-segmented, with the third segment enlarged and sometimes expanded laterally (there are many variations), and the fourth segment vestigial; tarsi of the first two pairs of legs are two-segmented; labial mentum more or less quadrate, and the anterior margin is never cleft .... **Gomphidae**

Antennae six or seven-segmented and filamentous; tarsi of all legs have three segments; labial mentum widest in the distal portion and narrowing towards the posterior with a cleft in the anterior margin .... **Aeshnidae**

15. Body elongate and covered with bristles or tufts of setae; distal margins of the palpal lobes of the labium with large irregular teeth which interlock with those on the corresponding lobe; anterior

margin of the mentum is cleft .... **Cordulegasteridae**

Body short and stout; distal margins of the palpal lobes of the labium smooth or rather evenly toothed; anterior margin of the mentum is never cleft... **16**

16. Legs very long giving the larvae a 'spidery' appearance; abdomen depressed and more-or-less circular in outline; a small 'horn' may be present between the antennal bases; rather long, regular teeth along the distal margins of the palpal lobes of the labium.........
.... **Macromiidae**

Legs rather short (apex of the hind femur does not extend beyond abdominal segment VIII); abdomen not markedly depressed or circular in outline; distal margins of the palpal lobes of the labium bear small teeth which *may* have associated setae ....
.... **Corduliidae** and **Libellulidae**

Zygoptera

Illustrations and descriptions of odonate larvae are scattered throughout the primary literature, but Ishida & Ishida (1985) provide an excellent series of figures of Japanese species which includes genera that are widely-distributed in Asia. Among the zygopteran families (Table 4.6), euphaeid larvae are unmistakable. They have lateral abdominal gills on segments II-VIII, and three large and swollen saccoid caudal gills. The former play a more important role in respiration (Norling, 1982) and — among Oriental Odonata — are confined to this family. The diagnostic features are shown clearly in *Euphaea* (Fig. 4.63); see also St Quentin, 1973; Matsuki & Lien, 1978), *Bayadera* (see Kumar, 1973a; Matsuki & Lien, 1978) and *Anisopleura* (Kumar, 1977). A *Euphaea* larvae is depicted — as *Indophaea* — by Fraser (1933: Fig. 1), and Needham (1930) figures Chinese *Anisopleura*, *Bayadera* and *Euphaea* (as *Taolestes*; see Needham, 1941). Euphaeid larvae are flattened as an adaptation for life under stones and cobbles in riffles; a univoltine life history with an extended flight period may be typical (e.g. Matsuki & Lien, 1978; Dudgeon, 1989b). There is a striking uniformity of appearance among larvae of different genera (Fraser, 1934: p. 74), but *Euphaea* larvae have a row of spines along the anteroventral margin of the compound eye which do not occur in *Bayadera* (and perhaps other euphaeids). Details of the spination on

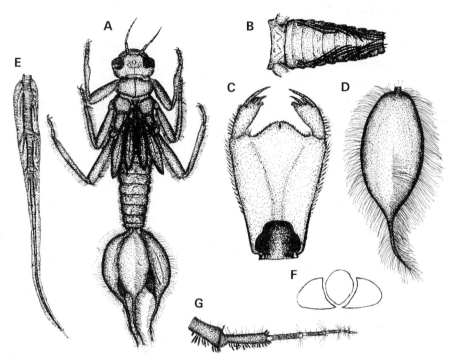

Fig. 4.63 *Euphaea decorata* (Zygoptera: Euphaeidae): A, dorsal view of larva; B; ventral view of abdomen showing lateral gills; C; labium; D, saccoid caudal gill; E; lateral abdominal gill; F; diagrammatic cross-section of caudal gills; G, antenna.

the palpal lobes of the labium may vary among genera (e.g. Needham, 1930: Plate VIII).

Among the Lestidae, *Orolestes* larvae have been described by Lieftinck (1939) and Lien & Matsuki (1983), and figures of *Lestes* can be found in publications dealing with Holarctic damselfly larvae (e.g. Ishida & Ishida, 1985). Lestids are uncommon in flowing water, and prefer lentic habitats with dense growths of macrophytes. The spoon-shaped labium is diagnostic for the family. Synlestid larvae occur in rocky streams. They are large and bear a superficial resemblance to Lestidae but are more robust. The labium is simple and lacks any long setae (*cf.* Lestidae). The rounded anterior margin has a median incision and is fringed with approximately 20 minute teeth. The palpal lobes have a long moveable hook and two robust spines. The caudal gills are short and broad, with oval apices and a smaller median lobe. *Megalestes* larvae are figured by Asahina (1985; see also Matsuki & Lien, 1978).

The Coenagrionidae are rather eurytopic damselflies that are mostly (although not exclusively) associated with standing water. Exceptions include *Archibasis* and *Agriocnemis* which occur in streams. Larvae of *Agriocnemis*, *Ischnura*, *Pseudagrion* and *Rhodischnura* have been described by Kumar (1973a), and *Ceriagrion* larvae are figured by Kumar (1973a) and Matsuki (1985). Needham (1930) provides illustrations of *Agriocnemis* and *Ceriagrion* also. The Protoneuridae includes *Prodasineura* which has been figured in the larval stage by Matsuki (1991). The caudal gills are somewhat ovoid in outline, and the labium has an intact anterior margin which lacks marginal teeth. The palpal lobes bear a long, moveable, anterior hook and a fixed spine, as well as three lateral setae on the outer margin near the base. The labium of *Disparoneura* is similar (see Kumar, 1973a) but has a toothed anterior margin, and the caudal lamellae have filamentous apices. In general, protoneurids resemble coenagrionids but the caudal gills are thick in the proximal half and membranous in the distal part. Their primary habitat is streams.

The Platycnemididae are often encountered in slow-flowing streams. Needham (1930) has figured the larva of *Platycnemis*; Lieftinck (1940) and Kumar (1973a) provide illustrations of *Copera* larvae; and, Matsuki & Lien (1984) describe the larvae of two species of *Coeliccia*. A *Calicnemia* larva described by Kumar (1977) is unusual in that it is semiterrestrial living among damp mosses and ferns. It superficially resembles the immature stages of the Megapodagrionidae which includes species that are terrestrial or live along the margins of forest streams. Such larvae have squat, rather robust bodies with short, stout legs and reduced gills. Other megapodagrionids (e.g. *Argiolestes* and *Podopteryx*) live in seeps, bogs, and 'container' habitats such as water-filled tree holes (Watson & Dyce, 1978). Platystictid larvae have saccoid caudal gills with attenuated apices, a flattened body, and a large head bearing bulbous eyes and a rectangular labium (Fig. 4.64). They tend to be somewhat pale and spindly in general appearance, and are found on and among cobbles in parts of stony streams with slow or moderate current. A detailed description of *Protosticta* (and a comparison with *Drepanosticta*) is given by Asahina & Dudgeon (1987); Lieftinck (1934) has figured *Drepanosticta*.

Like euphaeids and platystictids, the caudal gills of amphipterygid larvae are saccoid, although they have rather lanceolate tips. In genera such as *Devadatta*, the respiratory function of the caudal gills is taken over by a pair of retractable gill tufts between the anus and the non-respiratory caudal 'gills' (Norling, 1982). *Philoganga* larvae (Fig. 4.65; see also Asahina, 1967) are large (for a damselfly) and flattened (in a

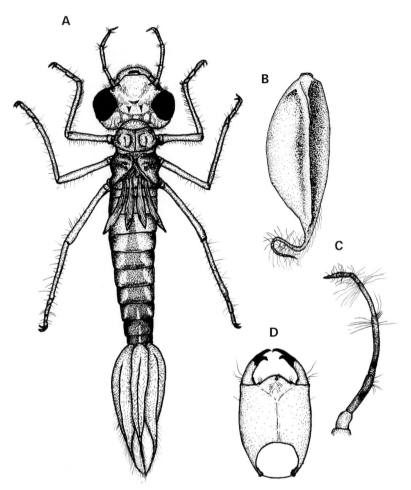

Fig. 4.64 *Protosticta taipokauensis* (Zygoptera: Platystictidae): A, dorsal view of larva; B; saccoid caudal gill; C; antenna; D, labium.

similar manner to Euphaeidae or perlid stonefly larvae). The antennae are unusually long (for a zygopteran) with seven segments. The labium is broad and ovate, with rows of tiny hooks along the lateral margins and a shallow median incision in the anterior margin. The palpal lobes bear three terminal, hooked spines, and a long moveable hook. There is a distinct postocular lobe, and a row of a dozen spines along the ventral margin of the eye. Amphiterygids are rather primitive odonates and adult *Philoganga* (but not *Devadatta*) hold their wings open when perching.

Chlorocyphid larvae are somewhat flattened and mainly confined to stony streams where they hide under small stones or submerged

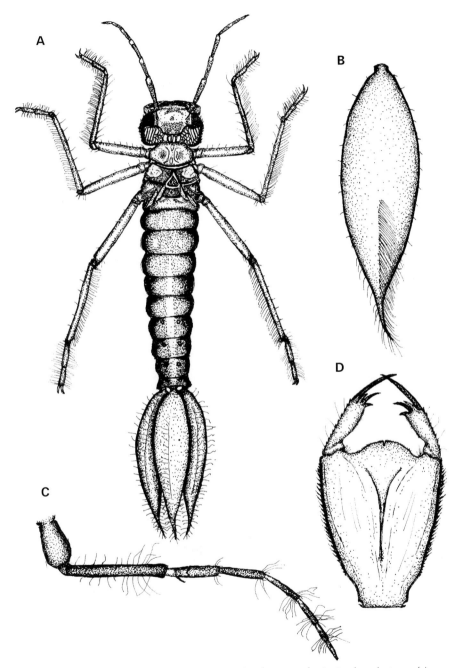

Fig. 4.65 *Philoganga vetusta* (Zygoptera: Amphipterygidae): A, dorsal view of larva; B; saccoid caudal gill; C; antenna; D, labium.

wood. Typically there are two, forceps-like caudal gills (the median gill is greatly reduced) as seen in *Rhinocypha* (Fig. 4.66; see also Needham, 1930; Kumar, 1973a; Kumar & Prasad, 1977). Needham (1930) has figured a Chinese *Libellago* (as *Micromerus*) larva, and the larvae of *Libellago* and *Rhinocypha* seem to be similar in all respects except minor details (Fraser, 1934: p. 4; see also Fig. 1 in Fraser, 1933).

Calopterygid larvae are slender with long legs and narrow caudal gills. The labium is elongate with a well-developed median cleft that

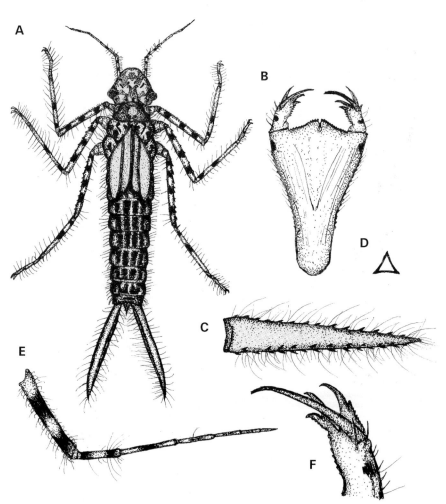

Fig. 4.66 *Rhinocypha perforata* (Zygoptera: Chlorocyphidae): A, dorsal view of larva; B; labium; C, caudal gill; D, diagrammatic cross-section of caudal gill; E; antenna; F, lateral lobe of labium.

can be over half the length of the mentum. Larvae of the widespread damselfly *Neurobasis chinensis* (figured by Lieftinck, 1965; Kumar, 1973a; see also Needham, 1930; Fraser, 1934: p. 120; St Quentin, 1973) are unusually long and slender with prominent antennae, and resemble a stick insect. They usually occur among trailing vegetation and roots along stream margins. *Mnias* larvae are likewise slender (Fig. 4.67; see also Matsuki & Lien, 1978), but tend to rest with their bodies flat upon the substrate (often leaf litter or wood). The blade-like caudal gills are relatively short and can be expanded like a fan or brought together to resemble a single structure. *Matrona* has unusually long caudal gills; the lateral gills are approximately the same length as the abdomen, but the median gill is shorter (see Matsuki & Lien, 1978); Needham (1930) has illustrated a Chinese *Matrona* larva. Like *Neurobasis*, they are often associated with trailing roots and vegetation.

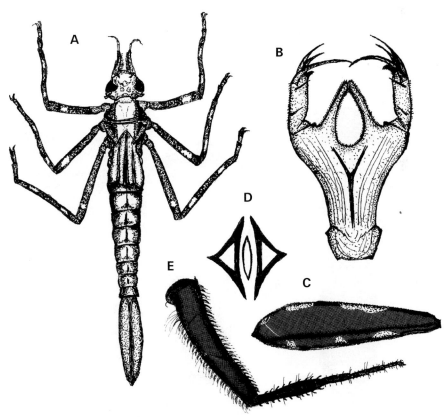

Fig. 4.67 *Mnias mneme* (Zygoptera: Calopterygidae): A, dorsal view of larva; B; labium; C, caudal gill; D, diagrammatic cross-section of caudal gills; E; antenna.

Psolodemus larvae have been figured by Matsuki & Lien (1978) and differ from other calopterygids in that the lateral caudal gills have toothed edges. These insects seem to favour slow-flowing streams. The larvae of *Vestalis* have been described by Lieftinck (1965). Fraser (1934: p. 135) reports that larvae of *Echo* (and *Vestalis*) are similar to *Neurobasis*.

Anisoptera

Anisoptera larvae are easily distinguished from the Zygoptera by the absence of caudal gills, as well as general facies which includes a rather squat body. The Aeshnidae are the most streamlined and elongated Anisoptera larvae (Figs 4.62 & 4.68); the majority live in standing water. Asahina (1974) provides diagnoses which allow separation of the final-instar larvae of some *Anax* species based on caudal appendages (see also Lieftinck, 1940; Fraser, 1943; Matsuki, 1987; Yoon & Dong, 1990). A diagnosis of the larval stages of

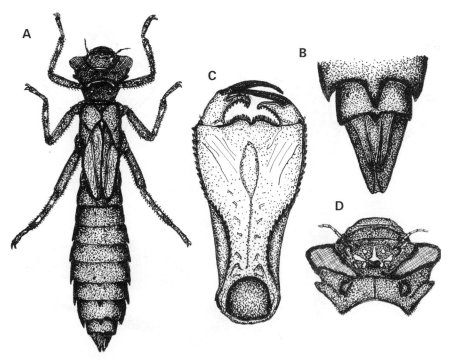

Fig. 4.68 *Tetracanthagyna* (Anisoptera: Aeshnidae); A, dorsal view of larva; B, anal pyramid; C, labium; D, dorsal view of larval head.

Planaeschna is given by Matsuki & Lien (1985), Matsuki (1989a) and Matsuki & Yamamoto (1990), while Matsuki (1986a; 1986b) and Matsuki & Kuwahara (1989) provide some information on the morphology of immature *Gynacantha* and *Polycanthagyna* respectively. Needham (1930) has figured Chinese larvae of *Cephalaeschna* and *Planaeschna*, in addition to *Aeschnophlebia* which may be confined to Palaearctic East Asia. Larvae of *Tetracanthagyna* — described by Matsuki (1988b; see also Fig. 4.68) — occur in forest streams which is an unusual habitat for aeshnids. They are dark brown with a twig-like habitus which renders them cryptic on their usual substrate of submerged wood. These insects withdraw their legs and feign death when disturbed.

The Gomphidae is a diverse family of dragonflies and includes a majority of species confined to streams. Within such habitats, larval microhabitats vary. Genera such as *Anisogomphus*, *Asiagomphus*, *Burmagomphus*, *Labrogomphus*, *Ophiogomphus* and *Onychogomphus* are burrowers in sand and mud, with *Labrogomphus* possessing an extension of the ninth abdominal segment which functions like a respiratory siphon while the animal is buried in the sediment. *Gomphidia* and *Megalogomphus* have broad, powerful legs that are clearly adapted for digging, while *Heliogomphus* and *Microgomphus* are flattened and leaf-like allowing them to nestle among detritus accumulations or between cobbles in streams. Unlike other gomphids, larvae of *Ictinogomphus* and *Sinictogomphus* are most common in standing water where they sprawl on mud surfaces. Fraser (1934: Fig. 45) has pictured the larvae of *Lamelligomphus*, *Macrogomphus*, *Microgomphus* and *Sieboldius*. *Ophiogomphus* larvae have been figured (as *Onychogomphus*) by Matsuki (1989b) and Dudgeon (1989c; see also Fig. 4.69), as have the larvae of *Heliogomphus* (Lieftinck, 1933; St Quentin, 1973; Matsuki, 1990a). *Heliogomphus scorpio* is shown in Fig. 4.70. In both genera, the morphology of the antennal segments are important identifying characters (although not all *Heliogomphus* have the distinctive fan-shaped antenna of *Heliogomphus scorpio*: e.g. Matsuki, 1978; Zhao, 1990: Fig. 5–1.8). Dumont *et al.* (1992) record the diagnostic features (principally, the degree of development of the dorsal and lateral abdominal spines) which can be used to separate larvae of *Onychogomphus* and *Ophiogomphus* (see also Matsuki, 1978; Yoon & Dong, 1990). Matsuki (1978; 1990b), Yoon & Dong (1990) and Wilson (1995b) have figured *Gomphidia* larvae in which the posterolateral margins of the rather flattened abdominal segments are drawn out into distinct spines. Other descriptions of gomphid larvae

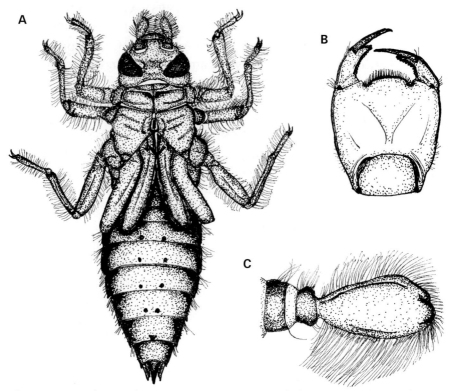

Fig. 4.69 *Ophiogomphus sinicus* (Anisoptera: Gomphidae); A, dorsal view of larva; B, labium; C, antenna.

include *Asiagomphus, Gastrogomphus* and *Trigomphus* (all as *Gomphus*; see Zhao, 1990), in addition to *Gomphidia, Ictinus* and *Sieboldius* by Needham (1930); *Microgomphus, Megalogomphus* and *Paragomphus* by Lieftinck (1940; 1941); *Anisogomphus, Burmagomphus* and *Paragomphus* (as *Mesogomphus*) by Kumar (1973b); *Anisogomphus, Burmagomphus, Fukienogomphus, Gomphus, Ictinogomphus* (see also Fig. 4.71), *Leptogomphus, Merogomphus, Sieboldius, Sinogomphus* and *Stylogomphus* by Matsuki (1978); *Anisogomphus, Burmagomphus, Davidius, Gomphidia, Gomphus, Ictinogomphus, Nihonogomphus, Sieboldius* and *Stylurus* — as well the mainly Palaearctic Asian genus *Trigomphus* — by Yoon & Dong (1990); *Anisogomphus, Asiagomphus, Burmagomphus, Davidius, Fukienogomphus, Gastrogomphus, Gomphidia, Heliogomphus, Ictinogomphus, Labrogomphus, Lamelligomphus, Leptogomphus, Melligomphus, Merogomphus, Nihonogomphus, Onychogomphus, Ophiogomphus, Shaogomphus, Sieboldius, Sinictogomphus,*

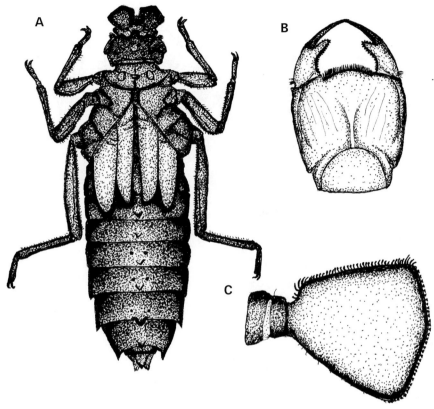

Fig. 4.70 *Heliogomphus scorpio* (Anisoptera: Gomphidae); A, dorsal view of larva; B, labium; C, antenna.

Sinogomphus, Stylogomphus, Stylurus and *Trigomphus* by Zhao (1990); and, *Lamelligomphus, Megalogomphus* and *Stylogomphus* by Wilson (1995b). Zhao (1990: pp. 82–83) gives key (in Chinese) to the larvae of some East Asian Gomphidae.

Like many of the Gomphidae, larvae of the Cordulegasteridae (sometimes incorrectly spelled Cordulegastridae) are burrowers but, unlike gomphids, they are usually only partly concealed within the sand or gravel bed of streams. Typically, the body and legs are extremely hairy. Asahina (1986) figures a larva of *Chlorogomphus* (Chlorogomphinae) which has a broad head (at least as wide as the broadest part of the thorax), with a strongly tapered and pointed abdomen. The body is covered with tufts of setae but they are less dense than in larval Cordulegasterinae. The front of the head is prominent, produced broadly between the antennae, and the palpal

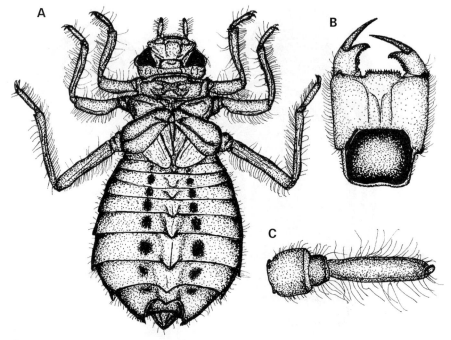

Fig. 4.71 *Ictinogomphus pertinax* (Anisoptera: Gomphidae); A, dorsal view of larva; B, labium; C, antenna.

lobes of the labium are well developed with sharply incised spines along the inner margin. *Neallogaster* (Cordulegasterinae) larvae have been described by Asahina (1982), and differ from other Cordulegasteridae in having wingsheaths that are parallel to the body axis; in *Anotogaster*, *Chlorogomphus* and *Cordulegaster* they diverge strongly from the mid-line. Fraser (1936: p. 13 & Fig. 13b) has remarked on the close resemblance of larvae of *Chlorogomphus* and *Cordulegaster* but there are differences among cordulegasterid genera in the details of the setation and spination of the labial mentum and palpal lobes. Kumar (1973b) figures an unidentified cordulegasterid larva (with parallel wing sheaths) from India.

The Macromiidae includes a number of genera which occur in tropical streams. *Macromia* larvae have long legs and the abdomen is ovate and rather flat ventrally imparting the general appearance of spider (see Lieftinck, 1940; 1950; Kumar, 1973b). They sprawl on the stream bed and are often partly covered by sand. *Epophthalmia* larvae are characterized by ' . . . their enormously long, spidery legs and comparatively small, quadrate head, armed with two small horns posteriorly. The mask is of great size and armed with a row of formidable teeth which resemble the branched antlers of a stag and is quite unlike any others in the order; the abdomen . . . is spined dorsally and laterally . . .' (Fraser, 1936: p. 193). A larva of this genus is figured (as *Azuma*) by Needham (1930). *Macromidia* larvae from Thailand have been described (as Corduliidae) by Matsuki (1989c; see also Fraser, 1936: Fig. 48A). All of these macromiids are spider-like in general appearance and hence rather easily recognized. This character sets them apart from the Corduliidae within which they have sometimes been treated as the subfamily Macromiinae.

The Libellulidae is a very large family with larvae that are stocky and rather short-bodied. Most species are associated with standing water, or even temporary habitats (e.g. *Diplacodes* and *Pantala*), although a number of genera may occur opportunistically in streams — especially in pools or among marginal or trailing vegetation. *Zygonyx* is, however, an exception and clings on to stone surfaces in torrents and even vertical rock faces in waterfalls. The larvae are moderately flattened with long, splayed legs (Fig. 4.72; see also St Quentin, 1973; Matsuki & Kitagawa, 1986; Matsuki, 1988a). Needham (1930) provides illustrations of the larvae of *Acisoma, Brachydiplax,*

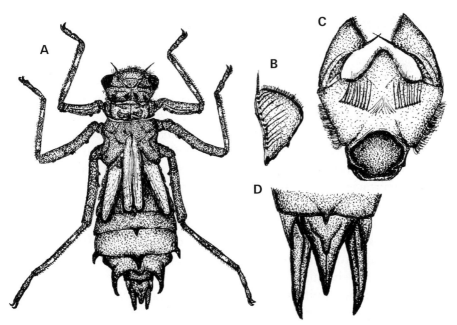

Fig. 4.72 *Zygonyx iris* (Anisoptera: Libellulidae); A, dorsal view of larva; B, lateral lobe of labium; C, labium; D, anal pyramid.

Crocothemis, Deielia, Hydrobasileus and *Lyriothemis*; *Cratilla* larvae are figured by Lien & Matsuki (1988); *Diplacodes* is described by Kumar (1984) and Rowe (1992); while Kumar (1973b) has figured the immature stages of *Brachythemis, Orthetrum, Pantala, Potamarcha, Sympetrum, Tholymis, Tramea, Trithemis* and *Zyxomma*.

Corduliidae larvae — which are infrequently encountered in tropical streams — strongly resemble libellulids in general appearance. However, the teeth borne on the distal margins of the palpal lobes of corduliids are more distinct than those in the majority of libellulids. In addition, the cerci extend over half of the length of the anal pyramid in Corduliidae, but are generally shorter in libellulids. The relatively strong development of dorsal and lateral spines on the ninth abdominal segment of corduliids may also be used as a character to separate them from the Libellulidae. Some Japanese corduliids are figured by Ishida & Ishida (1985), and Needham (1930) gives an illustration of a Chinese *Epitheca* larva — a Holarctic genus which does not extend far into the Oriental Region (if at all). The family becomes more diverse in northern Asia and Australia, but *Metaphya* occurs in New Guinea and Borneo, and the monotypic genus *Heteronaias* is confined to the Philippines.

PLECOPTERA

The Plecoptera (or stoneflies) is a small order of insects consisting of around 1800 species. They occur throughout most of the world but the highest species and family diversity is found in the streams and stony rivers of the temperate zone (Illies, 1965; Hynes, 1976). It is unlikely that there are any species of Asian Plecoptera confined to lentic habitats. Within streams, larval stoneflies have specific substrate requirements, and different species may be associated with cobbles, boulder surfaces, accumulations of detritus, and so on. Some taxa burrow into the stream substrate and spend most of their lives in the hyporheic zone. Diets vary considerably among taxa but many of the tropical Asian species are active predators with flat bodies. Phytophagous stoneflies tend to be more cylindrical in general body form and relatively small, while burrowing forms are long and slender. Adults are rather uniform in general appearance and habits. Most occur among rocks or plant litter along the stream margins, or are associated with riparian vegetation. Males of many species attract their mates by drumming on the substrate with the tip of the abdomen to produce a species-specific signal. Virgin females respond to the calls by producing their own signals (e.g. Maketon & Stewart, 1988; Stewart & Maketon, 1991; Hanada *et al.*, 1994; Stewart *et al.*, 1995). Next to nothing has been published on the ecology of stoneflies in the Old World tropics. A review of the literature by Brittain (1990) reveals that stoneflies usually complete their life cycles within one year in temperate latitudes, although larger species frequently take longer — possibly as much as three or four years (e.g. Ida, 1994; see also Hynes, 1976). Bivoltine or multivoltine life histories are rare.

There are two suborders of stoneflies, but only one — the Arctoperlaria — is found in the northern Hemisphere. This suborder is divided into two 'family groups': the Systellognatha and Euholognatha. Adults of both groups are winged (although aptery occurs in a few genera) with more-or-less flattened bodies, large heads, long antennae and well-developed eyes. The anal cerci vary in length, but are conspicuous. The Euholognatha feed as adults and the terrestrial phase of their life cycle may last several weeks (e.g. Zwick, 1990). Adult Systellognatha have reduced mandibles and cannot feed; they are relatively short-lived. The labial palps are short and blunt in the Euholognatha and long and filiform in the Systellognatha. In addition, the labial paraglossae (outer lobes) of the Systellognatha are large while the glossae (inner lobes) are reduced; those of the Euholognatha are of similar size. Labial characteristics also allow separation of larvae of these two groups. Reduction of the glossae in larval Systellognatha is accompanied by elongate mandibles, while euholognathan larvae with relatively 'normal' glossae have short, stout mandibles. Other larval characters used to differentiate among plecopteran taxa are the presence or absence of ocelli (as well as their number and position), and the location of gills on the coxa, prosternum or abdomen. The arrangement of hairs, bristles and stout setae on the body are also important distinguishing characters and, in families such as the Perlidae, larvae of some genera (*Kamimuria* and *Paragnetina*, for example) have characteristic markings which aid in species separation.

The Euholognatha comprises the Scopuridae plus the superfamily Nemouroidea which consists of the Nemouridae, Taeniopterygidae, Capniidae and Leuctridae. The Systellognatha includes the Pteronarcyidae, Styloperlidae and Pletoperlidae within the superfamily Pteronarcyoidea, and the Perlodidae, Perlidae and Chloroperlidae in the superfamily Perloidea. The Pteronarcyidae is one of the most primitive stonefly families (Illies, 1965). Larvae are large (up to 45 mm body length), and have branched gills on the two most anterior abdominal segments with a corresponding set of gills on all three thoracic sternites. Pteronarcyids are confined to North America and the far north of Palaearctic East Asia (e.g. Wu, 1936a; Illies, 1965; 1966). Larvae of the two Asian *Pteronarcys* species are figured by Nelson (1988) but the family is not considered further herein.

Plecoptera are particularly diverse in cool streams at high altitudes or in temperate latitudes, and most species are stenothermic (Hynes, 1976). Water temperature plays a preeminent role in longitudinal distribution (Hynes, 1976; Uchida, 1990; 1991), and governs egg hatching and adult emergence periods (e.g. Hynes, 1976; Brittain, 1990; Elliott, 1995; Zwick, 1996). Year-round emergence by tropical stoneflies has been reported in one instance (Zwick, 1976). It is uncertain whether this is a reflection of asynchronous development of univoltine species or if it represents a rare instance of multivoltinism which may be achievable by small stoneflies in warm streams (e.g. Alouf, 1991). At low temperatures, life cycles may be lengthened to obtain a larger body size, thereby increasing individual fecundity (Brittain, 1990). Egg hatching success is high at low temperatures (even below 5°C), but declines steeply above 20°C (Brittain, 1990, and references therein). This relationship goes some way to explain their relative abundance and diversity of stoneflies in temperate or high-altitude streams.

The taxonomic diversity of Plecoptera in Palaearctic Asia is apparent from an investigation of a medium-sized (80-km long) Japanese river system which yielded 87 species (Uchida, 1990). This diversity declines rapidly from temperate Asian latitudes (with at least nine families) to tropical latitudes (with four or fewer), where '. . . the only diverse tropical family of stoneflies is the Perlidae . . .' (Zwick, 1986a: p. 2). Together with the Nemouridae and, to a lesser extent, the Peltoperlidae and Leuctridae, perlids have penetrated tropical Asia as far as Borneo. The nemourids reach Bali (Baumann, 1975), but only the Perlidae extend further east. Even the widespread Perlidae have failed to colonize Australia (Illies, 1965) and they *appear* to be absent from New Guinea streams also (Dudgeon, 1990a; 1994b).

The following key can be used to identify the families and some of the common genera of Plecoptera larvae in tropical Asia. The couplets dealing with the Nemouridae are mainly derived from Baumann (1975) and those concerning the Perlinae are from Sivec *et al.* (1988). The Styloperlidae are incompletely known and determination of larvae using this key may be problematic. Accurate identification of stoneflies — especially to genus level — depends on the examination of well-grown larvae.

1. Larvae markedly flattened and cockroach-like in general appearance with the thorax much wider than the head and abdomen and head partly concealed beneath the prothorax; thoracic sternal plates overlapping; abdomen usually shorter than the thorax and body length (excluding cerci) generally < 10 mm .... **Peltoperlidae**

   Larvae not cockroach-like; head fully exposed and conspicuous; body may be flattened or cylindrical; thoracic sterna do not overlap; abdomen longer than the thorax, and body size variable but frequently > 20 mm 2

2. Glossae of labium as long (or almost as long) as paraglossae; mandibles short and stout; larvae *usually* neither strongly flattened nor particularly large (generally < 20 mm body length, excluding cerci); hind margins of the femora may bear scattered setae but do not usually have a setal fringe 3

   Glossae of labium much shorter than paraglossae; mandibles elongate and slender at the tip; larvae usually strongly flattened and may be quite large (often > 20 mm long); hind margin of the femora often fringed with long setae 11

3. Tarsus with each of the three segments (excluding the claw) as long or longer than the preceding one; mostly Palaearctic ....
   .... **Taeniopterygidae**

   Second segment of the tarsus shorter than the first segment; relatively common and widespread 4

4. Body cylindrical, elongate and sometimes delicate; hind leg when fully extended shorter than the abdomen (which may be rather long); relatively rare 9

   Body relatively stout; hind leg, when fully extended, longer than the abdomen (excluding cerci); common .... **Nemouridae** 5

5. Cervical gills present (on the 'neck'); whorls of spines absent from the forefemora .... **Amphinemurinae** 6

   Cervical gills usually very reduced or almost entirely absent; forefemora with a whorl or a group of long spines near the distal end .... **Nemourinae** 8

6. Cervical gills branched 7

   Cervical gills simple and reduced to stubby or triangular projections

.... *Mesonemoura* and probably *Indonemoura*

7. Cervical gills highly branched; common and widespread ....
    .... *Amphinemura*

    Cervical gills consisting of two or three branches only; rare, mainly
    Palaearctic Asia .... *Protonemura*

8. One small cervical gill present; distinct whorls of spines on the
    forefemora .... *Illiesonemoura*

    Cervical gills almost entirely lacking; spines present on the distal
    end of the forefemora, but not arranged in a whorl ..... *Nemoura*

9. Abdominal segments I-III (or IV) with discrete or separate terga
    and sterna (i.e. not fused, but with a distinct 'break' or 'fold'
    between the upper and lower surface of the abdomen); segments
    of the cerci usually increasing rapidly in length towards the tip;
    body long, thin and delicate, and usually white or pale yellow
    .... **Leuctridae**

    Abdominal segments not as above, the majority are either fused or
    separate; body elongate but not always thin or delicate and, *if*
    lightly pigmented, the body may be covered in short hairs or spines
    .... **10**

10. Abdominal segments I-IX with discrete terga and sterna; length of
    segments of the cerci increasing gradually toward the tip; antennae
    not exceptionally long (less than two-thirds of the body length);
    rare, probably confined to Palaearctic Asia .... **Capniidae**

    Abdominal sternum I fused to metasternum; abdominal sternum II
    separated from the tergum by a membrane but on other segments
    the terga and sterna are fused; antennae long (over two-thirds of
    the body length); southern China .... **Styloperlidae**

11. Thorax with branched lateral gills, and anal gills may be present;
    widespread and common .... **Perlidae** **13**

    Thorax lacks lateral gills (on the rare instances where they are
    present, the gills are unbranched), and anal gills are always absent;
    almost entirely Palaearctic **12**

12. Maxillary palp with terminal segment reduced, much smaller and
    narrower than the preceding one; legs rather short; cerci shorter
    than the abdomen; body colour more-or-less uniform; often quite

small (10 — 20 mm body length, excluding cerci) ....
....**Chloroperlidae**

Maxillary palp with terminal segment not markedly smaller than the preceding one; legs rather long; cerci at least as long as the abdomen; head and thorax may bear distinctive colour patterns; comparatively large (sometimes > 40 mm long) .... **Perlodidae**

13. Occiput with a transverse row of regularly-spaced spinules *or* with a distinctly-elevated transverse ridge .... **Perlinae** 14

    Occiput lacks spinules *or* bears a sinuate (= wavy) *or* an incomplete row of irregularly-spaced spinules .... **Acroneuriinae**

14. With two ocelli (biocellate).. 15

    With three ocelli (triocellate) 18

15. Posterior margin of mesosternum fringed with setae; a single bunch of lateral gills arising dorsal to the coxa of the metathoracic leg .... *Phanoperla* (in part)[28]

    Posterior margin of mesosternum lacks a setal fringe; two bunches of lateral gills arising dorsal to the coxa of the metathoracic leg 16

16. Occipital ridge with close-set, complete row of short, thick setae; body densely clothed with short black hairs................ *Tetropina*

    Occipital ridge with a few short, thick setae laterally; a few short brown hairs on the body surface 17

17. Lateral margins of pronotum completely fringed with thick setae; anal gills absent .... *Etrocorema*

    Lateral margins of pronotum with an incomplete fringe of thick setae; anal gills usually present...................................... *Neoperla*

18. Posterior margin of mesosternum fringed with setae; a single bunch of lateral gills arising dorsal to the coxa of the metathoracic leg .... *Phanoperla* (in part)

    Posterior margin of mesosternum lacks a setal fringe; two bunches

---

[28] *Chinoperla* — which has yet to be described in the larval stage — may key out here (see Zwick, 1982b).

of lateral gills arising dorsal to the coxa of the metathoracic leg 19

19. Anal gills present 20

Anal gills absent 22

20. Posterior setal fringe of abdominal segment VII incomplete on the underside (note that the last abdominal segment is X) ....
....*Paragnetina*

Posterior setal fringe of abdominal segment V.II complete on the underside 21

21. Fringe of setae on the hind margins of abdominal terga II-IV with (at least) a few conspicuously long setae intermixed with short thick ones; fewer than ten thick setae on the surface (excluding the margins) of abdominal tergum IX .... *Agnetina*

Fringe of setae on the hind margins of abdominal terga II-IV consists of relatively uniform setae; typically more than ten thick setae on the surface (excluding the margins) of abdominal tergum IX .....
.... *Claassenia*

22. Thorax and abdomen with a median row of long silky setae ....
.... 23

Thorax and abdomen without a median row of long silky setae .... *Togoperla*

22. Three single gills arising dorsal to the coxa of the mesothoracic leg .... *Tyloperla*

Three double gills arising dorsal to the coxa of the mesothoracic leg .... *Paragnetina* and *Kamimuria*

Early work on the Plecoptera of Asia (e.g. Banks, 1924; 1931; 1937; 1938; 1939; 1940; Wu, 1936a; 1936b; 1936c; 1936d; 1937a; 1937b; 1937c; 1937d; 1938; 1939; 1940; 1947; 1948a; 1948b; 1949; Wu & Claassen, 1934; Chao, 1946) was undertaken in circumstances where the global diversity of the group had been underestimated, and the literature contains many descriptions of species placed in the wrong genus or (in the case of authors such as Klapálek and Navás) where insufficient information is given to allow a correct reassignment without re-examination of the type specimens (Zwick, 1988a). In some instances, the situation has been resolved by more recent research but, unfortunately, the types of many of Asian Plecoptera — especially those described by the eminent entomologist Wu Chen-fu — have been lost or destroyed (Sivec, 1981a; Sivec & Zwick, 1987a; 1988a; I. Sivec, pers. comm.). Only the types of species named in Wu's last two papers (Wu, 1962; 1973) still exist (Sivec *et al.*, 1988). Accordingly, the status and geographic range of some genera remains uncertain. In other genera, even those which are apparently quite speciose, many more species exist than have actually been named to date. *Neoperla* (Perlidae) is a good example, and is probably the most diverse genus in the order Plecoptera (Zwick, 1986a). There is a great deal of local species-level endemicity among Oriental Plecoptera, especially among island faunas, which reflects the limitation of tropical Plecoptera to swift streams in mountainous areas. A lack of suitable habitat in lowland areas restricts their spread within river systems, with the consequence that typical stonefly habitats are widely separate and isolated (Zwick, 1986a).

Perlidae

Perlid larvae are characterized by the presence of thoracic gills, medium to large size, and labial glossae which are rounded and globular. They are predominately predaceous, although smaller larvae may ingest algae and/or detritus (Hynes, 1976). Dietary composition reflects the degree of overlap in microhabitat between predator and prey (which determine encounter rates), and its interaction with attack propensity and capture success. Slow-moving or sedentary prey with inefficient defenses usually make up the bulk of the diet (Tikkanen *et al.*, 1997 and references therein). The impact of predatory stoneflies on their prey may be significant, but studies which unequivocally demonstrate this point or estimate its magnitude are rare (Sih & Wooster, 1994; Kerans *et al.*, 1995; Kiffney, 1996; see Peckarsky *et al.*, 1997 for a discussion of this issue). Perlids may also influence community structure in a minor way by acting as attachment sites for ectosymbiotic chironomids (Bottorff & Knight, 1987).

Among the Oriental Perlidae, *Agnetina* (including *Phasganophora* and Asian *Marthamea* species) occurs in Vietnam, China and Taiwan (Banks, 1940; Zwick, 1984b; Stark & Sivec, 1991), and *Togoperla* extends from Thailand northward to Japan (Banks, 1940; Illies, 1966; Stark & Sivec, 1991). Illies (1965) states that *Formosita*, *Javanita*, *Oodeia*, *Neoperlops* and *Tylopyge* are established over the whole of the Oriental Region (see also Geijskes, 1952). *Kiotina* ranges from Japan southward through China to North Vietnam and Thailand (Illies, 1966; Kawai, 1968a; Yang & Yang, 1992; 1993), while *Ochthopetina* (which, together with *Oodeia*, Illies [1966] treats as a synonym of *Neoperla*) is widespread in Southeast Asia. Other perlids occur endemically in Taiwan (*Mesoperla* and *Schistoperla*; see also Kawai, 1968c), China (*Nirvania* and *Sinoperla*), Malaysia (*Etrocorema*, *Kalidasia* and *Neoeuryplax*), the Philippines (*Folga*) and Borneo (*Dyaperla*) (Illies, 1965). Many of these records must be treated with caution. For example, recent work has shown that *Etrocorema* occurs China, Thailand, Peninsula Malaysia, Borneo and Sumatra (Kawai, 1968a; Bishop, 1973; Zwick, 1982d; 1984a; Zwick & Sivec, 1985; Uchida & Yamasaki, 1989). Furthermore it is apparent that, for some time, the systematics of perlid stoneflies (especially the subfamily Perlinae and tribe Perlini) have been in an '. . . abysmal state . . .' and '. . . almost inextricably tangled . . .' (Sivec *et al.*, 1988: p. 3). This confusion is reflected in, for example, the transfer of certain species of *Tylopyge* to *Tyloperla* (Stark & Sivec, 1991) while others have been synonymised with *Paragnetina* (Zwick, 1988b). In addition, some species of *Javanita* have been reassigned to *Chinoperla* and the status of *Sinoperla* is questionable (Zwick, 1982). Other widespread perlid genera, such as *Kamimuria* (Banks, 1940; Harper, 1977; Yang & Yang, 1993) have been considered problematic also (Zwick & Sivec, 1980; Sivec, 1981b). Investigations of the full range of species within the region are needed to determine unambiguous generic limits between (for example) *Kamimuria*, *Paragnetina* and *Togoperla*, and among genera such as *Etrocorema*, *Neoperlops* (from China) and *Tetropina* (confined to Borneo) which have uncertain affinities and unknown or little-studied larvae (Zwick, 1984a).

The situation has been clarified recently by a revision of the World genera of Perlinae (Sivec *et al.*, 1988). It is now apparent that 19 genera in three tribes can be recognized; 13 genera occur in the Oriental Region (Figs. 4.73 & 4.74). The tribe Claasseniini includes the single genus *Claassenia* which occurs throughout China and into mainland Southeast Asia; Sivec *et al.* (1988) figure the larval stage. The tribe Perlini is represented by eight genera in Asia: *Tyloperla*, *Paragnetina* and *Agnetina* (Taiwan and mainland Southeast Asia), *Togoperla* (mainland Southeast Asia), *Kamimuria* (Taiwan and widespread over mainland Asia), *Etrocorema* (Southeast Asian mainland, Sumatra and Borneo), *Tetropina* (Borneo and Peninsular Malaysia) and *Neoperlops* (China and Vietnam). *Oyamia* occurs in Asia also but is confined to

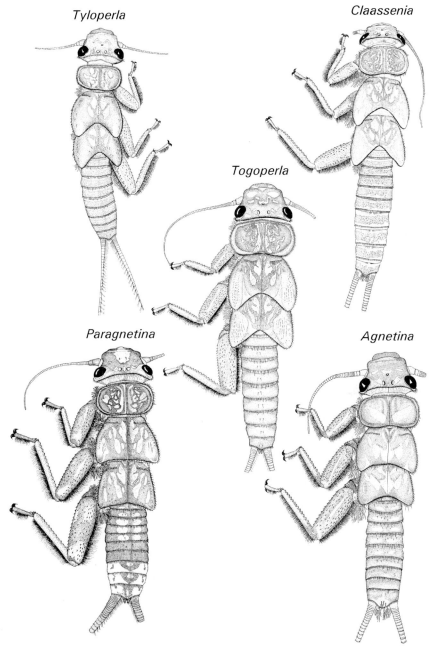

Fig. 4.73 Oriental Perlinae (Plecoptera: Perlidae) larvae (reproduced from Sivec *et al.* [1988] with permission).

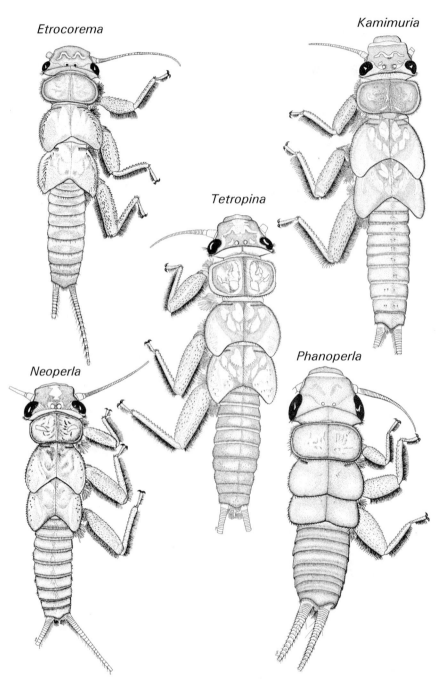

Etrocorema

Kamimuria

Tetropina

Neoperla

Phanoperla

Fig. 4.74 Oriental Perlinae (Plecoptera: Perlidae) — including Neoperlini — larvae (reproduced from Sivec *et al.* [1988] with permission).

the eastern Palaearctic and does not extent south into the Oriental Region. Sivec *et al.* (1988) provide illustrations of the larvae of all of these genera except *Neoperlops* which is unknown in the larval stage. Additional larval drawings of *Kamimuria* are given by Sivec (1981b) and Uchida & Isobe (1991), and Stark & Szczytko (1981a) figure *Paragnetina* larvae and note that the presence or absence of anal gills is a variable characteristic in this genus (see also Kawai, 1968a: Fig. 67). Illustrations of perlid larvae (e.g. Kawai, 1968a) published prior to the revision by Sivec *et al.* (1988) should be considered in the light of this more recent research.

Four genera in the tribe Neoperlini occur in tropical Asia (Fig. 4.74). *Furcaperla* contains a single Chinese species (originally placed in *Tylopyge*) from Fukien Province; the larval stages have yet to be described (Sivec *et al.*, 1988). *Neoperla* is found in North America and throughout the Oriental and Afrotropical regions (Banks, 1924; 1937; 1940; Geijskes, 1952; Jewett, 1958; Kawai, 1968a; 1969; 1973; Stark & Szczytko, 1979; Zwick, 1980; 1981; 1982; 1983a; 1986a; 1988a; Sivec, 1984; 1986a; Sivec & Zwick, 1987a; Zwick & Sivec, 1980; 1985; Stark, 1983; 1987; Uchida & Yamasaki, 1989; Yang & Yang, 1992; 1993; 1995a) The genus includes almost 200 species although the Chinese fauna is still inadequately known (Sivec *et al.*, 1988). Larvae of *Neoperla* — like all Perlidae — are largely or entirely carnivorous; adults are nocturnal and do not feed (Zwick, 1976).

Phanoperla is endemic to the Oriental Region (Zwick, 1982b) occurring from Sri Lanka and India (including Nepal), through Southeast Asia, Indonesia and the Philippines to Hainan (Jewett, 1958; Kawai, 1968a; 1969; 1973; Sivec, 1981b; Zwick, 1982b; 1982c; 1986b; Stark, 1983; 1987; Zwick & Sivec, 1985; Uchida & Yamasaki, 1989). This rather wide range is in stark contrast to Illies' (1965) view that *Phanoperla* is a Philippine endemic (see above), and reflects the extent to which our knowledge of Asian stoneflies has increased in recent years. The related genus *Chinoperla* (= some *Javanita* and *Ochthopetina*, and extant species of the fossil genus *Sinoperla*) ranges from southern China, Thailand and Peninsular Malaysia into Borneo and Sumatra and westward into India (Zwick, 1981; Zwick, 1982a; Sivec & Zwick, 1987b), although knowledge of the Chinese species is rather poor (Zwick & Sivec, 1980; Zwick, 1982a; Sivec & Zwick, 1987b). The larvae of *Chinoperla* have yet to be described. Larval illustrations for *Neoperla* and *Phanoperla* are given by Sivec *et al.* (1988), while Kawai (1973) and Zwick (1982b) figure larvae of *Neoperla* and *Phanoperla* (respectively). Zwick (1982b) notes that the posterior part of the head (anterior to the occipital ridge) of *Phanoperla* is pale and this is a distinctive feature of living specimens. In addition, the cerci tend to be short and stout, and anal gills are present. Like all Neoperlini, members of the genus *Phanoperla* are supposed to lack an anterior ocellus, but two species of *Phanoperla* are triocellate while two Perlini genera (*Tetropina* and *Neoperlops*) are apparently biocellate (Zwick, 1984a).

Among the perlid subfamily Acroneuriinae, the current situation is less clear than that in the Perlinae. *Acroneuria* (*sensu lato*: including *Calineuria*) ranges from Japan (Kawai, 1976) through China as far south as Hainan Island (Wu & Claassen, 1934; Wu, 1937c; Banks, 1940; Yang & Yang, 1995a) to Nepal (Harper, 1977; Sivec, 1981b). This genus is badly in need of revision (Uchida, 1983). Two species of *Perlesta* have been described from China and placed (incorrectly) within the Perlinae (Wu, 1938) but, because this acroneuriine genus appears to be an eastern Nearctic endemic (Stark, 1989b), the status of the Chinese species is uncertain. Wu (1937c; 1973) lists *Atoperla*, *Acroneuria*, *Gibosia* and *Kiotina* as among the Acroneuriinae of China, in addition to *Claassenia* which is now included within the Perlinae (see above). The Chinese *Atoperla* have since been reassigned to *Gibosia* which is known also from Taiwan and Japan but may be more widespread. *Kiotina* is found in Japan and China extending southward into Vietnam. Other Oriental genera include *Brahmana* (India and Nepal), *Kalidasia* (Peninsula Malaysia and India), *Mesoperla* (Taiwan), *Neoeuryplax* (Peninsular Malaysia) and *Nirvania* (Vietnam). A lack of information on the larval morphology of most of these genera precludes construction of a key to the subfamily, but larval Acroneuriinae form Japan are figured by Uchida (1983: *Calineuria*) and Kawai & Isobe (1985: *Acroneuria*, *Gibosia* and *Kiotina*).

Peltoperlidae and Styloperlidae

Although the Peltoperlidae extend into the Oriental Region, the distribution of the family as a whole is centred on Palaearctic East Asia and North America (Illies, 1965). The larvae are aberrant in their general appearance which is best described as 'cockroach-like'. They are detritivores and can be found among leaf packs and detritus accumulations where large larvae feed as shredders. A number of peltoperlid genera have been reported from tropical Asia, including *Peltoperla* (Banks, 1940; Kawai, 1961; 1968a; Wu, 1973), *Peltoperlodes* (Kawai, 1968a; 1969) and *Peltoperlopsis* (Kimmins, 1950; Kawai,

1973), as well as *Microperla, Neopeltoperla, Nogiperla* and *Cryptoperla* (Banks, 1938; 1940; Illies, 1966; Sivec, 1981b). The Holarctic genus *Yoraperla* is also known from Asia, but is restricted to Japan and Korea (Uchida & Isobe, 1988; Stark & Nelson, 1994). Until recently, understanding of the generic limits among Oriental Peltoperlidae was inadequate. This was partially rectified by Uchida & Isobe (1988) who reassigned all *Neopeltoperla* and *Nogiperla* species to the genus *Cryptoperla*. Although a full revision of the group must await data on larval morphology, Stark (1989) has done much to alleviate a confused situation: he has placed some genera in synonymy (e.g. *Peltoperlodes* with *Cryptoperla*), and established a new genus — *Peltopteryx*. Thus the current situation is that there are five peltoperlid genera in tropical Asia: *Cryptoperla* (widespread on the mainland; see also Sivec, 1995), *Microperla* (South China), *Peltopteryx* (northern India), *Peltoperlopsis* (the Philippines and Borneo) and *Peltoperla sensu lato* (Taiwan and China). The larvae of *Peltopteryx* have yet to be described, and the immature stages of *Peltoperla sensu lato* may not match the larval stages of Nearctic *Peltoperla sensu stricto* (Stark, 1989a). *Microperla* larvae are distinctive because they have three ocelli and lack gills, while *Peltoperlopsis* and *Cryptoperla* have two ocelli and gills on the thorax and tip of the abdomen. They can be distinguished on the basis of gill arrangement because *Peltoperlopsis* have single gills under the pro- and mesothoracic sternal plates, whereas *Cryptoperla* lack such gills (Stark, 1989a) although they have a pair of supracoxal gills on the meso- and metathorax (Uchida & Isobe, 1988). Uchida & Isobe (1989) established the subfamily Microperlinae to include *Microperla*, and have placed the remaining biocellate genera of peltoperlids within the Peltoperlinae.

The position of two genera — *Styloperla* and *Cerconychia* — which were incorrectly placed in the Perlidae upon initial description (Chao, 1946; Illies, 1965), is not addressed by Stark (1989a), but Illies (1966) includes them in a separate subfamily of the Peltoperlidae: the Styloperlinae. The matter has been resolved by Uchida & Isobe (1989) who elevated the subfamily to family rank. *Styloperla* (including *Nogiperla* described by Wu, 1973) appears to be widespread in southern China (Uchida & Isobe, 1989: Yang & Yang, 1990, 1992). *Cerconychia* was thought to occur in Taiwan only, but a new species was described recently from Hainan Island (Yang & Yang, 1995a). *Cerconychia* larvae lack the cockroach-like *habitus* of Peltoperlidae and have rather slender bodies (Uchida & Isobe, 1989). Gills are lacking. Larvae have two ocelli, long antennae (over two-thirds body length), and cerci which are approximately half of the body length. Mouthpart morphology indicates a phytophagous habit. *Styloperla* larvae as, as yet, unknown.

Nemouridae

Examination of the literature on Asian nemourids suggests that *Amphinemura* (Kimmins, 1950; Geijskes, 1952; Kawai, 1963; 1966; 1968c; 1969; zwick & Sivec, 1980; Sivec, 1981b; 1982; Zwick, 1983b), *Nemoura* (Kimmins, 1950; Kawai, 1968c; 1969; Zwick & Sivec, 1980; Sivec, 1981a; 1981b; 1981c) and *Protonemura* (Kimmins, 1950; Geijskes, 1952; Kawai, 1969; Harper, 1974) are widespread from Japan through China and Southeast Asia into the Indian subcontinent (Kimmins, 1950). *Indonemoura* is reported from Nepal (Sivec, 1981b; including a *Protonemura* described by Harper, 1974), and *Mesonemoura* occurs in Nepal also (Zwick & Sivec, 1980; Sivec, 1981b). The world fauna of Nemouridae has been reviewed by Baumann (1975) who provides diagnoses for the world genera. He cautions that the '. . . Asian Nemouridae have always been placed in one of the European genera even thought they did not belong there . . .' (Baumann, 1975: p. 1). In Baumann's view (see also Sivec, 1981b; 1981c), the Oriental genera comprise *Amphinemura* (China, Taiwan, Vietnam, Borneo, Java, Sumatra, India, Nepal and the Himalaya), *Mesonemoura* (China, northern India, Nepal and the Himalaya) and *Indonemoura* (Peninsular Malaysia, Java, Sumatra, Bali, India and probably Burma and Thailand) — placed in the subfamily Amphinemurinae — as well as *Illiesonemoura* (Taiwan, China, Nepal and the Himalaya) and *Nemoura* (Taiwan, China, Nepal and northern India) in the subfamily Nemourinae (Fig. 4.75). Note that Zwick & Sivec (1980) have suggested that *Illiesonemoura* may be synonymous with *Nemoura*. *Protonemura* is known from Nepal (Sivec, 1981c), but is otherwise confined to Palaearctic Asia (Baumann, 1975). Larval descriptions of nemourid genera are given by Geijskes (1952: *Amphinemura*; in addition, Fig. 6 is not *Protonemura* and should be relabelled *Indonemoura*), Baumann (1975: world genera) and Sivec (1981c: *Indonemoura, Mesonemura, Nemoura* and *Protonemura*). Sivec (1981c) notes that congeneric species of Nemouridae are not as uniform in some of their larval characters (such as gills and setal arrangement) as was presumed by Baumann (1975). In particular, the arrangement of the spines on the forelegs may be of limited use to distinguish

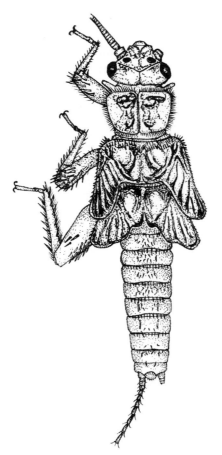

Fig. 4.75 *Nemoura* (Plecoptera: Nemouridae): dorsal view of larva (redrawn from Sivec, 1981c).

genera, and the cervical gills of some Asian nemourids do not match closely the larval diagnoses supplied by Baumann (1975). In essence, current knowledge of the Nemouridae is still inadequate to permit clear differentiation of Asian genera. A recent report of *Indonemoura* from Japan (Shimizu, 1994) indicates that the genus is more widespread than was previously supposed, and draws attention to our ignorance of these animals. Nemourid larvae can be quite abundant in collections from small streams draining forested catchments. They are phytophagous, subsisting on fine detritus and algae when small, but larger larvae may feed (at least in part) as shredders.

Leuctridae and other Plecoptera

Although Illies (1965) does not include the Leuctridae among the stonefly fauna of the Oriental Region, *Rhopalopsole* (Fig. 4.76) ranges from Japan southward through Taiwan and China to the Philippines, Vietnam, Peninsular Malaysia and Borneo (Jewett, 1958; Kawai, 1968c; 1969; Yang & Yang, 1993; 1995b). It is found in India and Nepal also (Harper, 1977; Zwick & Sivec, 1980; Sivec, 1981b). *Leuctra* is a primarily Palaearctic genus, but there are records of its occurrence in China along the northern fringes of the Oriental Region (Wu, 1937d; 1973). The picture is obscured by the tendency of some workers (e.g. Jewett, 1958) to treat *Rhopalopsole* as a subgenus of *Leuctra*. Illies

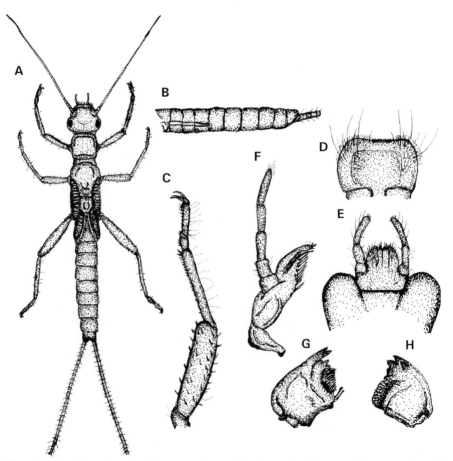

Fig. 4.76 *Rhopalopsole* (Plecoptera: Leuctridae): A, dorsal view of larva; B, lateral view of abdomen; C, foreleg; D labrum; E, labium; F, maxilla; G & H, mandibles.

(1966) reassigned some of these Chinese records to *Rhopalopsole*, but Yang & Yang (1995b) judge at least one of them (*Rhopalopsole orientalis*) to represent a member of the Palaearctic genus *Paraleuctra* (also reported from Nepal: Harper, 1977; Zwick & Sivec, 1980). Despite this uncertainly, the data suggest that *Rhopalopsole* is probably the only widespread genus of Leuctridae in tropical Asia. Leuctrids are not common in collections from the stream benthos as they dwell in the hyporheic zone for much of their larval life. The long slender body seems to be an adaption for a burrowing lifestyle, and most species are white or pale yellow as is often characteristic of animals which spend their days in stygian darkness.

Many Asian Plecoptera genera are confined to the Palaearctic Region or occur at high altitudes along the northern fringes of the tropics. Japan, for example, has 47 genera (Kawai, 1976) a number of which are endemic or (as in *Scopura*) shared with Korea and/or the Nearctic Region (Illies, 1965). This diversity of stoneflies is considerably greater than is found than Southeast Asia. Adults of the Scopuridae (comprising five species of *Scopura*) are apterous, and larvae have a rosette of gill tufts encircling the end of the abdomen. These insects are confined to boreal or alpine streams with mid-summer temperatures lower than 10°C (Kawai, 1974a; Uchida & Maruyama, 1987). They are not considered further herein. The Capniidae, as *Capnia* and *Allocapniella*, have been recorded at altitudes of over 3,400 m in the high Himalaya (Kawai, 1966; 1968b; see also Harper, 1977; Zwick & Sivec, 1980; Sivec, 1981b), and *Eucapnopsis* is known from Nepal (Zwick & Sivec, 1980). The larvae of *Allocapniella* is figured (as *Capnia*) by Kawai (1963; see also Kawai & Isobe, 1985). The Taeniopterygidae occurs in Palaearctic Asia and is represented by *Kyphopteryx*, *Mesyatsia* (including *Rhabdiopteryx*) and *Taenionema* in the Himalaya and/or Japan and the Soviet Far East (Kimmins, 1947; Jewett, 1970; Zwick & Sivec, 1980; Sivec, 1981b; Stanger & Baumann, 1993; Zhiltzova & Zwick, 1993). The family has not spread further south. Larval Taeniopterygidae can be recognized by the combination of gills at the bases of the legs and the subequal length of the tarsal segments. These and the related Capniidae, Leuctridae and Nemouridae always lack anal gills, but accessory gills may be present on the throat or leg bases.

Perlodes, *Isoperla* (Perlodidae) and *Alloperla* (Chloroperlidae) are known as larvae from the Himalaya (Kawai, 1963; 1966; see also Jewett, 1970), where *Xanthoperla* (Chloroperlidae) occurs also (Zwick & Sivec, 1980). *Alloperla* and *Chloroperla* have been reported from northern China (Wu, 1937c; 1973). Sivec (1981) records the perlodid

Neofilchneria from Nepal, and this genus occurs also in Tibet and northern China (Wu, 1936a). As far as is known, perlodid larvae in Asia do not have thoracic gills. Chloroperlids are smaller than perlodids, with shorter legs and cerci, and lack distinctive colour patterns on the body. Records of other perlodids in Palaearctic Asia (e.g. Wu, 1936a; 1973; Kawai, 1974b; Illies, 1966; Stark & Szczytko, 1981b) include *Perlodinella* (Tibet), *Protarcys* (northern China) and *Skobeleva* (northern China, Tibet, the Himalaya), as well as *Skwala* and *Sopkalia* (Japan). Although this list of genera is by no means comprehensive, it is evident that neither the Perlodidae nor Chloroperlidae should be considered as part of the Oriental fauna.

HETEROPTERA

The Hemiptera — or bugs — has been treated as comprising two suborders, only one of which — the Heteroptera — is aquatic or semiaquatic. There has been some disagreement over the years as to whether the Heteroptera warrants the status of an order, or should retain subordinal rank within the Hemiptera. For the remainder of this text I have decided to follow Henry & Froeschner (1988) and treat the taxon as an order, which makes ecological sense as all aquatic bugs are Heteroptera and none are Homoptera. A recent account of the group (Hilsenhoff, 1991) follows this arrangement also. The Heteroptera is a conspicuous component of the stream-associated fauna because members of the suborder Gerromorpha are neustic and walk or skate over the water surface. Like the whirligig beetles (Gyrinidae: p. 460), they forage for insects which have drowned or become trapped in the surface film, and are themselves protected from predators by chemicals produced from metasternal and abdominal scent glands. Unlike gyrinids, neustic bugs do not have aquatic larvae; both immatures and adults spend their entire lives above the water surface. This is a consequence of the combination of surface tension and hydrofuge hairs on the tibiae. At best, such Heteroptera can be described as semiaquatic: it is only their locomotory adaptations which show any modification for life in (or on) an aquatic milieu. A smaller number of taxa in the suborder Nepomorpha is fully aquatic, with adults and larvae which spend their lives submerged. Although aquatic, adults often have the ability to fly.

The combined total number of aquatic and semiaquatic species of

Heteroptera approaches 4,000. The Oriental Region is particularly rich, and in total number of species in most families is exceeded by no other region of comparable area; there is a high proportion of endemic genera and even subfamilies (Bishop, 1973; Andersen, 1982a; Spence & Andersen, 1994). The group is characterized by forewings (hemelytra, sometimes spelled hemielytra) which are hard and leathery in the basal (anterior) portion and membraneous posteriorly (although this part is thickened in the Pleidae and Helotrephidae). In addition, they all have beak-like mouthparts. In the majority of species they are modified into a cylindrical piercing and sucking structure (the rostrum), and feeding involves the injection of enzymes (and sometimes venom) into prey followed by extracorporeal digestion of tissue which is sucked out and swallowed. The forelegs of the Belostomatidae, Naucoridae, Nepidae and Gerridae are raptorial and variously modified for grasping prey; these adaptations are described by Gorb (1995). Some members of the Corixidae supplement feeding on small invertebrates with a secondarily herbivorous habit, and the mouthparts are adapted for rasping rather than sucking. Polymorphism in wing development (= alary) is common in many Heteroptera (e.g. Gerridae, Helotrephidae, Hydrometridae, Naucoridae and Veliidae). Adults may have fully-developed wings (= macroptery), short wings (= brachyptery), or may lack wings entirely (= aptery).

Life-history data on Oriental Heteroptera are fragmentary. Most have five larval instars (more rarely, four) which grow rapidly. They pass the dry season (usually the cooler months in the northern part of the tropics) in the adult stage which is relatively long lived. Ovarian diapause may occur during this dry period (Selvanayagam & Rao, 1988). Consequently, there are distinct seasonal fluctuations in population size (e.g. Tonapi, 1959; Nirmalakumari & Balakrishnan-Nair, 1984; Jebensan & Selvanayagam, 1994). Multivoltinism seems to be possible for Gerridae which complete larval development within about six weeks, allowing five or six generations per year (Hoffman, 1936a; 1936b; Selvanayagam, & Rao, 1988). The belostomatid *Diplonychus rusticum* may complete four generations per year (Su & Yang, 1992) and take little more than a month to mature (Saha & Raut, 1992), although the much larger *Lethocerus indicus* can manage only one or two generations annually (Hoffman, 1931). Altitude affects voltinism: a comparison of two Japanese *Diplonychus* species showed that one generation was completed each year in cooler, upland streams whereas two generations were possible in warmer lowland rivers (Okada & Nakasuji, 1993a; 1993b). Unfortunately, most life-history investigations of Asian Heteroptera have been carried out under laboratory conditions. Caution is needed when extrapolating from them to field situations where conditions are often suboptimal and life-cycle completion is slowed (Andersen, 1982a). An interesting facet of reproduction in some Heteroptera (including Belostomatidae, Gerridae, Notonectidae and Veliidae) is the generation of ripples to produce 'calling signals' during courtship (J.T. Polhemus, 1990a; Wilcox, 1995). Many taxa (e.g. Corixidae) have the ability to signal acoustically by stridulation also (J.T. Polhemus, 1994).

The following key to aquatic and semiaquatic Heteroptera has been modified from Hilsenhoff (1991); Andersen (1982a) gives a detailed key to the world genera of Gerromorpha. Successful identification requires the use of mature specimens. I have used faunal inventories for parts of the region (including Hoffman, 1933; Lundblad, 1933; Fernando, 1974; Polhemus, 1979; Yano *et al.*, 1981; Ameen & Nessa, 1985; Kovac & Yang, 1990; Polhemus & Polhemus, 1990; Nieser & Chen, 1991; 1992a; Chen & Andersen, 1993; and references therein), combined with data from Andersen (1982a) to describe the composition of the Nepomorpha and Gerromorpha of tropical Asia. The summary of the primary literature for the Gerromorpha draws heavily upon Andersen (1982a).

1. Antennae shorter than the head, inserted beneath the eyes and — in most instances — not visible from above.............................
.... suborder **Nepomorpha** 2

   Antennae markedly longer than the head, inserted in front of the eyes and visible from above **11**

2. Mouthparts short and broad, rather blunt and triangular or beak-like; the foretarsus is modified into a setaceous, scoop-like structure ....**Corixidae**

   Mouthparts elongate and cylindrical or cone-shaped, divided into segments and obviously adapted for piercing; foretarsi not as above (often raptorial) 3

3. Apex of abdomen terminates in a slender respiratory tube (over 4 mm long) made up of two grooved filaments .... **Nepidae**

   Posterior respiratory appendages either absent or very short.... 4

4. Mid- and hindlegs legs with fringes of swimming hairs; aquatic 5

Mid- and hindlegs legs without fringes of swimming hairs; riparian 10

5. Dorsoventrally flattened and somewhat oval insects; forelegs raptorial with somewhat broadened femora 6

   Elongate *or* hemispherical and distinctly ovid insects, not flattened; with rather slender forefemora 8

6. The tip of the abdomen bears a pair of short, strap-like appendages; eyes protrude from margin of the head; anterior margin of the pronotum is not concave; mid- and hindlegs somewhat flattened; often rather large (> 15 mm long and sometimes *much* larger) .... **Belostomatidae**

   Without appendages at the tip of the abdomen; eyes are flush with the margin of the head (i.e. they do not protrude); anterior margin of pronotum concave; mid- and hindlegs *often* more-or-less-cylindrical; relatively small (*usually* < 20 mm long) .... **Naucoridae** 7

7. The tip of rostrum extends back to the base of the hindlegs; forefemora broadened slightly; adults often apterous, rarely macropterous .... **Naucoridae: Aphelocheirinae**

   Rostrum relatively short and stout with the tip extending back to the base of the forelegs; forefemora broadened markedly; adults usually macropterous or brachypterous but sometimes apterous .... **Naucoridae** (in part, mostly **Cheirochelinae**)

8. Hemispherical and ovoid (the head and thorax are fused) with mid- and hindlegs of similar length; small, less than 5 mm long 9

   Elongate with long hind legs which are oar-like and fringed with hairs; usually over 5 mm long .... **Notonectidae**

9. Antennae with two segments; rounded anteriorly with the sides of the thorax tapering towards the head; small to minute (generally < 4 mm long) .... **Helotrephidae**

   Antennae with three segments; slightly convex anteriorly so that the 'face' appears rather flat and the sides of the thorax are almost parallel; up to 5 mm long .... **Pleidae**

10. Forelegs raptorial; short antennae not visible from above .... **Gelastocoridae**

Forelegs not conspicuously raptorial; short antennae visible from above .... **Ochteridae**

11. Mid- and hindlegs *often* unusually long; coxae of the hind legs small; body *sometimes* elongate or very slender; membranous part of hemelytra (if present) without veins or with dissimilar sized cells; common and widespread with neustic or riparian species .... suborder **Gerromorpha** 12

Mid- and hindlegs not unusually long; coxae of the hind legs large; body neither elongate or especially slender; membranous posterior part of hemelytra with veins forming four or five equal-sized cells; riparian .... Saldidae

12. Claws of tarsi inserted before the apex .... superfamily **Gerroidea** 13

Claws of tarsi situated at the apex 14

13. Hind legs conspicuously long (femora much longer than the abdomen); mesothorax more elongate (when viewed from the underside) than the other thoracic segments .... Gerridae

Hind legs not conspicuously long (femora not noticeably longer than the abdomen); thoracic segments of approximately equal length .... Veliidae

14. Head as long as thorax with eyes set halfway along; typically with an elongate and delicate stick-like body .... **Hydrometridae**

Without the above combination of characters 15

15. Ventral surface of head appears grooved with a pair of prominent vertical plates covering the base of the rostrum; tarsi with two segments; hindlegs lack spines .... Hebridae

Ventral surface of head not as above; tarsi with three segments; hind legs armed with spines .... **Mesoveliidae**

Nepomorpha

Typically, the Nepomorpha have short antennae which are often concealed from view beneath the head. One concept of phylogeny of the suborder (e.g. McCafferty, 1981: Chapter 10) is that with exception of the Corixidae — placed in the superfamily Corixoidea — all nepomorphan families are included within the superfamily Notonectoidea. An alternative opinion (e.g. Hutchinson, 1993) is that the Nepidae and Belostomatidae comprise the Nepoidea, the Notonectidae, Pleidae and Helotrephidae are contained within the Notonectoidea, and the Naucoridae is the sole member of the Naucoroidea. The Notonectidae — or backswimmers — live in stream pools. They swim with the long, oar-like hind legs which are fringed with hairs. The forelegs are variously modified and used to capture prey. These bugs always orientate with the ventral side uppermost, hence their common name. Respiration involves breaking the water surface with the tip of the abdomen to replenish a plastron on the underside of the abdomen. Notonectids often have colourful patterns on the dorsal surface. Of the two subfamilies, the Anisopinae (including only *Anisops*; see Brooks, 1951; Leong & Fernando, 1962) are small (less than 10 mm long), sleek-bodied, graceful swimmers with neutral buoyancy which can be adjusted by the uptake and release of oxygen from haemoglobin in the body. The antennae and rostrum have three segments, and there is a hair-lined pit situated anteriorly in the commissure (join) of the hemelytra. The ecology of lentic species of Anisopinae is discussed by Hutchinson (1993: pp. 621–633). The Notonectinae (comprising *Aphelonecta*, *Enithares*, *Notonecta* and *Nychia*) are larger (over 10 mm long) and more robust. They lack the hair-lined, hemelytral pit and often hang by the abdomen from the water surface. Notonectinae swim with a jerky movement, and the beak and antennae are four-segmented. *Anisops* (which is in need of generic revision) and *Enithares* (reviewed by Lansbury, 1968) occur throughout tropical Asia. *Notonecta* is widespread in the Palaearctic (e.g. Hoffman, 1933; Nieser & Chen, 1991), and the ecology of the genus has been reviewed by Hutchinson (1993: pp. 585–621). The Southeast Asian genera *Aphelonecta* and *Nychia* have fewer species and are less known.

The Pleidae (Fig. 4.77A) is represented in tropical Asia by the widely-distributed genus *Paraplea* (Polhemus & Polhemus, 1990). This family was formerly viewed as a subfamily of the Notonectidae (e.g. Hoffman, 1933), and is treated as *Plea* by some authors (e.g. Ameen & Nessa, 1985; Benzie, 1989). The Pleidae is a distinctive group because of their strongly convex, ovoid shape and small size (less than 5 mm long). The head is rather poorly defined and appears fused with the thorax. The antennae have three segments. Apart from the presence of a beak, pleids might be mistaken for beetles because the hemelytra appear uniformly hardened and lack the membraneous hind portion

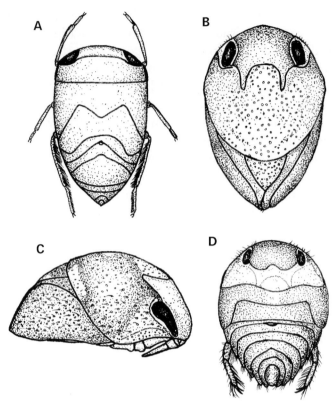

Fig. 4.77 A, *Paraplea* (Heteroptera: Pleidae) dorsal view; B, *Trephotomas* (Heteroptera: Helotrephidae) dorsal view of adult; C, lateral view of adult; D, dorsal view of larva (after Papácek *et al.*, 1988).

of other Heteroptera. The hind wings are vestigial. Although the legs have swimming hairs (and these bugs swim upside-down), the typical habitat is crawling among aquatic macrophytes in quiet water. Respiration is by a ventral plastron.

The Helotrephidae (Fig. 4.77B-D) is a pantropical family of minute bugs which do not exceed 4 mm long and are often much smaller. The head and thorax are fused into a cephalothorax, and the body is ovoid resembling the pleid genus *Paraplea*. However, pleids are only slightly convex anteriorly (so that the 'face' appears rather flat when seen from above, and the eyes are widely separated), whereas helotrephids are much more rounded and the sides of the thorax taper strongly toward the head (Fig. 4.77B). By contrast, the lateral margins of the prothorax are almost parallel in *Paraplea* (Fig. 4.77A). In addition, helotrephids have two-segmented antennae (*cf.* three segments in

Pleidae), and show considerably more generic diversity than the Pleidae. The Asia genera of Helotrephidae are *Distotrephes*, *Helotrephes*, *Heterotrephes* (southern Japan only), *Hydrotrephes*, *Idiotrephes*, *Limnotrephes*, *Tiphotrephes* and *Trephotomas*, plus *Mixotrephes* and *Paralimnotrephes* from Afghanistan (Lundblad, 1933; Papácek *et al.*, 1988; 1989; J.T. Polhemus, 1990b; 1997; Papácek, 1994; 1995). The group as whole is distinctly tropical, but does not extend into New Guinea (Polhemus & Polhemus, 1990). J.T. Polhemus (1990b) provides a world checklist of the family in which the Oriental fauna is subdivided into the Helotrephinae (*Distotrephes*, *Helotrephes*, *Hydrotrephes*, *Idiotrephes*, *Limnotrephes* and *Tiphotrephes*) and the Trephotomasinae (*Trephotomas*). The Helotrephinae have only a single tarsal segment on the first two pairs of legs, and two-segmented hind tarsi; the Trephotomasinae have two or (on the hind legs) three tarsal segments. Both macropterous and brachypterous morphs occur in (at least) *Idiotrephes* and *Mixotrephes*. Habitats probably vary among genera, but some (e.g. *Idiotrephes*) are quite eurytopic (Lundblad, 1993; Papácek, 1994). The biology of helotrephids is obscure, and nothing has been published on their life history or habits. A predatory diet is suggested by the morphology of the mouthparts (which are similar to those of most other aquatic Heteroptera), but the prey must be very small. Respiration involves a ventral plastron associated with dense pilosity on the abdomen.

The Nepidae are relatively large Heteroptera and may be up to 45 mm long (excluding the respiratory siphon). All nepids have raptorial forelegs (with a single claw) and a terminal respiratory siphon consisting of two grooved filaments. Oriental genera include *Borborophyes* (in the Philippines), *Cercotmetus*, *Laccotrephes*, *Montonepa*, *Ranatra* and *Telmatotrephes*. *Nepa* occurs in Palaearctic East Asia (Keffer *et al.*, 1990). There are two subfamilies with distinct body forms. The Nepinae (Fig. 4.78A) or waterscorpions (e.g. *Borborophyes*, *Laccotrephes*, *Montonepa* and *Telmatotrephes*) have rectangular and rather broad or more-or-less oval bodies; the slender and elongate Ranatrinae (e.g. *Cercotmetus* and *Ranatra*: Fig. 4.78B) are stick-like in appearance. Lansbury (1972a; 1973a) has revised the genera *Ranatra* and *Cercotmetus*, giving keys for the identification of their constituent species. Ranatrinae are usually found associated with trailing roots and vegetation along stream margins. Nepinae are benthic in shallow water with muddy sediments. They do not rely entirely on the siphon for respiration, as some taxa can store air beneath the hemelytra. Significantly, the metathoracic wings of *Telmatotrephes* are vestigial

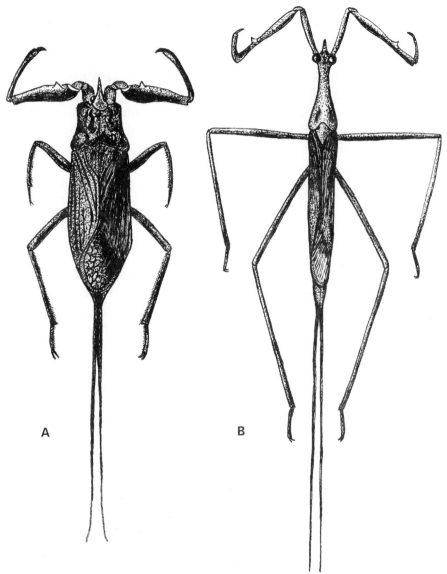

Fig. 4.78 Nepidae (Heteroptera): A, Nepinae; B, Ranatrinae (both from dorsal view).

(i.e. they are apterous), and those of *Borborophyes* are reduced; both genera have shorter respiratory siphons than other Nepidae (Keffer *et al.*, 1989; see also Lansbury, 1972b; 1973b). In general, the Nepinae move by slow crawling, and — unlike most other fully-aquatic Heteroptera — nepids are weak swimmers.

The Naucoridae (= creeping water bugs) are flattened, oval insects which may reach 25 mm in length but are usually smaller (around 15 mm). If wings are present, the hemelytra lack veins in the membraneous posterior part, but brachyptery or aptery are common (Figs. 4.79–4.81). Where wings are absent, a pair of scent glands on the mid-dorsum

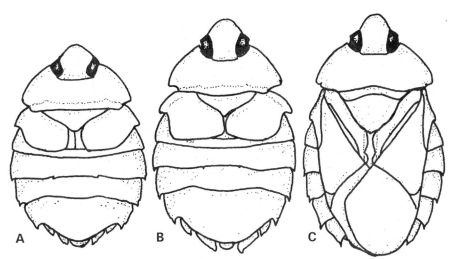

Fig. 4.79 *Aphelocheirus* spp. (Heteroptera: Naucoridae): A, brachypterous form of *Aphelocheirus dudgeoni*; B, brachypterous form of *Aphelocheirus malayensis*; C, macropterous form of *Aphelocheirus malayensis* (all from dorsal view; redrawn from D.A. & J.T. Polhmeus, 1988).

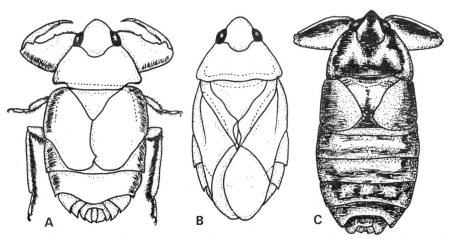

Fig. 4.80 Naucoridae (Heteroptera) from tropical Asia: A, brachypterous form of *Coptocatus*; B, macropterous form of *Coptocatus*; C, brachypterous form of *Idiocarus* (all from dorsal view; redrawn from D.A. Polhmeus [1986] and D.A. & J.T. Polhmeus [1986a]).

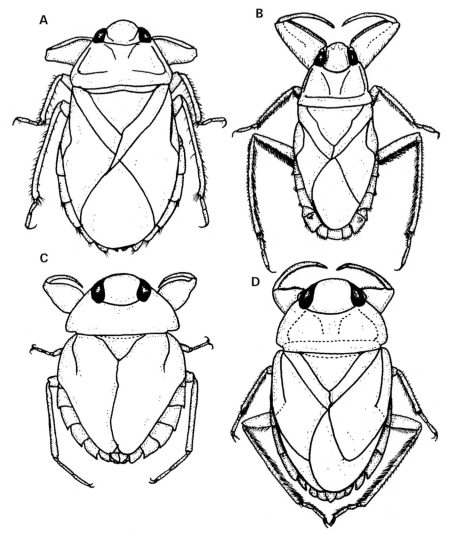

Fig. 4.81 Naucoridae (Heteroptera) from tropical Asia: A, *Nesocricos*; B, *Tanycricos*; C, *Cavocoris*; D, *Philippinocoris* (all macropterous forms from dorsal view; redrawn from D.A. & J.T. Polhmeus, 1985; 1986b; 1987; 1989).

of the abdomen may be apparent. The forelegs are conspicuously modified for holding prey, with expanded femora and a sickle-like distal segment. The mid- and hindlegs (especially the latter) are fringed with swimming hairs, but fast crawling is at least as important as swimming for these bugs. A diagnostic feature of the Naucoridae is the concave anterior margin of the pronotum which closely fits the hind margin of the head and usually extends along the sides to the eyes. As a result, the head appears to be part of the thorax rather than partially offset as in, for example, the Belostomatidae and Nepidae. Respiration in some genera is by a large, ventral plastron and because these insects are usually confined to well-oxygenated waters — especially stony streams — they rarely have to surface for air. If replenishment of the air store is necessary, it occurs through the tip of the abdomen.

The Naucoridae is particularly diverse in parts of tropical Asia and especially in New Guinea. The Oriental genera include *Aphelocheirus, Aptinocoris, Asthenocoris, Cavocoris, Cheirochela, Coptocatus* (confined to Borneo), *Ctenipocoris, Diaphorocoris, Gestroiella, Heleocoris, Idiocarus*[29], *Laccocoris, Naucoris, Nesocricos, Philippinocoris, Quadricoris, Sagocoris, Stalocoris, Tanycricos* and *Warisia* (La Rivers, 1970a; 1970b; D.A. Polhemus & J.T. Polhemus, 1985, 1986a; 1986b; 1987, 1988, 1989; J.T. Polhemus & D.A. Polhemus, 1990; Nieser & Chen, 1991; 1992a; Sites *et al.*, 1997). Eight of these genera are restricted to New Guinea (*Aptinocoris, Cavocoris, Idiocarus, Nesocricos, Quadricoris, Sagocoris, Tanycricos* and *Warisia*), and three are Philippines endemics (*Asthenocoris, Philippinocoris* and *Stalocoris*). Notwithstanding the comments of Nieser & Chen (1991; see also La Rivers, 1970b), *Sagocoris* is confined to New Guinea (D.A. Polhemus & J.T. Polhemus, 1987). *Aphelocheirus* is the only naucorid genus that New Guinea has in common with Asia or Australia (J.T. Polhemus & D.A. Polhemus, 1990). With that exception, the New Guinean naucorid fauna is entirely endemic, and appears to have radiated from a single ancestral stock (Polhemus & Polhemus, 1986b). *Ilyocoris* is known from Japan and northern China (Yano *et al.*, 1981) but does not penetrate tropical Asia. Nieser & Chen (1991) provide a key to 12 genera of Naucoridae from the region, eight of which are confined to New Guinea (see also Sites *et al.*, 1997).

The Aphelocheirinae (Fig. 4.79) — comprising *Aphelocheirus* (= *Tamopocoris*) only — are flattened, frequently wingless bugs which occur primarily in and among stones in riffles where the base substrate is gravel or coarse sand. They have less strongly-developed forefemora than other naucorids, and a relatively long rostrum. The species of *Aphelocheirus* occurring in tropical Asia have been reviewed by D.A. Polhemus & J.T. Polhemus (1988; see also Nieser & Chen, 1991;

---

[29] This genus is spelled incorrectly (as *Idiocoris*) in Table 1 of J.T. Polhemus & D.A. Polhemus (1990); the correct spelling (*Idiocarus*) is given by D.A. Polhemus & J.T. Polhemus (1986a).

Zettel, 1993; D.A. Polhemus, 1994a) and a key for their identification is given. There are two subgenera: *Aphelocheirus*, which is relatively speciose and, at 6–9 mm long, somewhat larger than *Micraphelocheirus* (< 5 mm long). The higher classification of the Naucoridae is in need of revision, and thus the limits and composition of the other subfamilies in tropical Asia are unclear. The Cheirochelinae (Figs 4.80 & 4.81) includes *Cheirochela*, *Gestroiella* and *Sagocoris* plus *Coptocatus* (reviewed by D.A. Polhemus, 1986; see also Nieser & Chen, 1991), *Idiocarus* (reviewed by Polhemus & Polhemus, 1986a) and *Tanycricos* (reviewed by Polhemus & Polhemus, 1986b). Most members of the subfamily *sensu stricto* have antennae recessed into grooves in the underside of the eyes; an extremely reduced labrum set in a deep cavity at the base of the rostrum; and, dense pads of hair at the apices of the mid- and hind tibiae. However, *Idiocarus* (which has a rather elongate body and resembles the New World Cryptocricinae: Fig. 4.80C) and *Tanycricos* have a well-developed labrum and slender, filiform antennae; all of the endemic New Guinea genera lack antennal grooves. Their inclusion within the subfamily Cheirochelinae is provisional pending a thorough revision of the group (Polhemus & Polhemus, 1989). The Naucorinae (which have foretarsi consisting of a single segment bearing a small claw) includes *Naucoris*, while *Ctenipocoris*, *Laccocoris* and *Heleocoris* are assigned to the Laccocorinae (with two-segmented foretarsi bearing two tiny claws [except in female *Heleocoris*]).

Little is known of the habits of naucorids in tropical Asia (but see Bisht & Das, 1982); all species are assumed to be predators and will attack prey larger than themselves. These insects bite if handled carelessly. Streams are the primary habitat of most naucorids, and many genera occur mainly in riffles (e.g. *Aphelocheirus*, *Diaphorocoris* and the Cheirochelinae *sensu lato*). Naucorids exhibit a strong tendency to aggregate in suitable microhabitats: for example, *Asthenocoris* and *Philippinocoris* are reported as sheltering in areas of relatively still water in the lee of boulders (Polhemus & Polhemus, 1988), whereas *Ctenipocoris*, *Laccocoris*, *Philippinocoris* and *Stalocoris* are associated with marginal roots and detritus away from the full force of the current (Bishop, 1973; D.A. Polhemus & J.T. Polhemus, 1987; Sites *et al.*, 1997). *Nesocricos* lives along the undercut banks of upland streams (altitudes > 800 m), where it shelters under rocks and logs (Polhemus & Polhemus, 1985). *Heleocoris* often occurs under algal mats on wet rock faces around cascades or in seeps (Mendis & Fernando, 1962; Bisht & Das, 1982). Exceptionally, *Naucoris* is most often encountered in standing water.

The Belostomatidae (or giant waterbugs) is represented by two widespread genera which occur throughout tropical Asia into New Guinea and Australia: *Diplonychus* (= *Sphaerodema*) and *Lethocerus* (sometimes incorrectly referred to as *Belostoma*: e.g. Tonapi, 1959). There are records of *Kirkaldyia* from Assam, China and Taiwan (Hoffman, 1933) — which actually to refer to *Lethocerus* — and a report of *Nectocoris* from Bangladesh (Ameen & Nessa, 1985). Like naucorids, the belostomatids are somewhat flattened and oval in shape, and the forelegs are robust and modified for grasping prey (Figs. 4.82 & Fig. 4.83). However, the mid- and hind legs are somewhat flattened (*cf.* more-or-less cylindrical in the Naucoridae) reflecting the relative importance of the swimming habit in the two families. Furthermore, the hemelytra of belostomatids have veins in the membranous part, the head is offset from — or protrudes in front of — the pronotum, and there are short, paired appendages at the end of the abdomen. These 'air straps' each bear a spiracular opening at the base. Belostomatids may reach a considerable size. The subfamily Lethocerinae includes very large bugs (*Lethocerus* are often well over

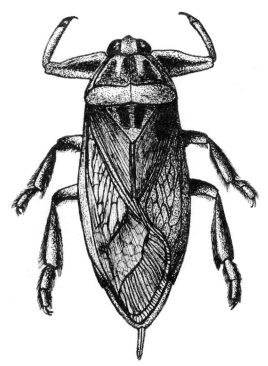

Fig. 4.82 *Lethocerus indicus* (Heteroptera: Belostomatidae): dorsal view of adult.

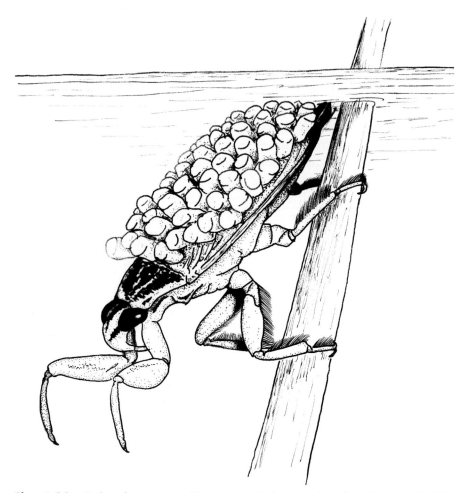

Fig. 4.83 *Diplonychus rusticum* (Heteroptera: Belostomatidae): lateral view of adult male carrying eggs.

60 mm long: Fig. 4.82) with paired claws on the raptorial forelegs; the smaller Belostomatinae (*Diplonychus* is usually < 25 mm long: Fig. 4.83) have single claws on the forelegs. Significantly, Venkatesan & Rhaghunatha-Rao (1981) ascribe all of the Indian Belostomatinae to *Diplonychus*.

Belostomatids are voracious predators, and can inflict a painful bite. They will attack almost any aquatic animal (even those larger than themselves) including pulmonate snails, insects, tadpoles and fish (e.g. Saha & Raut, 1992; Okada & Nakasuji, 1993b). In the laboratory, belostomatids are remarkably sensitive to the energetic benefits of different foraging modes (Dudgeon, 1990d; Cloarec, 1995 and references therein). The potential of belostomatids for the control of pestiferous mosquito larvae and snail parasite vectors has been mooted on a number of occasions (e.g. Sankaralingam & Venkatesan, 1989). These animals also find service as a source of human food: after they have been boiled in salt water or fried in oil, *Lethocerus* can provide a tasty snack. The head and hemelytra are discarded before the bugs are eaten (see Hoffman, 1931; Pemberton, 1988).

Belostomatids frequently fly to lights, but most often occur in slow-flowing streams where they are almost invariably associated with macrophytes and spend much time hanging from the water surface by the air straps. After mating, the female of *Diplonychus* lays her eggs on the back of the male whence they are carried around until hatching (Fig. 4.83). Male *Lethocerus* attend eggs laid above the water level on emergent vegetation. During the day, they remain in the water at the base of plant stems but, at night, climb up and cover the eggs with their wet bodies thereby moistening them (Ichikawa, 1988). The eggs fail to hatch if they are not supplied with water. When mature females encounter a brooding male in the water, they destroy his egg mass and take over the mate thereby gaining a nurse for their own egg mass (Ichikawa, 1995). Females are unable to detect brooding males when they are out of the water on the vegetation. Exclusive male parental investment of the type seen in *Diplonychus* and *Lethocerus* (where the female plays no part in parental care) is an extremely rare trait among animals.

The Corixidae — water boatmen — are small (no more than 10 mm long), somewhat flattened parallel-sided insects (Fig. 4.84) with a short,

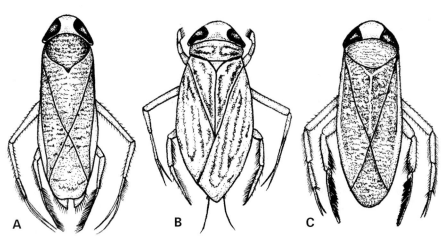

Fig. 4.84 Oriental Corixidae (Heteroptera): A, *Sigara*; B, *Micronecta*; C, Corixinae genus indet. (all from dorsal view).

blunt beak (*cf.* long and cylindrical in other groups). Although they resemble notonectids, they lack the deep bodies and long rostrum of those insects. In addition, the foretarsus of corixids is modified into a single-segment, scoop-like structure fringed with setae, and the hindlegs are oar-like and fringed with hairs. The tip of the abdomen is asymmetrical in the males of some species. Respiration involves an air bubble held by hydrofuge hairs on the underside of the body coupled to an air store beneath the hemelytra. Air-store replenishment occurs via the pronotum. Most species are able to stridulate. Corixids are most frequently collected in standing water or sandy-bottomed, slow-flowing streams where they are sometimes encountered in large schools or swarms. Hutchinson (1993: pp. 637–748) gives a detailed account of the ecology of lentic Corixidae. Feeding habits vary among taxa, and some are primarily predaceous (e.g. *Cymatia*) while others (such as the speciose genus *Sigara*) are exclusive herbivores. Many are collector-gatherers, a habit which is facilitated by the scoop-like, setaceous foretibiae. The Oriental genera (e.g. Fernando, 1964; 1974; Jansson, 1982; see Fig. 4.84) include *Agraptocorixa* (which is relatively large: up to 10 mm long), *Cnethocymatia* (New Guinea only), *Cymatia*, *Micronecta*, *Sigara*, *Synaptonecta*, *Trichocorixa* (in China) and *Tropocorixa* (which some treat as a subgenus of *Sigara*: e.g. Jansson, 1982). The subfamily Corixinae (e.g. *Agraptocorixa* and *Sigara*) have distinct transverse grooves on the beak, but they are lacking in the Cymatiinae (e.g. *Cymatia* and *Cnethocymatia*). Unlike these two subfamilies, the tiny Micronectinae have a conspicuous scutellum; this triangular area is an exposed portion of the mesonotum between the anterior margins of the wings and the posterior edge of the pronotum. At less than 3 mm long, the Micronectinae are smaller than other corixids which measure 5–10 mm. *Micronecta* is a speciose genus that is often found in pools along stream edges and in drying river beds (Fernando, 1964; see also Leong, 1961; Hutchinson, 1993: p. 637–748). Sri Lanka supports an unusually rich assemblage (over 20 species) of *Micronecta*. *Synaptonecta* contains only two species characterized by forelegs with fused tibiae and tarsi, and is an Asian endemic occurring through India, Sri Lanka and Indochina; *Synapotencta issa* has been introduced into Florida. Males are usually brachypterous (effectively apterous) and unable to fly, but the macropterous females often take to the wing.

There are two other families of Nepomorpha: both belong to the superfamily Ochteroidea and are semiaquatic. The Gelastocoridae (represented by the cosmopolitan genus *Nerthra*) and the Ochteridae live along water margins in moist, rocky sites. *Ochterus* occurs throughout

Asia. Both families have short antennae, as is typical of the Nepomorpha, and this sets them apart from the semiaquatic Gerromorpha which sometimes share the same habitat. Ochterids are fast runners and fly readily. They lack the raptorial forelimbs of gelastocorids.

Gerromorpha

The suborder Gerromorpha includes several families of semiaquatic Heteroptera which live on the water surface or along the margins. Andersen's (1982a) monographic treatment of the group is indispensable. The Gerromorpha can be distinguished easily from aquatic Heteroptera because their antennae are significantly longer than the head. The most striking and highly derived family is the Gerridae or waterskaters (also called waterstriders or pondskaters). Together with the Veliidae, they constitute the superfamily Gerroidea. Gerrids have very long mid- and hindlegs with retractible, preapical claws. The bases of these two pairs of legs are close together, while the shorter, raptorial forelegs are set relatively far forward. This reflects the fact that the mesothorax is the longest of the three thoracic segments. The retractability of the tarsal claws is important because, unlike the hydrofuge hairs on the legs, they are 'wettable'. As a result, they penetrate the surface film when extended, and can be used to provide extra traction during rapid movement over the water. The neustic habit has the particular advantage for gerrids that prey detection by visual means can be supplemented by responses to vibrations on the water surface; in some cases, visual stimuli only elicit a reaction when accompanied by vibrations. A variety of insect prey are taken, and cannibalism is common. Larvae are particularly vulnerable immediately after moulting.

Gerrids vary greatly in size, and body lengths vary from 3 to over 30 mm; some tropical species reach a handspan or more in size. *Gigantometra* (formerly *Limnometra*) *gigas* from Hainan, for example, has mid- and hindlegs that are around 100 mm long (Hoffman, 1936c). Many gerrids exhibit wing polymorphism and morphs with more or less shortened wings occur in almost all species. The picture in confused by the incidence of autotomy (or self-mutilation) in some species. This involves snapping the wings, usually with the aid of the legs, along preformed lines of breakage. Further complication is introduced by the incidence of seasonal polymorphism (in, for example, *Gerris*) whereby wings are only produced at certain times of the year. In general, the majority of species are effectively flightless (and apterous) because macropterous, flying morphs are rare in field populations. The immatures of gerrids which lack wings in the adult stage are easily recognized: the foretarsus consists of a single segment; it is two-segmented in adults. There is some evidence that the incidence of macroptery and other life-history traits are correlated with habitat characteristics (especially permanency) but the relationship is by no means simple (Spence, 1989; Spence & Andersen, 1994).

Gerrid subfamilies differ with respect to both morphology and habitat occupancy (Spence & Andersen, 1994). Seven subfamilies occur in the Oriental Region. Andersen (1982a) provides a detailed key to gerrid identification, but it is too complex to reproduce here (and has been updated by recent work cited below). Members of the Eotrechinae have peculiarly modified legs (in particular, reduced hind tibiae which are especially marked in *Amemboa*) — an apparent adaptation for clambering over seeping rocky surfaces or wet rocks along the side of waterfalls and cascades (i.e. madicolous or hygropetric habitats). Even in these relatively short-legged gerrids, the hind femora are about twice as long as the abdomen. To aid purchase on near-vertical surfaces the tarsal claws are either prolonged or — in *Eotrechus* — have shifted from a subapical to an subapical position. The subfamily Eotrechinae is confined to tropical Asia, and *Eotrechus* and *Onychotrechus* are typical of hygropetric habitats; *Eotrechus* is almost terrestrial. *Amemboa*, *Chimarrhometra* (the Himalaya only) and *Tarsotrechus* are known from the region also. *Chimarrohmetra* seems to prefer rock-pools along streams; *Amemboa* inhabits the boulder zone between solid substrate and water, spending much of its time on land (Andersen, 1982b; Polhemus & Andersen, 1984). The Rhagadotarsinae contains a single genus in Asia: *Rhagadotarsus*, which is mainly confined to standing water, has been reviewed by Polhemus & Karunaratne (1993). The Trepobatinae has been the subject of revision (Polhemus & Polhemus, 1993; 1994a; 1995a) following the discovery of a rich endemic fauna in New Guinea and surrounding archipelagos. The subfamily includes *Calyptobates*, *Cryptobates* and *Naboandelus* in southern Asia, as well as *Andersenella*, *Ciliometra*, *Isobates*, *Metrobatoides*, *Metrobatopsis* and *Stygiobates* in New Guinea. Several marine genera — *Pseudohalobates*, *Rheumatometroides* (see Lansbury, 1992), *Stenobates* and *Stenobatopsis* — have been revised recently by Polhemus & Polhemus (1996). Freshwater Trepobatinae occur in swarms or schools on rivers, and *Naboandelus* are nocturnal.

Of the Gerrinae, *Gerris* (subgenus *Gerriselloides*) is mainly Palaearctic but does occur in the northern fringes of the Oriental Region; the subgenus

Limnoporus, on the other hand, does not extend south of the Palaearctic (Andersen & Spence, 1992). Andersen & Chen (1993) review the *Gerris* of China, and Andersen (1993) gives a full account of the genus. *Aquarius*, formerly a subgenus of *Gerris*, occurs throughout most of tropical Asia and was raised to a full genus after a phylogenetic analysis by Andersen (1990). *Limnogonus* and *Neogerris* are pantropical. *Tenagogonus* (of which *Limnometra* has been treated as a subgenus [Nieser & Chen, 1992b] but see Polhemus & Polhemus [1997] for an alternative view of the status of *Limnometra*) is restricted to the Oriental Region, but the subgenus *Tenagogonus* extends into Africa; *Gigantometra* occurs in southern China and Vietnam. Most Gerrinae are associated with standing waters but may be encountered in slow-flowing streams and their pools. The Halobatinae includes coastal and oceanic genera (*Asclepios* and *Halobates*), but some freshwater species in the tribe Metrocorini occur in lotic habitats: *Esakia*, *Ventidiopsis* and *Ventidius* use relatively still water in the lee of boulders while *Metrocoris* (which was revised by D.A. Polhemus, 1990) lives mainly in riffles. The Halobatinae contains insects which are shorter and more stout-bodied than other Gerridae. A sixth gerrid subfamily in Asia — the Cylindrostethinae — is represented by *Cylindrostethus* only (reviewed by D.A. Polhemus, 1994b); they are rather large, stream-dwelling forms. The Ptilomerinae is primarily Oriental, and *Ptilomera* is widespread in fast-flowing, often turbulent reaches of streams, a habitat shared by other members of this subfamily (*Heterobates*, *Pleciobates*, *Potamometra*, *Potamometropsis*, *Rheumatogonus* and *Rhyacobates*: e.g. Polhemus & Zettel, 1997). All Ptilomerinae have extremely long, slender legs with tarsal claws that have been reduced or lost; in addition, the mid- and hindtarsi have only one segment (*cf.* two segments in Gerrinae, Halobatinae, etc.), and the foretarsi are very long.

The Veliidae are short, stout insects which are typically 6 — 8 mm long. Like gerrids, they have retractable preapical claws on the mid- and hindlegs. Wingless (apterous) forms are common but, again, the immatures and adults can be distinguished by the number of tarsal segments. Veliidae tend to accumulate in schools or swarms in areas where surface food may be brought by the current. Surface vibrations can be important in prey detection. Small live prey are readily attacked but large items must be helpless before feeding (often by several individuals) is initiated. Veliids occur at edges and on the surface of a variety of freshwater habitats, but the Rhagoveliinae are mainly associated with riffles. The tarsus of the mesothoracic legs of members of this subfamily is deeply cleft and contains a retractable fan of plumose hairs. When *Rhagovelia* is walking on fast-moving water this 'swimming fan' is expanded and serves as a paddle during the action stroke; the hairs are folded into the cleft during the recovery stroke. The Microveliinae and other veliid subfamilies are not found on open water in fast streams, and lack a cleft and associated swimming fans on the tarsi. They occur along stream banks, on emergent plants, or on the surface of slow-moving or standing waters. Some veliids produce a fluid which contains a surfactant (a chemical which reduces surface tension) from the tip of the rostrum — it may be secreted by salivary glands. The rostrum is flexible and fluid release can be regulated to propel the insect on a 'wave' of reduced surface tension in almost any direction. The resulting locomotion is called 'skimming' or 'expansion skating'. If used as an anti-predator device against another surface feeder it has the advantage of expelling the enemy in the opposite direction.

Among the Veliidae, the large subfamily Microveliinae has a cosmopolitan distribution — especially the species-rich genus *Microvelia*. Also known from tropical Asia are *Aphrovelia, Baptista, Lathriovelia, Neoalardus, Pseudovelia* (= *Perivelia*), *Velohebria* (which is terrestrial and restricted to New Guinea) and *Xiphovelia* (e.g. Andersen, 1982a; 1983; 1989). Polhemus & Polhemus (1994b) have established four additional genera (*Aegilipsicola, Neusterinsifer, Tanyvelia* and *Tarsovelia*) that are confined to New Guinea. The Microveliinae is characterized by foretarsi with a single segment only; the mid- and hindtarsi have two segments. Some genera (*Baptista* and *Lathriovelia*) live in cryptic and secluded habitats such as small holes or under turf along stream banks (Andersen, 1989), but an association with the margins of slow-moving or standing water may be more typical of the Microveliinae. The subfamily Perittopinae (in which the foretarsi have two segments, but the mid- and hindtarsi have three) contains the single genus *Perittopus* which occurs in India and Southeast Asia, and the Rhagoveliinae is represented by *Rhagovelia* and *Tetraripis*. J.T. Polhemus & D.A. Polhemus (1988) provide a monographic account of Southeast Asian *Rhagovelia* (including descriptions of 26 new species and a key) which supplements and updates Andersen's (1982a) report on the genus (see also Yang & Polhemus, 1994; Polhemus, 1996). They noted altitudinal zonation with habitat partitioning by elevation for different *Rhagovelia* species in Borneo and Sulawesi. Oriental Veliinae (with three-segmented tarsi on all legs) include *Angilia* (subgenus *Adriennella*) and *Angilovelia* (known only from Burma), with limited penetration of the Palaearctic genus *Velia* into southwestern China and the Himalaya. Members of the Haloveliinae have two-segmented tarsi on all legs. Most species are marine, but the freshwater genera *Entomovelia* and *Strongylovelia* are found in Indian and Southeast Asia; *Strongylovelia* extends into New Guinea.

The Hydrometridae (or watermeasurers) belong to the superfamily Hydrometroidea, and can be recognized easily by the distinctive head which is prolonged anteriorly with eyes set half-way along and antennae near the tip. Species of *Hydrometra* are delicate, elongate insects (up to 12 mm long) with slender — almost thread-like — legs bearing apical claws. They are resemble tiny twigs (reminiscent of the larger, fully-aquatic Ranatrinae; see above) and hide among emergent vegetation at the edge of streams and marshes. Some species are brachypterous. Of the two Oriental genera, *Hydrometra* (subfamily Hydrometrinae) is cosmopolitan and common; Polhemus & Polhemus (1995b) and Polhemus & Lansbury (1997) have revised part of this genus, including many Asian species. *Heterocleptes* (subfamily Heterocleptinae) is confined to Borneo. These animals have a relatively robust body (somewhat similar to a veliid), but the head is prolonged like that of *Hydrometra*. Hydrometrids subsist on dead prey which are approached carefully and inspected with the antennae for signs of life; predation on live prey is unusual. Several individuals may feed simultaneously from the same carcass.

The Mesoveliidae (superfamily Mesovelioidea) have three-segmented tarsi with apical claws and conspicuous spines on the hind leg. They are small (4 mm or less) and slender, occurring mostly along the edges and on the surface of standing waters or on floating vegetation (although the genus *Nereivelia* inhabits mangroves: Polhemus & Polhemus, 1989). Mesoveliids are scavengers, and do not attack live prey. Winged and apterous forms are known. Of the two subfamilies, the Madeoveliinae includes the cosmopolitan genus *Mesovelia*; the remaining genera are in the Mesoveliinae: *Nereivelia*, *Phrynovelia* (New Guinea only), and *Speovelia* from Japan. The Oriental Hebridae (velvet bugs: superfamily Hebroidea) comprises *Hebrus* (subgenus *Timasielloides*), *Merragata*, *Neotimasius* and *Timasius* (subfamily Hebrinae), as well as *Hyrcanus* (subfamily Hyrcaninae). Hebrids are small (< 3 mm long) with a pronotum that is wider than the rest of the body. They are covered with a short, dense pile of hydrofuge hairs. The tarsi comprise two segments and bear apical claws. The antennae have five segments, although four segments is more usual for the Gerromorpha. Hebrids lack wing venation, but aptery is rare. Most species have similar habits and are associated with aquatic vegetation but sometimes venture onto the water surface. Two *Hebrus* species are mangrove-associates (Polhemus & Polhemus, 1989), and *Timasius* is apparently hygropetric.

Other Heteroptera are associated with the marginal or riparian zone rather than the water surface. The Saldidae (shore bugs) belong to the suborder Leptopodomorpha and are small (no more than 8 mm long) oval, insects that occur along stream banks (e.g. Chen, 1987). The hemelytra have conspicuous veins (*cf.* Hebridae), and saldids fly readily when disturbed. They have large eyes which protrude laterally, and the coxa of each hindleg is modified into a flattened plate fused to the thorax. Oriental genera include *Macrosaldula*, *Metasalda*, *Orthosalda* (a Philippines endemic), *Pentacora* (mostly intertidal), *Rupisalda*, *Saldoida*, *Saldula* and *Salduncula*. Although they live near water, saldids cannot be considered aquatic in the strict sense. Some genera in the related Leptopodidae — *Leptopus*, *Patapsius* and the widespread genus *Valleriola* (see J.T. Polhemus & D.A. Polhemus, 1987) — occur in hygropetric habitats. Heteropteran genera from conventionally terrestrial families (for example, *Botocudo* in the Lygaeidae) sometimes share these environments (Slater & Polhemus, 1987).

TRICHOPTERA

The Trichoptera is a diverse order of holometabolous freshwater insects that occurs in all biogeographic regions. There are three suborders (Table 4.7): Spicipalpia (closed-cocoon makers), Annulipalpia (fixed-retreat makers) and Integripalpia (portable-case makers). The global species total is not known but there are over 600 extant genera. Schmid (1984) estimates that India alone may support 4,000 species with perhaps 50,000 species in the Oriental Region as a whole (see also Schmid, 1968a). Although it may be a little high, this figure seems to be within the right order of magnitude: for instance, Malicky (1989a) notes that the genus *Chimarra* (Philopotamidae) may contain 500 Southeast Asian species. Significantly, the available data suggest that the Oriental Region seems to be disproportionately rich in caddisflies when compared with the rest of the world (Schmid, 1984). Taiwan, for example, is host to at least 24 families of caddisflies whereas Poland, with more than eight times the area and a better-known fauna, has only 18 families (Chen & Morse, 1991). Certainly, the majority of the region's species have yet to be documented. Yang & Morse (1991) estimate that two-thirds of the caddisfly fauna of China await description.

Table 4.7 A checklist of tropical Asian Trichoptera genera. Caddisflies present in New Guinea are treated as part of the Oriental fauna (see Gressitt, 1982). Palaearctic genera which may penetrate the northern parts of the region, or which are present as a few isolated records, are indicated by *, and genera that seem to be entirely confined to Palaearctic Asia are indicated by **. I have omitted any doubtful records. Classification largely follows Wiggins (1996).

SPICIPALPIA
  Rhyacophilidae
    Fansipangana
    Himalopsyche
    Rhyacophila
  Hydrobiosidae
    Apsilochoreminae
      Apsilochorema (syn. *Bachorema*)
    Hydrobiosinae
      Tanorus (syn. *Ornatus*)
  Glossosomatidae
    Agapetinae
      Agapetus (syn. *Allagapetus*)
      Electragapetus **
      Synagapetus (syn. *Pseudagapetus, Myspoleo* and *Afragapetus*)
      Tagapetus
    Glossosomatinae
      Anagapetus **
      Glossosoma
        s.g. *Glossosoma* (syn. *Anseriglossa*)
        s.g. *Lipoglossa*
        s.g. *Muroglossa*
        s.g. *Synafophora* (syn. *Diploglossa, Eomystra, Mystroglossa* and *Mystropha*) *
      Poeciloptila
    Protoptilinae
      Nepaloptila
      Padunia (syn. *Uenotrichia*) *
      Temburongpsyche
  Hydroptilidae
    Hydrotilinae
      Catoxyethira
      Chrysotrichia
      Hydroptila (syn. *Oeceotrichia, Oxydroptila, Pasirotrichia* and *Sumatranotrichia*)
      Hellyethira
      Ithytrichia (syn. *Saranganotrichia*)
      Jabitrichia
      Macrostactobia
      Microptila
      Missitrichia
      Niuginitrichia
      Orthotrichia (syn. *Baliotrichia, Javanotrichia* and *Orthotrichiella*)
      Oxyethira (syn. *Gnathotrichia* and *Stenoxyethira*)
        s.g. *Dampfitrichia*
        s.g. *Oxyethira*
      Parastactobia

Table 4.7   *(cont.)*

  Plethus (syn. *Plethotrichia*)
  Scelotrichia (syn. *Madioxytheria* and *Pseudoxyethira*)
  Stactobia (syn. *Lamonganotrichia*)
  Stactobiella
  Tricholeiochiton (syn. *Synagotrichia*)
  Ugandatrichia (syn. *Moselyella*)
  Vietrichia
 Ptilocolepinae
  Palaeagapetus *
  Ptilocolepus[30]
ANNULIPALPIA
 Philopotamidae
  Philopotaminae
   Doloclanes
   Dolomyia
   Dolopsyche
   Dolophilodes (syn. *Sortosa*)
     s.g. *Dolophilodes*
     s.g. *Hisaura* **
   Edidiehlia
   Gunungiella
   Kisaura
   Wormaldia (syn. *Dolophiliella*)
  Chimarrinae
   Chimarra (syn. *Vigarrha*)
 Stenopsychidae
   Stenopsyche
 Psychomyiidae
  Psychomyiinae
   Eoneureclipsis
   Lype
   Padangpsyche
   Psychomyia (= *Khandalina* and *Psychomyiella*)
   Tinodes
  Paduniellinae
   Paduniella
 Xiphocentronidae
   Abaria
   Cnodocentron
   Drepanocentron
   Melanotrichia (syn. *Kibuenopsychomyia* and *Tsukushitrichia*)
   Proxiphocentron
 Dipseudopsidae
   Dipseudopsis (syn. *Bathytinodes*)
   Hyalopsyche (syn. *Hyalocentropus* and *Galta*)
   Hyalopsychella
   Phylocentropus
 Ecnomidae
   Ecnomus

---

[30] Malicky (1983b) and Malicky & Chantaramongkol (1996) place this genus in the Glossosomatidae.

Table 4.7   *(cont.)*

Polycentropodidae
   Kambaitipsychinae
      Kambaitipsyche
   Polycentropodinae
      Cyrnopsis
      Holocentropus
      Kyopsyche *
      Neucentropus **
      Nyctiophylax
      Pahamunaya
      Paranyctiophylax
      Plectrocnemia
      Polycentropus
      Polyplectropus
   Pseudoneureclipsinae
      Pseudoneureclipsis
Hydropsychidae
   Arctopsychinae
      Arctopsyche
      Arctopsychodes **
      Maesaipsyche
      Parapsyche
   Diplectroninae
      Diplectrona
      Diplectronella
      Diplex
      Sciops
   Hydropsychinae
      Abacaria
      Cheumatopsyche (= *Ecnopsyche*)
      Herbertorossia
      Hydatomanicus
      Hydatopsyche
      Hydromanicus
      Hydropsyche
      Hydropsychodes
      Occutanpsyche
      Potamyia
      Synaptopsyche
   Macronematinae
      Aethaloptera (syn. *Chloropsyche*)
      Amphipsyche (syn. *Amphipsychella* and *Phanostoma*)
      Baliomorpha
      Leptonema **
      Leptopsyche
      Hydronema **
      Macrostemum
      Oestropsyche
      Paraethaloptera
      Polymorphanisus
      Pseudoleptonema
      Trichomacronema

Table 4.7   *(cont.)*

INTEGRIPALPIA
  Limnocentropodidae
    Limnocentropus (syn. *Kitagamia*)
  Phryganaeidae
    Colpomera * *
    Dasytegia * *
    Eubasilissa * *
    Neurocyta * *
    Oligotrichia * *
    Oopterygia * *
    Semblis   * *
  Phryganopsychidae
    Phryganopsyche
  Brachycentridae
    Brachycentrus (syn. *Brachycentriella* and *Oligoplectrodes*) *
    Eobrachycentrus * *
    Micrasema
  Lepidostomatidae
    Lepidostomatinae
      Adinarthrella
      Adinarthrum
      Agoerodella
      Agoerodes
      Anacrunoecia
      Dinarthrella
      Dinarthrena
      Dinarthrodes (syn. *Crunoecia* [part] and *Maniconeura*) *
      Dinarthropsis
      Dinarthrum
        s.g. *Dinarthrum*
        s.g. *Indodinarthrum*
      Dinomyia * *
      Eodinarthrum *
      Goerodella
      Goerodes (syn. *Atomyiella, Crunobiodes, Crunoeciella, Dinarthrodes* [part],
        Goerinella, Paradinanthrodes and *Yamatopsyche*; Ito, 1984)
      Goerodina
      Hypodinarthrum
      Indocrunoecia
      Indodinarthrum
      Kodala
      Lepidostoma * * *(syn. Ayabeopsyche)*
      Mellomyia
      Metadinarthrum
      Neolepidostoma
      Neoseverinia * *
      Paraphlegopteryx (syn. *Neoseverinia* [part])
      Ulmerodes
    Theliopsychinae
      Zephyropsyche
  Limnephilidae
    Limnephilinae
      Aplatyphylax *

Table 4.7 *(cont.)*

  Asynarchus *
  Evanophanes **
  Glyphotaelius **
  Limnephilus *
  Mesophylax **
  Micropterna *
  Nothopsyche
  Philarctus **
  Psilopterna **
 Pseudostenophylacinae
  Astratodina **
  Phylostenax *
  Pseudostenophylax
Apataniidae
  Apatania
  Apataniana **
  Apatidelia *
  Moropsyche (syn. *Apatelina*)
  Notania
Uenoidae
 Thremmatinae
  Neophylax (syn. *Halesinus*) *
 Uenoinae
  Uenoa (syn. *Eothremma*)
Goeridae
  Gastrocentrella
  Gastrocentrides
  Goera
  Larcasia
Molannidae
  Indomolannodes
  Molanna
  Molannodes **
Calamoceratidae
  Anisocentropus (syn. *Kizakia*)
   s.g. *Anisokantropus*
   s.g. *Anisolintropus*
   s.g. *Anisomontropus*
  Ascalophomerus
  Ganonema (syn. *Asotocerus*)
  Georgium (syn. *Rhabdoceras*)
Leptoceridae
 Leptocerinae
  Adicella
  Allosetodes
  Athripsodes *
  Ceraclea
   s.g. *Athripsodina*
   s.g. *Ceraclea*
  Episetodes
  Erotesis **
  Lectrides
  Leptocella

Table 4.7   *(cont.)*

Leptocerus
>   *Mystacides*
>   *Nietnerella*
>   *Oecetis (syn. Oecetina)*
>   *Oecetodella*
>   *Parasetodes*
>   *Poecilopsyche*
>   *Setodes*
>   *Setodinella*
>   *Tagalopsyche*
>   *Trichosetodes*
>   *Triaenodella*
>   *Triaenodes*
>>   s.g. *Austotriaena*
>>   s.g. *Triaenodella*
>>   s.g. *Triaenodes*

>   Triplectidinae
>>   *Symphitoneuria* (syn. *Loticana* and *Notanatolica*)
>>   *Symphitoneurina*
>>   *Triplectides* (syn. *Tobikera*)

Odontoceridae
>   *Inthanopsyche*
>   *Lannapsyche*
>   *Marilia*
>   *Perissoneura* **
>   *Psilotreta*

Sericostomatidae
>   Asahaya
>   *Ceylanopsyche* (syn. *Noleca*)
>   *Gumaga*
>   *Karomana*
>   *Mpuga*
>   *Ngoya*
>   *Notidobia*

Beraeidae
>   *Ernodes*

Helicopsychidae
>   *Cochliophylax*
>   *Helicopsyche*

The higher classification of the Trichoptera was, for some time, based on the work of Ross (1956; 1967), but is now rather unsettled and characterized by '. . . fervent exchange of opinions . . .' about the benefits of competing phylogenies (Morse, 1997: p. 428). For the present purposes, I have based my arrangement of families largely on the phylogeny given by Weaver & Morse (1986) and Wiggins' (1996) review of North American caddisflies; the assignment of genera to families is after Higler (1981). Morse (1997) gives a recent account of the differing views on phylogeny held by Trichoptera workers.

Adult caddisflies are moth-like insects with wings that are covered by fine hairs. The antennae are long and filiform, and the wings are folded tent-like over the abdomen. Typically, adult caddisflies are dull coloured, but a few tropical taxa (such as the Macronematinae) may be boldly marked in yellow and black or black and white. The adults ingest liquid only (such as nectar; e.g. Nozaki & Shimada, 1997), although some of the mouthparts — the labial and maxillary palps — are well developed. They live for around a month (although this varies among species). Most are nocturnal or crepuscular but some brightly coloured Macronematinae (among others) are diurnal. Mating follows swarming or courtship behaviour (which can involve sex pheromones), and may take place among riparian vegetation or on the ground. Eggs are laid in or immediately above the water. Larvae can be found in almost every type of freshwater habitat, and are a major component of the stream benthos. There are generally five larval instars. Many species are univoltine although bivoltine and trivoltine life histories can occur. In Hong Kong, at least, adults of some species fly throughout the year, but others have a more restricted flight period (Dudgeon, 1988c; 1997; Wells & Dudgeon, 1990). Huisman (1991) reports that adults of most (but not all) families associated with streams in Borneo were least abundant in the dry season. Pupation almost invariably occurs under water, and the pupae in enclosed inside a cocoon within a pupal case that was built by the larva and fixed to the substrate. The pupae have large mandibles which are used to free the emerging pharate adult which is cloaked within the pupal integument. It nevertheless swims to the surface and moults to the winged adult stage. The pupal case is usually a modification of a moveable larval case or fixed shelter, and the ability of caddisfly larvae to build such structures out of organic or inorganic materials has been a source of fascination for many observers.

The functional significance of case-building and modes of building have been reviewed recently by Dudgeon (1994c) and so will not be repeated in detail here. In short, the moveable cases built by larvae in the suborder Integripalpia are for protection (some are crush resistant), defence (including camouflage) and/or may facilitate the production of respiratory currents of over the abdominal gills. Most larvae in the suborder Annulipalpia build shelters, tubes or other fixed structures, while those in the Spicipalpia are free-living (i.e. a shelter is made upon pupation only) or construct moveable cases. In the majority of caddisflies that construct a fixed shelter, the structure is associated with a silken net spun by the larvae and used to filter food from the current. In some cases, a tube is built which the grazing larvae extend periodically thereby gaining shelter while feeding. The diversity variety of larval constructions is considerable and there is no doubt that this has facilitated the occupation of a wide variety of niches. Note that some of the largest and most spectacular case builders — the Phyganaeidae and Limnephilidae — are largely absent from the Oriental Region, occurring only at high altitudes or in the north along the fringes of the boundary with Palaearctic Asia (e.g. Schmid, 1955; Dudgeon, 1987a).

All functional-feeding groups are represented among larval Trichoptera: one extraordinary Australasian hydroptilid is even a parasite on the pupae of other caddisflies (Wells, 1992). Filter-feeders are well represented in the order, while relatively few caddisflies are collector-gatherers. Filter-feeding based on silken nets occurs only in the Annulipalpia (especially the Philopotamidae, Stenopsychidae and Hydropsychidae), but larvae of certain Integripalpia (especially the Brachycentridae, Limnocentropodidae and some Leptoceridae) sieve food particles from the current with the aid of setae or spines on (usually) the legs. Most shredders are belong to the Integripalpia (e.g. Calamoceratidae, Lepidostomatidae and some Leptoceridae), while the Annulipalpia are mainly filter-feeders, and the Spicipalpia has a disproportionate share of the predators (notably the Rhyacophilidae and Hydrobiosidae). Grazers (e.g. Glossosomatidae, Xiphocentronidae, Psychomyiidae, Helicopsychidae, Molannidae and Odontoceridae) are represented in all three suborders (although the dependence upon algae in some families is not well established), but the habit of piercing plant cells and sucking out their contents is confined to the Hydroptilidae (Spicipalpia). Families which are feed mainly as collector-gatherers include the Dipseudopsidae, Xiphocentronidae and Psychomyiidae. Additional information summarising the ecology of each family will be given in the relevant sections below.

Research on tropical Asian caddisflies has been confined mainly to descriptions of the adult stages, and relatively few larvae have been characterized adequately (Higler, 1992). Unfortunately, identifications of larvae are required more frequently than identifications of adults, especially in biological surveys and biomonitoring. A detailed — and well illustrated — treatment of the biology of genera North American of caddisfly larvae is given by Wiggins (1996), and this monograph contains much information relevant to families of Trichoptera in Asia. Significant early work on caddisflies in Asia was undertaken by Ulmer

(1911; 1915; 1925; 1927; 1930; 1932b; 1932c), especially with regard to the Indonesian fauna (Ulmer, 1951; 1955; 1957); two of these papers (Ulmer, 1955; 1957) remain seminal sources for information on larval trichoptera in Asia. Important work on the Indian caddisfly fauna was undertaken by Martynov (1935; 1936) and Kimmins (1953a; 1953b; 1955a; 1957; 1963) who was among the first to document the Trichoptera of Borneo (Kimmins, 1955b) and New Guinea (Kimmins, 1962). Schmid (1958a; 1963; 1964; 1965a; 1968a; 1968b 1968c 1970a; 1971; 1972) has also studied the Indian and Sri Lankan caddisfly fauna, in addition to the Trichoptera of China (Schmid, 1959a; 1965b) and Pakistan (Schmid, 1958b; 1959b; 1960; 1961), and has completed global revisions of several caddisfly families and major genera (e.g. Schmid, 1968d 1968e 1969; 1970b; 1982; 1987; 1989). Other notable work on Chinese caddisflies includes Martynov (1914; 1931), Banks (1940), Mosely (1942), Hwang (1957; 1958) and Wang (1963). Banks (1937) also researched the Philippines Trichoptera. Among these studies, Ulmer (1955; 1957) and Wang (1963) are exceptional in considering larval stages; in all others the focus was on adult stages.

More recent investigations, which will be referred to below, have — on the whole — continued to emphasise adult stages, although some Japanese workers (e.g. Iwata, 1927; Ito & Kawamura, 1980; 1984; Ito, 1984; 1985a; 1985b; 1985c; 1985d; 1989; 1990a; 1990b; 1991a; 1991b; 1992a; 1992b; Ito & Hattori, 1986; Tanida, 1986a; 1986b; 1987a) have devoted their attention to larvae. However, there is still a tremendous gap between our capacity to identify adult stages and our ability to recognize larvae or associate them with a known adult. In part, this reflects the holometabolous life cycle of caddisflies, with a pupal stage which complicates larval rearing and ensures that none of the features of the adult are reflected in the final-instar larva (*cf.* mayflies or Hemiptera). As a result, reliable identification of species on the basis of larvae alone is not possible, and adult material is indispensable for the determination of species. This obstacle affects workers dealing with the relatively well-studied fauna of Europe and North America, but the situation is much worse in tropical Asia (Malicky, 1983a). For this reason, I have cited some of the recent literature dealing with adult caddisflies in the sections below; these citations are not exhaustive.

The following key to trichopteran families is modified from that of Ulmer (1957), and draws upon the key given by Wiggins (1996). Its use depends upon the availability of mature or well-grown (typically fifth-instar) larvae. Unfortunately, it is not practical to develop a key that will separate the all genera of Asian caddisflies, but some provisional guidance (and an incomplete key of the Hydroptilidae) is given below in the hope that they will encourage the development of improved, more comprehensive versions.

1. Larvae with spiral case of sand grains resembling a snail shell; body strongly curved; anal claw comb-like (with an inner row of pectinate spines) ....**Helicopsyche**

   Larvae free-living or within a case or shelter quite unlike a snail shell; anal claw generally hook-like... 2

2. Sclerotized plates covering the dorsal surface of all three thoracic segments (i.e. the pro-, meso- and metathoracic nota) 13

   Meso- *and/or* metathoracic nota largely membranous or fleshy and never entirely sclerotized (although they may have small sclerites) 3

3. Mesonotum partly or largely covered by sclerotized plates (although these may be lightly pigmented); larvae construct portable cases of various materials 18

   Mesonotum generally without sclerotized plates although a few small sclerites and setae *may* be present; larvae *usually* lack a portable case, but they may construct a shelter or live within a net 4

4. Sclerotized plate on the dorsal surface of abdominal segment IX; tracheal gills *may* be present on the sides of the abdomen....... 5

   Abdominal segment IX lacks a dorsal sclerotized plate (i.e. it is membranous); abdomen lacks gills 7

5. Tibia and tarsus of the forelegs are greatly modified — either chelate, or pincer-like with an attenuated tarsus; abdominal gills lacking .... **Hydrobiosidae**

   Forelegs not thus modified; limbs similar to each other in general form... 6

6. Anal prolegs short and broadly joined to abdominal segment IX; anal claws small with at least one dorsal accessory hook; forelegs not noticeably stouter than the other limbs; larvae live in a tortoise-like case constructed of small stones.................. **Glossosomatidae**

Anal prolegs free and rather long, with well-developed claws which may have accessory hooks on the inner margin; forelegs rather robust and raptorial (adapted for grasping prey); abdomen and thorax may bear lateral gills; free-living .... **Rhyacophilidae**

7. Labrum T-shaped and membranous, anterior margin densely fringed with fine setae; sclerotized parts usually yellow or orange and the posterior margin of the prothorax is rimmed with black; larvae spin sac-like nets of fine silk; may be locally abundant ....
.... **Philopotamidae**

   Labrum sclerotized, not T-shaped **8**

8. Larvae large (up to 45 mm long), and usually dark in colour with the blotches or other markings on the head and prothorax; head at least twice as long as wide; larvae spins an irregular net among stones and debris on the stream bed .... **Stenopsychidae**

   Larva usually rather small and pale; head never twice as long as wide; larvae spin fine-meshed nets or live in fixed tubes **9**

9. Body markedly elongated; forelimbs flattened with paddle-like tarsi covered by short, dense setae; labium rather long and drawn out beyond the anterior margin of the head; mandibles *may* appear massive, with setal brushes on the inner margin; larvae construct branching tubes that are partly buried within sandy substrates
.... **Dipseudopsidae**

   Tarsi of all legs normal; without the above combination of characters **10**

10. Fore-trochantin (a projection arising close to the base of the leg) is pointed (acute); the abdomen usually bears a lateral fringe of short, pale setae; tibiae with numerous long setae; anal proleg and claw well-developed; larvae may spin nets or build tubes on hard surfaces (rocks, submerged wood) **11**

    Fore-trochantin blunt or broad at the apex and somewhat hatchet-shaped; abdomen without lateral fringe of setae; forelegs usually more massive that the other limbs, and tibiae bear few setae; anal proleg short, claw not unusually developed; larvae construct tubular retreats on hard surfaces (rocks, submerged wood) **12**

11. Abdomen somewhat depressed and bearing a lateral fringe of setae; metanotum partially sclerotized along the lateral margins (Fig. 4.93A); abdomen with a lateral fringe of setae; labium

attenuated, extending beyond the maxillary palps; margins of the labrum densely fringed with setae; fore-trochantin relatively short; tube-dwelling .... **Polycentropodidae: Pseudoneureclipsinae**

Abdomen more-or-less cylindrical and lacking a lateral fringe of setae; metanotum entirely unsclerotized; labium rather short and not extending beyond the maxillary palps; margins of the labrum lack a dense fringe of setae; fore-trochantin relatively long; larvae spin nets or build tubes on hard surfaces ....
....**Polycentropodidae: Polycentropodinae**

12. Tibiae and tarsi of all legs fused into a single 'segment'; a broad lobate process of the mesothorax (the mesopleuron) extends anteriorly (Fig. 4.91A-C); anal claws lack teeth on the inner margin
.... **Xiphocentronidae**

Tibiae and tarsi separate; no anterior projection of the mesopleuron; anal claws *may* have spines or accessory hooks on the inner margin
.... **Psychomyiidae**

13. Abdomen with conspicuous ventral and (often) lateral gill tufts; anal prolegs with a distal brush of long setae (may be sparsely developed in some species); posterior margins of thoracic nota lobate; usually associated with a silken net and fixed retreat; very common and often locally abundant .... **Hydropsychidae** **14**

Ventral or lateral gills lacking; anal prolegs lack a distal brush; posterior margins of thoracic nota straight; often rather small (< 6 mm long) **17**

14. Mid-ventral line or suture on the underside of the head capsule lacking, replaced by a tongue-like ventral apotome running from the back to the front of the head separating the two halves completely; a pair of short, rather broad sclerites are present on the underside of abdominal segment VIII; relatively uncommon, usually at high altitudes .... **Hydropsychidae: Arctopsychinae**

Underside of head capsule not separated completely by the ventral apotome; sclerites on abdominal segment VIII not as above: common and locally abundant **15**

15. Ventral apotome divided into two pieces of approximately equal size (one anterior, the other posterior), and a mid-ventral suture in the central part of the head capsule connects these two parts; paired sclerites on the underside of abdominal segment VIII are

ovoid; meso- and metanota with transverse sutures; fore-trochantin not forked .... **Hydropsychidae: Diplectroninae**

Posterior portion of the ventral apotome (which is triangular and easy to see in Diplectroninae) minute or absent; paired sclerites on the underside of abdominal segment VIII are never ovoid and may be lacking; meso- and metanota lack transverse sutures; fore-trochantin forked or not... **16**

16. Abdominal gills feather-like, consisting of a central 'stalk' with filaments arising more or less uniformly along the entire length; paired sclerites on the underside of abdominal segment VIII lacking; fore-trochantin is never forked ....
.... **Hydropsychidae: Macronematinae**

Abdominal gills with the filaments arising near the apex resulting in a tufted appearance; paired sclerites on the underside of abdominal segment VIII are triangular or sometimes fused; fore-trochantin almost always forked ....
.... **Hydropsychidae: Hydropsychinae**

17. Minute larvae bearing a case of varying form in the final instar; long setae arising from the head capsule; abdomen usually swollen (sometimes to a considerable degree), and always much broader than the thorax; anal prolegs and fore-trochantin short ....
.... **Hydroptilidae**

Larvae — which are relatively large — do not build a portable case but may spin a net; the head lacks long setae; abdomen rather slim and not swollen; anal prolegs long; fore-trochantin rather long and pointed .... **Ecnomidae**

18. Pronotum is markedly thickened and robust, and the anterior margins are pointed and directed forward; mesonotum with four sclerites in addition to lateral projections that point forward; head can be retracted beneath the pronotum; case of rock fragments with lateral ballast stones .... **Goeridae**

Pronotum and mesonotum not as above; head not retractable **19**

19. Antennae rather prominent (at least six times longer than thick); sclerotized plates on the mesonotum *usually* lightly-pigmented and *sometimes* have a pair of dark curved lines on the posterior half; hind legs longer than other limbs with the femur and tibia (and,

in some cases, the tarsi) subdivided into two 'pseudosegments' .... **Leptoceridae** 20

Antennae short (less than three times longer than thick); mesonotum lacks dark curved lines on the posterior half **21**

20. Metanotum bears two small dorsal sclerites and two lateral sclerites; sclerotized parts elsewhere may be rather dark ....
.... **Leptoceridae: Tiplectidinae**

Metanotum lacks dorsal sclerites; sclerotized parts elsewhere usually rather pale (this subfamily includes most species of Leptoceridae) .... Leptoceridae: Leptocerinae

21. Dorsum of abdominal segment I bears a large, shield-like sclerite; meso- and metanota each have four sclerotized plates; meso- and metasterna and underside of abdominal segment I with sclerites and/or stout bristles; robust predatory larva in stalked case; rather rare, usually at high altitudes .... **Limnocentropodidae**

Dorsum of abdominal segment I is predominately membranous and lacking a shield-like sclerite, although setae or a dorsal hump or (rarely) a small sclerite may be present; lacking sclerites on the meso- and metasterna and underside of abdominal segment I; common **22**

22. Mesonotum *usually* without sclerotized plates (a pair of small sclerites *may* be present); a membranous, finger-like projection (the prosternal horn) is present on the underside of the prothorax prosternal horn; dorsal *and* lateral humps are borne on abdominal segment I; large larvae (often over 40 mm long) with conspicuously marked head capsules in cases of plant material; mostly Palaearctic Asia .... **Phryganeidae**

Mesonotum largely covered by sclerotized plates; without the above combination of characters **23**

23. Abdominal segment I lacks dorsal and lateral humps; pronotum with a distinct transverse ridge; most of the dorsal surface of the mesonotum is sclerotized; sclerites on the mesonotum widely separated; head *may* be flattened or *may* have dorsolateral ridges on either side; gills single filaments or absent .... **Brachycentridae**

Abdominal segment I always with lateral humps (although they may be rather inconspicuous) and sometimes with a dorsal hump;

without the above combination of characters **24**

24. Claw on the tip of the hind leg differs from those on the anterior limbs: either it is short and stump-like bearing short setae, or it is drawn out into a filament; flattened case of (mainly) sand grains with lateral flanges and a dorsal hood .... **Molannidae**

Tarsal claw of the hind leg resembles that on the other legs **25**

25. Mesonotal sclerites are well developed and have an anteromedian notch (or emargination); labrum with a membranous anterior edge bearing many short setae; case and larval body sometimes (not always) very slender; rather rare .... **Uenoidae**

Thoracic nota and labrum not as above; relatively common **26**

26. Labrum with a transverse row of 15 — 20 stout setae; abdominal segment I with dorsal and lateral humps; anterior corners of pronotum *usually* project forward; mesonotum with a pair of anteriolateral sclerites and a larger central plate; metanotum almost entirely lacking sclerotisation .... **Calamoceratidae**

Labrum not thus; without the above combination of characters **27**

27. Antennae situated close to the anterior margin of the eye; dorsal hump on abdominal segment I lacking; prosternal horn present; abdominal segment VIII with a fleshy lateral lobe on either side; case often square in cross-section and made of panels cut from leaves .... **Lepidostomatidae**

Antennae situated halfway between the eye and the anterior margin of the head capsule or even further forward; dorsal hump on abdominal segment I is usually present — if absent, the prosternal horn is present **28**

28. Antenna almost at anterior margin of the head capsule; prosternal horn lacking; head *usually* has dorsolateral ridges on either side (although they may be weakly developed) **29**

Antenna situated halfway between the eye and anterior margin of the head capsule; primarily Palaearctic or with a localised distribution at high altitudes (generally over 1,000 m) **31**

29. Sharp carina (ridge) extending obliquely across the pronotum, terminating in an rounded anterolateral lobes; highly localised and largely confined to Palaearctic Japan .... **Beraeidae**

   Pronotum lacking transverse ridge; widespread **30**

30. A cluster of 30 (or more) setae at the base of the proleg; lateral humps of abdominal segment partially encircled at the base by a narrow sclerotized band; fore-trochantin pointed and rather hook-like; prosternum lacks large sclerotized plates; rather uncommon .... Sericostomatidae

   Setal tuft at the base of proleg lacking almost entirely (and never more than five setae present); lateral humps of abdominal segment I without encircling sclerite; fore-trochantin short and blunt (and rather inconspicuous); prosternal plates *usually* present; relatively common .... **Odontoceridae**

31. Dorsal hump on abdominal segment I lacking; prosternal horn absent; metanotum largely membranous (small lateral sclerites only); surface of abdominal segment I with few setae; case of detritus .... **Phryganopsychidae**

   Dorsal hump and prosternal horn present; metanotum with small dorsal sclerites; surface of abdominal segment I with many scattered setae; case of various materials but often mineral **32**

32. A transverse row of setae present on the anterior portion of the metanotum (sclerites *usually* lacking in this position); mandibles lack teeth and are blade-like (adapted for scraping); case of rock fragments with an overhang above the anterior opening .... **Apataniidae**

   A pair of sclerites present on the anterior margin of the metanotum (setae between these sclerites are lacking); mandibles almost always toothed; cases of rock fragments or detritus.............................. .... **Limnephilidae** **33**

33. Gills consist of single filaments .... **Limnephilidae: Pseudostenophylacinae**

   Gills consist of multiple filaments .... **Limnephilidae: Limnephilinae**

Rhyacophilidae

The Rhyacophilidae includes two genera in tropical Asia: *Rhyacophila*, which is widespread and extremely speciose (especially in northern India and the Burma-Yunnan frontier area; Kimmins, 1953a:), and *Himalopsyche* which is much less diverse but Oriental in origin, occurring in China, Thailand, India, Nepal and Pakistan. A recently-established monotypic genus from northern Vietnam — *Fansipangana* (Mey, 1996a) — also belongs in the Rhyacophilidae. The genus *Rhyacophila* was reviewed by Schmid (1970b), and *Himalopsyche* was treated in detail but Schmid & Botosaneanu (1966). Banks (1940) established *Himalophanes* as a subgenus of *Himalopsyche*, but this arrangement not been adopted by later workers. The Rhyacophilidae as a whole decrease in diversity through the East Indies: species richness in Sulawesi is low, and the family is not present in New Guinea (Neboiss, 1986). Examples of recent publications that give descriptions of adult *Rhyacophila* include Hwang & Tian (1982: China), Tian & Li (1986; 1987a: China), Oláh (1987: Vietnam), Neboiss & Botosaneanu (1988: Sulawesi), Malicky (1989a: Sumatra; 1991: Thailand; 1993a: Pakistan; 1995a: Vietnam; 1995b: China; 1997: Nepal), Malicky & Chantaramongkol (1989a; 1993b; 1993c: Thailand), Sun & Yang (1995: China), Mey (1995a: the Philippines) and Mey (1996a: Vietnam); Tian & Li (1987a; China), Malicky & Chantaramongkol (1989a; Thailand), Malicky (1993a; Pakistan), Tian *et al.* (1993: China), Sun & Yang (1994: China) and Mey (1996a: Vietnam) also supply accounts of new *Himalopsyche*.

The larvae of Palaearctic Asian *Rhyacophila* have been figured by Iwata (1927), Kim (1974), Tanida (1985) and Yoon & Ki (1989a). Two examples from Hong Kong are shown in Fig. 4.85. Tanida (1985) provides illustrations of *Himalopsyche japonica*, while Iwata (1928) and Ulmer (1957) have described Oriental *Rhyacophila* and *Himalopsyche* species. The larvae of *Himalopsyche* (Fig. 4.86) are well illustrated by Schmid & Botosaneanu (1966) also. Rhyacophilid larvae are free-living and do not build a case until just before pupation. They are probably all predators which seek invertebrate prey on rock surfaces in fast-flowing streams and torrents. *Himalopsyche* seem to be restricted to habitats where torrential flows predominate, especially at high altitudes. Schmid & Botosaneanu (1966) refer to them as 'rithrobionts'. The adults of some species are brachypterous. The forelegs of the larvae in both *Rhyacophila* and *Himalopsyche* are raptorial and modified for grasping prey, while the apices of the

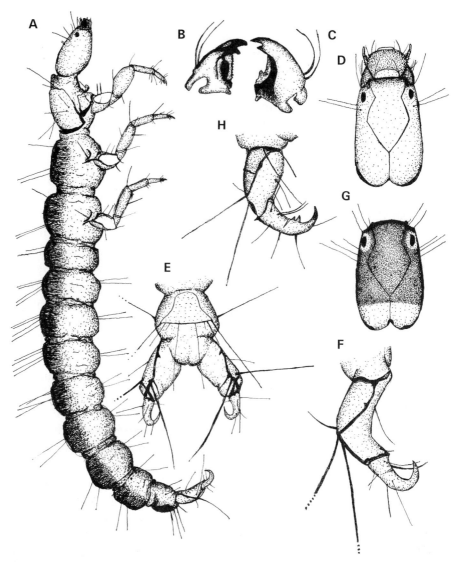

Fig. 4.85 *Rhyacophila* spp. indet. (Trichoptera: Rhyacophilidae): A, lateral view of larva of *Rhyacophila* sp. 1; B & C, mandibles; D, dorsal view of head capsule showing labrum; E, dorsal view of anal prolegs and abdominal segment IX with sclerite; F, proleg; G & H, head capsule and proleg of *Rhyacophila* sp. 2.

mandibles are narrow and rather pointed with cutting edges (e.g. Figs 4.85B & C). The head-capsule and pronotum are sclerotized, but the meso- and metanotum are membranous. The hooks on the anal prolegs are often well armed with accessory teeth, and the size and number of these, in addition to the arrangement of gills (if present),

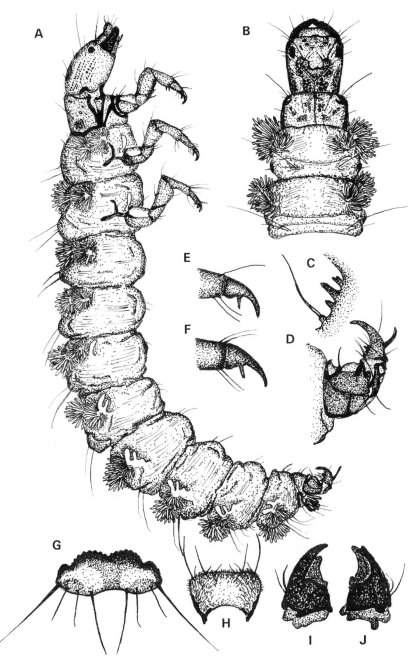

Fig. 4.86 *Himalopsyche* (Trichoptera: Rhyacophilidae): A, lateral view of larva; B, head and thorax; C, inner margin of anal claw (detail); D, anal proleg; E, tarsal claw of foreleg; F, tarsal claw of hindleg; G, dorsal sclerite on abdominal segment IX; H, labrum; I & J, mandibles.

patterns on the head capsule, and the form of the mandibular teeth, may be an aid to distinguishing species within *Rhyacophila* and *Himalopsyche*. Note that Asian *Rhyacophila* larvae often lack tufts of tracheal gills on the abdomen, but these seem to be invariably present and well developed on segments I-VIII in *Himalopsyche*. These larvae also have gills on the meso- and metathorax, and are generally larger and more robust than *Rhyacophila*. In addition, certain *Himalopsyche* have ovoid sclerites on the venter of some abdominal segments. The larvae of *Fansipangana* are unknown.

Hydrobiosidae

The Hydrobiosidae has been treated as a subfamily of the Rhyacophilidae but was raised to family status by Schmid (1970b). More recently, Schmid (1989) has argued that this family is more closely related to the Annulipalpi than the Rhyacophilidae. The matter is not settled (Morse & Yang, 1992; Morse, 1997), and a conservative classification which retains the Hydrobiosidae alongside the Rhyacophilidae within the Spicipalpia is adopted here. The Hydrobiosidae is most diverse in the Australian and Neotropical Regions, but a few species are found in tropical Asia (e.g. Schmid, 1970a; 1989). Schmid (1989) gives a valuable review of the family (although larval stages are not considered). There are approximately 50 genera in two subfamilies world-wide, but only *Apsilochorema* (Apsilochoreminae) occurs in the Oriental Region (Schmid, 1989: Fig. 253; Huisman, 1992). It is found throughout Southeast Asia to Sulawesi and New Guinea in the east, and through China and Taiwan toward Palaearctic Asia in the north. *Apsilochorema* also penetrates Pakistan, northern India and Sri Lanka (Schmid, 1958b). *Tanorus* (= *Ornatus*: Hydrobiosinae) is confined to New Guinea (Neboiss, 1984a; 1986). Although widely distributed, *Apsilochorema* is apparently restricted to montane or submontane habitats (> 950 m elevation) in forested areas (e.g. Huisman, 1992). I have collected larvae of *Apsilochorema* and *Tanorus* from similar habitats in New Guinea streams, but Chen & Morse (1991) report that hydrobiosids occur '. . . from low to high altitude . . .' in Taiwan. Studies of New Zealand hydrobiosid larvae suggest a preference for fast-flowing water (Collier *et al.*, 1995). Like rhyacophilids, larvae are free living, and do not construct a shelter until they are about to pupate. The abdomen lacks tracheal gills, and only the head-capsule and pronotum are sclerotized.

The limited data (e.g. P. Zwick quoted in Schmid, 1989: p. 107; Yule, 1996) suggest that the larvae are predators. This habit is reflected in the form of the prothoracic legs of *Apsilochorema*: the tibia, tarsus and claw are fused into a spike or sabre-like structure bearing a few large teeth on the inner basal margin. Like the blade of a pocket-knife, it can be folded back against the femur (which is greatly expanded with a toothed margin) thereby firmly grasping prey. The apex of the foreleg of *Tanorus* resembles the elongate chela of a portunid crab, in which the shortened tibia, tarsus and claw close against a concave extension of the femur thereby seizing and holding prey.

Recent descriptions of adult *Apsilochorema* include Neboiss & Botosaneanu (1988; Sulawesi), Huisman (1992: Sabah), Malicky & Chantaramongkol, 1993b: Thailand) and Malicky (1997: Nepal); Neboiss (1984a) also describes adult *Tanorus* from New Guinea, and considers that the female hydrobiosids figured by Kumanski (1979) as *Edpercivalia* probably belong to the genus *Tanorus*. Tian & Li (1985) report that *Psilochorema* occurs in China, but this record (originating in Hwang, 1957) of a New Zealand genus is certainly *Apsilochorema* (see Schmid, 1989). The larva of *Apsilochorema mancum* is illustrated and described in detail by Ulmer (1957), and *Apsilochorema* from Japan and Korea are figured by Tanida (1985) and Yoon & Ki (1989a).

Glossosomatidae

The Glossosomatidae includes about 450 described species, but the classification of the family is unsettled. For example, the status of the subgenera of *Agapetus*, which were established by Ross (1956) and adopted by Schmid (1959b), Kobayashi (1982), Oláh (1988), Nozaki *et al.* (1994) and others, is problematic with because some workers (e.g. Chantaramongkol & Malicky, 1986; Mey, 1990a; Morse & Yang, 1992) treat some of these subgenera (e.g. *Synagapetus*) as though they were genera (see also Kimmins, 1953b; 1962). There is a lack of agreement also over the subgeneric classification of *Glossosoma* (see, for example, Schmid, 1971; Kobayashi, 1982; Morse & Yang, 1992; Nozaki *et al.*, 1994). It seems that considerable revisionary work will be needed before the generic and higher classification of the Glossosomatidae can be resolved, and I have followed the pragmatic approach adopted by Morse & Yang (1992). On this basis, the Glossosomatidae comprises two subfamilies in tropical Asia — the Agapetinae and Glossosomatinae — with genera and subgenera set out as in Table 4.7. *Synagapetus* as used therein is *sensu lato*, and probably represents a genus-group rather than an single taxon (Morse & Yang, 1992). The subfamily Protoptilinae is represented in Palaearctic Asia by *Padunia*, a genus transferred to the Glossosomatidae from the Hydroptilidae by Marshall (1979). The only tropical Asian records are from northern Thailand (Malicky & Chantaramongkol, 1992), but *Nepaloptila* and *Temburongpsyche* are known from Thailand and Borneo respectively (Malicky & Chantaramongkol, 1992; Malicky, 1995b). Tian & Li (1986) have described a new species of *Mortoniella* from China, but this protoptiline genus is otherwise confined to South America and the generic placement is almost certainly incorrect (Morse & Yang, 1992).

Recent descriptions of adult glossosomatids from tropical Asia include *Agapetus* (*sensu stricto*) by Neboiss & Botosaneanu (1988), Oláh (1988: Vietnam), Malicky & Chantaramongkol (1992: Thailand), Malicky (1995a: Vietnam; 1995b; 1997: Malaysia, Nepal and China) and Mey (1996a: Vietnam); *Synagapetus* by Chantaramongkol & Malicky (1986: Sri Lanka), Oláh (1988: Vietnam), and Mey (1990a: the Philippines); and, *Glossosoma* by Malicky & Chantaramongkol (1992: Thailand) and Malicky (1994a: the Philippines; 1995; 1997: China and Nepal). In addition, Schmid (1991b) has erected a new genus — *Poeciloptila* — to accommodate two Indian glossosomatid species. Surprisingly, Malicky & Chantaramongkol (1996) have described a new species of the genus *Ptilocolepus* from Thailand as a glossosomatid. This genus is known from India and the Palaearctic (Ross, 1956; Marshall, 1979; Kumanski, 1980; Schmid, 1991b; Wiberg-Larsen *et al.*, 1991) where most workers have viewed it as a primitive hydroptilid (subfamily Ptilocolepinae) which has affinities with the Glossosomatidae, although Malicky (1983b) does not agree. Significantly, Wiberg-Larsen *et al.* (1991) report that the larvae resemble *Pseudagapetus* and show the typical hydroptilid hypermetamorphosis (see p. 381).

Larval glossosomatids make portable cases out of small stones and are usually found grazing algae from rock surfaces in moderate to fast currents. They are an important component of the scraper functional-feeding group in many streams. The case is small (typically less than 1 cm in length) and saddle-shaped, resembling the shell of a turtle. The stones used to make the case are of various sizes, but often include one or more larger 'ballast' stones along each side. The head and legs protrude from an anterior ventral opening in the case, and the anal prolegs contact the substrate through a similar aperture posteriorly. 'Anterior' in this instance is defined by the position of the head and not case architecture. Larvae can keep the head beneath the case while feeding so that they are protected from predators. Unlike most other case-building caddisflies, the larvae readily vacate their cases when collected. In addition, they to have leave the old case in order to construct a new one rather than simply extending the existing one.

Glossosomatid larvae (Fig. 4.87) have rather short, stout bodies, and lack abdominal gills. The mandibles are well adapted for scraping, with fused teeth and an inner brush-like array of stout setae (Figs 4.87D & E). The anal proleg is broadly joined with the abdomen in contrast to rhyacophilids, hydrobiosids and Hydropsychoidea where the proleg is free from the abdomen except at its base. However, the distal half is free, and this distinguishes glossosomatids from other case-makers (the Limnephiloidea) in which the anal proleg does not protrude from the case and is almost entirely fused to the abdomen. All glossosomatids have a dorsal hook on the claw of each proleg (Fig. 4.87G) which may help the larvae maintain their purchase on rock surfaces. The dorsum of the pronotum is sclerotized, but the meso and metanotum are more-or-less membranous. Reliable identification of larvae to genus is not possible in Asia, but recognition of subfamilies is straight-forward. The Agapetinae bear a small pair of median sclerites on both the meso- and metanotum which glossosomatine larvae lack. In addition, Agapetinae have more setae along the margins and on the surface of the pronotum than the Glossosomatinae, which are somewhat larger animals. Larvae of the Protoptilinae (*Nepaloptila*, *Padunia* and *Temburongpsyche*), which do not seem to penetrate far south into the Oriental Region, can probably be recognized by the presence of three sclerites on the mesonotum. Iwata (1937), Tanida (1985) and Yoon & Ki (1989a; 1989b) provides illustrations of *Glossosoma* and *Agapetus* larvae from Palaearctic Asia, and Ulmer (1957) figures tropical Asian *Glossosoma* and agapetine larvae which may be referable to *Synagapetus* and *Tagagapetus* (but see Morse & Yang, 1992). The larvae of *Poeciloptila* await description. Although ecological data from tropical Asia are limited, it is possible that Agapetinae favour lower-altitude and lower-latitude sites than the Glossosomatinae. Life-histories have yet to be investigated but, in Japan, *Glossosoma* may be able to complete three generations in a year; nevertheless, univoltine life histories seem more typical of Palaearctic glossosomatids (Sameshima & Satom, 1994).

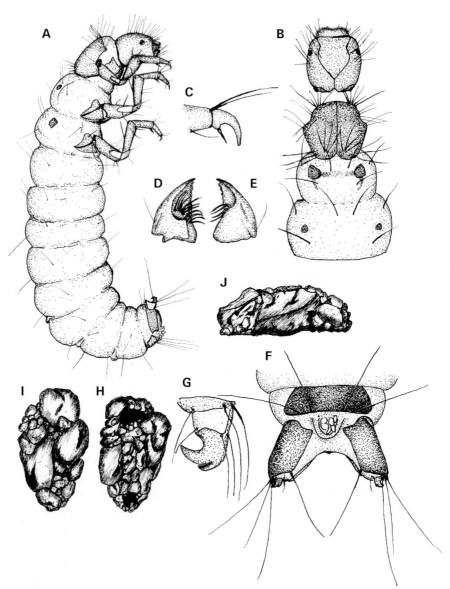

Fig. 4.87 *Agapetus* (Trichoptera: Glossosomatidae): A, lateral view of larva; B, head and thorax; C, tarsal claw of foreleg; D & E, mandibles; F, dorsal view of anal prolegs and abdominal segment IX with sclerite; G, anal claw; H, I & J, ventral, dorsal and lateral views of case (respectively).

Hydroptilidae

The Hydroptilidae is a diverse family of tiny caddisflies, the vast majority of which are no larger than 6 mm long. They are often referred to as microcaddisflies. Marshall (1979) has reviewed the genera, but concentrates on adult characteristics. The larvae are free-living for the first four instars (which are completed rapidly), and build a case or shelter in the fifth and final instar. These constructions are highly variable among species but, as in almost all caddisflies, are stereotyped within a species or genus. The cases may be bi-valved and purse-like, or resemble a flask or a barrel, or a flattened dome. They may be constructed entirely of silk secretions into which sand grains, diatoms or algal filaments are sometimes incorporated. Larvae change shape dramatically during the fifth instar, with the abdomen swelling greatly as pupation approaches, so the case must be enlarged to accommodate this increase. When fully mature, the larva fixes the case to the substrate, seals it and pupates. The term 'hypermetamorphosis' is applied to the hydroptilid life cycle. Fifth-instar larvae of different hydroptilid genera have distinctive body shapes, but early-instar larvae are rather similar. The group as a whole is characterized by small size, sclerotized plates which completely cover the dorsum of the thoracic nota, a lack of abdominal gills, and (in many cases) the presence of long setae on the head and thoracic nota; there is usually a sclerotized plate on the dorsum of abdominal segment IX. In other respects, however, the morphology of hydroptilid larvae is quite diverse. Their microhabitats are also variable. Most hydroptilids feed on algae (often sucking out the cell contents; e.g. *Hydroptila* and *Oxyethira*) and so are restricted to sunlit sections of streams but, within these habitats, species may be confined to seeps or thin films of water, cascades, rock surfaces, among filamentous algae, or on macrophytes. One member of the Australasian *Orthotrichia muscari* species group is a parasite upon the pupae of other caddisflies (Wells, 1992), while some other *Orthotrichia* species are predators. Ecological work on the family in tropical Asia is extremely limited. Wells & Dudgeon (1990) present some preliminary life-history data for Hong Kong hydroptilids which indicate that *Stactobia*, *Hydroptila*, *Oxyethira* and *Orthotrichia* species have long flight periods and may be bivoltine or multivoltine. Japanese *Hydroptila* and *Oxyethira* are apparently univoltine (Ito & Kawamura, 1980; 1984). Work on the larvae of a Thai species of *Ugandatrichia* — which reaches 10 mm in length and is a giant among microcaddisflies — by Hans Malicky and colleagues is ongoing, and it seems possible that these animals are filter-feeders.

A considerable amount of research has been directed towards the adult stages of tropical Asian hydroptilids, and many genera and species — most of them Hydroptilinae — are known from the region (Table 4.7). A number of these genera are known from only one or a small number of localities, and contain few species. Others, such as *Chrysotrichia, Hydroptila, Orthotrichia, Oxyethira* (which has been reviewed by Kelley, 1984), *Stactobia, Scelotrichia, Tricholeiochiton* and *Ugandatrichia*, are widespread and may include a large number of species. However, the range of species described so far surely under-represents the regional diversity of Hydroptilidae (Wells & Huisman, 1992), and the total number of genera and species continues to increase as investigation proceeds. For example, Schmid (1983) described 37 new *Stactobia* species from India; Oláh (1989) established the genus *Vietrichia* from Vietnam and documented a total of 35 new species of *Catoxyethira, Chrysotrichia, Hydroptila, Microptila, Orthotrichia, Oxyethira, Plethus, Scelotrichia, Stactobia* and *Ugandatrichia*; Mey (1990a; 1995a) reported seven new hydroptilids (in *Hydroptila, Orthotrichia, Stactobia, Sclerotrichia* and *Ugandatrichia*) from the Philippines; Chantaramongkol & Malicky (1986) described eight new species of the genera *Chrysotrichia, Hydroptila, Orthotrichia, Oxyethira* and *Tricholeiochiton* from Sri Lanka; Wells & Dudgeon (1990) figured 10 species of *Hydroptila, Oxyethira, Orthotrichia, Scelotrichia, Stactobia* and *Ugandatrichia* from Hong Kong; and, Wells (1990a; 1991) described a total of 41 hydroptilids from New Guinea including *Chrysotrichia, Hellyethira, Hydroptila, Orthotrichia, Oxyethira, Scelotrichia, Tricholeiochiton* and *Ugandatrichia* as well as two new monotypic genera, *Missitrichia* and *Niuginitrichia*. Wells (1990b) also reported 22 new species of *Chrysotrichia, Hellyethira, Hydroptila, Orthotrichia, Oxyethira, Parastactobia, Plethus* and *Scelotrichia* from Sulawesi; O'Connor & Ashe (1992) recorded a new *Jabitrichia* from Peninsular Malaysia; Wells & Huisman (1992; 1993) described 60 new species of *Chrysotrichia, Hellyethira, Hydroptila, Plethus, Oxyethira, Macrostactobia, Orthotrichia, Scelotrichia, Stactobia* and *Ugandatrichia* from Malaysia and Borneo; Wells (1993) figured eight new species of *Chrysotrichia, Microptila, Plethus* and *Stactobia* from Bali; Yang & Xue (1992; 1994) and Yang *et al.* (1997) gave accounts 12 new species of *Hydroptila, Orthotrichia* and *Oxyethira* from China; and, Mey (1996a) described two new *Scelotrichia* and a *Stactobia* from Vietnam. Wells & Malicky (1997) have recently described six new *Chrysotrichia*, two *Hydroptila*, four *Orthotrichia*, a *Stactobia*, two *Sclerotrichia* and a *Ugandatrichia* from Java and Sumatra, bringing the total hydroptilid fauna for these two islands to 21 and 32 species respectively.

Knowledge of larval hydroptilids is comparatively meagre, but has increased substantially during recent years. Ulmer (1957) figured Indonesian larvae of *Ithytrichia* (as *Saranganotrichia*), *Hydroptila* (as *Oeceotrichia*, *Pasirotrichia* and *Sumatranotrichia*), *Oxyethira*, *Orthotrichia* (as *Baliotrichia* and *Orthotrichiella*), *Plethus* (including *Plethotrichia*) and *Stactobia* (as *Lamonganotrichia*; all synonymies according to Marshall, 1979), in addition to four unidentified hydroptilid genera, while the immature stages of Taiwanese and Japanese *Hydroptila* and Japanese *Oxyethira* have been illustrated by Iwata (1928) and Ito & Kawamula (1980; 1984; see also Tanida, 1985). A diagnosis of larvae of the Holarctic and Oriental genus *Ithytrichia* is provided by Wiggins (1996). Valuable and definitive work on tropical members of the family has been undertaken by Alice Wells who has described larvae of many Australasian genera and figured their cases (e.g. *Hellyethira*, *Hydroptila*, *Orthotrichia*, *Oxyethira* and *Tricholeiochiton* in Wells, 1985). Wells & Huisman (1992) have figured the immatures of *Macrostactobia* (from Peninsular Malaysia) for the first time, and give a few details of *Stactobia* immatures also (Wells & Huisman, 1993). Wells (1991) provides some details of New Guinea *Orthotrichia* larvae and their cases, and Wells (1990a) describes immatures of *Niuginitrichia* and *Scelotrichia*. She also provides some illustrations of *Chrysotrichia*, *Hellyethira*, *Hydroptila*, *Plethus*, *Orthotrichia* and *Sclerotrichia* larvae from Sulawesi (Wells, 1990b). Despite such efforts, several genera (e.g. *Jabitrichia*, *Missitrichia*, *Parastactobia* and *Vietrichia*) are unknown as larvae and, citing the insufficiency of collecting effort, Wells & Huisman (1992) are reluctant to supply keys to even the adult stages. However, Wells (1990b) does give a key to final-instar larvae of those Sulawesi hydroptilids where the immatures are known. Although incomplete keys can be misleading, the following key to final-instar hydroptilid larvae is presented as an aid to further study of these unusual insects. Because of their small size, successful identification depends upon examination of larval morphology at high magnification. Each couplet includes a variety of characters (based largely on Wells, 1985; 1990b) which should reduce misidentifications. The Holarctic genus *Palaeagapetus* is also included in this key, since adults have been reported from Fujian Province, China (Tian & Li, 1985). Detailed descriptions of larval *Palaeagapetus* are given by Ito & Hattori (1986) and Ito (1991a; 1991b). The larvae of *Ptilocolepus* may also be encountered in tropical Asia (see above); they have been figured by Wiberg-Larsen *et al.* (1991).

1. Body rather flattened, and abdominal segments I-VIII with a truncate, fleshy tubercle on each side; legs rather short and similar in structure; case of two flattened valves covered with fragments of liverwort; probably confined to seeps and springs in Palaearctic Asia .... **Ptilocolepinae:** *Palaeagapetus*

    Abdominal segments I-VIII lack fleshy tubercles; body often somewhat compressed laterally; forelimbs shorter and usually stouter than the hindlegs; case as above (lacking liverwort fragments) .... **Hydroptilinae** 2

2. Mid- and hindlegs less than twice the length of the forelegs 3

    Mid- and hindlegs at least twice the length of the forelegs **11**

3. Claws on anal prolegs with at least two accessory dorsal hooks (on the outer margin behind the tip), or with a pair of stout setae arising from the same position; body slightly compressed laterally; abdominal segments enlarging gradually from I-IV or I-V; terminal segment (sensillum) of antenna discrete (not fused); case like a purse (ovoid or bean-shaped), comprising two parts (or valves), and including sand grains and/or algal filaments 4

    Anal claws lack accessory hooks or a pair of dorsal setae; body slightly flattened dorsoventrally; abdominal segments do not enlarge posteriorly; terminal segment of antenna *may* be fused; case may be fusiform, purse-shaped or otherwise 5

4. Anal claws with at least two accessory hooks or teeth on the outer margin; headcapsule and thoracic tergites *usually* pale; case usually includes sand or algae; common and widespread .... *Hydroptila*

    Anal claws with a pair of stout setae on the outer margin; headcapsule and thoracic tergites *usually* dark brown; case includes filamentous algae .... *Macrostactobia*

5. Single small tergites present on the dorsum of most abdominal segments, may be especially well developed on segments VIII-X 6

    Abdominal segments I-VIII without distinct dorsal tergites although dorsal and ventral projections may be present 7

6. Small tergites present on the dorsum of abdominal segments I-VIII; case bi-valved and flattened, with a smaller ventral valve .... *Plethus*

   Larvae *usually* have short rectangular sclerites on the dorsum of the abdominal segments, and they are *always* well developed (into plates) on segments VIII-X (at least); case flattened with ventral valve forming a median tube beneath a broader dorsal plate which may have fluted margins and/or anterior and posterior notches .... *Stactobia*

7. Body strongly compressed laterally; most abdominal segments with prominent dorsal and ventral lobe-like projections; case is a flattened pouch of secretions (silk) with a wide posterior opening and a small circular opening anteriorly; probably mainly Palaearctic Asia .... *Ithytrichia*

   Body and case not as above 8

8. One pair of dark sclerites on the dorsum of abdominal segment I, and a black T-shaped mark on tergite IX; case rather flat, round to ovoid or almost rectangular, with upper plate larger than the lower so that the structure is domed or tortoise-shaped, constructed of secretion and (sometimes) sand; New Guinea only .... *Niuginitrichia*

   Without the above combination of characters 9

9. Labrum with an asymmetrical beak-like projection (this may be hard to see): abdominal segment II is swollen to form paired lateral 'horns', and abdominal segment I has a median sclerotized ring in the form of a whorl of small sclerites; terminal segment of antenna fused; case fusiform and shaped like a wheat seed, rather flattened ventrally with some longitudinal dorsal ribs, and made of larval secretions .... *Orthotrichia*

   Labrum lacks a 'beak' and is more-or-less symmetrical; terminal segment of antenna discrete or fused; abdominal characteristics not as above; case bivalved and either purse-like or flattened **10**

10. Case laterally compressed and somewhat purse-like, comprising two equal valves and constructed of algal filaments, sand grains and/or tiny fragments of moss or other plants; a small spur-like seta borne beneath (subtending) the anal claw; terminal segment of antenna fused .... *Scelotrichia*

Case dorsoventrally flattened and made of larval secretions, with an upper dome which overlaps a smaller underlying plate, and an elongate ovoid shape; a fine seta subtends the anal claw; terminal segment of antenna discrete .... *Chrysotrichia*

11. Thoracic nota elongate, saddle-shaped (i.e. rather constricted posterior-laterally); abdominal segment I of a similar size to the metathorax; sensillum of antenna discrete and well-developed; case laterally compressed with two equal valves and generally much larger than the larva, shape variable but usually somewhat rectangular (the anterior and posterior margins may be expanded dorsally), sand or algal fragments may be incorporated into the structure .... *Hellyethira*

Thoracic nota short and broad; abdominal segment I considerably wider than the metathorax; case form variable but made entirely from larval secretions **12**

12. Mid- and hindlegs four times longer than the forelimbs, with unusually long tibiae; case rectangular with convex margins .... *Tricholeiochiton*

Mid- and hindlegs twice the length of the forelimbs, and tibiae not markedly longer than the femora or tarsi; case flask-shaped .... *Oxyethira*

Philopotamidae

The Philopotamidae is a rather homogenous family, both ecologically and morphologically. These Annulipalpia spin finger-shaped, fine-meshed silk nets within which the larvae dwell. Nets are usually situated in groups under stones, in seeps, or in thin films of water at the margins of waterfalls and cascades, where they collect fine organic material. Meshes are rectangular and very fine with dimensions in the order of 0.5 x 10 μm (e.g. Cartwright, 1990). Philopotamid larvae have a sclerotized pronotum (the mesonotum and metanotum are membranous) and lack abdominal gills. The sclerotized parts of the body are usually yellow or orange, while the abdomen and portions of the thorax are white. The labrum is membranous and T-shaped, with an expanded, brushlike apical margin (Fig. 4.88C); this feature clearly distinguishes members of the family from other Annulipalpia. Unlike larvae of the Spicipalpia, a sclerotized plate on the dorsum of

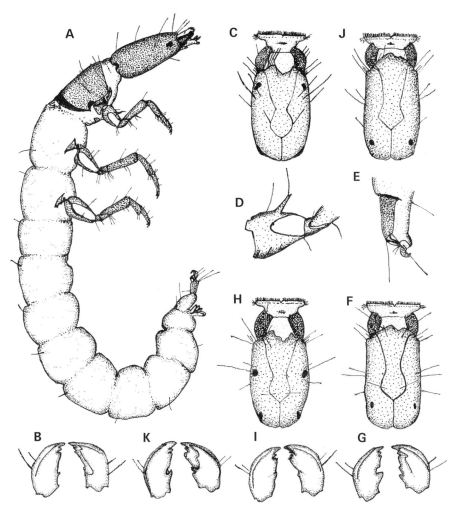

Fig. 4.88 *Chimarra* spp. indet. (Trichoptera: Philopotamidae): A, lateral view of larva of *Chimarra* sp. 1; B, mandibles; C, dorsal view of head capsule showing labrum and notching of the frontoclypeus; D, anterior view of fore coxa; E, anal proleg; F & G, head capsule and mandibles of *Chimarra* sp. 2; H & I, head capsule and mandibles of *Chimarra* sp. 3; J & K, head capsule and mandibles of *Chimarra* sp. 4.

abdominal segment IX is lacking. Despite such homogeneity, the Philopotamidae is quite speciose: for example, the genus *Chimarra* is estimated to contain around 100 species in Thailand (Chantaramongkol & Malicky, 1989) and about 500 species in Southeast Asia (Malicky, 1989a). This genus seems to dominate the philopotamid fauna of tropical Asia (Ross, 1956), and — in global terms — probably contains more species than all other genera in the family combined (Wiggins,

1996). In India (at least), however, *Gunungiella* is speciose also, comprising 35 described species compared to only 12 *Chimarra* (Higler, 1992); additions to both genera can be expected as research proceeds.

The philopotamid genera of tropical Asia are listed in Table 4.7. *Dolomyia* and *Dolopsyche* are recently-established genera from India (Schmid, 1991a), while *Edidiehlia* was erected by Malicky (1993a) and, to date, is confined to Sumatra. *Doloclanes* was established by Banks (1937) on basis of adults from the Philippines. Although Ross (1956) treated *Doloclanes* as a subgenus of *Wormaldia*, it was given full generic status by Schmid (1991a) and this view has received acceptance (e.g. Mey, 1993a; 1996; Malicky & Chantaramongkol, 1996). Kobayashi (1980) revised the Japanese Philopotamidae, but used the name *Sortosa* in preference to *Dolophilodes* (which has precedence), and established *Hisaura* (a subgenus of *Dolophilodes*) which seems to be confined to Japan. Recent work on the adults of Asian philopotamids include 43 new *Chimarra* species from Thailand (Chantaramongkol & Malicky, 1989; Malicky & Chantaramongkol, 1993b; 1993c; 1997); other new species of this genus have been documented from several Asian locations (including Nepal, Vietnam, China, Bali and the Philippines) by Malicky (1993b; 1994a; 1995b: 25 species) and Mey (1995b: five species), as well as from Sri Lanka (Chantaramongkol & Malicky, 1986: 2 species), Sumatra (Malicky, 1989a; nine species) and Vietnam (Malicky, 1994a; 1995a: 12 species). In addition, new species of *Gunungiella* have been described from Sri Lanka (Chantaramongkol & Malicky, 1986), Sabah (Huisman, 1993), Sumatra and the Philippines (Malicky, 1993a; Mey, 1995a), Thailand (Malicky & Chantaramongkol, 1993b; 1993c), Vietnam (Malicky, 1995a) and Peninsular Malaysia (Malicky, 1995b). Adult *Doloclanes* have been figured from India (Schmid, 1991a), Sumatra (Malicky, 1993a), Burma (Malicky, 1993b), Thailand (Malicky & Chantaramongkol, 1993b; 1996), China (Mey, 1993a), Nepal (Malicky, 1994a) and Vietnam (Malicky, 1995a; Mey, 1996a). New *Kisaura* (viewed by some workers as a subgenus of *Dolophilodes*) have been described from Nepal (Malicky, 1993a), Thailand (Malicky & Chantaramongkol, 1993b) and Vietnam (Malicky, 1995a; Mey, 1996a); new *Dolophilodes* from Nepal (Malicky, 1993a), Thailand (Malicky & Chantaramongkol, 1993b), China (Tian *et al.*, 1993) and Vietnam (Malicky, 1995a); and, new *Wormaldia* from India (Schmid, 1991a), Thailand (Malicky & Chantaramongkol, 1993b; 1993c), China (Tian *et al.*, 1993) and Vietnam (Malicky, 1995a; Mey, 1995b; 1996a).

Chimarra larvae (Fig. 4.88) are easily recognized by the presence of a distinct asymmetric notch in the anterior margin of the head capsule (e.g. Ulmer, 1957). This always occurs in combination with a finger-like process arising from the distal portion of the fore coxa, thus setting *Chimarra* apart from a few *Dolophilodes* species which show some asymmetry or notching of the front of the head. Cartwright (1990a) provides illustrations and an up-to-date diagnosis of larval *Chimarra*. Different species show slight variations in the shape of the anterior notch in the head capsule and in the form of the inner margin of the mandibles (Fig. 4.88). *Dolophilodes* larvae are characterized by a well-developed anterior projection — the fore-trochantin — which arises at the base of (but separate from) the foreleg. In the related genus *Wormaldia*, the fore-trochantin projects a relatively short distance forming a much shorter and more slender process. Larval *Doloclanes*, *Edidiehlia*, *Gunungiella* and *Kisaura* have yet to be described. The limited data suggest that *Dolophilodes* and *Wormaldia* may favour cooler, higher altitude/latitude streams than other philopotamids, especially *Chimarra*. Descriptions of Palaearctic Asian philopotamid larvae are given by Iwata (1927: *Wormaldia*), Kim (1974: *Chimarra* and *Dolophilodes*), Tanida (1985: *Dolophilodes* and *Wormaldia*), and Yoon & Ki (1989a: *Dolophilodes* and *Wormaldia*). Ulmer (1957) figures the larvae of four species of *Chimarra* from Indonesia.

Stenopsychidae

Stenopsychid caddisfly larvae are a distinctive element of the fauna of rivers and streams in Asia, although they also have a restricted distribution elsewhere. The family is made up of three genera: *Stenopsychodes* (nine species) from Australia, *Pseudostenopsyche* (two species) from Chile, and *Stenopsyche* found mainly in the Oriental and southeastern Palaearctic Regions (Martynov, 1926; Kuwayama, 1930; Hwang, 1963; Schmid, 1969; Neboiss, 1986; Weaver, 1987). *Stenopsyche* is the most speciose and widespread of the three genera (Schmid, 1969). It included 69 described species in 1969, but there have been a number of additions since then (e.g. Swegman & Coffman, 1980; Hwang & Tian, 1982; Tian, 1985; 1988; Kobayashi, 1987; Weaver, 1987; Tain & Zheng, 1989) bringing the current total to around 78. Although *Stenopsyche* penetrates the Oriental Region, species richness peaks in more northern latitudes, especially in the southeast Palaearctic. For example, a survey of the pre-1985 caddisfly literature (Dudgeon, 1987a) indicated that stenopsychids did not occur in Sri Lanka or Java, while one species each was known from Sumatra and Peninsula Malaysia; however, four species had been recorded from North Korea and 44 from China (see also Tian, 1988). Examination of (albeit incomplete) distributional data in a recent checklist prepared by Higler (1992) likewise shows that 11 of 14 Indian *Stenopsyche* species appear to be confined to latitudes north of the Tropic of Cancer.

Stenopsychid larvae are large (40 — 45 mm long), with a slender, elongated head that is slightly tapered anteriorly (Figs 4.89A, F & G). The sclerotized parts are generally yellowish or brown and mottled with darker markings, but may be almost black in some species; the membraneous parts of the thorax and abdomen are generally reddish-brown. The labrum (Fig. 4.89E) is short, broad and well sclerotized (unlike the Philopotamidae) with brush-like setae along the anterior margin. Abdominal segment IX lacks a dorsal sclerotized plate. Descriptions of larval *Stenopsyche* from Japan and Korea are given by Iwata (1927: as *Philopotamopsis*) and Yoon & Ki (1989a) respectively (see also Tanida, 1985), while Ulmer (1957) has figured *Stenopsyche ochripennis* from Indonesia, and Swegman & Coffman (1980) have described the larva of Indian *Stenopsyche kodikanalensis*. More recently, Rahim Ismail *et al.* (1996) have figured the larvae and pupae of *Stenopsyche siamensis* from Peninsular Malaysia in some detail.

Where they do occur, stenopsychids can be abundant, and Nishimura (1966) referred to them as '. . . dominant species in the climax . . .' in Japanese streams. *Stenopsyche* larvae build 'nests' between rocks or small boulders (or in places where pebbles have become lodged under larger stones) by spinning a coarse-meshed, irregular net structure which connects the two stones together and within which the larva rests and feeds (Nishimura, 1966). The structure includes a rather delicate cone-shaped net which is assumed to be the food-catching device (Rahim Ismail *et al.*, 1996). The 437 x 277 µm mesh measured from Malaysian *Stenopsyche siamensis* nets (Rahim Ismail *et al.*, 1996) is much coarser than most of other net-spinning caddisflies. For example, the rectangular mesh of nets spun by Hong Kong hydropsychids rarely exceeds 160 µm along the longest dimension (Dudgeon, 1992a: Figs 35 & 36), and the mesh of *Stenopsyche* nets is of a similar size to those spun by predatory arctopsychine Hydropsychidae (e.g. Malas & Wallace, 1977). Despite this, stenopsychid larvae consume detritus (mainly) and algae (predominately diatoms; Nishimura, 1966); animal food is rarely taken (Rahim Ismail *et al.*, 1996). They are most abundant in the middle of streams, where the current is strongest, and scarce close to the banks (Dudgeon, 1996b).

Most information on stenopsychid ecology comes from Japan where *Stenopsyche* species complete one or two generations per year (Nishimura, 1966; 1984; Gose, 1970b; Takemoto, 1983; Aoya & Yokoyama, 1987; 1990), and may shift between a univoltine and a

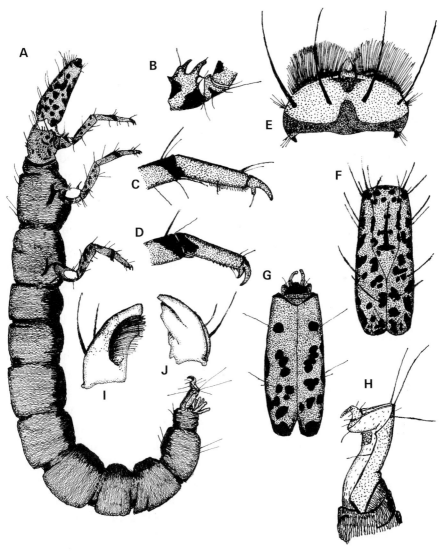

Fig. 4.89 *Stenopsyche angustata* (Trichoptera: Stenopsychidae); A, lateral view of larva; B, anterior view of fore coxa; C & D, tarsi of fore- and hindlegs; E, labrum; F, head capsule (dorsal view); G, head capsule (ventral view); H, anal proleg; I & J, mandibles.

bivoltine life history depending upon stream water temperatures (Gose, 1970b). In Hong Kong, *Stenopsyche angustata* seems to be univoltine (Dudgeon, 1996b), although adults are present throughout the year (Dudgeon, 1988c). All studies of Japanese species indicate a rather limited flight period, often associated with swarming behaviour (Nishimura, 1966; 1981; 1984; Gose, 1970b; Aoya & Yokoyama, 1987), and this seems to reflect rather synchronous larval growth. As adult *Stenopsyche marmorata* are presumed to live for only around 10 days (Nishimura, 1966), the presence of *Stenopsyche angustata* throughout the year in Hong Kong probably reflects warmer temperatures, allowing year-round emergence, and not longer-lived adults.

Psychomyiidae

Psychomyiids are small to minute (less than 10 mm body length) rather cryptic caddisflies. The larvae build silken tubes, covered with a thin layer of fine sand and detritus, which meander over wood and rock substrates in slow-flowing streams. They are confined to sites close to the banks in habitats where the mid-stream current is swift. Some species (e.g. in *Tinodes*) can also exploit well-oxygenated lake habitats. Larvae graze algae (especially diatoms: Mathis & Bowles, 1994) and ingest fine detritus, with the tube providing protection from predators and, perhaps, an aid to respiration (Wiggins, 1996). The tube is extended as areas around the anterior opening are depleted of food. The meso- and metathorax of psychomyiid larvae (Fig. 4.90) are membranous and more corpulent than the sclerotized prothorax, and the prothoracic legs tend to be relatively stout. In addition, the abdominal segments are sharply demarcated, and the anal prolegs are (relative to the Polycentropodidae: p. 397) rather short. A flange- or hatched-shaped fore-trochantin of the prothorax is the main distinguishing feature of this family (Fig. 4.90A & H). In addition, the tarsal claws have a stout process at the base from which arises a long seta (Fig 4.90C & D). The labium is well developed and extends beyond the anterior margin of the head; presumably this aids in the application of silk to the inner surface of their tubes. Unfortunately, the Psychomyiidae is quite incompletely known in Asia and, although many new species are being described in the adult stage, identification of the larvae to genus is problematic. Until recent revisionary work (see below), two subfamilies of Psychomyiidae were recognized. Of

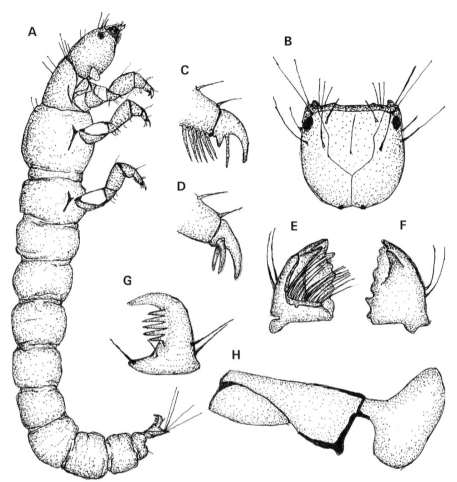

Fig. 4.90 *Psychomyia* (Trichoptera: Psychomyiidae): A, lateral view of larva; B, dorsal view of head capsule; C & D, tarsal claws of fore- and mesothoracic legs (respectively); E & F, mandibles; G, anal claw; H, fore-trochantin.

these, the Paduniellinae are occur throughout much of the Oriental Region. Recent descriptions of *Paduniella* adults include 16 new species from Bali, Sri Lanka, Nepal, Laos and Thailand (Chantaramongkol & Malicky, 1986; Malicky & Chantaramongkol, 1993c; 1996; Malicky, 1995b; 1997). The only account of immature *Paduniella* concerns a North American species which differed from *Psychomyia* in small details only (Mathis & Bowles, 1994; see also Wiggins, 1996).

Among the Psychomyiinae, new species of *Psychomyia* have been described recently by Oláh (1985: Bhutan), Tian *et al.* (1993: China) in addition to 20 species from Nepal, Laos, Thailand, Borneo, Sumatra,

Bali, Sulawesi, and the Philippines (Malicky, 1993b; 1994a; 1995b; 1997; Malicky & Chantaramongkol, 1997). Holarctic *Psychomyia* larvae are quite well known (see Iwata, 1928; Tanida, 1985; Wiggins, 1996), and a species from Korea is figured by Yoon & Ki (1989a). Diagnostic features include prolegs that are relatively stout and longer than the other legs, in combination with anal claws bearing four long teeth along the inner edge. Adults of *Psychomyiella* have been reported from Taiwan and Korea (Kobayashi, 1987; Kumanski, 1992), and the larva of an Indonesian species is illustrated by Ulmer (1957), but the genus is apparently unrecorded elsewhere in the region apart from India (under the junior synonym of *Khandalina*: Higler, 1992). Given the scarcity of material, it is difficult to define the limits of *Psychomyiella* on the basis of larval morphology. Like *Psychomyia*, the inner edge of the anal claws are toothed; indeed, *Psychomyiella* may be a junior synonym of this genus (see below). *Lype*, which occurs in Sri Lanka, India, Thailand and Vietnam (Schmid, 1972; Chantaramongkol & Malicky, 1986; Malicky & Chantaramongkol, 1993b; Mey, 1996a), extends into the Holarctic Region, and larvae from North America are illustrated by Wiggins (1996). The flattened silken tubes constructed by these larvae are shorter (typically less than 10 mm long) than those of other psychomyiids. *Tinodes* is one of the more speciose psychomyiid genera with, for example, at least 21 species in India (Higler, 1992). It is widespread throughout tropical Asia to New Guinea (e.g. Banks, 1937; Kimmins, 1963; Schmid, 1972; Tian & Li, 1985; Neboiss, 1986). Twenty-seven new species have been described recently from the Philippines, Bali, Thailand, Vietnam, Peninsular Malaysia, Burma and Nepal (Mey, 1990a; 1995a Malicky, 1993b; 1995a; 1995b; 1997; Malicky & Chantaramongkol, 1989b; 1993b; 1993c; 1996). The larvae have received relatively little attention but Taiwanese and Indonesian species has been figured by Iwata (1928) and Ulmer (1957) respectively. Illustrations of Holarctic *Tinodes* are provided by Tanida (1985) and Wiggins (1996). *Tinodes* and *Lype* lack the teeth on the inner margin of the anal claw which are present in *Paduniella*, *Psychomyia* and *Psychomyiella*. The larvae of the monotypic genus *Padangpsyche* from Sumatra (Malicky, 1993b). Likewise, the larvae of *Eoneureclipsis* which ranges from India through Burma, Thailand, Vietnam and Borneo (Kimmins, 1955; Schmid, 1972; Malicky & Chantaramongkol, 1989b; 1997; Malicky, 1995a) have yet to be reported in the literature.

A recent phylogenetic revision of the Psychomyiidae (Li & Morse, 1997a) changes the composition of the Psychomyiidae to include the genera *Paduniella* and *Psychomyia* in Asia. The Paduniellinae thus becomes a junior synonym of the Psychomyiinae. Li & Morse (1997a) establish a new subfamily — the Tinodinae — comprising the genera *Lype, Padangpsyche* and *Tinodes. Khandalina* and *Psychomyiella* are treated as junior synonyms of *Psychomyia*, and Li & Morse (1997a) do not recognize *Eoneureclipsis* despite recent descriptions of new species of the former genus from Thailand (e.g. Malicky & Chantaramongkol, 1997).

Xiphocentronidae

The archaic family Xiphocentronidae consists of seven genera of which five occur in the Oriental Region (Schmid, 1982). They have been included within the Psychomyiidae (as the Xiphocentroninae), but Schmid (1982) considered them so specialized as to warrant family status. Nevertheless, similarities in adult and larval morphology indicate that these families are sister groups (Barnard & Dudgeon, 1984). Xiphocentronids have a widely disjunct distribution in Asia, Central Africa, Mexico, Central and South America; *Cnodocentron*, for example, occurs in High Asia, Arizona and Mexico (Schmid, 1982; Moulton & Stewart, 1997). Barnard & Dudgeon (1982) synonymised *Kibuenopsychomyia* and *Tsukushitrichia* with *Melanotrichia*, and this genus (as *Kibuenopsychomyia*) has been reported from China (Hwang, 1957; Tian & Li, 1985). More recently, new species of *Abaria, Drepanocentron, Melanotrichia* and *Proxiphocentron* have been described from Thailand, Peninsular Malaysia and the Philippines (Malicky & Chantaramongkol, 1992; 1993c; Malicky, 1995b; Mey, 1995a). *Melanotrichia* larvae (described by Barnard & Dudgeon, 1984; see Fig. 4.91) are similar to *Abaria* (see Marlier, 1962), but the larvae of *Cnodocentron, Drepanocentron* and *Proxiphocentron* are unknown. Xiphocentronid larvae are somewhat similar to psychomyiids: for example, only the pronotum is sclerotized, and the labium is elongated. Unlike psychomyiids, the tibiae and tarsi of all legs are fused, and an anterolateral projection of the mesothorax (the mesopleuron: Figs 4.91B & C) can be used to differentiate xiphocentronids from other caddisflies. Like psychomyiids, the larvae build silken tubes covered with silt and sand (and sometimes a few small stones) at sites close to the banks in small streams; sometimes these tubes will loop above the water surface onto wetted surfaces. The gut contents of *Melanotrichia* from a Hong Kong stream consisted almost entirely (>99% by volume) of fine detritus, with diatoms making up the remainder. Schmid (1982) notes

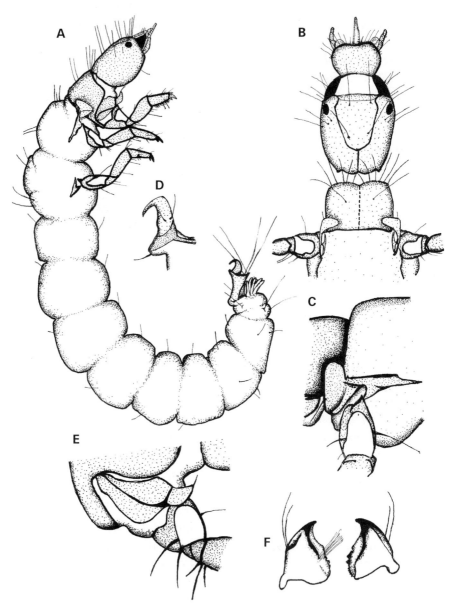

Fig. 4.91 *Melanotrichia serica* (Trichoptera: Xiphocentronidae): A, lateral view of larva; B, dorsal view of head and first two thoracic segments (note the mesopleuron projecting forward from the base of the mesothoracic legs); C, lateral view of the mesothorax showing the mesopleuron; D, anal claw; E, fore coxa and trochantin; F, mandibles (from Barnard & Dudgeon, 1984).

that, unusually for caddisflies, xiphocentronid adults are diurnal and active during periods of bright sunshine. In Hong Kong, adult *Melanotrichia serica* run around on the surface of leaves of riparian trees, and seem to favour well-lit sites.

Polycentropodidae

The Polycentropodidae is well represented in Asia but, until recently, knowledge of the family was rather limited. Many new species continue to be described, and a complete revision of the family is needed before the limits and species complement of some genera can be established with certainty (e.g. Neboiss, 1992). There are three subfamilies: one of them — the Kambaitipsychinae — consists of a single monotypic genus, *Kambaitipsyche*, established by Malicky (1991); the larvae are unknown. The Polycentropodinae is considerably more diverse. Genera include *Neucentropus* which is reported from the far North of China and Manchuria (Ulmer, 1932), but has yet to be recorded in tropical Asia. The genus *Holocentropus* is represented in the Oriental Region by a recently discovered Vietnamese species (Malicky, 1995a); it is present Palaearctic East Asia also (Ulmer, 1932). *Polycentropus* occurs in China, India, Borneo and Sulawesi (Ulmer, 1932; Schmid & Denning, 1979; Neboiss, 1987; 1989), while *Cyrinopsis* is present in India and (Neboiss, 1992; Malicky & Chantaramongkol, 1993c) but is poorly known. Hwang (1963) records *Kyopsyche* (a Japanese polycentropodid) from China.

The Polycentropodinae are particularly well represented in tropical Asia. In the case of *Polyplectropus*, for example, 48 new species of this genus have been described recently from Nepal, India, Burma, Thailand, Vietnam, Peninsular Malaysia, Borneo, Sumatra, Sulawesi and the Philippines (Neboiss, 1989a; Mey, 1990a; Malicky, 1993a; 1993b; 1994a; 1995a; 1995b; 1997; Malicky & Chantaramongkol, 1993b; 1993c; 1997); Ulmer (1957) has figured the larval stages. The genus *Plectrocnemia* is also widespread, and includes 15 recently-documented species from Nepal, Burma, Thailand, Vietnam, China, Sumatra and the Philippines (Malicky, 1993b; 1995b; 1997; Malicky & Chantaramongkol, 1993b; Tian *et al.*, 1993; Mey, 1996a). Among other Polycentropodinae, Neboiss (1992) has recently revised the genus *Nyctiophylax*, transferring some Palaearctic Asian species and several from New Guinea, Java, Sumatra, Sulawesi and Sri Lanka into *Paranyctiophylax*. Since that revision, 28 new *Nyctiophylax* have been described from Nepal, Burma, Thailand, Vietnam, Sumatra, Borneo and New Guinea (Malicky, 1993b; 1995a; 1997; Malicky & Chantaramongkol, 1993b; 1993c; 1997; Mey, 1995b); the genus is also known from China (Neboiss, 1992). *Pahamunaya* was established by Schmid (1958a) based on specimens from Sri Lanka; it is present also in Thailand and Vietnam (Schmid & Denning, 1979; Malicky & Chantaramongkol, 1993b; 1997; Malicky, 1995a; see also Kjaerandsen & Netland, 1997).

The larvae of Polycentropodidae have longest anal prolegs of any Trichoptera larvae, and a sharp, projecting fore-trochantin (Figs 4.92A & I). This combination of features separates them from the Psychomyiidae which they resemble owing to the sclerotized prothorax and overall larval *habitus*. Other distinguishing features include a lateral fringe of hair along the abdomen, well-developed setae on the legs, and tarsal claws which lack the basal process seen in psychomyiids. The labium is less pronounced because polycentropodids spin nets instead of constructing silken tubes. The net takes the form of a trumpet-like filtering device, or a short tube open at both ends from which arise a meshwork of associated threads that are attached to nearby objects (rather like an irregular spider's web). The predaceous larva resting in the tube is alerted when these threads are disturbed by unwary prey, whereupon the occupant rushes out and attempts to seize the unwary victim. Larvae occupy a range of habitats, but the flimsy nets break up if the flow becomes turbulent or too swift. When in static water, larvae undulate the abdomen to create a respiratory current.

Few polycentropodid genera are known as larvae, and the range of larval forms within genera has not been investigated. As a result, we know little of the extent of structural diversity within the group, and it is not possible to develop a workable key to larvae. Wiggins (1996) notes that the presence of teeth arising from the inner margin of the anal claw separates *Polyplectropus* and *Nyctiophylax* (and evidently *Paranyctiophylax*; see Ulmer, 1957: Fig. 358) from *Polycentropus*. However, an Indonesian *Polyplectropus* larva figured by Ulmer (1957: Fig. 347) appear to lack such teeth. *Polycentropus* larvae have two dark, sclerotized bands in the shape of an 'X' close to the base of the anal claw (Fig. 4.92H). Oriental larvae of thus type accord with descriptions of Holarctic *Polycentropus* but probably belong to one of the speciose polycentropine genera that occur in tropical Asia. Larvae of *Plectrocnemia* from Korea have been illustrated by Yoon & Ki (1989a), and Ulmer (1957) has figured the larval stages

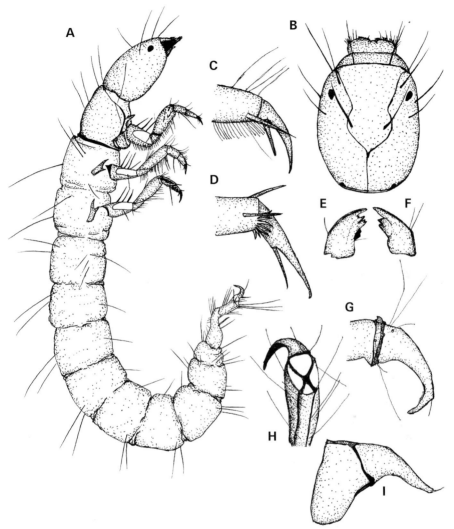

Fig. 4.92 *Polycentropus sensu lato* (Trichoptera: Polycentropodidae: Polycentropodinae): A, lateral view of larva; B, dorsal view of head capsule; C & D, tarsal claws of fore- and hindlegs (respectively); E & F, mandibles; G, anal claw; H, dorsal plate of anal proleg; I, fore-trochantin.

of an Indonesian 'Nyctiophylax' although this species probably belongs to *Paranyctiophylax* (Neboiss, 1992). Iwata (1928) has described larvae of *Neurecripsis* (*sic*) and *Holocentropus* from Japan, but Tanida (1987b) is doubtful about the identity of both. Other illustrations of polycentropodids from the Asian Palaearctic are provided by Iwata (1927) and Tanida (1985). The larvae of *Pahamunaya* are unknown.

The Pseudoneureclipsinae is represented by a single genus in tropical Asia: *Pseudoneureclipsis*. These animals occur throughout the region and, of late, more than 20 new *Pseudoneureclipsis* have been described from Nepal, Sri Lanka, Thailand, Vietnam, Laos, Sumatra, Burma and the Philippines (Chantaramongkol & Malicky, 1986; Malicky, 1993a; 1993b; 1995a; 1995b; 1997; Malicky & Chantaramongkol, 1993b; 1997). A *Pseudoneureclipsis* larva has been illustrated by Ulmer (1957). Superficially, these animals resemble psychomyiids with sharply demarcated abdominal segments, and have anal prolegs which are shorter than other polycentropodids with rather few setae on the legs (Fig. 4.93). In addition, the labium is attenuated. A distinctive feature of *Pseudoneureclipsis* is partial sclerotisation of the metanotum (i.e. not entire as in Ecnomidae: Figs 4.94A & B) resulting in a pair of lateral bar-like sclerites running along the length of the segment (Fig. 4.93A). Larvae live in short, straight silken tubes covered with fine detritus. They are small and largely detritivorous. Typically, the tube is attached to the underside of stones in shallow, fast-flowing water. According to Gibbs (1973), the tube has a wide, slanting opening directed upstream, and food may be concentrated around this aperture by eddy currents. However, the mandibles (e.g. Figs 4.93D & E) suggest a predaceous habit.

Ecnomidae

This family, previously treated as a subfamily of the Polycentropodidae, consists of one Asian genus. *Ecnomus* is widespread and known from China and the Palaearctic through Southeast Asia to New Guinea, and west to India, Pakistan and Sri Lanka (e.g. Ulmer, 1932; 1951; Hwang, 1963; Kimmins, 1963; Tian & Li, 1985; Neboiss, 1986; Higler, 1992; Tian *et al.*, 1992). Recent additions to *Ecnomus* in tropical Asia include 49 new species from Sri Lanka, Nepal, Burma, Thailand, Laos, Sumatra, Borneo, Bali, Sulawesi and the Philippines (Chantaramongkol & Malicky, 1986; Cartwright, 1992; 1994a; Malicky, 1993a; 1993b; 1995b; 1997; Malicky & Chantaramongkol, 1993b; 1993c; 1997). Twenty-two *Ecnomus* (including 18 new species) have been recorded from China (Li & Morse, 1997b), and New Guinea is estimated to support at least and additional 19 species (Cartwright, 1990b). These figures give an indication of the variety of *Ecnomus* species in the region and illustrate the limited extent of our knowledge of the group. Ecnomid larvae are unmistakable (Fig. 4.94). Although

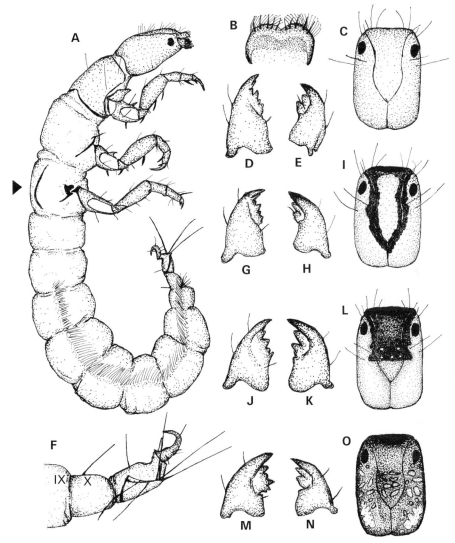

Fig. 4.93 *Pseudoneureclipsis* spp. indet. (Trichoptera: Polycentropodidae; Pseudoneureclipsinae); A, lateral view of larva of *Pseudoneureclipsis* sp. 1 with lateral sclerite on the metathorax; B, labrum; C, dorsal view of head capsule; D & E, mandibles; F, anal proleg and terminal abdominal segments; G, H & I, mandibles and head capsule of *Pseudoneureclipsis* sp. 2; J, K, & L, mandibles and head capsule of *Pseudoneureclipsis* sp. 3; M, N & O, mandibles and head capsule of *Pseudoneureclipsis* sp. 4.

they resemble polycentropodids, the dorsum of all three thoracic segments of *Ecnomus* are covered with sclerotized plates. Only the Hydroptilidae and Hydropsychidae show similar sclerotisation. However, *Ecnomus* does not construct a portable case and lacks the

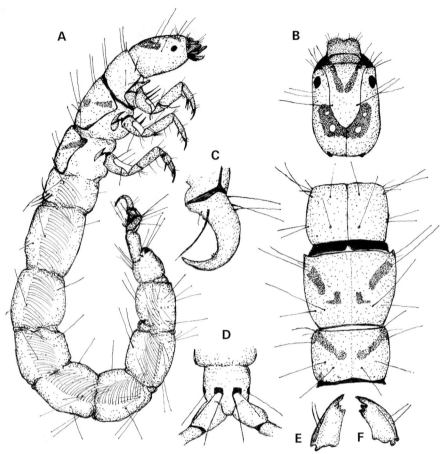

Fig. 4.94 *Ecnomus* (Trichoptera: Ecnomidae); A, lateral view of larva; B, head and thorax; C, anal claw; D, dorsal view of abdominal segment IX; E & F, mandibles.

dorsal sclerite on the abdominal segment IX which characterizes hydroptilids; the larvae are also relatively large. *Ecnomus* are easily distinguished from hydropsychids because (among other things) abdominal tracheal gills are absent. In addition, they have a row of tiny teeth along the inner margin of the anal claw (Fig. 4.94C) which hydropsychids lack. Palaearctic *Ecnomus* larva have been illustrated by Iwata (1928), Tanida (1985) and Yoon & Ki (1989a), an Ethiopian representative is described by Barnard & Clark (1986), and Ulmer (1957) has figured an Indonesian species. Most species have rather pale sclerites, but the head and prothorax usually bear distinctive dark markings. Larvae can occupy lentic and lotic habitats where they seem to prefer slow-flowing streams. Prey are captured with the aid of a polycentropodid-like net or 'web', and *Ecnomus* seems to feed unselectively on other invertebrates, eating smaller ones whole and tearing bigger prey into pieces (Barnard & Clark, 1986; Wiberg-Larsen, 1993). Often, only the fleshy tissue of large prey is consumed, leaving the heavily-sclerotized parts like legs and head capsules. Detailed observations on the ecology of these animals in tropical Asia are lacking.

Dipseudopsidae

Although previously treated either as a subfamily of the Psychomyiidae or placed within the Polycentropodidae, the prevailing consensus is that the Dipseudopsidae deserve family status (Weaver & Malicky, 1994). *Dipseudopsis* is the most diverse Asian genus: Weaver & Malicky (1994) recognize 36 species from Pakistan, India, Sri Lanka, Nepal, Bangladesh, Burma, Thailand, Cambodia, Vietnam, China, Japan, the Philippines, Peninsular Malaysia, Borneo and Indonesia (including Sumatra, Java, Borneo and Sulawesi). The larvae of Oriental *Dipseudopsis* are figured by Ulmer (1957), and a Japanese member of the genus is illustrated (as *Bathytinodes*) by Iwata (1928). Interestingly, some larvae have five lobe-like prolongations (possibly gills) arising ventrally at the junction of the metathorax and first abdominal segment (Iwata, 1928: Fig. 89). Other members of the Dipseudopsidae which occur in Asia include *Phylocentropus* known from Thailand, Vietnam and Malaysia (e.g. Mey, 1995b; Malicky & Chantaramongkol, 1997), and figured in the larval stage by Wiggins, 1996), as well as other genera that were treated as members of the family Hyalopsychidae (Ulmer, 1932; 1951; Kimmins, 1963; Neboiss, 1989a) or as members of the polycentropodid subfamily Hyalopsychinae (Neboiss, 1980; 1986): *Hyalopsyche* and *Hyalopsychella*. Recent investigations of the morphology of larval and adult *Hyalopsyche* (Fig. 4.95) reveal a close resemblance to North American *Phylocentropus* indicating that placement within the Dipseudopsidae is correct (Wells & Cartwright, 1993). That view will be followed here. Within tropical Asia, *Hyalopsyche* is reported from India, China and New Guinea (Ulmer, 1932; Neboiss, 1980; 1986; 1987), but *Hyalopsychella* occurs in Borneo, Sumatra and Sulawesi (Ulmer, 1951; Kimmins, 1963; Neboiss, 1987; 1989a). The larva of an Australia *Hyalopsyche* is figured by Wells & Cartwright (1993; see also Fig 4.95); *Hyalopsychella* larvae are unknown.

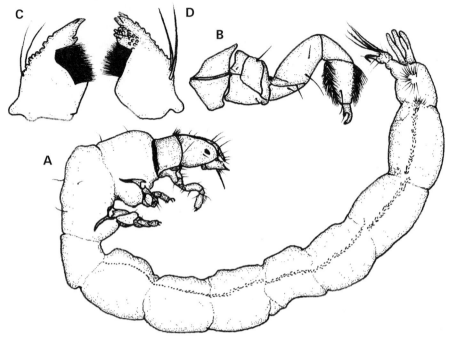

Fig. 4.95 *Hyalopsyche* (Trichoptera: Dipseudopsidae): A, lateral view of larva; B, foreleg; C & D, mandibles (redrawn from Wells & Cartwright, 1993).

Dipseudopsid larvae are long and slender (Ulmer, 1957; Gibbs, 1968; Wells & Cartwright, 1993; Wiggins, 1996), and the tip of the labium is attenuated which must facilitate maintenance of the tunnel system. The flattened limbs have paddle-shaped tarsi (which may bear reduced claws) with dense fringes of fine setae which are probably used to remove food from the filtering net. The anal prolegs are well developed and a group of elongated papillae or lobes arise from the end of the abdomen. *Dipseudopsis* larvae differ from *Hyalopsyche* and Phylocentropus in having a relatively short head capsule (i.e. it is shorter than wide) and well-developed lateral teeth on the rugose outer margins of the mandibles. Additional details of larval structure are given by Weaver & Malicky (1994). The typical habit of dipseudopsid larvae is to burrow in fine-grained stream sediments (although they also occur in lakes; Weaver & Malicky, 1994) where a silken tube covered with sand grains is constructed. By extending a portion of the tube above the sediment surface, water is diverted through the tunnel system and food particles are strained from the flow by a silken net before the water exits at a second opening downstream (Gibbs, 1968; Wallace *et al.*, 1976).

404

Hydropsychidae

The Hydropsychidae (comprising the subfamilies Arctopsychinae, Diplectroninae, Hydropsychinae and Macronematinae) is widely distributed and diverse, making up a conspicuous component of stream benthic communities (Wallace & Merritt, 1980). They are the most commonly encountered Trichoptera in Asian streams (Tsuda, 1960; Bishop, 1973; Tanida, 1986a). Hydropsychid larvae are distinguished from other Trichoptera by a combination of characters: sclerotisation of all three thoracic nota; branched gills on the meso- and metathorax and abdomen; and, usually, a tuft or 'fan' of long setae at the apex of each anal proleg (Fig. 4.96A). There are stridulatory files on the underside of the head (Fig. 4.96C) in most genera (except in some Macronematinae), with scrapers situated on the forelegs. These structures are used to generate sounds which may be important in interactions with conspecifics. All hydropsychids are filter-feeders, although the architecture of shelters and accompanying filtering nets varies somewhat. Larvae of the Arctopsychinae, Diplectroninae and Hydropsychinae attach fine-meshed silken nets to hard substrates in streams. Nets vary in mesh-size and location, which will affect dietary composition (see, for example, Malas & Wallace, 1977; Dudgeon, 1992a: pp. 87–93) but are always orientated upstream so that the current flows through perpendicularly through the capture mesh thereby sieving suspended particles efficiently. A silken retreat constructed adjacent to the net and covered with small stones and/or detritus provides concealment for the larva. The diet includes algae, fine detritus and fragments of leaf litter, as well as drifting invertebrates.

Many hydropsychid genera have been recorded from the Oriental Region (Table 4.7), including *Abacaria, Aethaloptera, Amphipsyche, Arctopsyche, Arctopsychodes, Baliomorpha, Ceratopsyche, Cheumatopsyche, Diplectrona, Diplectronella, Diplex, Ecnopsyche, Herbertorossia, Hydatomanicus, Hydatopsyche, Hydromanicus, Hydronema, Hydropsyche, Hydropsychodes, Leptopsyche, Macrostemum [= Macronema], Maesaipsyche, Mexipsyche, Occutanspsyche, Oestropsyche, Paraethaloptera, Parapsyche, Polymorphanisus, Potamyia, Pseudoleptonema, Sciops, Synaptopsyche* and *Trichomacronema* (e.g. Ulmer, 1932; Schmid, 1958a; 1959a; 1961; 1964; 1965b; Kimmins, 1962; 1963; Higler, 1981; 1992; Neboiss, 1986; Tian & Li, 1987b). Some were established on the basis of limited specimens from one or a few localities (e.g. *Ecnopsyche, Occutanspsyche, Paraethaloptera*) or animals collected outside their

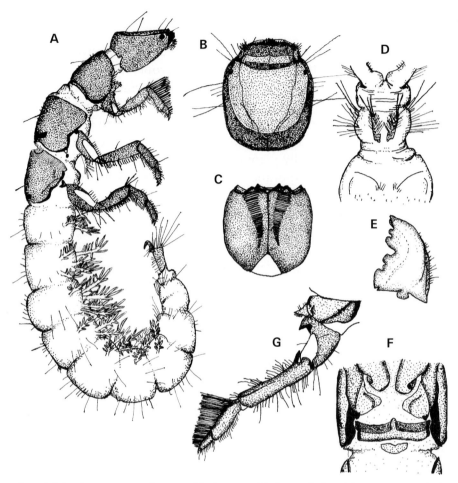

Fig. 4.96 *Macrostemum fastosum* (Trichoptera: Hydropsychidae: Macronematinae): A, lateral view of larva; B, dorsal view of head capsule; C, ventral view of head capsule; D, underside of abdominal segments VIII & IX; E, mandible; F, underside of pro- and mesothorax showing the arrangement of sclerites in the intersegmental fold; G, foreleg.

typical geographical range (e.g. *Mexipsyche*; see Tian & Li, 1987b). As a result, the status, limits and geographic distribution of some genera are unclear, and some changes (e.g. the recent synonymy of *Ecnopsyche* with *Cheumatopsyche*: Mey, 1997a) can be anticipated. However, certain valid genera — such as *Trichomacronema* which was established Schmid (1964) on the basis of a single Indian species — are either rather small or are simply under-collected. Similarly, *Pseudoleptonema* is recorded from Sri Lanka only (Schmid, 1958a; Malicky, 1973), and *Abacaria* and *Leptopsyche* are confined to New

Guinea (Kimmins, 1962; Neboiss, 1986). *Sciops* is apparently restricted to Sulawesi (Neboiss, 1987), and *Hydronema* records are limited to Pakistan (the Karakoram: Schmid, 1961). A second species of *Trichomacronema* has since been described from Thailand (Malicky & Chantaramongkol, 1991). The discovery of a new species of *Herbertorossia* in China (Li *et al.*, 1990) tends to support the view that the region as a whole has been under-collected, since records of this genus are otherwise confined to New Guinea (Kimmins, 1962; Neboiss, 1986).

The subfamily Arctopsychinae has been given full family rank by some workers (e.g. Schmid, 1968d; Malicky & Chantaramongkol, 1993b), but Morse (1997) does not treat the Arctopsychidae as separate from the Hydropsychidae (see also Givens & Smith, 1980; Tanida, 1987b; Higler, 1992; Wiggins, 1996) and that arrangement will be followed here. The subfamily is represented by *Arctopsyche* (in India, Vietnam, China and Taiwan; Chen & Morse, 1991; Higler, 1992; Mey, 1997c) and *Parapsyche* (in India, China, Thailand and Vietnam: Schmid, 1959a; 1964; Higler, 1992; Malicky & Chantaramongkol, 1992; Mey, 1996a; Malicky, 1997). Ulmer (1932) has recorded *Arctopsychodes* from China (Sichuan Province), although this genus is described as '. . . strictly Palaearctic . . .' by Givens & Smith (1980). More recently, Malicky & Chantaramongkol (1993b) have erected a new monotypic genus of Arctopsychinae (as Arctopsychidae) — *Maesaipsyche* — from Thailand. The larvae are unknown. *Arctopsyche* and *Parapsyche* larvae are characterized by the lack of a mid-ventral line or suture on the underside of the head (which tends to be rather square or quadrate). Instead, the tongue-like ventral apotome runs from the back to the front of the head completely separating the two halves of the head capsule (see Wiggins, 1996). Another distinguishing feature is the pair of short, rather broad sclerites on the underside of abdominal segment VIII. Arctopsychine larvae build coarse-meshed capture nets in fast current, and feed mainly on drifting insects. Most species seem to be confined to cool streams in mountainous areas, and the Arctopsychinae may not be truly 'tropical' (see Mey, 1997c). *Arctopsyche* and *Parapsyche* can be distinguished by details of the larval setation, and by the tendency for the ventral apotome of *Arctopsyche* to become narrower towards the back of the head; that of *Parapsyche* is usually more-or-less rectangular.

The Diplectroninae is represented in tropical Asia by *Diplectrona* which occurs throughout the region). New species have been described recently from Sri Lanka and the Philippines (Chantaramongkol &

Malicky, 1986; Mey, 1990a). *Diplectronella* is likewise widespread, occurring in India, Pakistan and Sri Lanka, Indonesia and China (Ulmer, 1951; Schmid, 1961; Malicky, 1973; Higler, 1992), *Diplex* is reported from Borneo (Kimmins, 1963), and *Sciops* from Sulawesi (Neboiss, 1987). Asian larvae of *Diplectrona* and *Diplectronella* are figured by Ulmer (1957; see also Iwata, 1927; Tanida, 1985), and African *Diplectronella* have been illustrated by Scott (1983), but immatures of *Diplex* and *Sciops* have not been described, Like the Arctopsychinae, the ventral apotome of Diplectroninae is distinctive. In this case it is divided into two pieces of approximately equal size: one anterior, the other posterior. A mid-ventral suture in the central part of the head capsule connects these parts of the apotome. The posterior portion (which is triangular and easy to see) is very small or absent in the Macronematinae and Hydropsychinae. The paired sclerites on the underside of abdominal segment VIII in Diplectroninae are ovoid. In addition, the meso- and metanota have transverse sutures (and break into two parts when the larva moults), and the fore-trochantin is not forked. The dorsum of the head in *Diplectrona* and *Diplectronella* may be distinctively-patterned (Ulmer, 1957; Malicky, 1973), but Hong Kong *Diplectrona* species from Hong Kong (at least) are uniformly pale. Filtering nets and the associated shelter are of the 'typical' hydropsychid structure. *Diplectronella* lacks a 'fan' of apical setae on the proleg (although a few setae are present), while rather more setae are seen in *Diplectrona* (see Wiggins, 1996: Fig. 7.3). A more satisfactory character separating these two genera is the combined length of the anterior and posterior ventral apotomes which are greater than the mid-ventral ecdysial line (or suture) in *Diplectrona*, but shorter in *Diplectronella* (Scott, 1983).

The larvae of Macronematinae are distinctive because the abdominal gills consist of a central 'stalk' with filaments arising more or less uniformly along the entire length (i.e. they are feather-like). In addition, the fore-trochantin is never forked and there are no sclerites on the underside of abdominal segment VIII. The genus *Macrostemum* (Fig. 4.96) is speciose and occurs throughout the Old World tropics; new species continue to be discovered (e.g. Chantaramongkol & Malicky, 1986; Mey, 1993b). Many Asian *Macrostemum* species were previously included in *Macronema* — among them, larvae illustrated by Iwata 1927) and Ulmer (1957) — but Flint & Bueno-Sorta (1982) restricted the use of *Macronema* to Neotropical species, and transferred Old-World *Macrostema* to the genus *Macrostemum*. In addition, five New Guinea species previously ascribed to *Macronema* have been transferred to *Baliomorpha* (Neboiss, 1984b); other species from the island have been placed in *Macrostemum*. *Macrostemum* larvae make specialized feeding chambers of fine sand and small stones, under or at the side of rocks. The structure is open at both ends and water flows through it; food particles are filtered from the current by an exceptionally fine-meshed net stretched across the chamber. The larvae rests beside the net in a small antechamber, using the setal fringes on the forelegs to collect food from the net (see Wallace & Merritt, 1980).

The monotypic genus *Oestropsyche* is widely distributed in Asia from India and Sri Lanka to China and through Indonesia to New Guinea (Barnard, 1980); the larva has been figured by Ulmer (1957). *Aethaloptera* has a similarly broad distribution (Barnard, 1980) although larvae from the Oriental Region have yet to be associated with adults; fortunately larval descriptions of African species are available (Gibbs, 1973; Scott, 1983; Statzner & Gibon, 1984). *Polymorphanisus* (Fig. 4.97) extends from India through Southeast Asia and Indonesia (but not New Guinea), and north into China. Barnard (1980) undertook a revision of the genus based on adult stages, and larvae of African species are illustrated by Scott (1983) and Statzner & Gibon (1984). Larvae of *Pseudoleptonema*, which occur in Sri Lanka (Schmid, 1958a; Malicky, 1973), are notable for producing nets which are typical of *Hydropsyche* (Malicky, 1973). Morphological characters match those of *Leptonema* (see Ulmer, 1957) — a genus which occurs in Africa, the Neotropics and parts of Palaearctic Asia (e.g. Mey, 1986), and has been described by Scott (1983). *Hydronema* is found in Pakistan and Central Asia (Schmid, 1961; Tian & Li, 1987b) but no further south; it is probably best considered as a Palaearctic genus. *Amphipsyche* occurs from Sri Lanka and India eastward to China and through Java, Sumatra, Borneo and the Philippines. Barnard (1984) has revised the adults of *Amphipsyche*, and Ulmer (1957) gives a detailed description of larvae from Java and Sumatra. *Amphipsyche* larvae are generally found in fast-flowing rivers and streams on a stony substrate, and are often associated with dams and impoundments where they reach high densities (e.g. Boon, 1979; 1984). Seshadri (1955) records the nuisance caused during the monsoon by swarms of adult *Amphipsyche* emerging from channels downstream of a dam in India. *Amphipsyche* nets consist of coarse strands that can withstand high current speeds; in addition, they do not collapse when flow velocities decline. At high densities, larvae exhibit unusual 'co-operative' behaviour and build large communal nets (Boon, 1979; 1984).

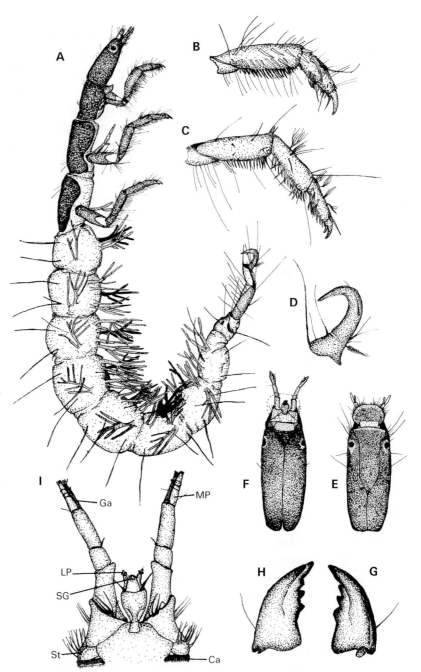

Fig, 4.97 *Polymorphanisus astictus* (Trichoptera: Hydropsychidae: Macronematinae): A, lateral view of larva; B & C, inner surfaces of the fore- and hindlegs (respectively); D, anal claw; E, dorsal view of head capsule; F, ventral view of head capsule; G & H, mandibles; I, details of the maxillae and labium (Ca, cardo; Ga, galea of maxilla; LP, labial palp; MP, maxillary palp; SG, opening of silk gland; St, stipes).

Based on Ulmer (1957), Gibbs (1973), Scott (1983), Dean (1984) and Statzner & Gibon (1984), it is possible to give larval diagnoses of most Asian macronematine larvae. *Pseudoleptonema* has a single row of two-branched lateral gills on either side of the abdomen which is densely covered with bristle-like setae; the fore-trochantin is stout. Other Macronematinae have at least two rows of gills (branched or unbranched) on either side of abdominal segments II-VI. *Baliomorpha* (which does not occur north of New Guinea) has feather-like gills on abdominal segments I-IX and long, slender anal prolegs. The mandibles are unusually short and robust, and the fore-trochantin is blunt (see Dean, 1984). The dorsum of the head is strongly flattened in *Macrostemum* (Figs 4.96A & B) and *Amphipsyche*, but the genera are easily separated because the former has two small sclerites at the base of the labrum and the latter has four. In addition, *Macrostemum* has a dense setal fringe on the tibia and tarsus of the proleg (Fig. 4.96G) which is not present in *Amphipsyche*. Like *Pseudoleptonema* and *Baliomorpha*, the head of *Aethaloptera* is not flattened but these larvae can be recognized by the fore-trochantin — which forms a long, straight spine — and the sparse setation of the abdomen. In common with *Polymorphanisus*, the head does not have stridulatory files. The head and thorax of *Polymorphanisus* (Fig. 4.97) are elongated and much narrower than the abdomen, and the prominent maxillary palps protrude beyond the mandibles. *Oestropsyche* resembles *Polymorphanisus* but lacks gills on the mesosternum (and has fewer abdominal gills on each segment). Furthermore, the tip of the fore-trochantin of *Oestropsyche* is blunt. The New Guinea macronematine *Leptopsyche* is not known in the larval stage, and *Trichomacronema* immatures have yet to be described.

The Hydropsychinae (Figs 4.98–4.101) have abdominal gills that are similar to the Macronematinae, but the filaments do not arise uniformly from the central stalk. Typically, most are near the apex (so that the gills appear tufted). In addition the fore-trochantin is forked (Figs 4.98C & 4.101E) — a condition never encountered among the Macronematinae. The paired sclerites on the underside of abdominal segment VIII are triangular or sometimes fused. The Oriental Hydropsychinae are very diverse, and probably includes several hundred species. Among them, *Hydropsyche* and *Cheumatopsyche* occur throughout the region and are highly speciose; *Hydromanicus* and *Hydatomanicus* are also diverse. *Hydropsychodes*, *Hydromanicus* and *Hydatomanicus* are widespread also, but these genera contain fewer species than *Hydropsyche et al.* Of these, new species of *Hydromanicus*,

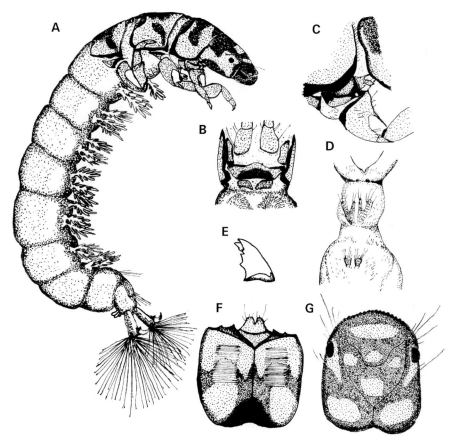

Fig. 4.98 *Hydropsyche sensu lato* (Trichoptera: Hydropsychidae: Hydropsychinae): A, lateral view of larva; B, underside of pro- and mesothorax showing the arrangement of sclerites in the intersegmental fold; C, fore coxa and trochantin; D, underside of abdominal segments VIII & IX; E, mandible; F, ventral view of head capsule; G, dorsal view of head capsule.

Hydatomanicus, Hydropsyche (Ceratopsyche) and *Hydropsyche (Mexipsyche)* from China have been described by Tian & Li (1985; 1988), Li & Tian (1990), Li *et al.* (1990) and Tian *et al.* (1993); the arrangement of *Ceratopsyche* as a subgenus of *Hydropsyche* follows Schefter *et al.* (1986). Kobayashi (1987), Mey (1990b; 1995a; 1995b; 1996a; 1996b) and Malicky (1997) have also described new *Hydropsyche* from Taiwan, the Philippines, Vietnam, Nepal and northern India. In addition, new *Cheumatopsyche* have been reported from China (Li & Dudgeon, 1988), Taiwan (Kobayashi, 1987), the Philippines (Mey, 1995a), and Vietnam (Mey, 1996a); new *Hydromanicus* from Burma, Vietnam, Thailand, Borneo and Sumatra

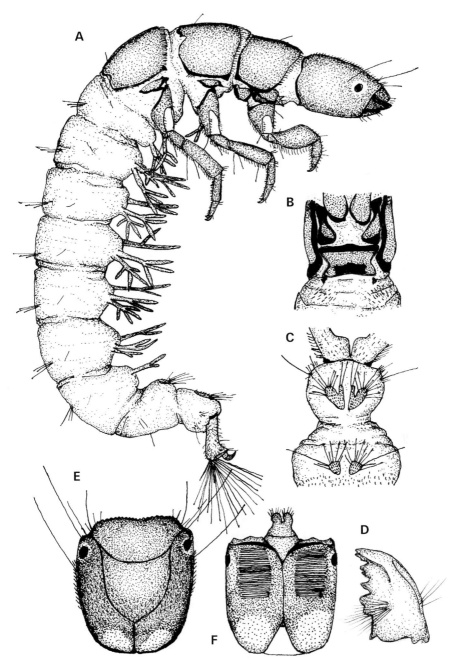

Fig. 4.99 *Cheumatopsyche criseyde* (Trichoptera: Hydropsychidae: Hydropsychinae): A, lateral view of larva; B, underside of pro- and mesothorax showing the arrangement of sclerites in the intersegmental fold; C, underside of abdominal segments VIII & IX; D, mandible; E, dorsal view of head capsule; F, ventral view of head capsule.

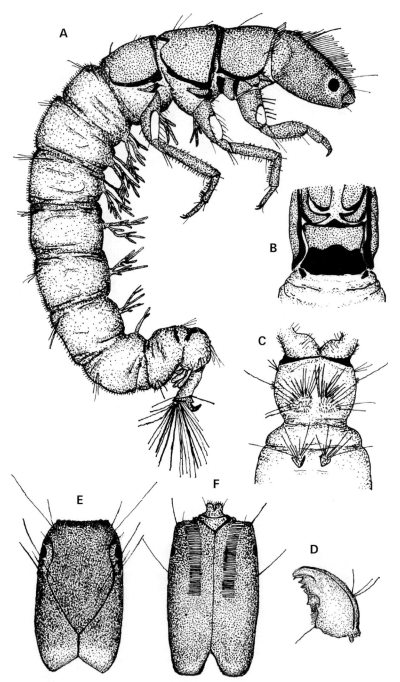

Fig. 4.100 *Cheumatopsyche (s.l.) ventricosa* (Trichoptera: Hydropsychidae: Hydropsychinae): A, lateral view of larva; B, underside of pro- and mesothorax showing the arrangement of sclerites in the intersegmental fold; C, underside of abdominal segments VIII & IX; D, mandible; E, dorsal view of head capsule; F, ventral view of head capsule.

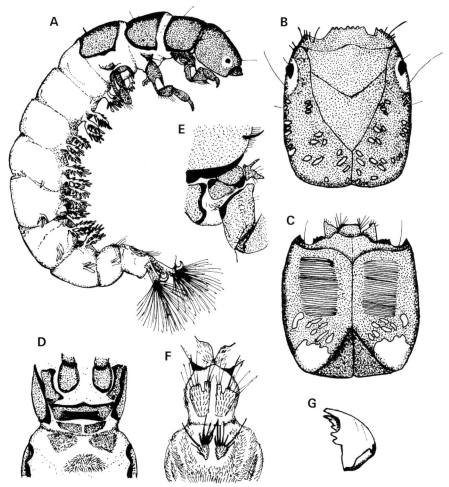

Fig. 4.101 *Herbertorossia quadrata* (Trichoptera: Hydropsychidae: Hydropsychinae): A, lateral view of larva; B, dorsal view of head capsule; C, ventral view of head capsule; D, underside of pro- and mesothorax showing the arrangement of sclerites in the intersegmental fold; E, fore coxa and trochantin; F, underside of abdominal segments VIII & IX; G, mandible.

(Malicky, 1993b; 1997; Malicky & Chantaramongkol, 1993b; 1996; Chantaramongkol & Malicky, 1995; Mey, 1996a; see also Rahim Ismail & Edington, 1990); and, new *Hydatomanicus* from Thailand (Malicky & Chantaramongkol, 1996), Vietnam (Mey, 1996a) and Sumatra (Malicky, 1997). Other hydropsychine genera (such as *Abacaria, Herbertorossia, Hydatopsyche* and *Synaptopsyche*) contain fewer species in Asia and — based on existing, incomplete data —

have either restricted or disjunct distributions. Mey (1995b) has recently described a new *Hydatopsyche* from Vietnam. *Potamyia* is reported from Mongolia (Schmid, 1965d), and apparently extends into southern China (Tian & Li, 1987) and Taiwan (Chen & Morse, 1991); it may be more widespread. Tian & Li (1987b) report a new monotypic genus — *Occutanopsyche* — from China (but do not give a description) as well as several species of *Mexipsyche* (a New World subgenus of *Hydropsyche*). They also refer to *Ecnopsyche* as part of the Indo-Malayan caddisfly fauna.

Larval descriptions of Asian Hydropsychinae include *Hydropsyche* from Korea (Kim, 1974), Japan (Tanida, 1986a; 1986b; 1987a) and Taiwan (Iwata, 1928); Ulmer (1957) has figured the larvae of *Hydropsyche*, *Hydropsychodes* and *Hydromanicus* from Indonesia. *Hydropsyche* larvae can be characterized by the presence of a pair of sclerites situated behind the sclerotized prosternal plate in the intersegmental fold on the underside of the prothorax (Fig. 4.98B). In addition, anterior margin of the head capsule (i.e. the anterior frontoclypeus) is *usually* slightly convex (and never has a median notch). In *Cheumatopsyche*, by contrast, the underside of prothorax has a tiny pair of inconspicuous sclerites in the intersegmental fold (Figs 4.99B & 4.100B), and the anterior margin of frontoclypeus usually has a small median notch or is crenulated (i.e. has a wavy outline: Fig. 4.99E). According to Ulmer (1957), *Hydropsychodes* larvae can be distinguished from *Hydropsyche* and *Cheumatopsyche* by the relative scarcity of hairs on the head capsule. The upper branch of the forked fore-trochantin of *Hydropsychodes* is longer than the lower one, but the tip is blunt (it is pointed in other Hydropsychinae). Despite these difference, Kimmins (1963) used the results of studies on adults as a basis for suggesting that *Hydropsychodes* might be a synonym of *Cheumatopsyche*. This view has not been widely accepted, and the genus is still seen as valid (e.g. Neboiss, 1986; 1987; see also Schmid, 1958a). Significantly, the *Hydropsychodes* larvae figured by Ulmer (1957) have strongly patterned head capsules which are not typical of the Asian *Cheumatopsyche* that I have seen. *Hydromanicus* larvae are very like *Hydropsyche*, and construct similar nets and shelters. The upper branch of the forked fore-trochantin of *Hydromanicus* is about 50% longer than the lower branch (which bears bristles); in *Hydropsyche* the upper branch is usually shorter (and certainly no longer) than the lower one (Fig. 4.98C). Furthermore, the head of *Hydromanicus* has few hairs or setae and a glossy sheen (i.e. it is glabrous), whereas *Hydropsyche* is conspicuously 'bristly'.

Hydromanicus larvae apparently lack gills on abdominal segment VII (Ulmer, 1957). The larvae of *Herbertorossia* are reported as similar to *Hydropsyche* by Ulmer (1957); however, the anterior margin of the frontoclypeus has distinctive, asymmetrical lateral notches in *Herbertorossia* from Hong Kong (Fig. 4.101B), and this is quite different to the entire frontoclypeal margin that seems typical of *Hydropsyche* (Fig. 4.98G) *Potamyia* larvae have rather variable characters: for instance, the fore-trochantin may be either forked, simple or weakly forked (Wiggins, 1996: Fig. 7.9). At least seven species occur in China (Tian & Li, 1987b), but none are known as larvae. The mandibles of *Potamyia* may have a prominent lateral flange, the side of the abdomen bear many scale-like setae, and there is a minute pair of sclerites in the intersegmental fold of the prothorax (as in *Cheumatopsyche*).

Within the genus *Hydropsyche*, species identification of larvae may be possible by comparison of markings on the head capsule and thoracic sclerites (see Fig. 4.102); it may be feasible in genrea such as *Cheumatopsyche* also (Fig. 4.103). However, the intensity of these patterns varies among individuals: those which have moulted recently tend to be rather pale, while those about to moult are relatively dark. Thus the size and intensity of pale markings on a darker background

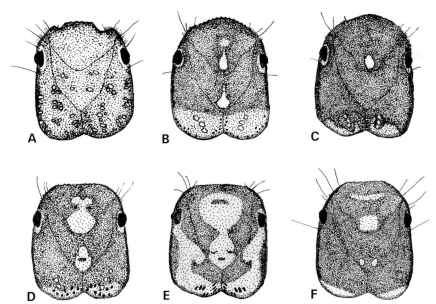

Fig. 4.102 Head capsule markings of Hydropsychinae (Trichoptera: Hydropsychidae): A, *Herbertorossia quadrata*; B, C, D, & F, *Hydropsyche* spp. indet. from Hong Kong; E, *Hydropsyche chekiangana*.

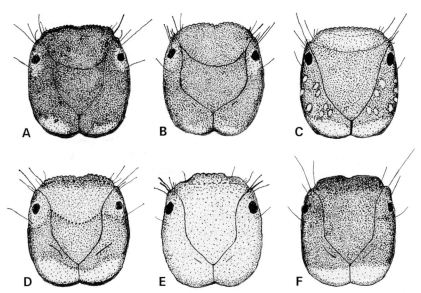

Fig. 4.103 Head capsule markings of Hydropsychinae (Trichoptera: Hydropsychidae): A - F, *Cheumatopsyche* spp. indet. from Hong Kong. Note distinctive frontoclypeal margins.

will change according to proximity to the next (or last) moult. Schefter & Wiggins (1986) regard head-capsule markings as an important guide to separating species of larval *Hydropsyche* (see also Malicky, 1973; Tanida, 1986a; 1986b; 1987a) but caution that, where there is variation and interspecific intergradation of this character (see also Tanida, 1986a), setal characters can be an effective aid to identification. These include primary setae (which are present in all five larval instars) and secondary setae (which arise during instars II-V). They may be hair-like or short and truncate, resembling spines, scales, brushes, bristles or pegs. An additional character of value is the shape of the anterior margin of the head capsule (the frontoclypeal apotome) which may distinctively shaped (e.g. concave or convex) or with notches and/or protuberances (e.g. Tanida, 1987a). Statzner (1984) has used similar characters (setae plus frontoclypeal apotome) to separate African species of *Cheumatopsyche* and *Hydropsyche*.

Ecological information on tropical Asian hydropsychids is scarce but, because of their abundance in most streams, I suspect we know more about these animals than all other caddisflies combined. Documented research includes studies of *Amphipsyche meridiana* in the outlet of a Javanese man-made lake (Boon, 1979; 1984), a report on altitudinal zonation of Thai caddisflies (Malicky &

Chantaramongkol, 1993), and limited data from Peninsular Malaysian and Hong Kong streams (Bishop, 1973; Edington *et al.*, 1984; Dudgeon, 1988a; 1988c; 1992a: pp. 87–93). These studies provide information mainly on the relationship between microdistribution, dietary composition and the pore size of the silk net used by the larvae to filter-feed (e.g. Edington *et al.*, 1984; Dudgeon, 1992a: pp. 87–93). By contrast, somewhat more research on hydropsychids has been undertaken in Palaearctic Japan, and information on life histories, microdistribution and longitudinal zonation are available for several *Hydropsyche* species (Tsuda, 1961; Tanida 1980; 1984; 1987b; 1989b). Data on the population dynamics, seasonality, microdistribution and productivity have recently become available for seven hydropsychids in a Hong Kong stream (Dudgeon, 1997): i.e. *Macrostemum fastosum* and *Polymorphanisus astictus* (Macronematinae), and *Cheumatopsyche criseyde (as C. spinosa), Cheumatopsyche ventricosa, Herbertorossia quadrata, Hydatopsyche melli* and *Hydropsyche chekiangana* (Hydropsychinae). They were not the only Hydropsychidae in the study stream (indicating the diversity of this family), but they made up > 95% of the individuals collected.

Total hydropsychid abundance in Hong Kong followed an annual pattern of wet-season decrease and dry-season increase, but this tendency was more apparent in some species (e.g. *Cheumatopsyche criseyde*) than others. *Macrostemum fastosum* was most numerous following recruitment during the wet season. Elsewhere in Asia, Tanida (1980) has reported declines in the abundance of three hydropsychid species in the River Kibune during summer floods, with densities of *Hydropsyche tsudai* (only) failing to recover to pre-spate levels. Even in tropical streams which receive aseasonal rainfall (e.g. Sungai Gombak, Malaysia) there is considerable temporal variation in hydropsychid abundance (Bishop, 1973: Fig. 52, p. 298). Densities build up during periods of stable flow, and decline due to washout after severe spates when substrate stability is reduced. In Hong Kong, *Cheumatopsyche* spp. and *Herbertorossia quadrata* had bivoltine life histories while *Polymorphanisus astictus, Macrostemum fastosum* and (probably) *Hydatopsyche melli* and *Hydropsyche chekiangana* were univoltine. *Macrostemum fastosum* grew rapidly to the final instar and then spent several months increasing in weight before emergence. One (or two) short, summer generations and a longer, overwintering generation can occur in some warm-temperate streams (e.g. Tanida, 1980); likewise, a relatively short, summer (wet-season) generation was seen also in bivoltine *Cheumatopsyche* spp. and *Herbertorossia quadrata* in Hong

Kong. Some degree of intraspecific life-history variation is also possible: for example, *Hydropsyche tsudai* in Japan switches between a univoltine life history in the upper River Kibune and bivoltinism in the lower course (e.g. Tanida, 1980; 1986b).

All Hong Kong hydropsychids (except *Polymorphanisus astictus*) were significantly more abundant in midstream microhabitats than close to the stream banks (Dudgeon, 1997), and the pattern was particularly marked in some species (*Hydatopsyche melli* and *Hydropsyche chekiangana*). While multivariate statistics pointed to a causal connection between sediment characteristics and hydropsychid abundance (Dudgeon, 1997), it was not possible to single out any individual variable that would have been responsible for the increased abundance of most hydropsychids in midstream. Some hydropsychids are confined to the surfaces of large rocks (see Bishop, 1973, p. 231; Edington *et al.*, 1984), but stone surface orientation and topography (e.g. the degree of pitting) also affect the abundance of certain species (Tanida, 1984; 1989b; Downes & Jordan, 1993). Current velocity can directly affect the distribution of hydropsychids (see, for example, Hideux *et al.*, 1991), but it also influences sediment particle-size distribution and the rate of supply of suspended food particles. Thus current velocity could affect hydropsychid microdistribution directly or indirectly. A recent view (Georgian & Thorp, 1992) is that most Hydropsychidae prefer microhabitats with large, stable substrate and high current velocity. When substrate type and size are held constant, larvae select microhabitats with high-flow rates which allow them to maximize their filter-feeding rate (Georgian & Thorp, 1992). Nevertheless, it is obvious that hydropsychids which spin rather fine-meshed capture nets cannot feed in fast current without damage to the net. In these species, high feeding rates will be enjoyed at intermediate current speeds which deliver large quantities of small organic particles to the surface of the capture net.

The apparent contradiction between microdistribution pattern (i.e. occupation of midstream sites with swift current) and feeding habit (i.e. fine-meshed capture nets that are damaged by fast current) arises in some species because of the scale of observation. At the scale of stream reach, most Hong Kong hydropsychids were found in midstream but, at a finer scale of resolution (i.e. within midstream patches), there were interspecific differences in larval positioning on stone surfaces which resembled those reported for Japanese *Hydropsyche* species (Tanida, 1984; 1989b). Species with relatively coarse-meshed capture nets (see Dudgeon, 1992a: pp. 87–93) — such as *Hydatopsyche melli* and *Hydropsyche chekiangana* — occurred on top of stones, while *Cheumatopsyche* spp. and *Herbertorossia quadrata* spun relatively fine-meshed nets on the sides or undersides. *Macrostemum fastosum*, which was confined to the underside of stones, ingests extremely fine organic material which is sieved from the slow-moving water by a delicate net with the smallest pores (approximately 5 x 40 µm) among all the Hydropsychidae (Dudgeon, 1992a: Fig. 37; see also Wallace & Merritt, 1980). The situation in Malaysian streams is similar: *Synaptopsyche* larvae spin finer nets than *Hydropsyche* and occur in microhabitats where the sediments have a relatively small particle size (Edington *et al.*, 1984). Unfortunately, a description of Malaysian *Synaptopsyche* larvae is not provided, although the diet is said to be mostly fine detritus with animals constituting a less important food than in *Hydropsyche* (Edington *et al.*, 1984). To summarise, within-patch microdistribution of hydropsychids probably reflects the varying efficiency of the species-specific net architecture at different current speeds. In turn, net architecture influences diet: species living on top of stones obtain larger food items (especially animal prey) than those dwelling on the sides or undersides which must ingest greater amounts of fine detritus (Dudgeon, 1992a: pp. 89–93).

Leptoceridae

The Leptoceridae is a large family that is well represented in the Oriental Region. In China, for example, the estimated species total increased from around 95 in 1989 (Yang & Morse, 1991) to 157 just three years later (Yang & Morse, 1992). Leptocerids are comparable to the Hydropsychidae and Hydroptilidae in terms of generic richness, although their importance declines northward into Palaearctic Asia (Dudgeon, 1987a). There are two subfamilies: the Triplectidinae — confined to southern Asia, Australia and the Neotropics — and the Leptocerinae which contains the majority of genera and species (Table 4.7). Wiggins (1996) notes that '. . . taxonomic problems in the genera of the Leptoceridae have made the biological literature difficult to interpret . . .'. This reflects, in part, the broad application of names to groups of species which should be more narrowly defined. Fortunately, substantial progress has been made with the Asian fauna — especially *Oecetis, Setodes, Trichosetodes* and related Leptocerinae — during recent years (e.g. Schmid, 1987; 1995a; 1995b; 1995c; Yang & Morse, 1988; 1989; 1991; 1992; Chen & Morse, 1991), although research on the larvae lags far behind that on adults. Some Leptocerine genera (e.g. *Ceraclea, Leptocerus, Oecetis, Setodes*) are particularly speciose (Schmid, 1987; 1995a; 1995b; 1995c; Neboiss, 1989b; Chen & Morse, 1991; Yang & Morse, 1988, 1989, 1991; Yang & Tian, 1987; 1989), and the Oriental Region seems to be a centre of radiation for *Setodes* and *Leptocerus* (Schmid, 1987; Mey, 1989). Indeed, Schmid (1987; 1994a; 1994b; 1995a; 1995b; 1995c) records 45 species of *Adicella*, 39 *Leptocerus*, 55 *Oecetis* and 103 *Setodes* in India alone, and most species of *Setodes* seem to be confined to the Oriental Region (Mey, 1989). Genera such as *Adicella, Leptocerus, Parasetodes, Setodes, Symphitoneuria, Triaenodes* and *Triplectides* occur throughout much of tropical Asia, while *Oecetis* is cosmopolitan (Neboiss, 1989b). Other genera (e.g. *Athripsodina, Nietnerella* and *Tagalopsyche*) are more restricted in distribution and diversity in Asia. *Allosetodes* includes only two species from Borneo and the Philippines (Yang & Morse, 1992), while *Lectrides* and *Symphitoneurina* do not occur north of New Guinea (Kimmins, 1962; Neboiss, 1986; 1987). Recent descriptions of adult caddisflies from the Oriental Region include 36 new species of *Adicella* from India (Schmid, 1994a; 1994b) and two from the Philippines (Mey, 1995a); new *Ceraclea* from China and northern Vietnam (Yang & Tian, 1987; 1989; Yang & Morse, 1988; 1997; Mey, 1997d); *Leptocerus* from Sri Lanka, India, Nepal, Vietnam, Laos, Thailand, Sumatra and the Philippines (Chantaramongkol & Malicky, 1986; Schmid, 1987; Malicky, 1993b; 1995a; Malicky & Chantaramongkol, 1996; Malicky, 1997); *Oecetis* from India, the Philippines, Sulawesi and New Guinea (Neboiss, 1989b; Mey, 1990a; 1995a; Schmid, 1995a; 1995b; 1995c); *Setodes* from India, Bhutan, Cambodia, Thailand, China, the Philippines (Oláh, 1985; Schmid, 1987; Mey, 1989; 1995a; Yang & Morse, 1989; 1997); *Triaenodes* from the Philippines and India (Mey, 1990a; 1995a; Schmid, 1994c); and, *Trichosetodes* from India, Thailand and China (Schmid, 1987; Yang & Morse, 1992; Malicky, 1995b).

Leptocerid larvae occupy a range of freshwater habitats. Some species are confined to the lower course of large rivers (e.g. certain *Setodes*; Mey, 1989), others are found in lakes (e.g. *Trichosetodes*; Schmid, 1987). *Leptocerus* species usually occur in cool, turbulent waters, especially mountain streams, but they are rarely abundant (Schmid, 1958). Despite their diversity, leptocerids do not approach the numerical or ecological importance of hydropsychids in streams. These case-bearing larvae can recognized by their long hind legs (which may bear rows of setae to aid swimming), and the division of both the femur and the tibia — and sometimes the tarsus — into two sections marked by a constriction near the middle. The mesothoracic notum is lightly sclerotized, and the metanotum is largely membranous but small sclerites are usually present. Dorsal and lateral humps occur on the first abdominal segment — although they may be inconspicuous — and gills may be present or absent. An important feature setting leptocerid larvae apart from other Integripalpia is their relatively long antennae (at least six times longer than wide). The length is relative to other caddisflies, and the antennae are short compared to most other insects. Leptocerids also have rather conspicuous unpigmented lines of weakness where the sclerites of the head and thorax split during moulting. Larval cases are varied in architecture and building materials, although the form is usually consistent within a genus.

Unfortunately, there are few descriptions of tropical Asian leptocerid larvae. The immature stages of many genera (e.g. *Episetodes, Nietnerella, Poeciliopsyche, Setodinella, Triaenodella*) are still unknown creating a major impediment to construction of a useful key. Ulmer (1955) has figured larvae of *Mystacides, Ocetodella, Setodes, Symphitoneuria* (as *Notanatolica*), *Tagalopsyche* and *Trichosetodes* from Indonesia, and Yang & Morse (1988) have illustrated larvae of two Chinese species of *Ceraclea*. Iwata (1927), Kim (1974) and Tanida (1985) give descriptions of *Leptocerus* and *Mystacides* larvae from Palaearctic Asia (see also Wiggins, 1996), and St Clair (1994a) figures Australian species of *Lectrides, Leptocerus, Oecetis, Symphitoneuria* and *Triaenodes*. All Leptocerinae lack sclerites on the metanotum. *Leptocerus* is distinct from other Leptocerinae (and, indeed, all leptocerids) because the mesothoracic legs have stout, hook-like tarsal claws and the tarsus is curved. The head is usually rather pale with dark blotches. The case is slender and delicate and constructed entirely of silk. Larvae of *Ceraclea* and the (mainly) Palaearctic genus *Athripsodes* can be recognized by the presence of longitudinal or oblique mesonotal bars on the weakly sclerotized plates; an abdomen which is stoutest anteriorly, tapering posteriorly; and, abdominal gills in clusters (Yang & Morse, 1988). *Ceraclea* has mesonotal bars that are curved or inflected at the midpoint (they are straight in *Athripsodes*); abdominal gills on at least segments II-VI (sometimes I-VII or I-VIII); abdominal segment IX lacks a dorsal sclerite, but a pair of gill-like projections are present posterolaterally; there are one or two dorsal accessory hooks on the anal claw; and, the dorsal lip of the case opening overhangs the ventral margin. The case may be made of silk or — more usually — sand grains, and is tapered and slightly curved posteriorly.

Among other Leptocerinae, *Setodes* larvae bear sclerotized plates with spiny posterior margins on either side of the anal prolegs; the larval case is made of small stones and is curved but not tapered. *Oecetis* have many setae scattered over the dorsal surface of the labrum (i.e. they are not arranged in rows); the maxillary palps are rather long, and the mandibles have a well-developed, sharp apical tooth. Cases vary in architecture and construction materials. *Mystacides* builds a distinctive case comprising a straight tube of small stones (or, more rarely, plant material) with small twigs or wood fragments attached parallel to the long axis. These are aligned along the long-axis of the case, and usually project beyond the anterior end. Larvae have patches of spines situated laterally at the base of the prolegs, and the head and sclerites are strongly marked with dark spots or blotches. *Triaenodes* has a elongated, slender case made up of spirally-arranged plant pieces. The tibiae and tarsi of the hind legs have a dense fringe of long setae, and the tarsus is divided. Head markings are usually dark bands or irregular markings contrasting with a light background. Larvae of the Triplectidinae have four sclerites on the metanotum (one lateral pair and a second pair situated middorsally); a fifth sclerite may occur behind the dorsal pair (in some *Triplectides*). The metanotal sclerites of *Triplectides* cover about half of the metanotum but, in *Symphitoneuria*, the two pairs of sclerites (and especially the lateral pair) are small. Both genera construct a variety of case types, although *Triplectides* larvae may hollow out a stem or twig to make a case.

Little is known of the ecology of tropical Asian leptocerids, and dietary studies are lacking. Data from elsewhere (St Clair, 1994b; Wiggins, 1996) suggest that most larvae are omnivorous. Mandibular specialization in some genera (e.g. *Oecetis*) suggests a predatory habit, and certain *Ceraclea* feed on freshwater sponges or Ectoprocta (including the genus *Lophopodella* which occurs in the Oriental Region; see p. 109) although this behaviour has yet to be confirmed in Asia. Studies of the life histories of Oriental leptocerids have yet to be undertaken, but information from Australia (St Clair, 1993) points to considerable variation in the degree of synchrony of larval development and in the duration of adult emergence. Some species complete two generations in a year, but others are semivoltine. An interesting feature of certain Chinese *Triplectides* is the capacity of adult females to give birth to first-instar larvae through ruptured intersegmental membranes of abdominal segment VIII (Yang & Morse, 1991).

Lepidostomatidae

The Lepidostomatidae is made up of two subfamilies: the Lepidostomatinae and the Theliopsychinae. The latter is represented in Asia by a single genus known only in the adult stage — *Zephyropsyche* (Weaver, 1992). Members of the Lepidostomatinae are common throughout the region. The generic classification of the Lepidostomatidae is confused, particularly in Asia, because secondary sexual characters (which can occur on the mouthparts, antennae, legs or wings) of the adult males have been used as a basis for proposing genera. Unfortunately, these features have little correlation with male or female genitalia (primary sexual characters) or the morphology of the larvae (Ito *et al.*, 1993). While the importance of larvae in establishing generic affinities has been established (Ito, 1984; 1990a; 1992a; Wiggins, 1996), Ito *et al.* (1993) note that assignment to genera has continued without consideration of larval characters (e.g. Weaver, 1985; 1992; Weaver & Huisman, 1992a). The fundamental problem is that species with almost identical larvae have been assigned to different genera, while the variation within a genus may be considerable (e.g. *Goerodes emarginatus*; Ito, 1985a: see below). Accordingly, there is some disagreement among workers as to the validity of lepidostomatid genera, and a few of those listed in Table 4.7 will almost certainly be relegated to synonymy as research proceeds. This process has already begun with the transfer of species from a number of genera into *Goerodes* (Ito, 1984; 1989; 1990a; 1992a; 1992b). On the other hand, some new species from Sulawesi attributed to *Goerodes* (by Neboiss, 1990) have been moved to *Lepidostoma* (by Weaver & Huisman, 1992b), and Weaver & Huisman (1992a) have transferred all lepidostomatids known previously from Borneo (including several *Goerodes*) to *Dinarthrum* and *Lepidostoma*. Significantly, a recent classification of the Lepidostomatidae (Weaver, 1992: Table 1) does not include some Indian genera (*Hypodinarthrum*, *Indodinarthrum* and *Metadinarthrum*) recorded by Higler (1992). Moreover, Weaver (1992) recognizes *Lasiocephala* — recorded from Japan by Iwata (1927) — as a valid lepidostomatid genus even though Tanida (1987) suggested that the identification was incorrect. It now seems that *Lasiocephala* is actually the odontocerid *Micrasema* (Ito *et al.*, 1993). Obviously, the situation is unstable, and a resolution of the status of certain genera will require a more complete knowledge of the Oriental lepidostomatid fauna.

Goerodes ranges from Palaearctic Asia southward through Indonesia to New Guinea and eastward through India and Sri Lanka to Africa (Schmid, 1958a; 1965; Kimmins, 1962; 1963; Ito, 1984; Weaver, 1985; Neboiss, 1986; Yang & Weaver, 1997). The genus includes some widely-distributed species: *Goerodes doligung*, for example, occurs in Taiwan, Hong Kong, Sumatra and southern India (Ito, 1992a). New species have been described recently from Nepal, the Philippines, Sumatra, Malaysia and China (Ito, 1986; Weaver, 1989; Mey, 1990a). *Dinarthropsis* occurs in Java and New Guinea (Kimmins, 1962; Weaver, 1985; Neboiss, 1986), and has recently been reported from Sumatra (Weaver, 1989); *Dinarthrum* is recorded from Japan, China, India and Pakistan (Ulmer, 1932; Hwang, 1951; 1958; Schmid, 1961; Tian & Li, 1985; Higler, 1992; Ito *et al.*, 1993; Yan, & Weaver, 1997), with new species described from Taiwan and Borneo (Ito, 1992a; Weaver & Huisman, 1992a). Ito *et al.* (1993) treat *Lepidostoma* as being confined to the Holarctic, but this distribution is extended by new *Lepidostoma* records from the Philippines, Thailand, Sumatra, Borneo, Sulawesi, Nepal and India (Weaver & Huisman, 1992a; 1992b; Malicky & Chantaramongkol, 1994). *Paraphlegopteryx* is found in southern China, northern Burma and throughout northern and eastern India at altitudes between 600 and 3,000 m (Kimmins, 1952; Schmid, 1965; Higler, 1992; Yang & Weaver, 1997). The genus is currently being reviewed by John Weaver (*in litt.*). New species of *Paraphlegopteryx* have been reported recently from Nepal (as *Neoseverinia*), India and Thailand (Ito, 1986; 1992a; Malicky & Chantaramongkol, 1994).

Other genera appear to have a relatively restricted range, including *Anacrunoecia, Dinomyia, Eodinarthrum* and *Mellomyia* in China (Ulmer, 1932, Martynov, 1931; Schmid, 1959; 1965; Yang & Weaver, 1997); *Adinarthrella, Adinarthrum, Agoerodella, Agoerodes, Anacrunoecia, Goerodella* and *Ulmerodes* in Burma and India (Kimmins, 1952; Higler, 1992); *Dinarthrena* in Burma (Kimmins, 1952); *Dinarthrella* in India, Nepal and Burma (Kimmins, 1952; Ito, 1986; Higler, 1992), *Dinarthrodes* in China and Japan (Hwang, 1957; Ito, 1978; 1989; Tian & Li, 1985; Yang & Weaver, 1997); *Dinanthropsis* in Indonesia (Ulmer, 1951); *Goerodina, Hypodinarthrum, Indocrunoecia, Indodinarthrum, Kodala* and *Metadinarthrum* in India (Higler, 1992); *Neolepidostoma* in Sumatra and New Guinea (Ulmer, 1951; Neboiss, 1986; although this distribution is regarded as dubious by Weaver & Huisman, 1992b); and, *Zephropsyche* from India and Thailand (Weaver, 1992; Malicky & Chantaramongkol, 1994). Again, it should be stressed that some of these taxa may eventually be synonymised with more widespread and speciose genera.

Lepidostomatid larvae have no median dorsal hump on the first abdominal segment, but there is a well-developed horn-like structure (of unknown function) on the prosternum. The pro- and mesonota are sclerotized, while the metanotum has four small mid-dorsal sclerites and one situated anterolaterally on either side (Figs. 4.104A & B). Gills are single or paired and arranged in dorsal and ventral rows on the abdomen, but may be lacking entirely. Abdominal segment VIII has a fleshy, lateral lobe on either side. Larvae of most genera make four-sided cases constructed of square panels cut from leaves or bark (Figs. 4.104F-I). In early instars the case is made of sand grains, and the characteristic four-sided case is built onto the anterior end of the sand case as the larva grows. While there have been a number of descriptions of Asian lepidostomatid larvae, robust generic diagnoses are not currently available. Detailed accounts of a number of *Goerodes* larvae are given by Ito (1984; 1985a; 1985b; 1985c; 1985d; 1986; 1989; 1990a; 1990b; 1992a; 1992b; see also Fig. 4.104). Most species are similar and differ only in arrangement and number of abdominal gills and the arrangement of setation on the meso- and metanota (Ito, 1985c). However, *Goerodes emarginatus* is distinctly different (Ito, 1985a): the dorsum of the head is flat, and the flattened are is bounded by a low, circular carina. In addition, the dark head capsule lacks the light-coloured spotting seen in other *Goerodes*. The difference between some *Dinarthrodes* and *Goerodes* larvae is indistinct, although the arrangement of abdominal gills may be important in species identification (Ito, 1978; 1989). The same situation applies in *Dinanthrella*: larval descriptions (Ito, 1986) are very similar to *Goerodes* but the arrangement of gill branching is slightly different and gills are lacking on segment VII although they are present in some *Goerodes*. This genus, like *Lepidostoma* (Wiggins, 1996), *Dinarthrodes* and *Goerodes*, makes a typical four-sided case of leaf squares. By contrast, *Dinarthrum* larvae build sand-grain cases which are cylindrical, slightly curved and tapered posteriorly (Ito, 1990b). Larval morphology is similar to *Goerodes*. *Paraphlegopteryx* larvae (described as *Neoseverinia* by Ito, 1986; 1992a) construct a case of irregular leaf fragments in early instars, but the definitive case is made of short pieces of twigs and bark arranged in a continuous spiral band. The abdominal gills are arranged in subdorsal and subventral rows on the posterior part of segments II-VII. A *Paraphlegopteryx* larvae which is to be described by John Weaver (*in litt.*) shares this distinctive case morphology, but can be distinguished by the presence of a small pair of sclerites on the underside of the first abdominal segment.

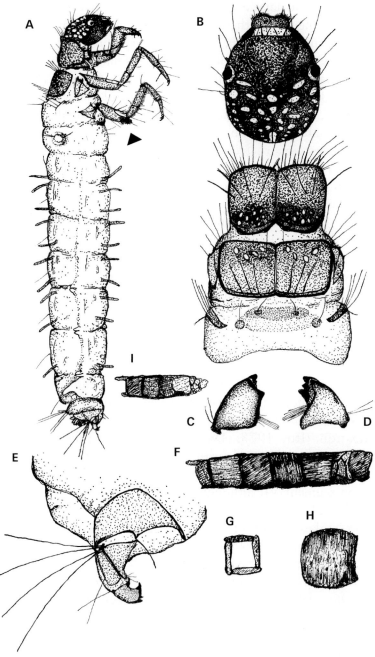

Fig. 4.104 *Goerodes doligung* (Trichoptera: Lepidostomatidae): A, lateral view of larva (a tuft of setae on the hind coxa is indicated by an arrow); B, head and thorax; C & D, mandibles; E, anal proleg; F, case of a fifth-instar larva (lateral view); G, case in cross-section; H, an individual leaf panel from the case; I, case of a third-instar larva constructed of leaf panels and sand grains.

Lepidostomatids inhabit various types of running water but usually small, often slow-flowing streams (and sometimes seeps) with abundant accumulations of terrestrial plant litter. They are shredders. Research in Japan reveals that some *Goerodes* are univoltine although others have a bivoltine or trivoltine life history (Ito, 1985b; 1985c; 1987). *Neoseverinia* (*sensu stricto*) are univoltine, while *Dinarthrum* larvae — which build a case of sand grains — take two years to complete their life cycle (Ito, 1987). Among Japanese Lepidostomatidae which exhibit interspecific variation in selection of case-building materials, species retaining a mineral case until emergence have a longer generation time and a smaller adult size than those which change to an organic case during larval life (Ito, 1987) In addition, the earlier the change occurs, the shorter the life cycle and the heavier are resulting adult females.

Calamoceratidae

The Calamoceratidae has a mainly tropical and subtropical distribution and two genera, *Anisocentropus* and *Ganonema*, are widespread in tropical Asia. *Anisocentropus* occurs throughout the Oriental Region and southward into Australia, while *Ganonema* (including *Asotocerus*) ranges from mainland Asia (and Sri Lanka) into Java and Sumatra (Ulmer, 1932; Schmid, 1958a; 1959; Kimmins, 1962; 1963; Neboiss, 1986; Higler, 1992; Malicky, 1995a; Mey, 1997d). A third genus, *Ascalaphomerus*, is present in China (Ulmer, 1932), while *Georgium* (= *Rhabdoceras*) appears to have a rather restricted range (Tanida, 1987). A recent revision of the adult stages of Asian Calamoceratidae (Malicky, 1994b) resulted in the division of *Anisocentropus* into three subgenera: *Anisokantropus*, *Anisolintropus* and *Anisomontropus*.

Calamoceratid larvae can be distinguished from other Integripalpia by the presence of a transverse row of at least a dozen (usually 15 – 20) stout setae on the labrum, and the presence of two anterolateral sclerites in addition to a larger dorsal plate on the mesonotum. The anterior corners of the pronotum are somewhat produced; this is especially noticeable in *Anisocentropus*. Larval cases are made from wood or dead leaves. Iwata (1928) and Ulmer (1955) have figured Oriental *Ganonema* larvae, and the latter also gives a description of *Anisocentropus*. The lateral humps on the first abdominal segment in *Ganonema* are flattened and bear a group of chitinous hooks. The plates of the mesonotum are more lightly-sclerotized than in

Anisocentropus, but all of the hard parts are deeply pigmented. The case of *Ganonema* consists of an excavated twig with a small posterior opening and a larger anterior opening on the underside. *Ganonema* larvae must change cases periodically as they grow because the twigs cannot be increased in size. *Anisocentropus* (Fig. 4.105) has long hind legs (twice the length of the middle legs); the tibiae are especially long (Fig. 4.105E). The larval body is depressed and the abdomen is wide and flat with a conspicuous lateral fringe of setae. The sclerotized parts are pale brown and the head bears lighter blotches. *Anisocentropus* constructs a case from two elongate and rather oval pieces of leaf that are cut to shape by the mandibles and bound together by silk. The upper, piece forms a dorsal shield over the smaller ventral portion thus enclosing the larva (Fig. 4.105K). The immature stages of *Ascalaphomerus* have not been reported in the literature and, aside from the original species descriptions, this genus is practically unknown. An undescribed calamoceratid larva from Hong Kong, which may be a possible candidate for *Georgium*, is shown in Fig. 4.106. The labral setae are characteristic of the family. The case consists of twigs arranged longitudinally and extending beyond the anterior and posterior ends (Fig. 4.106K). If this larva is *Georgium*, it represents a southern extension of the range of the genus. Calamoceratid larvae generally occur in stream pools where detritus accumulates. *Anisocentropus* and *Ganonema* are shredders (e.g. Nolen & Pearson, 1993) but, while *Anisocentropus* feeds entirely on leaf litter, *Ganonema* will gouge wood and is at least partially xylophagous. The presumed larva of *Georgium japonicum* is usually found on rocks within pools and appears to be a grazer/scraper of algae. Unpublished observations on the life-history of Hong Kong calamoceratids suggest that *Anisocentropus maculatus* completes two to three (or even more) generations annually, with asynchronous growth and adults present throughout the year. Larval densities were higher in the dry season when stream discharge was low and leaf litter accumulated in pools. *Ganonema extensum* and *Georgium japonicum* had similar life histories to *Anisocentropus*. *Anisocentropus kirramus* from tropical Australia also emerges over much of the year, but '. . . development appeared to take several months . . .' (Nolen & Pearson, 1992). *Anisocentropus* larvae in an aseasonal stream on Bougainville Island (Papua New Guinea) showed asynchronous development with continuous growth and recruitment (Yule, 1995b).

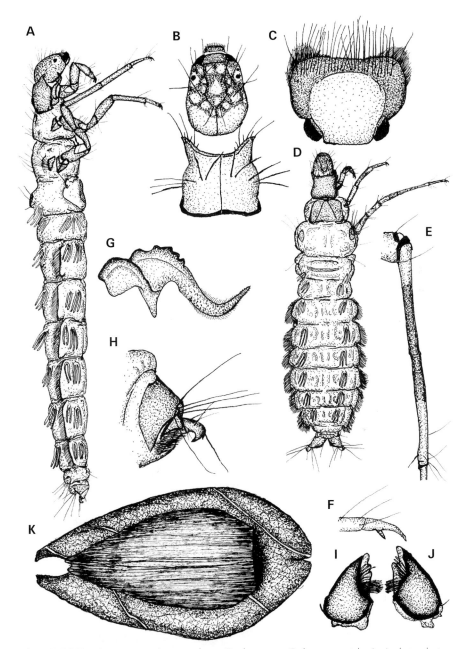

Fig. 4.105 *Anisocentropus maculatus* (Trichoptera: Calamoceratidae): A, lateral view of larva; B, head and prothorax; C, labrum; D, dorsal view of larva; E, hind tibia; F, tarsal claw of hind tibia; G, fore-trochantin 3 2; H; anal proleg and abdominal segment IX; I & J, mandibles; K, larval case.

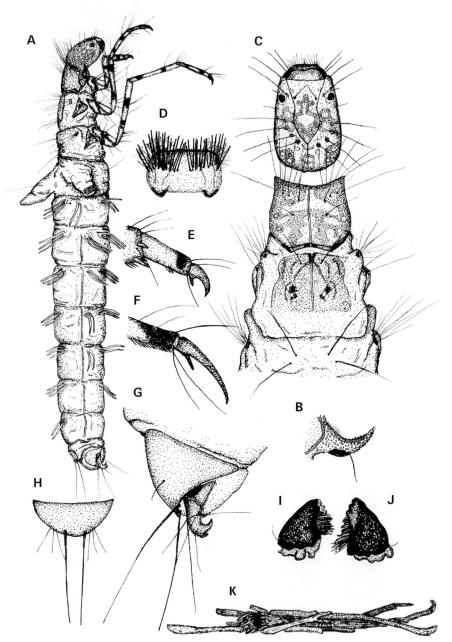

Fig. 4.106 Putative larvae of *Georgium japonicum* (Trichoptera: Calamoceratidae): A, lateral view of larva; B, fore-trochantin; C, head and thorax; D, labrum; E, fore tarsus; F, hind tarsus; G; anal proleg; H; dorsal sclerite of abdominal segment IX; I & J, mandibles; K, larval case.

Odontoceridae

The Odontoceridae includes five genera in Asia: *Ithanopsyche, Lannapsyche, Marilia, Psilotreta* and *Perissoneuria*. Records of *Odontocerum* from East Asia are incorrect (Tanida, 1987; Nozaki *et al.*, 1994). *Ithanopsyche* and *Lannapsyche* have, to date, been reported only from Thailand, Vietnam and China (Malicky, 1989a; 1995a; 1995b; Malicky & Chantaramongkol, 1993c; Mey, 1997c); their larvae are not known. *Perissoneuria* is a Japanese endemic (Nozaki *et al.*, 1994). *Psilotreta* occurs from Japan and Korea thorough China, Vietnam and Thailand eastward to India and Nepal (Ulmer, 1932; Schmid, 1959; Higler, 1992; Malicky, 1995a; 1995b; 1997; Mey, 1995b; 1997d; Malicky & Chantaramongkol, 1996). The genus has been revised by Parker & Wiggins (1987) who give a detailed larval diagnosis (see also Fig. 4.107). The head is shiny and often (but not invariably) marked with an inverted 'Y' (Fig. 4.107D). The pro- and mesonotum are well sclerotized, and the anterolateral margins of the pronotum are drawn out into points (Fig. 4.107D). The metanotum has two transverse sclerites, and there is a well-developed lateral sclerite at the base of the anal proleg. The case is cylindrical and curved; it tapers very slightly if at all (Fig. 4.107I). It is made of coarse sand grains with finer particles fitted into the interstices and reinforced by silk, producing a crush-resistant structure. Larvae inhabit stony hillstreams and are associated with sandy substrates where they sometimes burrow. Although rendered cryptic by their cases, the presence of *Psilotreta* is easily confirmed because larvae congregate in groups under stones at the onset of pupation. *Marilia* has a similar distribution to *Psilotreta* (Ulmer, 1932; Schmid, 1959; 1965; Tian & Li, 1985; Malicky & Chantaramongkol, 1996) but extends into Indonesia (Sumatra and Java; Ulmer, 1955; Malicky, 1989a) and Sri Lanka (Schmid, 1958a). Surprisingly, *Marilia* has yet to be recorded from New Guinea although it is in Australia (Neboiss, 1986; Drecktrah, 1990). An Indonesian larvae has been figured by Ulmer (1955), and Drecktrah (1990) compares this and other *Marilia* larvae with an Australian representative. Like *Psilotreta*, there is a conspicuous lateral sclerite at the base of the proleg, and a prosternal horn is lacking. However, the anterior corners of the pronotum are rounded, and the sclerotized mesonotum is divided into six plates (four centrally and two laterally). Two rectangular metanotal sclerites situated mesally are flanked by a smaller lateral pair. The cylindrical curved case is made of small sand grains and tapered posteriorly. The habits of odontocerids are obscure, but *Psilotreta* will attack and eat other invertebrates such as chironomid (Diptera) larvae.

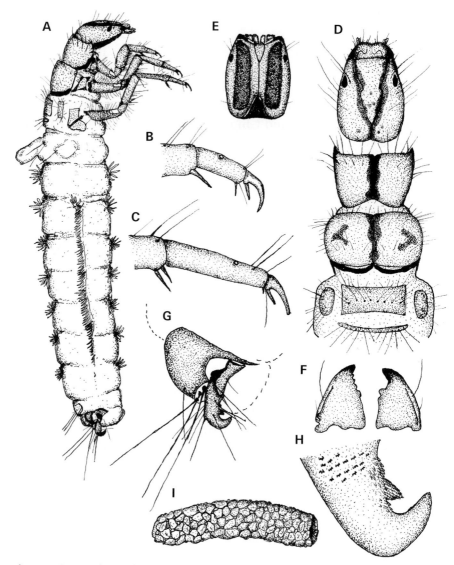

Fig. 4.107 *Psilotreta kwangtungensis* (Trichoptera: Odontoceridae): A, lateral view of larva; B, fore tarsus and claw; C, hind tarsus and claw; D, dorsal view of head and thorax; E, ventral view of head capsule; F, mandibles; G, anal proleg; H, anal claw; I, lateral view of larval case.

Helicopsychidae

The Helicopsychidae is distributed across most tropical and subtropical regions, and few species inhabit temperate latitudes. Many species occur in the Oriental Region, but only two have been described from

Palaearctic Asia (Korea and Japan: Johanson, 1994). *Helicopsyche* is the most speciose and widespread genus (e.g. Kimmins, 1952; Schmid, 1958a; Malicky, 1973; Higler, 1992), but does not extend into New Guinea (Neboiss, 1986), while *Cochliophylax* occurs in Thailand and India where it is represented by 12 species (Schmid, 1993b). Recent discoveries of *Helicopsyche* include 15 new species from India (Schmid, 1993b), as well as others from Sri Lanka, Thailand, Laos and Vietnam (Chantaramongkol & Malicky, 1986; Malicky & Chantaramongkol, 1992; 1993c; 1996; Schmid, 1993b; Johanson, 1994; Malicky, 1995a; 1997); many more can be expected. Johanson (1997) has recently reviewed the global diversity of Helicopsychidae. Larval *Helicopsyche* (Fig. 4.108) from Asia have been figured by Iwata (1927), Ulmer (1955), Kim (1974) and Tanida (1985), while the narrow range of case forms seen in the genus is illustrated by Malicky (1973). All species use sand grains to construct a case resembling a tightly-coiled snail shell (Fig. 4.108G & H); as a result, they are unmistakable. The larval body is modified in accordance with the spiral case architecture, and the abdomen is twisted (Fig. 4.108A). Diagnostic features include a long fore-trochantin, and a pectinate row of teeth on the inner margin of the anal claw (Fig. 4.108D). *Helicopsyche* larvae are scraper grazers and occur on stone surfaces in moderate current; they often share this habitat with glossosomatids such as *Agapetus*.

Limnocentropodidae, Phyrganopsychidae, Brachycentridae, Uenoidae, Goeridae, Molannidae and Sericostomatidae

The families treated in this section are grouped together for convenience and not because they are closely related. All contain few genera, and are rarely abundant in tropical Asian streams although most are widespread. The Limnocentropodidae contains a single genus — *Limnocentropus* — of less than 20 species which have been reported from India, Nepal, Thailand, Borneo, China and Japan (Ulmer, 1932; Kimmins, 1963; Wiggins, 1969; Tanida, 1987; Malicky & Chantaramongkol, 1989b; Higler, 1992). Ulmer (1955) figures a larva and case of *Limnocentropus*, but a complete diagnosis and detailed description (based on several species) is given by Wiggins (1969). Larvae have a robust and muscular appearance, and can be distinguished from all other caddisflies by a series of four sclerotized plates on the meso- and metanota, a large sclerite on the dorsum of the first

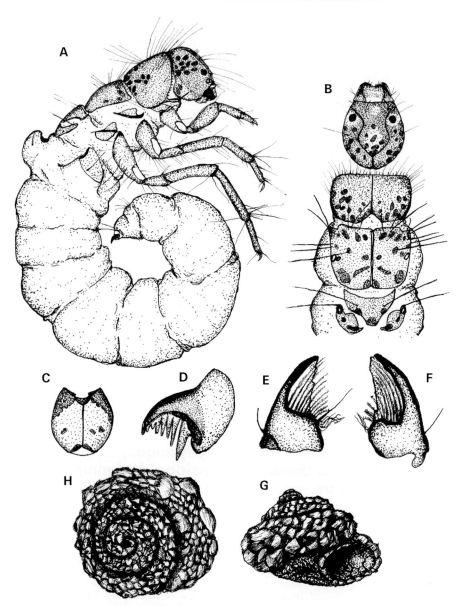

Fig. 4.108 *Helicopsyche* (Trichoptera: Helicopsychidae): A, lateral view of larva; B, dorsal view of head and thorax; C, ventral view of head; D, anal claw; E & F, mandibles; G & H, lateral and dorsal views (respectively) of larval case.

abdominal segment, and unusual patches of stout bristles on the meso- and metasterna and the underside of abdominal segment I. The case is cylindrical, curved and somewhat tapered, and may be made of organic or inorganic materials (possibly with variation among individuals of the same species). An unusual feature is the presence of a long stalk or 'mooring cable' on the larval and pupal cases. *Limnocentropus* are largely confined to stream rapids (Wiggins, 1969). They are predators, feeding on chironomids (Diptera) and other caddisflies. It appears that the larva fastens its case in position with the silken mooring cable thereby allowing it to maintain position in strong current while still permitting all of its legs to be devoted to capturing and handling prey. The legs are elongated and bear long stout spines which Wiggins (1969) speculates act like pincer traps and may also serve to 'sieve' drifting prey from the current. *Limnocentropus* favours very turbulent, rocky streams draining forests at altitudes between 550 and 2800 m. Pupae often occur in groups, and the exterior of the pupal cases may be covered by simuliid (Diptera) pupae and the tubes of chironomid larvae (Wiggins, 1969).

The Phryganopsychidae was erected in 1959 to hold several Asian caddisflies with an unusual but consistent combination of characters which had previously been placed within the Phryganeidae (Wiggins, 1959). The family consists of a single genus — *Phryganopsyche* — which occurs in India, Burma, China, Taiwan and Japan (Kimmins, 1952; 1952; Wiggins, 1959; Schmid, 1965; 1058a; Tian & Li, 1985). A larval diagnosis is given by Wiggins. The pro- and mesonota are heavily sclerotized, and the pronotum has a transverse fold (or furrow) near the posterior margin. The metanotum is membranous save for a small lateral sclerite on each side. There is a prosternal horn present, but no dorsal hump on the first abdominal segment. The case is cylindrical but rather flimsy and irregular. It is loosely constructed of detritus to which a few sand grains may be added. An unusual feature is the construction of an entirely new case in preparation for pupation. Note that *Phryganopsyche* larvae superficially resemble lepidostomatids but are easily distinguished by the structure of the thoracic nota and case architecture. Phryganopsychids are consistently referred to as occurring at high altitudes (e.g. Kimmins, 1952; Schmid, 1965; Chen & Morse, 1991), and the larvae live in stream pools or microhabitats with little or no flow. *Phryganopsyche* (as *Phryganopsis*) is recorded at elevations over 2,000 m in Burma (Kimmins, 1951), and Schimd (1968b) observed adults at altitudes between 950 and 3,000 m in northern India.

The Brachycentridae includes three Asian genera, but one (*Eobrachycentrus*) is confined to Japan and North America (Wiggins *et al.*, 1995). *Brachycentrus* occurs along the Himalaya, into northern China and Japan (Ulmer, 1932; Schmid, 1961; 1992; Wiggins *et al.*,

1995). *Micrasema* occurs throughout mainland Asia into Sumatra, Java and Borneo (Kimmins, 1952; 1963; Ulmer, 1951; Higler, 1992; Malicky & Chantaramongkol, 1992; Malicky, 1997; Mey, 1997d). The larva has been described by Ulmer (1955) and Ito (1995), and some aspects of the ecology of Japanese *Micrasema* are discussed by Kato (1997). Brachycentrid larvae are distinguished by a combination of characters. Most conspicuously, and unlike other Integripalpia, they lack both dorsal and ventral humps on the first abdominal segment. The head may have dorsolateral ridges and, in *Micrasema*, can be quite flat and/or strongly patterned (Ulmer, 1955: Fig. 51). There is a distinct transverse ridge on the pronotum. *Brachycentrus* and *Micrasema* are easily distinguished. The former has relatively long middle and hind legs, and the tibiae bear stout spurs and a row of short setae which may be used for filter-feeding. The foretibiae are densely fringed by longer setae. Gills (and sometimes a prosternal horn) are usually present. *Brachycentrus* constructs a four-sided case of tiny twigs arranged at right-angles to the long axis; more rarely, the entire structure is silk and circular in cross-section. *Micrasema* larvae have rather short legs, and gills and a prosternal horn are lacking. They construct a cylindrical case that tapers strongly and is usually curved; organic or inorganic building materials (but not both) can be employed. At least one species of *Micrasema* uses plant fibres which are wrapped around the long axis of case (Ulmer, 1951: Fig. 304). Although little is known of their ecology in Asia, brachycentrids are localised at high altitude in Taiwan (Chen & Morse, 1991), and this seems to be true of Indian *Brachycentrus* also (Schmid, 1992). All species are confined to streams or seeps, and *Brachycentrus* requires fast current if it is to supplement grazing with filter-feeding.

The Uenoidae, as originally conceived, contained only the Asian genus *Uenoa* (which had previously been placed in the Sericostomatidae; e.g. Iwata, 1927). Revisionary work (Wiggins *et al.*, 1985; Vineyard & Wiggins, 1988) has led to modification of this view so that the family is now made up of seven genera arranged in two subfamilies. The Thremmatinae includes *Neophylax* which was previously placed in the Limnephilidae (subfamily Neophylacinae). *Neophylax* occurs in Palaearctic Asia (Japan and China: Tanida, 1987; Tian *et al.*, 1993) but may penetrate further south since it has been recorded (as *Halesinus*, a junior synonym) from Sichuan Province (Ulmer, 1932; Banks, 1940). However, none of the other thremmatine genera have Asian representatives. Holarctic *Neophylax* larvae are figured by Vineyard & Wiggins (1988) and Wiggins (1996). A distinctive feature which separates them (and other Uenoidae) from the rest of the Integripalpia is the presence of an anteromedian notch in the well-developed mesonotal sclerites. Some *Neophylax* have a ventrolateral gill filament on the abdominal segment I which is not found in other ueonids. Additional distinctive features include a membranous anterior edge of the labrum; a rounded anterolateral margin on the pronotum; tiny metanotal sclerites; a dorsal accessory hook on the anal claw; and, a reduced prosternal horn. The larval case is rather short and broad reflecting the stout body of its occupant, and constructed of small stones flanked by a several larger ones along each side. *Uenoa* (= *Eothremma*) is known from Japan, Taiwan and China through Burma, Nepal, India and Pakistan (Iwata, 1927; Kimmins, 1952; Hwang, 1957; Schmid, 1961; Botosaneanu, 1976; Tian & Li, 1985; Chen & Morse, 1991; Higler, 1992). The larvae, which have been described by Botosaneanu and Wiggins *et al.* (1985), are long and slender. The pronotum has raised smooth areas, including one forming a Y-shaped ridge along the midline. The mesonotum bears two sclerites with a characteristic anteromedian notch, while there are four sclerites (a larger pair flanking a smaller median pair) on the metanotum. There is a prosternal horn and small paired sclerites on all thoracic sterna. The first abdominal segment has dorsal and lateral humps and, in some species, gills are confined to segment VIII. The larval case is slender (like a conifer needle) and made entirely of tough brown silk encircled by prominent annular ridges. *Uenoa* larvae are grazer-scrapers (Vineyard & Wiggins, 1988) and occur in small cold streams in montane areas; cascades and waterfalls are favoured habitats (Wiggins *et al.*, 1985). Larvae form large aggregations for pupation on the underside of stones in rapid water.

The Goeridae — and particularly the genus *Goera* — is widespread in Asia, from Korea and Japan in the north (Kim, 1974; Tanida, 1987) through China and Southeast Asia to Sulawesi, and eastward into India and Sri Lanka (Ulmer, 1932; Kimmins, 1952; 1958a; 1963; Hwang, 1957; Schmid, 1959; 1965; 1991c; Denning, 1982; Neboiss, 1990; Yang & Armitage, 1996). Although *Goera* has, at times, been placed among both the Sericostomatidae and Limnephilidae, most recent workers accord the goerids full family status. *Goera* is the largest genus, and most members of that genus are Oriental (Denning, 1982). Forty-nine new species have been described from the region recently, including nine from China and 21 from India (Schmid, 1991c; Malicky & Chantaramongkol, 1992; 1997; Malicky, 1995a; 1995b; Yang & Armitage, 1996; Mey, 1997d; Yang & Morse, 1997). In addition to

Goera, Gastrocentrides occurs in Asia but appears to be confined to Burma (Kimmins, 1952). *Larcasia* has an unusual disjunct distribution pattern: it is recorded from India (Kashmir and Assam only) and Thailand (Schmid, 1965c; Malicky & Chantaramongkol, 1996), as well as Europe (Malicky, 1983b). It seems likely that these animals will eventually be reported from some of the intervening territory. *Goera* larvae have been described from Indonesia (Ulmer, 1955) and Korea (Kim, 1974), but the larvae of *Gastrocentrides* and *Larcasia* are not known in Asia although de Jalón (1977) has figured a Spanish *Larcasia* larva. *Goera* build characteristic cases consisting of a central tube made of mineral fragments with a row of two or three larger pebbles along either side which serve as ballast stones. These cases are reminiscent of *Neophylax* (see above), but are more regular with fewer large stones. *Larcasia* cases may lack ballast stones. *Goera* larvae have distinctive anterolateral prolongations of the prothorax and mesothorax making them quite unmistakable (although the mesothoracic projections are apparently lacking in *Larcasia*: de Jalón, 1977). The prothorax is also thickened and rather robust. The sclerotized parts of the first two thoracic segments come together and block the mouth of the case when the larvae withdraws, thereby serving as an operculum. *Goera* has four mesonotal sclerites, and abdominal gills which (mostly) consist of three filaments. They are scraper grazers, and are confined to rock surfaces in flowing water.

The Molannidae is a small family comprising only three genera: *Indomolannodes*, *Molanna* and *Molannodes*. *Molanna* is widespread and is found throughout the Oriental Region (e.g. Ulmer, 1932; Neboiss, 1993; Malicky & Chantaramongkol, 1989b; Higler, 1992; Malicky, 1994a) although it does not reach New Guinea. The Holarctic genus *Molannodes* contains only two species which both occur in Palaearctic eastern Asia but have not been recorded south of Japan (Wiggins, 1968; Fuller & Wiggins, 1987; Nozaki *et al.*, 1994). *Indomolannodes* has its distribution centred in India (especially the Himalaya) and is also known from Burma (Wiggins, 1968), Vietnam (Malicky, 1995a) and Thailand (Chantaramongkol & Malicky, 1996). The larvae of *Indomolannodes* have yet to be described in the literature, but Wiggins (1996) gives diagnostic characters for *Molanna* and *Molannodes* larvae from North America. It is possible that a re-evaluation of the status of *Indomolannodes* (and possible synonymy with *Molannodes*) may follow description of the larval stage. *Molanna* larvae from Palaearctic Asia are figured by Iwata (1927), Kim (1974) and Tanida (1985). The larvae can be recognized easily by their cases. They are flattened and made of small rock fragments (although some detritus is often incorporated by *Molannodes*). The dorsal portion is larger than the ventral part so that the case has lateral flanges and a dorsal hood over the anterior end. The larvae have dorsal and lateral humps on abdominal segment I, abdominal gills and a lightly sclerotized mesonotum. In addition, the foretibia has a long distal process, and the hind legs are relatively long with the tarsus subdivided near the middle. The tarsal claws of the hind legs are different from those of the other legs and are diagnostic for the known genera: the claw of *Molanna* is reduced to a short, setose stub but is drawn out into a slender filament in *Molannodes*. In addition, the gills of *Molanna* each consist of two to four filaments; those of *Molannodes* are single. *Molanna* larvae occupy a range of lotic and lentic habitats from 60 — 2,300 m elevation; *Indomolannodes* adults, by contrast, are associated with turbulent, streams (770 — 2,200 m) in forested areas at higher altitudes (Wiggins, 1968).

The Sericostomatidae has, at various times, been defined so as to include genera that are now placed in the Lepidostomatidae, Uenoidae and Goeridae (see, for example, Tian & Li, 1985). The current, more limited, view of the Sericostomatidae encompasses *Gumaga* in Palaearctic East Asia, Vietnam and India (Tanida, 1987; Schmid, 1991b; Malicky, 1995a), *Notidobia* in China and Burma (Kimmins, 1952; Hwang, 1957) and *Asahaya* in Indian (Schmid, 1991b). The larvae of a Korean *Gumaga* is figured by Kim (1974; see also Wiggins, 1996), and illustrates the diagnostic features of the family: a dorsolateral ridge on each side of the head; the lack of a prosternal horn; broad and flat forefemora; a sclerotized mesonotum; a sclerotized ring around the base of the lateral humps on the first abdominal segment; a cluster of 30 or more setae at the base of the proleg; and, an accessory hook on the anal claw. The case, which is constructed of rock fragments, is cylindrical, curved and somewhat tapered posteriorly. *Gumaga* has only one pair of sclerites on the anterior part of the metathorax, but this feature may differ among genera. Some aspects of the life history of a North American *Gumaga* are reported by Resh *et al.* (1997).

Ceylanopsyche, which occurs in Sri Lanka, has been attributed to the Sericostomatidae but the precise family placement of this genus is unclear (Schmid, 1958a; Malicky, 1973; Chantaramongkol & Malicky, 1986). *Ceylanopsyche* was included within the Helicopsychidae by Schmid (1958a), but the larvae are quite unlike members of that family. According to Malicky (1973) the cases are long, slender and conical, and are usually covered with fine sand grains although in some instances they may be constructed of silk only. Larvae are robust with short, stout legs. They lack abdominal gills and a prosternal horn but have lateral humps on the first abdominal segment. There is a single large dorsal sclerite on each of the pro- and mesothoracic nota, and the metathorax has a narrow transverse sclerite situated posteriorly and two triangular sclerites anteriorly. Abdominal segment IX bears two small dorsal sclerites, and the anal claw is simple. In addition to the problematic *Ceylanopsyche*, Schmid (1993a) has described the adult stages of three new dravidian (South India and Sri Lankan) genera which cannot be assigned to any known family. They are *Karomana* (which is monotypic), *Mpuga* and *Ngoya*.

Other Integripalpia

Four other families of Integripalpia must now be considered briefly: the Beraeidae, Phryganeidae, Apataniidae and Limnephilidae. The Beraeidae is represented in East Asia by a single Japanese record: *Ernodes gracilis* (Nozaki *et al.*, 1994). Data on larval morphology are given by Nozaki & Kagaya (1994). There are as yet no reports of this genus from streams elsewhere in the region. The Limnephilidae and Phryganeidae are mainly Holarctic; they seem to lack warm-adapted species and cannot persist where temperatures regularly exceed 25°C. Indeed, the Limnephilidae is a fairly recent group of caddisflies that is almost entirely absent from southern Asia, Australia and Africa. Phryganeids occur in Palaearctic Japan and China, and there are a few scattered records from the Himalaya and elsewhere. Genera include *Colpomera*, *Dasystegia*, *Eubasilissa*, *Oligotricha*, *Oopterygia* and *Semblis* in China and Japan (Ulmer, 1932; Schmid, 1959; Tanida, 1987); *Oopterygia* occurs in the Himalaya and Hindukush of Pakistan also (Schmid, 1961). *Neurocyta* has a restricted distribution in the Indian Himalaya at high altitudes between 2,100 and 4,300 m (Schmid, 1968b), and *Eubasilissa* has been recorded at similar elevations in parts of India and the northern parts of Burma, Thailand and Vietnam (e.g. Kimmins, 1952; see also Higler, 1992). Many phryganeids have strikingly marked adults, and the genus *Eubasilissa* includes the largest caddisflies in the world. Larvae are predators and make cases of plant material which they readily abandon when disturbed. The distributional data on Asian phryganeids must be treated with caution as there may be many synonymies among the published names of these animals. Nevertheless, it is evident that the family is predominately Palaearctic.

An analysis of the Chinese caddisfly fauna (Dudgeon, 1987b) revealed only one phryganeid record (*Eubasilissa regina*) south of the Chang Jiang — i.e. the northern Boundary of the Oriental Region. Accordingly, these caddisflies will not be considered further.

Although previously treated as a subfamily of the Limnephilidae, the Apataniidae are now recognized as deserving of familial rank (e.g. Wiggins, 1996; Mey, 1997b; Morse, 1977). They inhabit cool montane steams (over 1,000 m elevation) in northern parts of the Oriental Region. Asian genera include *Apatania*, *Apataniana*, *Apatidelia*, *Moropsyche* and *Notania* which have been recorded from northern Burma and Vietnam, India, Nepal, Pakistan, Taiwan and the eastern Palaearctic (Ulmer, 1932; Mosely, 1942; Kimmins, 1952; Schmid, 1961; 1968c; Kobayashi, 1987; Mey & Levanidova, 1987; Malicky & Chantaramongkol, 1989b; Nishimoto, 1989; Tian *et al.*, 1992; Malicky, 1997; Mey, 1997b; 1997d). *Apatania* larvae can be recognized by the presence of a transverse row of setae on the anterior part of the metanotum and the absence of sclerites in this position. The mandibles are blade-like and usually lack teeth, suggesting that the larvae feed by scraping algae from hard surfaces. As is the case in limnephilids (see below), abdominal segment I bears a significant number of scattered setae. *Apatania* cases are made of rock fragments, and slightly curved with an overhang above the anterior opening. *Moropsyche* larvae are similar to *Apatania*, but differ in details of the setation (Nishimoto, 1989; see also Botosaneanu, 1968). The genus *Apatidelia* contains two Chinese species (Mey, 1997b) which are not known in the larval stage, while *Apataniana* is largely confined to the Karakoram and Central Asia (Schmid, 1961; 1968c; Mey & Levanidova, 1989). Larvae of the Indian genus *Notania* are likewise unknown, and a definitive larval diagnosis for the family is not yet feasible.

The Limnephilidae is a predominately Holarctic group (Schmid, 1955: Figs 5–8) and is the most diverse family of North American caddisflies (Wiggins, 1996). A few representatives of two subfamilies — the Pseudostenophylacinae and the Limnephilinae — occur in northern parts of the Oriental Region. Of particular note among the Pseudostenophylacinae are *Pseudostenophylax* in Burma, northern Vietnam, India, Pakistan and China (Ulmer, 1932; Banks, 1940; Kimmins, 1952; Hwang, 1958; Schmid, 1961; Tian & Li, 1988; Higler, 1992; Tian *et al.*, 1993; Mey, 1997d), and *Astratodina* and *Phylostenax* in India and Pakistan (Schmid, 1961; Higler, 1992). The Limnephilinae includes *Aplatyphylax* in India (Higler, 1992); *Asynarchus* in India and northern China (Ulmer, 1932; Higler, 1992); *Evanophanes* and *Glyphotaelius* in

China (Ulmer, 1932; Banks, 1940); *Limnephilus* in India, Pakistan and China (Ulmer, 1932; Schmid, 1961; Higler, 1992; Tian *et al.*, 1993); *Mesophylax* in Pakistan (Schmid, 1961); *Micropterna* in India and Pakistan (Schmid, 1961; Higler, 1992); *Nothopsyche* in China, North Vietnam and northern Thailand (Ulmer, 1932; Banks, 1940; Malicky & Chantaramongkol, 1989b; Mey, 1996c); *Philarctus* in Pakistan and northern China (Ulmer, 1932; Schmid, 1961); and, *Psilopterna* in Pakistan and China (Banks, 1940; Schmid, 1961). *Dicosmoecus* (Limnephilidae: Dicosmoecinae) is known from northern Japan (Nagayasu & Ito, 1997) but the genus is strictly Holarctic in distribution.

Asian limnephilids tend to be rather localised at high altitudes (1,200 — 4,800 m: e.g. Kimmins, 1952; Schmid, 1961; Malicky & Chantaramongkol, 1989b; Chen & Morse, 1991; Mey, 1996c), especially in the north of the region, and do not occur in New Guinea or most of Indonesia (Neboiss, 1986; 1987). This restricted distribution limits their ecological importance, although Schmid (1961) remarks on the local abundance of *Astratodina* between 2,000 and 3,400 m elevation in Pakistan (Schmid, 1961). Limnephilid larvae can be recognized by a combination of characters: antennae situated halfway between the eye and the anterior margin of the head capsule; the presence of a prosternal horn; and, extensive development of setae on abdominal segment I. Cases are made of small stones and sand in some genera, but others (e.g. *Limnephilus*) use organic building material also. The abdominal gills of pseudostenophylacine larvae consist of a single filament while those of the Limnephilinae usually have multiple gill filaments. Wiggins (1996) provides diagnoses for Holarctic larvae of *Pseudostenophylax*, *Asynarchus*, *Limnephilus* and *Philarctus* (see also Mey & Dulmaa, 1986), and other genera which occur in Palaearctic Asia; Tanida (1985) figures Japanese *Asynarchus*, *Limnephilus* and *Nothopsyche*. It is not clear whether the characteristics which define the genera of Holarctic limnephilid larvae can be applied without modification to Oriental representative of the family. Caution is advised.

COLEOPTERA

In terms of number of described species, the Coleoptera is the largest animal order. Although they are diverse in tropical Asian streams, aquatic Coleoptera — water beetles — have not been afforded the attention their frequency in collections warrants. Ecological research has been hampered by a lack of information on immature stages, and the majority of larvae cannot be identified to genus with any certainty. Adults are better known but, to date, most research has been directed towards cataloguing new genera and species (about 6,000 have been described). Virtually nothing is known about the habits or bionomics of the majority of these animals. In general, water beetles are aquatic as larvae, and may be terrestrial (in which case they tend to be rather short-lived) or aquatic (and relatively long-lived) as adults. A small number of genera (in the Hydraenidae and Dryopidae) have terrestrial larvae and aquatic adults. To quote Hinton (1955: p. 38): '. . . if the adults are aquatic the larvae are also aquatic, but terrestrial adults may have aquatic larvae . . .' (see also Hutchinson, 1981). Pupation almost invariably takes place on land, and the adults of many aquatic species undergo a dispersal flight before returning to the water. Feeding habits and habitat preferences vary greatly among beetle families: some are largely confined to running waters (e.g. Elmidae and Psephenidae), others are lentic. Certain families (such as the Scirtidae and Hydrophilidae) include species which vary in greatly in habitat occupancy, from rheophilic taxa to semiaquatic forms and beetles that are confined to standing waters.

Relative to our knowledge of adult water beetles (especially lentic forms), facts about the biology of larvae are scarce. This reflects, in part, the paucity of studies that have involved rearing larvae to the adult stage (Bertrand, 1983), and thus the association of larval and adult forms is often uncertain. Moreover, while some adults can be identified to species, the application of a correct name is infrequently correlated with an ability to state anything meaningful about the ecology of the animal concerned. For instance, we are able to identify many species of hydraenid beetle but have virtually no information about their habits or life histories. Obviously we cannot say much about the biology of these beetles if their larvae are unknown, but there is little incentive to learn more about the larval stages in cases where (because we lack information) the adults of all species within a particular family are treated as though they were ecologically equivalent to each other. Regrettably, there is little immediate prospect that this cycle will be broken.

There are four suborders within the Coleoptera but only two — the Adephaga and Polyphaga — are regularly encountered in streams (Table 4.8). The suborder Myxophaga has two families that are associated with water: the Microsporidae and Hydroscaphidae. Both are uncommon in benthic samples. The Microsporidae (formerly known

Table 4.8 Genera of aquatic Coleoptera from tropical Asian streams. The list includes taxa that may be associated with running water but because of taxonomic uncertainties and a paucity of ecological data this list cannot be regarded as definitive.

MYXOPHAGA
  Microsporidae
    Microsporus
  Hydroscaphidae
    Hydroscapha
ADEPHAGA
  Haliplidae
    Haliplus
    Peltodytes
  Hygrobiidae
    Hygrobia
  Amphizoidae
    Amphizoa
  Noteridae
    Canthydrus
    Hydrocanthus
    Neohydrocoptus
    Noterus
  Dytiscidae
    Laccophilinae
      Laccophilus
      Neptosternus
    Hydroporinae
      Allopachria
      Clypeodytes
      Deronectes
      Herophydrus
      Hydroglyphus
      Hydroporus
      Hydrovatus
      Hyphoporus
      Hyphovatus
      Hyphydrus
      Leiodytes
      Limbodessus
      Liodessus
      Methles
      Microdytes
      Neonectes
      Peschetius
      Pseuduvarus
      Uvarus
      Yola
    Colymbetinae
      Agabus
      Copelatus
      Hydronebrius
      Lacconectus
      Platambus
      Platynectes
      Rhantus

    Dytiscinae
      Cybister
      Eretes
      Hydaticus
      Rhantaticus
      Sandracottus
  Gyrinidae
    Enhydrinae
      Dineutus
    Gyrininae
      Aulonogyrus
      Gyrinus
      Metagyrinus
    Orectochilinae
      Orectochilus
POLYPHAGA
STAPHYLINOIDEA
  Hydraenidae
    Davidraena
    Gondraena
    Hydraena
    Laeliaena
    Limnebius
    Ochthebius
HYDROPHILOIDEA
  Hydrophilidae
    Hydrophilinae
      Agraphydrus
      Allocotocerus
      Ametor
      Amphiops
      Anacaena
      Berosus
      Chaetarthria
      Chasmogenus
      Crenitis
      Enochrus
      Helochares
      Helopeltarium
      Hydrobiomorpha
      Hydrobius
      Hydrocassis
      Hydrochara
      Hydrophilomima
      Hydrophilus
      Laccobius
      Oocylus
      Paracymus
      Pelthydrus
      Regimbartia
      Scoliopsis

Table 4.8   *(cont.)*

Sternolophus
Thysanarthria
Tylomicrus
Sphaeridiinae
Coelostoma
Helophoridae
Helophorus
Hydrochidae
Hydrochus
Spercheidae
Spercheus
Georissidae
Georissus
EUCINETOIDEA
Scirtidae
Cyphon
Elodes
Flavohelodes
Hydrocyphon
Mescrites
Ora
Prionocyphon
DRYOPOIDEA
Eulichadidae
Eulichas
Ptilodactylidae
Anchytarsinae
Epilichas
Psephenidae
Eubrianacinae
Eubrianax
Psepheninae
Mataeopsephus
Psephenus
Sinopsephenus
Psephenoidinae
Nematopsephus
Psephenoides
Eubriinae
Dicranopselaphus
Ectopria
Granuleubria
Homoeogenus
Macroeubria
Microeubria
Schinostethus
Dryopidae
Ceradryops
Dryops
Elmomorphus
Geoparnus
Helichus (sensu lato)
Malaiseanus

Pachyparnus
Praehelichus
Sostea
Uenodryops
Limnichidae
Byrrhinus
Limnichus
Pelochares
Heteroceridae
Heterocerus
Littorimus
Elmidae
Elminae
Aesobia
Ancyronyx
Aulacosolus
Austrolimnius
Cephalolimnius
Cuspidevia
Eonychus
Esolus
Graphelmis
Graphosolus
Grouvellinus
Haraldia
Hedyselmis
Homalosolus
Ilamelmis
Indosolus
Jilanzhunychus
Leptelmis
Limnius
Loxostirus
Macronevia
Macronychoides
Macronychus
Nesonychus
Ohiya
Ordobrevia
Paramacronychus
Podelmis
Podonychus
Prionosolus
Pseudamophilus
Rhopalonychus
Rudielmis
Simsonia
Sinonychus
Stenelmis
Taprobanelmis
Unguisaeta
Urumaelmis
Vietelmis

Table 4.8   *(cont.)*

        Zaiteviaria
        Zaitzevia
    Larinae
        Dryopomorphus
        Parapotamophilus
        Potamophilinus
        Potamophilus
CANTHAROIDEA
  Lampyridae
      Luciola
      Pyrophanes
CHRYSOMELOIDEA
  Chrysomelidae
    Donaciinae
CURCULIONOIDEA
  Curculionidae
    Molytinae
      Bagous

as Sphaeriidae) are semiaquatic and burrow in damp sand along stream margins. These minute beetles range from 0.5 to 0.9 mm body length. *Microsporus* (= *Sphaerius*) is the only genus. The Hydroscaphidae are small also (not exceeding 2 mm in length) and rather distinctive because the body is fusiform and the elytra are truncated and expose part of the dorsal surface of the abdomen. They have filiform antennae (of five or eight segments), short maxillary palps, and three-segmented tarsi. Hydroscaphids occur in streams or seepage pools; they respire by means of a plastron and eat algae. *Hydroscapha* is the only genus in Asia, and is known from China, Taiwan, Vietnam, Nepal, India and Sri Lanka (Jäch, 1984a; 1995d). These beetles are usually (but not always) associated with algae on submerged rocks in swift streams. The larvae are tiny fusiform creatures tapering posteriorly and sclerotized dorsally.

There are six aquatic families in the Adephaga (Amphizoidae, Gyrinidae, Haliplidae, Hygrobiidae, Noteridae and Dytiscidae), and 11 families in the Polyphaga have at least some species associated with fresh water. The latter suborder tends to be more varied in morphology than the rather uniform Adephaga. All adephagan water beetles have aquatic larvae and most have three larval instars, but the number is typically greater (and more variable) in the Polyphaga. Adult Adephaga which live under water respire by means of a reservoir of air held beneath the elytra (i.e. the hardened forewings which cover the abdomen). Aquatic adults of the suborder Polyphaga carry a film of air trapped by hydrofuge hairs and covering either the ventral surface or the entire body. It serves as a physical gill (a plastron) but, in some taxa, has to be renewed periodically at the water surface. Larval respiratory adaptations are quite varied, but are generally based upon tracheal gills (and independence from atmospheric air) or open spiracles (and access to the water surface). Because aquatic representatives of the Polyphaga occur in different superfamilies (the Hydrophiloidea, Staphylinoidea, Dryopoidea, Eucinetoidea, Chrysomeloidea and Curculionoidea; see Table 4.8), and as some of these superfamilies contain a majority of terrestrial species, it is reasonable to assume that the aquatic life style has evolved more than once. Obviously, therefore, the term 'water beetles' has no phylogenetic meaning. Moreover, because these animals are polyphyletic there is no single character (or combination of characters) which can be used to separate them unequivocally from their terrestrial ancestors.

The following key should allow the identification of families of larval and adult aquatic beetles, and draws from White & Brigham (1996). Identification of larvae to genus is, in most cases, not yet possible. Knowledge of adult stages is more complete but varies in extent among families. However, generic and specific identification of adults is feasible for some groups. For example, Hansen (1991a) provides an excellent key to the superfamily Hydrophiloidea including drawings of the adult *habitus* for each genus, and Pederzani (1995) has constructed a valuable key to the world genera of adult Dytiscidae (including Noteridae). Both are indispensable references but are too extensive to reproduce here. Partial or complete keys to species within some genera are available in the primary literature cited below and the interested reader should consult those publications for guidance. Identification of adult beetles usually depends upon details of the body form and markings or surface texture of the elytra. In addition, the morphology of the aedeagus (the male copulatory organ) is an important diagnostic feature. It is inconvenient to examine during routine ecological studies since it usually involves extraction and slide-mounting of the genitalia which can be a time-consuming process.

1. Wings present and enclosed beneath hardened wing covers (elytra) which overlay the abdomen .... **adult Coleoptera** 26

    Wings and elytra absent .... **Coleoptera larvae** 2

2. Legs absent or minute; thorax and abdomen short, obese and

without distinct sclerites; associated with aquatic macrophytes 3

Legs present (with three to six well-defined segments); body not as above; not usually confined to aquatic macrophytes 4

3. Legs absent .... **Curculionidae: Molytinae**

Legs minute; spiracles on abdominal segment VIII bearing large sclerotized hooks .... **Chrysomelidae: Donaciinae**

4. Legs with a single tarsal claw 5

Legs with two tarsal claws 6

5. Legs with five segments (excluding claw); body somewhat elongate and tapered posteriorly; relatively rare .... **Haliplidae**

Legs with three or four segments; body form varied; very common 11

6. Abdomen with two pairs of stout hooks at the tip (segment X) and lateral gills on segments I-IX .... **Gyrinidae**

Abdomen lacks terminal hooks and has only eight segments; lateral gills are absent 7

7. Abdominal segment VIII with a pair of large terminal spiracles; relatively common 8

Abdominal segment VIII without spiracles; rare .... **Hygrobiidae**

8. Cerci on the tip of the abdomen are slender and usually much longer than abdominal segment I .... **Dytiscidae** (in part)

Cerci either short and stout or absent 9

9. Thorax and abdomen distinctly flattened with tergites expanded laterally as thin, flat projections .... **Amphizoidae**

Body more-or-less cylindrical in cross section and not markedly flattened; tergites not expanded laterally 10

10. Legs long and slender, and may be adapted for swimming; mandibles sickle-shaped .... **Dytiscidae** (in part)

Legs short and stout and adapted for digging; mandibles not sickle-shaped .... **Noteridae**

11. Labrum separated from the front of the head (the clypeus) by a

distinct suture so it appears as a separate sclerite 14

Labrum not clearly differentiated from the rest of the head and not represented as a distinct sclerite 12

12. Body flattened, with large, rather wide thoracic and abdominal tergites and a membranous ventral surface; pronotum expanded anteriorly, and the retractable head is not usually visible when viewed from above; relatively rare ....**Lampyridae**

    Without the above combination of characteristics; relatively common 13

13. Abdomen with ten segments and very short cerci; legs short with three segments (excluding tarsal claw); relatively rare ....
    .... **Georissidae**

    Abdomen with eight segments (rarely with ten, in which case cerci are long and have two or three segments); legs long with five segments; widespread and abundant .... **Hydrophiloidea** 24

14. Abdominal segment VIII (and sometimes other segments) bears a pair of fleshy articulated finger-like lobes on the dorsal surface; antennae very short with only two segments; minute (< 2 mm long) larvae; rare 15

    Abdominal segments without pairs of articulated, finger-like lobes on the dorsal surface; antennae with at least three segments; body size varied; common 16

15. Finger-like dorsal lobes on abdominal segments I and VIII only
    .... **Hydroscaphidae**

    Finger-like dorsal lobes on abdominal segments I-VIII ....
    .... **Microsporidae**

16. Abdomen with ten segments; articulated cerci may be present
    17

    Abdomen with nine segments and lacking articulated cerci (stout upturned hooks — urogomphi — *may* be present) 18

17. Segment IX bears a pair of articulated cerci made up of two segments; segment X with a minute pair of recurved ventral hooks
    .... **Hydraenidae**

    Abdomen lacks cerci, gills or hooks .... **Heteroceridae**

18. Antennae long and conspicuous with many segments .... **Scirtidae**

Antennae short and inconspicuous, consisting of two or three segments only **19**

19. Body more or less cylindrical or fusiform; head and legs visible (at least in part) when viewed from above.................................. **20**

Body greatly flattened with thoracic and abdominal tergites expanded laterally giving the larva a more-or-less ovoid outline and concealing the legs and head when viewed from above.......
.... **Psephenidae**

20. Abdominal segment IX with a ventral, lid-like operculum concealing a chamber beneath; body entirely sclerotized and always lacking ventral gill tufts on abdominal segments I-VIII **22**

Abdominal segment IX lacks a ventral operculum, but ventral gill tufts may be present on segments I-VIII **21**

21. Ventral gill tufts on abdominal segments I-VIII; rather widespread but occurring at low densities .... **Eulichadidae**

Lacking ventral gills but with a tuft of gills in the anal region (segment IX); rare .... **Ptilodactylidae**

22. Tip of the abdomen distinctly pointed, bifid or notched with — in many cases — lateral ridges or longitudinal crests (carina); head capsule with groups of five lateral ocelli; relatively abundant ....
.... **Elmidae**

Tip of the abdomen rounded; head capsule with groups of six ocelli (five lateral and one ventral) or ocelli are absent; relatively scarce **23**

23. Opercular chamber (on abdominal segment IX) containing two retractile hooks and three tufts of retractile gills...... **Limnichidae**

Opercular chamber (on abdominal segment IX) lacks hooks or gills .... **Dryopidae**

24. Nine complete abdominal segments with the tenth reduced; abdomen may be noticeably sclerotized; rare ....
.... **Helophoridae** and **Sperchidae**

Eight abdominal segments; abdomen membraneous or fleshy; common **25**

25. Antennae inserted closer to the front of the head than are the mandibles; labium and maxillae are peculiarly positioned in a furrow beneath the head; rare .... **Hydrochidae**

Antennae arising behind the insertion point of the mandibles; labium and maxillae situated in the usual location (i.e. the ventral anterior margin of the head); common .... **Hydrophilidae**

26. Compound eyes divided into separate dorsal and ventral portions; forelegs long and raptorial, mid- and hindlegs short and paddle-like; antennae stout and club-shaped (slightly thicker toward the tip), with a small branch arising from the base (or pedicel); neustic .... **Gyrinidae**

Compound eyes undivided; form of legs and antennae various but not as above; aquatic or semiaquatic but never neustic 27

27. Head with a conspicuous anterior prolongation or rostrum (a drooping snout resembling an elephant's trunk); rare .... **Curculionidae**

Head not produced into a rostrum or snout 28

28. Elytra short or truncate, covering only part of the dorsal surface of the abdomen 29

Elytra covering all (or almost all) of the abdomen 30

29. Minute beetles (< 2 mm long) with eight antennal segments, the last being considerably longer than the preceding ones; aquatic and usually rather rare .... **Hydroscaphidae**

Body slender (usually > 5 mm long); antennae with more than eight segments, the last being no more than twice as long as the preceding one; semiaquatic and often shiny, metallic or iridescent beetles .... **Staphylinidae**

30. Hind coxae expanded posteriorly as plates overlying some or all of the underside of abdominal segments I-III 31

Hind coxae *sometimes* extended posteriorly along the midline but never expanded into plates 32

31. Beetles over 3 mm long; hind coxal plates greatly expanded to cover the first two or three abdominal segments and entirely concealing them from view; aquatic .... **Haliplidae**

Beetles only 1 mm long; hind coxal plates cover part of the first two or three abdominal segments leaving the lateral margins exposed; semiaquatic .... **Microsporidae**

32. Hind coxae extend posteriorly along the midline so that the sternum of the first abdominal segment is divided into a left and a right half; tarsi with five segments (excluding claw) 33

    Hind coxae not extended posteriorly to divide the first abdominal sternite; tarsi with three, four or five segments 36

33. Hind tarsi and (usually) tibiae flattened and bearing long stiff swimming hairs 34

    Hind tarsi and tibiae more-or-less cylindrical in cross section and lacking conspicuous swimming hairs .... **Amphizoidae**

34. Metasternum (i.e. underside of the metathorax) with a transverse suture; eyes strongly protuberant; body less than 12 mm long; uncommon .... **Hygrobiidae**

    Metasternum lacks a transverse suture; eyes flush with the margins of the streamlined body (i.e. they do not protrude); size variable, up to 40 mm body length; common, especially in slow-flowing water and marshes 35

35. Hind tarsus with a single claw; if two claws are present, the mesoscutellum (a triangular plate lying mid-dorsally on the mesothorax between the elytra) is large and exposed; dorsal and ventral surfaces almost equally convex .... **Dytiscidae**

    Hind tarsus with two claws; mesoscutellum concealed; dorsal surface more strongly convex than the ventral surface .... **Noteridae**

36. Tips of antennae form an abrupt globular or elongate club ... 37

    Antennae slender and elongate (lacking a clubbed tip), or very short and thick with the basal segment enlarged 41

37. Abdomen with five segments visible on the underside; antennal club made up of three segments 38

    Abdomen with six or seven segments visible on the underside; antennal club made up of five segments .... **Hydraenidae**

38. Fore tarsi with four segments (excluding claw); semiaquatic and rare .... **Georissidae**

Fore tarsi with five segments (excluding claw); common ....
.... **Hydrophiloidea** 39

39. Pronotum with five longitudinal grooves .... **Helophoridae**

Pronotum lacking grooves 40

40. Pronotum tapers posteriorly so that it is distinctly narrower than the bases of the elytra; eyes protruding; antennae with three segments before the club (or cupule); relatively rare ....
.... **Hydrochidae**

Pronotum broadens posteriorly and is not distinctly narrower than the bases of the elytra; eyes prominent or not; antennae with five segments before the 'club' (or cupule); common ... **Hydrophilidae**

41. Tarsi of all legs with five segments (excluding the claw) 42

Tarsi of all legs with three or four segments 49

42. Abdomen with five or six segments visible on the underside
.... 43

Abdomen with seven or eight segments visible on the underside
.... **Lampyridae** (= fireflies)

43. Sternum of the prothorax expanded as a prominent lobe beneath the head; head usually contracted into the prothorax concealing the antennae and eyes 44

Sternum of the prothorax not markedly expanded anteriorly; antennae clearly visible 46

44. Antennae rather thick and only as long as the head, with an enlarged basal segment 45

Antennae filiform and much longer than the head ....
.... **Elmidae** (in part)

45. Antennae with ten or fewer segments; hind coxae contiguous (i.e. not separate) .... **Limnichidae**

Antennae with 11 segments; hind coxae separate ....
.... **Dryopidae** (in part)

46. Tarsi with fourth segment bilobed; body weakly sclerotized ......
.... **Scirtidae**

Tarsi filiform and fourth segment not bilobed; body may be very hard or rather weakly sclerotized **47**

47. Antennae more-or-less filiform or concealed within the prothorax and never longer than the combined length of head and prothorax; body heavily sclerotized .... **Elmidae** (in part)

   Antennae serrate or pectinate (i.e. somewhat comb-like) and longer than the combined length of head and prothorax; **either** small beetles with weakly-sclerotized bodies **or** larger (> 10 mm long) hard-bodied beetles covered with a layer of short hairs **48**

48. Broadly oval and somewhat flattened weakly-sclerotized beetles generally much less than 10 mm body length; antennae inserted between the eyes .... **Psephenidae**

   Elongate and rather convex beetles with well-sclerotized bodies over 10 mm in length; antennae inserted below the eyes .... **Eulichadidae** (and **Ptilodactylidae**)

49. Antennae thickened apically and shorter than the combined length of head and prothorax; mandibles rather conspicuous, projecting horizontally in front of the head; flattened, rectangular beetles covered with fine hair; semiaquatic .... **Heteroceridae**

   Antennae slender and longer than the combined length of head and thorax; mandibles inconspicuous, directed ventrally; may be rather brightly coloured, metallic or iridescent beetles; semiaquatic and associated with macrophytes .... **Chrysomelidae**

Dytiscidae

The Dytiscidae (Figs 4.109A,B & 4.110) is a large family of around 150 genera and over 3,000 species globally. Among aquatic insects, it is probably second only to the Chironomidae in terms of species richness. Adults range in size from minute to substantial (2 — 40 mm body length). They are streamlined and rather flattened, and are sometimes conspicuously patterned (usually in combinations of black, brown, yellow and red). Dytiscids secrete a range of defensive compounds when squeezed or grasped by fish. Some species produce toxic alkaloids, while others release steroids which anaesthetize fish temporarily (e.g. Sipahimalani *et al.*, 1970). Almost all types of freshwater habitat from lowland to montane — including temporary water bodies — are used by dytiscids. Despite the existence of some

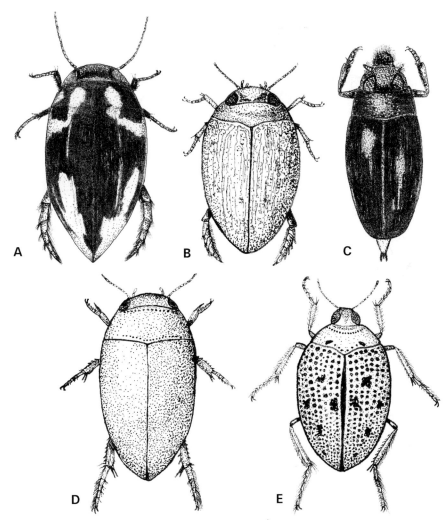

Fig. 4.109 Adult aquatic beetles (Coleoptera: Adephaga): A, *Neptosternus* (Dytiscidae); B, *Laccophilus* (Dytiscidae); C, *Orectochilus* (Gyrinidae); D, *Hydrocanthus* (Noteridae); E, *Peltodytes* (Haliplidae).

stream specialists, it must be stressed that — among rheophilic Coleoptera — the Dytiscidae rank far behind the Elmidae, or even the Psephenidae and Hydrophilidae, in terms of diversity and abundance. Although a minority of dytiscid species are confined to running waters, many others are opportunists living along stream margins and in floodplain pools and marshes. Adults and larvae usually occur within the same habitat and, because they are aeropneustic, must visit the surface periodically to obtain a new air supply. Dytiscid adults are

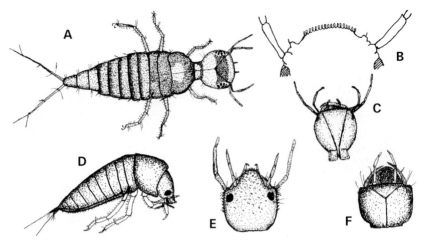

Fig. 4.110 Unasociated dytiscid larvae (Coleoptera: Adephaga): A, *Neptosternus*-type larva (dorsal view); B, anterior margin of head capsule; C, ventral side of head capsule showing division by the gular suture; D, *Hydrovatus*-type larva (lateral view); E, dorsal view of head capsule; F, ventral view of head capsule showing anterior bifurcation of the gular suture.

strong swimmers with the mid- and hind legs (which are often fringed with long hairs) functioning as flattened paddles, but larvae are benthic and have limited swimming ability. Both life stages are voracious predators; adults will also eat carrion. The diversity and pleasing appearance of dytiscids has attracted considerable attention, and spawned a vast literature on their distribution and habitat preferences in Europe and North America. Publications concerning adult beetles have been reviewed by Nilsson *et al.* (1989), who provide a checklist of the valid generic names now in use, and Pederzani (1995) who has constructed a key to the world genera of Dytiscidae. The latter is an essential starting point for students of the group.

Ecological information on tropical Asian dytiscids is hard to come by despite an extensive taxonomic literature dealing with the adults (e.g. Brancucci, 1986; 1988; Holmen & Vazirani, 1990; Balke, 1992; 1993; Wewalka & Biström, 1994; Balke & Sâto, 1995; Biström, 1995; Nilsson, 1995; Balke & Heinrich, 1997; Heinrich & Balke, 1997; Wewalka, 1997). This reflects the fact that most larvae (Fig. 4.110) have yet to be associated with adults. Dytiscid life cycles involve pupation on land followed by flights in search of new habitats after eclosion. However, adult flights do not take place in every case because some species contain both flightless morphs and individuals with fully-developed wings. After mating and oviposition, eggs take no more than two weeks to hatch, and there is rapid development through the first two larval instars so that the third and final instar is reached in about two weeks. The duration of this stage is longer (up to six weeks), and the mature larvae leaves the water to pupate in the soil, or under rocks and plant litter. The precise duration of the life cycle will vary among species, and in response to temperature and food supply, but there is the potential for several generations each year in the tropics — especially in perennial streams. Where habitats dry up for part of the year (on river floodplains, for instance), adults may leave to search for new water bodies. Adults of many species pass the dry season buried in the substrate, remaining quiescent until the next monsoon.

Among the Dytiscidae which occur in running water (e.g. *Agabus*, *Deronectes* and *Platambus*), the subfamily Laccophilinae (Table 4.8) contains the most specialized stream-dwelling taxa. Of these, *Neptosternus* seems to prefer stony streams (e.g. Bertrand, 1973; Jäch, 1984a; Heinrich & Balke, 1997). The Southeast Asian *Neptosternus* have been revised by Hendrich & Balke (1997), who provide illustrations of the 50 species are known from this part of the Oriental Region. The distinctive black and yellow markings on the dorsal surface of *Neptosternus* adults are an aid to identification (Fig. 4.109A). *Laccophilus*, which is also widespread in Southeast Asia, likewise has species-specific patterns on the dorsal surface (Fig. 4.109B); a revision of this genus is ongoing (Balke & Hendrich, 1997). Other rheophilic dytiscids include the Oriental genus *Microdytes* (subfamily Hydroporinae) which occurs in springs and small streams throughout the region; many species are restricted to areas of primary forest. The elytra of most species bear combinations of ferruginous, dark brown and black markings. Wewalka (1997) gives a key to the adults of the 30 known *Microdytes* species.

Although they may not be confined to running waters, dytiscid genera (subgenera have been omitted for brevity) in listed in faunal inventories of tropical Asian countries include the following: *Agabus*, *Allopachria*, *Clypeodytes*, *Copelatus*, *Cybister*, *Deronectes*, *Eretes*, *Herophydrus* (= *Hyphoporus*), *Hydaticus*, *Hydroglyphus* (= *Guignotus*), *Hydronebrius*, *Hydrovatus*, *Hygrotus* (including *Coelambus* which is sometimes treated as a subgenus or synonym: Nilsson *et al.*, 1989), *Hyphydrus*, *Lacconectus*, *Laccophilus*, *Leiodytes*, *Limbodessus*, *Liodessus*, *Methles*, *Microdytes*, *Neonectes* (Taiwan only), *Neptosternus*, *Peschetius*, *Platynectes*, *Pseuduvarus*, *Rhantaticus*, *Rhantus*, *Sandracottus*, *Uvarus* and *Yola* (e.g. Feng, 1932a; 1932b;

1933; Gschwendtner, 1932; Mendis & Fernando, 1962; Fernando, 1964; 1969; Rocchi, 1976; 1982; 1986; Vazirani, 1977 and references therein; Nilsson, 1995). Additional genera such as *Achilus, Bidessus, Colymbetes, Dytiscus, Graphoderus, Hydroporus* (= *Hydrocoptus*), *Ilybius, Nebrioporus* (= *Potamodytes* and *Potamonectes* in part) and *Stictotarsus* (= *Potamonectes* in part) may penetrate the northern fringes of the Oriental Region but should be considered as Holarctic or Palaearctic rather than Oriental (Nilsson, 1995). New species of Dytiscidae continue to be described from tropical Asia (e.g. Holmen & Vazirani, 1990; Biström & Wewalka, 1994; Nilsson & Wewalka, 1995; Wewalka & Brancucci, 1995; Balke & Hendrich, 1997; Wewalka, 1997) but, relative to other water beetles, the family is quite well known. It is probable that relatively few new genera have yet to be discovered, although Wewalka & Biström (1994) recently established the genus *Hyphovatus* based on specimens from streams in Thailand and Sumatra.

Gyrinidae

The Gyrinidae (or whirligig beetles: Fig. 4.109C) is a cosmopolitan family of somewhat more than 1,000 described species. The adult stage is neustic, while larvae are typically benthic. Adults vary considerably in size from 2 to over 30 mm in body length. They are highly modified for life on the water surface, and the body is highly streamlined. The antennae are short and lie flush alongside the head in front of the eyes which are divided into upper and lower facets. when in the normal swimming position, the upper set is above the water surface while the lower set is submerged giving a comprehensive field of vision. The forelegs are usually tucked beneath the thorax, but the mid- and hind legs are paddle-like and fringed with swimming hairs allowing these beetles dart over the surface at top speeds of between 0.5 and 1.0 m s^{-1}. They can also swim well under water and dive when startled. Other adaptations to a neustic existence include the ability to 'echo-locate' prey (or other food items) by detecting waves reflected from objects trapped on the surface film. Dytiscids are conspicuous insects which require an effective defense. It is afforded by noxious secretions of norsesquiterpenes from the pygidial glands which deter predatory fish and bats. Many gyrinids form large single- or mixed-species aggregations (schools or rafts) on the water surface. The advantage of such groupings to each individual is probably a reduced risk from predators; surprisingly, feeding rates do not seem to be significantly reduced by the presence of conspecifics (Wilkinson *et al.*, 1995).

Gyrinids occur in unpolluted lotic and lentic habitats from sea level to mountain streams, although individual species may be highly stenotopic. The Orectochilinae are mainly found in streams, but the other two subfamilies (Enhydrinae and Gyrininae) are more frequently encountered in standing water. Where a single species occupies a range of sites, there is evidence of habitat-related morphological variation among populations but such variation is smaller within than between species (Svensson, 1991). In continental Asia, at least, gyrinids show a latitudinal gradient in diversity, and tropical regions contain more species (Mazzoldi, 1995). Most are associated with standing water, but some taxa are associated with pools and undercut banks (or other shaded localities) in streams. The limited data suggest that the abundance of adults declines during the monsoon (Wilkinson *et al.*, 1995), but information on life history and phenology is generally lacking. Some lotic gyrinids are inactive during the day but forage widely over the stream surface at night. Brinck (1981) considers that lowland gyrinids are apt to occur in schools on the surface of rivers and streams during daylight, while species in montane streams occupy sheltered microhabitats and are more wary and likely to fly (rather than dive) when disturbed. The upland species are descended from lowland ancestors, and may be rather specialized. They include large, sexually-dimorphic forms with powerful forelegs and distinctively marked dorsal surfaces (Brink, 1984). The larval habits of Gyrinidae have received little or no attention in the literature, but they are likely to be carnivorous. Larvae (Fig. 4.111) have paired tracheal gills on abdominal segments I-VIII, with two pairs on segment IX. There are four hooks at the end of the abdomen (4.111B). Some larvae (e.g. *Aulonogyrus*) have distinct cuticular ornamentation. Pupation takes place on land. Bertrand's (1972) preliminary key to the gyrinid larvae is an essential starting point for students of these beetles.

Of the three gyrinid subfamilies, the Enhydrinae contains *Dineutus* which occurs throughout tropical Asia and comprises several subgenera: *Cyclous, Dineutus, Merodineutus* (which is restricted to New Guinea), *Porrorhynchus, Rhombodineutus, Spinosodineutes* and Afrotropical *Gyrinodineutus* (e.g. Wu, 1931; Ochs, 1937; Mendis & Fernando, 1962; Bertrand, 1973; Brinck, 1981a; 1981b; 1983; 1984; Vazirani, 1984; Mazzoldi, 1995, and references therein). The Gyrininae comprises *Aulonogyrus* (known from China to Sri Lanka: e.g. Jäch, 1984a;

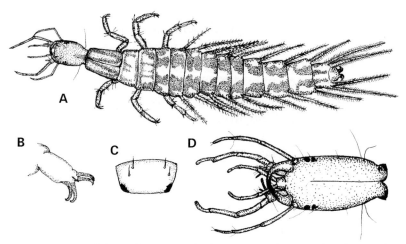

Fig. 4.111 *Orectochilus* (Coleoptera: Adephaga: Gyrinidae): A, dorsal view of larva; B, terminal hooks on abdomen; C, labrum; D, head capsule.

Mazzoldi, 1995), *Gyrinus* (subgenera *Gyrinus* and Palaearctic *Gyrinulus*) which is quite widespread in Asia, and *Metagyrinus* which is endemic to China. The Orectochilinae consists of *Orectochilus* (Fig. 4.109C: subgenera *Orectochilus* and *Patrus*), including many species limited to China (Mazzoldi, 1995; see also Vazirani, 1984). Members of the Orectochilinae have a ventral ridge of hairs at the apex of the abdomen which serves as a 'rudder' during swimming.

Noteridae and other Adephaga

Although once included within the Dytiscidae (as the Noterinae), the Noteridae is regarded by most workers as being sufficiently distinctive to warrant familial rank. They differ from the Dytiscidae, and almost all other Asian water beetles, by having a fully-aquatic pupal stage. Noterids usually occur in shallow lentic habitats but may be encountered in pools and oxbow lakes on river floodplains. Adults are fusiform (i.e. tapered at both ends of the body) and have a more convex dorsal surface than dytiscids. They are active swimmers. Noterid larvae are likewise fusiform and have short, rather broad forelegs enabling them to burrow in mud. Genera occurring in tropical Asia include *Canthydrus*, *Hydrocanthus* (Fig. 4.109D) and *Neohydrocoptus* (including some species formerly placed in the dytiscid genus *Hydroporus* [= *Hydrocoptus*]; Nilsson *et al.*, 1989); *Noterus* is known from the northern fringes of the region (Gschwendtner, 1932; Feng, 1933; Mendis & Fernando, 1962; Fernando, 1969; Rocchi, 1976; 1986; Vazirani, 1967; Wewalka, 1992; Nilsson, 1995). Pederzani (1995) has produced a useful key to adults of the world genera of Dytiscidae *sensu lato* which includes the Noteridae.

The Amphizoidae occur in northern China and Xizang (Tibet), but have yet to be recorded from the Oriental Region. *Amphizoa* is the only genus. Adults live under rocks or in gravel along the edges of cool, fast-flowing streams but, until recently, the larvae of Asian species were unknown (Yu & Stork, 1991). Ji & Jäch (1995) have figured the larva of *Amphizoa sinica* which is spindle-shaped and rather flat, being broadest in the middle of the body. The dorsal surface is relatively well sclerotized, and the abdomen is rather short (equal to the combined length of the head and abdomen. The terga of the thorax and abdomen are expanded and project laterally; there is a pair of spiracles on abdominal segment VIII. Larvae cannot swim and probably feed on drowned insects. They crawl on partially submerged rocks and logs which allow them easy access to the surface for respiration (Ji & Jäch, 1995).

The Haliplidae are aquatic as larvae and adults, and are typically less than 10 mm long. Adults can be recognized by a conspicuous pair of ventral plates formed by posterior expansion of the hind coxae. A store of air is sandwiched between these postcoxal plates and the underside of the abdomen. Haliplid larvae can be recognized by the presence of nine (sometimes ten) abdominal segments and five-segmented legs with single claws (Bertrand, 1972). Haliplidae in the Oriental Region include *Peltodytes* (Fig. 4.109E) and *Haliplus* (with the subgenera *Haliplus* and *Liaphlus*) which occur throughout most of China southward into Indonesia. *Haliplus* (*Liaphlus*) is also distributed eastward into India and Sri Lanka (Wu, 1932a; Falkenström, 1933; Mendis & Fernando, 1962; Vazirani, 1984; Van Vondel, 1992; 1993; 1995; Mazzoldi & Van Vondel, 1997). Larvae of *Haliplus* are elongate and tapered from the head to the tip of the abdomen which ends in a long process (resembling a blunt spine). The cuticle is rather rough and rigid, and gills are lacking. *Peltodytes* larvae have long hair-like gills on the thorax and abdomen: the prothorax has three pairs; the remaining thoracic segments and all but the last two abdominal segments have two pairs; and, abdominal segments IX and X bear one pair each. Larvae of both genera feed on algae (especially macroalgae such as *Chara* or *Spirogyra*), and are most common in standing water — less often along stream margins.

The remaining Adephagan family is the Hygrobiidae (= Pelobiidae) which represented by one genus and five species world-wide. The only Oriental species — *Hygrobia davidi* — is recorded from Jiangxi Province, southeastern China (Wu, 1932c; Jäch, 1995c). The larvae and adults of these predatory beetles live in stagnant water, and the adults stridulate loudly to produce squeaks when handled.

Hydraenidae

Hydraenid beetles are found along stream margins, in seeps, and in the splash zone of cascades. Many of them are probably best characterized as semiaquatic. They are small (typically less than 3 mm long), slow-moving and inconspicuous. There have been few accounts of hydraenid larvae, but recent research by Delgado & Soler (1997) is directed towards rectifying this situation. Like the adults, larvae seem to be semiaquatic are probably herbivorous.

Over 1,000 species of Hydraenidae have been described to date but many of them have restricted distributions so this total is certainly far short of the true value. Our knowledge of Asian hydraenids is largely derived from species descriptions, taxonomic revisions and faunal inventories (e.g. Orchymont, 1934; Pu, 1942; 1951; 1956; 1958; Jäch, 1982; 1994a; 1995a, and references therein; Hansen, 1991b) rather than ecological work. Hansen (1991b) provides a valuable key to 22 recognized genera, although additional genera (*Davidraena* and *Gondraena*: Jäch, 1994a) have been established since then and others await description. To date, six hydraenid genera are known from the Oriental Region: *Davidraena*, *Gondraena*, *Hydraena* (Fig. 4.112A), *Laeliaena* and *Limnebius* and *Ochthebius* (Fig 4.112B). *Davidraena*, *Gondraena* and *Laeliaena* contain few species and have been reported from India and (in the case of *Laeliaena*) Sichuan Province in China only (Hansen, 1991b; Jäch, 1994a; 1995a). The remaining three genera (and their constituent subgenera) are speciose and widely distributed. *Neochthebius* is confined to marine rock pools in Japan and has not been reported further south (Hansen, 1991b).

Hydrophiliodea

The superfamily Hydrophiloidea consists of five families. Of these, the Hydrophilidae is the most diverse containing approximately 140 genera

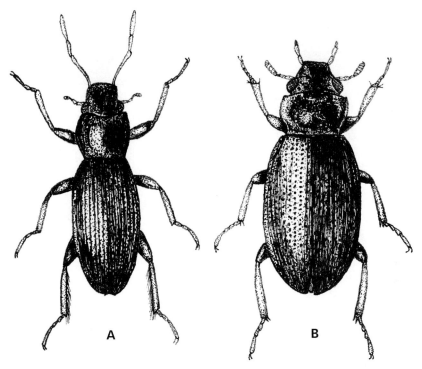

Fig. 4.112 Adult Hydraenidae (Coleoptera: Polyphaga): A, *Hydraena*; B, *Ochthebius* (both redrawn from Hansen, 1991b)

and around 2,000 species (Hansen, 1991a). The group is incompletely known — especially in the immature stages. There are two subfamilies in Asia. The Sphaeridiinae are predominately terrestrial or associated with dung and carrion. They will not be considered here, although one genus (*Coelostoma*) may occur in seeps and other aquatic habitats. Members of the Hydrophilinae — termed water scavenger beetles — are aquatic. The adults feed on a variety of foods, including plant and animal tissue, while most larvae are predaceous. Larvae and adults of the majority of species are aquatic, although adults of a few genera (e.g. *Chaetarthria* and *Thysanarthria*) are semiaquatic. Both stages are rather poor swimmers and most adults crawl rather than swim. They are, however, strong fliers. Hydrophilids vary greatly in size from around 2 mm to over 40 mm body length. Adults (Fig. 4.113) have distinctive clubbed antennae and well-developed maxillary palps that are longer than the antennae. The mid- and hindlegs may be flattened and/or fringed with hairs to aid swimming, and the subfamily Hydrophilinae have a distinctive sternal keel which distinguishes them from the Sphaeridiinae (and other beetles).

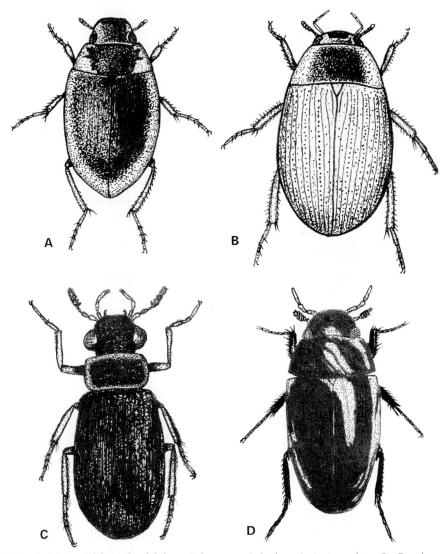

Fig. 4.113 Adult Hydrophilidae (Coleoptera: Polyphaga): A, *Laccobius*; B, *Enochrus*; C, *Berosus* (redrawn from Schödl, 1997); D, *Hydrophilomima* (redrawn from Hansen & Schödl, 1997).

Most hydrophilids are found in lentic habitats, but some (e.g. *Hydrophilomima* and *Pelthydrus*) occur principally in streams where they can be quite abundant. These lotic taxa have been almost entirely neglected by ecologists, and the details of their biology remain unknown. Life-history investigations from the tropics are lacking although, in common with temperate hydrophilids, it appears that the larval stage is of short duration. Pupation takes place in damp soil along the water margins. Bertrand (1972) provides the most comprehensive account of hydrophilid larvae from tropical latitudes, but this monograph is far from complete. The larvae have eight abdominal segments (Figs 4.114A & D), and are distinguished from most other families of Coleoptera by their five-segmented legs and single tarsal claw (the latter being a feature shared with the Haliplidae). In addition, the mandibles have distinctive toothed or serrated inner margins (Fig. 4.114E). Some

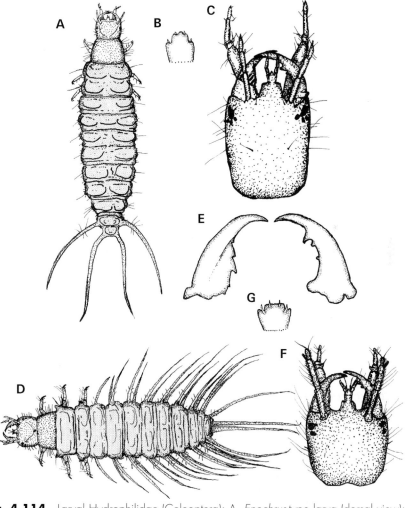

Fig. 4.114   Larval Hydrophilidae (Coleoptera): A, *Enochrus*-type larva (dorsal view); B, labrum; C; head capsule; D, *Berosus*-type larvae (dorsal view); E, mandibles; F, head capsule (dorsal view); G, labrum.

hydrophilid larvae have lateral tracheal gills — notably *Berosus* (Fig. 4.114D) and allied genera; these gills occur on all abdominal segments and are associated with closed spiracles. Nevertheless, the vast majority of hydrophilid larvae have a respiratory chamber (or atrium) and a pair of well-developed (open) spiracles on abdominal segment VIII. Diets seem to consist of animal prey, and Hutchinson (1993: p. 772) remarks that the larvae '. . . have the interesting habit of feeding with the head pushed through the surface film, presumably to avoid diluting the saliva involved in external digestion'.

By combining the genera included in an early checklist of the Chinese Hydrophilinae (Orchymont, 1934) with an updated version of that list (Gentili *et al.*, 1995) and records from other parts of the region (e.g. Pu, 1964; 1981; Satô & Chûgô, 1961; Mendis & Fernando, 1962; Bertrand, 1973; Jäch, 1984a), it is possible to produce a list of hydrophiline genera known from tropical Asia: *Agraphydrus, Allocotocerus* (= *Globaria*), *Amphiops, Anacaena, Berosus, Chaetarthria, Chasmogenus, Crenitis, Enochrus, Helochares, Helopeltarium, Hydrobiomorpha* (= *Neohydrophilus*), *Hydrophilus* (= *Hydrous*), *Laccobius, Oocylus, Paracymus, Pelthydrus, Regimbartia, Scoliopsis, Sternolophus* and *Thysanarthria*. Synonymies follow Hansen (1991a), and subgenera have not been included. Additional genera such as *Ametor, Hydrochara, Hydrobius* and *Hydrocassis* are primarily Palaearctic, but some species occur in the north of tropical Asia (see, for example, Gentili *et al.*, 1995; Schödl & Ji, 1995).

Recent taxonomic studies of the Hydrophilinae have increased the number of genera and species. For example, Hansen & Schödl (1997) have erected the genus *Hydrophilomima* (Fig. 4.113D) to include three new species of hydrophilid from Thailand, Vietnam and southern China; adults live among detritus in running water. In addition, Schödl (1995) has established a new, monotypic genus — *Tylomicrus* — for a recently-discovered Malaysian hydrophilid. As part of a revision of the genus *Berosus* (Fig. 4.113C), Schödl (1992; 1993) gives keys to the known Oriental representatives, although new species continue to be discovered (Schödl, 1997). The taxonomic situation has become more stable after recent revisions of *Pelthydrus* (Schönmann, 1994), *Hydrocassis* and *Ametor* (Schödl & Ji, 1995), and a partial review of the widely-distributed (but predominately) lentic genus *Laccobius* (Gentili, 1995; see Fig. 4.113A).

Four families of beetles (Helophoridae, Hydrochidae, Georissidae and Spercheidae) which comprise the rest of the Hydrophiloidea were formerly treated as subfamilies of the Hydrophilidae but are now viewed as distinct by most workers (e.g. Hansen, 1991a). The Helophoridae consists of a single genus — *Helophorus* (and constituent subgenera). These beetles are largely Holarctic in distribution but penetrate the northern fringes of the Oriental Region in China (e.g. Orchymont, 1934) and there is a single non-Chinese record of the *Helophorus* (subgenus *Thaumhelophorus*) from Assam in India (Angus, 1995). Some *Helophorus* are terrestrial or semiaquatic, and aquatic members of the group are not usually found in streams. Larvae resemble hydrophilids but have nine instead of the usual eight abdominal segments. The abdominal spiracles are open (i.e. functional). Larvae of the Georissidae (= Georyssidae) look like hydrophilids also (van Emden, 1956), but have ten abdominal segments. *Georissus* occurs in Asia. These tiny (less than 2 mm-long) beetles have a strongly convex body shape and are semiaquatic inhabitants of muddy stream banks. The Hydrochidae are small (1.5–5.5 mm long), elongate beetles that live mainly in stagnant water where they burrow in mud. The family is almost cosmopolitan, and *Hydrochus* seems to be widely distributed in Asia (e.g. Orchymont, 1934; Satô & Chûgô, 1961; Mendis & Fernando, 1962; Makhan, 1995; Jäch, 1995b). The Spercheidae — consisting of the singe genus *Spercheus* — have a rather globular body shape resembling some Hydrophilidae. Adults can be recognized by the presence of dense hairs on the antennal club. Moreover, there are no more than three segments before the antennal 'club' in spercheids as compared to five in most Hydrophilidae. As in the Helophoridae, larvae have nine abdominal segments. Little is known of the biology of *Spercheus* but both larvae and adults seem mainly confined to lentic habitats.

Scirtidae

The Scirtidae (= Helodidae) are small (up to 4 mm long) ovoid beetles with a strongly deflexed head, protuberant eyes, weakly-sclerotized elytra, and (in many cases) swollen hind femora. They are usually found close to water. Larvae (Fig. 4.115) are elongate and rather flattened with a large head bearing complex mouthparts. All body segments are moderately sclerotized, and there are retractable anal gills on abdominal segment IX. A pair of functional spiracles are located within a concealed ventral chamber on abdominal segment VIII. The presence of long, multi-segmented antennae sets scirtids apart from other beetle larvae. They occupy a variety of lotic and lentic habitats, ranging from marshes to 'container' habitats such as treeholes,

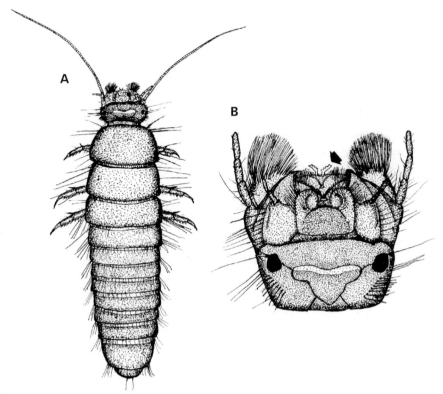

Fig. 4.115 *Hydrocyphon* (Coleoptera: Scirtidae): A, dorsal view of larva; B, dorsal view of head (antennae omitted for clarity), with the distinctive hypopharyngeal teeth indicated by an arrow.

and larvae have been reported from ground water in Europe (Klausnitzer & Popsil, 1991). Scirtid larvae are herbivorous (Bertrand, 1972), and fine detritus collected and sorted by comb-like hairs on the maxillae is probably the main food item.

Klausnitzer (1995) lists the Chinese species of Scirtidae but, apart from that inventory, almost all studies of the family in tropical Asia consist of descriptions of the adults of new species in genera that include *Cyphon* (Klausnitzer, 1979; 1980a; 1980b; 1980c; 1980d), *Elodes* (Yoshitomi & Satô, 1996a), *Flavohelodes* (Klausnitzer, 1980b; 1980e; Yoshitomi & Satô, 1996b), *Hydrocyphon* (Klausnitzer, 1976 [as *Cyphon*]; 1980b; 1980c; Nyholm, 1981), and *Prionocyphon* (Klausnitzer, 1980c). In addition, Fernando (1964) records *Mescrites* and *Ora* (a semiaquatic genus) from Sri Lanka. Of the Asian Scirtidae, *Cyphon* is particularly speciose and distributed throughout the Oriental Region but, like *Prionocyphon*, is mainly found in standing water.

Hydrocyphon seems to be common in flowing water (e.g. Bertrand, 1973). A study of the larval ecology of *Hydrocyphon* in a Hong Kong stream (Dudgeon, 1992c) revealed rather asynchronous growth, sporadic occurrence of adults and almost continuous recruitment with indications that four generations were produced each year. Population densities close to the stream banks were low but larvae were relatively numerous in midstream. Scirtids can be abundant in intermittent reaches of Hong Kong streams, where they may comprise over 60% of benthos densities during the dry season due to an ability to survive among the damp stones and gravels of the stream bed when surface flows cease (Dudgeon, 1992d). The larval morphology of European scirtids has been the subject of close investigation directed towards constructing a phylogeny for the group (Hannappel & Paulus, 1987), and although the results have limited direct application for our fragmentary knowledge of the ecology of these animals in tropical Asia, European representatives of all of the tropical Asian genera mentioned above have been studied.

Eulichadidae and Ptilodactylidae

The Eulichadidae (formerly treated by some workers as part of the Ptilodactylidae) are easily distinguished from other beetle larvae. They are elongate, more-or-less circular in cross section, and the body ends in a pair of upturned hooks (urogomphi). The cuticle is rather granulate in appearance. The larvae are large (up to 50 mm body length) and have paired, branched abdominal gills on segments I-VII (Hinton, 1955 gives details; see also Bertrand, 1972). The spiracles are non-functional. *Eulichas* is the only genus in tropical Asia. It contains two subgenera (*Eulichas* and *Forficulichas*) and 20 species, ten of which were described recently (Jäch, 1995e). Although widely distributed throughout the Oriental Region, the Eulichadidae is one of the least known families of aquatic Coleoptera. Larvae are found buried in sand or among detritus in stony streams where they feed on leaf litter and wood. Pupation occurs in just above the water level. The terrestrial adults are elongate (reaching 35 mm in length), pubescent beetles with long antennae and well-developed mandibles. Jäch (1995e) gives a detailed diagnosis of the genus. Although male *Eulichas* are slightly smaller than females, they have longer, serrate antennae. The Ptilodactylidae are closely allied to the Eulichadidae and members of one subfamily — the Anchytarsinae — have aquatic larvae. *Epilichas* is known from Taiwan and Japan (Nakane, 1985; 1996), but are more widespread than literature records indicate. I have collected larvae (which are have highly localised distributions) from streams in Hong Kong and New Guinea. *Epilichas* larvae resemble *Eulichas* but lack the paired abdominal gills and upturned urogomphi. Practically nothing is known of the larval habits of the Anchytarsinae. They are likely to be detritivores.

Psephenidae

At present, the Oriental Psephenidae are represented by 13 genera and four subfamilies (Table 4.8). They are known as water pennies because of the flattened almost spherical shape of the larvae (Figs. 4.116–4.118). The limpet-like form — which resembles a trilobite — and the

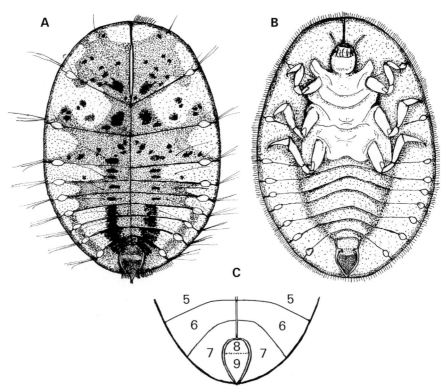

Fig. 4.116 *Psephenoides* (Coleoptera: Psephenidae: Psephenoidinae): A, dorsal view of larva; B, ventral view of larva; C, diagrammatic representation of the extensions of the abdominal segments and associated operculate cavity.

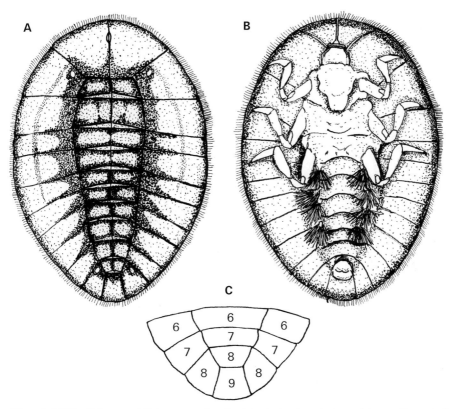

Fig. 4.117 *Eubrianax* (Coleoptera: Psephenidae): A, dorsal view of larva; B, ventral view of larva; C, diagrammatic representation of the extensions of the abdominal segments.

ventral head which is hidden when viewed from above make these larvae unmistakable. Adults may be found crawling on rocks close to the water line, but they fly readily when disturbed. They are usually less than 6 mm long, with rather flattened bodies and weakly sclerotized elytra. The antennae of the males are pectinate or flabellate. Psephenid adults seem to be rather short lived and are seldom collected even around habitats where larvae are abundant (e.g. Dudgeon, 1995c). In Asia, adults are nocturnal and are attracted to lights (Lee *et al.*, 1990; Dudgeon, 1995c), although this is apparently not the case for Nearctic psephenids. Larvae are found clinging to stone surfaces (usually the underside) in riffles, or are associated with wet rocks in and around waterfalls and seeps (e.g. *Schinostethus*: Lee *et al.*, 1993), but members of the subfamily Eubriinae seem to be associated with leaf packs (Lee & Yang, 1994b). Most psephenid larvae are active at night, when they

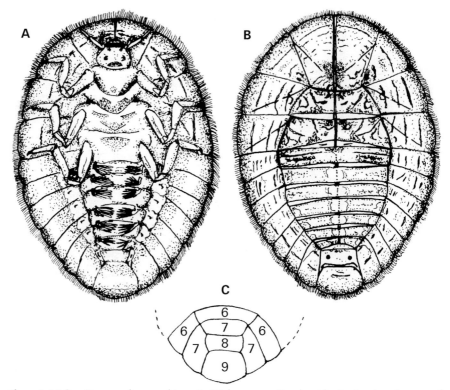

Fig. 4.118 *Sinopsephenus chinensis* (Coleoptera: Psephenidae): A, dorsal view of larva; B, ventral view of larva; C, diagrammatic representation of the extensions of the abdominal segments.

move to the upper surface of rocks and stones to graze periphyton. *Nipponeubria* in Japan is thought to have a life cycle lasting two to three years (Lee & Satô, 1996), but Taiwanese *Ectopria* are apparently univoltine (Lee & Yang, 1994b). Life-history investigations in a Hong Kong stream revealed that *Sinopsephenus chinensis* was bivoltine, while other psephenids (*Eubrianax* sp. and *Psephenoides* sp.) showed asynchronous growth and almost continuous recruitment (Dudgeon, 1995c). Microdistribution patterns also varied among these psephenids: *Eubrianax* sp. and (especially) *Sinopsephenus chinensis* were most abundant in midstream microhabitats but *Psephenoides* sp. was more numerous close to the stream banks.

Psephenids are mostly confined to lotic habitats but *Homoeogenous*, recorded from standing waters in India and Taiwan, is an exception (Lee & Yang, 1993). Larval Psephenidae have a variety of respiratory adaptations (Hinton, 1955). The Psephenoidinae (Fig. 4.116) have anal tracheal gills that can be retracted into an operculate cavity on the dorsum of abdominal segments VIII and IX (Fig. 4.116C). The Eubriinae have retractile anal gills also (on abdominal segment IX) but, unlike the Psephenoidinae which pupate under water, they have a pair of functional spiracles with an associated brush of setae situated at the end of the abdomen (segment VIII). These brushes are thought to keep the spiracles unobstructed when larvae leave the water to pupate (Hinton, 1955). The remaining two subfamilies lack anal gills. Instead, larvae of the subfamily Eubrianacinae (Fig. 4.117B) have paired ventral gills on the first four abdominal segments. These comprise branched, unsclerotized prolongations of the body wall, each containing an extension of the tracheal system. The gills of the Psepheninae are similar (Fig. 4.118B), but occur on the first six abdominal segments. Both of these subfamilies usually pupate on land just above water level, but *Eubrianax* will sometimes pupate under water (Lee & Yang, 1990).

In addition to differences in gill arrangement noted above, larvae of the Eubrianacinae have lateral extensions on abdominal segment VIII (Fig. 4.117C), and maxillary palps with two segments. Psepheninae larvae lack lateral extensions on abdominal segment VIII, and have maxillary palps with four segments. In addition to the absence of functional spiracles and associated brushes, Psephenoidinae differ from Eubriinae in the absence of lateral extensions on abdominal segment VIII (Fig. 4.118C). However, the extensions of segment VII are prolonged backwards and meet at the midline thereby enclosing segments VIII and IX.

In a global review of the genera that had been described by that date, Brown (1981a) recorded *Eubrianax* (Eubrianacinae), *Mataeopsephus* and *Psephenus* (Psepheninae), *Psephenoides* (= *Microeubrianax*: Psephenoidinae), plus *Cophaesthetus*, *Ectopria*, *Eubria* and *Schinostethus* (Eubriidae). Of these, *Eubrianax* is particulary speciose and widespread in Oriental streams (e.g. Bertrand, 1973; Jäch, 1982; 1984a; Satô, 1983; Lee & Yang, 1990). The larvae are figured by Bertrand (1972; 1973; see also Fig. 4.117). More recent work has added to this list of genera. For example, *Sinopsephenus* has been recorded from Hong Kong (Dudgeon, 1995c; see also Fig. 4.118), and a new genus of Psephenoidinae — *Nematopsephus* — has been erected based on adult beetles from Thailand (Jäch & Jeng, 1995). The eubriine genera *Spineubria* and *Macroeubria* are in Asia also (Sâto, 1985; Lee & Yang, 1994a), as is *Homoeogenus* which was described in the larval stage by Lee & Yang (1993). An account of the immature stages of *Schinostethus* is given by Lee *et al.* (1993), but eubriine larvae are generally poorly known.

The taxonomy of the Eubriinae is in a state of flux. Of the genera recognized by Brown (1981a), *Cophaesthetus* is a junior synonym of *Schinostethus* (Lee *et al.*, 1993), and Sâto (1985) treats *Grammeubria* as a synonym of *Ectopria*. An additional genus — *Granuleubria* (= *Drupeus*) — has been described by Jäch & Lee (1994) on the basis of adult beetles from Pakistan, Nepal and India. Further changes involve a revision of the genus *Eubria* by Lee & Jäch (1996) leading to the transfer of some species of *Eubria* to *Microeubria* — a genus which occurs in Peninsular Malaysia, Thailand, Java and Nepal and was established by Lee & Yang (1994a). As a result, *Eubria* — which now includes a single European species — can no longer be considered part of the Asian fauna. More recently, Lee & Yang (1996) have revised the genus *Dicranopselaphus* — which is widespread in the Oriental Region from China through Southeast Asia (including Indonesia and the Philippines) to Nepal — describing 21 new species and reducing several species of Eubriinae (in *Ectopria* and *Grammeubria*, and the entire genus *Spineubria*) to junior synonyms or placing them in new combinations within *Dicranopselaphus*. Lee & Sâto (1996) have established a new genus — *Nipponeubria* — which is, so far, known only from Japan, but these authors give a useful comparison of the larval morphology of this genus and that of *Ectopria* — a genus which occurs in Taiwan and has been reviewed by Lee & Yang (1994b). While the situation may change subsequently, Table 4.8 lists the present generic composition of the Eubriinae — and other psephenid subfamilies — in tropical Asia.

Elmidae

The Elmidae (= Elminthidae or Helminthidae) are known as riffle beetles because of their preference for fast-flowing streams where they are associated with stones and submerged wood. The group is diverse (for example, at least nine genera and 20 species are known from Hong Kong alone), and there are approximately 1200 species in 130 genera described globally. Tropical streams appear to be richer in riffle beetles than equivalent temperate habitats, and variety declines with altitude (Brown, 1981a). Elmid larvae and most adults are aquatic whereas pupation takes place on land. There are usually seven larval instars, but the number varies between five and eight in different genera according to the size of the adult beetle (Brown, 1987). Intraspecific variation in adult size in response to stream temperature has been reported also, with smaller-than-normal individuals occurring in warmer streams (Shepard, 1992). After emergence, adults may disperse by flying (e.g. Jäch, 1997) before entering the water, whereupon the fight muscles apparently atrophy. There are two elmid subfamilies: the Elminae and the much less diverse Larinae. Many members of the subfamily Larinae are semiaquatic as adults, crawling on rocks and wood just above the water surface. Elmine adults are aquatic and remain permanently submerged after entering the water. They are quite long lived, surviving for several weeks or months, and may be present in streams for much of the year. Brown (1987) has summarised much of the available information on elmid life histories and ecology. Respiration by adults is accomplished with a plastron which is formed by hydrofuge hairs which cover most of the body and envelope the beetle in a layer of air. In addition, there is an air store beneath the elytra. Oxygen is replenished by diffusion from the water into the air held by the plastron.

Most elmids are rather small with an adult length that rarely exceeds 5 mm and, more usually, is only 2–3 mm (Fig. 4.119). Diets of both stages include periphyton and bacteria in addition to detritus. Neither stage can swim, and adults in particular have stout claws allowing them to maintain their purchase in fast current. Some genera (e.g. *Ancyronyx*; Jäch, 1993) have long, spider-like legs. Most elmids are associated with rocks and stones but certain genera (including *Ancyronyx* and *Graphelmis*) seem to prefer submerged wood. Larvae and adults are heavily sclerotized and the larval cuticle is covered with small tubercles. Colours are usually dull, but adults of some species have warning colouration (generally red, yellow or cream markings contrasting with a darker background) which advertises a chemical defense against fishes that seems to be concentrated on or within the elytra (White, 1989). This pigmentation can be used as an aid for the identification of adult elmids, but the shape and sculpturing of the thorax and elytra are more useful characteristics for the majority of dull-coloured species. The relative lengths of the legs and size of the tarsal claws can be helpful also, but cannot be substituted for extraction of the male genitalia and microscopic examination of the aedeagus.

Mature elmid larvae are usually less than 8 mm long, with an elongate body that is somewhat triangular in cross section and tapers towards the end (Figs 4.120 & 4.121). Each of the first five abdominal segments (at least) have a pair of ventral pleura (i.e. lateral areas

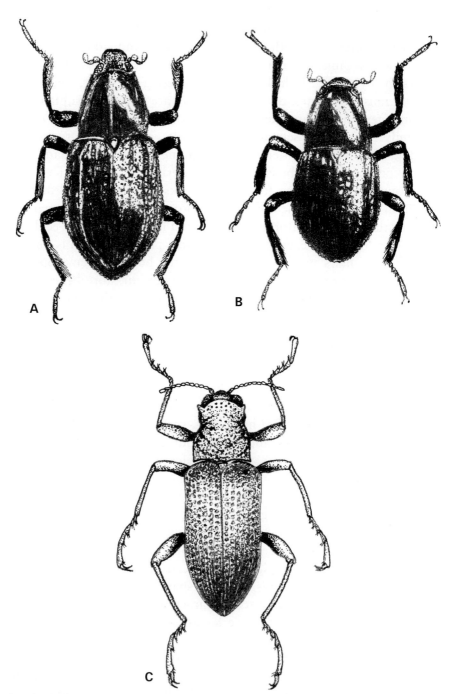

Fig. 4.119 Adult Elmidae (Coleoptera): A, *Aulacosolus*; B, *Nesonychus* (both redrawn from Jäch & Boukal, 1997a); C, *Leptelmis* (redrawn from Jeng & Yang, 1993).

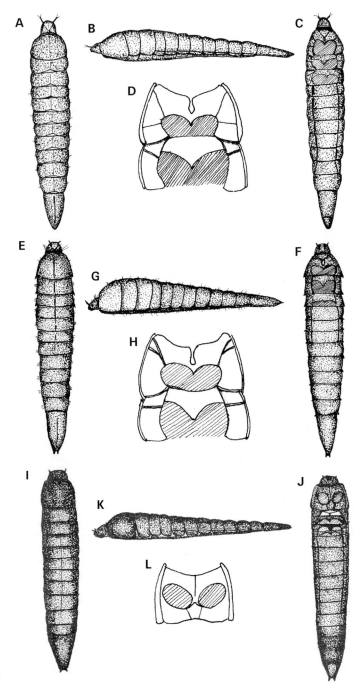

Fig. 4.120 Larval Elmidae (Coleoptera) showing dorsal (A, E, I), lateral (B, G, K) and ventral (C, F, J) views of three Hong Kong species which have not yet been associated with named adults. Diagrams of the arrangement of pleura on the prothoracic and mesothoracic sterna (after removal of the legs) are shown also (D, H, L).

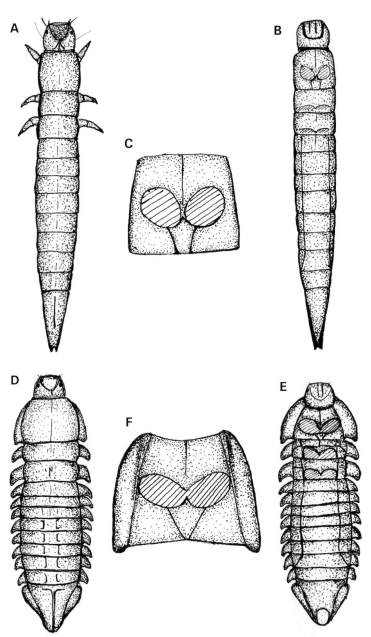

Fig. 4.121 Larval Elmidae (Coleoptera) showing dorsal (A, D) and ventral (B, E) views of two Hong Kong species which have not yet been associated with named adults. The arrangement of pleura on the prothoracic sterna (after removal of the legs) are shown also (C, F).

demarcated by sutures) as well as a tergum and sternum. Respiration is effected by a group of tracheal gills than can be retracted into an operculate chamber on the underside of abdominal segment IX. There is a pair of hooks or claws on the inner surface of this chamber. In larvae of some Asian genera (e.g. *Ilamelmis*) the lateral margins of the tergites are expanded, and may be more or less quadrate or rounded (Fig. 4.121D & E). This expansion may be associated with the development of carinae (dorsal longitudinal crests) on the tergites. The carinate tergites may be restricted to the posterior-most segment (as in *Ilamelmis*) or almost the entire body (as in *Taprobanelmis*, *Grouvellinus* and *Ancyronyx*). Bertrand (1973) has figured Sri Lankan larvae of *Ilamelmis*, *Podelmis*, *Potamophilinus* and *Taprobanelmis*, and (together with Bertrand, 1972) this is the most important source of information on elmid larvae in Asia. Moubayed (1983) has figured the larva of *Grouvellinus* from Lebanon — a genus which is widely distributed in the Oriental Region (Jäch, 1984b) — and Jäch (1994b) gives a very brief account of *Ancyronyx* larvae. Bertrand (1972) figures the larvae of some Holarctic elmids, including genera (e.g. *Stenelmis*, *Zaitzevia*) which occur in the Oriental Region.

The two elmid subfamilies differ greatly in terms their representation in tropical Asia. Brown (1981a; 1981b) lists three genera of Larinae as present in the Oriental Region: *Dryopomorphus*, *Potamophilinus* and *Potamophilus* (see also Delève, 1970; Jäch, 1985; Spangler, 1985). Brown (1981b) also gives a well-illustrated key to the world genera of adult Larinae, and establishes a new genus — *Parapotamophilus* — from New Guinea. The Oriental Elminae are much more diverse than the Larinae, and many genera are listed in the literature: *Aesobia*, *Ancyronyx*, *Austrolimnius* (New Guinea and the Moluccas only), *Cephalolimnius*, *Esolus*, *Graphelmis* (including some species ascribed to *Cylloepus*), *Grouvellinus*, *Hedyselmis*, *Ilamelmis*, *Leptelmis* (see Fig. 4.119C), *Limnius*, *Macronychoides*, *Macronychus*, *Paramacronychus*, *Podelmis*, *Pseudamophilus*, *Ohiya*, *Ordobrevia*, *Simsonia*, *Stenelmis*, *Taprobanelmis*, *Unguisaeta*, *Urumaelmis*, *Vietelmis*, *Zaitzevia* and *Zaiteviaria* (Delève, 1968; 1970; Bertrand, 1973; Brown, 1981a; Jäch, 1982; 1983; 1984a; 1984b; 1985; 1993; 1994b; Brown & Tobias, 1984; Jeng & Yang, 1991; 1993; Kodada, 1992; 1993; Jäch & Boukal, 1995a; Jäch & Kodada, 1995; Boukal, 1997). The list is incomplete because numerous genera of Elmidae in tropical Asia streams await description (Jäch & Boukal, 1995a). This is evident from the number of recent additions to the fauna of the region. For example, Jäch & Boukal (1997a) established two new genera based on adults: *Aulacosolus* and *Nesonychus* (Macronychini: see Fig. 4.119A & B). *Aulacosolus* is widely distributed in the Oriental Region, and seems to prefer shallow, slow-flowing streams. *Nesonychus* is known from running waters in Borneo, Sumatra and Java. Other recent discoveries of Elmidae include four new genera from China (*Cuspidevia*, *Eonychus*, *Jilanzhunychus* and *Sinonychus*; Jäch & Boukal, 1995a); *Homalosolus*, *Loxostirus* and *Rhopalonychus* from Borneo (Jäch & Kodada, 1996a); *Rudielmis* from southern India (Jäch & Boukal, 1995b); *Haraldia* and *Macronevia* (which includes some species previously placed in *Zaitzevia*) from Peninsular Malaysia (Jäch & Boukal, 1996c); the genus *Graphosolus* from the Philippines, Borneo and Java (Jäch & Kodada, 1996b); and, *Podonychus* and *Prionosolus* from Southeast Asia (Jäch & Boukal, 1997b). *Indosolus*, which is widespread from India through Malaysia to China, was redescribed recently by Jäch & Boukal (1995a) who elevated it from a subgenus of *Esolus* to full generic rank. In addition, a recent revision of *Austrolimnius* (Boukal, 1997) added a total of 22 new species to the genus. Descriptions of the larval stage are lacking for the majority of elmid genera in Asia, giving an indication of the amount of research on riffle beetles that is still outstanding. This is due, in part, to the difficulty of rearing elmid larvae in captivity.

Other Dryopoidea

The Dryopidae are inconspicuous reddish-brown or black beetles typically no more than 7 mm in body length. Adults cannot swim and are rather similar to elmids in general appearance. The family includes a significant proportion of species that is entirely terrestrial. However, adults of genera such as *Helichus* and *Ceradryops* live on or under wet or submerged rocks and wood in streams, and adult *Dryops* occur along stream margins sometimes in vegetation overhanging the water. Unlike elmids (to which the dryopids are closely allied) and most other aquatic beetles, the larvae appear to be terrestrial living in damp soil or rotting wood, but little ecological information on the group has been published and some *may* be fully aquatic. Dryopid larvae (see Bertrand, 1972) are reminiscent of elmids but have a more cylindrical body and a smooth tubercle-fee cuticle. In addition, abdominal segment IX lacks retractile tracheal gills, although there is a flat operculum on the ventral side. Eyes are made up of scattered groups of stemmate (simple eyes or ocelli: *cf.* a single group in elmids).

Brown (1981a) recorded seven Oriental genera in a global review of the Dryopidae: *Ceradryops, Dryops, Elmomorphus, Geoparnus, Helichus, Malaiseanus* and *Sostea* (see also Delève, 1970; 1973). An additional dryopid genus — *Uenodryops* (Sâto, 1981) — has been established since, based on adults from Nepal, and *Praehelichus* (formerly a subgenus of *Helichus*; Nelson, 1990) has been added from the northern fringe of the Oriental Region in China (Kodada & Jäch, 1995a). *Sostea* includes over half of the known species in the region and, when combined with *Helichus* and *Elmomorphus*, makes up 95% of the 1981 regional total. Additional research has added further species to *Dryops* (Olmi, 1986), *Helichus* (Kodada & Jäch, 1995b) and *Elmomorphus* (Jäch, 1985; Sâto, 1981; 1992), and some generic reassignment will probably be needed after thorough revision of the Old-World species of *Helichus* (Kodada & Jäch, 1995b). Significantly, Nelson (1989) considers that *Helichus* is restricted to North America and that Asian records of this genus are erroneous. He has raised several subgenera of *Helichus* to full genera (Nelson, 1990), and recognizes 12 species of *Pachyparnus* from the Oriental Region.

The Limnichidae is a family of small (0.5 — 5 mm body length) pubescent beetles represented by genera such as *Byrrhinus, Limnichus* and *Pelochares*. They are not truly aquatic, occurring in splash zones close to cascades and waterfalls and in the riparian zone of streams. Larvae resemble elmids and some may be terrestrial. The related Heteroceridae are also confined to riparian habitats. They live along banks of slow-flowing streams or marshes where they excavate galleries of tunnels in the mud just above the water line. Adults do not exceed 8 mm body length, and are flattened, rectangular beetles covered with a fine pubescence. The genus *Heterocerus* is widespread, occurring in China, Vietnam, Thailand, Burma, Cambodia and Nepal (Wu, 1932b; Mascagni, 1990; 1991; 1995a; 1995b); *Littorimus* is also known from the region (Mascagni, 1995a; 1995b). Heterocerid larvae are elongate, tapered towards the hind end, with a ten-segmented abdomen that lacks hooks, gills or an operculum. The body is clothed in dark, erect setae and the antennae are minute.

Lampyridae and other Polyphaga

A few terrestrial families of Polyphaga include a small number of aquatic or semiaquatic representatives. The most striking of these are the Lampyridae. Adult lampyrids — or fireflies — are luminescent, using light as a signal during courtship displays. Almost all lampyrid larvae are terrestrial with a predilection for moist microhabitats, but some (especially in the widespread genera *Luciola* and *Pyrophanes*) are semiaquatic and a few are benthic in streams (Annandale, 1900; Bertrand, 1973; Jäch, 1984a). Larvae are elongate, tapered at both ends, flattened and dorsally sclerotized. The head is small and retractable (and hence usually concealed when viewed from above) with sickle-shaped mandibles. The dorsal surface of the body consists of a series of medially-divided, sclerotized plates; the underside is generally pale and membraneous although sternal plates cover part of the abdominal segments. Some lampyrid larvae lack gills, while others have a pair of branched or unbranched lateral gill filaments on abdominal segments I-VII (Bertrand, 1973). There may be a weak light-emitting organ on the underside of the abdominal segment VIII. The larvae are predators upon snails which they paralyse with glandular secretions injected through their perforate mandibles. Digestion occurs outside the body and the larva imbibes liquified snail tissue.

Other beetles associated with freshwater habitats include certain Chrysomelidae (especially in the subfamily Donaciinae) and Curculionidae (e.g. *Bagous*) that live on floating and emergent plants (e.g. *Nuphar*, *Nymphaea*, *Sagittaria* and *Trapa*). In most cases, larvae feed inside the roots, stems and leaves of the aquatic host but do not come into contact with water and so are not obviously adapted for a submerged existence. These beetles are generally confined to lentic habitats and will not be considered here. Caldara & O'Brien (1995) provide a useful summary of the current state of knowledge in Asia. Some Staphylinidae are associated with streams, but do not enter the water. Instead they patrol the banks or forage over rock surfaces in search of insect prey. The slender body and extremely short elytra of these beetles distinguish them from most other semiaquatic Coleoptera. A shiny metallic or iridescent appearance is not uncommon. Members of the staphylinid subfamily Steninae stalk prey along the water margins but can zoom across the surface at high speed due to a chemical secreted by glands at the end of the abdomen. It acts as a surfactant and reduces the surface tension, pushing the beetle forward as it spreads rapidly over the surface (see also p. 354).

DIPTERA

Among the major orders of insects in streams, the Diptera (= true flies) have received the least attention from ecologists. Leaving aside certain biting flies such as the mosquitoes (Culicidae) and certain blackflies (Simuliidae) which have medical importance, a common approach has been to identify dipteran larvae to the level of family or subfamily and then treat groups within the higher taxon as being ecologically equivalent and/or taxonomically intractable. There are a number of reasons for this state of affairs. Firstly, most Diptera larvae have elongate bodies and all lack segmented legs. Accordingly, many of them will pass through a 500-μm or even finer mesh net of the type used to collect macroinvertebrates; however, this net will retain (say) a stonefly larvae of equivalent body length. In addition to the absence of legs, many dipterans are grub-like and do not have conspicuous external gills, cerci and the like. Even a sclerotized head capsule is lacking in certain groups. The shortage of conspicuous visual cues for separation of different taxa enhances the tendency of many ecologists to lump Diptera larvae into a few convenient categories, and to label them 'difficult'. Even where identification to genera has been accomplished (species separation is rarely possible), the inefficiency of routine sampling methods ensures that the abundance and biomass of Diptera — and hence their importance — is underestimated. An extra complication is that species identification depends upon examination of the terrestrial adults, and these are rarely associated with the larvae. While it will not be possible to rectify the paucity of ecological and taxonomic data herein, I will attempt to summarise relevant information on the major dipteran taxa that occur in tropical Asian streams so that the reader is made aware of the current state of knowledge.

There are two suborders of Diptera: Nematocera and Brachycera (Table 4.9). The latter includes taxa which were formerly referred to as the suborder Cyclorrhapha. Nematoceran larvae have a sclerotized head capsule (although this may be retracted into the body and difficult to see clearly). The mandibles — which have subapical teeth — move laterally. Prolegs are usually present. Brachyceran larvae lack a head capsule entirely or it is partially formed only. The mandibles have no subapical teeth (often resembling hooks) and move vertically. Prolegs are sometimes absent (or reduced to pseudopods or welts), and the body is often pale and maggot-like. Both suborders have aquatic representatives, but the majority of Brachycera are terrestrial and most aquatic members of the suborder pupate on land. The nematocerans are the most diverse aquatic dipterans and, within this suborder, the Chironomidae (or non-biting midges) are arguably the most species-rich and abundant group of stream macroinvertebrates. Our inadequate

Table 4.9   Families of aquatic (or semiaquatic) Diptera in the Oriental Region.

Suborder Nematocera	Suborder Brachycera
Tanyderidae	Tabanidae
Tipulidae	Athericidae
Blephariceridae	Stratiomyidae
Deuterophlebiidae	Empididae
Nymphomyiidae	Dolichopodidae
Psychodidae	Phoridae
Ptychopteridae	Sciomyzidae
Dixidae	Syrphidae
Culicidae	Ephydridae
Chaoboridae	Muscidae
Thaumaleidae	
Simuliidae	
Ceratopogonidae	
Chironomidae	

knowledge of the larval stages makes it impossible to estimate the number of species with any accuracy, but it is likely that chironomids comprise about 30% of all stream invertebrate species. Nematocerans are particularly diverse in streams, and some families (e.g. the Blepharoceridae, Deuterophlebiidae and Simuliidae) are confined to flowing waters. Others (e.g. the Culicidae) occur in almost all aquatic habitats, and some also inhabit damp soil, plant litter and other terrestrial environments (e.g. the Tipulidae and Ceratopogonidae). Delfinado & Hardy (1973) provide a useful catalogue to the Oriental genera of Nematocera that were recognized in 1970. The Brachycera has fewer aquatic species than the Nematocera, and most of them live in standing water. Again, the Oriental genera have been catalogued by Delfinado & Hardy (1975; 1977). Because of their relatively high diversity in running water, most attention will be focused on Nematocera larvae.

The following key can be used to distinguish the main taxa of larval Diptera likely to be encountered in streams. Keys to genera within families are not feasible with the current state of knowledge, but subfamilial identification for Chironomidae is both possible and desirable on ecological grounds (see below). The key characters for identification of subfamilies of Ceratopogonidae are taken from Elson-Harris (1990).

1. Body *appears* to be divided into seven segments; first six 'segments' with a conspicuous ventral sucker ....
.... **suborder Nematocera: Blephariceridae**

Body not as above (with more than seven segments and lacking ventral suckers) 2

2. Larvae with a sclerotized head capsule (although this may be retracted into the body and may not be fully sclerotized posteriorly); mandibles move laterally (i.e. in a horizontal plane) and have subapical teeth; usually abundant... **suborder Nematocera** 3

   Larvae lack a head capsule entirely or it is partially formed; mandibles move vertically and lack subapical teeth so that they are hook-like; prolegs sometimes (but not always) absent and body *may* be pale and maggot-like; generally much less abundant than Nematocera .... **suborder Brachycera** 16

3. Head capsule strengthened only by sclerotized rods; body segments secondarily segmented (i.e. divided up by annuli); prolegs absent; rare .... **Ceratopogonidae: Leptoconopinae**

   Head capsule well sclerotized or, if strengthened only with sclerotized rods, then body segments not secondarily segmented; prolegs present or absent 4

4. Head capsule with incomplete posterior margins and retracted into body cavity giving a truncated appearance to the anterior end of the body; prolegs usually lacking; often rather dark in colour and may be quite large (body length > 20 mm) .... **Tipulidae**

   Head capsule fully formed and not retracted into the body cavity; at least one proleg *usually* present 5

5. Prolegs absent although ventral suckers sometimes present; thoracic and abdominal segments divided up by one or more annuli; dorsal sclerites may be present; usually black or grey; relatively rare .... **Psychodidae**

   Prolegs present; no annuli or secondary segmentation 6

6. Prolegs on anterior and/or posterior segments only; common 7

   Prolegs on many body segments; relatively rare 15

7. A single proleg on the prothorax, and no apparent posterior proleg(s); body club-shaped, posterior end swollen with a ring of tiny hooks at the tip .... **Simuliidae**

   Posterior prolegs *usually* present; body not club-shaped or swollen posteriorly 8

8. Only posterior prolegs present 10

   Anterior and (usually) posterior prolegs present 9

9. Body segments bear either small spines or long processes ....
   .... **Ceratopogonidae: Forcipomyiinae**

   Body segments lack spines or processes; common and often
   abundant **12**

10. Three pairs of long filaments arising from the last two abdominal
    segments .... **Tanyderidae**

    Last two abdominal segments lacking filaments ....
    .... **Ceratopogonidae** **11**

11. Anal segment with retractile posterior prolegs; larval movements
    slow and cumbersome .... **Dasyheleinae**

    Anal segments with a circlet of bristles surrounding the anal
    opening; larvae usually swim with a quick, serpentine motion ...
    .... Ceratopogoninae

12. Posterior and anterior prolegs paired (although the anterior pair
    may be fused partially); widespread and abundant ....
    .... **Chironomidae** **13**

    Prolegs single (never paired); prothorax with dorsolateral spiracles
    on short stalks; thoracic and abdominal segments with dorsal
    sclerotized plates; found in rocky streams but never abundant ...
    .... **Thaumaleidae**

13. Anterior and posterior prolegs strongly developed; head capsule
    elongated and conspicuously longer than wide; prothorax often
    rather swollen and wider than the head or succeeding segments; a
    single eyespot (ocellus) on either side of the head; antennae retractile
    into head capsule; lacking a broad, toothed plate on the underside
    of the head .... **Tanypodinae**

    Anterior and posterior prolegs less-strongly developed (and the
    posterior pair may be greatly reduced or lacking almost entirely);
    head capsule not markedly longer than wide and not especially
    elongated; prothorax not usually swollen or wider than the
    succeeding segments; *may* have two ocelli situated close together
    on either side of the head; antennae not retractile; a broad, toothed
    plate (the hypostoma) almost always present on the ventral anterior

margin of the head **14**

14. With two ocelli of similar size on either side of the head situated so that the lower ocellus lies immediately below or slightly behind the other; some genera have one or two pairs of 'blood gills' on abdominal segment VIII .... **Chironominae**

    Either with one ocellus on either side of the head *or* with two ocelli of different sizes in which case the anterior one is the smaller of the two; without 'blood gills' on abdominal segment VIII; the posterior prolegs are *sometimes* lacking almost entirely ....
    .... **Orthocladiinae**

15. Eight pairs of slender prolegs projecting ventrally; body somewhat compressed laterally, minute (< 3 mm long) and delicate; antennae short and rather inconspicuous .... **Nymphomyiidae**

    Seven pairs of stout prolegs projecting ventrolaterally; body rather flat and robust; long, branched antennae .... **Deuterophlebiidae**

16. Head capsule partially formed and visible; body flattened, broadened anteriorly with a circlet of long setae at the tip; exoskeleton hardened with calcium carbonate giving it a rough, granular appearance .... **Stratiomyidae**

    Head enclosed within the thorax and poorly developed; body rather soft (without calcium carbonate deposits) and *usually* broadened posteriorly, lacking a posterior ring of setae **17**

17. Larvae with an extendable, posterior respiratory tube which is well over half the length of the body; anterior end of the body rather blunt .... **Syrphidae**

    Posterior respiratory tube lacking or, if present, it is short (less than one third of the body length) and not extendable; anterior end of the body somewhat tapered **18**

18. Body ends in a short respiratory tube that is divided at the apex
    .... **Ephydridae**

    Respiratory tube lacking **19**

19. Abdomen without distinct prolegs but with a girdle of at least six pseudopods around each segment .... **Tabanidae**

    Prolegs may be present or absent, but girdles of pseudopods are lacking **20**

20. Body with ventral prolegs, and terminating in a pair of caudal, filamentous projections fringed with setae; the body surface is covered with many finger-like prolongations .... **Athericidae**

    Body not as above; prolegs present or absent **21**

21. Seven or eight pairs of prolegs present, the posterior-most pair being longer than the anterior ones .... **Empididae**

    Abdomen may have ventral pseudopods but lacks prolegs; relatively rare in streams .... **Muscidae**

The major Nematoceran families are the Ceratopogonidae, Chaoboridae, Chironomidae, Culicidae, Deuterophlebiidae, Dixidae, Psychodidae, Simuliidae and Tipulidae (Table 4.9). The Chaoboridae and Culicidae (the mosquitoes) are neither primarily stream-dwelling nor benthic (although the latter family is extremely diverse), whereas the Ceratopogonidae, Dixidae and Psychodidae have relatively limited representation in streams or are confined to particular microhabitats. However, the Simuliidae, Tipulidae and especially the Chironomidae are important components of stream benthic communities. The Blephariceridae and Deuterophlebiidae are rithrobionts but are hardly ever abundant, and the Nymphomyiidae, Tanyderidae and Thaumaleidae — although primarily lotic — are likewise uncommon. Certain small families have been omitted. For example, the Ptychopteridae, represented by only two genera (*Bittacomorphella* and *Ptychoptera*) in tropical Asia, are not considered herein. Their larvae are semiaquatic or aquatic in stagnant or slow-flowing water. Knowledge of the ecology of most nematocerans is extremely limited, and usually confined to preliminary information on diet or feeding mode. Most Nematocera have four larval instars, and the majority of aquatic families pupate in water. Life-history investigations are in their infancy, but it is probably a safe assumption that the majority of small Nematocera complete several generations each year. Larger nematocerans, such as some tipulid larvae, are likely to be univoltine or bivoltine.

Nematocera: Tipulidae and Tanyderidae

Tipulid or cranefly larvae differ from other Nematocera in that the head capsule is usually retracted into the thorax and is therefore not apparent at first glance. The posterior part of the head is weakly sclerotized, but the mandibles and front of the head are typical of the suborder. The body is cylindrical and elongated, often clothed with short pubescence which sometimes resembles a brown pelt so that the larvae look like furry cigars. Body lengths range from around 5 to 60 mm. Although the Tipulidae is one of the largest dipteran families (approximately 15,000 described species world-wide), most genera are not aquatic and the habits of many of the Oriental species catalogued in Delfinado & Hardy (1973) are obscure.

Lotic tipulids often have complicated respiratory organs associated with a pair of functional spiracles at the end of the abdomen. However, these structures (the spiracular disc and its associated lobes) can be retracted, and are not always conspicuous. *Antocha*, and a few other genera, are exceptional among the Tipulidae in that functional spiracles are absent (i.e. they are hydropneustic). These larvae live entirely submerged, but most other aquatic tipulids must come to the surface for oxygen (i.e. they are aeropneustic), and usually move to land or dry stream margins to pupate. Cutaneous respiration may be important for some aeropneustic larvae, and a few genera (e.g. *Pseudolimnophila*) have elongated and membranous anal gills. Tipulid larvae lack prolegs, but some have creeping welts on the underside of the body that are used for crawling over rocks. However, the majority of taxa is associated with detritus or burrow in sand. The group is very incompletely known in tropical Asia (especially in the immature stages), and would provide a fertile research area.

Aquatic genera of Tipulidae include *Antocha* (which is rheophilic: see Fuller & Hynes, 1987), *Dicranota*, *Dicranomyia*, *Elliptera*, *Gonomyia* (probably confined to water margins), *Helius*, *Hesperoconopa*, *Hexatoma* (including *Eriocera* which some workers have treated as a separate genus), *Erioptera*, *Limnophila*, *Limonia*, *Lipsothrix*, *Molophilus*, *Orimarga*, *Ormosia*, *Paradelphomyia*, *Pedicia*, *Pilaria*, *Polymera*, *Pseudolimnophila*, *Rhabdomastix*, *Trentepohlia* (which may be confined to water-filled leaf axils) and *Thaumastoptera* — all in the subfamily Limoniinae — as well as *Ctenacroscelis*, *Dolichopeza*, *Holorusia*, *Leptotarsus*, *Megistocera*, *Nephrotoma*, *Tipula* and *Tipulodina* — in the Tipulinae. An Oriental record of the Palaearctic tipuline genus *Prionocera* cited by Alexander (1931) is undoubtedly incorrect (see, for example, Brodo, 1987). At least two genera of Cylindrotominae — *Phalacrocera* and *Triogma* — are present also. *Limonia* and *Tipula* have several subgenera and are the major tipulid genera in the region, although not all species live in streams.

The most important single reference concerning Oriental tipulid larva is Alexander (1931), but Yoon & Kim (1992) provide an illustrated key to the larvae of some Korean crane flies. Byers (1996) gives a key to Palaearctic genera which may be helpful for the identification of widely-distributed genera (of which there are at least 30). Unfortunately, there is a paucity of information on the biology of larval Tipulidae in general and the Limoniinae — which is the most diverse subfamily — in particular (Pritchard, 1983). Ecologies vary widely (see Pritchard, 1983) and, even within a genus (e.g. *Tipula*), different species may be strictly aquatic or entirely terrestrial. A variety of feeding modes (except filtering) is represented in the family. Members of some genera — especially in the Limoniinae — are quite large and feed by engulfing other invertebrates. Table 4.10 gives a summary of the larval habits of some tropical Asian genera.

The Tanyderidae are sometimes referred to as primitive craneflies and, although lacking prolegs, they resemble chironomids more closely than they look like tipulids. The elongate body is approximately 20 mm long and the most distinctive feature is three pairs of long caudal

Table 4.10 The larval habits of aquatic tipulid genera recorded from tropical Asia. Data derived mainly from Alexander (1931) and Byers (1996).

Taxa	Habitat	Habit	Feeding behaviour
Tipulinae			
Ctenacroscelis	Lotic (among moss and algae on rock faces)	clinger/sprawler	?
Dolichopeza	Aquatic (among moss and algae on wet faces)	sprawler/? burrower?	
Holorusia	Water margins (in detritus & silt)	burrower	detritivore/ shredder
Leptotarsus	Small streams (depositional areas)	burrower	?
Megistocera	Lentic margins (among macrophytes)	neustic	chewing herbivore
Nephrotoma	Aquatic (among moss algae on wet surfaces)	sprawler/ burrower?	?
Tipula	Lotic depositional (also lentic and semiaquatic)	burrower	variable (shredder, collector, and scraper; predaceous?)
Tipulodina	Lotic and lentic (various including margins)	burrower	?

Table 4.10   *(cont.)*

Cylindrotominae

Phalacrocera	Lentic margins (among macrophytes)	burrower/ sprawler	chewing herbivore/ shredder
Triogma	Water margins (semiaquatic)	burrower/ sprawler	herbivore?

Limoniinae

Antocha	Lotic (rheophilic)	clings onto rocks within silken tubes	collector-gatherer
Dicranota	Lotic (also lentic and semiaquatic)	burrower/ sprawler	predator (engulfer)
Dicranomyia	Lotic (among moss and algae on rock faces)	clinger/burrower	? chewing herbivore/ shredder
Elliptera	Lotic and semiquatic	burrower	?
Erioptera	Water margins (semiaquatic)	burrower	collector-gatherer
Gonomyia	Water margins (semiaquatic)	burrower	?
Helius	Water margins (semiaquatic)	burrower	?
Hesperoconopa	Lotic (in sand)	burrower	?
Hexatoma	Lotic and lentic (among detritus and moss)	burrower/ sprawler	predator (engulfer)
Limnophila	Lotic and lentic (margins and fine sediments)	burrower	predator (engulfer)
Limonia	Lotic and lentic (mainly margins; semiaquatic)	burrower/ sprawler	chewing herbivore/ shredder
Lipsothrix	Lotic (in wood)	burrower	xylophagous shredder
Molophilus	Water margins (semiaquatic)	burrower	?
Orimarga	Water margins (in wood; semiaquatic)	burrower	? xylphagous shredder
Ormosia	Water margins (among detritus; semiaqautic)	burrower	? collector-gatherer
Paradelphomyia	Water margins	burrower	?
Pedicia	Water margins (among detritus; semiaquatic)	burrower	predator (engulfer)
Pilaria	Water margins	burrower	? predator (engulfer)
Polymera	Lotic margins (among detritus)	burrower	?
Pseudolimnophila	Water margins	burrower	?
Rhabdomastix	Lotic (in sand)	burrower	?
Trentepohlia	Water-filled leaf axils	sprawler	?
Thaumastoptera	Lotic (in fine sediment)	burrower	? predator (engulfer)

filaments which arise from the last two body segments. Although rare, these animals — represented by *Protanyderus* in tropical Asia — can be locally common in sandy streams where they burrow in the bottom sediments. Almost nothing is known about their ecology.

Nematocera: Blephariceridae, Deuterophlebiidae and Nymphomyiidae

Blepharicerid larvae are unmistakable: the body is flattened and divided into seven apparent 'segments', with a sucker disc on the underside of the first six of them. These discs allow the larvae to maintain a hold on substrates in fast current where they scrape diatoms and other algae from rock surfaces. Genera found in the Oriental Region include *Blepharicera* (apparently widespread from Japan and Taiwan to Fukien Province and Hainan Island in China, and through Thailand, Peninsular Malaysia and India), as well as *Diotopsis*, *Horaia* (= *Manaliella*) and *Philorus* from northern India (Delfinado & Hardy, 1973; Zwick, 1990b; 1991) and *Hammatorrhina* from Sri Lanka (Fernando, 1964; Delfinado & Hardy, 1973). *Agathon* and *Bibiocephala* (= *Amika*) occur in Palaearctic Asia (Zwick, 1990b), but do not extend further south. *Curupirina* is known from New Guinea only, but the closely-related genus *Apistomyia* is more widespread (Zwick & Hortle, 1989). While research on blepharicerid larvae is limited, it appears that *Blepharicera* can be separated from other Oriental genera by the possession of seven (rather than five) gills per tuft, and an extended membraneous section between their two antennal segments (Zwick & Hortle, 1989).

Like blepharicerids, deuterophlebiid larvae (mountain midges) have a rather specialized morphology that is associated with living on rock surfaces in swift currents (generally over 1 m s^{-1}) at high altitudes. *Deuterophlebia* is the only genus in the family, and most species are Holarctic (Courtney, 1994a). They are small (no more than 6 mm long), rather flattened insects with seven pairs of stout, eversible, crotchet-tipped prolegs on the underside of the abdomen. These prolegs, combined with the possession of long branched antennae, set deuterophlebiids apart from other Diptera. The original description of the family Deuterophlebiidae was based on specimens collected from Asia (Kashmir) in the 1920s. These animals have received little attention since then, although Courtney (1994a) has recently described some new species from the region. It is likely that larvae feed on algae scraped from the rocks in fast current, but '. . . there is a significant lack of biological data about these flies . . .' (Courtney, 1990). Courtney (1991) gives a useful account of deuterophlebiid ecology, noting that a univoltine life history and asynchronous growth seems typical of *Deuterophlebia* (see also Yie, 1933).

The Nymphomyiidae is a small family, and only seven species within the genus *Nymphomyia* (= *Felicitomyia* and *Palaeodipteron*) are known worldwide. Courtney (1994b) has reviewed the group, reporting one species (*Nymphomyia holoptica*) from Hong Kong, another from the Himalaya, and three from Palaearctic East Asia. The larvae are minute (often less than 2 mm long), slender insects that can be locally abundant in some streams. They are easily recognized by the presence of paired, crotchet-tipped prolegs on the underside of abdominal segments I-VII and IX. Nymphomyiid larvae are associated with rock surfaces, particulary where they are covered with mosses such as *Amblystegium*, *Fontinalis* or *Rhynchostegium*, although Takemon & Tanida (1994) consider that the interstices among pebbles and cobbles is also a likely microhabitat. The larvae are collector-gatherers or grazers, and most species complete at least two generations each year (Courtney, 1994b). Apterous adults *in copula* are often encountered under water (Courtney, 1994b; Takemon & Tanida, 1994) but this follows a winged terrestrial stage, and the mechanism by which their wings are lost is mysterious.

Nematocera: Simuliidae

Simuliid or blackfly larvae have a distinctive club-like body shape (Fig. 4.122), and have been dubbed '. . . little snub-nosed being(s) . . .' by Crosskey (1990: p. 3). The posterior is swollen and attached to the substrate by a ring of tiny hooks. (These actually represent a terminal proleg which is not demarcated from the body.) There is a single ventral proleg on the thorax, and the large head capsule bears a pair of hemispherical labral fans (= cephalic fans) that are used to strain particles of food from the current. A mucosubstance of uncertain composition that is secreted by glands associated with the mouthparts coats the fan rays and their fringes of microtrichia, and facilitates the retention of fine particles including those of colloidal size. Filter-feeding is the main mode of nutrition for most larval simuliids, but it may be supplemented by scraping algae from the surroundings and even by carnivory. Coprophagy is common in dense larval populations. Adult females of many species in the genus *Simulium* are bloodsucking ectoparasites of

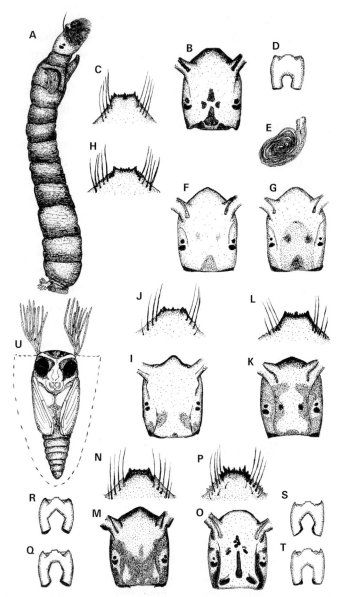

Fig. 4.122 *Simulium* larvae (Diptera: Nematocera: Simuliidae) from Hong Kong: A, lateral view of *Simulium* sp. 1 (with cephalic fans folded); B, dorsal view of head capsule; C, hypostomium (= anteroventral margin of the head capsule); D, postgenal cleft (in the posteroventral margin of the head capsule); E, wing 'pad'; F, G, H & Q, head-capsule markings (two specimens illustrated to show variation), hypostomium and postgenal cleft of *Simulium* sp. 2; I & J, head-capsule markings and hypostomium of *Simulium* sp. 3; K, L & R, head-capsule markings, hypostomium and postgenal cleft of *Simulium* sp. 4; M, N, S, head-capsule markings, hypostomium and postgenal cleft of *Simulium* sp. 5; O, P & T, head-capsule markings, hypostomium and postgenal cleft of *Simulium* sp. 6; U, pupa of *Simulium* sp. 6 (ventral view).

birds and mammals. Some are vectors of human diseases and are of significant veterinary importance also. As a result, biting adults have social and economic impacts in the vicinity of streams where larvae are abundant. For this reason their ecology (especially in the adult stage) has received a great deal of attention.

The natural history of simuliids has been well reviewed by Crosskey (1990), and that volume is essential reading for students of the group. Simuliid larvae are confined to running waters. They tend to congregate on rock surfaces or on the leaves of trailing vegetation in areas of swift laminar flow where the cephalic fans function most efficiently. Larvae sometimes attach to the bodies of other arthropods (e.g. Hora, 1930) and, in Africa at least, may be epizootic on freshwater crabs and (with lesser frequency) on atyid shrimps. Unlike other Nematocera which have only four larval instars, simuliids may have between six and nine. Pupation occurs under water in the neighbourhood of the larval feeding site. The pupae is sheathed in a cocoon, and has a pair of branched thoracic gills (Fig. 112U) projecting anteriorly (although in nature they trail behind the pupae which are typically orientated with the head facing downstream). Tropical simuliids are multivoltine and with larval development taking around one month, although up to 16 generations per year may be possible in warm lowland rivers (Crosskey, 1990).

The Simuliidae contains around 1,600 described species. The most diverse and widespread genus is *Simulium* (with various subgenera). The fauna of tropical Asia is incompletely known, but almost all strictly Oriental simuliids can be referred to *Simulium*, with the majority of species belonging to the subgenera *Simulium* (around 100 species and about half of the regional fauna) or *Gomphostilbia* (over 40 species). *Morops* and *Nevermannia* each contain around 20 Oriental species, and a few more are known in the subgenera *Byssodon*, *Eusimulium*, *Himalayum*, *Montisimulium* and *Wallacellum* (Crosskey, 1990 and references therein; see also Smart & Clifford, 1969; Datta, 1983; Takaoka & Suzuki, 1984; Takaoka & Davies, 1995: pp. 159–168). New species continue to be described from the region (e.g. Takaoka & Hadi, 1991; Zhang & Wang, 1991a; 1991b; Davies & Györkös, 1992; Takaoka & Sigit, 1992; Takaoka & Davies, 1995; 1996; Takaoka *et al.*, 1995) and, among them, Takaoka & Davies (1995) recently discovered the genus *Sulcicnephia* in Peninsular Malaysia. These larvae differ from *Simulium* (Figs 4.122; see also Smart & Clifford, 1969; Zhang & Wang, 1991b) in the form of the hypostomium (which has reduced teeth) and the deep postgenal cleft.

Nematocera: Ceratopogonidae

The larvae of the Ceratopogonidae (= Heleidae) or biting midges inhabit a wide range of aquatic habitats and moist terrestrial environments. The adults of some ceratopogonids are bloodsucking ectoparasites of vertebrates, and may serve as vectors for human and animal diseases. Their biology has been reviewed by Kettle (1977). There are four subfamilies: Ceratopogoninae, Dasyheleinae, Forcipomyiinae and Leptoconopinae (Wirth *et al.*, 1974). Ceratopogonine larvae are markedly vermiform, and larvae lack prolegs. They rarely exceed 10 mm in length and many are no more than 5 mm long. The head capsule is strongly prognathous and usually well-sclerotized (although there are a few exceptions). The mouthparts are adapted for predation: chironomid larvae are a major food item (Wirth & Grogan, 1988). Body setae are inconspicuous but there is often a group of setae at tip of the terminal segment. Larvae of the Ceratopogininae usually frequent the open-beach like margins of rivers and large streams, but others are associated with algal mats or damp soil at stream edges and are sometimes included in benthic samples. Even within a genus, there is considerable variation in larval habits: *Culicoides*, for example, occurs along a variety of water margins, rotting plant material and animal dung (Howarth, 1985). Although the immature stages and habits of the majority of species are unknown, genera reported as adults from the Oriental Region include *Allohelea, Alluaudomyia, Brachypogon, Bezzia, Camptopterohelea, Clinohelea, Culicoides, Downeshelea, Echinohelea, Johannsenomyia, Mallochohelea, Monohelea, Nannohelea, Nilobezzia, Paeabezzia, Palpomyia, Pseudostilobezzia, Serromyia, Sphaeromias* and *Stilobezzia* (Wirth & Grogan, 1988). Several of these genera are very large: *Culicoides*, for example, includes 168 Southeast Asian species (Wirth & Hubert, 1989). Descriptions and a key to the immature stages of some Indonesian Ceratopogoninae are given by Mayer (1934), while Elson-Harris (1990) provides a key to known larvae of Australian Ceratopogoninae which includes some genera found in Asia.

The Forcipomyiinae contains only two speciose genera: *Atrichopogon* (including *Dolichohelea*: Wirth & Ratanaworabhan, 1993) and *Forcipomyia*. The bodies of forcipomyiine larvae bear spines or elongate processes (the number and arrangement of which are important characters in species determination), and the prolegs bear hook-like crotchets at the tip. The head is hypognathous. Typical *Atrichopogon* larvae are more flattened than *Forcipomyia* (which have a cylindrical body) owing to some lateral extension of the thoracic and abdominal segments. Habits are varied, and larvae may occur in streams but are also found along water margins and may be associated with aquatic macrophytes or algal mats. Some *Forcipomyia* are terrestrial. Aquatic larvae are probably scrapers and collector-gatherers. Life-history studies suggest rapid growth, and the immature stages of some Forcipomyiinae can be completed in two to three weeks which is the most rapid of any Ceratopogonidae (Chan & Linley, 1989). The Leptoconopinae has only one genus in tropical Asia: *Leptoconops*. The head capsule is prognathous but incompletely sclerotized, and most of the body segments (which lack setae) are secondarily divided. Larvae are probably restricted to lentic water margins and beaches, and are unlikely to be encountered in streams. The Dasyheleinae likewise consists of the single genus *Dasyhelea*. Larvae, which are herbivorous and semiaquatic, are elongated with a prognathous head. There is a ring of curved hooks and elongate papillae around the anus on the terminal segment. *Dasyhelea* may be encountered along stream margins, as well as in a variety of other situations (Mayer, 1934).

Nematocera: Chironomidae

The Chironomidae — or non-biting midges — are abundant in all freshwaters and are of great ecological importance. Globally there may be around 20,000 species, and a total of at least 3,000 species in the Oriental Region seems reasonable but rather conservative. 'Chironomid larvae are responsible for the turnover of large quantities of many kinds of food materials . . .' and '. . . nearly every macroinvertebrate predator and most vertebrate predators, in most aquatic systems, utilize chironomid larvae (and, to a lesser extent, pupae and adults) at some point in their life cycle . . .' (Coffman, 1995; p. 446). However, because adult chironomids drink only water or nectar and are not blood-feeders, they have not received the same attention as those Nematocera (e.g. Ceratopogonidae, Culicidae and Simuliidae) with biting adults. Larvae (Fig. 4.123) are elongate, cylindrical and slender, and are rarely over 10 mm long although a few species reach 30 mm. There is a well-developed, sclerotized head capsule, paired anterior and posterior prolegs bearing apical hooks, and (usually) anal tubules and tufts of anal setae on the terminal segments. There are no tubercles or projections elsewhere on the body segments. Sometimes the posterior prolegs and setae and/or tubules are lacking, and the anterior pair of prolegs may be fused.

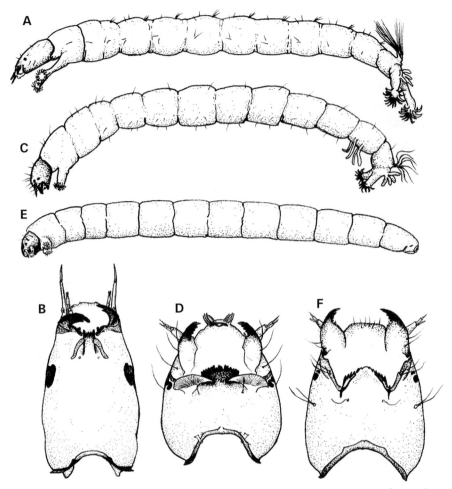

Fig. 4.123 Chironomidae (Diptera: Nematocera) larvae (mainly redrawn from Oliver & Roussel, 1983): A & B, Tanypodinae (lateral view of larva and ventral view of head capsule); C & D, Chironominae (lateral view of larva and ventral view of head capsule); E & F, Orthocladiinae (lateral view of larva and ventral view of head capsule).

The key on pp. 486–490 includes only the three most important chironomid subfamilies (Tanypodinae, Orthocladiinae and Chironominae: Fig. 4.123) but they comprise the vast majority (at least 95%) of species in the Oriental Region; Cranston (1995c) gives a more complete (and more complex) key to subfamilies. The cold-adapted Diamesinae occur in northern China and the Himalaya (e.g. Ashe *et al.*, 1987; Wang & Zheng, 1993), and so may be encountered in mountain streams along the boundary of the Oriental and Palaearctic Regions. Larvae of some diamesine genera are omnivorous or largely predatory (e.g. *Pseudodiamesa*, and *Protanypus* which appears to be confined to mountain lakes). The Usambaromyiinae, Chilenomyiinae and Aphroteniinae do not occur in tropical Asia. The Telmatogetoniinae are exclusively marine. The Buchonomyiinae, Podonominae and Prodiamesinae are rare in streams (e.g. Chaudhuri & Ghosh, 1981) and are of little ecological importance; together they make up only about 5% of the chironomid genera (and a smaller proportion of the species) recorded from the Oriental Region.

The feeding habits of larval chironomids vary widely, from predation through grazing, detritivory, wood-boring, leaf-mining and filter-feeding. A few are ectoparasitic (Kobayashi, 1995), and some species are facultative or obligate ectosymbionts or commensals of other aquatic insects including ephemerid mayflies, odonate larvae and perlid stoneflies (Parker & Voshell, 1979; Matěna & Soldán, 1986; Bottorff & Knight, 1987; Dudgeon, 1989f; Tokeshi, 1993 and references therein). Predation by chironomids is frequently associated with a free-living lifestyle, but most chironomids build tubes or cases of fine particles bound together by salivary secretions. Herbivores and detritivores usually collect food from the area around their cases, and filter-feeders construct a web or a net that strains water passing through or over their retreats. Subfamily provides a guide to chironomid larval ecology: the Tanypodinae (roughly 20% of chironomid species) are free-living predators (although the diet may be supplemented by algae and detritus), while the Chironominae (approximately 50%) and Orthocladiinae (around 25%) usually construct some sort of case or tube and are rarely predaceous (Berg, 1995). Tanypodine larvae occur in both flowing and standing waters, but some genera (e.g. *Rheopelopia* and *Nilotanypus*) are strongly rheophilic. Orthoclad larvae are well represented in streams. They are frequently rather specialized in terms of diet and/or microhabitat including larvae that are ectosymbiotic or phoretic (e.g. *Epoicocladius* and *Nanocladius*) or ectoparasitic (e.g. *Cardiocladius*) on other insects. The Chironominae are diverse in tropical latitudes but much of this diversity seems to be associated with standing waters. Unfortunately, information on the habitat preferences of most Asian genera are in short supply but some (e.g. filter-feeding *Rheotanytarsus* and *Sublettea*) are confined to streams. *Harnischia* and related genera (*Cryptochironomus*, *Demicryptochironomous* and *Paracladopelma*) are probably predators (Ashe *et al.*, 1987), and *Xenochironomus* burrows into and feeds on freshwater sponges. Notwithstanding, carnivory is not typical of the majority of Chironominae. Life-history data on chironomids from tropical latitudes are sparse. A review of the global literature (Tokeshi, 1995a) suggests that multivoltine species are abundant in warmer regions, and Masheshwari (1989) reports generation times of less than one month for *Tanytarsus ipei* in India. However, even in temperate latitudes, larval development is rapid (in the order of two to three weeks: Hauer & Benke, 1991). Such findings have important implications for the contribution of chironomids to secondary production in streams (see also Tokeshi, 1995b).

As the wide range of feeding modes suggests, chironomids occupy all types of freshwater habitat, and some representatives (in, for example, the *Chironomus* and *Polypedilum* species-complexes) possess haemoglobin which allows them to tolerate the low oxygen concentrations associated with organic pollution (see p. 200) or the burrowing habit (as in wood-boring *Stenochironomus*). The red tints imparted by haemoglobin can be used as a means of recognizing certain Chironominae larvae, since the pigment is largely confined to some members of this subfamily (it occurs in a few Tanypodinae and one Palaearctic genus of Orthocladiinae also). The biology of the Chironomidae has been reviewed by Oliver (1971) and Pinder (1986), and Hudson (1987) gives a synopsis of some more unusual larval life styles. Two recent books (Armitage *et al.*, 1995; Cranston, 1995a) discuss many aspects of chironomid natural history. Armitage *et al.* (1995), in particular, provides valuable reviews of many aspects of chironomid biology, and this volume is an essential starting point for students of the group. Interestingly, these books (with a combined total of 1054 printed pages) say virtually nothing about the ecology of Oriental (or, indeed, tropical) chironomids, and the fauna of this region is described as '. . . poorly known' (Cranston, 1995b: p. 65). Indeed, reiterating the view of Ashe *et al.* (1987), Saether & Wang (1993: p. 185) state that ' . . . with respect to chironomids the Oriental Region is probably the least investigated of all zoogeographical regions . . .'.

Inventories of Oriental chironomids have been prepared for China (Wang & Zheng, 1993), India (Chaudhuri & Guha, 1987; Coffman *et al.*, 1988; Roback & Coffman, 1989), Thailand (Moubayed, 1989; 1990 and references therein), and Sulawesi (Ashe, 1990). The small proportion of the Chinese fauna that is known (expected to comprise around 1200 species in total; Wang, S., 1987) consists mainly of Palaearctic taxa (see Wang & Zheng, 1993) reflecting a scarcity of collections from south of the Chang Jiang. There is no doubt that many, perhaps most, species of tropical Asian Chironomidae are still undescribed. This is demonstrated by the rate at which new species are reported in the recent literature: for example, from India (Bhattacharyay & Chaudhuri, 1990; Chaudhuri & Datta, 1991; Chattopadhyay & Chaudhuri, 1991; 1993; Chaudhuri *et al.*, 1992; Datta *et al.*, 1992a; 1992b; 1992c; 1996a; 1996b; Bhattacharyay *et al.*, 1991; 1993; Datta & Chaudhuri, 1995), China (Reiss, 1988; Wang, S., 1990; Wang & Zheng, 1990a; 1990b; 1990c; 1990d; 1991; 1990f; Saether & Wang, 1992; 1993; Wang & Saether, 1993a; 1993b; 1993c; Wang, 1995), Sumatra (Kikuchi & Sasa, 1990), and Sulawesi (Ashe & O'Connor, 1995; Murray, 1995). Although the vast majority of such papers deal with adults only, there are a few which describe larval stages (e.g. Yan & Ye, 1977; Matëna & Soldán, 1986; Chaudhuri & Chattopadhyay, 1988; Wang, S., 1990; Chaudhuri *et al.*, 1992; Murray, 1995).

Table 4.11, which lists the Oriental genera of freshwater Chironomidae, is based on an analysis of the global distribution of chironomids by Ashe *et al.* (1987) but updates it with information from more recent publications (e.g. Ashe, 1990). In particular, records of genera not previously known from Asia have been added: *Epoicocladius* (phoretic on ephemerid mayflies) from Vietnam (Matëna & Soldán, 1986); *Kloosia, Tokyobrilla, Pseudobrilla* and *Antillocladius* from Oriental China (Reiss, 1988; Saether & Wang, 1992; Wang & Saether, 1993b); *Thalassosmittia* from Xizang (Wang & Saether, 1993a; all other species in this genus are marine); *Demicryptochironomus* and *Kribiocosmus* from India (Datta & Chaudhuri, 1995; Datta *et al.*, 1996a); and, *Cryptotendipes* and *Pseudosmittia* from Sumatra (Kikuchi & Sasa, 1990). Other additions comprise new genera described from tropical Asia in recent years, such as *Xiaomyia, Shangomyia* and *Zhouomyia* from Oriental China (Saether & Wang, 1993), and *Asachironomus, Sumatendipes* and *Tobachironomus* from Sumatra (Kikuchi & Sasa, 1990). Chironomid genera with (at least) some species which occur in streams are highlighted in Table 4.11, but information on the habitat occupancy of many genera is incomplete or lacking entirely.

To date, a key to the tropical Asian genera of chironomid larvae is lacking. Sasa & Kikuchi (1995) provide a key to the adult males of the Japanese chironomids and, although this is an essentially Palaearctic fauna of over 700 species, many of the genera illustrated inhabit the Oriental Region. Unfortunately, larvae receive brief consideration only. Wang & Zheng (1993: p. 247) caution that a ' . . . high percentage . . .' of published work dealing with Chinese chironomid larvae '. . . are dubious determinations and require further study . . .'. Keys to chironomid larvae of the Holarctic Region have, unfortunately, limited

Table 4.11 Subfamilies and genera of Chironomidae in tropical Asia. Marine taxa (such as the Telmatogetoninae and the orthoclad *Clunio*) have been omitted. Genera that occur in streams are marked *. For sources, see text.

Tanypodinae
  Ablabesmyia *
  Clinotanypus *
  Conchapelopia *
  Djalmabatista *
  Fittkauimyia
  Krenopelopia *
  Labrundinia *
  Larsia *
  Macropelopia *
  Monopelopia
  Nilotanypus *
  Paramerina *
  Procladius *
  Rheopelopia *
  Tanypus
  ? Telmatopelopia
  Thienemannimyia *
  Zavrelimyia *
Buchonomyiinae
  Buchonomyia *
Podonominae
  Boreochlus *
  Neopodonomus
Diamesinae
  Boreoheptagyia *
  Diamesa *
  Potthastia *
  Protanypus
  Pseudodiamesa *
  Sympotthastia *
Prodiamesinae
  Monodiamesa *
  Odontomesa *
Orthocladiinae
  Acricotopus *
  Allotrissocladius
  Antillocladius
  Asclerina
  Brillia *

Bryphaenocladius *
Cardiocladius *
Chaetocladius *
Corynoneura *
Cricotopus *
Epoicocladius *
Eukiefferiella *
Eusmittia
Heleniella *
Heterotrissocladius *
Krenosmittia *
Limnophyes *
Metriconemus *
Nanocladius *
Orthocladius *
Paracricotopus *
Parakiefferiella *
Parametriocnemus *
Paratrichocladius *
Psectrocladius *
Pseudosmittia *
Rheocricotopus *
Semiocladius
Smittia *
Thalassosmittia
Thienemanniella *
Tvetenia *
Xylotopus *
Chironominae
  Asachironomus
  Baeotendipes
  Chernovskiia *
  Chironomus *
  Cladopelma
  Cladotanytarsus *
  Cryptochironomus *
  Cryptotendipes *
  Demicryptochironomus *
  Dicrotendipes

Einfeldia
Endochironomus
Glyptotendipes (*)
Harnischia
Kiefferulus
Kloosia
Kribiocosmus
Lauterborniella
Microchironomus
Micropsectra *
Microtendipes *
Neozavrelia *
Nilodorum
Parachironomus
Paracladopelma *
Paratanytarsus *
Paratendipes *
Polypedilum
Pontomyia
Pseudobrilla
Rheotanytarsus *
Robackia *
Shangomyia
Stempellina *
Stempellinella
Stenochironomus
Stictochironomus
Sublettea *
Sumatendipes
Tanytarsus *
Tobachironomus
Tokyobrilla
Trichotendipes
Virgatanytarsus
Xenochironomus *
Xiaomyia
Zavrelia *
Zavreliella
Zhouomyia

applicability to the Asian fauna, but Oliver & Roussel (1983) give a well-illustrated guide to the Canadian genera which includes a number of cosmopolitan forms. Wiederholm (1983; 1986) also provides useful keys to — and diagnoses of — those genera of Holarctic chironomids where immature stages are known. They complement a more recent, valuable key to North American genera Coffman & Ferrington (1996). Examination of chironomid larvae for identification to genus generally requires slide mounting of the head capsule and scrutiny of the mouthparts and antennae with a compound microscope. This rather tedious procedure has probably contributed to the neglect of chironomids in many ecological studies.

Other Nematocera

Larval Psychodidae (moth flies) are rather uncommon in streams and are more typical of marshes or moist terrestrial habitats. *Horaiella*, *Pericoma*, *Psychoda* (an extremely speciose genus), *Telmatoscopus* and *Trichopsychoda* may be associated with lotic habitats, but most species are probably confined to stream margins or stagnant backwaters. However, larvae of *Neotelmatoscopus* (Fig. 4.124) — a genus which is confined to the Oriental Region (Ilango, 1994) — occur on rock surfaces in streams and have ventral suckers to aid attachment. Psychodid larvae are easily recognized by the secondary annulations (usually two or three) that occur within body segments and, in some genera, the presence of dorsal sclerites on some 'segments'. Larvae cannot usually be identified to species and little is known of their ecology. They are often associated with organically-polluted habitats.

Larvae of Dixidae occur at the water surface along stream margins where they often adopt a U-shaped resting position. They feed on algae and organic matter trapped on the surface film and collected with the aid of setal brushes on the labrum. Asian dixids have two pairs of prolegs on the first two abdominal segments, and the body terminates in two pairs of fringed processes. A pair of caudal spiracles are present also. Although they are not members of the stream benthos, they are occasionally taken in benthic samples. The family is rather small. *Dixa* is the most widespread and speciose genus (e.g. Nowell, 1980; Wagner, 1982), but a new genus — *Metadixa* — was erected recently by Peters & Savary (1994) based on larval material collected from the Philippines. These authors give a key to the larvae of known genera of world Dixidae.

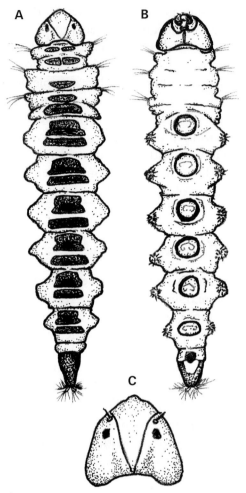

Fig. 4.124 *Neotelmatoscopus* (Diptera: Nematocera: Psychodidae); A, dorsal view of larva; B, ventral view of larva; C, head capsule.

The larval morphology of Thaumaleidae (= Orphnelidae) tends to be quite varied, and some resemble chironomids superficially. All are small (less than 12 mm long) and have single — not paired as in chironomids — anterior and posterior prolegs. The head is hypognathous so that the mouthparts are arranged more-or-less ventrally. A pair of short, stalked spiracles are situated dorsolaterally on the prothorax. Larvae have dorsally sclerotized thoracic and abdominal segments, and may have bristles or bunches of stout setae projecting from the body. The ecology and habits of thaumaleids are obscure. Most are associated with films of water flowing over stone surfaces where the feed on algae (especially diatoms). *Thaumelea* occurs in tropical Asia (Schmid, 1958c), but may not be the only genus in the region.

Brachycera

The Brachycera (Table 4.9) have received far less attention from stream ecologists than the Nematocera — in part, because the group are less diverse and have many fewer medically-important species (*cf.* blackflies and ceratopogonids). Larval stages of members of this suborder are particularly hard to identify to species or genus because most of them have yet to be associated with a named adult. Even within a single genus, species may have terrestrial, semiaquatic or fully-aquatic larvae. Some have aquatic larvae that do not occur in streams: the Phoridae, for example, are confined to 'container habitats' such as pitcher plants (e.g. *Nepenthes*: Nepenthaceae) or the internodes of large bamboos (see Disney, 1991; 1994). Taxonomic difficulties have also hindered ecological studies of Brachycera since, if larvae cannot be identified, life-history investigations are futile. Accordingly, the following discussion of Oriental Brachycera is brief.

Unlike most other Brachycera, all Athericidae (recently excised from the Rhagionidae) have aquatic larvae. Abdominal segments I-VII have paired ventral abdominal prolegs; segment VIII has only one proleg and a pair of caudal, filamentous projections fringed with setae. In addition, the body surface bears many finger-like prolongations, although these are shorter than the caudal projections and lack setae. Athericid larvae attain a body length of around 20 mm. They are predators (e.g. Hunter & Maier, 1994) with canaliculate mandibles that are used to pierce and suck the body fluids from other invertebrates. Some information on the biology of *Atherix* in Palaearctic Asia is given by Nagatomi (1962; see also Thomas, 1993). Athericid genera encountered in Oriental streams are likely to include *Atrichops*, *Asuragina*, *Ibisia* and *Suragina* (Nagatomi, 1984; 1985; Yang & Nagatomi, 1991; 1992); *Atherix* occurs in the north of the region (e.g. Thomas, 1993). Descriptions of larval stages of *Atherix* and *Suragina* are given by Thomas (1993) and Webb (1994) respectively.

The Empididae are relatively small (< 7 mm long) and have seven or eight pairs of ventral abdominal prolegs. The posterior-most pair is longer than the anterior ones, and there is a pair of rather short caudal appendages (which lack the setal fringes of athericid larvae).

Larvae of some species are terrestrial and, like other Brachycera, their ecology has received little or no attention in Asia. Aquatic larvae are likely to be predaceous. Lotic genera include *Clinocera*, *Dolichocephala*, *Hemerodromia* and (probably) *Chelipoda* (e.g. Yang & Yang, 1990; 1991; Horvat, 1994), but the association of other genera (such as *Aclinocera*, *Syneches* and *Hybos*) with streams is unclear.

Aquatic Ephydridae (shore flies) are usually found in lentic habitats or brackish water, but may be associated with submerged macrophytes or occur along stream margins. The body rarely exceeds 10 mm in length, and the abdomen ends in a pair of respiratory tubes that usually have a dark sclerotized ring around the tip. Some — but not all — larvae have short ventral prolegs. Diets include algae and fine detritus. Foote (1995) gives a useful review of the biology of the family. Like ephydrids, the majority of Muscidae are not found in streams, but they are occasionally taken in benthic samples. Larvae have a pair of short respiratory tubes at the end of the abdomen, and the same segment bears a pair of relatively long prolegs that are tipped with a scatter of small hooks. Prolegs are lacking elsewhere on the body although ventral welts which aid creeping movements are sometimes present.

Larvae of Stratiomyidae (including *Odontomyia* which is widespread in the Oriental region: Yang, 1995) differ substantially from other Brachycera. The body is flat, broadened anteriorly, and hardened by deposits of calcium carbonate. The head is partly sclerotized, and the body terminates in a pair of caudal spiracles surrounded by a ring of long setae. Diets probably consist of algae and fine detritus obtained by filter-feeding. Although not all species are aquatic, larval habits include floating on the surface or sprawling on mud; they are thus confined to sites along stream margins where there is little or no flow. Larvae of Syrphidae — 'rat-tailed maggots' — have a very long extendible respiratory tube or siphon and, relative to other Brachycera, are blunter and more-rounded anteriorly. They are found mainly in standing water, and are tolerant of polluted conditions where oxygen is scarce. Tabanidae are likewise rarely encountered in streams, but larvae are easily recognized by the absence of prolegs and the presence of girdles of six or more pseudopods on most abdominal segments. Teng (1990) has characterized the larval habitats of some Chinese tabanids, reporting that species of *Atylotus*, *Chysops*, *Haematopota* and *Tabanus* sometimes occur in running water. These four genera (and *Hybomitra*) also contain species with larvae which live in lakes, marshes, paddy fields and temporary ponds. Other Brachycera with aquatic or semiaquatic larvae include the Sciomyzidae (among them *Pherbellia*, *Sepedon* and possibly the Palaearctic genus *Colobaea* which are predators of pulmonate snails) and the Dolichopodidae (including the cosmopolitan genera *Dolichopus*, *Hercostomus* and *Paraclius*: Yang, 1996a; 1996b; 1996c), but they are unlikely to be encountered in most Oriental streams.

MINOR AQUATIC INSECT ORDERS

In addition to the seven major aquatic insect orders considered above, representatives of six additional orders live in or around streams. The Collembola or springtails are small, primitively wingless insects (almost always < 5 mm long and generally < 3 mm) with simple eyes. They lack marked adaptations to aquatic life and although they may be found at the water's edge or on the surface film, most live on land in soil and among leaf litter where they consume decaying organic material. Neustic species are encountered in 'rafts' comprising many hundreds of individuals resembling ground black pepper sprinkled upon the water surface. The order is distinguished by a short, ventral tube on the first abdominal segment; in addition, a pair of spring-like, fused appendages (the furca or furcula) on the fourth abdominal segment is usually present. When retracted under the body, the furca is held by the retinaculum or 'catch' on the third abdominal segment. Upon release, the furcula suddenly moves downwards and backwards throwing the insect into the air. In consequence, springtails can be difficult to catch because they jump when disturbed. *Podura aquatica* (Poduridae) and *Sminthurides viridis* (Sminthuridae: Fig. 4.125) are cosmopolitan species that occur in slow-flowing water at stream margins. Their small size and hydrophobe (water-repelling) integument keep them on the water surface (McCafferty, 1981: Chapter 21; Christiansen & Snider, 1996;), but their ecology has received little attention and their habits remain obscure. Diets probably include diatoms and other unicellular algae trapped on the surface film. Christiansen & Snider (1996) give a key the families and genera of Collembola in North America.

The order Orthoptera (grasshoppers and their allies) include some hydrophilous species which live near water and others (especially in the family Acrididae) which feed on emergent or floating aquatic

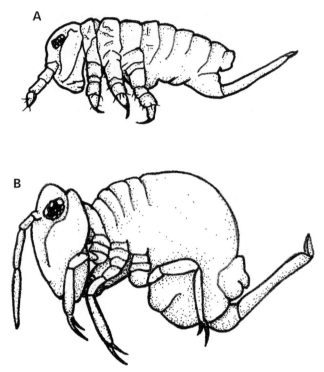

Fig. 4.125 Aquatic springtails (Collembola): A, *Podura aquatica* (Poduridae); B, *Sminthurides viridis* (Sminthuridae).

macrophytes. The grouse locusts (family Tetrigidae) are characterized by the possession of a pronotum that is prolonged posteriorly beyond the apex of the abdomen (Figs. 4.126A & B); they occur on rocks and boulders along stream margins (e.g. Kaltenbach, 1973; Roonwal, 1979; Bhalerao & Paranjape, 1986). Tetrigids leap or crawl into the water when disturbed and, in common with many Orthoptera, are competent swimmers (e.g. Roonwal, 1979). Alternatively they crawl into the water and remain submerged until danger passes. Although their biology is not well known, grouse locusts appear to feed on decaying plant tissue, algae and mosses growing in damp conditions at stream margins (Bhalerao & Paranjape, 1986; Reynolds *et al.*, 1988). Sand is also consumed for its nutritional coating of microorganisms and possibly for the aid to digestion of its grinding action in the crop of the insect. Genera which occur in the region include *Acrydium*, *Cassitettix*, *Clivitettix*, *Criotettix*, *Euscelimena*, *Gavialidium*, *Hedotettix*, *Hyboella*, *Mazarredia*, *Paratettix*, *Saussurella*, *Scelimenia* and *Tetrix*, but there are many others and it is not clear which are confined to lotic habitats.

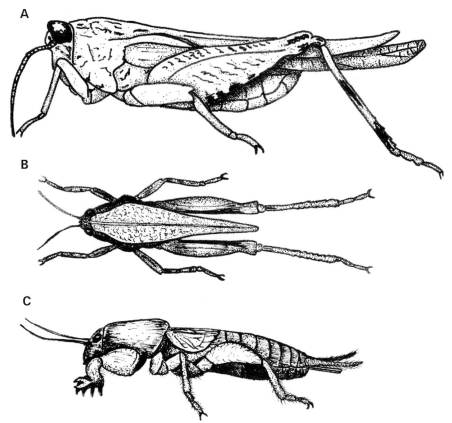

Fig. 4.126 Semiaquatic Orthoptera: A & B, lateral and dorsal views of a grouse locust (Tetrigidae); C, lateral view of a mole cricket (Gryllotalpidae).

Mole crickets (family Gryllotalpidae; *Gryllotalpa* spp.) burrow into wet muddy substrates around freshwater habitats, but do not voluntarily enter the water (Fig. 4.126C). By contrast, the poorly-known pygmy mole crickets (family Tridactylidae; *Tridactylus* spp.) burrow in the loose, saturated sand bordering water, and are able to swim by means of lamellae or plates borne on the end of the hind tibiae (La Rivers, 1956). Tridactylids may jump into water and burrow into the bottom sediments when disturbed (Bishop, 1973). Certain Gryllidae ('true' crickets in the subfamily Nemboiinae: e.g. *Paranemobius*) are able to swim and are tunnel into the banks of streams (Kaltenbach, 1973).

Several species of amphibious cockroaches (order Blattodea) in the family Blaberidae (subfamily Epilamprinae: e.g. *Epilampra*, *Rhabdoblatta* and *Rhicnoda*) are known from streams in Southeast Asia, India and Sri Lanka (Shelford, 1909; Bishop, 1973; Kaltenbach,

1973; Hutchinson, 1993: p. 569). Although able to remain submerged for some time, they are similar to related terrestrial species and do not have specific modifications for aquatic life. In Hong Kong streams, cockroaches occur close to the banks among accumulations of detritus or sheltering between stones. The same aquatic habit is reported in Peninsular Malaysia (Bishop, 1973). Because they are found in damp areas on land as well as in the water (Shelford, 1909), these cockroaches can be considered semiaquatic. The significance, if any, of the aquatic phase of their lives warrants investigation.

The Neuroptera includes three families with aquatic or semiaquatic representatives. Larvae of the family Sisyridae are parasites on freshwater sponges (family Spongillidae: see p. 99), hence their common name of spongillaflies (McCafferty, 1981: chapter 12). The mouthparts are modified into a needle-like sucking apparatus and these, plus seven pairs of ventral tracheal gills on the abdomen, are diagnostic of the family (Fig 4.127). Like all aquatic Neuroptera, sisyrids have three larval instars. A single genus of spongillaflies occurs in the Old World; *Sisyra aurorae* is known from China and I have collected larvae in New Guinea also. At least some members of the related Osmylidae are semiaquatic and lack the abdominal gills of sisyrids. They are predators of larval Diptera and small insects, occurring either in or near the water along the borders of clear streams (Richards & Davies, 1977b; Kawai, 1985a). Asian genera include *Osmylus*, *Heliosmylus* and *Spilosmylus*, the latter being relatively speciose (Banks, 1937), but other genera occur in the region. The family is in need of revision, and larvae are not yet known for many genera. Some may not be aquatic. The Neurorthidae is an obscure group of Neuroptera represented by *Nipponeurorthus* in Taiwan (Nakahara, 1958). The larvae are presumably aquatic but their morphology and habits have yet to be described.

The Megaloptera (alderflies, dobsonflies and fishflies) includes the Sialidae and Corydalidae. Both families have aquatic larvae with terrestrial pupae and adults. *Sialis* is the sole Asian sialid genus. It is mainly restricted to temperate latitudes such as Japan and Korea, and it has been reported from Sichuan Province, China (Banks, 1940). The Corydalidae are widespread and include *Acanthacorydalis*, *Anachauliodes*, *Ctenochauliodes*, *Neochauliodes*, *Neoneuromus*, *Neurhermes*, *Neuromus*, *Parachauliodes* and *Protohermes* in Asia (e.g. Banks 1940; Ghosh, 1981; Yang & Yang, 1986; 1991; 1992; 1993). The immature stags of several of these genera are not yet known. Corydalid larvae are generally large (up to 60 mm in length) with

512

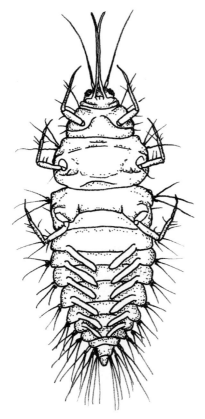

Fig 4.127 A *Sisyra* larva (Neuroptera: Sisyridae), ventral view (after McCafferty, 1981).

development times of some species likely to be in excess of 12 months (Yoshida *et al.*, 1985; Hayashi, 1988a; 1996), although univoltine life cycles are possible at high temperatures (Hayashi, 1996) and are typical of smaller members of this family (e.g. Dolin & Tarter, 1981) including *Neochauliodes boweringi* in Hong Kong which shows rather synchronous growth after recruitment at the start of the wet season. Pupation takes place on land; usually within a chamber constructed in damp soil. Corydalid larvae are predaceous with well-developed strong mandibles, and consume a variety of aquatic invertebrates which they catch by ambushing them (Hayashi & Nakane, 1989). Maximum width of prey eaten coincides approximately with mandible length and therefore mouthpart dimensions affect prey selection (Hayashi, 1988b). Abdominal segments I-VIII bear a pair of two-segmented lateral filaments which may have a respiratory function. *Protohermes* possesses gill tufts at the bases of lateral filaments I-VII (Fig. 4.128), but *Neochauliodes* (which is much smaller) lacks these tufts. In addition, all corydalids have functional abdominal spiracles which allow aeropneustic respiration and may enhance survival in streams which dry out periodically. There is a pair of anal prolegs, each bearing two terminal hooks, at the end of the abdomen. Small corydalids may sometimes be confused with certain hydrophilid beetle larvae (e.g. *Berosus*: see Fig 4.114D) which they resemble superficially, but the

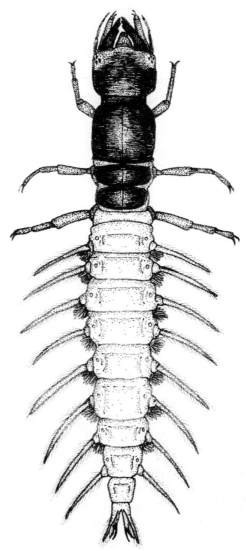

Fig. 4.128   Dorsal view of a *Protohermes* larva (Megaloptera: Corydalidae).

presence of hooked prolegs and a clearly-demarcated labrum should be sufficient distinguishing characters.

Adult corydalids have elongate, net-veined wings patterned with dark blotches; that are held flat over the body when the insect is at rest (Fig. 4.129). The complex pattern of wing venation is thought to be primitive, and megalopterans are generally regarded as the most ancient endopterygote order. The adults are nocturnal and are rather weak fliers, although they are sometimes attracted to lights. Plant sap and small insects are eaten by adults of some species (Yoshida *et al.*, 1985). The eggs are deposited on vegetation overhanging potential larval habitats. Hatchlings drop from the egg masses into the water.

The caterpillars of certain pyralid moths are the only strictly aquatic Lepidoptera in tropical Asia; of these, the subfamily Nymphulinae is most conspicuous. It is treated by some authors as belonging to the Pyraustidae (e.g. Rose & Pajni, 1987) of Crambidae (e.g. Habeck & Solis, 1994; Munroe, 1995), but the more conventional placement within the Pyralidae is followed here. In addition to the Nymphulinae,

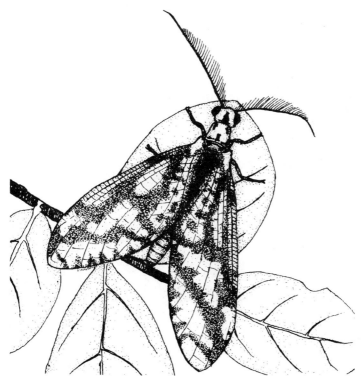

Fig. 4.129 Dorsal view of an adult *Neochauliodes* (Megaloptera: Corydalidae).

a few other pyralids and members of other families may be associated with aquatic or semiaquatic macrophytes. For example, larvae of the pyralid family Schoenobiinae (e.g. *Neoschoenobia, Niphadoses, Schoenobius* and *Scirpophaga*) are rice-stem borers (and may attack other emergent plants also). They are economically important crop pests over much of Asia but do not occur in streams. Nymphuline larvae can be recognized by a combination of characters: presence of a distinct head capsule bearing stemmata; three pairs of thoracic legs; prolegs on the abdominal segments III-VI plus a pair of anal prolegs which bear curved hooks or crotchets at the tip; and, spiracles on the prothorax and abdomen. Filamentous tracheal gills may be present, and vary in form and arrangement among species (Reichholf, 1973) or may be lacking entirely. Benthic taxa dwell under silken tent-like retreats on rocks in fast current (tribe Argyractini: *cf. Eoophyla*; Figs. 4.130A & B), or in tunnels reinforced by sand grains and small stones, beneath

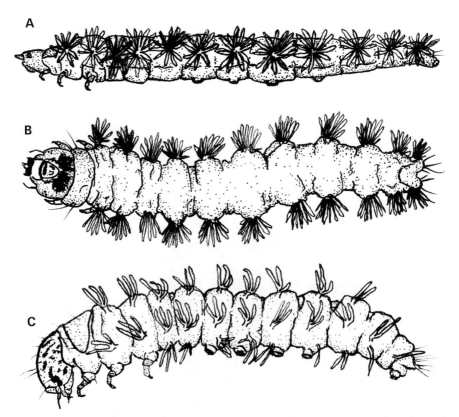

Fig. 4.130 Larval Nymphulinae (Lepidoptera: Pyralidae): A & B, lateral and dorsal views of *Eoophyla*; C, lateral view of *Parapoynx*.

which they graze algae from the rock surface. Upon pupation, the silken retreat is replaced by a tougher-textured cocoon with holes along the periphery that permit water circulation. The pupa cuts a slit in the cocoon, allowing the adult to escape and swim to the water surface. Aquatic caterpillars can be locally abundant in streams on the Asian mainland, where the genus *Aulocodes* seems to be widespread (e.g. Hora, 1927; Reichholf, 1973). These insects may be the dominant freshwater invertebrate on some oceanic island in the Pacific where competition is minimal because the older aquatic groups have not dispersed to these isolated habitats (Hynes, 1984).

In addition to rock-dwelling species, larvae of some Nymphulinae in the tribe Nymphulini (e.g. *cf. Parapoynx* [= *Paraponyx*]: Fig. 4.130C) are associated with macrophytes such as *Hydrilla verticillata* (Hydrocharitaceae) and usually occur in standing water. Larvae live within portable retreats constructed of several longitudinally-arranged leaves (Buckingham & Bennett, 1996), while Asian *Elophila* build a case of leaf fragments or live between two pieces of leaves bound together by silk (Nagasaki, 1992). North American Nymphulinae have also been recorded as associated with *Ceratophyllum* (Ceratophyllaceae), *Nymphaea* (Nymphaeaceae), *Vallisneria* (Hydrocharitaceae) and *Lemna* (Lemnaceae) (McCafferty, 1981: Chapter 15; Lange, 1996) in lentic habitats. These genera are widespread in the Oriental Region, and it is likely that nymphuline caterpillars use them as host plants. Although aquatic moths have received little attention in the ecological literature, nymphuline genera such as *Agassizia, Ambia, Arxama, Bradina, Camptomastix, Eoophyla, Eristina, Hymenoptychis, Massepha, Metoeca, Nymphicula, Oligostigma, Paracymoriza, Parthenodes, Stegothryris* and *Strepsinoma* (plus, possibly, the Palaearctic genera *Nymphula* and *Cataclysta sensu lato*) may be encountered in Oriental streams (e.g. Rose & Pajni, 1987; Yoshiyasu, 1987). Furuya (1989) and Yoshiyasu (1985) describe the larvae of several genera of aquatic pyralids from Japan (*Cataclysta, Elophila, Neoschoenobia, Nymphula, Paracymoriza, Parapoynx* and *Schoenobius*), and Reichholf (1973) figures some unidentified larvae from Sri Lankan streams. Life history data are scant, but pyralids from standing waters in Japan complete two or there generations each year (Nagasaki, 1992). *Parapoynx diminutatis* larvae accidentally introduced from Asia which have become naturalised in Florida (USA) take between 32 and 50 days to develop from egg through seven larval instars to the adult (Buckingham & Bennett, 1996).

Certain members of the Order Hymenoptera (suborder Apocrita, section Parasitica) are parasitoids of aquatic insects (Fig. 4.131), the adult wasps entering the water to locate and oviposit on or in their aquatic hosts (Hagen, 1956; 1996; McCafferty, 1981: Chapter 20; Elliott, 1982). In most cases, these diving wasps do not show any external modifications in either the endoparasitic larval stage or the free-living adult as compared with those parasitizing terrestrial hosts. In streams, cased caddisfly larvae are parasitised by Agriotypidae (*Agriotypus* spp.; Kawai, 1985b). Final-instar larvae of this genus construct a ribbon-like gas-filled cocoon which extends outside the host's case. The ribbon probably functions as a plastron, and its removal kills the parasitoid (Thorpe, 1950; Hagen, 1956). The influence of these parasitoids on caddisfly — or aquatic insect — population dynamics in Asian streams is unknown. Care is needed in the identification of these parasitic Hymenoptera, because a few species in all major superfamilies of the section Parasitica have aquatic insect hosts (see Hagen, 1996 for more details). In addition to parasitic wasps, at least one species of Pompilidae (the spider-hunting wasps) in the section Aculeata can be considered semiaquatic also, as it is a specialized predator of pisaurid spiders (*Dolomedes*: see pp. 188–190).

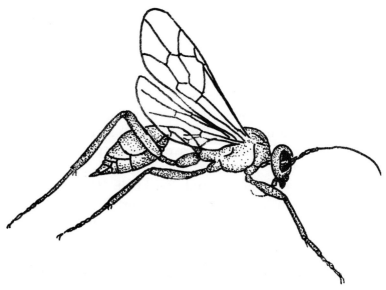

Fig. 4.131 Adult *Argiotypus* (Hymenoptera: Apocrita): several species of Argiotypidae are parasitoids of aquatic insects such including caddisfly larvae.

5

Anthropogenic Threats

Human influences on tropical Asian rivers are all-pervasive, and reflect the development of ancient civilizations around the great Asian rivers — for example, the Harappa and Mohenjodaro cultures along the banks of the Indus — coupled with increasing use of the region's extensive freshwater resources (summarized by Ali *et al.*, 1987). There are now probably no large, tropical Asian rivers in pristine condition (Hynes, 1989), and a map of the extent of pollution in the region would have to include most of the major rivers (Fig. 5.1). Generally, we do not know what these rivers were like in their original state. Studies which have been undertaken show little continuity, and the only data that are fairly good (and that is only true for some rivers) are fisheries statistics; however, they are at best only a crude estimate of the totality of conditions in a large water body (Hynes, 1989). Attention was drawn to declines in fisheries stocks forty years ago (Hora, 1952). Today, erosion, deforestation, urbanization, channelization, flow regulation, water transfers, irrigation, salinization, domestic and industrial pollution, and many other human activities have altered rivers in ways that we do not understand fully.

Stream biologists may deplore the despoliation and degradation of lotic habitats that can arise from human activities such as damming or

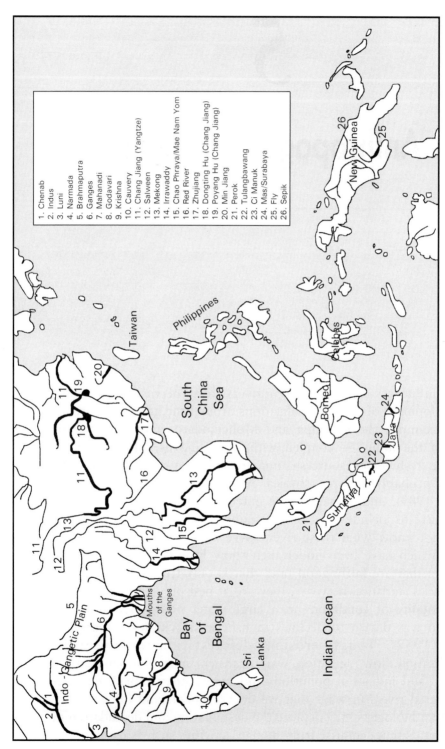

1. Chenab
2. Indus
3. Luni
4. Narmada
5. Brahmaputra
6. Ganges
7. Mahanadi
8. Godavari
9. Krishna
10. Cauvery
11. Chang Jiang (Yangtze)
12. Salween
13. Mekong
14. Irrawaddy
15. Chao Phraya/Mae Nam Yom
16. Red River
17. Zhujiang
18. Dongting Hu (Chang Jiang)
19. Poyang Hu (Chang Jiang)
20. Min Jiang
21. Perok
22. Tulangbawang
23. Ci Manuk
24. Mas/Surabaya
25. Fly
26. Sepik

Fig. 5.1 The major rivers of tropical Asia, showing the extent of pollution: modified from Lean *et al.* (1990) and Dudgeon (1992b). The thick lines indicate severely polluted stretches or sites which exceed GEMS (United Nations Global Environmental Monitoring System) median values in more than one category of pollutant.

impounding rivers, but it is a fact that — because of the monsoonal climate — tropical Asia is a region of floods and droughts which cause considerable human misery. While the concerns of stream biologists and conservationists have yet to carry much weight when matters of flood control, water supply and hydroelectric power generation are considered by governments, special pleading for attention to ecological issues is unlikely to meet with much success if relevant and up-to-date information on the consequences of anthropogenic modification of ecosystems is not made available to those in decision-making positions. Unfortunately, no such compendium of information exists for tropical Asian rivers. Clearly, we need a better understanding of these changing and despoiled ecosystems so that we can appreciate the consequences of our past activities, and thus manage rivers and their drainage basins more effectively. To achieve this goal, we must first summarize the major threats to these ecosystems. That is one objective of this book. Because of the variety of investigations — especially primarily-descriptive pollution studies — undertaken in recent years, it is impossible for a review of this type to be exhaustive. The extent of the region also rules out the possibility of summarizing all relevant information in a sufficiently brief fashion without sacrificing a good deal of precision. Herein, some 'case studies' of the potential effects of human impacts — especially river regulation — are described in order to highlight the main factors contributing to habitat degradation. A further obstacle to effective conservation and management of tropical Asian rivers is the shortage of basic ecological knowledge which adds a degree of uncertainty to predictions of changes that will be brought about by on-going and planned human activities. While the uncertainty should not be down-played, the following account of threats to these ecosystems is probably correct in general essence even if estimates of the magnitude and extent of deleterious changes are imprecise.

RIVER REGULATION

The management of water resources has a long history in Asia, and the importance of streams as sources of irrigation water cannot be overestimated. Diversion and containment of rivers is a centuries-old tradition, and has influenced the culture of the region profoundly. Examples include the Kauvery Delta canals in India constructed during the second century AD; the 4,000 year-old Ifugao rice terraces of the

Philippines; the 10th century AD Barrai irrigation dam in Burma; and, early attempts to dam the Mahaweli River in Sri Lanka (3rd century AD), combined with the extent of ancient irrigation canals on the island which point to the existence of an early 'hydraulic civilization' dating back to around 600 BC. Dam and reservoir construction associated with irrigation in China has strongly influenced settlement patterns for over 2,000 years, and major schemes include the Dujiangyan irrigation system (256 BC), the Zhengguo (246 BC), Lingqu (214 BC) and Longshou (128 BC) Canals, and the Grand Canal (6th century A.D.). More recently, there have been detrimental effects also, and it is estimated that over 10 million people have been displaced by dam building since the 1950s (Ryder, 1991). The antiquity and pervasiveness of river regulation, in one form or another, confirm the impression that present-day riverine ecology in tropical Asia must be viewed against a backdrop of existing and past anthropogenic interactions in which humans influence rivers and *vice versa*. Nowhere is this been more apparent than in China (Dudgeon, 1995e), and detailed consideration of cases from southern China can serve as an example of the causes and consequences of river regulation. While aspects of the irrigation systems of China have been reviewed elsewhere (Nickum, 1982; Ministry of Water Resources & Electric Power, 1987), ecological effects upon the source rivers is not considered in these publications.

River regulation in China

China receives a mean annual rainfall 680 mm but, despite this apparent plenitude, about half of the country, especially in the west and northwest, is arid, semiarid or desert. The region north of the Chang Jiang makes up approximately 60% of the land mass but has only about 20% of the water resources (World Resources Institute, 1994: p. 64; see also Chen & Wu, 1987), while water is abundant in the Chang Jiang and Zhujiang basins in the south. As a consequence of the large population of 1.15 billion and a shortage of water in much of the country, annual per capita water use is less than 500 t or approximately 25% of that (1,868 t) consumed in the United States — a country of comparable area (World Resources Institute, 1994: pp. 346–347; see also Chen & Wu, 1987). This relative shortage of water has meant that all river resources in China are harnessed to some degree, thus contributing to the extent of river regulation in the country. Approximately 45% (48 x 10^6 ha) of all farmland is irrigated (Ministry of Water Resources & Power, 1987; Smil, 1993), and annual water use (withdrawals) constitutes 16% of total surface water resources (World Resources Institute, 1994: p. 346) although, in some areas, up to 68% of surface water is used. Most of this (87%) is attributable to agriculture (i.e. irrigation). In the north, over 300 cities experience water shortages which can be severe (Smil, 1993), and aggressive use of groundwater has led to falling water tables (Smil, 1993; World Resources Institute, 1994: p. 73).

In essence, not only is the total amount of water far from plentiful when divided by the huge population, but the water resources are unevenly distributed in space. There is also marked temporal variation in the water supply as a result of the Chinese climate, and year-to-year variations in cereal production are strongly dependent upon weather conditions (Hulme *et al.*, 1992). Joseph Needham, who was responsible for the 15-volume treatise *Science and Civilization in China*, stated in a recent interview (Vines, 1993) that hydrological factors had played an pivotal role in the development of Chinese civilization. Because China is a continental country, bound to agriculture, its highly seasonal and variable rainfall necessitated large hydrological schemes — including works of irrigation and water conservation, river control, drainage and inland navigation — on a scale unprecedented in the West. For example, by 1985 there were 83,219 reservoirs in China (gross capacity 430 x 10^9 m^3), 177,000 km of dykes and 24,816 km of sluices, and approximately 48 x 10^6 ha of land under irrigation (Chen & Wu, 1987; Ministry of Water Resources & Electric Power, 1987). Most significantly, the many great water engineering works, inter-basin transfers and interconnected canals can be thought of as analogous to a 'hydraulic state' which eroded the autonomy and power of individual feudal lords; that power was absorbed by the imperial authority in Beijing (Vines, 1993; for a more detailed analysis of the 'hydraulic state' see Will, 1985). The importance of river control and irrigation in the agricultural production and social economy of China is indisputable and one result of this, according to Needham, is that probably no country in the world has so many legends about heroic engineers and their deeds. In a review of the history of hydrology, Biswas (1972) describes the work of the legendary Chinese hero-emperor Yu the Great who lived around 2200 BC, and was instrumental in building dams, dykes and great waterworks. Hirth (1911: p. 34) reports that Yu '. . . cut canals through the hills in order to furnish outlets to the floods; . . . that he traced each river to its source and back again to its mouth, in order to clear its spring, regulate its course, deepen its bed, raise embankments, and change its direction'. Such was his success that Yu subsequently became Emperor of China (2205–2198 BC). The importance of river regulation in the history of China can be seen also from the fact that emperors were classified as 'good dynasty' or 'bad dynasty' according to whether waterworks were maintained carefully or allowed to fall into disrepair (Biswas, 1972). The devastating effects of floods and droughts also had the consequence that the Chinese were among the earliest (13th century) users of rain gauges.

Floods and droughts are inescapable features of the Chinese environment. Drought affects between 1 — 18 x 10^6 ha per year while, in recent years, the average area disastrously affected by floods was approximately 5.5 x 10^6 ha (Smil, 1993). Particularly serious floods in the Chang Jiang have occurred in 1931, 1954 and 1980: in the former case, a total area of over 80 x 10^6 ha was inundated, and the death toll was estimated at one million. The severity of these floods will be apparent from the observation that water levels at the height of these floods were 27 — 30 m above normal (Zhang & Lin, 1992: p. 91). Although the historical data are incomplete, it has been estimated (Ministry of Water Resources & Electric Power, 1987) that 1,092 flood disasters and 1,056 major droughts occurred between 206 BC and 1949 AD. The importance of floods as an incentive for the regulation of Chinese rivers in the present day (e.g. Chen & Wu, 1987) can be appreciated from the fact that about 10% of the territory, inhabited by around 65% of the population and responsible for approximately 70% of the agricultural and industrial output, is below the flood level of major rivers (Smil, 1993). Accordingly, it is not surprising that river control has been an important goal of the Chinese government, and millions of citizens were involved in water conservancy and flood control projects in the decade following the revolution in 1949. One characteristic of these developments has been long-distance transfers of water, from one river catchment to another (inter-basin transfers) or to distant storage reservoirs (Biswas *et al.*, 1983 and references therein), and work is proceeding on one route which will involve northward diversion of Chang Jiang water on a massive scale through the ancient Grand Canal to northern Jiangsu, Shandong, Hebei and Tianjin Provinces and Beijing (see pp. 534–537). The flood and water-supply problem in China has also been approached in a smaller-scale piecemeal fashion through construction of dykes, embankments, dams, and so on. In 1992 the Chinese National People's Congress endorsed a massive scheme to tame the waters of the Chang Jiang: the

Three Gorges Project is the most ambitious of any flood-control measure so far undertaken, and will include the largest dam — and the biggest engineering project — in the world (Van Slyke, 1988).

The Three Gorges Scheme

Over a 200-km stretch between Baidicheng in Sichuan to Nanjinguan in Yichang Hubei, the Chang Jiang cuts deeply into limestone and forms many gorges. The area is named after the three most spectacular ones. The Chinese government is in the process of constructing a high dam, begun in 1996, which will flood the Three Gorges (Fig. 5.2). The objectives are to generate hydroelectricity and increase irrigation, plus improved navigation and flood prevention. It will take 12 years to put the first generating unit in the Three Gorges High Dam into commission and an estimated 18 years to complete the project. The new dam would lie upstream of the huge Gezhouba or Chang Jiang Low Dam sited at Yichang, which serves the joint purposes of navigation, hydroelectric power generation, and partial regulation of water level. Planning for the Three Gorges Project began in 1956 when the intention was '. . . settling once and for all the Chang Jiang flood problem that has a history of about a thousand years' (Carin, 1962). The need for flood prevention should not be overestimated given that there are over 32,000 km of embankments (3,748 km of them major) along the main river channel and tributaries (Luk & Whitney, 1993b). Many of the embankments are rather unstable and frequent maintenance is required; a programme of raising the main embankments by 1–2 m is ongoing (Luk & Whitney, 1993a). Together with several flood diversion projects on the Chang Jiang floodplain these embankments reduce the effects of small flood peaks, but the Three Gorges Project is required to prevent the devastation that can be caused by severe floods when the river level may rise by over 10 m (Wang, C., 1993). While flood control is not the only benefit projected to arise from the Three Gorges Scheme (see below) — and even that benefit has been questioned (Fang, 1993) — the frequency and extent of damaging floods in southern China (most recently in 1991, 1994 and 1996) draws attention to the point that enhanced river regulation can be related directly to a reduction in the scale of misery visited upon the Chinese populace. In summer 1996, for example, 2,700 people were killed when the Chang Jiang rose to its highest levels in 65 years (SCMP, 1996a; 1996b)

The Three Gorges Dam will be 185 m high and store almost 40 x 10^9 m^3 water; estimated capacity (13 — 19 x 10^6 kilowatts)

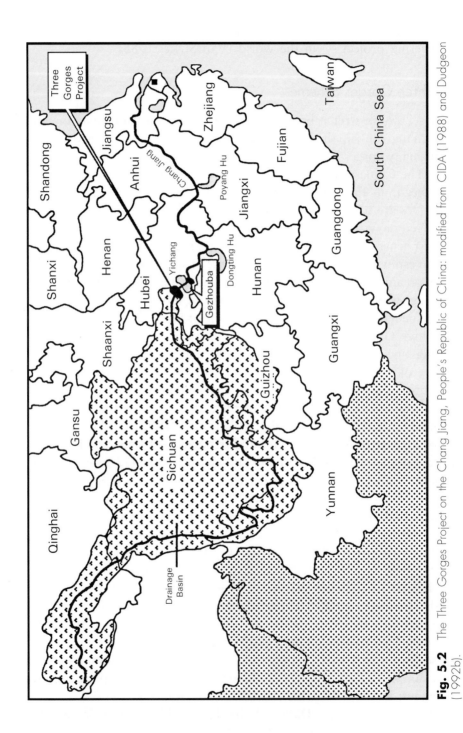

Fig. 5.2 The Three Gorges Project on the Chang Jiang, People's Republic of China: modified from CIDA (1988) and Dudgeon (1992b).

would exceed the largest dams in the world now operating by 40%, and projected annual power generation is 68–89 x 10^9 kilowatt hours — equivalent to 50 x 10 t coal.yr^{-1} (Shen, 1992; Li, 1994). The project has perceived environmental advantages because severe air pollution is a direct consequence of China's reliance upon coal for over 75% of its commercial energy supply; the completion of the Three Gorges Dam would reduce SO_2 and CO_2 emissions substantially. By lowering demand for coal, the project would also relieve pressure on the overburdened railway network in China (Wang, C., 1993; Li, 1994). Considerable discussion as to the costs and benefits of the project has ensued both within China and elsewhere; a range of points of view from both sides have are given in a recent volume dealing with the Three Gorges Project edited by Luk & Whitney (1993b).

Ecological impacts of the Three Gorges Project are not yet known, but some hold that they will be minor (CIDA, 1988; Wang, 1993; Li, 1994) because of the small operating volume of the reservoir in relation to the large volume of river discharge. It is projected that storage for seasonal flow regulation will be only 2.6% of the average annual flow. However, it is notable that the 1988 CIDA report concentrates largely on engineering aspects of the Three Gorges Project, and contains very little substantial data about the ecology of the river environment. Wang Chaojun (1993: p. 106), nevertheless, reports that '. . . the development of fisheries in the reservoir area and downstream of the dam is very promising . . .' and that '. . . the overall impact is that the positive effects will outweigh the negative ones' (Wang, C,. 1993: p. 107). However, there are alternative views of the potential for ecological damage (Larson, 1991), and — despite Wang Chaojun's (1993) optimistic view — impacts seem inevitable because the riverine habitat of the Three Gorges will be transformed into a lacustrine environment to which the existing biota are not adapted. The volume edited by Ryder (1991) contains a particularly trenchant criticism of the CIDA assessment of the Three Gorges Project, and calls into account all of the supposed benefits of the dam. Other workers consider also that '. . . serious conceptual and data shortcomings remain . . .' (Luk & Whitney, 1993a: p. 36) in connection with the Project. One point which has been stressed by proponents of the dam is that the environmental benefits to be gained from the huge amount of clean energy to be generated will more than offset ecological impacts or problems arising from relocation and resettlement of the 1.1 million inhabitants of areas which will be inundated following impoundment (Li, 1992; Tian & Lin, 1993; Wang, C., 1993; Li, 1994; for alternative views see Fang & Wang, 1993; Freeberne, 1993; Luk & Whitney, 1993a). This is despite the fact that the impounded water will submerge some of China's finest scenery and an important source of tourist revenue (Li, 1992).

Concerns about the severity of the ecological impact of the Three Gorges Project may be lessened by the fact that it will affect an area of the Chang Jiang which has been used intensively by man for centuries. Little of the original environment remains undisturbed; natural habitats and wildlife populations have been replaced by dense rural settlement with intensive use of land for farming and aquaculture. Deforestation has taken place throughout most of the Three Gorges and along the downstream reaches of the Chang Jiang where the river and its tributaries are confined within dykes. Recent urbanization and industrialization has affected water quality increasingly (CIDA, 1988). Nevertheless, the Chang Jiang basin — particularly Dongting Hu and Poyang Hu — provides habitat for commercial as well as rare and endangered species including certain fishes and waterfowl, as well as two endemic and highly-endangered vertebrates: the Chinese river dolphin or *baiji* (*Lipotes vexillifer* — also known as the whitefin dolphin) and the Yangtze alligator (*Alligator sinensis*). Appropriate mitigation and monitoring programmes will be essential to avoid habitat damage and species loss and possible global extinction. The extent to which these will be implemented has yet to become apparent. In March 1992, the Three Gorges Scheme was approved by the Standing Committee of the Chinese National People's Congress (CNPC) and was passed by the CNPC (after some dissent) in April 1992 (Sun & Wang, 1992). Construction of the dam began in late 1994, and is due to be completed in 2009; the first turbine should go into operation in 2003. The rate of progress will depend on the generation of adequate funding, and there are widely varying estimates as to the total cost of the project (Wang, J., 1993) from RMB 57 billion (US$10.6 billion; Shen, 1992) to RMB 100 billion (US$18.6 billion; Li, 1994). At the time of writing, the World Bank and the US Import-Export Bank had refused funding for the dam on environmental grounds (SCMP, 1996c), although these decisions could be reversed subsequently.

Obviously, the 'ecological acceptability' of any river regulation project will vary according to the type of engineering work proposed, the status and biology of the species involved, and so on. It is impossible to predict what the precise impacts of the Three Gorges Scheme on species in the drainage basin will be, but the possible detrimental effects will vary above and below the dam. Upstream of the proposed dam, rare and endangered species include the giant salamander (*Andrias davidianus*) and Yangtze sturgeon (*Acipenser dabryanus*). The former species inhabits mountain brooks (at altitudes above 1,000 m) in central and southwest China. It is the largest living salamander (reaching 180 cm) and is intensively hunted for its flesh. Concern has also been expressed also over the fate of populations of endangered *Adiantum reniforme* ferns which will be submerged by water impounded behind the dam (Xie, 1993). Downstream effects of the dam are likely to be felt in the huge lateral lakes associated with the Chang Jiang floodplain. Dongting Hu and especially Poyang Hu (which has been a Chinese national nature reserve since 1988) are important wintering grounds for waterfowl including migrants from the north in winter, such as white-fronted geese (*Anser albifrons*), the rare eastern white stork and the black stork *(Ciconia boyciana* and *C. nigra*), the Japanese crane (*Grus japonensis*), and 95% of the world population of the endangered Siberian white crane (*Grus leucogeranus*) (Zhao *et al.*, 1990; Melville, 1994). Poyang Hu also has the largest known concentration of swan geese (*Anser cygnoides*) in the world, as well as a several species of birds and two mammals (the Chinese water deer, *Hydropotes inermis*, and the black finless porpoise, *Neophocaena phocaenoides*) of national conservation importance.

The Three Gorges Project will alter inundation patterns of the lateral lakes and floodplain and may restrict biological production through an influence on land-water interactions. One projection is that — after the dam is completed — the water levels in the Chang Jiang during the dry season will be 1–2 m higher than before; investigators nevertheless concluded that the operation of the dam and reservoir would '. . . not influence the hydrogeological environment of the Poyang Lake area and river mouth' (Guo, 1993; see also Wang, C., 1993; Qian *et al.*, 1993). However, despite Wang Chaojun's (1993: p. 108) opinion that the Three Gorges Project would '. . . protect the ecological environment . . .' of the Chang Jiang floodplain it seems unlikely that there will be no detrimental ecological effects. Indeed, other investigators have predicted that feeding habitats for waterfowl at Poyang Hu may be reduced by one third to one half as a result of the Three Gorges Project (Zhu *et al.*, 1987; Melville *et al.*, 1992; Melville, 1994). A possible scenario associated with higher river levels can be extrapolated from observations made by Wang (1987). He noted that land reclamation from Poyang Hu maintained the water in the main course of the Chang Jiang at higher levels, decreasing fluctuations and reducing the development of grass 'flushes' on exposed shores. In consequence, food availability and breeding sites for cyprinids were reduced and fish populations declined greatly.

The Chinese river dolphin, the Chinese sturgeon (*Acipenser sinensis*), and Chinese alligator occur downstream of the Three Gorges region. The *baiji* is highly endangered: the estimated world population in the Chang Jiang was 200 in 1990 (Zhao *et al.*, 1990), and 100 or fewer in 1994 (SCMP, 1995a; SCMP, 1996d). Continued declines in numbers reflect hook fishing and accidents caused by the propellers of river traffic (Chen *et al.*, 1979; Zhou *et al.*, 1979; Han, 1982; Perrin *et al.*, 1989) and changes in habitat caused by the construction of dams and floodgates that interrupt flow between the Chang Jiang and its lateral lakes (Reeves & Leatherwood, 1994). The *baiji* has abandoned former habitat in Dongting Hu as a result of increasing sedimentation (Chen *et al.*, 1979), and the lake has become increasingly eutrophic in recent years (Liu, 1984; Jin *et al.*, 1990). Last-ditch conservation efforts include a scheme to capture all remaining wild *baiji* and maintain them in an enclosed and protected 21-km long section of the river (the 'Swan Oxbow') in Hunan Province (Ames, 1991; Dikkenberg, 1992). A second smaller reserve is planned, but funding is extremely limited. Location and capture of dolphins to stock the reserves has proved difficult also, and only one of two captive individuals has survived (SCMP, 1995a; 1996d). The endemic Yangtze alligator (estimated population 300–500 individuals) is found only in the region between Anhui, Zhejiang, and Jiangsu, where it inhabits streams, irrigation pools and rice fields bordering the river (Huang *et al.*, 1986). Although the wild population is decreasing, artificial propagation of this species has been successful. At present, it is not possible to predict what effect the proposed dam may have on endangered species in the Chang Jiang, but any habitat change could lead to further population reductions.

The Three Gorges High Dam will have no facilities for upstream or downstream migration of fish. Such movements were blocked by the Chang Jiang Low Dam at Gezhouba in 1981; fish passages were not provided as they were deemed inefficient for the species involved (CIDA, 1988). This has had the effect of reducing populations of the Chinese sturgeon, partly through alteration of downstream flows and changes in sediment characteristics that reduce spawning success. Nevertheless, some spawning by sturgeons and other fishes does take place in the changed habitat (Yu *et al.*, 1985a; 1985b; 1986; Liu *et al.*, 1990), although the Chinese sturgeon may still be at risk from commercial fishing because these fish take a decade or more to become sexually mature. Based on the successful spawning of major carp (*Ctenopharyngodon idella*, *Mylopharyngodon piceus*, *Hypophthalmichthys molitrix* and *Aristichthys nobilis*) in the Chang Jiang after dam construction at Gezhouba, Yu *et al.* (1985a) concluded that the expense of constructing a fishway as part of the Three Gorges High Dam was unjustified. In a comparable study of the Han Shui River — a tributary of the Chang Jiang in Hubei Province — Yu *et al.* (1981) reported that construction of a dam had no significant deleterious effect on fish resources, and that population densities of some species had increased. In their view, the reservoir behind the dam provided a habitat in which many of the river fish could flourish, and similar conclusions were reached by Liu & Yu (1992) following construction of the Danjiangkou Dam on a Chiang Jiang tributary — despite markedly changed conditions and the absence of a fish pass. Liu & Yu (1992) did, however, note that spawning grounds of rheophilous fishes downstream of the dam were much diminished. Likewise, Yu *et al.* (1985a) reported a decline in the abundance of fry in the Chang Jiang after dam construction at Gezhouba; there was also some modification of the river habitat attributable to the dam but they judged the impacts on the fish that they studied to be rather minor. The fishes investigated in both of these studies were mostly major carp which are highly-adaptable aquaculture species farmed widely in China. Their responses cannot be considered as representative of the fish fauna as a whole which, in the region of the Gezhouba and Three Gorges Dams, comprises 109 species (Yu *et al.*, 1985b).

Some of the potamodromous species which are most at risk from blocked migration routes (e.g. *Myxocyprinus asiaticus* and both sturgeons found in the Chang Jiang) can be cultured artificially (Yu *et al.*, 1985b; 1986; 1988), thereby offsetting some of the potential economic loss arising from blockage of migration routes, but it is debatable as to whether this can be said to compensate for the interference with the natural reproductive behaviour of these fishes. Moreover, the Chinese sturgeon is now considered as endangered while the Yangtze sturgeon is extremely endangered (Birstein, 1993), and their survival in the Chang Jiang is by no means assured. The threatened Chinese paddlefish (*Psephurus gladius*) also occurs in the Chang Jiang, where it can reach 7 m in length (Birstein, 1993). Although routes are not well known (LaBounty, 1984; Liu & Zheng, 1988), paddlefish are reported to migrate from the Chang Jiang into the East China and Yellow seas (Liu & Zheng, 1988). The anadromous Reeves' shad (*Tenualosa reevesii*: Clupeidae) has declined greatly in the Chang Jiang during recent years and, in 1987, was classified as a protected species by the Chinese Ministry of Agriculture (Wang, 1996). Yu (1983) has expressed concern about possible impacts of large-scale water transfers from the Chang Jiang (see pp. 534–537) on this species, and Liu & Zheng (1988) and Birstein (1993) have drawn attention to the dangers posed to paddlefish and sturgeons by overfishing and pollution of the Chang Jiang. The paddlefish is now protected in China (Liu & Zheng, 1988) but legal protection may offer little security from the effects of large-scale habitat alteration resulting from river regulation. A recent assessment of the status of the Chinese paddlefish and Yangtze sturgeon is that '. . . both seem to face extinction' (Birstein, 1993: p. 779).

Regulation of the Zhujiang

Flooding problems associated with southern Chinese rivers have been met largely with embankments. There are difficulties in the Zhujiang (Fig. 3.1) — the largest river in China south of the Chang Jiang — as rainfall in the catchment is the highest in China (three times the annual average of northern China) with uneven precipitation throughout the year. Construction of dykes, dams, and reservoirs has been undertaken to mitigate floods, with 40 large reservoirs (and almost 200 smaller ones) constructed between 1949 and 1960 (Carin, 1962); nevertheless, monsoonal rains and typhoons still caused losses in the Zhujiang basin during the 1960s. Construction of further drainage and irrigation schemes has continued, and over 3,000 dams of various sizes have been built along the Zhujiang and its tributaries during the last 30 years. Despite these efforts, the three southern provinces regularly experience serious flooding and, in summer 1994, over 1,400 people were killed when 5.2×10^6 ha were inundated during the worst flooding to hit the region for 45 years. Some of the severity of these floods has been blamed upon local authorities who are reported to have diverted funds earmarked for flood-control into projects which would yield short-term economic gains (SCMP, 1994).

The ecological effects of river regulation on southern Chinese rivers has received less attention than has been the case for the Chang Jiang, because of the lack of a 'high-profile' project equivalent to the Three Gorges Dam or the south-to-north water-transfer scheme (see pp. 534–537). Indeed, there have been relatively few published reports on the Zhujiang in the primary scientific literature despite the fact that, with an annual discharge of approximately 20% (11,000 $m^3.s^{-1}$) that of the Chang Jiang (Zhang *et al.*, 1987), the Zhujiang ranks second in terms of discharge in China and is — by any standards — one of the world's great rivers. There is a notable lacuna in information about fisheries stocks or their protection in the Zhujiang (Liao *et al.*, 1989). Surveys undertaken between 1981 and 1983 (Liao et al., 1989) showed that the river contained 381 fish species, of which 262 were freshwater species, the remainder being brackish water or diadromous species including the endangered Chinese sturgeon. Phytoplankton (219 genera), zooplankton (410 species), zoobenthos (268 species) and macrophytic hydrophytes (132 species) were likewise diverse.

Historical analysis of fish catches from the Zhujiang reveal a peak during the 1950s (10,367 t.yr^{-1}) but, since the 1960s, catches have declined and in the early 1980s had fallen by 38% to 6,463 t.yr^{-1} (Liao *et al.*, 1989). In the case of the clupeid *Tenualosa reevesii*, which has also declined in the Chang Jiang, landings in the 1980s were only one-fifth of those in the 1960s, and this fish has been almost eliminated from the upper tributaries (Wang, 1996). Significant declines in the recruitment of major carp were noted, which has been attributed particularly to the use of ever-finer meshed nets and overfishing of spawning grounds, combined with the use of poisons, explosives and electric shocking to increase catches. Thus overfishing — combined with increasing pollution loads in the Zhujiang (Liao *et al.*, 1989; see also Qi *et al.*, 1982; Qi & Lin, 1985; Qi, 1987) — have led to declining fisheries yields, and regular fish kills have been attributed to untreated industrial waste water and pesticide pollution (Liao *et al.*, 1989).

River regulation is the third factor contributing to fishery declines in the Zhujiang because of the large number (3,300) of dams that have been constructed on the river since the late 1950s. Most were built without fishways and have caused reductions in populations of migratory fishes both through blockage of migration routes and transformation of lotic spawning sites into lacustrine habitats. Changes in the thermal characteristics of rivers downstream of impoundments may also have been a factor contributing to declines in fish populations. For example, construction of the Sijin Hydroelectric Station in 1958 caused a reduction in major carp populations which was attributed to a combination of reduced river flow, blockage of migration routes and lowered water temperatures (Liao, 1980). Liao *et al.* (1989) report that fish passes have been effective in facilitating fish migration past more than 20 dams on streams in Jiangsu Province (eastern China), and have called for their incorporation into new dams built on the Zhujiang. Damming has also produced changes in the Dongjiang (East

River), a major tributary of the Zhujiang. Although only fragmentary data are available, it appears that clupeids such as *Tenualosa reevesii* and *Clupanodon thrissa* which once migrated into the upper reaches of the Dongjiang to breed were eliminated by the construction of five dams in the lower reaches of the river. These dams, combined with construction of a series of reservoirs, have reduced the abundance of carp, especially *Cirrhinus molitorella*, to levels where there is no longer an economically-viable fishery in the river (Liao *et al.*, 1989).

While it seems clear that dams on the Zhujiang block migration routes of riverine fishes and have caused declines in the abundance of several species (Liao *et al.*, 1989), elsewhere in China increases in fish resources are reported as a result of impounding rivers (Yu *et al.*, 1981). While it may be unwise to extrapolate from a few particular cases, it seems unlikely that dams will be beneficial to potamodromous or diadromous fishes. Probably rather few species are capable of adjusting to existence in a reservoir, and thus where fisheries yields have increased this is likely to have occurred at the expense of species diversity. Moreover, it must be stressed that, as in the Zhujiang, flow regulation is often accompanied by other human impacts (pollution, over-fishing) which act synergistically to impact fishes and the riverine biota.

Long-distance water transfers

Possible responses to the shortage of clean water in many parts of China could include more effective water conservation and a reduction in the extent of pollution through tough legislation and compulsory waste treatment. An alternative approach would be to transfer water from the south where, at least during the summer monsoon it is relatively plentiful, to the more arid north. Inter-catchment transfer or movement of water to distant storage reservoirs is practised widely in China (Biswas *et al.*, 1983 and articles therein), and the Territory of Hong Kong, on the coast of southern China, depends upon transfer of water over a distance of 83 km from the Dongjiang in Guangdong Province; 1.1×10^9 m^3 of water is transferred each year, amounting to 8% of the total discharge of the river, and this meets 75% of Hong Kong's water needs.

Of the many water transfer schemes in China, the largest is that intended to transfer water northwards from the Chang Jiang. The initial scheme (proposed in 1978, and based on an idea conceived in 1952 by Mao Zedong) involved construction of three south-to-north canals to divert 30 gigatonnes of water each year, but the investment required in pumping equipment and modernization of the Grand Canal was judged to be excessive (Smil, 1993). Despite opposition to this scheme in the south on the grounds of cost, possible impacts on ports and harbours, and problems associated with salt water intrusion which could enter city water supplies, a decision was made in 1983 to undertake a reduced (but nevertheless massive) project. Recent discussions suggest that the original scheme may be revived (SCMP, 1995b) but, currently, work is proceeding on the first stage of the scaled-down development which will involve northward diversion of Chang Jiang water through the ancient Grand Canal to northern Jiangsu, Shandong, Hebei and Tianjin Provinces and Beijing (Fig. 5.3). This 650-km diversion will divert Chang Jiang water through a series of pumping stations along the Grand Canal, raising the water by 40 m before it passes into Dongpinghu (Dongping Lake) immediately south of the Huang He (Yellow River). Upon completion, the capacity of this system would be 10 gigatonnes.yr^{-1}. The second stage involves the construction of a tunnel under the Huang He to deliver some (approximately 3.2 gigatonnes.yr^{-1}) of Chang Jiang water further northwards to Hebei, Tianjing and Beijing (Smil, 1993). Water transfer may not take place during periods of low flow in the Chang Jiang, and this should reduce the possibility of salt water intrusions into the lower course of the river (Shen *et al.*, 1983).

Before the decision was made to scale down the initial (1978) scheme, the whole water-transfer project was the subject of an edited volume (Biswas *et al.*, 1983) considering the costs and benefits of the project. Most contributors to the volume addressed the impacts and implications of the three separate parts of the initial proposal so that the potential impacts of the part of the development which was actually approved in 1983 (initially termed the 'Eastern Route') are apparent from the published reports. Some (Liu & Liu, 1983; Huang, 1983; Yao & Chen, 1983; Zuo, 1983; Caulfield, 1986; Wang, 1987) emphasise the great benefits that would be derived from the scheme, while others considered that the project would be damaging and — in particular — would cause increased salinization in irrigated areas in the north (Herrmann, 1983; Stone, 1983; see also Xu & Hong, 1983); still others (Liu & Zuo, 1983; Yao & Chen, 1983; Zhu *et al.*, 1983) were cautiously optimistic. A report considering the benefits of reviving the initial scheme by the Chinese government is under active consideration and construction may begin before 2000 if the project wins approval (SCMP, 1995b). Work may go ahead because, as

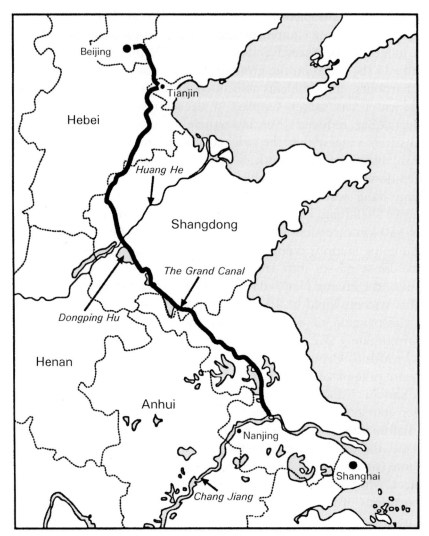

Fig. 5.3 South-to-north transfer of water from the Chang Jiang to Beijing, using the ancient Grand Canal after modification and installation of pumping stations (redrawn from Smil, 1993).

Caulfield (1986) states, '. . . inter-provincial water transfers from the Chang Jiang through Central and North China may be justifiable to "drought-proof" and "flood-proof" the area'.

Impacts of the water transfer on aquatic organisms have received relatively little attention in China or elsewhere (Davies *et al.*, 1992), but it is clear that the water levels in the lakes along the transfer route will become more stable which will reduce the magnitude and extent of land-water interactions causing a decline in primary and secondary aquatic productivity (Xu & Hong, 1983). Some concern has been expressed also over the possible northward spread of fish diseases (Stone, 1983) and parasitic schistosomes or blood flukes (Stone, 1983). Yao & Chen (1983) consider that the snail intermediate hosts (*Oncomelania* spp.; see p. 18 and p. 138) of the schistosome could not establish themselves far north, which Xu & Hong (1983) attribute to unfavourable temperatures along the water-transfer route. Yu (1983) has drawn attention to possible impacts of removal of water from the Chang Jiang on fish stocks; a particular problem was that extraction of water in the summer wet season might remove the stimulus required for breeding migrations by diadromous fishes (especially Clupeidae, Engraulidae and the Chinese sturgeon) and reduce the extent of the inundated floodplain which is an important feeding site for young fishes. In this context, Yu (1983) highlighted the paucity of data on migratory species in the Chang Jiang, while Zhu *et al.* (1983) have expressed general concern over the limited information upon which many assessments of ecological impacts are based. Other damaging effects of the water transfer were projected to include the mortality of larger carps and eels as they pass through the hydraulic turbines of the pumping stations where the pump blades would cut them to pieces (Yu, 1983). There are to be 13 such stations along the first part of the transfer route (Smil, 1993). Changes in conditions in lakes along the transfer route would reduce foraging and spawning areas for fishes, and lead to the loss of large areas of submerged and emergent plants (especially *Phragmites communis* and *Zizinia latifolia*). Furthermore, alterations in lake nutrient status caused by inflowing Chang Jiang water combined with increased suspended-sediment loads would reduce photosynthesis and zoobenthos production with deleterious impacts upon fish stocks, especially major carp (Yu, 1983; see also Xu & Hong, 1983).

River regulation elsewhere in tropical Asia

China is not the only nation planning ambitious river-control and water-transfer schemes and a few of them will be mentioned here. Almost all rivers of the Indian Subcontinent are impounded or regulated to serve as a source of irrigation water (Chitale, 1992): for example, over 13,000 km of canals originate from the Ganges and irrigate 7 x 10^6 ha (Natarajan, 1989), while the state of Punjab (Pakistan) has the largest canal irrigation system in the world (Firdaus-e-Bareen & Iqbal,

1994). One effect of the removal of irrigation water from the Ganges has been a gradual salinity increase in downstream reaches because the marine influence of the estuary is less diluted by freshwater discharge (Roy, 1955; Basu, 1965; Morton, 1977a). Attempts have been made to reverse this trend in the Hooghly River by diverting water from the main stream of the Ganges via the Farakka Barrage. Additional, large-scale river regulation projects in India are planned or under construction, including the 120-m high Sardar Sarovar Dam on the Narmada River (Wood, 1993). This is part of the Narmada Valley Development Project which involves plans for 30 big dams, 135 medium-sized ones and 3,000 smaller impoundments in the valley. The scheme has been the subject of heated debate over its projected benefits and potential environmental and social impacts (e.g. Fisher, 1995). Construction was halted in 1994 when the dam had reached a height of 82 m, and World Bank funding ceased. Resumption of work was delayed pending a ministerial review of the effects of the project (SCMP, 1996e), and protests against the different components of the scheme have continued (Vidal, 1998). The extent of regulation of Indian rivers reflects the fact that about 80% of their annual discharge passes through them during a four-month period, and thus the construction of dams is necessary to store water for use throughout the year (Chitale, 1992). Proposals for inter-basin transfers of water within India are being investigated by the National Water Development Agency (Chitale, 1992). In recent years, the Indian Government has been attempting to integrate environmental issues into the planning process for water resource projects (Chitale, 1988), but it not clear to what extent consideration is being given to the potentially serious ecological impacts of inter-basin transfers (Davies *et al.*, 1992). Iyer (1994) considers that basin-wide planning for Indian rivers — especially those crossing state boundaries where disputes arise frequently (Shah, 1994) — is needed urgently, and that more attention needs to be paid to the environmental consequences of flow regulation and water extraction (see also Sinha, 1984).

Some of the largest rivers in the Indian Subcontinent flow across international boundaries, and thus schemes for water transfer and sharing are needed. Thus agreement had to be reached between India and Pakistan to ensure equitable sharing of the water resources of the Indus (Caponera, 1987; Kirmani, 1990). Disputes over water rights have arisen over the Ganges and Brahmaputra, the major rivers of Bangladesh. They, like almost all of the rivers of that country, have their headwaters outside the borders, and derive only a negligible proportion of their flow from run-off within the country. Consequently, the discharge of many rivers in Bangladesh is affected by river regulation within India (e.g. Bandyopadhyay & Gyawali, 1994), and negotiations on water sharing are a regular feature of these countries' international relations (e.g. Crow, 1985; Abbas, 1987; Chaube, 1990; Quinn & Harrington, 1992; see also Biswas, 1993 for a general account of hydropolitics and international river basins). Difficulties arise because the existing dry-season flows are insufficient to meet the needs of both countries, and aggressive extraction of water from upstream reaches can seriously deplete water supplies in Bangladesh. By contrast, calamitous floods can occur in Bangladesh during the monsoon — in 1987 and 1988, for example, as well as in 1996 when Dhaka was flooded — inundating almost half of the country and giving rise to extensive human suffering (Anwar Khan, 1987; Rahman Khan, 1987; Khalil, 1990; Khan, 1991). While the harm caused by unusually severe floods is regrettable, it must not be forgotten that the development of agriculture (especially deep-water rice cultivation) in Bangladesh has been based upon anticipation of some annual flooding and silt deposition. A massive scheme to manage floods and droughts — the Bangladesh National Flood Action Plan — has been commissioned (Brammer, 1990). This multi-billion-dollar engineering project involves construction of a vast network of embankments, dykes and canals as well as dams and associated storage reservoirs. Undoubtedly, it will have significant impacts on the hydrology, ecology, agriculture and fisheries of the Bangladesh floodplain; indeed, there has been a long-term declining trend in inland capture fisheries since 1970 (Sultana & Thompson, 1997). Significantly, the Flood Action Plan provides an opportunity to develop a plan for the management and conservation of capture fisheries in Bangladesh through controlled flooding. Potential strategies include the protection of wetlands by building bunds to retain water during the dry season, and designing flow regulators that are operated to give adults of commercially important (but declining) major carp access to spawning grounds. Appropriate management and operation of flood control and drainage measures could also regulate the passage of water containing carp spawn and fry in a 'fish friendly' manner (Sultana & Thompson, 1997 give details).

Other major river control schemes in tropical Asia include the construction of a chain of dams along the Mahaweli River, the largest (10,750 km^2 basin; annual discharge 953 x 10^6 m^3) and longest (315 km) river system in Sri Lanka (Bandara, 1985; Hewavisenthi, 1992). Also in Sri Lanka, there are plans to divert the Menik River so that it will flow into the Kirindi Oya Reservoir (Amarasekara, 1992). It is of interest that the earliest attempts to dam the Mahaweli River for irrigation were made 1,700 years ago by King Mahasena (274–301 AD), and the extent of ancient irrigation canals in Sri Lanka points to the existence of an early 'hydraulic civilization' dating back to around 600 BC.

A recent proposal to dam Malaysia's largest river indicates some disturbing trends in water-resources development in the region. The 205-m Bakun Dam on the Balui River in the upper Rajang basin, Sarawak, provides an outstanding example of the conflicts that can arise among different interest groups involved in dam projects, and the important influence of government policy tin this regard. The Bakun Dam is intended to provide cheap power to fuel development and to enhance opportunities for investment. Although the project is backed strongly by the Malaysian Government, there is opposition to it from Malaysian non-government organizations (NGOs) and residents along the Balui River. An environmental impact assessment prepared for the developers notes that (among other things) there will be a degradation of fish habitat and a loss of fishery resources both downstream and within the inundation area. However, the consequences of these changes for biodiversity and subsistence fishery resources, and the effects of reduced frequency and duration of inundation of the floodplain, are not addressed in detail. In June 1996, the Kuala Lumpur High Court ruled that residents of the area to be submerged had been deprived of their right under federal environmental law to be consulted before official approval was given for the dam. In mid-July, however, project proponents persuaded the High Court to suspend the ruling pending an appeal, and work at the site by a Swiss-Swedish engineering multinational company was set to begin in March 1997. Press statements by Malaysian Prime Minister Dr Mahathir Mohamad, who visited Sarawak in August 1996, and his ministers have made it abundantly clear that the Federal Government is in favour of the dam (Anon, 1996a; SCMP, 1996f), asserting also that the government would ensure that construction would have no adverse effect on the environment. Interestingly, this statement follows a 1990 decision to cancel dam construction on environmental grounds that was reversed when the government revived the project in 1993. In the meantime, the project developer has stated that no one can enter the area around the dam site without permission from the government or state police (SCMP, 1996f). The 1997 economic downturn in Asian economies has lead to the cancellation or postponement of a number of large infrastructural projects, including dams. This may delay the onset of anthropogenic impacts on streams and rivers, but will not change the perceived need to undertake such projects. Issues of conservation policy and practice relating to the rivers of tropical Asia lie beyond the scope of this book, and the matter is discussed in detail by Dudgeon *et al.* (in press).

The Mekong River

Planned and ongoing water-resource developments in the Mekong River basin illustrate the extent and range of threats to the ecology of tropical Asian rivers. This example demonstrates also the conflict between the pressures for economic development, which are usually driven by governmental imperatives, and the ecological consequences of such development which are most often felt by local communities some distance from centres of political power and policy (as epitomised by the Bakun Dam discussed above).

Plans for the Mekong basin date back to the 1950s when Raymond Wheeler, a retired general of the United States Army Corps of Engineers, headed a mission to study the hydropower potential of the Mekong. The annual discharge of the Mekong is over 475×10^6 m^3, yielding a potential energy capacity of 58,000 MW and the possibility of irrigating some 6,000,000 ha (Chomchai, 1987). In 1957, Wheeler's report to ECAFE, an Economic Commission for Asia and the Far East created by the United Nations (renamed ESCAP — the Economic and Social Commission for Asia and the Pacific — in 1974), identified seven sites on the river that were suitable for the multi-purpose development of water resources. Also in 1957, the riparian states formed a coordinating committee to function under the auspices of the United Nations as represented by ECAFE. Thus the Committee for the Coordination of Investigations of the Lower Mekong Basin came into being.

The Mekong Committee (serviced by the Mekong Secretariat comprising an array of water-resource specialists) assigned first priority to dams at Pa Mong and Sambor (Thailand), and to a barrage which would control the movement of waters into and out of Tonlé Sap River and Le Grand Lac in Cambodia. Work on these was envisaged to proceed in parallel with projects on smaller tributaries of the Mekong in Cambodia, Laos, Vietnam and Thailand. By 1967, however, only two small tributary projects (both in Thailand) had been completed. War and political instability in the region (combined with more recent economic constraints) made the construction of large infrastructural projects and work of an international nature impossible, and for nearly three decades the Mekong Committee sat more-or-less idle, unable to fulfil its mandate. Recently, the political situation has stabilized and, by 1986, there were 14 dams on tributaries in the lower basin (Pantulu, 1986a). Following completion of a major dam on a tributary of the Mekong at Nam Ngum in Laos (Fig. 5.4), further feasibility studies were launched (Chomchai, 1987). Despite the potential for detrimental impacts on river biota, the reservations of former staff of the Mekong Committee regarding the acceptability of impacts on fish and fisheries (Usher, 1996), and some criticism by the World Bank, in December 1994 the Mekong Secretariat published the *Mekong Mainstream Run-of-River Hydropower* study document which identified 12 potential dam sites on the mainstream, 11 of them with generating capacities of over 1000 megawatts. Five months later, the Mekong Committee was replaced by the Mekong Commission formed by the four lower riparian states. At the same time, the Swedish International Development Authority strengthened its previous support for the authority (Usher, 1996). The Commission was set up as an autonomous, intergovernmental organization to develop policies on the Mekong basin, one of the first being that member states were allowed to use the river waters without seeking the approval of other members — except during the dry season (SCMP, 1995c). The Mekong Commission has stressed its commitment to dam construction at 12 sites along the mainstream of the river in Laos, Thailand and Cambodia (Roberts, 1995). The environmental impacts of these dams on, for example, fish spawning and migrations, have yet to be assessed fully, but are hardly likely to be positive (Pantulu, 1986b; Lohmann, 1990; Roberts, 1992; 1993a; 1993b; 1995: Baird, 1994). This has important socioeconomic implications given that the annual fisheries yield from the lower Mekong basin is approximately 0.5 million t (Lohmann, 1990), and estimates for the Vietnamese portion alone are 100,000 — 160,000 t (Tran Thanh Xuan, Ministry of Fishery — Vietnam, pers. comm.). Adequate data for sound management decisions are still lacking (Pantulu, 1986b; Roberts, 1993a), insufficient or unreliable (Hori, 1993), although the Mekong Committee has been attempting to rectify this deficiency (Petersen & Sköglund, 1990; Choowaew, 1993).

It should be stressed here that although plans for mainstream dams in the lower Mekong basin (Fig. 5.4) have been stalled for many years, this is not the case in the upper basin. For example, China is not a member of the Mekong Committee, but nevertheless has its own schemes for the Mekong: the 1,500-MW Manwan Dam in Yunnan

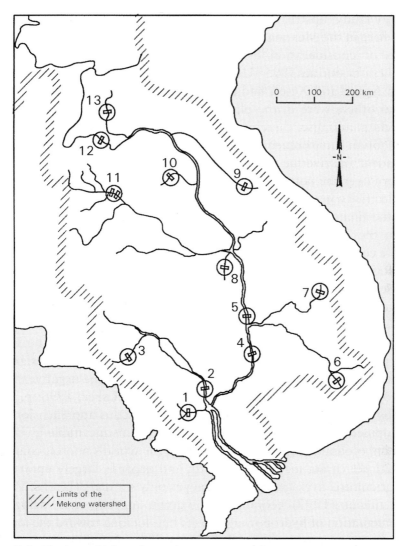

Fig. 5.4 The lower Mekong basin showing some of the major river regulation projects that have been proposed or constructed: 1, Prek Thonot; 2, Tonlé Sap; 3, St. Battambang; 4, Kratie; 5, Stung Treng; 6, Upper Srepok; 7, Upper Se San; 8, Lam Dom Noi; 9, Upper Nam Theun; 10, Nam Pung; 11, Nam Pong; 12, Pa Mong; 13 Nam Ngum. From various sources, including Takenouchi, 1966; Pantulu, 1970; Chomchai, 1987 (see Lang [1996] and Usher [1996] for further information and more detailed maps).

Province (completed in 1995) is only one of at least 14 dams that China intends to build on the river (SCMP, 1993; 1995). China and Burma have recently agreed to participate in an subcommittee of the Mekong Commission which is designed to facilitate information sharing and to study upstream hydropower development (Anon, 1996b). Elsewhere in the Mekong basin, over sixty dam projects were in various stages of consideration by the Laotian Ministry of Industry and Handicrafts during 1995 (Usher, 1996). Although only two projects (Nam Ngum and Xeset) had reached the construction stage by 1995, several others were in the planning stage. If even a fraction of these projects materialise, there are expected to be serious implications for the Laotian environment and for the people whose livelihoods are dependent on riverine ecosystems (Usher, 1996). There is particular concern over the potentially damaging effects of these projects because the electricity generated will be sold abroad, mostly to Thailand, and thus the financial benefits are likely accrue to those far from the sites where the impacts of dam construction are felt. Secondly, foreign private companies — especially from Nordic countries — are investing in Laotian hydro-power development, and it seems that there is an inverse relationship between the costs of avoiding or mitigating impacts in Laos and the profits to be gained abroad by these firms (Usher, 1996). There has been concern also over the role of private consultant firms in the hydropower development in Laos, since they are hired not only to assess social and economic interests, but also to design the projects. This dual role '. . . practically ensures that the negative effects of projects are glossed systematically over . . .' (Usher, 1996: p. 125) thereby leading to 'appraisal optimism'. Despite this apprehension, the development of hydropower and irrigation seems inevitable given the potential economic benefits: Laos is one of the world's poorest countries (annual per capita income <US$300), and depends largely upon rain-fed agriculture; average life expectancy is only 49 years (Jacobs, 1994). The current (1997) economic slowdown in Asia may delay the implementation of hydropower projects but, looking toward the longer-term, it seems most unlikely that they will be abandoned.

Among other dams to be completed on Mekong tributaries in Laos is the Nam Theun Hinboun, a 210-MW hydropower project on the Theun River. Construction — funded by companies in Sweden and Norway — began in 1994 and is expected to be complete in 1997. In another example of 'appraisal optimism', the interests of the rural villagers, who depend upon fish stocks and river transport routes, were given scant consideration by foreign consultants who concluded that there would be no negative effects on fish and significant beneficial environmental impacts (Usher, 1996). This is despite a projection that the river bed will be reduced to a series of pools downstream of the dam during the three-month dry season. It is evident that the dam will block the migration routes and destroy the downstream habitat of some of the 140 species of fish that inhabit the Theun basin and must inevitably decrease this source of food and income for local people. In this and other dam projects in Laos, northern interests and agencies, especially from the Nordic countries, have provided both the means and incentive to harness the hydropower potential of Laotian rivers. Though the pervasive 'appraisal optimism' which exudes from the proponents and evaluators of dam schemes (as well as aid donors and financiers), the negative environmental and social consequences are down-played, leaving an impression of a 'win-win' solution that will generate hard currency and a 'green' environmentally-benign energy (Usher, 1996).

There is a persistent view that '. . . water resources development may create positive effects especially on fisheries through the integration of freshwater aquaculture in reservoir projects . . .' (ESCAP, 1992: p. 177; see also Djuangsih, 1993), and that '. . . the nutritional status of rural people is further improved by the availability of animal protein through . . . the development of inland fisheries in the newly created reservoirs' (Biswas & El-Habr, 1993: p. 121). The project consultant's view in the Nam Theun Hinboun case was that increasing the water level behind the dam would provide an ideal environment for rearing fish, and would improve the situation for local people (Usher, 1996). This statement is, however, based upon a lack of understanding of fish ecology (see Roberts, 1993b) and reflects the conflict of interest that (in this case) foreign consultants face if they can benefit from one outcome over another (Usher, 1996). An increased fisheries yield at Nam Theun Hinboun might have to depend on exotic species, and would certainly be accompanied by a loss of indigenous biodiversity.

Hydropower development in Vietnam includes the Yali Falls Dam which is part of a six-dam scheme proposed by the Mekong Committee for the Se San River. Although the environmental impact assessment of the project states that the general pressure on wildlife habitat will be increased by the project, the prevailing Government view seems to be that hydropower will decrease the use of wood and charcoal for fuel thereby reducing deforestation and resulting in a net ecological benefit (Lang, 1996). This ecological equation seems rather simplistic. After completion of the Hao Binh Dam in northern Vietnam — which is the largest dam in Southeast Asia — displaced people moved to the upstream part of the drainage basin where they cleared forest to plant crops, thereby increasing runoff and sedimentation (Lang, 1996). It is possible that this situation will occur again around the Yali Falls Dam, unless there is an increase in the government will and wherewithal to preserve forested catchments upstream (and so, incidentally, prolonging the useful life of the dam). Even if this does occur, downstream impacts of the dam on the river biota are likely to such that the projected net ecological gain of the scheme disappears.

Important environmental consequences of dam construction along the Mekong will include changes in inundation patterns which have, for example, affected aquatic productivity of waters at Nam Ngum (de Bont & Kleijn, 1984; Beeckman & de Bont, 1985). Roberts (1993b; 1995) reports that the recently-completed Pak Mun Dam on the Mekong in Thailand has been extremely damaging to fish populations, which declined from 1991 (when dam construction began) until 1994 (when it was finished). Affected fishers received a one-off US$3,600 per family compensation payment. Dam building on the Mekong mainstream will lead to changes in flow and temperature regime and may remove important directive factors which stimulate fish migratory and breeding behaviour. Such dams will obstruct the passage of several long- and medium-range whitefish migrants. These include Pangasidae catfish (especially *Pangasius* spp. which contribute 10–20% of total landings from the river; Pantulu, 1970), cyprinids such as *Cirrhinus auratus, P. jullieni,* and other fisheries species. Of these, the giant 3-m Mekong catfish, *Pangasius* (= *Pangasianodon*) *gigas* is believed to undergo spectacular long-range migrations for spawning, travelling as far upstream as Yunnan Province (Smith, 1945; Pantulu, 1970; but see Bhukaswan, 1983). Attention was drawn to declining populations of this poorly-known species earlier this century (Smith, 1945), when was the basis of an important fishery in Cambodia wherein the fish as caught for its flesh and the extraction of oil due to its high fat content. It has declined greatly in recent years (Nandeesha, 1994). *Pangasius siamensis* may be extinct already. *Pangasius krempfi* is an anadromous species, with a life history resembling that of salmon; it appears to comprise a single population in the Mekong River (Roberts & Baird, 1995). These large fish, reaching up to 15 kg and over a meter in length, are of particular economic importance to fishers along the river (Baird, 1994).

Roberts (1995) is of the opinion that the series of mainstream dams proposed by the Mekong Committee in 1994 will cause '. . . extirpation or extinction of many fish species, including strongly migratory species that are the main basis for Mekong wild-capture fisheries . . .'. Some of these species are confined to the Mekong: for example, *Pangasius gigas* and *P. krempfi,* the carps *Aaptosyax grypus* and *Probarbus labeamajor*

(Cyprinidae), and the freshwater herring *Tenualosa thibaudeaui* (Clupeidae) which is already endangered (see also Roberts, 1992; Roberts & Baird, 1995). They could be driven to extinction by a single mainstream dam (Roberts, 1995). Fish ladders — which have been proposed as a means of mitigating the impacts of dams on migratory fishes — are unlikely to be successful in the Mekong, because the majority of species in the river do not jump (Roberts, 1993b; 1995). Moreover, both upstream and downstream passage must be provided (*cf.* salmon). As Hill (1995) points out, research and development of passage facilities for tropical fish species has rarely been addressed and there is virtually no relevant information to draw upon. Meisner & Shuter (1992) raise an the additional issue of the effects of global climate change on the annual flow regimes and breeding migrations of floodplain fishes in the Mekong. The extent to which possible changes might magnify or ameliorate the effects of mainstream dams is unknown.

The impacts will extend beyond fish to other taxa. The Irrawaddy dolphin, *Orcaella brevirostris*, enters rivers throughout Southeast Asia (Sigurdsson & Yang, 1990; Baird & Mounsouphom, 1994), and is in need of protection in some rivers (Wirawan, 1986; Baird, 1994). Recent research suggests dolphin numbers in the Mekong have fallen because of declines in fish prey and death of animals trapped in gill nets, but the large-scale dams proposed for the basin will be greater threats to *O. brevirostris* in Laos and Cambodia (Baird & Mounsouphom, 1994). Considering possible impacts on invertebrate taxa, we do not know what effect the proposed dams along the Mekong will have on the endemic species-flocks of stenothyrid and pomatiopsid gastropods (over 100 species; Davis, 1979) in the river. Attwood (1995b) considers that modified flow regimes caused by dams may lead to changes in food availability and substrate characteristics which will favour some snails — in particular, *Neotricula aperta* which is a host of the human parasite *Schistosoma mekongi* (Trematoda: Schistosomatidae). Such changes, together with the translocation of riparian peoples and the influx of casual labour from endemic areas in Laos, may result in an epidemic of human schistosomiasis in northeast Thailand (Attwood, 1995b). Significantly, Srivardhana (1987) reports that there has been a great increase in flukes and intestinal parasites among residents of the irrigated area associated with the Nam Pong Dam in northeast Thailand.

In addition to the impacts which may arise from implementation of the Mekong Secretariat's proposals for the lower basin, and various hydropower plants in Laos and Vietnam, there may be cumulative impacts on river ecology from other projects. The Thai Government has been formulating a master plan for dam building and drainage-basin development covering the north and northeast parts of the country where rainfall is low and unreliable (Office of the National Economic and Social Development Board, 1994). It is likely that the latter project will go ahead because of the urgent need to improve soil conditions, agriculture and forest management, and to enhance the economic development of this, the poorest region of Thailand (Krishna, 1983; Hori, 1993; Jacobs, 1994). Significantly, the World Bank has been critical of the scale of the developments proposed by the Mekong Secretariat for the lower basin, and considers that a more appropriate approach to development in the riparian states should, in the first instance, be centred on improvements in agricultural productivity through small-scale irrigation projects, rural development and related hydropower projects (Kirmani, 1990). The Thai Government plans seem more in line with World Bank thinking in this regard and, because of their smaller scale, are likely to be less damaging — and more 'sustainable' — than the large mainstream dams that have been proposed for the Mekong. Indeed, the Mekong Committee and Secretariat has recently broadened its programme of work to take greater account of environmental and social concerns (Jacobs, 1994).

Significantly, Usher (1996: p. 130) reports that '. . . some 30 Thai NGOs and local water basin groups issued a statement . . . opposing the influence of the dam-building industry in the creation of the new Commission'. Similarly, Kirmani (1990: p. 203) considers that the lower Mekong basin project is a '. . . classic example of external effort, external management and external planning with little involvement of the beneficiaries . . .' and '. . . the assistance of donor countries was utilized mostly to finance their own experts and consulting firms . . .' while the '. . . . Mekong Secretariat was managed by foreign experts'. While such comment and criticism is well founded (see also Lohmann, 1990; Roberts, 1993b) and will probably delay the construction of mainstream dams on the lower Mekong for the foreseeable future, Jacobs (1994) points out that without dam construction for hydropower (and the concomitant impacts on river ecology) to fuel economic development, Thailand will burn more lignite while Cambodia and Laos will continue to clear their forests for fuelwood. 'To reject outright the prospect of hydropower project construction denies these countries the opportunity for development and keeps them upon a path of environmental degradation and poverty . . .' (Jacobs, 1994: p. 50). McNeely (1987) suggests that drainage-basin protection could have been used to justify reserves and preservation of riparian habitats along the Mekong, while the economic impetus arising from water-resource development might stimulate government control and enforcement of environmental protection. It is not yet clear whether such hopes for the integration of conservation with development along the Mekong will bear fruit, but there is little room for optimism over the fate of migratory river fishes.

DRAINAGE-BASIN MANAGEMENT

The need for river regulation is to a large degree a response to rapid population growth in Asia combined with the drive towards economic development, both of which have been accompanied by significant environmental problems in drainage basins. Forests have been cleared to make way for agriculture, but also for fuel wood: excluding China, South Asia contains 30% of the world's population but only 2% of the known reserves of fossil fuels (Revelle, 1980). Rapid loss of forest cover in drainage basins (figures vary, but estimates of annual deforestation in tropical Asia are around 2%: Lanly, 1982; Lean *et al.*, 1990; World Resources Institute, 1994) leads to an increase in sediment-rich surface run-off and stream-flood flows (Sutadipradja & Hardjowitjitro, 1984; Bartarya, 1991; Haigh & Krecek, 1991; Tejwani, 1993). For example, forest clearance has affected fish populations of Le Grand Lac on the Mekong, leading to a decline in catches attributed to erosion, siltation and the lessened availability of allochthonous food arising from the reduced area of forest (Welcomme, 1979). The silting threatens the existence of the lake which has an average depth of only about 1 m during the dry season (Pantulu, 1986a). Siltation has also increased turbidity through resuspension of particles by wave action in the increasingly shallow water. This, in turn, has led to a drop in primary productivity which is reflected in lower fish populations.

Drainage-basin degradation exacerbates the seasonal floods and droughts characteristic of tropical Asian rivers (Tejwani, 1993), and thus contributes to the perceived need for flow regulation. This cause and effect link has been long been acknowledged in China where deforestation is thought to precipitate floods and droughts (e.g. Carin, 1962; Smil, 1993), including the 1991 deluge of the Chang Jiang floodplain when approximately 2,500 people died and 18.4 million were seriously affected. Chinese researchers have repeatedly drawn attention to the connection between the loss of tree cover and the incidence of destructive floods and stream siltation. Significantly, the average area of China disastrously affected by floods was approximately 5.5 x 10⁶ ha during the 1980s (Smil, 1993), nearly a 2.5 fold increase compared to the 1970s, and the risks of catastrophic flooding have been growing almost everywhere in China (Smil, 1993). Deforestation probably also contributes to mudflow disasters in mountainous areas of China and elsewhere (Rapp *et al.*, 1991; Ahmad, 1993; but see Hofer, 1993; Bandyopadhyay & Gyawali, 1994).

The transport of silt and soil during floods in Chinese rivers is of major national concern. The once-forested upper reaches of the Chang Jiang have been denuded in the last 30 years, leaving only 3% of forest cover. Spectacular erosion has resulted, and flood waters may bear more than 50% by weight of suspended material (Pereira, 1981; 1989). The situation is particularly serious in loess (red soil) areas of Sichuan Province and, in the early 1980s, an estimated 67.3% of the area of the Province was affected by soil erosion (Luk & Whitney, 1993a). In the 1970s it was estimated that significant erosion affected 20% of the Chang Jiang basin and, a decade later, the figure had risen to over 30% or 560,000 km² (Smil, 1993). Luk & Whitney (1993a: p. 15) give an even higher estimate of 41% (739,376 km²) of the river basin being affected by soil erosion in the '. . . early 1980s'.

Heavy loads of sand and silt have led to a deterioration of riverine environments, and deposited sediments block river channels and spill floods across the plains. Extensive flooding in the Chang Jiang valley which has occurred on several occasions during the past 15 years is probably attributable to increasing siltation which has led to elevation of the river bed and the disappearance of many small floodplain lakes in Hubei and Hunan Provinces. The situation is serious in Dongtinghu (Hunan) and Poyanghu (Jiangxi) — especially the former where the lake bed is rising by 2.5 cm.yr⁻¹ and sedimentation has caused the lake area to shrink from 4,350 km² in 1949 to 2,740 km² in 1987 (Wang, 1987; Jin *et al.*, 1990; Smil, 1993). Consequently, the lake's ability to buffer flood flows has decreased substantially, and it has been estimated (Wang, 1987) that the lakes of the Chang Jiang floodplain have lost a storage capacity of 35,000 x 10⁶ m³. Aside from the damage caused by floods, economic consequences of increased silt loads include the clogging of irrigation systems leading to reduced grain yields. In Guangxi Province, for example, over 20% of the irrigation network has silted up (World Resources Institute, 1994: p. 72). Li (1989) has estimated that the increased accumulation of silt has eliminated approximately 25% of the reservoir cavity built between 1960 and 1989. While small check dams (low dams built across gullies and streams to store flood run-off) have been used with some success to reduce sediment inflow to reservoirs (Biswas, 1990; see also Das, 1991), the problem of reservoir sedimentation remains serious.

In parts of the Chang Jiang valley where erosion has not taken place, Smil (1993: p. xvii) notes that '. . . the slopes are covered mostly with grasses. The absence of large trees, groves, and forests seems unnatural in such humid places. There are . . . no larger forest patches even in higher elevations. Occasional aligned plantings of thin-stemmed saplings or a few trees left standing . . . actually accentuate the sweep of deforestation'. In response to such damage, the Chinese Government instituted a reforestation programme with a target of 70×10^6 ha by the year 2000 (Pereira, 1981). The programme has had variable success, but over 30×10^6 ha of trees have been planted in China since 1949, an area exceeding the total for 76 other developing countries (Lanly, 1982; World Resources Institute, 1994: p. 69). Despite increased planting in recent years, the situation remains serious (SCMP, 1996g). Mixed agricultural practises have been employed to improve degraded uplands in southern China: hilltops are reforested to reduce erosion and run-off, upper slopes are managed as plantations while lower slopes are given over to orchards and mixed agriculture. Low-lying floodplain areas (often termed 'wastelands' in China) are drained and converted to grain production or the cultivation of livestock, ducks and fishes in traditional dike-pond systems.

Reforestation has a significant, but limited, ability to moderate flood flows because — relative to natural forests — protection forests and orchards have less biomass and less diversity and therefore generally less ground cover and poorer absorptive capacity (Kumar & Verma, 1991; Pande, 1991; Ambasht *et al.*, 1994). Reforestation moderates the destructive effects of floods by damping peak flows, strengthening river banks, and reducing erosion and transport of soil and detritus. However, even large areas of forest can become saturated with water and may be unable to ameliorate flood flows (Hsia, 1987). For instance, in the eastern Himalaya and Burma up to 5,000 mm rain may fall during the six-month monsoon, and annual peak discharges of southern Chinese rivers may be over an order of magnitude greater the low-flow (dry season) values (see p. 24 *et seq.*).

Elsewhere in Asia, drainage-basin degradation and resulting effects of sediment on receiving waters are just as critical as those recorded in China (see, for example, Chanphaka, 1982; Gupta, 1983; Sutadipradja & Hardjowitjitro, 1984; Henderson & Rouysungnern,

1985; Pereira, 1989; Bartarya, 1991; Das, 1991; Melkania, 1991; Chakrapani & Subramanian, 1993; Ambasht *et al.*, 1994). Over 20 years ago, Thiemmedh (1970) pointed out problems of river degradation in Thailand, where patterns of stream flow were altered as a consequence of forest clearance in drainage basins in the north (see also Chanphaka, 1968; 1982). Douglas (1968, 1970) recorded similar effects of forest removal on suspended-sediment loads in Malaysian rivers at around the same time, and the situation has not improved in either country since then as continued development in upland areas has lead to soil erosion, silting and flash floods (Chanphaka, 1982; Henderson & Rouysungnern, 1985; Sun, 1985; Jahi & Sani, 1987; Charoenphong, 1991). Logging practises influence the degree to which sediment loads of Malaysian streams are elevated (Yusop & Suki, 1994), although the association of forest clearance with increased turbidity and suspended solids remains.

Heavy loads of sand and silt have led to a deterioration of riverine environments in many parts of Asia. Deposited sediments block river channels and spill floods across the plains. Deforestation, with a consequent upsetting of the hydrological balance, has been considered a major cause of this deterioration (Henderson & Rouysungnern, 1985; Bartarya, 1991; Das, 1991; Melkania, 1991; Ahmad, 1993; Chakrapani & Subramanian, 1993) so precipitating floods, droughts and mudflows (Charoenphong, 1991; Rapp *et al.*, 1991), although some workers find no systematic correlation between upland deforestation and severe lowland flooding (Hofer, 1993; Bandyopadhyay & Gyawali, 1994). A confounding environmental factor is the intensity and variability of natural processes in the Himalaya which may mask anthropogenic changes in the hydrologic regime of rivers. Nevertheless, it is regrettable that clear-felling has been official forest policy in India for many years (Bandyopadhyay, 1988). Slopes too steep for cultivation (e.g. in the hill districts of the Indian Himalaya) are officially designated as forest, but are in a degraded state because of overgrazing by unproductive, undersized animals kept mainly to produce manure for fuel (Melkania, 1991; Ahmad, 1993). In consequence, an estimated 33% of forests administered by Indian federal or state forest services has been degraded severely (Pereira, 1989).

Hill-slope degradation and resulting effects of sediment on receiving waters are critical in Nepal (Thapa & Weber, 1994; 1995). This mountainous territory has a monsoon climate (1,500–3,500 mm annual rainfall) and its major natural resource is a vast potential for hydroelectric power (Shibusawa, 1987; Rao & Prasad, 1994). By the late 1980s, a mere 110 MW have been commissioned from Nepal's estimated 20,000 MW of generating potential. Obstacles to the development of the enormous hydroelectric potential include slope instability, high sediment discharge regimes, extremes in flow, and vulnerability of dams to seismic activity (Bandyopadhyay & Gyawali, 1994). In particular, funding the construction of major dams and reservoirs is only practical if there are stable soil and water conditions in the drainage basins, but they are generally lacking. A shortage of fuelwood, overgrazing resulting from competitive overstocking of a common fodder resource, and a rapidly-growing (over 40 million) population have had the combined effect of causing severe soil erosion (Shrestha *et al.*, 1983; Pereira, 1989) which has contributed to a great increase in both dissolved and suspended loads in the Ganges River. The high turbidity has reduced the effective photic zone in the river and diminished benthic productivity (Natarajan, 1989); undoubtedly, it has had other ecological effects.

India would not only provide the market for Nepalese electricity, but would benefit also from dams which would reduce havoc caused by annual floods on the Gangetic plains where over 400 million people dwell (Shibusawa, 1987; Rao & Prasad, 1994). Bandyopadhyay & Gyawali (1994) refer to this as the '. . . plains bias . . .' whereby large dams are built upstream or at high altitude to control floods in the foothill zone; when the additional purposes of hydroelectric power development and irrigation are taken into consideration, the construction of high dams becomes an attractive investment objective. In this context, the interests of the indigenous mountain dwellers (whether in Nepal or India) are largely ignored (Bandyopadhyay & Gyawali, 1994), and new approaches to sustainable development in the region are needed urgently (Ahmad, 1993; see also Shrestha & Paudyal, 1992). Bandyopadhyay & Gyawali (1994) list examples of schemes showing the 'plains bias' including the controversial Tehri Dam Project on the Bhagirathi River in Garhwal Himalaya (see also Das, 1991; Nautiyal *et al.*, 1991), the multipurpose high (270 m) dam across the Chisapani Gorge on the Karnali River, and the Three Gorges Project in China.

The extent of dam construction in the Himalaya is apparent from the following breakdown according to major river basins (dating from 1992; figures from Bandyopadhyay & Gyawali, 1994): 59 dams on the Indus in India; 56 dams in the Ganges basin; and 37 dams in the Brahmaputra basin. The sustainability and economic viability of such dam projects are questionable, and relevant ecological and hydrological data are lacking almost entirely (Bandyopadhyay & Gyawali, 1994). There huge areas of eroded land in the catchments of Indian reservoirs (e.g. 45.8 x 10^6 ha in the catchments of 21 reservoirs of flood-prone rivers), causing high rates of sedimentation (Das *et al.*, 1981; Gupta, 1983; Das, 1991: Melkania, 1991). In Bihar, which contributes a major part of the discharge of the Ganges River, forest cover is only 2.4% (Natarajan, 1989). The problem was highlighted by the largest flood ever recorded on the Yamuna River which swept through New Delhi in 1978 (see also Sharma & Vangani, 1982). Despite the use of embankments, there has been progressive increase in the area affected by floods in India each year (Pereira, 1989): this may reflect the inefficiency of massive embankments as a permanent flood defense against rivers carrying heavy silt loads. Significantly, the amount of land in India classified as 'flood prone' has increased from 20 x 10^6 ha in 1970 to 40 x 10^6 ha in 1980; by 1984, the figure had reached 60 x 10^6 ha (Bandyopadhyay, 1988). The Indian government's response to a deteriorating situation was to appoint a National Commission of Floods in 1976 to conduct an in-depth study of flood control and floodplain management for the whole of India. This monumental task is still in progress, and a lack of successful management of Himalayan drainage basins may to contribute to devastating floods in Bangladesh during the monsoon (Anwar Khan, 1987; Rahman Khan, 1987). The Bangladesh National Flood Action Plan has been devised to mitigate the effects of these calamities (see p. 539).

Problems arising from the misuse (largely overgrazing) of steep uplands in important drainage basins in Pakistan are exemplified by the mountainous Hazara district of the Northwest Frontier where major reservoirs are located: the Tarbela Dam on the Indus River and the Mangla Dam on the Jhelum. Together they provide about five days supply of irrigation water. Sedimentation rates in both reservoirs are high because of erosion in the drainage basin (Khattak, 1983; Khan, 1985), and this is critical given that the economy of Pakistan depends upon irrigation of the alluvial plain of the Indus with water from these reservoirs (Pereira, 1989).

The examples given above outline the scope of the problem of drainage-basin degradation, and the matter is discussed at more length by Dudgeon *et al.* (in press). Accounts of attempts to manage Asian forest catchments so as to reduce soil erosion and degradation of streams draining damaged drainage basins are given by van Bronckhorst (Java; 1985), Flemming (1983; Nepal), Gunatilleke & Gunatilleke (1983; Sri Lanka), Khattak (1983; Pakistan), Pant (1983; India),

Bartarya (1991; India), Singh *et al.* (1991; India), Reyes & Mendoza (1983; Philippines) and Nykvist *et al.* (1994; Malaysia). Overviews of major river valley projects are provided by Chakravarty (1969; India) and Petersen & Sköglund (1990; lower Mekong basin), while drainage-basin management problems are described by Chanphaka (1968; 1982) and Sabhasri (1968; Thailand), Balchand (1983; India), Huang & Ferng (1990a, 1990b; Taiwan), and Bandyopadhyay & Gyawali (1994; eastern Himalaya) who question the accepted wisdom regarding drainage-basin management and large-scale flow-regulation projects.

River drainage basins in overcrowded tropical lands are not all misused. Java has a population of 91 million (average 600 individuals.km^{-2}), but the steep upper slopes of many volcanoes remain under protective forest cover, with 21% of the land area forested. Here, catchments are protected because of a lack of free-range grazing, a tradition which has arisen because the land is held individually (Pereira, 1989), and produce is derived from mixed, multistorey tree gardens. Similar tree gardens are also a feature of the intensively-settled slopes of Sabah, Bali, and parts of Sri Lanka. They represent a modification of the natural forest ecology that (as an incidental consequence) reduces impacts upon rivers. Where the Javanese population has spread out onto thinner soils and steeper slopes, and cleared the forest for agriculture, increased sediment loads have been recorded (van Bronckhorst, 1985; Pereira, 1989).

Tropical deforestation and river fisheries

Southeast Asia has lost approximately 67% of its original wildlife habitat (Braatz *et al.*, 1992), and, as mentioned above (see p. 549), the rate and extent of deforestation in the region is disturbing. Most of the loss has been recent: forest cover in Thailand decreased from 53% of the total area in 1961 to 26% in 1993. Likewise, 90% of lowland forest in the Philippines has disappeared during the last three decades, and only 5% of the land area remains under natural forest. Rates of loss there and in Vietnam are over 1.5% per year, while forest clearance in Indonesia proceeds at a rate of 10,000 km^2 per annum (Braatz *et al.*, 1992). The general pattern over the last three decades has '. . . been for forests to be degraded and destroyed, for much native biodiversity to be eroded and lost, and for any distinctive local cultures to be erased . . .' (Caldecott, 1996: p. 141). This has important implications for river conservation since a significant proportion of the rich fish fauna of the region exploits inundated forest and allochthonous foods (see p. 82). Moreover, the economic pressures for logging which cause deforestation often have national or international origins whereas the effects are felt by local communities, especially fishers. There is thus considerable potential for conflict among different interest groups.

As an example, an issue that came to the fore in Sarawak during the late 1980s is considered. It involved the side effects of logging on the lifestyles of indigenous people who were mainly shifting cultivators. Shifting cultivation requires access not only to land, but to a wide range of materials for use or sale which are derived from the surrounding forests and rivers: fish were of particular importance and accounted for up to one third of the food intake of forest-inhabiting Kenyah and Iban communities (Caldecott, 1996; Parnwell & Taylor, 1996). Prawns and turtles were harvested also. Fishes are especially significant as a source of protein because hunting land animals carries a greater risk of failure than fishing (Parnwell & Taylor, 1996). In the Baram River drainage basin, mechanised logging at high intensities was detrimental to fish population in the river through muddying, increased sedimentation and diesel pollution. This disturbance led to a decline in average ration harvested from 54 to 18 kg per person per year during the first decade of logging, and a local perception that logging was primarily to blame for a significant increase in hardship in rural areas (Caldecott, 1996).

While the Sarawak Forest Department was sympathetic to the aim of preserving biodiversity and drainage-basin integrity by managing the forests of the Baram drainage to a high professional standard, the department's position has been conflicted (Caldecott, 1996). The role of the Forestry Department was to supervise a forestry system which functioned by awarding and exploiting industrial logging concessions for profit in order to support the state's growing economy. As a result, conflicts arose between the demands of this system on natural resources, and the needs of wildlife and local communities in the affected areas.

One possible solution to conflicts such as these is maintenance of protected areas or drainage basins in which logging is not permitted. Sustainable harvests of fish from protected rivers are feasible as long as the ecological requirements of the target population continue to be met and exploitation is monitored so that over-harvesting is prevented. However, a particular danger arises when a management strategy devised for one set of harvesting techniques is upset by the introduction of a new, more effective means of capture. The impact of fishing by hooks, harpoons, traps and plant toxins is very different from rates of exploitation that can be achieved when they are replaced by more efficient nets, explosives, electricity and artificial toxins. For example, more damaging fishing practices were adopted by the by the Iban in Sarawak because logging led to a scarcity of fish and prawns, and spurred an increased use of pesticides (replacing natural phytochemicals) and fishing by electrocution (using power generators) to increase catches (Parnwell & Taylor, 1996). It also initiated a change in community practices. Traditional fish poisoning by the Iban, for example, was a rather labour intensive social activity involving extracting the poison from plants, damming rivers, and harvesting fish (Caldecott, 1996; Parnwell & Taylor, 1996). Such poisons (e.g. rotenone from *Derris*) are usually biodegradable and non-persistent. The use of agricultural pesticides and artificial toxins can have disastrous effects, since the concentrations of commercial insecticide needed to disable fish will be extremely damaging to aquatic communities. By contrast, rotenone is efficacious against fish, degrades rapidly in bright sunlight, and has relatively minor effects on benthic invertebrates (Dudgeon, 1990at). Commercial markets for fish also encourage more frequent use of poisons, and the intensity of fishing can also be affected by transport and trading patterns. During the 1970s, for example, unharvested fish in remote locations in interior Borneo were exploited by urban markets because of the increased availability of speedboats and cold-storage facilities (Caldecott, 1996). Obviously, continued monitoring and flexibility must be built into systems for managing river fisheries if they are to be responsive to changes such as these.

In addition to logging for timber, plantation agriculture for the growing pulp and paper industry in Thailand and Indonesia, poses a significant threat to the river biota. Natural forests have been transformed to monocultures of acacia and eucalyptus and huge pulp mills have been constructed along major waterways. Plantations in Indonesia '. . . are a direct cause of legal and illegal deforestation, devastating biodiversity, agricultural watersheds, soils and fisheries' (Lohmann, 1993: p. 28); pollution from pulpmills has resulted in '. . . poisoning of drinking and bathing water and in large-scale fish kills which in some areas have devastated one of the rural villagers' most important protein sources and marketable products'. Lohmann (1993: p. 34) also reports that rivers in Kalimantan became sediment filled and 'useless for transport to market or fishing' as a result of logging and plantation establishment. Interestingly, Tejwani (1993: p. 513) states that there is '. . . no direct evidence of an adverse impact of heavy loads of sediment in water on inland fish and fish breeding grounds . . . however, it is reasonable to conclude that siltation of . . . rivers and streams . . . adversely influence the breeding and multiplication of fish'.

POLLUTION

Water quality varies considerably throughout tropical Asia (Alabaster, 1986). However, on the basis of data drawn from the Global Environmental Monitoring System water project (GEMS/WATER), initiated in 1976 by WHO, UNESCO, WMO and UNEP, the region's rivers are perhaps the most degraded in the world (Lean *et al.*, 1990; see also Fig. 5.1). Many rivers — particularly in India — have a low percentage of saturation by dissolved oxygen (< 70%), accompanied by high BOD and COD values (e.g. Alabaster, 1986; Meybeck, 1987; Trivedy, 1988; Meybeck *et al.*, 1989: pp. 70–73 & 288; Bhargava, 1992; Dudgeon, 1992b; Gopal & Sah, 1993). Increasing population pressures account for much of the region's poor water quality. As cities and industries expand without adequate waste-treatment facilities, rivers and streams are used increasingly as receptacles for discharged effluents. This is reflected in levels of faecal pollution in Asian rivers: median concentrations of faecal coliforms are $10,000.L^{-1}$ (compared with levels of less than a quarter of that in North and Central America); 17% of Asian GEMS rivers have levels more than $100,000.L^{-1}$ and 10% over $1,000,000.L^{-1}$ (Meybeck, 1987). Even in rivers such as the Purari in New Guinea, mean faecal coliform concentrations of $>30,000.L^{-1}$ have been recorded (Petr, 1983). Nevertheless, tropical Asian rivers are, in general, less mineralized than the global average (mean conductivity 170 $\mu S.cm^{-1}$) because of relatively limited pollution through industry and mining (Meybeck, 1987). Malaysia is an exception to this generalization, and effluents from tin mining have depleted the fish fauna (Alfred, 1968), although other pollutants have contributed to more recent declines (Khoo *et al.*, 1987); fish communities of the Fly River in Papua New Guinea are at risk from mining wastes also (Smith *et al.*, 1990; Smith & Hortle, 1991; Smith & Morris, 1992; Jaensch, 1994).

Chinese rivers face significant qualitative problems due to pollution from a variety of sources (Chen & Wu, 1987; Smil, 1993; Zhang *et al.*, 1994; Dudgeon, 1995e). Out of 78 rivers monitored by the

GEMS/WATER initiative in the People's Republic of China, 54 are seriously polluted with untreated sewage, while industrial wastes (such as heavy metals) are significant in some reaches (Dou *et al.*, 1987; Zhang *et al.*, 1987; Lean *et al.*, 1990). Over 80% of sewage and 85% of industrial waste water are discharged without any treatment and, as a result, nearly 13% of the total length of major rivers is polluted (Liu & Wang, 1987). Wang (1989) states that 82% of 878 Chinese rivers surveyed in the early 1980s were polluted to some degree, and fish had been eliminated from more than 5% of the total river length. Regular fish kills are caused by untreated industrial waste water and pesticides in Zhujiang tributaries and other rivers in China (Liao *et al.*, 1989; Wang, 1989). In Taiwan, also, rivers pollution by heavy metals, fertilizers, and pesticides is a significant problem (Hung, 1987; Lo & Chen, 1991; Yu *et al.*, 1995)

In the same context, Lean *et al.* (1990) have written that 'India's rivers are little more than open sewers, carrying untreated wastes from urban and rural areas to the sea'. They present data indicating that out of some 3,119 Indian towns and cities, only 217 have partial or complete sewage treatment facilities (see also Mahajan, 1988). Pollution of Indian rivers by industrial wastes dates from (at least) the 1940s and 1950s when fish kills, caused by paper-mill effluent, were reported from the River Godavari (Hickling, 1961: pp. 12–13). The Ganges — tropical Asia's third largest river by discharge volume and probably the most polluted large Asian river (Meybeck *et al.*, 1989; p. 288) — exemplifies the extent of the problem (Natarajan, 1989): almost 600 km (about 25%) of its length is heavily polluted with human and animal wastes as well as increasing amounts of effluents from industries and agriculture. Summer fish kills are common. Only 12 out of 132 industrial plants discharging directly into the river are said to have waste treatment plants in working order. Population densities in the Ganges catchment are high, and some cities along the river (e.g. Varanasi) lack sewage treatment facilities completely. Efforts are being made to alleviate the situation, and in 1989 the Indian Government allocated US$195 million to a five-year programme aimed to clean up the Ganges. Among other measures, this involved new or improved sewage treatment plants, and tighter regulation of pesticide and fertilizer use in the river catchment (Natarajan, 1989). River pollution is also a significant and increasing problem in the Indian Himalaya (Gautam, 1990; Das, 1991; Joshi, 1991; Melkania, 1991), and in Nepal where domestic wastes, sewage and industrial effluents are discharged directly into water courses (Ram Vidya, 1995; Ranjit, 1995; Shresta, 1995). A similar situation prevails in Bangladesh (Ghosh & Konar, 1993; Khondker, 1994; Ullah *et al.*, 1995) and parts of Sri Lanka (Costa, 1994; de Silva & de Silva, 1994; Pethiyagoda, 1994). Additional, detailed data from India are given by Ghose & Sharma (1989), Trivedy (1988, and references therein) and Gopal & Sah (1993).

Measures have now been taken in many Asian countries to contain pollution, including enactment of laws to control point-source pollution in China, Indonesia and Malaysia (Y.C. Ho, 1987; Jalal, 1987; Lim, 1987; Smil, 1993; Ho, 1994; 1996; Nontji, 1994; Zhang *et al.*, 1994) although pollution from non-point sources has been neglected and is in urgent need of attention (Zhang *et al.*, 1994). There are moves in China to make damage to the environment a crime punishable by imprisonment or even death, and the charge of 'jeopardising the environment' will soon be incorporated into China's criminal law (SCMP, 1995d). Despite vigorous efforts, however, pollution levels in Asian rivers have not been reduced substantially. In some cases they have actually risen (Pantulu, 1986a; Ali *et al.*, 1987; Jalal, 1987; Lim, 1987; Shen *et al.*, 1987; Welch & Lim 1987; Wang, 1989; Baconguis *et al.*, 1990; Djuangsih, 1993; Low, 1993; Bukit, 1995; Dudgeon *et al.*, in press), and may continue to do so. For example, despite enforcement efforts, 40 of the 119 principal rivers of the Philippines are 'biologically dead' because of pollution by industrial, domestic and mining wastes (Villavicencio, 1987). In a recent incident, a toxic tailings from a copper mine caused mass mortality of fish and shrimp and poisoned livestock along 27 km of the Boac River on Marinduque Island (SCMP, 1996h; see also Low, 1993). Saeni *et al.* (1980) concluded that almost all of Indonesia's rivers and streams were polluted by domestic wastes, but industrial effluents have become increasingly important (Djuangsih, 1993; Nontji, 1994; Bukit, 1995; Palupi *et al.*, 1995). Conditions in the Citarum River in West Java have been described as 'supercritical' because of pollution by industrial wastes, organochlorines, pesticides and fertilizers (Djuangsih, 1993). In parts of northeast Thailand, as well as in the Chao Phraya River and the lower Mekong (predominately in the delta), pollution by domestic wastes, agricultural run-off, and recent industrial developments is becoming acute (Srivardhana, 1984; Pantulu, 1986a: Hunt, 1992; Vadhanaphuti *et al.*, 1992; Roberts, 1993b; Muttamara & Sales, 1994). Likewise, many of Malaysia's major rivers have been contaminated with industrial and agricultural wastes (including oil palm and rubber processing residues), and over 40 of them have been classified as 'biologically dead' (Khoo *et al.*, 1987; Phang, 1987; Y.C. Ho, 1987;

Lim, 1990; Ho, 1994). This contrasts with the situation 30–40 years ago when inorganic sediments from tin mines were the major pollutant in most of the country's rivers (Johnson, 1957, 1968; Alfred, 1968; Prowse, 1968). More recently, fertilizers, pesticides, herbicides and a range of industrial wastes have become important (Lim, 1990).

There are significant weaknesses in the ability or willingness of Southeast Asian governments to enforce pollution-control legislation. Common problems include inadequate enforcement of laws and regulations, ineffectiveness of water-quality standards, regulations and penalties, and a general lack of treatment of domestic and industrial effluents (e.g. Srivardhana, 1984; Y.C. Ho, 1987; Vadhanaphuti *et al.*, 1992; Low, 1993; Muttamara & Sales, 1994; Choowaew, 1995; Afsah *et al.*, 1996; Ho, 1996). Centralised collection and disposal of urban-generated sewage is inadequate over most of the region, and much of what is collected continues to be discharged into streams and rivers without treatment (Hufschmidt, 1993). Even in the few cases (e.g. Singapore) where point source pollution is treated, non-point sources contribute to gross water pollution during periods of high rainfall and associated run-off. The capital investment required for sewage collection and treatment are far beyond the immediate financial capacity of many cities — especially fast growing ones. With few exceptions (again, Singapore), waste-water treatment to the secondary level is not a realistic option, at least over the next decade or two (Hufschmidt, 1993; see also Hjorth & Nguyen, 1993). According to Low (1993: p. 534), '. . . water pollution monitoring and enforcement are particularly weak as they depend on adequate manpower and funding, as well as the will to act . . . which many countries lack. The financial and economic gains from unfettered development are too attractive to be hampered by enforcement of anti-pollution legislation . . .'. Likewise, Setamanit (1987: p. 207) writes with regard to river pollution in Thailand that '. . . the difficulty appears to lie more with the enforcement of laws rather than with the lack of them . . .', and '. . . there appears to be a lack of urgency in formulating clear policies and developing strategies to deal with the problem ..'. Similarly, while the Philippines has 'one of the most comprehensive environmental legislation, regulation and enforcement of anti-pollution laws remain weak . . . (and) purely regulatory solutions to environmental management have not produced the expected results' (Villavicencio, 1987: p. 107).

One consequence of the ubiquity of pollution is the plethora of studies describing the general effects of a range of pollutants (sewage, industrial effluents, pesticides, mine wastes) on individual rivers or drainage systems (e.g. Ho, 1975; 1976; Verma & Dalela, 1975a; 1975b; Konar, 1977; Rama Rao *et al.*, 1978; 1979a; Patil *et al.*, 1986; Arya *et al.*, 1987; Chattopadhyay *et al.*, 1987; Ajmal & Razi-Ud-Din, 1988; Dash *et al.*, 1988; Nazneen & Begum, 1988; 1994; Palharya & Malviya, 1988; Somashekar, 1988a; Tandon, 1988; Gautam, 1990; Salagar & Hosetti, 1990; Singh & Nautiyal, 1990; Khan, 1992; Abdullah & Nainggolan, 1991; Khan & Lim, 1991; Kulshrestha *et al.*, 1991; Smith & Hortle, 1991; Bhatt & Pathak, 1992; Klepper *et al.*, 1992; Shukla *et al.*, 1992; Smith & Morris, 1992; Kannan *et al.*, 1993; Tariq *et al.*, 1993; 1994; Ambasht & Srivastava, 1994; Hameed *et al.*, 1995; Suryadiputra & Suyasa, 1995; Unni, 1995; Yule, 1995a). A variety of methods for water-quality evaluation has been considered (Rama Rao *et al.*, 1979b; Verma *et al.*, 1987), involving — for example — the use of algae (e.g. Venkateswarlu & Jayanti, 1968; Rai, 1978; Reddy & Venkateswarlu, 1986; Hiremath & Shetty, 1988; Rana & Palria, 1988; Somashekar, 1988b; Kulshrestha *et al.*, 1989; Joy, 1990; Venkateswarlu *et al.*, 1990; Khan, 1990b; 1991; Shaji & Patel, 1991; Yu *et al.*, 1995) or benthic invertebrates (e.g. Ho & Hsu, 1977; Rama Rao *et al.*, 1978; Qureshi *et al.*, 1980; Lin, 1982; Qi *et al.*, 1982; Shen & Qi, 1982; Qi & Erseus, 1985; Qi & Lin, 1985; Qi, 1987; Su & Li 1988; Datta Munshi *et al.*, 1989; Jhingran *et al.*, 1989; Kulshrestha *et al.*, 1989; Sharma *et al.*, 1990; Yang *et al.*, 1990; Augustine & Diwan, 1991; Rao *et al.*, 1991; Khan & Kulshrestha, 1993; Tong *et al.*, 1995; Trihadiningrum *et al.*, 1995; Yu *et al.*, 1995; Mustow *et al.*, 1997) as pollution indicators and biomonitors. These and a host of other studies show that the biology of pollution in tropical Asia differs only in detail from accounts based upon research undertaken in north-temperate latitudes. The subject will not be considered further here, except to note two things. Firstly, that biologists have a reasonably good understanding of the effects of pollution; pollution control, however, is more often within the bailiwick of the legislator than the scientist. A second, important, point relates to the research strategies employed to assess the impact of pollutants on streams and rivers. Generally, they do not involve adequate controls or replication with the consequence that it may be impossible to separate the suspected effects of a pollutant from effects that are due to other, independently-varying environmental factors. This is a critical point which hampers our ability to assess environmental impacts, and will be discussed in detail in Chapter 6.

EXTINCTIONS

The biota of many rivers in Southeast Asia are at risk of impoverishment or — in extreme cases — extinction. The overall pattern throughout the region is one of decline and species loss, reflecting widespread urbanization, pollution, river regulation and land-use changes (Zakaria-Ismail, 1994; Kottelat & Whitten, 1996). This loss of fish biodiversity has parallels in streams and rivers in all countries (Maitland, 1995), and 20% of the world's freshwater fish fauna is already extinct or in danger of extinction in the foreseeable future (Moyle & Leidy, 1992). A particular problem that is obvious from three recent reviews of threatened fishes of the world (Moyle & Leidy, 1992; Bruton, 1995; Maitland, 1995) is the critical shortage of information on the conservation status of Asian freshwater fishes. A report on the biodiversity of the region published at the time of writing (Kottelat & Whitten, 1996) underscores this point. Indeed, it is obvious that fish biodiversity has been declining throughout the region for over 30 years. In 1965, for example, only 35 of 54 indigenous Singaporean fish species still survived in the wild (Alfred, 1968); of the survivors, eight were rare. In addition, 11 species of exotics had become established. At that time, the giant cyprinid *Probarbus jullieni* was endangered in Malaysia (Alfred, 1968), as several species of Thai river fish (Suvatti & Menasveta, 1968). The decline of *P. jullieni* reflected losses of gravid individuals to fisherman during upstream spawning migrations, and obstruction of migration routes by the Chenderoh Dam on the Sungai Perak — one of the only two rivers that the species inhabits in Malaysia (Alfred, 1968; Tan, 1980; Khoo *et al.*, 1987). Between 1971 and 1980 the availability of commercially important fishes in this river had fallen from nine to three species (Khoo *et al.*, 1987; Ho, 1996). Reductions in *P. jullieni* catches, apparently a result of overfishing, have been reported also from Laos, Cambodia and Thailand (Nandeesha, 1994; Roberts & Warren, 1994; Roberts & Baird, 1995), where *Catlocarpio siamensis* — one of the world's largest cyprinids, and formerly the basis of an important fishery — has now become rare (Nandeesha, 1994). Similarly, overfishing in east Sumatran rivers poses a threat to the freshwater 'shark' (*Wallago attu*) which has been exterminated over part of its range (Claridge, 1994). Recently, Ng & Lim (1992) and Ng *et al.* (1994) concluded that an array of stenotopic Singaporean and Malaysian fishes confined to soft, acidic blackwater streams and peat swamps was severely threatened, and stated that '. . . conservation of the remaining blackwater biotopes is

critically important if extinction of many species is to be prevented'
(Ng *et al.*, 1994).

Recent data from Singapore (Lim & Ng, 1990), which update
Alfred's (1968) investigation, show that only 29 of the original 54
primary freshwater species survived in 1990; 18 of them were
endangered. The number of established exotic species had risen to 14.
Concern has also been expressed over the future of fishes in Sri Lankan
forest streams, which are endangered by overfishing, pollution caused
by agricultural chemicals, and deforestation (Kortmulder *et al.*, 1978;
de Silva, 1994; de Silva & de Silva, 1994; Pethiyagoda, 1994). The
situation is particularly serious because the known range of many Sri
Lankan stream fishes (especially the 26 endemic species) are exceedingly
small and almost all are habitat specialists (Pethiyagoda, 1994; see also
p. 82 *et seq.*). Similar threats to fish biodiversity — resulting from
overfishing and pollution by tin-mine effluents — were highlighted in
Malaysia almost 30 years ago (Johnson, 1968; Prowse, 1968), but have
persisted and been exacerbated in recent years (Zakaria-Ismail, 1987;
1994). Elsewhere, river pollution has caused declines of freshwater
flatfish (*Synaptura panoides*: Soleidae) — among many other species —
in Thailand (Wongratana, 1988), and the frequency of fish kills due to
pollution has increased in rivers throughout the region (Hunt, 1992; see
pp. 553, 557 and 559–560). An additional threat to fishes is climate
change (Meisner & Shuter, 1992). A rise in global temperatures due to
the increasing 'greenhouse effect' would influence precipitation patterns
and the seasonality of flow regimes in rivers which would, in turn, affect
inundation of the floodplain and spawning migrations of fishes. Meisner
& Shuter believe that the potential for negative effects of climate change
on fishes warrants detailed investigation.

Reference has been made above (pp. 541–549) to the possible
effects of impoundments along the Mekong River on fish spawning
migrations, the plight of sturgeons and paddlefish in the Chang Jiang
(pp. 525–532), and declines in fish stocks in the Zhujiang (pp. 532–
543). In India, overfishing has reduced the stocks of major carp in the
Ganges and Godavari Rivers (Jhingran, 1980), and throughout much
of the Himalaya (Joshi, 1991). The irrigation system associated with
the Indus River has contributed significantly to dwindling fish stocks
because of inadequate consideration for the needs of fish to pass dams,
weirs and barrages (Jhingran, 1980). Hydraulic structures have all but
eliminated the fishery for anadromous *Tenualosa* (= *Hilsa*) *ilisha*
(Clupeidae) in riverine stretches of the Ganges (Chandra *et al.*, 1990)
despite warnings of harmful effects on the fisheries during the 1950s

(Hickling, 1961: p. 88). In addition, there have been considerable declines in mahseer stocks and the major carp fishery has virtually disappeared as a result of habitat degradation, river control and pollution (Jhingran & Ghosh, 1978; Sehgal, 1983; Natarajan, 1989; Joshi, 1991; Shreshta, 1991; Singh *et al.*, 1991). The same factors, combined with overfishing in spawning grounds, have caused declines in Zhujiang fish stocks, and catches of some commercially-important species have declined to less than 15% of the average for the 1960s (Liao *et al.*, 1989). Natarajan (1989), Liao *et al.*, (1989) and Melkania (1991) discuss depletion of fish stocks in relation to man-made changes, and outline river-basin management strategies to rehabilitate and enhance affected populations.

Concern has been expressed in Thailand over the decline of some long-distance migratory fishes due to over-fishing, blockage of migration routes, and destruction of feeding or breeding grounds. They include *Catlocarpio siamensis* (which can exceed 120 kg), three *Probarbus* spp. (including *P. jullieni*, and *P. labeamajor* which reaches 80 kg) and *Pangasius gigas* (300 kg: Nandeesha, 1994). Two of them (*P. jullieni* and *Pangasius gigas*) are listed in the appendices of CITES and the Migratory Species Convention (Kottelat & Whitten, 1996). The Thailand Fisheries Department has collected mature individuals of all three species from the wild in order to induce spawning. The hatchlings were reared and resulting juveniles released into the wild. Success with induced breeding of captive *Pangasius gigas* (Pholprasith, 1983; 1993; Chang, 1992) has averted, at least temporarily, the listing of the giant catfish as an endangered species in Thailand (Nandeesha, 1994) although the effort that been devoted to this species is as much an attempt to exploit its potential for aquaculture as it is a conservation imperative. More recently, the Wildlife Fund of Thailand (WFT) has been working to increase awareness of the need to protect *P. gigas* and, in 1996 began a high-profile programme to buy live specimens and return them to the wild. WFT figures show a decline in commercial catches of *P. gigas* from 69 fish in 1990, 48 in 1993, and 18 in 1995. In Cambodia, too, attention has been paid to *Probarbus* spp. and *P. gigas* which disappeared from commercial catches in the Mekong during the early 1990s. As a result, the government outlawed fishing for these species. Collection of *Pangasius* spp. seed from the Mekong supports a major aquaculture industry (especially in Vietnam), but is believed by fishers to be damaging to natural populations (Nandeesha, 1994). Accordingly the Cambodian Government banned *Pangasius* seed collection in 1994, and emphasized the importance of re-establishing natural stocks by a means of well-publicised release of wild-caught *Pangasius* seed back into the Mekong.

The aquarium fish trade has been accused of driving species to extinction, because of selective overfishing (especially of stenotopic blackwater fishes). However, there is direct little evidence of this since declines in important trade species have always been associated with other causes (Kottelat & Whitten, 1996). An exception to this is may be the endangered golden dragon fish (*Scleropages formosus*: Osteoglossidae), as there has been little effective regulation of its capture in Malaysia (Zakaria-Ismail, 1994). It is protected by law in Thailand, as is another aquarium fish, *Botia sidthimunki*, although in this case habitat destruction rather than exploitation has caused its decline in the wild (Kottelat & Whiten, 1996). *Scleropages formosus* is protected in Indonesia also (Kottelat & Whitten, 1996), but these animals are nevertheless fished intensively in Sumatra during the wet season when they breed in shallow floodwaters adjacent to rivers (Claridge, 1994). Recent success with the propagation of *S. formosus* in fish farms should reduce pressures on wild stocks (Andrews, 1990) and, in theory, a proportion of the animals bred under artificial conditions in West Kalimantan have to be returned to the Kapuas River to restore a much exploited population (Kottelat & Whitten, 1996). While the aquarium trade has been largely responsible for the harvest of *S. formosus* in the wild, it is possible that the trade may help save species from extinction. For instance, the popular aquarium fish *Epalzeorhynchos* (= *Labeo*) *bicolor* may no longer exist in the wild in Thailand, but captive-bred specimens are in abundant supply within the trade (Kottelat & Whitten, 1996). Pethiyagoda (1994) believes that '. . . given existing trends, no means of protection is likely to ensure their survival in the wild in the medium term . . .' and concludes that '. . . the strategy most likely to succeed is the maintenance of . . . captive populations with a view to reintroduction once the pressures on wild populations have been controlled'. *Ex situ* protection of this type can effectively conserve only a small — usually critically-endangered — proportion of total biodiversity. These efforts usually involve 'charismatic' species — probably the most ornamental or unusual small fishes. The number of species which can be maintained in living collections is limited and, for large river fish, will be constrained by the size of the facilities and the high maintenance cost per species.

Many Asian fish species are poorly known and their conservation status cannot be determined accurately. Among these are several species of freshwater dasyatid stingrays (Ahmad & Sonoda, 1979; Compagno

& Roberts, 1982; Roberts & Karnasuta, 1987; Monkolprasit & Roberts, 1990); one of them is indigenous to the Fly River (New Guinea) which has been polluted by mining wastes (Smith *et al.*, 1990; Smith & Morris, 1992). Another, *Dasyatis* (= *Himantura*) *chaophraya*, has declined due to pollution of the Chao Phraya River and may soon be included on the protected species list under the 1992 Wildlife Conservation Act. Like other species of obligate freshwater elasmobranchs, these stingrays are vulnerable to extinction because they occupy a restricted range of habitats which '. . . strictly limit their opportunities to evade pollutants, habitat modifications, or directed and incidental capture in local fisheries . . .' (Compagno & Cook, 1995). Their vulnerability may be increased by low fecundity: *D. chaophraya* gives birth to only a single offspring at a time. Pethiyagoda (1994) points out that few Asian countries have made systematic assessments of their native fish resources (or faunal diversity in general), and emphasizes that recently-published faunal inventories are based more on dated literature or old museum collections than on recent surveys. In such circumstances, it is unlikely that extinctions will become known until long after they have occurred and '. . . the state of taxonomic knowledge of the fishes of the Asian region in general is so poor that many extinctions may never be known' (Pethiyagoda, 1994). Problems with fish taxonomy and correct identifications have been highlighted recently by Kottelat & Whitten (1996), and can give rise to situations where protected status is sometimes granted to species which do not exist in the country or are misidentified.

Reptiles and amphibians, as well as fish, are at risk from anthropogenic modification of Asian streams and rivers. In Thailand, for example, the long-nosed crocodile (or Indonesian false gharial, *Tomistomus schlegelii*) and giant frog (*Rana macrodon*) were identified as being threatened with extinction three decades ago (Suvatti & Menasveta, 1968). A variety of Thai river turtles are also in need of protection in Asia (Belsare, 1994; Thirakhupt & Van Dijk, 1994) — which is of particular importance given that western Thailand contains the worlds most diverse turtle community (Thirakhupt & Van Dijk, 1994). Elsewhere in the region, the Himalayan newt (*Tylototriton verrucosus*) of Nepalese hillstreams is threatened by pollution and land-use changes (Shrestha, 1994). Dussart (1974) urged that more research be undertaken on Asian amphibians to inform conservation measures. This plea was repeated recently by Belsare (1994) who considers that many of the endemic anurans of India are on the verge of extinction as a result of habitat destruction, human exploitation and pollution. It has become clear that hunting and man-made changes in habitat characteristics have caused a general decline in crocodilian populations throughout tropical Asia (Ahmad, 1986). *Crocodylus porosus*, which inhabits river estuaries but can penetrate over 1000 km upstream (Pernetta & Burgin, 1983), is now rare in parts of tropical Asia where it was once abundant (Whitaker & Whitaker, 1978; Pernetta & Burgin, 1983; Whitten *et al.*, 1984, 1987), and the population structure of remaining populations (e.g. in the Klias River, Sabah) shows evidence of pressure from hunting (Stuebing *et al.*, 1994). The mugger (*Crocodylus palustris*) which once occurred through eastern India to Sri Lanka, living in streams and rivers as well as lakes and saline lagoons, has likewise been exterminated over much of its range (Whitaker & Whitaker, 1984). Attempts have been made to reintroduce both of these crocodiles to parts of India (Choudhury & Bustard, 1982; Choudhury & Chowdhury, 1986) but, despite the success of captive breeding of muggers, the reintroduction programme has been stopped (Sebastian, 1992). The gharial (*Gavialis gangeticus*) is critically endangered in the Indian Subcontinent (Belsare, 1994) although viable populations persist in some national parks and reintroduction of captive-raised gharials to the wild has met with some success (Choudhury & Chowdhury, 1986; Dodd & Seigel, 1991). The New Guinea crocodile (*Crocodylus novaeguineae*) inhabits rivers and floodplain swamps (especially when young); it is under threat from hunting (Pernetta & Burgin, 1983) which could account for the high level of nocturnal activity in this species (Montague, 1983). The Philippines crocodile (*Crocodylus mindorensis*), too, is endangered (Baconguis *et al.*, 1990), and populations of the Indonesian false gharial (*Tomistomus schlegelii*) have been greatly depleted by hunting (Whitten *et al.*, 1984). Urgent action is needed to protect these species — along with the endangered Siamese crocodile (*Crocodylus siamensis*) and Chinese alligator (*Alligator sinensis*; see p. 528 and p. 530) — in the wild (Scott & Poole, 1989).

Asia is home to three of the four extant river dolphins (Reeves *et al.*, 1991), including the *susu* or Gangetic dolphin (*Platanista gangetica*), which occurs in the Ganges, Brahmaputra, Meghna and Karnapuli River systems of India, Bangladesh, Bhutan and Nepal. It is endangered because of habitat degradation (including pollution) and population fragmentation by dykes and dams (Dussart, 1974; Pelletier & Pelletier, 1980; Kannan *et al.*, 1993; Smith, 1993; Reeves & Leatherwood, 1994), and its decline is liable to be hastened by the ongoing Bangladesh National Flood Action Plan (Reeves & Leatherwood, 1994; see p. 539 and p. 554). Indus dolphins (of which about 500 remain; Reeves *et al.*,

1991) are generally treated as a separate species, *Platanista minor* (= *P. indi*; Pilleri & Pilleri, 1987). Like the *susu* they have been hunted for their oil which is said to have medicinal value (Pelletier & Pelletier, 1980) and, despite legal protection in since the 1970s, numbers are declining as a result of accidental captures in fishing nets and habitat loss (Chaudhry & Khalid, 1990; Reeves *et al.*, 1991). Aggressive water extraction and barrage construction are particular threats to Indus dolphins, and populations have been fragmented into four or five sub-populations dams and barrages thereby increasing the risks of extinction (Reeves *et al.*, 1991; Reeves & Leatherwood, 1994). Remaining individuals of the endemic and endangered Chinese river dolphin or *baiji* (*Lipotes vexillifer*) are gravely threatened (see p. 528 and p. 530) and the species may soon become extinct. The Irrawaddy dolphin, *Orcaella brevirostris* (the *pesut*), enters rivers throughout Southeast Asia from India to Vietnam and south through the Indonesian archipelago. It has been found over 1,000 km from the sea in the Irrawaddy (Sigurdsson & Yang, 1990), and is in need of protection in some areas (Wirawan, 1986; Baird, 1994; Baird & Mounsouphom, 1994) while in others it is relatively secure (Freeland & Bayliss, 1989). They are protected by law (although designated as 'fish species') in Laos (Baird & Mounsouphom,1994). The Irrawaddy dolphin has disappeared from rivers (such as the Chao Phraya) in recent years, while declines in numbers in the Mekong have been attributed to a reduction in fish prey and the death of dolphins trapped in gill nets (Baird & Mounsouphom, 1994). These workers suggest that large-scale dams proposed for the Mekong further threaten *O. brevirostris* in Laos and Cambodia.

Riparian vegetation and gallery forest along tropical Asian rivers are under threat (Cox, 1987; Johnsingh & Joshua, 1989; Le, 1994). This loss would affect terrestrial animals which depend upon the riverine habitat during the dry season (Cox, 1987), and would cause the demise of plant and animal species which are associated exclusively with the riparian zone. For example, certain Araceae (*Schismatoglottis okadae*) and bamboos (*Racemobambos setifera*) are confined to riparian zones in Sumatra and Malaysia respectively (Okada & Hotta, 1987; Wong, 1987); the latter is endangered because of an inability to tolerate habitat alteration resulting from logging. Riparian habitats have diverse plant communities — for example, 78 species of Cyperaceae have been found along the banks of the River Ganges and its tributaries (Bakshi *et al.*, 1977) — and are also sites of unusually high herpetofauna abundance (Inger *et al.*, 1987). Johnsingh & Joshua (1989) give an account of the diversity of plants and vertebrates of a gallery forest along the River Tambiraparani (India) which is threatened by river control projects. Such forest supports the endangered Niligiri langur (*Presbytis johnii*). Elsewhere in Southeast Asia the riparian zone sustains the flat-headed cat (*Felis planiceps*), the otter civet (*Cynogale bennettii*) (Cranbrook & Furtado, 1988) and fishing cat (*Felis viverrina*) (Cox, 1987). All are endangered or of uncertain status. Four species of otter are known from rivers in the region (Medway, 1978; Payne *et al.*, 1985; Sivasothi & Nor, 1994), and Southeast Asia is quite exceptional in the degree of range overlap of otter species (Kruuk *et al.*, 1994). *Lutra lutra* (common or Eurasian otter), *Lutra sumatrana* (hairy-nosed otter), *Lutra* (= *Lutrogale*) *perspicillata* (smooth otter) and *Aonyx cinereus* (Oriental small-clawed otter) are quite widespread but their status in the region is either uncertain or causing serious concern (Kruuk *et al.*, 1994;). Populations of some species (especially those of the hairy-nosed otter) are small and declining (McKay, 1984; Sivasothi & Nor, 1994; see also Scott & Poole, 1989).

The status of birds associated with Asian wetlands has been reviewed recently by Scott & Poole (1989; see also Belsare, 1994; Shulka, 1994) and will not be considered here. However, it should be noted that genera such as *Ichthyophaga* (fishing eagles), *Alcedo* (kingfishers) and *Enicurus* (forktails) show longitudinal separation of species along their riverine habitat (Cranbrook & Furtado, 1988). This has implications as to the extent of riverine corridors which must be protected to avoid loss of avifauna. For endangered vertebrates in general it seems that species confined to continental Asia — or those with restricted distributions — are in greater danger of extinction than those which range across the Malesian islands where, for example, a vast extent of wetland habitat (including riparian forest) remains in Sumatra and Borneo (Scott & Poole, 1989). Hunting pressure and habitat alteration because of burgeoning human populations seem to be the main cause of declines at mainland sites.

Riparian forests provide habitat for wetland vertebrates and are major contributors of food (flowers, fruits, leaves, terrestrial insects, etc.) to aquatic consumers, including fishes (see pp. 82–85), so that vegetation clearance could result in the loss of stenophagous or specialized species (see, for example, Pethiyagoda, 1994). Submerged roots of riparian plants serve as habitats for aquatic invertebrates, and their significance as a nursery area has been emphasized in conservation plans for the endangered Sri Lankan shrimp, *Caridina singhalensis* (Benzie & de Silva 1988). This species is occurs in streams above

1,500 m (de Silva & de Silva, 1988), and is the only atyid in the world found at such high altitudes (de Silva, 1982). It is endangered because of habitat changes resulting from forest clearance and the introduction of predatory rainbow trout (*Oncorhynchus mykiss*). The shrimp is now confined to a 10-km stretch of single stream at 2,200 m (de Silva, 1982; de Silva & de Silva, 1994). Although this is the only documented case of population decline in a tropical Asian stream invertebrate, the extent of habitat alteration in the region suggests that the phenomenon may not be unusual. As an example, it is reasonable to ask what effect the proposed dams along the Mekong will have on the endemic species-flocks of gastropods (see p. 547) in the river. Attwood (1995b) considers that modified flow regimes caused by dams may lead to changes in food availability and substrate characteristics which will favour some snails — in particular, *Neotricula aperta* which is a host of the human parasite *Schistosoma mekongi* (Trematoda: Schistosomatidae). Such changes, together with the translocation of riparian peoples and the influx of casual labour from endemic areas in Laos, may result in an epidemic of human schistosomiasis in northeast Thailand (Attwood, 1995b). There is a complete lack of information on the possible impacts of dams and other planned changes along the Chang Jiang upon the many endemic pomatiopsid gastropods reported from that river (Davis *et al.*, 1986; 1992).

Floodplain habitats in India are threatened by overgrazing, deforestation, and land reclamation (Gopal, 1983; 1988). Seasonally inundated grasslands on such floodplains, which have been reduced in extent by flood-control dykes and levees, are important habitat for the few remaining wild Indian rhinoceros (*Rhinoceros unicornis*) (Chowdhuary & Ghosh, 1984). Excessive loads of suspended solids in the Ganges, because of deforestation and soil erosion in the upper catchments, have increased rates of siltation downstream in oxbow lakes, backwaters and on the floodplain, degrading these habitats (Natarajan, 1989). Newby (1989) records 'crocodiles' (probably gharials), Gangetic dolphins, and freshwater sharks (*Carcharias gangeticus*) in the Ganges; all are under threat as modification and degradation of the river and floodplain habitat continue (Scott & Poole, 1989). Similar problems of silt deposition have arisen in the Kashmir Himalaya wetlands where many species — especially waterfowl but also macrophytes and fishes — have vanished (Pandit & Qadri, 1990; see also Belsare, 1994).

Flushes of vegetation which grow up after flood waters recede provide dry-season grazing for cattle and wildlife on Asian floodplains, including elephants and rhinoceroses. They provide habitat for a variety of Cervidae, including the swamp deer, Schomburgk's deer and brow-antlered deer *(Cervus duvauceli, C. shomburgki* and *C. eldi* respectively), and species or subspecies may be confined to particular river systems. These deer avoid dense covets and thick forests and are restricted to open floodplains (Boonsong & McNeely, 1977). Habitat destruction has fragmented swamp deer populations into small remnants (Scott & Poole, 1989; Henshaw, 1994) while Schomburgk's deer, which was once abundant in swamps along the Chao Phya River (Thailand), is now extinct because of hunting and conversion of floodplain for wet rice cultivation (Boonsong & McNeely, 1977). In Sri Lanka, use of floodplains by elephants is limited by agricultural activities but, after the rice harvest, paddyfields become important dry-season grazing sites (Ishwaran, 1993). Belsare (1994) has summarized the status of wetland wildlife in Southeast Asia, and gives additional information of some of the species mentioned above. He and Choudhury (1994) highlight the plight of the wild Asiatic water buffalo (*Bubalus bubalus*) which is confined to a few isolated populations on floodplains and in riverine wetlands of India, Bhutan and Kampuchea. Choudhury (1994) states that '. . . unless protection measures are intensified, the species could disappear in India within a few decades'. A bleaker view is taken by Belsare (1994) who considers that, without strenuous efforts, wild water buffalos may become extinct by the year 2000.

Declines in many species associated with riverine wetlands are due to hunting and destruction or degradation of habitats, but river control schemes have an effect on species loss also. The precise outcome will vary according to the type of engineering work proposed, the status and biology of the species involved, and so on. The example of the possible consequences of the proposed Three Gorges High Dam on the Chang Jiang drainage basin has been considered on pp. 525–532. They may involve declines in waterfowl (including winter migrants and endangered species) and mammals associated with lateral lakes on the floodplain (Zhao *et al.*, 1990; Melville, 1994), as well as reductions in fish populations with severe impacts upon — or extinctions of — rare and endangered sturgeons and paddlefish (Birstein, 1993). The wetlands of Vietnam and Kampuchea around the Mekong delta have been changed markedly by drainage, channelization and diking such that the seasonal inundation of the floodplain has been prevented and extensive areas of habitat have been degraded (Beilfuss & Barzen, 1994; Le, 1994); undoubtedly, large-scale impoundment of the Mekong will alter inundation patterns further. In this context, recent attempts to restore areas of the floodplain (the Tram Chim Reserve) to their original hydrological state are welcome but, as most of the lower Mekong wetlands have been disturbed by channelization (Beilfuss & Barzen, 1994), the scale of restoration that will be needed on this and other floodplains is vast.

PROSPECTS

There is a great diversity of running-water habitats in tropical Asia, from small blackwater streams in rainforest to floodplain rivers fed by snowmelt and monsoonal rains. This diversity is more than matched by the variety of uses to which man has put these habitats, and the harvest of fish (and palaemonid shrimps in some areas) provides a source of animal protein throughout the region. Schemes of river regulation and control, often involving long-distance transfers of water, have been developed over centuries and are likely to become more frequent (Chitale, 1992). In consequence, few large rivers now flow unmodified from headwaters to mouth. As the Three Gorges Scheme aptly demonstrates, as engineering techniques have advanced, man's ability to block and divert river flows has increased. So, too, has the capacity to wreak environmental havoc. Increasing population, urbanization and industrial effluents have put new pressures on tropical Asian rivers, and the prognosis for biological diversity in some systems — the Ganges, for example — is not good.

In the modified and often highly-disturbed rivers of tropical Asia, a further threat has materialized. The spread of exotic species in these ecosystems is a matter for great concern because, in many cases, we do not know how the invaders will affect the native biota. Species introductions or translocations (especially of fish) are not a new phenomenon (Hickling, 1961: pp. 237–248; Welcomme, 1988; Fernando, 1991), and are often associated with aquaculture which tends to have negative effects on biodiversity (Beveridge *et al.*, 1994). One estimate is that two thirds of species introduction into tropical freshwaters have established self-sustaining populations (Welcomme, 1988). The number of unplanned introductions is increasing and, in some cases, their ecological effects — as well as those arising from deliberate introductions — have been detrimental (e.g. Welcomme, 1988; Ramakrishnan, 1991; Pethiyagoda, 1994). The issue of species introductions is complex, because food fish may be introduced into regions where human populations experience protein shortages (West & Glucksman, 1976); under such circumstances, the ecological case for maintaining a pristine ecosystem is open to argument. In some circumstances, such as the recent deliberate introduction of *Tilapia rendalii* (Cichlidae) to the Sepik River, Papua New Guinea, (where it has joined another exotic 'Tilapia', *Oreochromis mossambicus*), the added species appears to fill a niche left vacant by the native assemblage of fishes (Coates, 1987; 1993) and may ultimately support an important fishery. It seems probable also that the Java carp, *Puntius goniotus*, will be introduced into highland tributaries of the Sepik.

By no means all introductions are planned or involve species which can be exploited by humans. Many introductions involve invertebrates or macrophytes (see, for example, Dudgeon, 1992a: pp. 103–109; Ng *et al.*, 1993), and considerable efforts have been devoted to attempts to control exotic weeds (e.g. Nanadeesha *et al.*, 1989; Gopal, 1987; Barrett, 1989; Room, 1990). Clearly, the habitats where exotic species establish themselves are those which are most susceptible to invasion. The main criteria (Elton, 1958; Diamond & Case, 1986) determining such susceptibility are the species richness of the receiving community (invasion success declines steeply with species richness of the extant flora and fauna), and the number of recent extinctions (the more recent extinctions there have been, the greater the chance of an invader's success). Species introductions appear to be most successful in disturbed, man-modified habitats (as discussed by Mooney & Drake, 1989; see also Ng *et al.*, 1993). This has important implications for many far-from-pristine tropical Asian rivers. Establishment of exotic species will reduce the chances of restoring disturbed or degraded habitats to their original state. There are engineering and technological solutions to most problems arising from river regulation and pollution but, as yet, no infallible methods for eliminating exotic species once they have become established.

We must fill some of the gaps in our knowledge of tropical Asian rivers. Descriptive investigations of many endangered species or poorly-known habitats are still required. Long-term studies over several years are especially needed to provide a knowledge of ecological 'crunch' times — for example, a year when the monsoonal rains are insufficient to inundate the floodplain — which will inform predictions about the consequences of management plans (such as controlling water flow) on river productivity. Long-term studies will be more valuable if they deal with entire river systems, from headwaters to mouth, rather than involving only a single tributary or reach. Investigations carried out in the context of an existing paradigm — such as the RCC — will be particularly useful as they will allow comparison between rivers in tropical Asia and elsewhere. Attention should be drawn to the fact that experimental approaches (hypothesis testing) to problems in stream biology have yet to be applied widely in tropical Asia, even though they are recognized widely as yielding less ambiguous results than some traditional methods and need be no more costly to undertake. In particular, rigorous procedures for assessing environmental impacts must be adopted to replace the descriptive and correlative studies which are undertaken widely at present. It is this topic which will comprise the subject matter of the next chapter.

6

Experimental Design and Detection of Anthropogenic Impacts in Streams

Our ability to predict and ameliorate or mitigate the effects of human activities on stream and rivers depends upon an understanding of the ecology of these systems. If we do not know what the ecological 'rules of existence' might be for the biota of tropical Asian rivers and streams, we are in no position to formulate the conservation and management strategies required to ensure habitat integrity and maintain biodiversity. There is also a need to apply existing information and research strengths in the most appropriate manner, and in the context of an appropriate legal framework which will deal effectively with those whose actions are shown to cause environmental degradation. While such a framework does not exist in many parts of Asia, legislation has been put in place in some countries (Jalal, 1987; Lim, 1987; Smil, 1993; Nontji, 1994; Zhang *et al.*, 1994) and is likely to be strengthened. In anticipation of an appropriate legal framework, stream biologists must design and undertake studies which unambiguously address the issues at hand to the exclusion of any confounding variables. Unfortunately, studies aimed at investigating the environmental impact of pollutants in tropical Asian streams and rivers have failed to adopt appropriate, statistically-rigorous designs such as those advocated by, for example, Green (1979), Stewart-Oaten *et al.* (1986) and Underwood (1991, 1993).

The purpose of this and the following chapter is to consider how research strategies might be improved so as to enhance our abilities to understand, conserve and manage tropical Asian streams. Strategies for the design and execution of ecological and environmental investigations are stressed, rather than statistical analysis of the resulting data. This reflects the fact that sophisticated data analysis cannot salvage the results of a study which has weaknesses in sampling or experimental design. In particular, it is intended to highlight those strategies by which the following two objectives can be achieved:

(a) Identify unambiguously the effect and magnitude of human impacts on streams.

(b) Enhance mechanistic understanding of ecological patterns and processes in the wetted channel, riparian zones and drainage basins of running waters, and thereby improve our ability to conserve and manage stream ecosystems in the face of anthropogenic change.

Objective (a) can be achieved more easily than objective (b) since, in the former case, it is clear what kind of information is required; the main question is 'how can impacts be assessed unambiguously?' Objective (b) is more complicated, since there may be disagreement over the type of information needed to underpin conservation and management strategies. Even when there is agreement on what we need to know, there may be dissent over how best to achieve the agreed goals. Because of the relative complexity of these two matters, objective (a) will be dealt with first and will form the subject matter of this chapter. Objective (b) will be addressed in Chapter 7.

CHOICE OF BIOMONITORS

Detection of environmental impacts in streams depends upon the use of biomonitors in combination with physical (e.g. temperature, suspended solids) and chemical (e.g. nutrient levels, concentrations of potential toxins) data. Biomonitors are a crucial component of environmental-impact assessments (EIAs) and monitoring programmes that aim to determine how effective management strategies have been in keeping impacts within acceptable limits, since it is only studies on living organisms that can define, or asses, the overall effects of pollution or ecosystem modification (see also Cullen, 1990; Camargo, 1994). All EIAs or monitoring programmes are selective with regard to which components of the biota should be included as biomonitors, and which attributes (population density, biomass, diversity, etc.) should be measured. Usually, abundance, species richness or some combination of these (e.g. diversity) is measured, although functional organization (Rabeni *et al.*, 1985), energy pathways, secondary production and ecosystem function deserve more attention (Benke, 1993). The choice of taxa included is often rather arbitrary, depending upon local expertise or the existence of a historical data set for the putatively-impacted stream; it is less frequently based upon an impartial decision as to what would be the most appropriate type of biological data to test for the presence of an impact or would be most useful for the problem in question (Hellawell, 1986). Stream biologists have frequently employed fishes, macroinvertebrates or algae as biomonitors, although changes in populations of microbes, macrophytes and zooplankton (in large rivers) have sometimes been used to infer environmental impacts. Macroinvertebrates — generally zoobenthic forms — are especially widely used as biomonitors (Hellawell, 1986; Abel, 1989; Metcalfe, 1989; Rosenberg & Resh, 1993 and references therein), and are therefore the main focus herein. Accordingly, it is worthwhile to describe briefly the advantages of benthic macroinvertebrate communities (or populations or assemblages within these communities) for biomonitoring and EIAs in streams; the list below follows Humphrey & Dostine (1994).

1. A diversity of forms and, accordingly, a wide range of sensitivity to many kinds of anthropogenic change and stresses.

2. Relatively limited mobility and life cycles of weeks to months in duration, so that animals collected at a site accurately reflect prevailing conditions at that site. By contrast, fish are mobile and are more likely to reflect overall conditions throughout the stream in which they move. Some macroinvertebrates (e.g. unionid bivalves, prosobranch gastropods, decapods) are sufficiently long-lived for use in bioaccumulation studies.

3. Sampling techniques (Chapter 8) and methods of data analysis that are well established.

4. High diversity and abundance coupled with relative ease of identification. While it remains an open question as to what level of taxonomic resolution is needed to detect changes in ecological conditions most effectively (Hart, 1994; Humphrey & Dostine, 1994), in some cases at least species-level determinations increase sensitivity (Humphrey & Dostine, 1994).

CONFOUNDING EFFECTS OF ENVIRONMENTAL VARIABILITY IN STREAMS

A major obstacle to stream conservation and management in tropical Asia (and elsewhere) is the difficulty of distinguishing changes due to human impacts from changes resulting from natural variability (i.e. distinguishing the 'signal' due to impact from the environmental 'noise'). The types of natural variability in streams include those which occur along the length of a river (e.g. hillstream *versus* floodplain reaches), those that arise in the same place due to season (e.g. wet season *versus* dry season) and those that occur in the same place in different years (e.g. a drought year *versus* a year with higher-than-average rainfall). Almost all studies that purport to investigate the effects of human impacts on tropical Asian streams (e.g. Dudgeon, 1984a; 1984b; 1990b; 1994a; 1995b which are discussed below; see also pp. 558–562) fail to distinguish between natural and anthropogenic change. If, for example, samples of benthic invertebrates are taken from one stream (the putatively-impacted site) that receives effluent from a factory and compared with samples from a second stream which does not receive such effluent, the effects of the pollutant on the benthos cannot be distinguished from the consequences of natural inter-site variation in stream characteristics. In other words, a lower species diversity at the putatively-impacted site might reflect the consequences of pollution, but could be due to a slower current speed, finer-grained bottom sediments, or any number of other causes. Under these circumstances, natural variability confounds attempts to identify an environmental impact unequivocally.

Poor design and confounded sampling in EIAs often rule out unambiguous and objective interpretation of the results because natural variation cannot be separated from that caused by human impacts. Separation is possible with appropriate sampling strategies or study designs, but the specific strategy that should be employed depends upon the nature or source of the confounding factors. The latter can be divided into four categories (pp. 580–589) which are set out below; some possible solutions to the confounding effects of each category are given also.

Longitudinal variation

There is an obvious gradient of physical and biological conditions along streams, from headwaters to estuary, which is well known to stream biologists (Hynes, 1970). Despite this knowledge, many studies that purport to test for environmental impacts compare conditions at an upstream 'control' or unimpacted with those at a putatively-impacted site further downstream. For example, Dudgeon (1994a; 1995b) attempted to monitor the impacts of increased suspended-sediment loads in a Hong Kong stream by comparing a putatively-impacted site with a second site upstream; the riparian zone between the two sites had been cleared during stream channelization thereby contributing suspended sediments to the downstream site. Increased suspended loads and proportions of fine particles in the interstitial sediments at the downstream site were associated with a decline in species richness and abundance of benthic macroinvertebrates but, in the absence of other information, the difference between the two study sites cannot be attributed to channelization. This uncertainty arises because we do not know if the two sites were the same prior to the 'impact', and there is every reason to believe that the longitudinal gradients in physical conditions that exist along any stream would give rise to some natural differences in benthos community structure between the two sites.

Similarly, studies of longitudinal zonation of benthic macroinvertebrates along Lam Tsuen River (Hong Kong) have been correlated with anthropogenic effects; in this case, organic enrichment (Dudgeon, 1984a; 1984b; 1990b). Increased zoobenthos densities — but lower species diversity — were associated with elevated nutrient loads in the lower course arising from input of wastes from pig and chicken farms. However, the downstream changes in community structure cannot be unequivocally ascribed to increased nutrients, because natural variations in the physical environment along the stream course will alter the benthic community in the absence of any longitudinal change in nutrient loads. The correlation between nutrients and benthic community structure cannot, therefore, be assumed to be causal.

An additional problem arises in any design that purports to test for an environmental impact by employing upstream (control site) *versus* downstream (putatively-impacted site) comparisons of stream communities; the sites or sample replicates are not statistically independent — they are pseudoreplicated (*sensu* Hurlbert, 1984). There are two problems here:

1. The impact (analogous to an experimental 'treatment') cannot be interspersed or assigned randomly to upstream or downstream sites; the upstream site is always the control (unimpacted) 'treatment' and the downstream one is inevitably the experimental 'treatment'.

Moreover, because the sites are segregated (as is the 'treatment' effect) replicate samples taken at one site may share properties not found at the other site. In other words, replicates taken at a particular site are not independent but linked (pseudoreplicated) because they are all derived from the same site. They are subsamples and not replicates (Hurlbert, 1984; Eberhardt & Thomas, 1991), and misleading conclusions may arise if the results of an unreplicated study are analysed under the mistaken assumption that subsamples can be regarded as replicates.

2. Because the control site is, of necessity, upstream of the downstream one, then the latter is influenced by the former. The one-way flow of water along a stream means that processes in the upstream site may affect those downstream; for example, a reduction in the rate of processing of allochthonous leaf litter at the upper site will alter the magnitude of transport of fine particulate organic matter to downstream reaches. Obviously, then, the two sites (experimental 'treatments') are not truly independent of each other, but such independence is required for statistical analysis. It is important to stress that the lack of independence of downstream from upstream sites is not a trivial matter. The prevailing paradigm in stream ecology — the RCC — stresses the downstream transport of material and linkage of sites along a stream course. Moreover, recent research in North America (Cushing *et al.*, 1993) has shown that organic matter from headwater reaches can be transported large distances along a stream, and strongly supports the notion that sites along a stream system are strongly connected.

Even if pseudoreplication were not a problem, and it was possible to test differences between the putatively-impacted and control sites for statistical significance, the analysis is confounded because there is only a single replicate of each type of site ('treatment'). As mentioned above, under such circumstances it is not possible to separate the effects of an impact from differences that arise simply because the two sites vary with respect to some environmental factors.

Pseudoreplication of the type outlined above is encountered commonly in EIAs in streams in tropical Asia and elsewhere, and appear to rule out the application of inferential statistics in data analysis. However, such concerns are rarely addressed specifically (Humphrey & Dostine, 1994). One logical and valid response to these difficulties is to abandon statistics altogether (Carpenter, 1990). Some environmental impacts may cause changes that are so large that they are convincing in the absence of statistical analysis; however, the conservationist or environmental manager would be unwise to assume that all observers will respond to the evidence in the same way (see pp. 592–593). Some optimism is possible, however, given the recent application of novel statistical techniques to analysis of the effects of unreplicated large-scale changes in ecosystems (Frost *et al.*, 1988; Carpenter, 1990; Reckhow, 1990).

Where an upstream *versus* downstream comparison is the only option available to a scientist undertaking an EIA (as was the case described in Dudgeon, 1994a), the authority of any conclusions drawn can be greatly enhanced by the inclusion of 'before-impact' information. If the putatively-impacted site is sampled before the impact, and demonstrated to be closely similar to the unimpacted site, then the hypothesis that (as a consequence of natural variability) the two sites were different before the impact can be ruled out. In the example given by Dudgeon (1994a), before-impact data showed that macroinvertebrate species richness at the downstream site was the same as that at the upstream control site after impact, suggesting the observed post-impact difference between the two sites was, indeed, a consequence of the increased suspended-sediment loads. Conclusions about an impact effect can be strengthened still further by continuing the evaluation of the impact over time until natural recovery has progressed and convergence of the control and impacted site can be demonstrated. The opportunity to monitor such recovery will occur only when the impact ·is in the form of a 'pulse disturbance' (*sensu* Bender *et al.*, 1984); i.e. a short-term disturbance (such as a chemical spill into the stream) which is discrete in time with a begining and an end. The converse — an impact due to a 'press disturbance' — is a sustained disturbance that impacts the stream continuously (e.g. effluent from a factory or pig farm) and offers no opportunity for recovery. Unfortunately, most anthropogenic impacts on tropical Asian streams are of this type.

Before- and after-impact comparisons are, of course, only available if prior knowledge of the timing of the impact is available. If it is known that a potentially-polluting factory is to begin discharging into a stream in (say) 24 months, then collection of before-impact data is possible. If the putative impact is unpredictable, as in the case of an accidental spill of toxic chemical into a stream, the EIA is seriously constrained. The only option available is a comparison of the upstream and putatively-impacted downstream site (where differences may reflect natural variability or, perhaps, the effect of the accidental impact), accompanied by a longer-term study that follows the process of recovery at the downstream site. The demonstration of recovery is crucial because it strengthens the argument that initial conditions at the upstream and downstream sites did not differ (Peterson, 1993). Monitoring recovery may involve novel methods of time-series analysis (e.g. Green, 1993), some of which do not depend upon site replication (e.g. Jassby & Powell, 1990).

Small-scale spatial heterogeneity

Natural environments show great variability from place to place over several spatial scales. This variation necessitates replicate sampling within study sites along (or among) streams; the question is, how much sampling effort should be allocated to this replication? Clearly, the answer depends upon how variable stream reaches within a study site might be, and how much of that variability is of interest to the investigator. A simple answer is that where there are natural features or microhabitats in the study site (e.g. riffles, pools and runs) they should each be the subject of replicated sampling and treated as separate parts of the site. This is necessary so that variation among microhabitats is not confounded with variability from one site to another. For example, a sample from a pool microhabitat at one stream site would differ from the contents of a sample from a riffle microhabitat at another site, even if the two stream sites were identical (e.g. Logan & Brooker, 1983). In other words, if localities within a site differ, then — in the absence of an environmental impact — single samples from a several sites are likely to differ also and the inter-site comparison is confounded.

The confounding effects of small-scale spatial heterogeneity can be reduced by comparing 'like with like'. Most stream invertebrates attain peak abundance and diversity in riffles and, unless a complete species list for each stream site is needed, comparison between sites can be based upon riffle inhabitants only. Thus replicated (or, strictly speaking, pseudoreplicated) samples from riffles in two or more stream sites can be used as a basis for comparison. An alternative approach depends upon the fact that EIAs always involve some sort of comparison (control *versus* impacted site, before *versus* after impact). This can provide a way of reducing the confounding effects of small-scale environmental heterogeneity because an inter-site comparison rarely requires a knowledge of the abundance of every species at each stream site, but instead a reliable estimate of the abundance of some of those species (preferably the common ones). Usually such data can be obtained by the use of introduced or artificial substrates (e.g. Hester-Dendy multiplate samplers used by the United States Environmental Protection Agency; see Chapter 8). Thus if several artificial samplers are deployed in comparable microhabitats (similar current speed, bottom sediments, etc.) at each study site, we would expect the mean density of macroinvertebrate colonists at each site to be similar unless there is some difference among the sites. The uniform surface area and form of each homogenous substrate unit reduces the magnitude of variance around the sample mean that is so typical of samples taken from natural substrate units (rocks, boulders, sand, etc.) in streams, and thus improves the precision of estimates of population size at each site. The greater precision increases an investigator's ability to detect impacts, because small changes in abundance due to the impact are less likely to be masked by variation among replicates (i.e. the ratio between the impact 'signal' and the 'noise' of natural variation is increased). For example, if mean population densities of benthic macroinvertebrates on artificial substrates can be estimated with a precision of $\pm$ 10% (at the 95% confidence level), then an impact which causes mean densities to decline by 30% should be readily detectable with an appropriate sampling design. By contrast, if mean population densities using benthic Surber samplers can be estimated with a precision of only $\pm$ 30% (at the 95% confidence level), then it will not be possible to separate the effects of an environmental impact causing a 30% reduction in population densities (the impact 'signal') from the effects of inter-sample variability (environmental 'noise'). Note that the use of artificial substrates will not yield a perfectly-representative sample of the benthic macroinvertebrates present in a stream, but (with appropriate replication) it does produce reliable estimates of the population sizes of those taxa which will colonize artificial substrates and thereby provides a robust basis for assessing environmental impacts on those taxa.

Inter-stream variability

This point is rather straightforward, and echoes the problems that arise within a single stream when a single upstream site is compared with one downstream. If we extend the case to one in which a single 'control' stream is sampled and compared with a single putatively-impacted stream, it is clear that we cannot use the comparison to make any inference about the effects of the impact. This is because the two streams may have been intrinsically different (with different species complements and population densities) before the impact occurred, or may be different notwithstanding the effect of the impact. Clearly, replicated control streams are necessary so that the extent of variability among streams can be clarified. For example, if a stream receiving chemical pollutant has a mean benthic macroinvertebrate density of 250 individuals.m^{-2} and densities in a single control stream are 400 individuals.m^{-2}, then an investigator might (unwisely) conclude that the reduced densities in the impacted stream were due to the impact of the chemicals. There are examples of many such investigations in the literature on Asian streams, and such studies are published because they concur with our biological intuition and preconceptions of the effects of chemical on benthic macroinvertebrates. If a second control stream with densities of 100 individuals.m^{-2} were added to the imaginary comparison, then it would be apparent that densities in the putatively-impacted stream were well within the range of natural variability (250 *versus* 100–400 individuals.m^{-2}). Moreover, a comparison of the second control stream with the putatively-impacted one would have led to the bizarre conclusion that chemical effluent had the effect of increasing macroinvertebrate densities from 100 to 250 individuals.m^{-2}. This conclusion would most likely have been rejected by the investigator because it did not match expectations; in fact, the appropriate response would have been to increase the number of replicate control streams.

There is an inherent asymmetry here, because the investigator is often in a position to establish replicate control streams, but rarely (if ever) has more than one stream experiencing the same environmental impact. Nevertheless, an asymmetric design incorporating the presence of multiple control streams can help to reduce the confounding effects that occur when a comparison involves only two streams (control *versus* putatively-impacted).

The spatial scale of variation or patchiness in the environment is generally not known before assessment of environmental impacts is undertaken. Does, for example, patchiness occur at the scale of the replicate samples (e.g. within a riffle reach), at a higher scale (e.g. among riffles) within the same stream, or among streams? This lack of information creates the problem that patchiness at any intermediate spatial scale between that of the sampling units (small scale) and the locations sampled (large scale) will not be revealed by the sampling design, and consequently valid comparisons among locations are impossible. In other words, the comparison is confounded (i.e. the data are pseudoreplicated) because the within-location variance or patchiness has not been adequately represented by the replicate samples. A method of detecting the spatial scales of patchiness has been described by Morrisey *et al.* (1992a) in which a nested sampling design was employed to investigate patchiness of marine benthos. Each of a series of successively smaller spatial scales was nested within the scale above it, and partitioning of the variances associated with each scale (using analysis of variance) provided an estimate of the contribution of each scale to the total variation among samples within the largest scale of comparison. For example, if the focus of attention had been comparison of streams on two sides of a mountain range, then the hierarchical (or nested) sampling design to examine variation of benthic macroinvertebrates at a range of spatial scales would be as follows: three riffle reaches could be nested within each of three streams nested within each of two sides of the mountain range; three (or more) replicate multiplate samples could be taken in each riffle reach. Sampling designs of this type (described in more detail by Morrisey *et al.*, 1992a) have yet to be used in studies of tropical Asian streams, but investigation of the scale of spatial variation in these environments would yield data of great value to the design and execution of EIAs.

Temporal variation in stream communities

The choice of the frequency with which impacted and control sites are sampled is not a simple matter, but often seems to be made in an arbitrary fashion (Underwood, 1991). There are two main issues. The first is that whatever the interval at which samples are taken (monthly, seasonally, yearly, etc.), there must be replication within each interval so that comparisons among sampling periods are not confounded by short-term temporal changes (Underwood, 1991, 1993). For example, if a stream biologist takes samples of benthic macroinvertebrates on one occasion in the wet season and on one occasion during the dry season and finds that densities are higher during the dry season, it is not possible to ascribe the density difference as being due to season, because the magnitude of short-term within-season fluctuations in density are not known. Would the difference between wet and dry seasons have remained consistent if samples had been taken on different days during the two seasons? Clearly, an appropriate sampling design in this case would include temporal replication of samples within each season. This could be achieved in a straightforward manner by visiting each site on three or four randomly-selected occasions during each season, with the constraint that the dates chosen should be near the middle of each season. Two additional points need to be made:

1. Spatial replication of samples on each visit to the site will not compensate for a lack of temporal replication since, if they were all taken on the same occasion they cannot represent any variation in the temporal dimension.
2. Where two or more sites are to be compared, sampling of each should be undertaken on the same day. This minimizes the chances of inter-site differences which might arise if there are regular and synchronous short-term cycles in the benthic communities among study sites.

The use of replicate samples within each season illustrates the general point that replicated temporal sampling at smaller time scales is essential to demonstrate that an apparent trend at a larger (in this case, seasonal) time scale is not caused by shorter-term fluctuations. The nested (hierarchical) sampling designs which have been used to examine variation at a range of spatial scales (Morrisey *et al.*, 1992a) can be applied to variation at different temporal scales (Morrisey *et al.*, 1992b), whereby each of a series of samples taken at a progressively shorter time intervals is nested within the scale above it (e.g. days nested within months nested within seasons nested within years). However, while replication within seasons will not place a great additional burden upon the investigator, if the time scale of interest is months rather than seasons, replication within months combined with the collection of samples at monthly intervals will add considerably to the sampling effort expended. This raises the question as to whether sampling at a monthly time interval would be an appropriate focus for an EIA, and is related to the second major point to be made with regard to the effects of temporal variation. It is important to ensure that sampling occurs with a frequency which is related to the duration of the life history or turnover rate of the benthic community because — for the purposes of an EIA — samples in each time interval need to be independent of each other or not serially correlated (Stewart-Oaten *et al.*, 1986; Underwood, 1993). Indeed, such independence is a requirement of most statistical procedures used to test for environmental impacts (Underwood, 1981). The main issue here is that successive sampling intervals should be sufficiently spaced in time to ensure that members of a population collected during the first set of samples are not included in the second set of samples, or that different components of the same cohort are not sampled in successive sampling periods. If this is not the case (i.e. if the population has not 'turned over') then the size or composition of the population collected by the second sampling will be correlated with that in the first and, if sampling involves observation and counting rather than removal sampling, the same individuals may be counted on both sampling occasions. Under such circumstances the statistical assumption of independence cannot be met. Stewart-Oaten *et al.* (1986) have also pointed out that however many times and at whatever intervals populations are sampled, the occasions must be chosen randomly and should not be regular. This will minimize the concordance of sampling dates with regular fluctuations of the environment of which the investigator is unaware.

Serial correlation and non-independence of benthic samples from streams on successive time intervals is likely to be less of a problem in tropical Asia than in, for example, north-temperate streams. This is simply because the higher water temperatures at lower latitudes allow rapid completion of life cycles, with some species breeding throughout the year (Bishop, 1973; Maheshwari, 1989; Dudgeon, 1992b; see also Gupta *et al.*, 1993). It should be stressed here that although the non-independence of successive samples presents a problem for the statistical analyses that are an essential part of EIAs, this obstacle does not arise in studies of the life history or production of stream invertebrates in tropical Asia. For such studies, non-independence is essential as the goal is to monitor the growth and development of a particular cohort of individuals. However, it is worth stressing here that the rapid growth of tropical stream invertebrates will require more frequent population sampling than the monthly intervals that are frequently employed in studies of north-temperate taxa.

BACI AND BEYOND-BACI DESIGNS

To avoid the confounding factors, pitfalls and obstacles to environmental-impact assessment, an appropriate study design is needed. There is some consensus on what this design should involve, and it is based on a proposal by Green (1979). The essentials of the design are a before- and after-impact comparison of the abundance of organisms at a control site and a putatively-impacted site. This BACI (before-after, control-impact) design was modified by Stewart-Oaten *et al.*

(1986) who realized that temporal variability dictated that a single sampling occasion before the impact followed by another one subsequent to the impact would be confounded by the lack of temporal replication (see above) and would be insufficient to represent the possible natural fluctuations in population size. They extended the design to include simultaneous sampling at both sites on two or more occasions both before and after the impact, with the sampling times chosen randomly to avoid possible confounding effects caused by cyclical processes not known to the observer. Each time of sampling is characterized by some measure of the difference between control and putatively-impacted sites, rather than by the data for individual samples at the two sites. The set of differences before and after impact are then compared statistically. A further design modification was conceived by Underwood (1991) who proposed the beyond-BACI design, whereby the lack of spatial replication in the original BACI design is dealt with by the addition of more control (unimpacted) sites. Underwood (1991) argued that a BACI design which compared only single impacted and control sites was inadequate because there was no valid reason to assume that a difference between populations at the control and impacted sites would be unchanged in the absence of any impact. The inclusion of replicate control sites is the most practical solution to this problem because impacted sites generally cannot be replicated, but the outcome is an asymmetric design consisting of replicated (repeated-measures) before- and after-impact data from several control sites and one impacted site. Methods for statistical analysis to detect environmental impacts using an asymmetric sampling design are given by Underwood (1993). Note that the BACI approach employs population-level or univariate data; it has yet to be applied to community or multivariate data. Attempts are being made to incorporate multivariate data by using dissimilarity indices (e.g. the Bray-Curtis measure) to reduce the differences between control and putatively-impacted communities over many different species to a single value (Faith *et al.*, 1991; Humphrey & Dostine, 1994; Faith *et al.*, 1995; Humphrey *et al.*, 1995).

SIGNIFICANCE OF CONFOUNDING EFFECTS

Is it always necessary to use the beyond-BACI design in EIAs, and is it essential for us to avoid all of the confounding factors discussed above? Some would argue that, in certain cases, biological intuition — or the opinion of an experienced biologist — will be enough to persuade an unbiased audience that an effluent or other anthropogenic modification of a stream has caused the changes in community structure that we observe. However, the issue is not usually one of persuading like-minded colleagues, but rather tough-minded industrialists, government officials, or officers from regulatory agencies and the like, who may have an economic or political stake in underplaying the importance of any environmental impact. In cases where a dispute over the significance of an environmental impact arises between scientists and other interested parties, resolution may involve a court hearing or tribunal. Under these circumstances, an inadequate design for detecting environmental impacts prevents compelling and rigorous scientific assessment of effects and therefore actually promotes dispute (Peterson, 1993). The uncertainty of the effect of a putative impact implies that reasonable arguments can be made on each side because of a lack of the unequivocal and unambiguous scientific data needed to solve the dispute. Under such circumstances, rigorous and unbiased scientists may disagree and, in a courtroom where resolution of disputes may be based upon an adversarial legal system, there is every likelihood that the findings of an EIA will be challenged. Accordingly, the interests of conservation and environmental management are best served if the design of EIAs are sufficiently rigorous to give rise to unambiguous conclusions.

Statistical power

One situation in which ambiguity may arise occurs when there is uncertainty as to the power of the statistical tests used to detect an impact during an EIA. If information on the power of these tests is not presented, it is not possible to determine whether an impact might have occurred but the test was too weak to detect it (a Type-II error) or if there really was no impact effect. Traditionally, biologists and other scientists have been concerned about making Type-I errors, and falsely concluding that an effect existed when it did not. Conclusions that an effect existed are based on the fact that a statistical test is significant at some chosen level of probability, usually $P = 0.05$. That is, the chances of making a Type-I error, and concluding that the impact had an effect when in fact there was none, is 0.05 (or 5%). In EIAs, it would seem appropriate to consider the alternative Type II error; that of falsely concluding that there was no impact when, in fact, one had taken place. The importance of this point has been emphasized by Peterman (1990) and Fairweather (1991), and has begun to become more widely recognized recently (Cooper & Barmuta, 1993; Peterson, 1993; Underwood, 1993; Humphrey & Dostine, 1994). This reflects the fact that failure to detect a real impact has serious consequences for conservation and environmental management.

Power is the probability of detecting a real difference of a certain magnitude, or the probability that the null hypothesis of 'no effect of impact' is rejected; thus it is obvious that a given statistical test will be more powerful when the effect caused by an impact is great or when many replicate samples are collected (Fairweather, 1991). By contrast, the test is weakened if there is great variability in the response variable (e.g. benthic population densities) and the impact effect is small relative to sample variability. If the power of any statistical test used in an EIA is not given, then definitive conclusions about impact are unjustified. Furthermore, it is necessary to decide what magnitude of effects are to be detected, and to ensure that the test has appropriate power; this is not a trivial issue, as the investigator has to decide whether, say, a 5% reduction in population size is ecologically significant. If it is, then the test must have the power to detect such a change if a Type-II error is to be avoided. The calculation of power to detect environmental change is described in detail by Underwood (1993).

Statistical errors and the 'burden of proof'

Type-I and Type-II errors vary inversely in any given test. There is thus good reason to reconsider the widespread approach of setting the probability of committing a Type-I error at 0.05, especially in EIAs, because the more we reduce the chances of a Type I error, the greater is the possibility that a Type-II error — and concomitant environmental damage — will result. Since the consequences and costs of making a Type-II error (environmental damage and the need for restorative measures) are generally greater than those of a Type-I error (treatment of effluent and modification of engineering projects), and because Type-II errors are rarely in the public interest (although they may serve the interests of private developers and industry), then an increase in the P value used in EIAs to 0.10 or even higher seems appropriate (Peterman, 1990). This is simply the greater acceptance of the chances of making a Type-I error (i.e. concluding that an effect existed when it did not) but is in the public interest in cases where the consequences of failing to detect a real impact (a Type-II error) are irreversible degradation of the affected stream. Note that, in contrast to a Type-II error, a Type-I error incurs no biological cost. This could be combined with a practice of shifting the 'burden of proof' (Peterman, 1990) from the usual situation where developments are allowed to continue if 'proof' of impact is missing (although a Type-II error may be made), to one where statistical tests with adequate power are employed to ensure that impacts exceeding a prescribed amount would not go undetected (i.e. the chances of a Type-II error are minimized). The main point here is that the 'burden of proof' needs to be shifted so that the industrialist, developer, or those seeking to change stream environments in some way are required to demonstrate that their proposed modifications will not have an impact greater than a predetermined amount. This is in contrast to the usual practice whereby environmental modifications are permitted unless biologists can demonstrate unequivocally that significant environmental degradation will result. In other words, there should be a shift towards proof that proposed changes *will not* have an impact, rather than the conventional approach requiring proof that changes *will* have an impact.

RECOMMENDED STRATEGIES

Ecological research on tropical Asian streams and rivers is increasingly concerned with predicting, assessing and mitigating the effects of various types of human impacts. There is a particular need to improve our ability to detect environmental impacts, and to distinguish the effects of natural change from those caused by human activities. To this end, a variety of strategies are proposed.

1. Longitudinal variation must be taken into account in designs that purport to test for an environmental impact by employing upstream (control site) *versus* downstream (putatively-impacted site) comparisons of stream communities, because the sites or sample replicates are pseudoreplicated. Under such circumstances, conclusions can be enhanced by the inclusion of before-impact information. However, a better approach is adoption of BACI designs (Green, 1979; Stewart-Oaten *et al.* 1986), with sampling at control and putatively-impacted sites on two or more occasions

before and after impact to reduce the confounding effects of a lack of temporal replication.

2. Small-scale spatial variation or patchiness in streams confounds inter-stream or intersite comparison because, if localities within a site differ, then — in the absence of an environmental impact — single samples from several sites are likely to differ also and the inter-site comparison is confounded. The confounding effects of small-scale spatial heterogeneity can be reduced by comparing 'like with like', and by the use of artificial or introduced substrates.

3. Problems of inter-stream spatial variability arise when a single control stream is sampled and compared with a single putatively-impacted stream, because we cannot use the comparison to make any inference about the effects of the impact. An asymmetric design, where a single putatively-impacted stream is compared with several control streams, can help to reduce this effect by exposing the extent of natural variability, and is a feature of beyond-BACI sampling designs (e.g. Underwood, 1991).

4. The scale of natural variability within and among streams is rarely known *a priori*, but such knowledge is essential for planning an appropriate sampling strategy. A nested sampling design for detecting the spatial scales of patchiness (e.g. Morrisey *et al.*, 1992a), in which each of a series of successively smaller spatial scales is nested within the scale above it, can provide an estimate of the contribution of each scale to the total variation among samples within the largest scale of comparison. Such designs have yet to be applied in tropical Asian streams, but would yield the information needed to design sampling strategies that avoid the pitfalls described above.

5. The confounding effects of temporal variation can pose difficulties for ecological studies and EIAs in streams because, whatever the interval at which samples are taken (seasonally, yearly, etc.), there must be replication within each interval so that comparisons among sampling periods are not confounded by short-term temporal changes. Replicated temporal sampling at smaller time scales is essential to demonstrate that an apparent trend at one time scale is not caused by shorter-term fluctuations. Nested or hierarchical sampling designs can be applied to variation at different temporal scales (e.g. Morrisey *et al.*, 1992b), whereby each of a series of samples taken at a progressively shorter time intervals is nested within the scale above it. Serial correlation and non-independence of samples pose difficulties for impact assessment, as independence

is an assumption of most statistical procedures used in EIAs. Accordingly, sampling frequency must be less than the turnover rate of the community so as to ensure that samples in each time interval are independent of each other and not serially correlated (Stewart-Oaten *et al.*, 1986; Underwood, 1991; 1993).

6. Most EIAs give no information about the power of the statistical tests used to detect an impact. Power is the probability of detecting a real difference of a certain magnitude, or the probability that the null hypothesis of 'no effect of impact' is rejected. If information on the power of statistical tests is not presented, it is not possible to determine whether an impact might have occurred but the test was too weak to detect it or if there really was no impact effect. If the power of any statistical test used in an EIA is not given, then definitive conclusions about impact are unjustified; Underwood (1993) has described the calculation of power to detect environmental change.

7. Failure to detect a real impact has serious consequences for stream conservation and environmental management. Accordingly, use of power analysis in EIAs should be combined with a shift in the 'burden of proof'. The usual situation where a proposed environmental modification is allowed to continue if proof of impact is missing (which may arise if statistical power is low) should be changed to one where developers are required to demonstrate that proposed modifications will not have an impact greater than a predetermined amount; i.e. there should be a requirement to prove that anthropogenic modifications *will not* have a detrimental impact.

7

Process-Orientated Studies in Stream Ecology

Assessment of environmental impacts is an important part of the activities of many stream biologists, but empirical studies of impacts need to be buttressed by process-orientated studies that investigate the mechanisms underlying the changes caused by the impact. Obviously, if we are to make well-founded predictions of the effects of anthropogenic activities, we need to understand the processes and mechanisms by which stream ecosystems are structured. We may never understand stream ecosystem structure and function fully, but this should not prevent us from identifying the priority areas in which knowledge of underlying mechanisms (or even identification of fundamental patterns) is needed urgently. It is therefore reasonable to ask: what do we need to know about stream ecology in Asia? Process-orientated studies (*sensu* Peterson, 1993) which examine the mechanisms underlying observed patterns or changes will be essential for understanding stream ecology, and would have the additional advantage of strengthening the rather weak links that exist between developments in 'pure' ecology and the practice of biomonitoring (Hart, 1994). Functional understanding of ecological patterns and processes is largely lacking for Asian streams, and process studies of population-, community- and ecosystem-level responses are needed desperately to improve the science of environmental assessment, and so advance stream conservation and management. Again, therefore, we can ask: which process-oriented studies need to be undertaken most urgently? In this chapter, an attempt will be made to identify the necessary process-orientated studies, and then the relative merits of examples of research strategies for obtaining the requisite basic information on ecological processes will be discussed.

Process-orientated studies of streams can be grouped into three general categories: descriptive studies; experimental studies; and, studies of impact mitigation and restoration management. Although these categories are not entirely exclusive, for convenience they will be considered in turn.

DESCRIPTIVE STUDIES THAT REVEAL PATTERNS

A useful descriptive study of a stream system should include the following:
1. Characterization of the major habitat types.
2. A catalogue of the species (or species-groups) present in (or characterizing) each habitat type (i.e. an inventory of patterns of biodiversity).
3. Elucidation of seasonal patterns of distribution and abundance of the biota.
4. Basic ecological information — such as feeding, reproduction, microhabitat use — for (at least) the major species in each habitat type.
5. Information on the magnitude and seasonality of the flow of materials or energy through the main habitat types (e.g. the input of allochthonous material, downstream transport of organic matter).
6. Estimates of primary and secondary production in the major habitat types, together with some indication of the importance of autochthonous and allochthonous material in supporting secondary production.
7. Identification of potentially endangered, endemic or vulnerable species, and the threats to their existence.

While *some* of this information is available for a few Asian rivers (the Ganges, Chang Jiang and Mekong are possible examples), the data are incomplete and concerned largely with fish stocks or large, charismatic vertebrates. Little or no attention has been paid to investigations of fishes which have no direct economic value, and very little is known of the composition, distribution or abundance of lotic invertebrate taxa and primary producers. Certainly, the vast majority of studies of tropical Asian streams are fragmentary, focusing on very few parameters or taxa (Gopal & Sah, 1993; Dudgeon, 1995a), and ecological data of any sort are wanting for most systems.

A key issue, which perhaps should be the starting point for research and management agendas, is to ascertain how much is actually known about the taxonomic relationships, diversity and conservation status of endemic aquatic organisms in particular countries. These data are lacking for most of tropical Asia (Pethiyagoda, 1994; see also p. 563 and p. 567), and only in Sri Lanka can the stream biota be described as reasonably well known (see Fernando, 1990; Costa, 1994; de Silva, 1994; de Silva & de Silva, 1994 and references therein) although, even here, knowledge of the conservation status and taxonomic status of stream fishes is considered to be inadequate (Pethiyagoda, 1994). With the nature and status of endemic biota categorized, the next step towards the development of management plans and rehabilitation strategies would be accumulation of essential biological and ecological information that could be used to establish management plans for the protection of particular species and their habitats. Knowledge of habitat requirements and environmental tolerances, reproductive biology, patterns of dispersal and interactions among species in tropical Asian streams will be critical here, but such information is generally inadequate. There is a particular need for greater knowledge of the biota of many large river systems and, as an example of the type of research needed, a recent study of the use of floodplain habitats by the fishes of the Fly River, Papua New Guinea (Smith & Bakowa, 1994) can be cited. This example draws attention to the fact that even the migration patterns of commercially-important fish species up and down rivers, in response to varying food supply and to reach breeding grounds, are almost unknown in most tropical Asian countries (Lowe-McConnell, 1987; Dudgeon, 1992b; Roberts, 1993a). Smith & Bakowa (1994) highlight several important methodological issues, including the difficulties of quantitative sampling in large river systems and the need for sophisticated multivariate analyses of biological and environmental data sets. Clearly, execution of the types of studies needed in many tropical Asian streams will challenge our ingenuity and resources in many ways.

How can biologists make an effective contribution to the description of tropical Asian streams, when such huge gaps exist in our knowledge?

One answer is that description must take place in the context of some paradigm or model of what the system *might* be like, which can then be taken as a null model against which predictions about system characteristics may be tested. As discussed in Chapter 2, the RCC, which attempts to link patterns of productivity and community structure to gradients in the physical environment, can serve as such a paradigm or null model because it makes predictions about longitudinal patterns of ecological change from headwaters to estuary which can be used as a basis for description and comparison (see pp. 63–70). To date, these predictions have received scant attention from workers in tropical Asia (Arunachalam & Arun, 1994; Dudgeon, 1995a).

An important feature of the RCC is the integration of in-stream and catchment processes and an emphasis on the importance of inputs of material from the land. The significance of allochthonous energy sources and organic matter in stream energy budgets is now widely recognized from studies in most major continents, although most of the available information has been derived from research in temperate systems (Lake & Barmuta, 1986). One technique which has been used by stream biologists in the temperate region (e.g. Rounick *et al.*, 1982) involves the measurement of carbon stable-isotope ratios of consumer organisms. Aquatic plants have higher stable carbon isotope ratios than terrestrial plants, and the relative ^{13}C-depletion or enrichment of consumer issues can be used to indicate dependence on allochthonous or autochthonous energy sources. This approach has yet to be employed in tropical Asian streams, but seems to be a rather straight-forward means of examining the contribution of allochthonous detritus to in-stream secondary productivity. While we need to know more about the importance of allochthonous food sources in tropical Asian streams, the limited data suggest that they may constitute the basis for production in some streams — especially those draining forest (pp. 46–49). Tropical forests are being cleared at an alarming rate and, by 1990, over 40% of the total area had been destroyed (Lean *et al.*, 1990). We must increase understanding of the roles of natural allochthonous energy sources in streams if management practises for streams and their catchments are to be placed on a sound footing.

If, and there is little reason to doubt this, the links between tropical Asian streams and their valleys are of similar importance to those reported for temperate streams (Hynes, 1975), it is sensible to assume that management practises employed elsewhere might be applicable to tropical Asia. For example, the maintenance of buffer strips of riparian vegetation along stream banks may help to protect aquatic communities from the effects of land-use changes within the drainage basin. Unfortunately, we do not know how wide such buffer strips would need to be to ensure effective protection. Retention of riparian buffer strips of at least 10 m width along each bank does reduce the effects of clear-cutting and soil erosion in north-temperate latitudes (Burns, 1972; Newbold *et al.*, 1980; Ahtiainen, 1992; Holopainen & Huttunen, 1992), but it is not yet clear what width of strips is necessary to maintain the structural and functional integrity of stream communities. This is an important management question for streams in Asia, since there are not even the most basic data pertaining to the question of what buffer strip widths are both necessary and sufficient to ensure stream protection. A straight-forward comparative study of streams which differ only with respect to the width of the buffer strip would help to resolve this issue. An additional consideration is the width of buffer strip (or riparian corridor) needed to conserve populations of terrestrial, streamside animals. Research in North America suggests minimum widths of 10–30 m and 75–175 m to include 90% of the streamside plants and birds respectively (Spackman & Hughes, 1995), but use of a standard corridor width to conserve flora and fauna is a poor substitute for individual, stream-specific assessments of species distributions.

A note of caution must be inserted here. While we cannot — and should not — ignore prevailing paradigms such as the RCC, stream biologists in tropical Asian must question the axioms and generalizations derived from research in temperate systems where catchments are dominated by deciduous forest species and climatic factors impose fundamentally different regimes on aquatic biota and ecological processes (Williams, 1988, 1994). Rigorous testing of the RCC will be needed to determine whether this can be used as a predictive model for tropical Asian ecosystems. There are, for instance, considerable doubts as to whether the Concept can be applied in any wholesale fashion to Australian and New Zealand streams (e.g. Winterbourn *et al.*, 1981; Bunn, 1986; Lake & Barmuta, 1986). At present, there are no complete studies of any tropical Asian river system of any size which would allow us to test fully the predictions of the RCC, nor have the aquatic, semiaquatic and terrestrial habitats on the floodplain — a major site of land-water interactions and allochthonous inputs into the river (Welcomme, 1979; Junk *et al.*, 1989; Sedell *et al.*, 1989) — been adequately integrated into the RCC (see pp. 90–91).

There are two additional concepts of lotic ecology which are related to the RCC and have yet to receive any attention from stream biologists in tropical Asia, despite the fact that they may have implications for effective conservation and management. Spiralling is used to refer to the process of reuse (ingestion — egestion — reingestion) of allochthonous material in streams (Wallace *et al.*, 1977), and it may contribute significantly to the 'tightness' of nutrient cycles and retentiveness of the lotic community (Newbold *et al.*, 1983a, 1983b). The spiralling concept may have utility in the development of models for understanding the dynamics of nutrients and carbon in stream ecosystems, but we lack even the most basic data on this process in tropical Asia. Secondly, the serial discontinuity concept (Ward & Stanford, 1983) seems to be a useful construct for estimating the impacts of impoundments on streams in the north-temperate zone (with particular reference to the predictions of the RCC), but has yet to be applied to the many regulated streams of tropical Asia.

EXPERIMENTAL STUDIES

A striking feature of ecological science over the past two decades has been the increased use of experiments to answer questions about pattern and process in nature (reviewed by Hairston, 1989). This has arisen because conclusions derived from purely descriptive studies lack predictive ability, and because of the uncertainty that can arise when observations of a natural phenomenon are used as a basis for subsequent interpretation of that pattern (so-called 'weak inference'; Peters, 1976). Many scientists (Hairston, 1989 and references therein) question the extent to which natural situations can be used to interpret the processes that brought them about and, in a review of methods for sampling stream benthos, Peckarsky (1984) advocated a problem-orientated, experimental approach over the traditional, descriptive approach to stream ecology. Cooper & Barmuta (1993) have also stressed the importance of field experiments in biomonitoring efforts, because they allow greater control over relevant variables and can reveal the proximal mechanisms which produce the observed responses. Despite the greatly increased use of field experiments in other regions, I have been able to discover only six published studies of stream ecology in Asia that involve manipulative experiments (Benzie, 1984; Dudgeon, 1990a; 1991a; 1991b; 1993; Dudgeon & Chan, 1993). The scarcity of experimental studies is an important issue: although association techniques or interpretation may generate hypotheses when applied to large sets of data, causal relationships are least-ambiguously elucidated through manipulative experiments with adequate controls and replication.

The practice of stream ecology in tropical Asia generally involves a series of observations followed by formulation of some plausible explanation to account for patterns revealed by the observations. Several alternative hypothesis can usually be called upon to account for the patterns observed, and thus this approach to science can generate explanations which are incorrect; at best, we have little assurance that they are right! However, an observational study can be used to generate hypotheses about causative factors that are tested subsequently by experiment. For example, Dudgeon (1988a; 1989a) examined changes in macroinvertebrate community structure and abundance in four streams which differed in the degree of shading by riparian vegetation. Inter-stream variations in species complement and abundance appeared to be related to the degree of shading and the standing stocks of allochthonous detritus and algae in each stream, and Dudgeon (1988a; 1989a) suggested that the change in food base caused the modification of macroinvertebrate communities. While this interpretation may have been correct, it was possible that some other factor (e.g. intensity of predation by fishes, disturbance history) could have caused the observed patterns. In a subsequent experimental study, Dudgeon & Chan (1992) manipulated shading intensity in a small Hong Kong stream, and found that shaded patches of the stream bed had less algae, as well as fewer species and lower densities of macroinvertebrates. This experiment unambiguously supported the initial interpretation — arrived at by weak inference — that shading altered the food base of streams and caused changes in the macroinvertebrate communities. The point here is not that the interpretation of the observed pattern turned out to be correct, but that a manipulative experiment was *required* to test it; had the initial interpretation been incorrect, then the results of the experiment would have shown that it was wrong.

The greater explanatory power of experimental studies when compared to weak inference strongly suggests that there should be more emphasis placed on field experiments in those process-orientated studies where environmental variables of interest can be manipulated. Inferences about the importance of predation or competition as structuring processes in stream communities are of limited value unless they are backed up by experimental studies which manipulate the variables of interests (i.e. populations of predators and prey; the relative abundance of putative competitors); this may involve the use of cages or enclosures to exclude certain species or to enclose populations with densities changed from those prevailing in the habitat. Such manipulations are crucial because descriptive studies of niche segregation (or partitioning), for instance (e.g. Edington *et al.*, 1984; Moyle & Senanayake, 1984; Dudgeon, 1987d; 1989e; Wikramanayake & Moyle, 1989; Wikramanayake, 1990), tell us merely that animals can exploit various dimensions of the stream habitat in different ways; without experimental manipulation, they reveal nothing about the mechanisms underlying the observed patterns. While ecomorphological relationships among species *may* reflect evolutionary adjustments to facilitate resource partitioning and reduce competition (e.g. Kortmulder *et al.*, 1990; Wikramanayake, 1990), the evidence for this (however persuasive) is no more than circumstantial in the absence of manipulative studies demonstrating the occurrence of competition.

The experimental approach could also play a role in biological monitoring programs because they allow prediction of the effects of anthropogenic manipulation of the environment and can indicate ecological mechanisms controlling responses of organisms to such changes (see also Courtemanch, 1994; Hart, 1994). Dudgeon (1990a) gives an example of the link between biomonitoring and experimental studies. Rotenone piscicide was introduced into two Papua New Guinea streams, and the response of macroinvertebrates in a treated reach was compared with that in an untreated reach of each stream. The design permitted unambiguous identification of the influence of rotenone on benthic communities because it involved replicated control (unimpacted) and treatment (impacted) stream sections. In a similar type of investigation, Dudgeon (1991b) demonstrated that periodic ('pulse') chemical disturbances had rather minor effects on macroinvertebrate community structure and abundance, but caused measurable reductions in community functioning (leaf-litter processing). Note that an intriguing result of the latter experiment was that community function seemed to be more sensitive than community structure to disturbance effects. If this finding has any generality, it could have important implications for biomonitoring programs.

A major reason for advocating the experimental approach is that environmental disturbances are going to exist anyway - therefore why not treat all proposed developments that might have environmental impacts as experiments? This approach has been proposed in the literature several times (e.g. Hilborn & Walters 1981; Walters & Holling, 1990; Peterson, 1993; Underwood, 1995) and can be termed 'experimental management', wherein predictions about the effects of a disturbance (e.g. stream channelization) are followed up by monitoring through all phases of the development from pre-impact to post-impact recovery (if any). All such disturbances must include replicated control areas and, if possible, parallel or similar developments in the same region should be treated as replicates of the experimental treatments. The results of the 'experiment' will enable similar future developments to be better planned so as to minimise environmental impact.

RESTORATION AND MITIGATION OF ENVIRONMENTAL IMPACTS

Restoration

An important goal of process-orientated ecological studies is provision of information that can be used to manage ecosystems with the aim of mitigating human impacts; in extreme cases, environmental management may be directed at enhancing restoration of severely-impacted systems. Peterson (1993) points out that an attempt to reconstruct a damaged ecosystem represents the ultimate challenge to ecology: do we know enough about how tropical Asian stream systems work to be able to engineer a desired ecosystem? The answer to that question is 'no', but by extrapolation from what is known about streams elsewhere we should be able to suggest interventions which change the system in the right direction (i.e. make things better and not worse). Information derived from process-orientated studies should improve our ability to intervene and produce beneficial consequences. However, it is worth stressing here that there appears to be a complete lack of published studies on the rate and trajectory of recovery of any tropical Asian stream following the cessation of anthropogenic disturbance. This is a significant gap in our knowledge because it has the inevitable consequence that it is impossible to predict the duration of any active management strategy for streams recovering from disturbance. Since we do not know how long a tropical Asian stream takes to recover from anthropogenic disturbance, all that can be said is that (presumably) a more severe disturbance will have a longer recovery trajectory. Moreover, we do not know whether a degraded stream will eventually regain its original community structure and species diversity, or whether a radically different combination of species and population densities will be attained. More information on this subject is required urgently because, with increasing pressures being placed upon tropical Asian streams, there will be an ever- greater need for positive intervention to restore damaged or degraded ecosystems.

Mitigation of stream-flow regulation

As discussed above (pp. 598–602), knowledge of links between streams and their valleys indicate that the maintenance of buffer strips of riparian vegetation may mitigate the effects of land-use changes in the drainage basin upon the aquatic biota. What other measures can be employed to mitigate human impacts? A unifying characteristic of tropical Asian streams is the dominant influence of the monsoonal climate which, typically, includes distinct dry and wet seasons giving rise to stream discharge peaks and seasonal flow minima with associated floods and droughts (see pp. 24–31). Impoundment of rivers for flood control, water storage and irrigation has, in consequence, been practised since ancient times (see p. 521 *et seq.*), and few rivers are now unregulated. Nonetheless, little attention has been given in the literature to the water (or in-stream flow) requirements of aquatic species and stream ecosystems in the region. Arthington (1994) has described a case history using the In-stream Flow Incremental Methodology (IFIM), and a range of other approaches, to the development of a water allocation strategy for an impounded subtropical system in Australia. This region has a wet-dry climate which is comparable to parts of tropical Asia, and may therefore have some relevant lessons for monsoonal areas. The IFIM, developed by the United States Fish and Wildlife Service (Bovee, 1982, 1986), is considered by many to be the most sophisticated methodology for assessing the environmental water requirements of streams. The physical habitat component of IFIM (termed PHABSIM) involves transect analysis at various points along the stream course and is based on the simulation and evaluation of habitat availability (expressed in terms of 'weighted useable area') for particular aquatic species at different levels of stream discharge.

It is possible that with appropriate modifications and care in project execution (e.g. Kwak *et al.*, 1992), IFIM methods will be of value as techniques for fish and habitat management in tropical streams. However, Arthington (1994) concludes that while IFIM and related methods may be useful for assessing the requirements of economically-important fish species (e.g. salmonids) in relatively-stable, homogeneous rivers, they are seldom useful for assessing the water needs of entire fish communities, and are far from adequate when the protection of functional aquatic ecosystems is required. Moreover, it is unlikely that the application of transect-based models applied at a few points along stream channels can realistically characterize geomorphologically-heterogeneous and hydrologically-variable stream systems (Arthington, 1994; Dudgeon *et al.*, 1994). Finally, there are practical and theoretical problems inherent in the application of PHABSIM, and IFIM in its entirety (e.g. Mathur *et al.*, 1985; Bain & Boltz, 1989; Gan & McMahon, 1990; Arthington, 1994). Jowett (1997) provides a useful comparison of different approaches to the problem of providing a certain level of protection for the aquatic environment under conditions of lower-than-natural flows.

There is growing recognition that the utility of IFIM requires further investigation in areas with strongly-seasonal rainfall. Strategies for water allocation must place much greater emphasis on the maintenance of the entire stream ecosystem, including such components as the source area, stream channel, riparian zone, floodplain, groundwater, wetlands and estuary, as well as any particularly important features such as rare and endangered species. The holistic approach to water allocation discussed by Arthington (1994) calls for greater emphasis to be given to water allocation for the environment, especially in regions with markedly-seasonal flow regimes; i.e. much of tropical Asia. The current knowledge base suggests that the environmental response to flow is not linear; the relative change in width and habitat with flow is greater for small rivers than for large (Jowett, 1997). Small rivers are therefore more 'at risk' than large rivers and require a higher proportion of the average flow to maintain similar levels of environmental protection. Unfortunately, there is no Asian research which can be called upon to inform water-allocation policies, and it is likely that water extraction — as practised currently — significantly degrades the ability of downstream reaches to support aquatic biota.

The need for development of ecologically-sound water-allocation strategies is urgent given the extent and magnitude of water transfers and river control projects in tropical Asia, as well as some of their known deleterious effects (see p. 521 *et seq.*). The proposals for large-scale projects on the Mekong River and the Chang Jiang Three Gorges Dam are conspicuous examples but, in densely-populated areas, almost every stream is dammed or impounded, and reduced flows or even dewatering of reaches downstream is a frequent occurrence during the dry season (Dudgeon, 1992c). Human impact on the potamon sections of major rivers in Europe, North America and Asia is particularly severe (Benke, 1990; Zwick, 1992; Dudgeon, 1992b), perhaps because they are separated by distances beyond the average dispersal capacity of the fauna so that recolonization after local extinction of populations cannot occur. In other words, potamon habitats are widely-separated ecological 'islands' with known risk of species losses (Zwick, 1992). Clearly, research effort needs to be directed towards estimating minimum in-stream flow requirement for the tropical Asian stream biota. However, given that development in many parts of the region is taking place under conditions of water scarcity (and occasional devastating floods) which necessitates manipulation of stream flows, water extraction, and so on (Bandyopadhyay, 1988; Falkenmark & Suprapto, 1992), significant conflicts over the use of water (stream conservation *versus* economic growth) seem inevitable.

INTEGRATED APPROACHES TO STREAM AND DRAINAGE BASIN MANAGEMENT

It is axiomatic that (catchment) processes play a significant role in defining the character and functioning of river ecosystems, and this notion constitutes a core component of the RCC. The close links between the stream and its valley might imply that effective stream conservation and management can only be achieved if the entire drainage basin is allowed to remain in a pristine state. This is unrealistic, and socioeconomic pressures towards development have resulted in many modifications of tropical Asian streams and their drainage basins, leading to irreversible degradation of some of them. A more balanced perspective is needed, where there is proper consideration of ecological, recreational, scientific and cultural uses of natural resources. It is clear that to achieve this balance, not only a better understanding of the scientific issues is needed, but also improved approaches to the integration of these broader concerns. The expertise of many disciplines must be drawn upon, and multi-scale (*cf.* Dudgeon, 1994c), multidisciplinary studies will be an essential component of effective conservation and management strategies for tropical Asian streams and their valleys. This will entail cooperative research by large (sometimes multinational) teams (the cross-sectoral approach of Dugan [1994a]). One example of this type of approach is the Wetlands Management Programme of the Mekong Secretariat for the lower Mekong basin (described on pp. 541–542; see also Choowaew, 1993), which provides a model of the type of structure and organization that is required for coordination of conservation and management efforts for a large river.

RECOMMENDED STRATEGIES

Descriptive, process-orientated studies of stream ecosystems in tropical Asia are needed because little is known of the patterns of biodiversity and biological production in the systems that we seek to conserve and manage. Accordingly, the following approaches are recommended:

1. Descriptive studies comprise the basis of biodiversity inventories and form an essential foundation for experimental studies. A useful descriptive study should include characterization of the major habitat types; a catalog of the species present in each habitat type (including endangered, endemic or vulnerable species); elucidation of seasonal patterns of distribution and abundance of the biota, as well as studies of their feeding habits and reproduction; information on the magnitude and seasonality of energy and materials; estimates of primary and secondary production, and an estimate of the importance of autochthonous and allochthonous material in supporting secondary production.

2. There is a need for greater use of experimental techniques in studies of tropical Asian streams because causal or mechanistic relationships are least-ambiguously elucidated through manipulative experiments with adequate controls and replication. Studies of the importance of interspecific interactions (competition, predation) in streams would benefit greatly from the inclusion of experimental manipulation of populations. Proposed anthropogenic modifications that might have impacts on stream ecosystems should be treated as experiments. With the inclusion of replicated control sites, predictions about the effects of a modification can be investigated by monitoring pre-impact to post-impact recovery (if any), and the results used to plan subsequent developments in a manner which will minimise environmental impacts.

3. Two topics where there are significant lacunae in our knowledge are highlighted. Flow regulation, impoundment and water extraction — as practised currently — significantly degrade the ability of downstream reaches to support aquatic biota. Much greater emphasis must be placed on water allocation to the environment

so as to ensure the maintenance of the entire stream ecosystem, but research is needed to determine the appropriate balance between water allocation for human use and environmental protection. In addition, the rate and trajectory of recovery of streams from disturbance is not known. We do know whether stream communities eventually return to their pre-disturbance state, nor how this can be enhanced. Clearly our ability to mitigate environmental impacts and restore damaged ecosystems is hampered severely by this lack of information.

4. A variety of strategies for improving ecological research and impact assessment in tropical Asian streams have been advocated in herein. Knowledge of the ecology of tropical Asian streams is far from complete, but the rate of ecosystem degradation is such that stream biologists will continue to be asked to assess possible environmental impacts, and advise on ways in which they can be mitigated. A pragmatic approach must be adopted. While we should maintain scientific rigour, a refusal to participate in decision-making processes because we lack some of the data will merely contribute to the irreversible despoliation of tropical Asian streams and rivers.

8

Concluding Remarks

Even if the research strategies advocated in Chapters 6 and 7 of this book were taken up by biologists, stream conservation will be possible only when they are combined with a move beyond the bailiwick of science into the political arena. If they are to succeed, ecologically-viable management strategies for tropical Asian streams must take account of socioeconomic contexts; in particular, the pressure of increasing human populations and the desire for economic growth which drive development and urbanization. While the 1997 depression of Asian economies may slow growth temporarily, it will do nothing to lessen the forces driving such growth. Socioeconomic advancement in many tropical countries must proceed as a way of reducing poverty, disease and deprivation, and the '. . . fact that the safeguarding of biodiversity is a desirable goal for the whole world is meaningless to the tens of millions who live on the edge of starvation . . .' (Zuckerman, 1992) is inescapable. Unfortunately, development almost inevitably brings in its train new environmental problems. There is thus every likelihood that ecosystems will continue to be degraded and biodiversity will decline until such a time that they can be assigned an economic value and factored into the costs of developments (Dudgeon *et al.*, 1994). This means that the 'indirect use value' of environmental functions provided by stream ecosystems needs to be estimated (Aylward & Barbier, 1992; Dugan, 1994b); i.e. some economic value must be put on streams. A stress can be placed upon high-value functions such as fisheries support, since river fisheries provide the principal source of animal protein for many rural societies (Dugan, 1994b).

Putting an economic value on the biota and its functions will not, on its own, prevent further degradation of tropical Asian stream ecosystems. Neither will publication of more and better research, or a more frequent adoption of experimental approaches to stream ecology. Inevitably, scientists will have to address political issues in order to influence policy-makers and the populace at large. If we are to be effective in this regard, we must ensure that the scientific data and predictions about ecological change and environmental impacts are rigorous. If we lack understanding of the processes and mechanisms underlying observed patterns or changes in streams, or if we use inadequate designs to test for environmental impacts, then compelling and rigorous scientific assessments or predictions are not possible, and disputes are promoted. Any uncertainty arising from poor design or inadequate functional understanding means that reasonable arguments can be made for and against the incidence of a 'real' effect because of the lack of unequivocal evidence. Ecologists and conservationists cannot be effective in the political arena unless they have at hand rigorous and unbiased data — based upon appropriate monitoring designs and research strategies — which can be analysed unequivocally thereby avoiding disputes among interested parties. If we fail to rise to this challenge, then we will witness a dilution of the scientific input into public debates that incorporate the wider socioeconomic or political issues relating to stream conservation and management.

To be effective advocates of stream ecosystems, we must enhance (and maintain) the rigour of our research strategies. Decision-makers in government and state organizations must be informed of our findings and made aware of a number of key facts (after Williams, 1994):

1. That stream ecosystems (indeed, all inland waters) are of considerable economic, cultural, aesthetic, educational and scientific value.
2. Water resources can be (and are being) significantly degraded in value, sometimes irreparably so.
3. The conservation and management of stream ecosystems both for present and future generations is manifestly in the interests of the government or state.
4. Effective conservation and management of stream ecosystems is

possible only when based upon a thorough scientific knowledge of them.

To conclude, we must assess impacts and undertake ecological process-orientated studies in as rigorous a manner as is possible. Obviously, better management of ecological resources will occur if there is an improved understanding of the organization and function of aquatic systems — especially if 'ecological health' can be invoked as a goal of environmental management (Courtemanch, 1994). In this regard it is important that we address the nature of the link between aquatic biodiversity and ecosystem functioning. The existence of such a link is implicit in the RCC, yet we do not know the form of the relationship between these two variables, nor do we know how much biodiversity (if any) can be lost before ecosystem functioning is impaired. Perhaps there is much redundancy in functional roles among different taxa so that replacement of one by another is possible. If so, it might be argued that species- or generic-level identification of (say) benthic invertebrates is irrelevant in many instances. If, however, ecosystem functioning and biodiversity are inextricably linked, then our options and approaches to environmental management must take full account of this fact. We do not know which of these possibilities matches reality more closely and, while wise counsel dictates that we assume that all components of biodiversity have functional significance, in the longer term we must determine whether this is, in fact, the case. The relationship between biodiversity and ecological function is fundamental, and one which we should understand sufficiently well to allow prediction of the wider consequences of species loss. Decision-makers and the public are certain to expect us to be able to accomplish this much.

Our knowledge of the ecology of tropical Asian streams is nowhere near as complete as we would like, but stream biologists will nevertheless be asked for their professional judgement about possible environmental impacts on habitats, and ways by which impacts might be mitigated. Clearly, we will have to give advice in situations were data are ambiguous, or even lacking, because a delay in offering an opinion could lead to irreversible changes in ecosystems. It is, for example, most unlikely that industrial development or civil-engineering projects along the banks of tropical Asian rivers will cease until scientists have at hand all of the data they need to render an informed opinion about the consequences of a proposed development. Moreover, this assumes that research funding to obtain such information will become available; past experience gives us little confidence that funding will ever be anything but extremely limited. Pollution and river regulation are often irreversible processes and a delay in proffering advice may mean that streams are degraded beyond their capacity to recover. It thus behooves us to provide scientific advice whenever and wherever possible, notwithstanding the uncertainty with which we make our judgements. The information provided should be appropriately focused, achieving a balance between the ideal and the achievable, recognizing the functioning of river basins and not merely river channels (Boon, 1992). Clearly, it will be helpful if the input is unequivocal, authoritative and simple (but not simplistic). A pragmatic approach is essential (and has been advocated to marine biologists by Warwick, 1993), although we must be mindful of the words of Slobodkin (1988) that '. . . it is by no means obvious that crudely imperfect advice is more valuable than none at all'. We should, of course, maintain the highest levels of scientific rigor and objectivity, but a refusal to participate in public debate and decision-making processes because we lack some of the data will do nothing to prevent the on-going despoliation of streams and rivers throughout tropical Asia.

Notwithstanding the need for more research, decisions taken by legislators and politicians concerning tropical Asian rivers will have a greater impact on the future of these ecosystems than any amount of ecological work. The biological solutions to many environmental problems are within reach, given the political will and suitable legislation. To quote McNeely (1987): '. . . environmentalists often behave as if the progression from knowledge to action should somehow be automatic, but experience has taught the hard lesson that conduct of governments is not directly guided by ecological science, or even objective logic . . .'. Moreover, ecologically-viable management strategies will fail if they do not address socioeconomic and cultural contexts, or are considered in isolation from the aspirations of the local populace. The implications are clear: we must continue to contribute to the academic development of stream biology, but make greater efforts to disseminate our knowledge of river ecosystems and communicate with those planning large-scale developments as well as those whose activities have a direct effect on the ecosystems at issue. As Caldwell (1985) puts it: '. . . what is needed, but is not present, is a popular movement, fundamentally political, to translate the 'oughts' and 'shoulds' of environmental findings and declarations into workable and widely acceptable programmes of action . . . .'. Development of effective dialogue which will influence or affect politicians will not be easy and may divert efforts away from research. Nevertheless, further habitat degradation and loss of biodiversity will result from a failure to engage in wide-ranging discourse. We must now weigh our priorities. Should we increase output of learned papers, while ignoring the *realpolitik* of conservation and management in tropical Asia? Or, should we devote more effort to communicating the relevance and importance of our science, even if this means we spend less time publishing research papers? It may be argued that this choice is a false dichotomy, but I believe that a more focused approach to our science is required. It is no longer appropriate (if, indeed, it ever was) to devote ourselves *solely* to spawning manuscripts which begin with statements like 'Not much is known about the biology of . . . .'. There must be a greater attempt to generate ideas, hypotheses or theories which will have direct application to stream conservation and management. Perhaps this change will come about only when we recognize many of the engineering projects along Asian rivers for what they are: giant, unreplicated, experiments in environmental management. They should be treated as such, and studied so that we can learn from them how to better understand and ameliorate their environmental effects. This approach may provide our best chance of preserving many stream ecosystems into the next century and beyond.

References

Abbas, B.M., 1987 Agreement on the Ganges. In: Ali, M., Radosevich, G.E. & Ali Khan, A. (Eds), *Water Resources Policy for Asia*. A.A. Balkema Publishers, Boston: 517–538.

Abbot, T.R., 1952. A study of an intermediate snail host (*Thiara scabra*) of the Oriental lung fluke (*Paragonimus*). *Proc. U.S. Nat. Mus.* **102**: 71–116.

Abdullah, M.D.P. & Nainggolan, H., 1991. Phenolic water pollutants in a Malaysian river basin. *Environmental Monitoring & Assessment* **19**: 423–431.

Abel, P.D., 1989. *Water Pollution Biology*. Ellis Horwood Limited, Chichester: 231 pp.

Abele, L.G. & Blum, N., 1977. Ecological aspects of the freshwater decapod crustaceans of the Perlas Archipelago, Panama. *Biotropica* **9**: 239–252.

Aditya, A.K. & Mahapatra, S., 1991. Notes on the biology of the freshwater planarian *Dugesia bengalensis* (Platyhelminthes, Turbellaria, Tricladida). *Hydrobiologia* **227**: 145.

Afsah, S., Laplante, B. & Makarim, N., 1996. Programme-based pollution control management: the Indonesian PROKASIH programme. *Asian Journal of Environmental Management* **4**: 75–93.

Afser, M.R., 1990. Food and feeding habit of a teleostean fish *Clupisoma gaura* (Ham.) in the Ganga River system. *J. Freshwat. Biol.* **2**: 159–167.

Agarwal, G.D., 1990. Water-quality in some Himalayan rivers. In: Agarwal,

V.P. & Das, P. (Eds), *Recent Trends in Limnology.* Society of Biosciences, Muzaffarnagar (India): 411–418.

Agarwal, N.K., Bahuguna, S.N. & Badola, S.P., 1990. Seasonal variation in the gut contents of a snow-trout (*Schizothoraichthys progastus* McClelland) from the River Ganga in Garhwal Himalaya. *Indian J. Anim. Sci.* **60**: 750–752.

Agrawal, H.P., 1980. Some observations on the seasonal variation in the gonads of *Indonaia caerulea* (Lea) (Mollusca: Unionidae). *Bull. Zool. Surv. India* **3**: 87–92.

Ahktar, S., 1978. On a collection of freshwater molluscs from Lahore. *Biologia (Lahore)* **24**: 437–447.

Ahmad, A., 1986. The distribution and population of crocodiles in the provinces of Sind and Baluchistan (Pakistan). *J. Bombay Nat. Hist. Soc.* **83**: 220–223.

Ahmad, A., 1993. Environmental impact assessment in the Himalayas: an ecosystem approach. *Ambio* **22**: 4–9.

Ahmad, M. & Sonoda, S., 1979. The ecology of freshwater elasmobranch of the Indragiri River, Riau. *Oseanol. Indones.* **12**: 13–19 (in Indonesian).

Ahmad, S.H. and Singh, A.K., 1989. A comparative study of the benthic fauna of lentic and lotic water bodies around Patna (Bihar) India. *J. Environ. Biol.* **10**: 283–292.

Ahmed, M. & Tana, T.S., 1996. Managemetn of freshwater capture fisheries of Cambodia: issues and approaches. *Naga, the ICLARM Quarterly* **19**: 16–19.

Ahtiainen, M., 1992. The effects of forest clear-cutting and scarification on the water quality of small brooks. *Hydrobiologia* **243/244**: 457–464.

Ajmal, M. & Razi-Ud-Din, 1988. Studies on the pollution of Hindon River and Kali Nadi (India). In: Trivedy, R.K. (Ed.), *Ecology and Pollution of Indian Rivers.* Ashish Publishing House, New Delhi: 87–112.

Aki, K. & Berthelot, R., 1974. Hydrology of humid tropical Asia. In: *Natural Resources of Humid Tropical Asia.* Natural Resources Research, XII, Unesco, Paris: 145–158.

Alba-Tercedor, J. & Zamora-Munoz, C., 1993. Description of *Caenis nachoi* sp. n., with keys for the identification of the European species of the *Caenis macrura* group (Ephemeroptera: Caenidae). *Aquatic Insects* **15**: 239–247.

Alexander, C.P., 1931. Deutsche Limnologische Sunda-Expedition. The Crane-flies (Tipulidae, Diptera). *Arch. Hydrobiol., Suppl.* **2**: 135–191.

Alfred, E.R., 1968. Rare and endangered fresh-water fishes of Malaya and Singapore. In: Talbot L.M. & Talbot, M.H. (Eds), *Conservation in Tropical South East Asia.* IUCN Publications New Series No. 10, IUCN, Morges, Switzerland: 325–331.

Alfred, E.R., 1969. The Malayan cyprinoid fishes of the family Homalopteridae. *Zool. Meded., Leiden* **43**: 213–237.

Alabaster, J.S., 1986. Review of the state of water pollution affecting inland fisheries in Southeast Asia. *FAO Fisheries Technical Paper* **260**: 1–25.

Ali, A.B. & Kathergany, M.S., 1987. Preliminary investigation on standing stocks, habitat preference and effects of water level on riverine fish population in a tropical river. *Trop. Ecol.* **28**: 264–273.

Ali, M., Radosevich, G.E. & Ali Khan, A., 1987. *Water Resources Policy for Asia*. A.A. Balkema Publishers, Boston: 628 pp.

Ali, S.R., 1971. The nymphs of new species of genus *Ephemerella* (Order: Ephemeroptera). *Pakistan Journal of Forestry* **1971**: 359–366.

Alikunhi, K.H., 1953. Notes on the bionomics, breeding and growth of the murrel, *Ophicephalus striatus* Bloch. *Proc. Indian Acad. Sci. (B)* **38**: 10–20.

Allan, J.D. & Flecker, A.S., 1989. The mating biology of a mass-swarming mayfly. *Anim. Behav.* **37**: 361–371.

Allen, G.R., 1987. *Melanotaenia iris*, a new freshwater rainbowfish (Melanotaeniidae) from Papua New Guinea with notes on the fish fauna in head waters. *Japan. J. Ichthyol.* **34**: 15–20.

Allen, G.R., 1991. *Field Guide to the Freshwater Fishes of New Guinea*. Publication No. 9, Christensen Research Institute, Madang: 268 pp.

Allen, G.R. & Coates, D., 1990. An ichthyological survey of the Sepik River, Papua New Guinea. *Rec. West. Aust. Mus., Suppl.* **34**: 31–116.

Allen, R.K., 1971. New Asian Ephemerella with notes (Ephemeroptera: Ephemerellidae). *Can. Ent.* **103**: 512–528.

Allen, R.K., 1980. Geographic distribution and reclassification of the subfamily Ephemerellinae (Ephemeroptera: Ephemerellidae). In: Flannagan, J.F. & Marshall, K.E., *Advances in Ephemeroptera Biology*. Plenum Press, New York: 71–91.

Allen, R.K., 1984. A new classification of the subfamily Ephemerellinae and the description of a new genus. *Pan-Pacific Entomologist* **60**: 245–247.

Allen, R.K., 1986. Mayflies of Vietnam: *Acerella* and *Drunella* (Ephemeroptera: Ephemerellidae). *Pan-Pacific Entomologist* **60**: 245–247.

Allen, R.K. & Edmunds, G.F., 1963. New and little known Ephemerellidae from southern Asia, Africa and Madagascar. *Pacific Insects* **5**: 133–137.

Allen, R.K. & Edmunds, G.F., 1976. *Hyrantella*: a new genus of Ephemerellidae from Malaysia (Ephemeroptera). *Pan-Pacific Entomologist* **52**: 133–137.

Alouf, N.J., 1991. Développment de Plecopteres (Insecta) d'un ruisseau permanent du Liban. *Annl Limnol.* **27**: 133–140.

Alstad, D.N., 1987. Particle size, resource concentration, and the distribution of net-spinning caddisflies. *Oecologia* **71**: 525–531.

Amarasekara, N., 1992. Environmental consequences of the Menik Ganga diversion. *Water Resources Development* **8**: 286–291.

Ambasht, R.S. & Srivastava, N.K., 1994. Restoration strategies for the degrading Rihand River and reservoir ecosystems in India. In: Mitsch, W.J. (Ed.), *Global Wetlands Old World and New*. Elsevier, Amsterdam: 483–492.

Ambasht, R.S., Kumar, R. & Srivastava, N.K., 1994. Strategies for managing the Rihand River riparian ecosystem deteriorating under rapid industrialization. In: Mitsch, W.J. (Ed.), *Global Wetlands Old World and New*. Elsevier, Amsterdam: 725–728.

Ameen, M. & Nessa, S.K., 1985. A preliminary identification key to the aquatic Hemiptera of Dhaka city. *Bangladesh J. Zool.* 13: 49–60.

Ames, M.H., 1991. Saving some cetaceans may require breeding in captivity: work on bottlenose dolphin may be applied to the baiji. *BioScience* 41: 746–749.

Andreev, S., 1982a. Sur une nouvelle espec cavernicole du genre *Cyathura* (Isopoda, Anthuridae). *International Journal of Speleology* 12: 55–62.

Andreev, S., 1982b. Une *Cyathura* cavernicole nouvelle de Sarawak — Kalimantan du Nord. *Bulletin Zoologisch Museum Universiteit van Amsterdam* 8: 149–155.

Andersen, N.M., 1982a. *The Semiaquatic Bugs (Hemiptera, Gerromorpha). Phylogeny, Adaptations, Biogeography and Classification*. Entomonograph Volume 3, Scandinavian Science Press Ltd, Klampenborg, Denmark: 455 pp.

Andersen, N.M., 1982b. Semiterrestrial water striders of the genus *Eotrechus* Kirkaldy and *Chimarrhometra* Bianchi (Insecta, Hemiptera, Gerridae). *Steenstrupia* 9: 1–25.

Andersen, N.M., 1983. The Old World Microveliinae (Hemiptera: Veliidae) I. The status of *Pseudovelia* Hoberlandt and *Perivelia* Poisson, with a review of Oriental species. *Ent. Scand.* 14: 253–268.

Andersen, N.M., 1989. The Old World Microveliinae (Hemiptera: Veliidae). II. Three new species of *Baptista* Distant and a new genus from the Oriental region. *Ent. Scand.* 19: 363–380.

Andersen, N.M., 1990. Phylogeny and taxonomy of water striders, genus *Aquarius* Schellenberg (Insecta, Hemiptera, Gerridae) with a new species from Australia. *Steenstrupia* 16: 37–81.

Andersen, N.M., 1993. Classification, phylogeny, and zoogeography of the pondskater genus *Gerris* Fabricius (Hemiptera: Gerridae). *Can. J. Zool.* 71: 2473–2508.

Andersen, N.M. & Chen, P., 1993. A taxonomic review of the pondskater genus *Gerris* Fabricius in China, with two new species (Hemiptera: Gerridae). *Ent. Scand.* 24: 147–166.

Andersen, N.M. & Spence, J.R., 1992. Classification and phylogeny of the Holarctic water strider genus *Limnoporus* Stål (Hemiptera, Gerridae). *Can. J. Zool.* 70: 753–785.

Andrews, C., 1990. The ornamental fish trade and fish conservation. *J. Fish. Biol.* 37 (**Suppl. A**): 53–59.

Anger, K., 1991. Effects of temperature and salinity of the larval development of the Chinese mitten crab *Eriocheir sinensis* (Decapoda: Grapsidae). *Mar. Ecol. Prog. Ser.* 72: 103–110.

Angus, R.B., 1995. Helophoridae: the *Helophorus* species of China, with notes on the species from neighbouring areas (Coleoptera). In: Jäch, M.A. & Ji, L. (Eds), *Water Beetles of China*. Zoologisch-Botanische Gesellschaft & Wiener Coleopterologenverein, Vienna: 185–206.

Annandale, N., 1900. Observations on the habits and natural surroundings of insects made during the 'Skeat Expedition' to the Malay Peninsula, 1899–1900. VI. Insect luminosity. An aquatic lampyrid larva. *Proc. Zool. Soc. Lond.* **70**: 862–865.

Annandale, N., 1911. *The Fauna of British India including Ceylon and Burma. Freshwater Sponges, Hydroids and Polyzoa.* Taylor & Francis, London: 251 pp.

Annandale, N., 1912. *Caridinicola*, a new type of Temnocephaloidea. *Rec. Indian Mus.* **7**: 232–243.

Anon, 1996a. Lies, statistics and dam lies. *Guardian Weekly* **June 30, 1996**: 13.

Anon, 1996b. Mekong nations to share information. *The Nation (Bangkok)* **July 27, 1996**: 5.

Ansari, Z.A., Parulekar, A.H. & Matondkar, S.G.P., 1981. Seasonal changes in meat weight and biochemical composition in the black clam *Villorita cyprinoides* (Grey). *Indian Journal of Marine Sciences* **10**: 128–131.

Anwar, S. & Siddiqui, M.S., 1992. Observation on the predation by *Mystus seenghala* (Sykes) and *Wallago attu* (Bloch and Schneider) of the River Kali in North India. *J. Environ. Biol.* **13**: 47–54.

Anwar Khan, T., 1987. The water resource situation in Bangladesh. In: M. Ali, M., Radosevich, G.E. & Ali Khan, A. (Eds), *Water Resources Policy for Asia*. A.A. Balkema Publishers, Boston: 139–164.

Anwar, S. & Siddiqui, M.S., 1988. On the distribution of macro-invertebrate fauna of the River Kali in northern India. *J. Environ. Biol.* **9**: 333–341.

Ao, M., Alfred, J.R.B. & Gupta, A., 1984. Studies on some loic systems in the north-eastern hill regions of India. *Limnologia* **15**: 135–141.

Aoya, K. & Yokoyama, N., 1987. Life cycles of two species of *Stenopsyche* (Trichoptera: Stenopsychidae) in Tohoku District. *Jpn. J. Limnol.* **48**: 41–53 (in Japanese).

Aoya, K. & Yokoyama, N., 1990. Production of two species of *Stenopsyche* (Trichoptera: Stenopsychidae) in Tohoku District. *Jpn. J. Limnol.* **51**: 249–260 (in Japanese).

Arain, R., 1987. Persisting trends in carbon and mineral transport monitoring of the Indus River. *Mitt. Geol.-Palaont. Inst. Univ. Hamburg, SCOPE/ UNEP Sonderbd.* **64**: 417–421.

Armitage, P., Cranston, P.S. & Pinder, L.C.V., 1995. *The Chironomidae: Biology and Ecology of Non-biting Midges.* Chapman & Hall, London: 572 pp.

Arnqvist, G., 1993. Courtship behaviour and sexual cannibalism in the semi-aquatic fishing spider, *Dolomedes fimbriatus* (Clerck) (Araneae: Pisauridae). *J. Arachnol.* **20**: 222–226.

Arthington, A.H., 1994. A holistic approach to water allocation to maintain the environmental values of Australian streams and rivers: a case history. *Mitt. Internat. Verein. Limnol.* **24**: 165–178.

Arumuga Soman, A.K., 1991. A new species of *Thraulus* (Ephemeroptera: Leptophlebiidae: Atalophlebiinae) from Nilgiris, South India. *J. Bombay Nat. Hist. Soc.* **88**: 99–102.

Arunachalam, M. & Arun, L.K., 1994. Input, transport and storage of organic matter in two south Indian Rivers. *Mitt. Internat. Verein. Limnol.* **24**: 179–188.

Arunachalam, M., Madhusoodanan Nair, K.C., Vijverberg, J., Kortmulder, K. & Suriyanarayanan, H., 1991. Substrate selection and seasonal variation in densities of invertebrates in stream pools of a tropical river. *Hydrobiologia* **213**: 141–148.

Arya, S.C., Mudgal, S. & Shrivastava, P., 1987. Effects of sewage and industrial waste on river ecosystem. *Indian J. Limnol.* **15**: 49–56.

Asahina, S., 1967. Notes on two amphipterygid dragonflies from Southeast Asia. *Deutsche Entomologische Zeitschrift* **14**: 323–326.

Asahina, S., 1974. Diagnostic notes on the ultimate instar larvae of some *Anax* species. *Tombo* **17**: 10–16.

Asahina, S., 1981. Seasonal variation in *Neurothemis tullia* (Drury). *Tombo* **24**: 12–16.

Asahina, S., 1982. The larval stage of the Himalayan *Neallogaster hermionae* (Fraser) (Anisoptera: Cordulegastridae). *Odonatologica* **11**: 309–315.

Asahina, S., 1985. Contributions to the taxonomic knowledge of the *Megalestes* species of continental South Asia (Odonata, Synlestidae). *Chô Chô* **8**: 2–18.

Asahina, S., 1986a. A list of the Odonata recorded from Thailand Part XVI. Cordulegastridae. *Proc. Jap. Soc. Syst. Zool.* **34**: 39–45.

Asahina, S., 1986b. Revisional notes on Nepalese and Assamese dragonfly species of the genus *Chlorogomphus*. *Chô Chô* **9**: 11–26.

Asahina, S., 1993. *A List of the Odonata of Thailand (Parts I — XXI)*. Bosco Offset, Bangkok (each part is individually paginated).

Asahina, S. & Dudgeon, D., 1987. A new platystictid damselfly from Hong Kong. *Tombo* **30**: 2–6.

Ashe, P., 1990. Ecology, zoogeography and diversity of Chironomidae (Diptera) in Sulawesi with some observations relevant to other aquatic insects. In: Knight, W.J. & Holloway, J.D. (Eds), *Insects and the Rain Forests of South East Asia (Wallacea)*. The Royal Entomological Society of London, London: 261–268.

Ashe, P. & O'Connor, J.P., 1995. A new species of *Sublettea* Roback from Sulawesi, Indonesia. (Diptera: Chironomidae). In: Cranston, P., (Ed.), *Chironomids: From Genes to Ecosystems*. CSIRO Publications, Melbourne: 431–435.

Ashe, P., Murray, D.A. & Reiss, F., 1987. The zoogeographical distribution of Chironomidae (Insecta: Diptera). *Ann. Limnol.* **23**: 27–60.

Askew, R.R., 1978. *The Dragonflies of Europe*. Harley Books, Colchester: 291 pp.

Attwood, S.W., 1995a. The effect of substratum grade on the distribution of the freshwater snail *y-Neotricula aperta* (Temcharoen), with notes on the sizes of particles ingested. *J. Moll. Stud.* **61**: 133–138.

Attwood, S.W., 1995b. A demographic analysis of *y-Neotricula aperta* (Gastropoda: Pomatiopsidae) populations in Thailand and southern Laos, in relation to the transmission of schistosomiasis. *J. Moll. Stud.* **61**: 29–42.

Audley-Charles, M.G., 1981. Geological history of the region of Wallace's line. In: T.C. Whitmore (Ed.), *Wallace's Line and Plate Tectonics*. Oxford Science Publications, Clarendon Press, Oxford: 24–35.

Audley-Charles, M.G., 1987. Dispersal of Gondwanaland: relevance to evolution of the angiosperms. In: T.C. Whitmore (Ed.), *Biogeographical Evolution of the Malay Archipelago*. Oxford Science Publications, Clarendon Press, Oxford: 5–25.

Augustine, S. & Diwan, A.P., 1991. Population dynamics of midge worms with relation to pollution in River Kshipra. *J. Environ. Biol.* **12**: 255–262.

Aylward, B. & Barbier, E.B., 1992. Valuing environmental functions in developing countries. *Biodiversity & Conservation* **1**: 34–50.

Babu, N., 1963. Observations on the biology of *Caridina propinqua* De Man. *Indian J. Fish.* **10**: 107–117.

Baconguis, S.R., Cabahug, D.M. Jr & Alonzo-Pasicolan, S.N., 1990. Identification and inventory of Philippine forested-wetland resource. *For. Ecol. Manag.* **33/34**: 21–44.

Badola, S.P. & Singh, H.R., 1981. Hydrobiology of the River Alaknanda of the Garhwal Himalaya. *Indian J. Ecol.* **8**: 269–76.

Bae, Y.J., 1995. *Ephemera separigata*, a new species of Ephemeridae (Insecta: Ephemeroptera) from Korea. *Korean J. Syst. Zool.* **11**: 159–166.

Bae, Y.J. & McCafferty, W.P., 1991. Phylogenetic systematics of the Potamanthidae (Ephemeroptera). *Trans. Am. Entomol. Soc.* **117**: 1–143.

Bae, Y.J. & McCafferty, 1994. Microhabitat of *Anthopotamus verticis* (Ephemeroptera: Potamanthidae). *Hydrobiologia* **288**: 65–78.

Bae, Y.J. & McCafferty, W.P., 1995. Ephemeroptera tusks and their evolution. In: Corkum, L. & Ciborowski, J. (Eds), *Current Directions in Research on Ephemeroptera*. Canadian Scholars' Press, Toronto: 377–405.

Bae, Y.J., McCafferty, W.P. & Edmunds, G.F., 1990. *Stygifloris*, a new genus of mayflies (Ephemeroptera: Potamanthidae) from Southeast Asia. *Ann. Entomol. Soc. Am.* **83**: 887–891.

Bae, Y.J., Yoon, I.B. & Chun, D.J., 1994. A catalogue of the Ephemeroptera of Korea. *Ent. Res. Bulletin* **20**: 31–50.

Baer, J.G., 1953. Zoological results of the Dutch New Guinea Expedition 1939. No. 4. Temnocephalida. *Zool. Meded., Leiden* **32**: 119–139.

Bain, M.B. & Boltz, J.M., 1989. *Regulated Streamflow and Warmwater Stream Fish: a General Hypothesis and Research Agenda*. Biological Report 89 (18), US Department of the Interior, Fish and Wildlife Service, Washington, D.C.: 28 pp.

Baird, I.G., 1994. Community management of Mekong River resources in Laos. *Naga, the ICLARM Quarterly* **17**: 10–12.

Baird, I.G. & Mounsouphom, B., 1994. Irrawaddy dolphins (*Orcaella brevirostris*) in southern Lao PDR and northeastern Cambodia. *Nat. Hist. Bull. Siam Soc.* **42**: 159–175.

Bakshi, D.N.G., Mondol, S.K. & Sen, S., 1977. Contributions to the study of Cyperaceae in West Bengal, India. *Bull. Bot. Soc. Bengal* **31**: 90–97.

Balasubramanian, C., Venkataraman, K. & Sivaramakrishnan, K.G., 1991. Life stages of a South Indian burrowing mayfly, *Ephemera (Aethephemera) nadinae* McCafferty and Edmunds 1973 (Ephemeroptera: Ephemeridae). *Aquatic Insects* **13**: 223–228.

Balasubramanian, C., Venkataraman, K. & Sivaramakrishnan, K.G., 1992. Bioecological studies on the burrowing mayfly *Ephemera (Aethephemera) nadinae* McCafferty & Edmunds 1973 (Ephemeroptera: Ephemeridae) in Kurangani Stream, Western Ghats. *J. Bombay Nat. Hist. Soc.* **89**: 72–77.

Balasuriya, S.I., 1979. An unusual home. *Loris* **15**: 30–31.

Balchand, A.N., 1983. Kuttanad: a case study on environmental consequences of water resources mismanagement. *Water International* **8**: 35–41.

Balete, D.S. & Holthuis, L.B., 1992. Notes on the cave shrimp *Edoneus atheatus* Holthuis, 1978, with an account of its type locality and habits (Decapoda, Caridea, Atyidae). *Crustaceana* **62**: 98–101.

Balke, M., 1992. Systematische und faunistische Untersuchungen an paläarktischen, orientalischen und afrotropischen Arten von *Rhantus* Dejean (Coleoptera: Dytiscidae). *Mitt. schweiz. ent. Ges.* **65**: 283–296.

Balke, M., 1993. Taxonomische Revision der pazifischen, australischen und indonesischen Arten der Gattung *Rhantus* DEJEAN (Dytiscidae). *Koleopterologische Rundschau* **60**: 39–84.

Balke, M. & Hendrich, L., 1997. A new species of *Laccophilus* LEACH, 1815 from Vietnam (Coleoptera: Dytiscidae). *Koleopterologische Rundschau* **67**: 99–100.

Balke, M. & Sâto, M., 1995. *Limbodessus compactus* (Clark): a widespread Austro-Oriental species, as reveled by its synonymy with two other species of Bidessini (Coleoptera: Dytiscidae). *Aquatic Insects* **17**: 187–192.

Ball, I.R., 1970. Freshwater triclads (Turbellaria, Tricladida) from the Oriental Region. *Zool. J. Linn. Soc.* **49**: 271–294.

Baloni, S.P. & Grover, S.P., 1987. Bio-ecological notes on *Pseudecheneis sulcatus* (McClelland) in Garhwal Himalaya. *Indian J. Phy. Nat. Sci. Sect. A* **7**: 21–24.

Baloni, S.P. & Tilak, R., 1985. Ecological observations on *Schizothorax richardsonii*. *J. Bombay Nat. Hist. Soc.* **82**: 581–585.

Ban, R., 1988. The life cycle and microdistribution of *Ephemera strigata* Eaton (Ephemeroptera: Ephemeridae) in the Kumogahata River, Kyoto Prefecture, Japan. *Verh. Internat. Verein. Limnol.* **23**: 2126–2134.

Ban, R. & Kawai, T., 1986. Comparison of the life cycles of two mayfly species between upper and lower parts of the same stream. *Aquatic Insects* 8: 207–215.

Banarescu, P., 1972. The zoogeographical position of the east Asian freshwater fish fauna. *Rev. Roum. Biol., Zool.* 17: 315–323.

Banarescu, P. & Coad, B.W., 1991. Cyprinids of Eurasia. In: Winfield, I.J. & Nelson, J.S. (Eds), *Cyprinid Fishes: Systematics, Biology and Exploitation.* Chapman & Hall, London: 127–155.

Bandara, M., 1985. The Mahaweli strategy of Sri Lanka — great expectation of a small nation. In: Lundqvist, J., Lohm, U. & Falkenmark, M. (Eds), *Strategies for River Basin Management.* D. Reidel Publishing Co., Dordrecht: 265–277.

Bandhavong, P., 1990. Bottom fauna in the Nam Ngum River, station: Tha Ngone. In: Petersen, R.C. & Sköglund, E. (Eds) *Wetlands Management Programme. Study to Formulate Plans for the Management of the Wetlands in the Lower Mekong Basin (1.3.13/88/SWE).* Mekong Secretariat, Bangkok, Thailand: 181–186.

Bandyopadhyay, J. 1988. The ecology of drought and water scarcity. *The Ecologist* 18: 88–95.

Bandyopadhyay, J. & Gyawali, D., 1994. Himalayan water resources: ecological and political aspects of management. *Mountain Research & Development* 14: 1–24.

Banks, N., 1924. Descriptions of new neuropteroid insects. *Bull. Mus. Comp. Zool.* 65: 412–455.

Banks, N., 1931. Some neuropteroid insects from the Malay Peninsula. *J. Fed. Malay States Mus.* 16: 377–409.

Banks, N., 1937. Philippine neuropteroid insects. *Philippine J. Sci.* 63: 125–174.

Banks, N., 1938. Further neuropteroid insects from Malaya. *J. Fed. Malay States Mus.* 18: 220–235.

Banks, N., 1939. New genera and species of neuropteroid insects. *Bull. Mus. Comp. Zool.* 85: 439–504.

Banks, N., 1940. Report of certain groups of neuropteroid insects from Szechwan, China. *Proc. U.S. Natn. Mus.* 88: 173–220.

Barnard, J.L. & Barnard, M., 1983. *Freshwater Amphipoda of the World* (volumes 1 & 2). Hayfield Associates, Mt. Vernon, Virginia: 830 pp.

Barnard, P.C., 1980. Macronematine caddisflies of the genus *Amphipsyche* (Trichoptera: Hydropsychidae). *Bull. Br. Mus. nat. Hist. (Ent.)* 48: 71–130.

Barnard, P.C., 1984. A revision of the Old World Polymorphanisini (Trichoptera: Hydropsychidae). *Bull. Br. Mus. nat. Hist. (Ent.)* 41: 59–106.

Barnard, P.C. & Clark, F., 1986. The larval morphology and ecology of a new species of *Ecnomus* from Lake Naivasha, Kenya (Trichoptera: Ecnomidae). *Aquatic Insects* 8: 175–183.

Barnard, P.C. & Dudgeon, D., 1984. The larval ecology and morphology of a new species of *Melanotrichia* from Hong Kong (Trichoptera: Xiphocentronidae). *Aquatic Insects* 6: 245–252.

Barrett, S.C., 1989. Waterweed invasions. *Scientific American* 261 (4): 66–73.

Bartarya, S.K., 1991. Watershed management strategies in Central Himalaya: the Gaula River Basin, Kumaun, India. *Land Use Policy* 8: 177–184.

Bartarya, S.K., 1993. Hydrochemistry and rock weathering in a subtropical Lesser Himalayan river basin in Kumaun, India. *J. Hydrol.* 146: 149–174.

Basu, A.K., 1965. Observations on probable effects of pollution on primary productivity of the Hooghly and Mutlah estuaries. *Hydrobiologia* 25: 302–316.

Baumann, R.W., 1975. Revision of the stonefly family Nemouridae (Plecoptera): a study of the world fauna at the generic level. *Smithsonian Contributions to Zoology* 211: 1–74.

Bayley, P.B., 1989. Aquatic environments in the Amazon Basin, with an analysis of carbon sources, fish production, and yield. *Can. Spec. Publ. Fish. Aquat. Sci.* 106: 399–408.

Beeckman, W. & de Bont, A.F., 1985. Characteristics of the Nam Ngum reservoir eco-system as deduced from the food of the most important fish-species. *Verh. Internat. Verein. Limnol.* 22: 2643–2649.

Begum, A., Bashar, M.A. & Biswas, V., 1990a. On the life history of *Neurothemis tullia* (Drury) from Dhaka, Bangladesh (Anisoptera: Libellulidae). *Indian Odonatol.* 3: 11–20.

Begum, A., Bashar, M.A. & Ahmed, M.U., 1990b. Biology and larval morphology of *Pantala flavescens* (Fabricius) (Anisoptera: Libellulidae). *The Dhaka University Studies Part E* 5: 41–48.

Begum, F. & Rizvi, S.N., 1987. Biological studies on freshwaters of Pakistan, 13. Some freshwater bivalves of Sind. *Biologia (Lahore)* 33: 1–14.

Beilfuss, R.D. & Barzen, J.A., 1984. Hydrological wetland restoration in the Mekong Delta, Vietnam. In: Mitsch, W.J. (Ed.), *Global Wetlands Old World and New*. Elsevier, Amsterdam: 453–468.

Belfiore, C., 1994. Taxonomic characters for species identification in the genus *Electrogena* Zurwerra and Tomka, with a description of *Electrogena hyblaea* sp. n. from Sicily (Ephemeroptera, Heptageniidae). *Aquatic Insects* 16: 193–199.

Belov, V.V., 1982. Systematic position and synonomy of *Cinygma tibiale* ULMER, 1920 (Ephemeroptera: Heptageniidae). *Entomol. Mitt. Zool. Mus. Hamburg* 115: 193–194.

Belov, V.V., 1983. Further remarks on the type-specimens of the genus *Cinygma* EATON, 1885 (Ephemeroptera: Heptageniidae). *Entomol. Mitt. Zool. Mus. Hamburg* 118: 387–389.

Belsare, D.K., 1994. Inventory and status of vanishing wetland wildlife of Southeast Asia and an operational management plan for their conservation. In: Mitsch, W.J. (Ed.), *Global Wetlands Old World and New*. Elsevier, Amsterdam: 841–856.

Bender, E.A., Case, T.J. & Gilpin, M.E. 1984. Perturbation experiments in community ecology: theory and practise. *Ecology* **65**: 1–13.

Benke, A.C., 1993. Concepts and patterns of invertebrate production in running waters. *Verh. Internat. Verein. Limnol.* **25**: 15–38.

Benke, A.C. & Jacobi, D.I., 1986. Growth rates of mayflies in a subtropical river and their implications for secondary production. *J. N. Amer. Benthol. Soc.* **5**: 107–114.

Benke, A.C., 1990. A perspective on America's vanishing streams. *J. N. Amer. Benthol. Soc.* **9**: 77–88.

Benzie, J.A.H., 1982. The complete larval development of *Caridina mccullochi* Roux, 1926 (Decapoda, Atyidae) reared in the laboratory. *J. Crust. Biol.* **2**: 493–513.

Benzie, J.A.H., 1984. The colonization mechanisms of stream benthos in a tropical river (Menik Ganga: Sri Lanka). *Hydrobiologia* **111**: 171–179.

Benzie, J.A.H., 1989. The immature stages of *Plea frontalis* (Fieber, 1844) (Hemiptera, Pleidae), with a redescription of the adult. *Hydrobiologia* **179**: 157–171.

Benzie, J.A.H. & de Silva, P.K., 1983. The abbreviated larval development of *Caridina singhalensis* Ortmann, 1894 (Decapoda, Atyidae). *J. Crust. Biol.* **3**: 117–126.

Benzie, J.A.H. & de Silva, P.K., 1988. The distribution and ecology of the freshwater prawn *Caridina singhalensis* (Decapoda, Atyidae) endemic to Sri Lanka. *J. Trop. Ecol.* **4**: 347–359.

Berg, M.B., 1995. Larval food and feeding behaviour. In: Armitage, P., Cranston, P.S. & Pinder, L.C.V., (Eds), *The Chironomidae: Biology and Ecology of Non-biting Midges*. Chapman & Hall, London: 136–168.

Berra, T.M., Moore, R. & Reynolds, L.F., 1985. The freshwater fishes of the Laloki River system of New Guinea. *Copeia* **1985**: 316–326.

Berry, A.J. & Kadri, A.B.H., 1974. Reproduction in the Malayan freshwater cerithiacean gastropod *Melanoides tuberculata*. *J. Zool., Lond.* **172**: 369–381.

Bertrand, H., 1972. *Larves et nymphes des Coléoptères aquatiques de globe.* H. Paillart, Paris: 804 pp.

Bertrand, H.P.I., 1973. Results of the Austrian-Ceylonese hydrological mission, 1970. Part XI: larvae and pupae of water beetles collected from the island of Ceylon. *Bull. Fish. Res. Stn, Sri Lanka (Ceylon)* **24**: 95–112.

Beveridge, M.C.M., Ross, L.G. & Kelly, L.A., 1994. Aquaculture and biodiversity. *Ambio* **23**: 497–502.

Bhalerao, A.M. & Paranjape, S.Y., 1986. Studoes on the bioecology of a grouse locust *Euscelimena harpago* Serv. (Orthoptera: Tetrigidae). *Geobios* **13**: 145–150.

Bhargava, D.S., 1992. Why the Ganga (Ganges) could not be cleaned. *Environmental Conservation* **19**: 170–172.

Bhatnagar, G.K. & Karamchandani, B.J., 1970. Food and feeding habits of

Labeo fimbriatus (Bloch) in River Narbada near Hoshangabad (MP). *J. Inland Fish. Soc., India* **2**: 30–50.

Bhatt, S.D. & Pathak, J.K., 1991. Streams of the great mountain arc: Physiography and physico-chemistry. In: Bhatt, S.D. & Pande, R.K. (Eds), *Ecology of the Mountain Waters*. Ashish Publishing House, New Delhi: 43–58.

Bhatt, S.D. & Pathak, J.K., 1992. Assessment of water quality and aspects of pollution in a stretch of River Gomti (Kumaun, Lesser Himalaya). *J. Environ. Biol.* **13**: 113–126.

Bhatt, S.D., Bisht, Y. & Negi, U., 1984. Ecology of the limnofauna in the River Kosi of the Kumaun Himalaya, Uttar Pradesh, India. *Proc. Indian natn. Sci. Acad.* **B50**: 395–405.

Bhatt, S.D., Bisht, Y. & Negi, U., 1985. Hydrology and phytoplankton populations in River Kosi of the western Himalaya (Uttar Pradesh). *Indian J. Ecol.* **12**: 141–146.

Bhattacharyay, S. & Chaudhuri, P.K., 1990. Two new Indian species of *Pseudorthocladius* Goetghebuer (Diptera: Chironomidae). *Rev. Bras. Entomol.* **34**: 331–335.

Bhattacharyay, S., Ali, A. & Chaudhuri, P.K., 1991. Orthoclads of tribe Orthocladiini (Diptera: Chironomidae) from India. *Beitr. Entomol.* **41**: 333–349.

Bhattacharyay, S., Chattopadhyay, S. & Chaudhuri, P.K., 1993. Four new species of *Chaetocladius* (Diptera: Chironomidae) from India. *Eur. J. Entomol.* **90**: 87–94.

Bhuiyan, A.S. & Islam, M.N., 1991. Observation on the food and feeding habits of *Ompok pabda* (Hamilton) from the River Padma (Siluridae: Cypriniformis). *Pak. J. Zool.* **23**: 75–77.

Bhukaswan, T., 1983. *Pla buk* (*Pangasianodon gigas*) in Chaing Khong. *Thai Fisheries Gazette* **36**: 339–346.

Bij de Vaate, A. & Greijdanus-Klaas, M., 1990. The Asiatic clam, *Corbicula fluminea* Müller, 1774 (Pelecypoda, Corbiculidae), a new immigrant in The Netherlands. *Bull. zool. Mus. Univ. Amsterdam* **12**: 173–178.

Birstein, V.J., 1993. Sturgeons and paddlefishes: threatened fishes in need of conservation. *Conservation Biol.* **7**: 773–787.

Bishop, J.E., 1973a. *Limnology of a small Malayan river Sungai Gombak*. Dr W. Junk Publishers, The Hague: 485 pp.

Bishop, J.E., 1973b. Observations on the vertical distribution of the benthos in a Malaysian stream. *Freshwat. Biol.* **3**: 147–156.

Bisht, R.S. & Das, S.M., 1982. Observations on the ecology and biology of a saxicoline naucorid, *Heleocoris vicinus* Montandon (Hemiptera: Naucoridae) of Kumaon lakes. *Proc. Natl Acad. Sci. India* **52B**: 301–307.

Biström, O., 1995. Taxonomic revision of the genus *Hydrovatus* Motschulsky (Coleoptera, Dytiscidae). *Entomologica Basiliensia* **20**: 57–142.

Biström, O. & Wewalka, G, 1994. *Hydrovatus asymmetricus* sp. n.

(Coleoptera, Dytiscidae) an enigmatic species from Southest Asia. *Entomomlogica Fennica* **5**: 103–104.

Biswas, A.K., 1972. *History of Hydrology.* North-Holland Publishing Company, Amsterdam: 336 pp.

Biswas, A.K., 1990. Watershed management. In: Thanh, N.C. & Biswas, A.K. (Eds), *Environmentally-sound Water Management.* Oxford University Press, New Delhi: 155–175.

Biswas, A.K., 1993. Management of international waters: problems and perspective. *Water Resources Development* **9**: 167–188.

Biswas, A.K. & El-Habr, H.N., 1993. Environment and water resources management: the need for a holistic approach. *Water Resources Development* **9**: 117–125.

Biswas, A.K., Zuo, D., Nickum, J.E. & Liu, C., 1983. *Long-distance Water Transfer: a Chinese Case Study and International Experiences.* Tycooly International Publishers Limited, Dublin: 417 pp.

Blache, J., 1951. Aperçu sur le placton des eaux douces du Cambodge. *Cymbium* **6**: 62–94.

Bleckmann, H. & Barth, F.G., 1984. Sensory ecology of a semi-aquatic spider (*Dolomedes triton*). 2. The release of predatory behaviour by water surface waves. *Behav. Ecol. Sociobiol.* **14**: 303–312.

Bleckmann, H. & Bender, M., 1987. Water surface waves generated by the male pisaurid spider *Dolomedes triton* (Walckenaer) during courtship behaviour. *J. Arachnol.* **15**: 1988.

Bleckmann, H. & Lotz, T., 1987. The vertebrate-catching behaviour of the fishing spider *Dolomedes triton* (Araneae, Pisauridae) *Anim. Behav.* **35**: 641–651.

Boon, P.J., 1979. Adaptive strategies of *Amphipsyche* larvae (Trichoptera: Hydropsychidae) downstream of a tropical impoundment. In: Ward, J.V & Stanford, J. (Eds), *The Ecology of Regulated Streams.* Plenum Press, New York: 237–255.

Boon, P.J., 1992. Channelling scientific information for the conservation and management of rivers. *Aquatic Conservation* **2**: 115–123.

Boon, P.J., 1995. The conservation of freshwaters: tepmerate experience in a tropical context. In: Timotius, K.H. & Göltenboth, F. (Eds), *Tropical Limnology Volume I. Present Status and Challenges.* Satya Wacana University Press, Salatiga, Indonesia: 149–159.

Boonsom, J., 1976. Studies on Benthic Animals in the Chao Phya River using multiple-plate samplers. In: *Symposium on the Development and Utilization of Inland Fishery Resources.* Colombo, Sri Lanka: 132–137.

Boonsong, L. & McNeely, J.A., 1977. *Mammals of Thailand.* Association for the Conservation of Wildlife, Sahakarnbhat C., Bangkok: 758 pp.

Botosaneanu, L., 1976. Les jeunes stades des Moropsychini: *Moropsyche vanegudha* n. sp. et *krichnaruna* n. sp. *Can. Ent.* **100**: 1269–1277.

Botosaneanu, L., 1976. Une collection de stades aquatiques de Trichoptères du Népal, réalisée par le Professeur H. Janetschek. *Khumbu Himal* 5: 187–200.

Bott, R., 1970. Die Süsswasserkrabben von Europa, Asien, Australien, und ihre Stammesgeschichte. Eine Revision der Potamoidea und Parathelphusoidea (Crustacea, Decapoda). *Abhandlungen Senckenbergischen naturforschenden Gesellschaft* 526: 1–338.

Bott, R., 1974. Die Süsswasserkrabben von Neu Guinea. *Zool. Verh. (Leiden)* 136: 1–36.

Bottorff, R.L. & Knight, A.W., 1987. Ectosymbiosis between *Nanocladius downesi* (Diptera: Chironomidae) and *Acroneuria abnormis* (Plecoptera: Perlidae) in a Michigan stream, USA. *Entomol. Gen.* 12: 97–113.

Boukal, D.S., 1997. A revision of the genus *Austrolimnius* CARTER & ZECK, 1929 (Insecta: Coleoptera: Elmidae) from New Guinea and the Moluccas. *Ann. Naturhist. Mus. Wien* 99B: 155–215.

Bovee, K.D., 1982. *A Guide to Stream Habitat Analysis Using the Instream Flow Incremental Methodology.* Instream Flow Information paper No. 12, U.S. Fish and Wildlife Service, Washington, D.C.: 248 pp.

Bovee, K.D., 1986. *Development and Evaluation of Habitat Suitability Criteria for Use in the Instream Flow Incremental Methodology.* Instream Flow Information Paper No. 21, U.S. Fish and Wildlife Service, Washington, D.C.: 235 pp.

Boyden, C.R., Brown, B.E., Lamb, K.P., Drucker, R.F. & Tuft, S.J., 1978. Trace elements in the upper Fly River, Papua New Guinea. *Freshwat. Biol.* 8: 189–205.

Braasch, D., 1979. Neue Heptageniidae aus Asien. *Reichenbachia* 17: 261–272.

Braasch, D., 1980a. *Ecdyonurus cristatus* n.sp. aus Nepal (Ephemeroptera, Heptageniidae). *Reichenbachia* 18: 227–32.

Braasch, D., 1980b. Eine neue *Cinygmula*–Art aus der Mongolei (Ephemeroptera, Heptageniidae). *Reichenbachia* 18: 161–163.

Braasch, D., 1980c. Eintagsfiegen (Gattungen *Epeorus* und *Iron*) aus Nepal und Indien (Ephemeroptera, Heptageniidae). *Reichenbachia* 18: 56–65.

Braasch, D., 1981a. Zum Status einiger Eintagsfliegenarten der Ausbeute der japanischen Himalaya-Expeditionen 1952–53 und 1960, bearbeitet von M. UENO (I). *Reichenbachia* 19: 31–32.

Braasch, D., 1981b. Beitrag zur Kenntnis der Heptageniidae des Himalaya (Ephemeroptera). *Reichenbachia* 19: 127–132.

Braasch, D., 1981c. *Epeorus gilliesi* n.sp. aus Indien. *Reichenbachia* 19: 117–118.

Braasch, D., 1981d. Zum Artstatus des *Epeorus psi* EATON, 1885 von Taiwan. *Deutsche Entomologische Zeitschrift* 28: 113–115.

Braasch, D., 1981e. Eintagsfliegen (Gattungen *Epeorus* und *Iron*) aus Nepal (II) (Ephemeroptera, Heptageniidae). *Reichenbachia* 19: 105–110.

Braasch, D., 1981f. Beitrag zur Kenntnis der Ephemerellidae des Himalaya (Ephemeroptera). *Reichenbachia* 19: 85–88.

Braasch, D., 1983a. Neue Baetidae von Nepal (Ephemeroptera). *Reichenbachia* 21: 147–155.

Braasch, D., 1983b. Eintagsfiegen (Gattungen *Epeorus* und *Iron*) aus Nepal und Indien (Ephemeroptera, Heptageniidae). *Reichenbachia* 21: 195–196.

Braasch, D., 1984a. Beitrag zur Kenntnis der Heptageniidae des Himalaya (II) (Ephemeroptera). *Reichenbachia* 22: 45–50.

Braasch, D., 1984b. Beitrag zur Kenntnis der Heptageniidae des Himalaya (III) (Ephemeroptera). *Reichenbachia* 22: 65–74.

Braasch, D., 1986a. Eintagsfliegen aus der Mongolischen Volksrepublik. *Entomologishce Nachrichten und Berichte* 30: 77–79.

Braasch, D., 1986b. Zum status der Gattung *Heptagenia* WALSH, 1863 in Indien (Ephemeroptera, Heptageniidae). *Reichenbachia* 23: 131–134.

Braasch, D., 1986c. Zur Kenntnis der Gattung *Notacantnurus* TSHERNOVA, 1974 aus dem Himalaya (Ephemeroptera, Heptageniidae). *Reichenbachia* 23: 117–125.

Braasch, D., 1990. Neue Eintagsliegen aus Thailand, nebst einigen Bemerkungen zu deren generischem Status (Insecta, Ephemeroptera: Heptageniidae). *Reichenbachia* 28: 7–14.

Braasch, D. & Soldán, T., 1979. Neue Heptageniidae aus Asien. *Reichenbachia* 17: 261–272.

Braasch, D. & Soldán, T., 1980. *Centroptella* n. gen., eine neue Gattung der Eintagsfliegen aus China. *Reichenbachia* 18: 123–127.

Braasch, D. & Soldán, T., 1982. Neue Heptageniidae (Ephemeroptera) aus Asien III. *Entomologische Nachrichten und Berichte* 26: 25–28.

Braasch, D. & Soldán, T., 1983. Baetidae in Mittelasien III (Ephemeroptera). *Entomologishe Nachrichten und Berichte* 27: 266–271.

Braasch, D. & Soldán, T., 1984a. Zwei neue Arten der Gattung *Cinygmina* KIMMINS, 1937 aus Vietnam (Ephemeroptera, Heptageniidae). *Reichenbachia* 22: 195–200.

Braasch, D. & Soldán, T., 1984b. Beitrag zur Kenntnis der Gattung *Thalerosphyrus* EATON, 1881 im Hinblick auf die Gattung *Ecdyonuroides* THANH, 1967. *Reichenbachia* 22: 201–206.

Braasch, D. & Soldán, T., 1984c. Eintagsfliegen (Gattungen *Epeorus* und *Iron*) aus Vietnam (Ephemeroptera, Heptageniidae). In: Landa, V. (Ed.), *Proceedings of the 4th International Conference on Ephemeroptera*. CSAV, Czechoslovakia: 109–114.

Braasch, D. & Soldán, T., 1986a. *Asionurus* n. gen., eine neue Gattung der Heptageniidae aus Vietnam (Ephemeroptera). *Reichenbachia* 23: 155–159.

Braasch, D. & Soldán, T., 1986b. Die Heptageniidae des Gombak River, Malaysia (Ephemeroptera). *Reichenbachia* 24: 41–52.

Braasch, D. & Soldán, T., 1986c. Zur kenntnis der Gattung *Compsoneuria* EATON, 1881 von den Sunda-Inseln (Ephemeroptera). *Reichenbachia* 24: 59–62.

Braasch, D. & Soldán, T., 1986d. *Rhithogena diehliana* n. sp. von Sumatra (Ephemeroptera, Heptageniidae). *Reichenbachia* 24: 91–92.

Braasch, D. & Soldán, T., 1987a. Neue *Cinygmina*-Arten aus Vietnam (Ephemeroptera, Heptageniidae). *Reichenbachia* **24**: 123–126.

Braasch, D. & Soldán, T., 1987b. Neue Heptageniidae von Indien. *Reichenbachia* **24**: 131–134.

Braasch D. & Soldán, T., 1988. *Trichogenia* gen. n., eine neu Gattung der Eintagsfliegen aus Vietnam (Insecta, Ephemeroptera, Heptageniidae). *Reichenbachia* **25**: 119–124.

Braatz, S., Davis, G., Shen, S. & Rees, C., 1992. Conserving biological diversity. A strategy for protected areas in the Asia-Pacific Region. *World Bank Technical Paper* **193**: 1–66.

Brammer, H., 1990. Floods in Bangladesh. II. Flood mitigation and environmental aspects. *Geographical Journal* **156**: 158–165.

Brancucci, M., 1986. A revision of the genus *Lacconectus* Motschulsky (Coleoptera, Dytiscidae). *Entomologica Basiliensia* **11**: 81–202.

Brancucci, M., 1988. A revision of the genus *Platambus* Thomson (Coleoptera, Dytiscidae). *Entomologica Basiliensia* **12**: 165–239.

Brandt, R.A.M., 1974. The non-marine aquatic Mollusca of Thailand. *Arch. Mollusck.* **105**: 1–423.

Brandt, R.A.M. & Temcharoen, P., 1971. The molluscan fauna of the Mekong at the foci of schistosomiasis in South Laos and Cambodia. *Arch. Mollusk.* **101**: 111–140.

Brewin, P.A. & Ormerod, S.J., 1994. Macroinvertebrate drift in streams of the Nepalese Himalaya. *Freshwat. Biol.* **32**: 573–583.

Brinck, P., 1981a. *Spinosodineutes* (Coleoptera: Gyrinidae) in New Guinea and adjacent islands. *Ent. Scand., Suppl.* **15**: 353–364.

Brink, P., 1981b. Taxonomic status and evolutionary trends in the whirligig beetle taxon *Spinosodineutes*. *Entomol. Gen.* **7**: 17–32.

Brinck, P., 1983. A revision of *Rhombodineutus* Ochs in New Guinea (Coleoptera: Gyrinidae). *Ent. Scand.* **14**: 205–223.

Brinck, P., 1984. Evolutionary trends and specific differentiation in *Merodineutus* (Coleoptera: Gyrinidae). *Int. J. Entomol.* **26**: 175–189.

Brinkhurst, R.O. & Gelder, S.R., 1991. Annelida: Oligochaeta and Branchiobdellidae. In: Thorp, J.H. & Covich, A.P. (Eds), *Ecology and Classification of North American Freshwater Invertebrates*. Academic Press, San Diego: 401–435.

Brinkhurst, R.O. & Jamieson, B.G.M., 1971. *Aquatic Oligochaeta of the World*. Toronto University Press, Toronto: 860 pp.

Brinkhurst, R.O. & Wetzel, M.J., 1984. Aquatic Oligochaeta of the world: supplement. A catalogue of new freshwater species, descriptions and revisons. *Canadian Technical Report of Hydrography and Ocean Sciences* **44**: 1–101.

Brinkhurst, R.O., Qi, S. & Liang, Y., 1990. The aquatic oligochaeta from the People's Republic of China. *Can. J. Zool.* **68**: 901–916.

Brittain, J.E., 1990. Life history strategies in Ephemeroptera and Plecoptera.

In: Campbell, I.C. (Ed.), *Stoneflies and Mayflies: Life Histories and Biology*. Kluwer Academic Publishers, Dordrecht: 1–12.

Brodo, F., 1987. A revision of the genus *Prionocera* (Diptera: Tipulidae). *Evolutionary Monographs* 8: 1–93.

Brooks, G.T., 1951. A revision of the genus *Anisops* (Notonectidae, Hemiptera). *Kansas University Science Bulletin* 34: 301–519.

Brown, H.P., 1981a. A distributional survey of the world genera of aquatic dryopoid beetles (Coleoptera: Dryopidae, Elmidae, and Psephenidae *sens. lat.*). *Pan-Pacific Entomologist* 57: 133–148.

Brown, H.P., 1981b. Key to the world genera of Larinae (Coleoptera, Dryopoidea, Elmidae), with descriptions of new genera from Hispaniola, Columbia, Australia, and New Guinea. *Pan-Pacific Entomologist* 57: 76–104.

Brown, H.P., 1987. Biology of riffle beetles. *Ann. Rev. Entomol.* 32: 253–273.

Brown, H.P. & Thobias, M.P., 1984. World synopsis of the riffle beetle genus *Leptelmis* Sharp, 1888, with a key to Asiatic species and description of a new species from India (Coleoptera, Dryopoidea, Elmidae). *Pan-Pacific Entomologist* 60: 23–29.

Brown, K.M., 1991. Mollusca: Gastropoda. In: Thorp, J.H. & Covich, A.P. (Eds), *Ecology and Classification of North American Freshwater Invertebrates*. Academic Press, San Diego: 285–314.

Bruce, A.J., 1989. Reestablishment of the genus *Coutierella* Sollaud, 1914 (Decapoda: Palaemonidae), with a redescription of *C. tonkinensis* from the Mai Po Marshes, Hong Kong. *J. Crustacean Biol.* 9: 176–187.

Bruton, M.N., 1995. Have fishes had their chips? The dilemma of threatened fishes. *Environ. Biol. Fish.* 43: 1–27.

Buckingham, G.R., & Bennett, C.A., 1996. Laboratory biology of an immigrant Asian moth, *Parapoynx diminutalis* (Lepidoptera: Pyralidae), on *Hydrilla verticillata* (Hydrocharitaceae). *Florida Entomologist* 79: 353–363.

Bukit, N.T., 1995. Water quality conservation for the Citarum River in West Java. *Water Science & Technology* 31: 1–10.

Bunn, S.B., 1986. Spatial and temporal variation in the macroinvertebrate fauna of streams of the northern jarrah forest, Western Australia: functional organization. *Freshwat. Biol.* 16: 621–632.

Burns, J.W. 1972. Some effects of logging and associated road construction on northern California streams. *Trans. Amer. Fish. Soc.* 101: 1–17.

Bushnell, J.H., 1973. The freshwater Ectoprocta: a zoogeographical discussion. In: Larwood, G.P. (Ed.), *Living and Fossil Bryozoa. Recent Advances in Research*. Academic Press, London: 503–521.

Bushnell, J.H., 1974. Bryozoa (Ectoprocta). In: Hart, C.W. & Fuller, S.L.H. (Eds): *Pollution Ecology of Freshwater Invertebrates*. Academic Press, New York: 157–194.

Butt, J.A. & Khan, K., 1988. Food of freshwater fishes of North West Frontier Province, Pakistan. In: Ahmad, M. (Ed.), *Proceedings of the Seventh*

Pakistan Congress of Zoology. The Zoological Society of Pakistan, Quetta: 217–233.

Byers, G.W., 1996. Tipulidae. In: Merritt, R.W. & Cummins, K.W. (Eds), *An Introduction to the Aquatic Insects of North America. Third Edition)*. Kendall/Hunt Publishing Company, Iowa: 549–570.

Caldara, R. & O'Brien, C.W., 1995. Curculionidae: aquatic weevils of China. In: Jäch, M.A. & Ji, L. (Eds), *Water Beetles of China*. Zoologisch-Botanische Gesellschaft & Wiener Coleopterologenverein, Vienna: 389–408.

Caldecott, J., 1996. *Designing Conservation Projects*. Cambridge University Press, Cambridge: 312 pp.

Caldwell, L., 1985. Science will not save the biosphere but politics might. *Environmental Conservation* **12**: 195–197.

Calow, P., 1978. The evolution of life-cycle strategies in freshwater gastropods. *Malacologia* **17**: 351–364.

Camargo, J.A., 1994, The importance of biological monitoring for the ecological risk assessment of freshwater pollution: a case study. *Environment International* **20**: 229–238.

Caponera, D.A., 1987. International water resources law in the Indus basin. In: M. Ali, M., Radosevich, G.E. & Ali Khan, A. (Eds), *Water Resources Policy for Asia*. A.A. Balkema Publishers, Boston: 509–515.

Carin, R., 1962. *River Control in Communist China*. Communist China Problem Research Series, EC 31, Union Research Institute, Hong Kong: 124 pp.

Carle, F.L., 1986. The classification, phylogeny and biogeography of the Gomphidae (Anisoptera). I. Classification. *Odonatologica* **15**: 275–326.

Carpenter, A., 1978 Protandry in the freshwater shrimp, *Paratya curvirostris* (Heller, 1862) (Decapoda: Atyidae), with a review of the phenomenon and its significance in the Decapoda. *J. Roy. Soc. New Zealand* **8**: 343–358.

Carpenter, S.R., 1990. Large-scale perturbations: opportunities for innovation. *Ecology* **71**: 2038–2043.

Cartwright, D.J., 1990a. Taxonomy of the larvae, pupae and females of the Victorian species of *Chimarra* Stephens (Trichoptera: Philopotamidae) with notes on biology and distribution. *Proc. Roy. Soc. Victoria* **102**: 15–22.

Cartwright, D.J., 1990b. The Australian species of *Ecnomus* McLachlan (Trichoptera: Ecnomidae). *Memoirs of the Museum of Victoria* **51**: 1–48.

Cartwright, D.J., 1992. Descriptions of four new species of *Ecnomus* McLachlan (Trichoptera: Ecnomidae) from North Sulawesi. *Bull. Zool. Mus. Univ. Amsterdam* **13**: 101–108.

Cartwright, D.J., 1994. New secies and new records of *Ecnomus* McLachlan (Trichoptera: Ecnomidae) from Indonesia. *Memoirs of the Museum of Victoria* **54**: 447–459.

Caulfield, H.P., 1986. Viability of interbasin, interstate/international transfers of water. *Water International* **11**: 32–37.

Chace, F.A., 1983. The *Atya*-like shrimps of the Indo-Pacific region (Decapoda: Atyidae). *Smithsonian Contributions to Zoology* **384**: 1–54.

Chace, F.A., 1992. On the classification of the Caridea (Decapoda). *Crustaceana* **63**: 70–80.

Chace, F.A., 1997. The caridean shrimps (Crustacea: Decapoda) of the Albatross Philippine Expedition 1907–1910, Part 7: families Atyidae, Eugonatonotidae, Rhynchocinetidae, Bathy-palaemonellidae, Processidae, and Hippolytidae. *Smithsonian Contributions to Zoology* **587**: 1–106.

Chace, F.A. & Bruce, A.J., 1993. The caridean shrimps (Crustacea: Decapoda) of the Albatross Philippine Expedition 1907–1910, Part 6: superfamily Palaemonoidea. *Smithsonian Contributions to Zoology* **543**: 1–152.

Chacko, P.I. & Ganapati, S.V., 1952. Hydrobiological survey of the Suruli River in the Highwavys, Madurai District, India, to determining its suitability for the introduction of the Rainbow Trout. *Arch. Hydrobiol.* **46**: 128–141.

Chakrabarty, R.D., Ray, P. & Singh, S.B., 1959. A quantitative study of plankton and the physico-chemical conditions of the River Yamuna at Allahbad in 1954–55. *Indian J. Fish.* **6**: 186–208.

Chakrapani, G.J. & Subramanian, V., 1993. Rates of erosion and sedimentation in the Mahanadi River Basin, India. *J. Hydrol.* **149**: 39–48.

Chakravarty, R.B., 1969. Utility of river valley projects in India. *Bull. Natn. Inst. Sci. India* **40**: 1–9.

Chan, K.L. & Linley, J.R., 1989. Laboratory studies on the immature stages of *Atrichopogon wirthi* (Diptera: Ceratopogonidae). *Florida Entomologist* **72**: 85–88.

Chan, T.Y., Hung, M.S. & Yu, H.P., 1995. Identity of *Eriocheir recta* (Stimpson, 1858) (Decapoda: Brachyura), with description of a new mitten crab from Taiwan. *J. Crustacean Biol.* **15**: 301–308.

Chandra, R., Saxena, R.K. & Tyagi, R.K., 1990. *Hilsa ilisha* (Ham.) of Ganaga — its glory and downfall — a retrospect. In: Agrawal, V.P. & Das, P. (Eds), *Recent Trends in Limnology*. Society of Biosciences, Muzaffarnagar (India): 365–378.

Chandrashekar, K.R., Sridhar, K.R. & Kaveriappa, K.M., 1990. Periodicity of water-borne hyphomycetes in two streams in Western Ghat forests (India). *Acta Hydrochim. Hydrobiol.* **18**: 187–204.

Chang, H., 1989. Six synnematous hyphomycetes new for Taiwan. *Bot. Bull. Acad. Sinica (Taipei)* **30**: 161–166.

Chang, K.M. & Lin, C.C., 1978. Description of a new subspecies of *Pisidium, taiwanense* from Tiench, Taiwan (Bivalvia; Spaheriidae). *Bulletin of the Malacological Society of China (Taipei)* **5**: 23–27.

Chang, W.Y.B., 1992. Giant catfish (*pla beuk*) culture in Thailand. *Aquacult. Mag.* **18**: 54–58.

Chaniotis, B.N., Butler, J.M., Ferguson, F.F. & Jobim, W.R., 1980. Presence of males in Puerto Rican *Thiara (Tarebia) granifera* (Gastropoda: Thiaridae) thought to be parthenogenetic. *Caribbean J. Sci.* **16**: 81–90.

Chanphaka, U., 1968. The Royal Forest Department program for watershed management and its problems. In: Talbot, L.M. & Talbot, M.H. (Eds), *Conservation in Tropical South East Asia*. IUCN Publications New Series No. 10, IUCN, Morges, Switzerland: 100–102.

Chanphaka, U., 1982. Catchment management for optimum use of land and water resources in Thailand. *New Zealand Water and Soil Misc. Publ.* **45**: 203–211.

Chantaramongkol, P. & Malicky, H., 1986. Beschreibung von neuen Köcherfliegen (Trichoptera, Insecta) aus Sri Lanka. *Ann. Naturhist. Mus. Wien* **88/89**: 511–534.

Chantaramongkol, P. & Malicky, H., 1989. Some *Chimarra* (Trichoptera: Philopotamidae) from Thailand. *Aquatic Insects* **11**: 223–240.

Chantaramongkol, P. & Malicky, H., 1995. Drei neue asiatische *Hydromanicus* (Trichoptera: Hydropsychidae). *Ent. Z.* **105**: 92–95.

Chao, H., 1946. Two new species of stoneflies of the genus *Styloperla* Wu (Perlidae, Plecoptera). *Biological Bulletin, Fukien Christian University* **5**: 93–96.

Charoenphong, S., 1991. Environmental calamity in southern Thailand's headwaters: causes and remedies. *Land Use Policy* **8**: 185–188.

Chatterji, A., Ansari, Z.A. & Parulekar, A.H., 1984. Growth of the black clam, *Villorita cyprinoides* (Grey), from Colvale River (Goa). *Indian Journal of Marine Sciences* **13**: 119–120.

Chatterji, A., Siddiqui, A.Q. & Khan, A.A., 1977. Food and feeding habits of *Labeo gonius* (Ham.) from the River Kali. *J. Bombay. Nat. Hist. Soc.* **75**: 104–109.

Chattopadhyay, D.N., Saha, M.K. & Konar, S.K., 1987. Some bio-ecological studies of the River Ganga in relation to water pollution. *Environ. Ecol.* **5**: 494–500.

Chattopadhyay, S. & Chaudhuri, P.K., 1991. *Kiefferulus inciderus* sp. nov. from West Bengal, India (Diptera: Chironomidae). *Hexapoda* **3**: 24–26.

Chattopadhyay, S. & Chaudhuri, P.K., 1993. *Paratendipes unimaculipennis*, a new chironomine midge from India (Diptera: Chironomidae). *Proc. Zool. Soc. (Calcutta)* **46**: 51–53.

Chao, H.,[1] 1984. Reclassification of Chinese gomphid dragonflies, with the establishment of a new genus and species (Anisoptera: Gomphidae). *Odonatologica* **13**: 71–80.

Chaube, U.C., 1990. Water conflict resolution in the Ganga-Brahmaputra basin. *Water Resources Development* **6**: 79–85.

Chaudhry, A.A. & Khalid, U., 1990. Indus dolphin population in the Punjab. In: Ahmad, M. (Ed.), *Proceedings of the Ninth Pakistan Congress of Zoology*. Zoological Society of Pakistan, Quetta: 291–296.

---

[1] Chao Hsiu-fu is the same person as Zhao, X.

Chaudhuri, P.K. & Chattopadhyay, S., 1988. Studies on the juveniles of *Microchironomus* from India (Diptera: Chironomidae). *Oriental Insects* **22**: 175–183.

Chaudhuri, P.K. & Datta, T., 1991. One new species and one new record of the genus *Microspectra* Kieffer (Diptera: Chironomidae) from India. *Proc. Zool. Soc. (Calcutta)* **44**: 29–32.

Chaudhuri, P.K. & Ghosh, M., 1981. A new genus of Podomine midge (Chironomidae) from Bhutan. *Systematic Entomology* **6**: 373–376.

Chaudhuri, P.K. & Guha, D.K., 1987. A conspectus of chironomid midges (Diptera: Chironomidae) of India and Bhutan. *Ent. Scand., Suppl.* **29**: 23–33.

Chaudhuri, P.K., Ali, A., Majumdar, A., 1992. Life stages of *Tanytarsus vanderwulp* with description of a new species from India (Diptera: Chironomidae). *Deutsche Entomologische Zeitschrift* **39**: 381–396.

Chen, G.X. & Chang, J., 1982. A new species of *Parapotamon* from China (Crustacea: Decapoda). *Acta Zootaxonomica Sinica* **7**: 42–46 (in Chinese).

Chen, J. & Wu, G., 1987. Water resources development in China. In: Ali, M., Radosevich, G.E. & Ali Khan, A. (Eds), *Water Resources Policy for Asia*. A.A. Balkema Publishers, Boston: 51–60.

Chen, P., 1987. Three species of saldid bugs recorded as new to China (Hemiptera: Saldidae). *Entomotaxonomia* **9**: 163–166 (in Chinese).

Chen, P. & Andersen, N.M., 1993. A checklist of Gerromorpha from China. *Chinese J. Entomol.* **13**: 69–75.

Chen, P., Liu, P., Liu, R., Li, K. & Pilleri, G., 1979. Distribution, ecology, behavior and conservation of the dolphins of the Chanjiang (Yangtze) River, Wuhan — Yueyang, China. *Invest. Cetacea* **10**: 87–104.

Chen, Q., 1979. A report on Mollusca in Lake Hwama, Hubei Province. *Oceanol. Limnol. Sinica* **10**: 46–62 (in Chinese).

Chen, Q. & Wu, T., 1983. Ecological aspects of Mollusca in the lower reaches (Nanjing to Jiangyin) of Changjiang River (the Yangtze River). *Trans. Chinese Soc. Malacol.* **1**: 103–114 (in Chinese).

Chen, T.C., Liao, K.Y. & Wu, W.L., 1992. The research and evaluation on *Corbicula fluminea* in Taiwan (Bivalvia: Corbiculidae). *Bulletin of Malacology, Republic of China* **17**: 37–49.

Chen, Y., 1988. The freshwater snails of Poyang Lake and its surrounding waters, Jiangxi Province. *Sinozoologia* **6**: 69–75 (in Chinese).

Chen, Y.E. & Morse, J.C., 1991. A preliminary examination of caddisflies from Taiwan, with special reference to Leptoceridae. In: Tomaszewski, C. (Ed.), *Proceedings of the 6th International Symposium on Trichoptera*. Adam Mickiewicz University Press, Lodz-Zakopane, Poland: 377–380.

Chew, Y.F., 1995. Tasek Bera: Malaysia's first Ramsar site. *Asian Wetland News* **8** (**1**): 20–21.

Chitale, M.A., 1988. Environmental management in water resources projects: Indian experiences of irrigation and power projects. *Water Resources Development* **4**: 108–116.

Chitale, M.A., 1992. Development of India's river basins. *Water Resources Development* **8**: 30–44.

Chitramvong, Y.P., 1992. The Bithyniidae (Gastropoda: Prosobranchia) of Thailand: comparative external morphology. *Malacol. Rev.* **25**: 21–38.

Chomchai, P., 1987. The Mekong project: an exercise in regional cooperation to develop the lower Mekong basin. In: Ali, M., Radosevich, G.E. & Ali Khan, A. (Eds), *Water Resources Policy for Asia*. A.A. Balkema Publishers, Boston: 497–508.

Choowaew, S., 1993. Inventory and management of wetlands in the Lower Mekong Basin. *Asian Journal of Environmental Management* **1** (3): 1–10.

Choowaew, S., 1995. Sustainable agricultural development in Thailand's wetlands. *TEI Quarterly Environmental Journal* **3**: 2–13.

Chopra, A.K., Madhwal, B.P. & Singh, H.R., 1990. Abiotic variables and primary productivity of River Yamuna at Naugaon, Uttarkashi-Garhwal. *Indian J. Ecol.* **17**: 61–64.

Choudhary, S.K., Singh, R.B., Nayak, M. & Choubey, S., 1992. Diurnal profile of some physico-chemical and biological parameters in certain perennial pond and River Ganga at Bhagalpur (Bihar). *J. Freshwat. Biol.* **4**: 45–51.

Choudhury, A., 1994. The decline of the wild water buffalo in north-east India. *Oryx* **28**: 70–74.

Choudhury, B.C. & Bustard, H.R., 1982. Restocking mugger crocodile *Crocodylus palustris* (Lesson) in Andhra Pradesh: evaluation of a pilot release. *J. Bombay Nat. Hist. Soc.* **79**: 275–289.

Choudhury, B.C. & Chowdhury, S., 1986. Lessons from crocodile reintroduction projects in India. *Indian For.* **112**: 881–890.

Chowdhary, S.K., 1984. Studies on the bioecology of aquatic insects of Sind and Lidder streams of Kashmir. *Indian J. Ecol.* **11**: 160–165.

Chowdhuary, M.K. & Ghosh, S., 1984. Operation Rhino, Jaldapara Sanctuary, India. *Indian For.* **110**: 1098–1108.

Chowdhury, M.I., Safiullah, S., Iqbal Ali, S.M., Mofizuddin, M. & Enamul Kabir, S., 1982. Carbon transport in the Ganges and Brahmaputra: preliminary results. *Mitt. Geol.-Palaont. Inst. Univ. Hamburg, SCOPE/UNEP Sonderbd.* **52**: 457–468.

Choy, S.C., 1991. The atyid shrimps of Fiji with description of a new species. *Zool. Meded. (Leiden)* **65**: 343–362.

Choy, S.C., 1992. *Caridina bruneiana*, a new species of freshwater shrimp (Decapoda, Caridea, Atyidae) from Negara Brunei Darussalam, Borneo. *Zoologica Scripta* **21**: 49–55.

Choy, S.C. & Ng, P.K.L., 1991. A new species of freshwater atyid shrimp, *Caridina temasek* (Decapoda: Caridea: Atyidae) from Singapore. *Raffles Bulletin of Zoology* **39**: 265–277.

Christensen, M.S., 1992. Investigations on the ecology and fish fauna of the Mahakam River in East Kalimantan (Borneo), Indonesia. *Int. Rev. ges. Hydrobiol.* **77**: 593–608.

Christensen, M.S., 1993a. The artisanal fishery of the Mahakam River floodplain in East Kalimantan, Indonesia. I. Composition and prices of landings, and catch rates of various gear types including trends in ownership. *J. Appl. Ichthyol.* **9**: 185–192.

Christensen, M.S., 1993b. The artisanal fishery of the Mahakam River floodplain in East Kalimantan, Indonesia. II. Catch, income and labour requirements of fisher households. *J. Appl. Ichthyol.* **9**: 193–201.

Christensen, M.S., 1993c. The artisanal fishery of the Mahakem River floodplain in East Kalimantan, Indonesia. III. Actual and estimated yields, and their relationship to water levels and management options. *J. Appl. Ichthyol.* **9**: 202–209.

Christiansen, K.A. & Snider, R.J., 1996. Aquatic Collembola. In: Merritt, R.W. & Cummins, K.W. (Eds), *An Introduction to the Aquatic Insects of North America. Third Edition.* Kendall/Hunt Publishing Company, Iowa: 113–125.

Chuang, C.T.N. & Ng, P.K.L., 1991. Preliminary descriptions of one new genus and three new species of hymenosomatid crabs from Southeast Asia (Crustacea: Decapoda: Brachyura). *Raffles Bull. Zool.* **39**: 363–368.

Chuang, C.T.N. & Ng, P.K.L., 1994. The ecology and biology of Southeast Asian false spider crabs (Crustacea: Decapoda: Brachyura: Hymenosomatidae). *Hydrobiologia* **285**: 85–92.

Chuensri, C., 1974. Freshwater crabs of Thailand. *Kasetart University Fishery Research Bulletin* **7**: 12–40.

Chookajorn, S., 1990. The proper generic name of the Mekong giant catfish. *Thai Fisheries Gazette* **43**: 299–301 (in Thai).

CIDA (Canadian International Development Agency), 1988. *Three Gorges Water Control Project Feasibility Study, People's Republic of China. Volume 1: Feasibility Report.* Canadian International Development Agency & People's Republic of China Ministry of Water Resources and Electric Power (CIPM Yangtze Joint Venture), Montreal (no pagination).

Claridge, G., 1994. Management of coastal ecosystems in eastern Sumatra: the case of Berbak Wildlife Reserve, Jambi Province. *Hydrobiologia* **285**: 287–302.

Clark, W.H., 1986. A note on predation of caddisflies (*Cheumatopsyche logani* and *Hydropsyche wineman*, Trichoptera: Hydropsychidae) by the wolf spider, *Paradosa steva* (Lycosidae). *Pan.-Pac. Entomol.* **62**: 300.

Clifford, H.F., 1980. Numerical abundance values of mayfly nymphs for the Holarctic region. In: Flannagan, J.F. & Marshall, K.E. (Eds), *Advances in Ephemeroptra Biology.* Plenum Press, New York: 503–509.

Cloarec, A., 1995. General activity and foraging tactics in a water bug. *Journal of Ethology* **13**: 31–39.

Coates, D., 1985. Fish yield estimates for the Sepik River, Papua New Guinea, a large floodplain system east of 'Wallace's line'. *J. Fish. Biol.* **27**: 431–443.

Coates, D., 1987. Consideration of fish introductions into the Sepik River, Papua New Guinea. *Aquacult. Fish. Manag.* **18**: 231–241.

Coates, D., 1990a. Biology of the rainbowfish, *Glossolepis multisquamosus* (Melanotaeniidae) from the Sepik River floodplains, Papua New Guinea. *Environ. Biol. Fish.* **29**: 119–126.

Coates, D., 1990b. Aspects of the biology of the perchlet *Ambassis interupta* Bleeker (Pisces: Ambassidae) in the Sepik River, Papua New Guinea. *Aust. J. Mar. Freshwat. Res.* **41**: 267–274.

Coates, D., 1991. Biology of fork-tailed catfishes from the Sepik River, Papua New Guinea. *Environ. Biol. Fish.* **31**: 55–74.

Coates, D., 1993. Fish ecology and management of the Sepik-Ramu, New Guinea, a large contemporary tropical river basin. *Environ. Biol. Fish.* **38**: 345–368.

Coates, D. & Van Zwieten, P.A.M., 1992. Biology of the freshwater halfbeak *Zenarchopterus kampeni* (Teleostei: Hemiramphidae) from the Sepik and Ramu River basin, northern Papua New Guinea. *Ichthyol. Explor. Freshwat.* **3**: 25–36.

Coffman, W.P., 1995. Conclusions. In: Armitage, P., Cranston, P.S. & Pinder, L.C.V., (Eds), *The Chironomidae: Biology and Ecology of Non-biting Midges*. Chapman & Hall, London: 436–447.

Coffman, W.P. & Ferrington, L.C., 1996. Chironomidae. In: Merritt, R.W. & Cummins, K.W. (Eds), *An Introduction to the Aquatic Insects of North America. Third Edition*. Kendall/Hunt Publishing Co., Iowa: 635–754.

Coffman, W.P., Yurasits, L.A. & de la Rossa, C., 1988. Chironomidae of South India. 1. Generic composition, biogeographical relationships and descriptions of two unusual pupal exuviae. *Spixania, Suppl.* **14**: 155–165.

Coleman, J.M., 1969. Brahmaputra River: channel processes and sedimentation. *Sed. Geol.* **3**: 131–139.

Collier, K.J., Croker, G.F., Hickey, C.W., Quinn, J.M. & Smith, B.S., 1995. Effects of hydraulic conditions and larval size on the microdistribution of Hydrobiosidae (Trichptera) in two New Zealand rivers. *N.Z. J. Mar. Freshwat. Res.* **29**: 439–451.

Collins, N.M., 1980. The habits and populations of terrestrial crabs (Brachyura: Gecarcinucoidea and Grapsoidea) in the Gunung Mulu National Park, Sarawak. *Zool. Meded. (Leiden)* **55**: 82–85.

Compagno, L.J.V. & Roberts, T.R., 1982. Freshwater stingrays (Dasyatidae) of Southeast Asia and New Guinea with description of *Himantura signifer* (new species) and reports of unidentified species. *Environ. Biol. Fish.* **7**: 321–340.

Compagno, L.J.V. & Cook, S.F., 1995. The exploitation and conservation of freshwater elasombranchs: status of taxa and prospects for the future. *Journal of Aquariculture & Aquatic Sciences* **7**: 62–90.

Cook, D.R., 1967. The water mites of India. *Memoirs of the American Entomological Institute* **9**: 1–411.

Cooper, S.D. & Barmuta, L. 1993. Field experiments in biomonitoring. In:

Rosenberg, D.M. & Resh, V.H. (Eds), *Freshwater Biomonitoring and Benthic Macroinvertebrates*. Chapman & Hall, New York: 399–441.

Corbet, P., 1980. Biology of Odonata. *Ann. Rev. Entomol.* **25**: 189–217.

Corbet, P.S., 1981. Seasonal incidence of Odonata in light-traps in Trinidad, West Indies. *Odonatologica* **10**: 179–187.

Corner, E.J.H., 1978. The freshwater swamp-forest of south Johore and Singapore. *Gardens' Bull. Singapore, Suppl.* **1**: 1–266.

Costa, H.H., 1974. Limnology and fishery biology of the streams at Horton Plains, Sri Lanka (Ceylon). *Bull. Fish. Res. Stn, Sri Lanka (Ceylon)* **25**: 15–26.

Costa, H.H., 1984. The ecology and distribution of free-living meso- and macrocrustacea of inland waters. In: C.H. Fernando (Ed.), *Ecology and Biogeography in Sri Lanaka*. Dr. W. Junk Publishers, The Hague: 195–213.

Costa, H.H., 1967. A systematic study of freshwater Oligochaeta from Ceylon. *Ceylon J. Sci. (Bio. Sci.)* **7**: 37–51.

Costa, H.H., 1994. The status of limnology in Sri Lanka. *Mitt. Internat. Verein. Limnol.* **24**: 73–85.

Costa, H.H. & Fernando, E.C.M., 1967. The food and feeding relationships of the common meso- and macrofauna in the Maha Oya, a small mountainous stream at Peradeniya, Ceylon. *Ceylon J. Sci. (Biol. Sci.)* **7**: 74–90.

Costa, H.H. & Starmühlner, F., 1972. Results of the Austrian-Ceylonese hydrobiological mission, 1970. Part I. Preliminary report: Introduction and description of the stations. *Bull. Fish. Res. Stn, Sri Lanka (Ceylon)* **23**: 47–76.

Courtemanch, D.L., 1994. Bridging the old and new science of environmental monitoring. *J. N. Amer. Benthol. Soc.* **13**: 117–121.

Courtney, G.W., 1990. Revision of Nearctic mountain midges (Diptera: Deuterophlebiidae). *J. Nat. Hist.* **24**: 81–118.

Courtney, G.W., 1991. Life history patterns of Nearctic mountain midges (Diptera: Deuterophlebiidae). *J. N. Am. Benthol. Soc.* **10**: 177–197.

Courtney, G.W., 1994a. Revision of Palaearctic mountain midges (Diptera: Deuterophlebiidae), with phylogenetic and biogeographic analyses of world species. *Systematic Entomology* **19**: 1–24.

Courtney, G.W., 1994b. Biosystematics of the Nymphomyiidae (Insecta: Diptera): Life history, morphology, and phylogenetic relationships. *Smithsonian Contributions to Zoology* **550**: 1–41.

Covich, A.P., 1987. Atyid shrimps in the headwaters of the Luquillo Mountains, Puerto Rico: filter feeding in natural and artificial streams. *Verh. Internat. Verein. Limnol.* **23**: 2108–2113.

Covich, A.P. & Thorp, J.H., 1991. Crustacea: Introduction and Peracarida. In: Thorp, J.H. & Covich, A.P. (Eds), *Ecology and Classification of North American Freshwater Invertebrates*. Academic Press, San Diego: 665–689.

Cox, B.S., 1987. Thailand's Nam Chaon dam: a disaster in the making. *The Ecologist* **17**: 212–219.

Cranbrook, Earl (of) & Furtado, J.I., 1988. Freshwaters. In: Cranbrook, Earl (of) (Ed.), *Key Environments: Malaysia*. IUCN & Pergamon Press, Oxford: 225–250.

Cranston, P., 1995a. *Chironomids: From Genes to Ecosystems*. CSIRO Publications, Melbourne: 482 pp.

Cranston, P., 1995b. Biogeography. In: Armitage, P., Cranston, P.S. & Pinder, L.C.V., (Eds), *The Chironomidae: Biology and Ecology of Non-biting Midges*. Chapman & Hall, London: 62–84.

Cranston, P., 1995c. Systematics. In: Armitage, P., Cranston, P.S. & Pinder, L.C.V., (Eds), *The Chironomidae: Biology and Ecology of Non-biting Midges*. Chapman & Hall, London: 31–61.

Crisman, T.L. & Streever, W.J., 1995. The legacy and future of tropical limnology. In: Timotius, K.H. & Göltenboth, F. (Eds), *Tropical Limnology Volume III. Tropical Rivers, Wetlands and Special Topics*. Satya Wacana University Press, Salatiga, Indonesia: 235–248.

Crosskey, R.W., 1990. *The Natural History of Blackflies*. John Wiley & Sons, Chichester: 711 pp.

Crow, B., 1985. The making and breaking of agreement on the Ganges. In: Lundqvist, J., Lohm U. & Falkenmark, M. (Eds), *Strategies for River Basin Management*. D. Reidel Publishing Co., Dordrecht: 255–264.

Cullen, P., 1990. Environmental monitoring and environmental management. *Environ. Monit. Assess.* **14**: 107–114.

Cummins, K.W., 1973. Trophic relations of aquatic insects. *Ann. Rev. Entomol.* **18**: 183–206.

Cummins, K.W., 1974. Structure and function of stream ecosystems. *BioScience* **24**: 631–641.

Cummins, K.W. & Klug, M.J., 1979. Feeding ecology of stream invertebrates. *Ann. Rev. Ecol. Syst.* **10**: 147–172.

Cushing, C.E., 1963. Filter-feeding insect distribution and planktonic food in the Montreal River. *Trans. Amer. Fish. Soc.* **92**: 216–219.

Cushing, C.E., Minshall, G.E. & Newbold, J.D., 1993. Transport dynamics of fine particulate organic matter in two Idaho streams. *Limnol. Oceanogr.* **38**: 1101–1115.

Dai, A.Y, 1983. Crayfish — a kind of fishery resource. *Chinese J. Zool.* **3**: 48–50 (in Chinese).

Dai, A.Y., 1990. Zoogeographical analysis of the freshwater crab from Hubei Province, with descriptions of new subspecies (Malacostraca: Decapoda). *Acta Zootaxonomica Sinica* **15**: 417–426 (in Chinese).

Dai, A.Y. & Chen, G.X., 1979. On the freshwater crabs from Fujian Province. *Acta Zoologica Sinica* **25**: 243–249 (in Chinese).

Dai, A.Y. & Chen, G.X., 1985. A preliminary report of the freshwater crabs of Hengduan Mountains area. *Sinozoologia* **3**: 39–72 (in Chinese).

Dai, A.Y. & Chen, G.X., 1987. A study on the genus *Nanhaipotamon*

(Decapoda: Isolapotamidae). *Acta Zootaxonomica Sinica* **12**: 31–35 (in Chinese).

Dai, A.Y. & Chen, G.X., 1990. A study of freshwater crabs of Sichuan Province. *Acta Zootaxonomica Sinica* **15**: 282–297 (in Chinese).

Dai, A.Y. & Naiyanetr, P., 1994. A revision of the genus *Tiwaripotamon* Bott, 1970. The freshwater crabs from China (Decapoda: Brachyura: Potamidae). *Sinozoologia* **11**: 47–72 (in Chinese).

Dai, A.Y. & Song, Y.C., 1982. New species of *Malayopotamon* from Guangxi. *Acta Zootaxonomica Sinica* **7**: 372–373 (in Chinese).

Dai, A.Y. & Yuan, S.L., 1988. A study of freshwater crabs from Chisui County of Guizhou province. *Acta Zootaxonomica Sinica* **13**: 127–130 (in Chinese).

Dai, A.Y., Chen, G.X., Song, Y.Z., Fan, F.P., Lin, Y.g. & Zeng, Y.Q., 1979. On new species of freshwater crabs harbouring metacercariae of lung flukes. *Acta Zootaxonomica Sinica* **4**: 122–131 (in Chinese).

Dai, A.Y., Chen, G.X., Zhang, S.Q. & Lin, H., 1986. A study of the freshwater crabs from Hubei Province. *Sinozoologia* **4**: 55–71.

Dai, A.Y., Song, Y.Z., He, L.Y., Cao, W.J., Xu, Z.B. & Zhong, w.L., 1975. Description of several new species of freshwater crabs belonging to the intermediate hosts of lung flukes. *Acta Zoloogica Sinica* **21**: 257–264 (in Chinese).

Dai, A.Y., Song, Y.C., Li, M.G., Chen, Z.Y., Wang, P.P. & Hu, Q.X. 1984. A study of freshwater crabs from Guizhou Province 1. *Acta Zootaxonomica Sinica* **9**: 257–267 (in Chinese).

Dai, A.Y., Song, Y.C., Li, M.G., Chen, Z.Y., Wang, P.P. & Hu, Q.X. 1985. A study of freshwater crabs from Guizhou Province 2. *Acta Zootaxonomica Sinica* **10**: 34–44 (in Chinese).

Dai, A.Y., Song, Y.C., Li, L.L. & Liang, P.X., 1980. New species and new records of freshwater crabs from Guangxi. *Acta Zootaxonomica Sinica* **5**: 369–376 (in Chinese).

Dai, A.Y., Zhou, X.M. & Peng, W.D., 1995. On seven new species of freshwater crabs of the genus *Huananpotamon* Dai & Ng, 1994 (Crustacea: Decapoda: Brachyura: Potamidae) from Jiangxi Province, southern China. *Raffles Bull. Zool.* **43**: 417–433.

Datta, T. & Chaudhuri, P.K., 1995. First record of *Kribiocosmus* Kieffer from the Oriental Region, with description of *K. tumulus* spec. nov. from India (Insecta: Diptera: Chironomidae). *Reichenbachia* **31**: 93–96.

Datta, T., Majumdar, A. & Chaudhuri, P.K., 1992a. *Cladotanytarsus* Kieffer (Diptera: Chironomidae) from India with description of three new species. *Rev. Bras. Entomol.* **36**: 671–684.

Datta, T., Majumdar, A. & Chaudhuri, P.K., 1992b. Indian species of genus *Paratanytarsus* Thienemann and Bause (Diptera: Chironomidae). *Aquatic Insects* **14**: 129–135.

Datta, T., Majumdar, A. & Chaudhuri, P.K., 1992c. Taxonomic studies on *Tanytarsus* V.D. Wulp (Diptera: Chironomidae) in India. *Oriental Insects* **26**: 39–66.

Datta, T., Ali, A., Majumdar, A. & Chaudhuri, P.K., 1996a. Chironomid midges of *Harnischia* complex (Diptera: Chironomidae) from the Duars of the Himalayas, India. *Eur. J. Entomol.* **93**: 263–279.

Datta, T., Majumdar, A. & Chaudhuri, P.K., 1996b. Two new species of *Paratendipes* Kieffer (Diptera: Chironomidae) from the Duars of the Himalayas of West Bengal, India. *Entomon* 21: 49–54.

Davies, D.M. & Györkös, H., 1992. The Simuliidae (Diptera) of Sri Lanka. Descriptions of additional species of *Simulium (Simulium)*, with a key for Sri Lankan species in the genus and a checklist for the country. *Can. J. Zool.* **70**: 1029–1046.

Dang, Ngoc-Thanh, 1967. Nouveaux genres, nouvelles espèces de la faune des invertébrés des eaux douces et saumâtres du Nord Vietnam. *Tap san Sinh Vat — Dia Hoc* **6**: 155–165 (in Vietnamese).

Dang, Ngoc-Thanh & Tran, Ngoc-Lan, 1992. Two new species of freshwater crab from Vietnam. *Tap Chi Sinh Hoc* **14**: 17–21 (in Vietnamese; not seen by author).

Dang, L.V., 1970. Contribution to a biological study of the lower Mekong. *Regional Meeting of Inland Water Biologists in Southeast Asia. Proceedings.* Unesco Field Science Office for Southeast Asia, Djakarta: 65–90.

Das, D.C., Bali, Y.P. & Kaul, R.N., 1981. Soil conservation in multi-purpose river valley catchments. Problems, Programme approach and effectiveness. *Indian J. Soil Conservation* **9**: 5–26.

Das, D.N., Mitra, K., Mukhopadhyay, P.K. & Chaudhuri, D.K., 1994. Periphyton of the deepwater rice field at Pearapur village, Hoogly, West Bengal, India. *Environment & Ecology* **12**: 551–556.

Das, S.M., 1991. Water conservation, development, hill erosion and Himalayan mini dams. In: Bhatt, S.D. & Pande, R.K. (Eds), *Ecology of the Mountain Waters*. Ashish Publishing House, New Delhi: 383–385.

Dash, M.C., Mishra, P.C., Choudhury, K. & Das, R.C., 1988. Bioaccumulation of mercury in River Ib. In: Trivedy, R.K. (Ed.), *Ecology and Pollution of Indian Rivers*. Ashish Publishing House, New Delhi: 287–292.

Datta, M., 1983. A review of the Diptera (Simuliidae) from the Oriental Region. *Oriental Insects* **17**: 215–267.

Datta Munshi, J.S. & Srivastava, M.P., 1988. *Natural History of Fishes and Systematics of Freshwater Fishes in India*. Narendra Publishing, Delhi, India: 421 pp.

Datta Munshi, J.S., Singh, O.N. & Singh, D.K., 1989. Ecology of freshwater polychaetes of River Ganga. *J. Freshwat. Biol.* **1**: 103–108.

Datta Munshi, J.S., Singh, O.N. & Singh, D.K., 1990. Food and feeding relationships of certain aquatic animals in the Ganga ecosystem. *Trop. Ecol.* **31**: 138–144.

Davies, B.R. & Walker, K.F., 1986. River systems as ecological units. An introduction to the ecology of river systems. In: Davies, B.R. & Walker, K.F. (Eds), *The Ecology of River Systems*. Dr W. Junk Publishers, The Hague: 1–8.

Davies, B.R., Thoms, M. & Meador, 1992. An assessment of the ecological impacts of inter-basin water transfers, and their threats to river basin integrity and conservation. *Aquatic Conservation* 2: 325–349.

Davies, R.W., 1991. Annelida: leeches, polychaetes, and acanthobdellids. In: Thorp, J.H. & Covich, A.P. (Eds), *Ecology and Classification of North American Freshwater Invertebrates.* Academic Press, San Diego: 437–479.

Davis, G.M., 1971. Systematic studies of *Brotia costula episcopalis*, first intermediate host of *Paragonimus westermani* in Malaysia. *Proc. Acad. Natur. Sci. Philad.* 123: 51–86.

Davis, G.M., 1979. The origin and evolution of the gastropod family Pomatiopsidae with emphasis on the Mekong River Triculinae. *Monogr. Acad. Natur. Sci. Philad.* 20: 1–120.

Davis, G.M., Chen, C.E., Wu, C., Kuang, T.F., Xing, X.G., Li, L., Liu, W.J. & Yan, Y.L., 1992. The Pomatiopsidae of Hunan, China (Gastropoda: Rissoacea). *Malacologia* 34: 143–342.

Davis, G.M., Chen, C.E., Zeng, X.P., Yu, S.H. & Li, L., 1994. Molecular, genetic and anatomical relationships among pomatiopsid (Gastropoda: Prosobranchia) genera from southern China. *Proc. Acad. Natur. Sci. Philad.* 145: 191–207.

Davis, G.M., Guo, Y.H., Hoagland, K.E., Chen, P.L., Zheng, L.C., Yang, H., Chen, D.J & Zhou, Z.F., 1986. Anatomy and systematics of triculine snails (Prosobranchia: Trichulinae) freshwater snails from Yunnan, China, with descriptions of new species. *Proc Acad. Natur. Sci. Philad.* 138: 466–575.

Davis, G.M., Kitikoon, V. & Temcharoen, P., 1976. Monograph on 'Lithogyphopsis' aperta, the snail host of Mekong River schistosomiasis. *Malacologia* 15: 241–287.

Dayakar, Y. & Ramana Rao, K.V., 1992. Breeding periodicity of the paddy field crab *Oziotelphusa senex senex* (Fabricius) (Decapoda: Brachyura), a field study. *J. Crustacean Biol.* 12: 655–660.

de Beauchamp, P., 1973. Results of the Austrian-Ceylonese hydrological mission, 1970. Part X: freshwater Triclads (Turbellaria, Tricladida) from Ceylon. *Bull. Fish. Res. Stn, Sri Lanka (Ceylon)* 24: 89–93.

de Bont, A.F. & Kleijn, L.J.K., 1984. Limnological evolution of Lake Nam Ngum (Lao P.D.R.). *Verh. Internat. Verein. Limnol.* 22: 1562–1566.

de Jalón, D.G., 1977. The larva of *Larcasia partita* Navas (Trichoptera). *Annl Limnol.* 13: 221–226.

de Rosayro, R.A., 1974. Vegetation of humid tropical Asia. In: *Natural Resources of Humid Tropical Asia.* Natural Resources Research, XII, Unesco, Paris: 179–195.

de Silva, K.H.G.M., 1982. Aspects of the ecology and conservation of Sri Lanka's endemic freshwater shrimp *Caridina singhalensis*. *Biological Conservation* 24: 219–231.

de Silva, K.H.G.M., 1988. Studies on Atyidae (Crustacea: Decapoda: Caridea) of Sri Lanka. 5. pH tolerance of three species of *Caridina*. *Ceylon Journal of Science, Biological Sciences* **19/20**: 18–24.

de Silva, K.H.G.M., 1989. Temperature tolerances and geographic distribution of three species of *Caridina* (Decapoda, Atyidae) in Sri Lanka. *Int. Rev. ges. Hydrobiol.* **74**: 95–107.

de Silva, K.H.G.M., 1991a. Population ecology of the paddy field-dwelling fish *Channa gachua* (Günther) (Perciformes, Channidae) in Sri Lanka. *J. Fish Biol.* **38**: 497–508.

de Silva, K.H.G.M., 1991b. Population dynamics and production of the rocky stream-dwelling fish *Garra ceylonensis* (Cyprinidae) in Sri Lanka. *J. Trop. Ecol.* **7**: 289–303.

de Silva, K.H.G.M. 1994. Diversity, endemism and conservation of three major freshwater groups in Sri Lanka: Atyidae (Decapoda), Gastropoda and Teleostei. *Mitt. Internat. Verein. Limnol.* **24**: 63–71.

de Silva, K.H.G.M. & de Silva, P.K. 1994. The effects of human modification of lotic habitats on the freshwater fauna of Sri Lanka. *Mitt. Internat. Verein. Limnol.* **24**: 87–94.

de Silva, P.K. & de Silva, M., 1980. An ecological study of *Dugesia nannophallus* (Ball) (Turbellaria, Tricladida) in Sri Lanka. *Arch. Hydrobiol.* **88**: 363–377.

de Silva, P.K. & de Silva, K.H.G.M., 1988. Temperature, salinity, and pH tolerance, and the influence of other ecological factors in the geographic isolation of a freshwater atyid shrimp (Decapoda, Caridea) in Sri Lanka. *Arch. Hydrobiol.* **111**: 435–448.

de Silva, S.S. & Kortmulder, K., 1977. Some aspects of the biology of three species of *Puntius* (= *Barbus*) (Pisces, Cyprinidae), endemic to Sri Lanka. *Neth. J. Zool.* **27**: 182–194.

de Silva, S.S., Schut, J. & Kortmulder, K., 1985. Reproductive biology of six *Barbus* species indigenous to Sri Lanka. *Environ. Biol. Fish.* **12**: 201–218.

De Sousa, S.N., Sen Gupta, R., Sanzgiri, S. & Rajagopal, M.D., 1981. Studies on nutrients of Mandovi and Zuari River systems. *Indian J. Mar. Sci.* **10**: 314–321.

Dean, J.C., 1984. Immature stages of *Baliomorpha pulchripenne* (Tillyard) from Australia, with comments on generic placement (Trichoptera: Hydropsychidae). *Proc. R. Soc. Victoria* **96**: 141–145.

Degens, E.T., Kempe, S. & Richey, J.E., 1991. Summary. In: Degens, E.T., Kempe S. & Richey, J.E. (Editors), *Biogeochemistry of Major World Rivers*. SCOPE/John Wiley & Sons, Chichester: 323–347.

Delagado, J.A. & Soler, A.G., 1997. Morphology and chaetotaxy of larval Hydraenidae (Coleoptera) I: genus *Limnebius* Leach, 1815 based on a description of *Limnebius cordobanus* d'Orchymont. *Aquatic Insects* **19**: 37–49.

Delève, J., 1968. Dryopidae et Elminthidae (Coleoptera) du Vietnam. *Annales Historico-Naturales Musei Nationalis Hungarici* **60**: 149–181.

Delève, J., 1970. Contribution a l'étude des Dryopoidea XXI. Elminthidae (Coleoptera) peu ou mal connus de l'Indonésie et du Vietnam. *Bulletin et Annales de la Société royale belge de l'Entomologie* **106**: 235–273.

Delève, J., 1973. Coleoptera: Dryopidae et Elminthidae. *Ent. Scand., Suppl.* **4**: 5–24.

Delfinado, M.D. & Hardy, D.E., 1973. *A Catalog of the Diptera of the Oriental Region. Volume 1. Suborder Nematocera.* The University Press of Hawaii, Honolulu: 618 pp.

Delfinado, M.D. & Hardy, D.E., 1975. *A Catalog of the Diptera of the Oriental Region. Volume 2. Suborder Brachycera.* The University Press of Hawaii, Honolulu: 459 pp.

Delfinado, M.D. & Hardy, D.E., 1977. *A Catalog of the Diptera of the Oriental Region. Volume 3. Suborder Cyclorrhapha.* The University Press of Hawaii, Honolulu: 854 pp.

Denning, D.G., 1982. A review of the Goeridae (Trichoptera). *Can. Ent.* **114**: 637–642.

Desai, V.R., 1970. Studies on the fishery and biology of *Tor tor* (Hamilton) from River Narbada. 1. Food and feeding habits. *J. Inland Fish. Soc., India* **2**: 101–112.

Diamond, J. & Case, T.J., 1986. Overview: introductions, extinctions, exterminations, and invasions. In: Diamond, J. & Case, T.J. (Eds), *Community Ecology.* Harper & Row, New York: 65–79.

Diaz, H. & Rodriguez, G., 1977. The branchial chamber of some terrestrial and semiterrestrial crabs. *Biol. Bull.* **153**: 485–504.

Diemont, W.H., Smiet, A.C. & Nurdin, 1991. Re-thinking erosion on Java. *Neth. J. Agric. Sci.* **39**:213–224.

Dikkenberg, J., 1992. Can the pink dolphin be saved? *Asia Magazine* **18 September 1992**: 9–15.

Disney, R.H.L., 1991. The aquatic Phoridae (Diptera). *Ent. Scand.* **22**: 171–191.

Disney, R.H.L., 1994. *Scuttle Flies: the Phoridae.* Chapman & Hall, London: 467 pp.

Djajasasmita, M., 1982. The occurrence of *Anodonata woodiana* Lea, 1937 in Indonesia (Pelecypoda: Unionidae). *Veliger* **25**: 175.

Djajasasmita, M. & Budiman, A., 1984. Population density of *Batissa violacea* (Lamarck, 1818) in the Pisang Tiver, Lampung, Sumatra (Mollusca, Bivalvia: Corbiculidae). *Treubia* **29**: 179–183.

Djuangsih, N., 1993. Understanding the state of river basin management from an environmental toxicology perspective: an example from water pollution at Citarum River Basin, West Java, Indonesia. *Science of the Total Environment, Suppl.* **1**: 283–292.

Dobriyal, A.K., 1985. Ecology of limnofauna in the small steams and their importance to the village life in Garhwal Himalaya, India. *Uttar Pradesh J. Zool.* **5**: 139–144.

Dobriyal, A.K. & Kumar, N., 1988. Fish and fisheries of the River Mandakini.

In: Khulbe, R.D. (Ed.), *Perspectives in Aquatic Biology*. Papyrus Publishing House, New Delhi: 337–340.

Dobriyal, A.K. & Singh, H.R., 1987. The reproductive biology of a hillstream minor carp *Barilius bendelisis* (Ham.) from Garhwal Himalaya, India. *Vest. Cs. Spolec. Zool.* **51**: 1–10.

Dobriyal, A.K. & Singh, H.R., 1989. Ecology of rithrofauna in the torrential waters of Garhwal Himalaya, India: fecudity and sex ration of *Glyptothorax pectinopterus* (Pisces). *Vestn Cesk. Spol. Zool.* **53**: 17–25.

Dobriyal, A.K. & Singh, H.R., 1990. Ecological studies on the age and growth of *Barilius bendelisis* (Ham.) from India. *Arch. Hydrobiol.* **118**: 93–103.

Dobriyal, A.K., Bahuguna, A.K., Kumar, N. & Kotnala, C.B., 1993. Ecology & seasonal diversity of plankton in a spring-fed stream Khandagad of Garhwal Himalaya. In: Singh, H.R. (Ed.), *Advances in Limnology*. Narendra Publishing House, Delhi: 175–180.

Dobriyal, A.K., Kumar, N., Kotnala, C.B., Bahuguna, A.K. & Singh, H.R., 1992. Preliminary observations on seasonal trends in macro-zoobenthic diversity in the River Nayar of Garhwal Himalaya. In: Sehgal, K.L. (Ed.), *Recent Researches in Coldwater Fisheries*. Today & Tomorrow's Printers & Publishers, New Delhi: 119–127.

Dodd, C.K. Jr & Seigel, R.A., 1991. Relocation, repatriation and translocation of amphibians and reptiles: are they conservation strategies that work? *Herpetologica* **47**: 336–350.

Dolin, P.S. & Tarter, D.C., 1981. Life history and ecology of *Chauliodes rastricornis* Rambur and *C. pectinicornis* (Linnaeus) (Megaloptera: Corydalidae) in Greenbottom Swamp, West Virginia. *Brimleyana* **7**: 111–120.

Domrös, M. & Peng, K., 1988. *The Climate of China*. Springer Verlag, Berlin: 360 pp.

Donnelly, T.W., 1993. Two new genera of isostictid damselflies from New Britain, Bougainville, and the Solomon Islands (Odonata: Zygoptera). *Tijdschrift voor Entomologie* **136**: 125–132.

Dou, Y., Liu, Y., & Lu, G., 1987. Distribution of heavy metals in the upper Changjiang (Jinsha River, Dukou reach). *Mitt. Geol.-Palaont. Inst. Univ. Hamburg, SCOPE/UNEP Sonderbd.* **64**: 233–241.

Douglas, I., 1970. Measurements of river erosion in Malaysia. *Malay. Nat. J.* **23**: 78–83.

Douglas, I., 1968. Erosion in the Sungai Gombak catchment, Selangor, Malaysia. *J. Trop. Geog.* **26**: 1–16.

Douglas, I., 1969. The efficiency of humid tropical denudation systems. *Trans. Br. Geogr. Inst.* **46**: 1–16.

Downes, B.J. & Jordan, J., 1993. Effects of stone topography on abundance of net-building caddisfly larvae and arthropod diversity in an upland stream. *Hydrobiologia* **252**: 163–174.

Drecktrah, H.G., 1990. Larval and pupal descriptions of *Marilia fusca*

(Trichoptera: Odontoceridae). *Ent. News* **101**: 1–8.

Dudgeon, D., 1980. A comparative study of the Corbiculidae of southern China. In: Morton, B. (Ed.), *Proceedings, First International Malacological Workshop on the Malacofauna of Hong Kong and Southern China*. Hong Kong University Press, Hong Kong: 37–60.

Dudgeon, D., 1982a. Aspects of the hydrology of Tai Po Kau Forest Stream, New Territories, Hong Kong. *Arch. Hydrobiol., Suppl.* **64**: 1–35.

Dudgeon, D., 1982b. Spatial and temporal changes in the sediment characteristics of Tai Po Kau Forest Stream, New Territories, Hong Kong, with some preliminary observations upon within-reach variations in current velocity. *Arch. Hydrobiol., Suppl.* **64**: 36–64.

Dudgeon, D., 1982c. Spatial and seasonal variations in the standing crop of periphyton and allochthonous detritus in a forest stream in Hong Kong, with notes on the magnitude and fate of riparian leaf fall. *Arch. Hydrobiol., Suppl.* **64**: 189–220.

Dudgeon, D., 1982d. Aspects of the microdistribution of insect macrobenthos in a forest stream in Hong Kong. *Arch. Hydrobiol., Suppl.* **64**: 221–239.

Dudgeon, D., 1982e. An investigation of physical and biotic processing of two species of leaf litter in Tai Po Kau Forest Stream, New Territories, Hong Kong. *Arch. Hydrobiol.* **96**: 1–32.

Dudgeon, D., 1982f. The life history of *Brotia hainanensis* (Brot 1872) (Gastropoda: Prosobranchia: Thiaridae) in a tropical forest stream. *Zool. J. Linn. Soc.* **76**: 141–154.

Dudgeon, D., 1983a. Preliminary measurements of primary production and community respiration in a forest stream in Hong Kong. *Arch. Hydrobiol.* **98**: 287–298.

Dudgeon, D., 1983b. Spatial and temporal changes in the distribution of gastropods in the Lam Tsuen River, New Territories, Hong Kong, with notes on the occurrence of the exotic snail *Biomphalaria straminea*. *Malacol. Rev.* **16**: 91–92.

Dudgeon, D., 1983c. An investigation of the drift of aquatic insects in Tai Po Kau Forest Stream, New Territories, Hong Kong. *Arch. Hydrobiol.* **96**: 434–447.

Dudgeon, D., 1983d. The utilization of terrestrial plants as a food source by the fish stock of a gently sloping marginal zone of Plover Cove Reservoir, Hong Kong. *Environ. Biol. Fish.* **8**: 73–77.

Dudgeon, D., 1984a. Seasonal and long-term changes in the hydrobiology of the Lam Tsuen River, New Territories, Hong Kong, with special reference to benthic macroinvertebrate distribution and abundance. *Arch. Hydrobiol., Suppl.* **69**: 55–129.

Dudgeon, D., 1984b. Longitudinal and temporal changes in functional organization of macroinvertebrate communities in the Lam Tsuen River, Hong Kong. *Hydrobiologia* **111**: 207–217.

Dudgeon, D., 1985a. The population dynamics of some freshwater carideans (Crustacea: Decapoda) in Hong Kong, with special reference to

Neocaridina serrata (Atyidae). *Hydrobiologia* **120**: 141–149.

Dudgeon, D., 1985b. *Anodonta woodiana* (Bivalvia: Unionacea): the egg repository of *Rhodeus sinensis* (Pisces: Cyprinidae). *Malacol. Rev.* **18**: 110.

Dudgeon, D., 1986. The life cycle, population dynamics and productivity of *Melanoides tuberculata* (Müller, 1774) (Gastropoda: Prosobranchia: Thiaridae) in Hong Kong. *J. Zool., Lond.* **208**: 37–53.

Dudgeon, D., 1987a. Preliminary investigations on the faunistics and ecology of Hong Kong Trichoptera. In: Bournaud, M. & Tachet, H. (Eds), *Proceedings of the 5th International Symposium on Trichoptera*. Dr W. Junk Publishers, The Hague: 111–117.

Dudgeon, D., 1987b. The ecology of a forest stream in Hong Kong. *Arch. Hydrobiol. Beih., Ergebn. Limnol.* **28**: 449–454.

Dudgeon, D., 1987c. The larval development of *Neocaridina serrata* (Stimpson) (Crustacea: Decapoda: Atyidae) from Hong Kong. *Arch. Hydrobiol.* **110**: 339–355.

Dudgeon, D., 1987d. Niche specificities of four fish species (Homalopteridae, Cobitidae, Gobiidae) from a Hong Kong forest stream. *Arch. Hydrobiol.* **108**: 349–364.

Dudgeon, D. 1988a. The influence of riparian vegetation on macroinvertebrate community structure in four Hong Kong streams. *J. Zool., Lond.* **216**: 609–627.

Dudgeon, D., 1988b. Hong Kong freshwaters: seasonal influences on benthic communities. *Verh. Internat. Verein. Limnol.* **23**: 1362–1366.

Dudgeon, D., 1988c. Flight periods of aquatic insects from a Hong Kong forest stream I. Macronematinae (Hydropsychidae) and Stenopsychidae (Trichoptera). *Aquatic Insects* **10**: 61–68.

Dudgeon, D., 1989a. The influence of riparian vegetation on the functional organization of four Hong Kong stream communities. *Hydrobiologia* **179**: 183–194.

Dudgeon, D., 1989b. Life cycle, production, microdistribution and diet of the damselfly *Euphaea decorata* (Odonata: Euphaeidae) in a Hong Kong forest stream. *J. Zool., Lond.* **217**: 57–72.

Dudgeon, D., 1989c. Resource partitioning among Odonata (Insecta: Anisoptera and Zygoptera) larvae in a Hong Kong forest stream. *J. Zool., Lond.* **217**: 381–402.

Dudgeon, D., 1989d. Gomphid (Odonata: Anisoptera) life cycles and production in a Hong Kong forest stream. *Arch. Hydrobiol.* **114**: 531–536.

Dudgeon, D., 1989e. Ecological strategies of Hong Kong Thiaridae (Gastropoda: Prosobranchia). *Malacol. Rev.* **22**: 39–53.

Dudgeon, D., 1989f. Phoretic Diptera (Nematocera) on *Zygonyx iris* (Odonata: Anisoptera) from a Hong Kong river: incidence, composition and attachment sites. *Arch. Hydrobiol.* **115**: 433–439.

Dudgeon, D., 1990a. Benthic community structure and the effect of rotenone piscicide on invertebrate drift and standing stocks in two Papua New Guinea streams. *Arch. Hydrobiol.* **119**: 35–53.

Dudgeon, D., 1990b. Determinants of the distribution and abundance of larval Ephemeroptera (Insecta) in Hong Kong running waters. In: Campbell, I.C. (Ed.), *Mayflies and Stoneflies: Life Histories and Biology*. Kluwer Academic Publishers, Dordrecht: 221–232.

Dudgeon, D., 1990c. Seasonal dynamics of invertebrate drift in a Hong Kong stream. *J. Zool., Lond.* **222**: 187–196.

Dudgeon, D., 1990d. Feeding by the aquatic heteropteran *Diplonychus rusticum*: an effect of prey density on meal size. *Hydrobiologia* **190**: 93–96.

Dudgeon, D., 1991a. An experimental study of the effects of predatory fish on macroinvertebrates in a Hong Kong stream. *Freshwat. Biol.* **25**: 321–330.

Dudgeon, D., 1991b. An experimental study of abiotic disturbance effects on community structure and function in a tropical stream. *Arch. Hydrobiol.* **122**: 403–420.

Dudgeon, D., 1992a. *Patterns and Processes in Stream Ecology: A Synoptic Review of Hong Kong Running Waters*. Schweizerbart'sche Verlagsbuchhandlung, Stuttgart: 147 pp.

Dudgeon, D., 1992b. Endangered ecosystems: a review of the conservation status of tropical Asian rivers. *Hydrobiologia* **248**: 167–191.

Dudgeon, D., 1992c. Effects of water transfer on aquatic insects in a stream in Hong Kong. *Regulated Rivers: Research & Management* **7**: 369–377.

Dudgeon, D., 1993. The effects of spate-induced disturbance, predation and environmental complexity on macroinvertebrates in a tropical stream. *Freshwat. Biol.* **30**: 189–197.

Dudgeon, D., 1994a. Functional assessment of the effects of increased sediment loads resulting from riparian-zone modification in a Hong Kong stream. *Verh. Internat. Verein. Limnol.* **25**: 1790–1792.

Dudgeon, D., 1994b. The influence of riparian vegetation on macroinvertebrate community structure and functional organization in six New Guinea streams. *Hydrobiologia* **294**: 65–85.

Dudgeon, D., 1994c. The need for multi-scale approaches to the conservation and management of tropical inland waters. *Mitt. Internat. Verein. Limnol.* **24**: 11–16.

Dudgeon, D., 1994d. The functional significance of selection of particles by aquatic animals during building behaviour. In: Wotton, R.S. (Ed.), *The Biology of Particles in Aquatic Systems, Second Edition*. Lewis Publishers, Boca Raton, Ann Arbor, London & Tokyo: 289–312.

Dudgeon, D., 1995a. The ecology of rivers and streams in tropical Asia. In: Cushing, C.E., Cummins, K.W. & Minshall, G.E. (Eds), *Ecosystems of the World 22: River and Stream Ecosystems*. Elsevier, Amsterdam: 615–657.

Dudgeon, D., 1995b. Environmental impacts of increased sediment loads caused by channelization: a case study of biomonitoring in a small Hong

Kong river. *Asian J. Environ. Manag.* **3**: 69–77.

Dudgeon, D., 1995c. Life histories, secondary production and microdistribution of Psephenidae (Coleoptera: Insecta) from a tropical forest stream. *J. Zool., Lond.* **236**: 465–481.

Dudgeon, D., 1995d. Life history, secondary production and microdistrbution of *Hydrocyphon* (Coleoptera: Helodidae) in a tropical forest stream. *Arch. Hydrobiol.* **133**: 261–271.

Dudgeon, D., 1995e. River regulation in southern China: ecological implications, conservation and environmental management. *Regulated Rivers: Research & Management* **11**: 35–54.

Dudgeon, D., 1996a. The life history, secondary production and microdistribution of *Ephemera* spp. (Ephemeroptera: Ephemeridae) in a tropical forest stream. *Arch. Hydrobiol.*: **135**: 473–483.

Dudgeon, D., 1996b. Life history, secondary production and microdistribution of *Stenopsyche angustata* (Trichoptera: Stenopsychidae) in a tropical forest stream. *J. Zool., Lond.* **238**: 679–691.

Dudgeon, D., 1996c. Life histories, secondary production and microdistrbution of heptageniid mayflies (Ephemeroptera) in a tropical forest stream. *J. Zool., Lond.* **240**: 341–361.

Dudgeon, D., 1997. Life histories, secondary production and microdistrbution of hydropsychid caddisflies (Trichoptera) in a tropical forest stream. *J. Zool., Lond.*: in press.

Dudgeon, D. & Bretschko, G., 1995a. Land-water interactions and stream ecology: comparison of tropical Asia and temperate Europe. In: Timotius, K.H. & Göltenboth, F. (Eds), *Tropical Limnology Volume I. Present Status and Challenges*. Satya Wacana University Press, Salatiga, Indonesia: 69–108.

Dudgeon, D. & Bretschko, G., 1995b. Allochthonous inputs and land-water interactions in seasonal streams: tropical Asia and temperate Europe. In: Schiemer, F. (Ed.), *Tropical Limnology, Past and Present*. SPB Academic Publishing, The Hague: in press.

Dudgeon, D. & Chan, I.K.K., 1993. An experimental study of the influence of periphytic algae on invertebrate abundance in a Hong Kong stream. *Freshwat. Biol.* **27**: 53–63.

Dudgeon, D. & Cheung, C.P.S., 1990. Selection of gastropod prey by a tropical freshwater crab. *J. Zool., Lond.* **220**: 147–155.

Dudgeon, D. & Corlett, R.T., 1994. *Hills and Streams: an Ecology of Hong Kong*. Hong Kong University Press, Hong Kong: 234 pp.

Dudgeon, D. & Lam, P.K.S., 1985. Freshwater gastropod foraging strategies: interspecific comparisons. In: Morton, B. & Dudgeon, D. (Eds), *Proceedings of the Second International Workshop on the Malacofauna of Hong Kong and Southern China, Hong Kong, 1983*. Hong Kong University Press, Hong Kong: 591–600.

Dudgeon, D. & Lam, P.K.S., 1994. *Inland Waters of Tropical Asia and Australia: Conservation and Management*. Schweizerbart'sche

Verlagsbuchhandlung, Stuttgart: 386 pp.

Dudgeon, D. & Morton, B., 1983. The population dynamics and sexual strategy of *Anodonta woodiana* (Bivalvia: Unionacea) in Plover Cove Reservoir, Hong Kong. *J. Zool., Lond.* **201**: 161–183.

Dudgeon, D. & Morton, B., 1984. Site selection and attachment duration of *Anodonta woodiana* (Bivalvia: Unionacea) glochidia on fish hosts. *J. Zool., Lond.* **204**: 355–362.

Dudgeon, D. & Wat, C.Y.M., 1986. Life cycle and diet of *Zygonyx iris insignis* (Insecta: Odonata: Anisoptera) in Hong Kong running waters. *J. Trop. Ecol.* **2**: 73–85.

Dudgeon, D. & Yipp, M.W., 1983. A report on the gastropod fauna of aquarium fish farms with special reference to an introduced human schistosome host species, *Biomphalaria straminea* (Pulmonata: Planorbidae). *Malac. Rev.* **16**: 93–94.

Dudgeon, D. & Yipp, M.W., 1985. The diets of Hong Kong freshwater gastropods. In: Morton, B. & Dudgeon, D. (Eds), *Proceedings of the Second International Workshop on the Malacofauna of Hong Kong and Southern China, Hong Kong, 1983*. Hong Kong University Press, Hong Kong: 491–509.

Dudgeon, D., Arthington, A.H., Chang, W.Y.B., Davies, J., Humphrey, C.L., Pearson, R.G. & Lam. P.K.S., 1994. Conservation and management of tropical Asian and Australian inland waters: problems, solutions and prospects. *Mitt. Int. Verein. Limnol.* **24**: 369–386.

Dudgeon, D., Choowaew, S. & Ho, S.C., in press. River conservation in Southeast Asia: conflicts, constraints and compromise. In: Boon, P.J., Petts, G.E. & Davies B.R., *Global Perspectives on River Conservation: Science, Policy and Practice*. John Wiley & Sons, Chichester.

Dugan, P.J., 1994a. Constraints and opportunities for training in wetland management. *Mitt. Internat. Verein. Limnol.* **24**: 365–368.

Dugan, P.J., 1994b. The role of ecological science in addressing wetland conservation and management in the tropics. *Mitt. Internat. Verein. Limnol.* **24**: 5–10.

Dumont, H.J., 1994. The distribution and ecology of the fresh- and brackish-water medusae of the world. *Hydrobiologia* **272**: 1–12.

Dumont, H.J., Haritonov, A.Y. & Borisov, S.N., 1992. Larval morphology and range of three West Asiatic species of the genus *Onychogomphus* Selys, 1854 (Insecta: Odonata). *Hydrobiologia* **245**: 169–177.

Dussart, B.H., 1974. Biology of inland waters in humid tropical Asia. In: *Natural Resources of Humid Tropical Asia*. Natural Resources Research, XII, Unesco, Paris: 331–353.

Dutta, S.P.S. & Malhotra, Y.R., 1986. Seasonal variations in the macrobenthic fauna of Gadigarh Stream (Miran Sahib) Jammu. *Indian J. Ecol.* **13**: 138–145.

Eberhardt, L.L. & Thomas, J.M., 1991. Designing environmental field studies. *Ecol. Monogr.* **61**: 53–73.

Edds, D.R., 1993. Fish assemblage structure and environmental correlates in Nepal's Gandaki River. *Copeia* **1993**: 48–60.

Edington, J.M., Edington, M.A. & Dorman, J.A., 1984. Habitat partitioning amongst hydropsychid larvae of a Malaysian stream. In: Morse, J.C. (Ed.), *Proceedings of the 4th International Symposium on Trichoptera.* Dr W. Junk Publishers, The Hague: 123–129.

Edmunds, G.F., 1959. Subgeneric groups within the mayfly genus *Ephemerella* (Ephemeroptera: Ephemerellidae). *Ann. Entomol. Soc. Am.* **52**: 543–547.

Edmunds, G.F. & Edmunds, C.H., 1980. Predation, climate, and emergence and mating of mayflies. In: Flannagan, J.F. & Marshall, K.E. (Eds), *Advances in Ephemeroptera Biology.* Plenum Press, New York: 277–285.

Edmunds, G.F. & Polhemus, D.A., 1990. Zoogeographical patterns among mayflies (Ephemeroptera) in the Malay Archipelago, with special reference to Celebes. In: Knight, W.J. & Holloway, J.D. (Eds), *Insects and the Rain Forests of South East Asia (Wallacea).* The Royal Entomological Society of London, London: 49–56.

Edmunds, G.F. & McCafferty, W.P., 1996. New field observations on burrowing in Ephemeroptera from around the world. *Entomological News* **107**: 68–76.

Edmunds, G.F., Jensen, S.L. & Berner, L., 1976. *The Mayflies of North and Central America.* University of Minnesota Press, Minneapolis: 330 pp.

Elliott, J.M., 1982. The life cycle and spatial distribution of the aquatic parasitiod *Agriotypus armatus* (Hymenoptera: Agriotypidae) and its caddis host *Silo pallipes* (Trichoptera: Goeridae). *J. Anim. Ecol.* **51**: 923–941.

Elliott, J.M., 1995. Egg hatching and ecological partitioning in carnivorous stoneflies (Plecoptera). *Comptes Rendus de l'Academie des Science Serie III Sciences de la Vie* **318**: 237–243.

Elpers, C. & Tomka, I., 1994. Structure of the mouthparts and feeding habits of *Potamanthus luteus* (Linné) (Ephemeroptera: Potamanthidae). *Arch Hydrobiol./Suppl.* **99**: 73–96.

Elson-Harris, M.M., 1990. Keys to the immature stages of some Australian Ceratopogonidae (Diptera). *J. Aust. ent. Soc.* **29**: 267–275.

Elton, C.S., 1958. *The Ecology of Invasions by Animals and Plants.* Chapman & Hall, London: 181 pp.

Emden, F. van, 1956. The *Georyssus* larva — a hydrophilid. *Proc. R. Ent. Soc. Lond. A* **31**: 20–24.

Enckell, P.H., 1970. Isopoda Asellota and Flabellifera from Ceylon. *Arkiv för Zoologi* **22**: 557–570.

Eriksen, C.H., Resh, V.H. & Lamberti, G.A., 1996. Aquatic insect Respiration. In: Merritt, R.W. & Cummins, K.W. (Eds), *An Introduction to the Aquatic Insects of North America. Third Edition.* Kendall/Hunt Publishing Co., Iowa: 29–73.

Erseus, C. & Qi, S., 1985. Two aberrant Tubificidae (Oligochaeta) from Pearl River in the People's Republic of China. *Hydrobiologia* **127**: 193–196.

ESCAP, 1992. *Towards and Environmentally Sound and Sustainable Development of Water Resources in Asia and the Pacific.* Water Resources Series No. 71, Economic & Social Commission for Asia & the Pacific, Bangkok, Thailand, and United Nations, New York: 214 pp.

Fairweather, P.G., 1991. Statistical power and design requirements for environmental monitoring. *Aust. J. Mar. Freshwat. Res.* **42**: 555–567.

Faith, D.P., Humphrey, C.L. & Dostine, P.L., 1991. Statistical power and BACI design in environmental monitoring: comparative evaluation of measures of community dissimilarity based on benthic macroinvertebrate communities in Rockhole Mine Creek, Northern Territory, Australia. *Aust. J. Mar. Freshwat. Res.* **42**: 589–602.

Faith, D.P., Dostine, P.L. & Humphrey, C.L., 1995. Detection of mining impacts on aquatic macroinvertebrate communities: results of a disturbance experiment and the design of a multivariate BACIP monitoring program at Coronation Hill, Northern Territory. *Aust. J. Ecol.* **20**: 167–180.

Falkenmark, M. & Suprapto, R.A., 1992. Population-landscape interactions in development — a water perspective to environmental sustainability. *Ambio* **21**: 31–36.

Falkenström, G., 1933. Schwedisch-chinesiche wissenschaftliche Expedition den nordwestlichen Provinzen Chinas. Coleoptera. 1. Haliplidae und Dytiscidae. *Arkiv för Zoologi* **27A**: 1–23.

Fang, J.Q., 1993. Lake evolution during the last 300 years in China and its implications for environmental change. *Quaternary Research* **39**: 175–185.

Fang, Z., 1985. Some strategic principles for long-term river basin development — the case of the Han River. In: Lundqvist, J., Lohm U. & Falkenmark, M. (Eds), *Strategies for River Basin Management.* D. Reidel Publishing Co., Dordrecht: 141–150.

Fang, Z., 1993. The flood prevention function of the Three Gorges Project — disadvantages outweigh advantages. In: Luk, S. & Whitney, J. (Eds), *Megaproject. A Case Study of China's Three Gorges Project.* M.E. Sharpe Inc., New York: 161–175.

Fang, Z. & Wang, S., 1993. Resettlement problem of the Three Gorges Project. In: Luk, S. & Whitney, J. (Eds), *Megaproject. A Case Study of China's Three Gorges Project.* M.E. Sharpe Inc., New York: 3–39.

Farrelly, C.A. & Greenaway, P., 1993. Land crabs with smooth lungs: Grapsidae, Gecarcinidae, Sundathelphusidae ultrastructure and vasculature. *J. Morphol.* **215**: 245–260.

Feng, H.T., 1932a. Aquatic insects of China. Article II. Catalogue of Chinese Dytiscidae. *Peking Nat. Hist. Bull.* **7**: 17–37.

Feng, H.T., 1932b. Aquatic insects of China. Article VIII. Additions and corrections to the catalogue of Chinese Dytiscidae. *Peking Nat. Hist. Bull.* **7**: 323–333.

Feng, H.T., 1933. Classification of Chinese Dytiscidae. *Peking Nat. Hist. Bull.* **7**: 323–333.

Fernando, C.H., 1960. The Ceylonese freshwater crabs (Potamonidae). *Ceylon J. Sci. (Biol. Sci.)* **3**: 191–222.

Fernando, C.H., 1964. A preliminary account of the water bugs of the family Corixidae in Ceylon. *J. Bombay Nat. Hist. Soc.* **61**: 603–613.

Fernando, C.H., 1969. A guide to the freshwater fauna of Sri Lanka. Supplement 3. *Bull. Fish. Res. Stn, Ceylon* **20**: 15–25.

Fernando, C.H., 1974. A guide to the freshwater fauna of Sri Lanka, Supplement 4. *Bull. Fish. Res. Stn, Sri Lanka (Ceylon)* **25**: 27–81.

Fernando, C.H., 1980. The freshwater invertebrates fauna of Sri Lanka. *Spolia Zeylanica* **35**: 15–42.

Fernando, C.H., 1980a. The freshwater zooplankton of Sri Lanka, with a discussion of tropical freshwater zooplankton composition. *Int. Rev. ges. Hydrobiol.* **65**: 85–125.

Fernando, C.H., 1980b. The species and size composition of tropical freshwater zooplankton with special reference to the Oriental Region (Southeast Asia). *Int. Rev. ges. Hydrobiol.* **65**: 411–426.

Fernando, C.H., 1984a. Reservoirs and lakes of Southeast Asia (Oriental Region). In: Taub, F.B. (Ed.), *Ecosystems of the World 23: Lakes and Reservoirs.* Elsevier, Amsterdam: 411–446.

Fernando, C. H., 1984b. Freshwater invertebrates: some comments. In: Fernando, C.H. (Ed.), *Ecology and Biogeography in Sri Lanaka.* Dr. W. Junk Publishers, The Hague: 145–148.

Fernando, C.H., 1990. *Freshwater Fauna and Fisheries of Sri Lanka.* Natural Resources, Energy & Science Authority, Colombo: 444 pp.

Fernando, C.H., 1991. Impacts of fish introductions in tropical Asia and America. *Can. J. Fish. Aquat. Sci.* **48**: 24–32.

Fink, T.J., Soldán, T., Peters, T.J. & Peters, W.L., 1991. The reproductive life history of the predacious, sand-burrowing mayfly *Dolania americana* (Ephmeroptera: Behningiidae) and comparisons with other mayflies. *Can. J. Zool.* **69**: 1083–1093.

Firdaus-e-Bareen & Iqbal, S.H., 1994a. Seasonal occurrence of freshwater hyphomycetes on submerged fallen leaves in canal waters. *Can. J. Bot.* **72**: 1316–1321.

Firdaus-e-Bareen & Iqbal, S.H., 1994b. Freshwater hyphomycetes of the Karakorum Range. *Annales Botanici Fennici* **31**: 147–154.

Fisher, S.G. & Likens, G.E., 1973. Energy flow in Bear Brook, New Hampshire: an integrative approach to stream ecosystem metabolism. *Ecol. Monogr.* **43**: 421–439.

Fisher, W.F., 1995. *Toward Sustainable Development?: Struggling over India's Narmada River.* M.E. Sharpe Inc., New York: 481pp.

Flecker, A.S., Allan, J.D. & McClintock, N.L., 1988. Male body size and mating success in swarms of the mayfly *Epeorus longimanus. Holarct. Ecol.* **11**: 280–285.

Flemming, W.M., 1983. Phewa Tal catchment management program: benefits and costs of forestry and soil conservation in Nepal. In: Hamilton, L.S.

(Ed.), *Forest and Watershed Development and Conservation in Asia and the Pacific*. Westview Press, Boulder, Colorado: 217–288.

Flint, O.S. & Bueno-Soria, J., 1982. Studies of Neotropical caddisflies XXXII: the immature stages of *Macronema variipenne* Flint & Bueno, with the division of Macronema by the resurrection of *Macrostemum* (Trichoptera: Hydropsychidae). *Proc. Biol. Soc. Wash.* **95**: 358–370.

Flowers, R.W. & Pescador, M.L., 1984. A new *Afronurus* (Ephemeroptera): Heptageniidae) from the Philippines. *Int. J. Entomol.* **26**: 362–365.

Foote, B.A., 1995. Biology of shore flies. *Ann. Rev. Entomol.* **40**: 417–442.

Fraser, F.C., 1933. *The Fauna of British India including Ceylon and Burma. Odonata. Vol. I.* Taylor & Francis, London: 432 pp.

Fraser, F.C., 1934. *The Fauna of British India including Ceylon and Burma. Odonata. Vol. II.* Taylor & Francis, London: 398 pp.

Fraser, F.C., 1936. *The Fauna of British India including Ceylon and Burma. Odonata. Vol. III.* Taylor & Francis, London: 461 pp.

Fraser, F.C., 1943. New Oriental odonate larvae. *Proc. R. Ent. Soc. Lond.* B **12**: 81–93.

Freeberne, M., 1993. The Three Gorges project and mass resettlement. *Water Resources Development* **9**: 337–351.

Freeland, W.J. & Bayliss, P., 1989. The Irrawaddy River dolphin (*Oracella brevirostris*) in coastal waters of the Northern Territory: distribution, abundance and seasonal changes. *Mammalia* **53**: 49–58.

Frost, T.M., 1991. Porifera. In: Thorp, J.H. & Covich, A.P. (Eds), *Ecology and Classification of North American Freshwater Invertebrates*. Academic Press, San Diego: 95–124.

Frost, T.M., DeAngelis, D.L., Bartell, S.M., Hall, D.J. & Hurlbert, S.H., 1988. Scale in the design and interpretation of aquatic community research. In: Carpenter, S.R. (Ed.), *Complex Interactions in Lake Communities*. Springer Verlag, New York: 229–260.

Frusher, S.D., 1983. The ecology of juvenile penaeid prawns, mangrove crab (*Scylla serrata*) and the giant freshwater prawn (*Macrobrachium rosenbergii*) in the Purari delta. In: Petr, T. (Ed.), *The Purari — Tropical Environment of a High Rainfall River Basin*. Dr W. Junk Publishers, The Hague: 341–353.

Fuller, E.R. & Wiggins, G.B., 1987. A new species of *Molannodes* McL. from Hokkaido, Japan (Trichoptera: Molannidae). *Aquatic Insects* **9**: 39–43.

Fuller, R.L. & Hynes, H.B.N., 1987. The life cycle, food habits and production of *Antocha saxicola* O.S. (Diptera: Tipulidae). *Aquatic Insects* **9**: 129–135.

Funk, D.H. & Sweeney, B.W., 1994. The larvae of eastern North American *Eurylophella* Tiensuu (Ephemeroptera: Ephemerellidae). *Trans. Am. Entomol. Soc.* **120**: 209–286.

Furtado, J.I., 1969. Ecology of Malaysian odonates: biotope and association of species. *Verh. Internat. Verein. Limnol.* **17**: 863–887.

Furtado, J.I. & Mori, S., 1982. *Tasek Bera — the Ecology of a Freshwater Swamp*. Dr W. Junk Publishers, The Hague: 413 pp.

Furuya, Y., 1980. On aquatic lepidopterous larva, *Cataclysta* sp. CA, collected in the Hijikawa River. *Biology of Inland Waters* **1**: 17–18 (in Japanese).

Gan, W. & Kempe, S., 1987. The Changjiang: its longterm changes in PCO_2 and carbonate mineral saturation. *Mitt. Geol.-Palaont. Inst. Univ. Hamburg, SCOPE/UNEP Sonderbd.* **64**: 207–215.

Gan, K. & McMahon, T., 1990. Variability of results from use of PHABSIM in estimating habitat area. *Regulated Rivers: Research & Management* **5**: 233-239.

Gangotri, M.S., Vasantha, N. & Venkatachari, S.A.T., 1978. Sexual maturity and breeding behaviour in the freshwater crab *Barytelphusa guerini* Milne Edwards. *Proc. Indian Acad. Sci., Sect. B* **87**: 195–201.

Gautam, A., 1990. *Ecology and Pollution of Mountain Waters (a Case Study of Bhagirathi River)*. Ashish Publishing House, New Delhi: 209 pp.

Geisler, R., Schmidt, G.W. & Sookvibul, S., 1979. Diversity and biomass of fishes in three typical streams of Thailand. *Int. Rev. ges. Hydrobiol.* **64**: 673–697.

Gelder, S.R. & Brinkhurst, R.O., 1990. An assessment of the phylogeny of the Branchiobdellida (Annelida: Clitellata) using PAUP. *Can. J. Zool.* **68**: 1318–1326.

Gentili, E., Hebauer, F, Jäch, M.A., Ji, L. & Schödl, S., 1995. Hydrophilidae: 1. Checklist of the Hydrophilinae recorded from China (Coleoptera). In: Jäch, M.A. & Ji, L. (Eds), *Water Beetles of China*. Zoologisch-Botanische Gesellschaft & Wiener Coleopterologenverein, Vienna: 207–219.

Gentili, E., 1995. Hydrophilidae: 3. The genus *Laccobius* ERICHSON in China and neighbouring areas (Coleoptera). In: Jäch, M.A. & Ji, L. (Eds), *Water Beetles of China*. Zoologisch-Botanische Gesellschaft & Wiener Coleopterologenverein, Vienna: 245–286.

Georgian, T. & Thorp, J.H., 1992. Effects of microhabitat selection on feeding rates of net-spinning caddisfly larvae. *Ecology* **73**: 229–240.

Gettmann, W.W., 1978. Untersuchungen zum Nahrungsspektrum von Wolfsspinnen (Lycoside) der Gattung *Pirata*. *Mitteilungen der Deutschen Gelleschaft für allgemeine und angewandte Entomologie* **1**: 63–66.

Ghose, N.C. & Sharma, C.B., 1989. *Pollution of Ganga River (Ecology of Mid-Ganga Basin)*. Ashish Publishing House, New Delhi: 262 pp.

Ghosh, M. & Gaur, J.P., 1990. Aplication of algal assay for defining nutrient limitation in two streams at Shillong (India). *Proc. Indian Acad. Sci. Plant Sci.* **100**: 361–368.

Ghosh, M. & Gaur, J.P., 1991a. Sructure and interrelation of epilithic and epipelic algal communities in two deforested streams at Shillong, India. *Arch. Hydrobiol.* **122**: 105–116.

Ghosh, M. & Gaur, J.P., 1991b. Regulatory influence of water current on algal colonization in an unshaded stream at Shillong (Meghalaya, India). *Aquat. Bot.* **40**: 37–46.

Ghosh, M. & Gaur, J.P., 1994. Algal periphyton of an unshaded stream in relation to *in situ* nutrient enrichment and current velocity. *Aquat. Bot.* **47**: 185–189.

Ghosh, S.K., 1981. Notes on the Indian species of *Neochauliodes* Weele (Neuroptera: Corydalidae). *Bull. Zool. Surv. India* **4**: 199–202.

Ghosh, T.K. & Konar, S.K., 1993. Mass mortality of fish in the River Mathabhanga-Churni, West Bengal. *Environment & Ecology* **11**: 833–838.

Gibbs, D.G., 1968. The larva, dwelling-tube, and feeding of a species of *Protodipseudopsis* (Trichoptera: Dipseudopsidae). *Proc. R. ent. Soc. Lond. (A)* **43**: 73–79.

Gibbs, D.G., 1973. The Trichoptera of Ghana. *Dtsch. Ent. Z.* **20**: 363–424.

Gibbs, R.J., 1970. Mechanisms controlling world water chemisty. *Science* **170**: 1088–1090.

Gibson, R. & Moore, J., 1976. Freshwater nemerteans. *Zool. J. Linn. Soc.* **58**: 177–218.

Gibson, R. & Qi, S., 1991. A new freshwater heteronemertean from the Zhujiang (Pearl River), People's Republic of China. *Hydrobiologia* **220**: 167–178.

Giesen, W., 1994. Indonesia's major freshwater lakes: a review of current knowledge, development processes and threats. *Mitt. Internat. Verein. Limnol.* **24**: 115–128.

Gijskes, D.C., 1952. Die Plekopteren der Deutschen Limnologischen Sunda-Expedition, nebst einigen Neubeschreibungen. *Arch. Hydrobiol., Suppl.* **21**: 275–297.

Gillies, M.T., 1951. Further notes on Ephemeroptera from India and Southeast Asia. *Proc. R. Ent. Soc. Lond.* **B20**: 121–130.

Gillies, M.T., 1990. A revision of the African species of *Centroptilum* Eaton (Baetidae, Ephemeroptera). *Aquatic Insects* **12**: 97–128.

Gillies, M.T., 1991. A diphyletic origin for the two-tailed baetid nymphs occurring in East African stony streams with a description of the new genus and species *Tanzaniella spinosa* gen. nov. sp. nov. In: Alba-Tercedor, J. & Sanchez-Ortega, A. (Eds), *Overview and Strategies of Ephemeroptera and Plecoptera*. Sandhill Crane Press, Gainesville: 175–187.

Girardi, H. & Ledoux, J.C., 1989. Presence d'*Anodonta woodiana* (Lea) en France (Mollusques, Lamellibranches, Unionoidae). *Bulletin Mensuel de la Societét Linneenne de Lyon* **58**: 285–291.

Givens, D.R. & Smith, S.D., 1980. A synopsis of the western Arctopsychinae (Trichoptera: Hydropsychidae). *Melanderia* **35**: 1–24.

Gopal, B., 1987. *Water Hyacinth*. Elsevier, Amsterdam: 471 pp.

Gopal, B., 1988. Wetlands: management and conservation in India. *Water Quality Bull.* **13**: 3–6.

Gopal, B. & Sah, M., 1993. Conservation and management of rivers in India: case study of the River Yamuna. *Environmental Conservation* **20**: 243–254.

Gorb, S.N., 1995. Design of the predatory legs of water bugs (Hemiptera:

Nepidae, Naucoridae, Notonectidae, Gerridae). *J. Morphol.* **223**: 289–302.

Gose, K., 1970a. Life history and production rate of *Ephemera strigata* (Ephemeroptera). *Jap. J. Limnol.* **31**: 21–26 (in Japanese).

Gose, K., 1970b. Life history and instar analysis of *Stenopsyche griseipennis* (Trichoptera). *Jpn. J. Limnol.* **31**: 96–106 (in Japanese).

Gose, K., 1985. Ephemeroptera. In: Kawai, T. (Ed.), *An Illustrated Book of Aquatic Insects of Japan.* Tokai University Press, Tokai: 7–32 (in Japanese).

Grant, P.M. & Peters, W.L., 1993. Description of four new genera of *Thraulus* group mayflies from the Eastern Hemisphere and a redescription of *Simothraulus* and *Chiusanophlebia* (Ephemeroptera: Leptophlebiidae: Atalophlebiinae). *Trans. Am. Entomol. Soc.* **119**: 131–168.

Grant, P.M. & Sivaramakrishnan, K.G., 1985. A new species of *Thraulus* (Ephemeroptera: Leptophlebiidae: Atalophlebiinae) from southern India. *Florida Entomologist* **68**: 424–432.

Green, R.H., 1979. *Sampling Design and Statistical Methods for Environmental Biologists.* Wiley Interscience, New York: 257 pp.

Green, R.H., 1993. Application of repeated measures designs in environmental impact and monitoring studies. *Aust. J. Ecol.* **18**: 81–98.

Gressitt, J.L., 1982. Zoogeographical summary. In: Gressitt, J.L. (Ed.), *Biogeography and Ecology of New Guinea. Volume 2.* Dr W. Junk Publishers, The Hague: 897–918.

Grondejs, K., 1978. Once again *Rhodeus ocellatus* from Taiwan. *Aquarium (Den Haag)* **49**: 78–79 (in Dutch).

Gschwendtner, L., 1932. Aquatic insects of China. Article V. Neue Dytiscidae aus China. *Peking Nat. Hist. Bull.* **7**: 159–164.

Gui, H. & Zhang, J., 1992. A new species of genus *Epeorus* Eaton (Ephemeroptera, Heptageniidae). *Acta Zootaxonomica Sinica* **17**: 61–62.

Gunatilleke, C.V.S. & Gunatilleke, I.A.U.N., 1983. A forestry case study of the Sinharaja rainforest in Sri Lanka. In: Hamilton, L.S. (Ed.), *Forest and Watershed Development and Conservation in Asia and the Pacific.* Westview Press, Boulder, Colorado: 289–357.

Guo, J.Y., Ng, N.K., Dai, A. & Ng, P.K.L., 1997. The taxonomy of three commercially important species of mitten crabs of the genus *Eriocheir* de Haan, 1835 (Crustacea: Decapoda: Brachyura: Grapsidae). *Raffles Bull. Zool.* **45**: 445–476.

Guo, X., 1993. *The Three Gorges Project (TGP) on the Yangtze River: Main Geoenvironmental Problems.* Unpublished manuscript, Geotechnics & the Environment '93 Conference, The Hong Kong Institution of Engineers Geotechnical & Environmental Division, The University of Hong Kong, September 1993.

Guo, Z. Choy, S. & Gui, Q., 1996. *Caridina semiblepsia,* a new species of troglonic shrimp (Crustacea: Decapoda: Atyidae) from Hunan Province, China. *Raffles Bull. Zool.* **44**: 65–75.

Guo, Z., Jiang, H. & Zhang, M., 1992. A new species of *Caridina* from

Hunan, China (Decapoda: Atyidae). *Sichuan Journal of Zoology* **11**: 4–6 (in Chinese).

Gupta, A., 1983. High-magnitude floods and stream channel response. *Spec. Publs int. Ass. Sediment.* **6**: 219–227.

Gupta, A., 1993. Life histories of 2 species of *Baetis* (Ephemeroptera, Baetidae) in a small north-east Indian stream. *Arch. Hydrobiol.* **127**: 105–114.

Gupta, A. & Michael, R.G., 1983. Seasonal differences and relative abundance among populations of benthic aquatic insects in a moderately high altitude stream. In: Tikader, B.K. (Ed.), *Proceedings of a Workshop on High Altitude Entomology & Wildlife Ecology*. The Zoological Survey of India, Calcutta: 21–28.

Gupta, A. & Michael, R.G., 1992. Diversity, distribution, and seasonal abundance of Ephemeroptera in streams of Meghalaya State, India. *Hydrobiologia* **228**: 131–139.

Gupta, L.P., 1981. The setae of the subterranean isopod *Nichollsia menoni* Tiwari 1955 (Crustacea, Isopoda, Phreatoicoidea, Nichollsidae). *Bulletin of the Zoological Survey of India* **7**: 93–101.

Gupta, L.P., 1985. On some anomolies in the abdominal region of *Nichollsia menoni* Tiwari 1955 (Isopoda: Phreatoicoidea). *Bulletin of the Zoological Survey of India* **4**: 7–8.

Gupta, P.D. & Ghosh, R.K., 1979. Catalogue of type specimens in the Zoological Survey of India Part 3. Helminths. *Records of the Zoological Survey of India* **74**: 243–331.

Gupta, R.K., 1983. Land use regulations for flood control and watershed management in the Himalayas. *Indian J. Soil Conservation* **11**: 10–29.

Gupta, S., Michael, R.G. & Gupta, A. 1993. Laboratory studies on the life cycle and growth of *Cloeon* sp. (Ephemeroptera: Baetidae) in Meghalaya State, India. *Aquatic Insects* **15**: 49–55.

Haas, F., 1969. Superfamily Unionacea Fleming, 1828. In: Moore, R.C. (Ed.), *Treatise on Invertebrate Palaeontology, Part N, Volume 1 (of 3), Mollusca 6: Bivalvia*. The Geological Society of America and The University of Kansas, Lawrence: N411–N467.

Habe, T., 1985. Freshwater arcid mussel, *Scaphula pinna* Benson, from Thailand. *Chiribotan* **16**: 47 (in Japanese).

Habeck, D.H. & Solis, M.A., 1994. Transfer of *Petrophila drumalis* (Dyar) to *Argyractis* based on immature and adult characters with a larval description of *Argyractis subornata* (Hampson) (Lepidotera: Crambidae: Nymphulinae). *Proc. Entomol. Soc. Wash.* **96**: 726–734.

Hagen, K.S., 1996. Aquatic Hymenoptera. In: Merritt, R.W. & Cummins, K.W. (Eds), *An Introduction to the Aquatic Insects of North America. Third Edition)*. Kendall/Hunt Publishing Company, Iowa: 474–483.

Haigh, M.J. & Krecek, J., 1991. Headwater management: problems and policies. *Land Use Policy* **8**: 171–176.

Haines, A.K., 1983. Fish fauna and ecology. In: Petr, T. (Ed.), *The Purari —*

Tropical Environment of a High Rainfall River Basin. Dr W. Junk Publishers, The Hague: 367–384.

Hairston, H.G., 1989. *Ecological Experiments: Purpose, Design and Execution.* Cambridge University Press, Cambridge: 370 pp.

Hall, C.A.S., 1972. Migration and metabolism in a temperate stream ecosystem. *Ecol. Monogr.* **43**: 421–439.

Hameed, P.S., Sherief, Y.H. & Amanullah, B., 1995. Study of zooplankton in Kaveri River system (Tiruchirapalli, India) with reference to water quality assessment. In: Timotius, K.H. & Göltenboth, F. (Eds), *Tropical Limnology Volume III. Tropical Rivers, Wetlands and Special Topics.* Satya Wacana University Press, Salatiga, Indonesia: 111–119.

Han, C., 1982. The status of the research on the Chinese river dolphin *Lipotes vexillifer* in China. *Acta Theriol. Sinica* **2**: 245–252 (in Chinese).

Hanada, S., Isobe, Y., Wada, K. & Nagoshi, M., 1994. Drumming behaviour of two stonefly species *Microperla brevicauda* KAWAI (Peltoperlidae) and *Kamimuria tibialis* (PICTET) (Perlidae), in relation to other behaviours. *Aquatic Insects* **16**: 75–89.

Hannappel, U. & Paulus, H.F., 1987. Arbeiten zu einem phylogenetischen System der Helodidae (Coleoptera): Feinstrukturuntersuchungen an europaischen Larven. *Zool. Beitr.* **31**: 77–150.

Hansen, M., 1991a. *The Hydrophiloid Beetles. Phylogeny, Classification and a Revision of the Genera (Coleoptera: Hydrophiloidea).* Biologiske Skrifter No. 40, The Royal Danish Academy of Sciences and Letters, Copenhagen: 367 pp.

Hansen, M., 1991b. A review of the genera of the beetle family Hydraenidae (Coleoptera). *Steenstrupia* **17**: 1–52.

Hansen, M. & Schödl, S., 1997. Description of *Hydrophilomima* from Southeast Asia (Coleoptera: Hydrophilidae). *Koleopterologische Rundschau* **67**: 187–194.

Harding, W.A., 1927. Rhynchobdellae. In: Shipley, A.E. (Ed.), *The Fauna of British India, including Ceylon and Burma. Hirudinea.* Taylor & Francis, London: 13–96.

Harper, P.P., 1974. New *Protonemura* (S.L.) from Nepal (Plecoptera; Nemouridae). Psyche **81**: 367–376.

Harper, P.P., 1977. Capniidae, Leuctridae and Perlidae (Plecoptera) from Nepal. *Oriental Insects* **11**: 53–62.

Harikumar, S., Padmanabhan, K.G., John, P.A. & Kortmulder, K., 1994. Dry-season spawning in a cyprinid fish of southern India. *Environ. Biol. Fish.* **39**: 129–136.

Harker, J.E., 1953. An investigation of the distribution of the mayfly fauna of a Lancashire stream. *J. Anim. Ecol.* **22**: 1–13.

Hart, D.D. 1994. Building a stronger partnership between ecological research and biological monitoring. *J. N. Amer. Benthol. Soc.* **13**: 110–116.

Hart, R.C., 1980. Embryonic duration and post-embryonic growth rates of the tropical freshwater shrimp *Caridina nilotica* (Decapoda: Atyidae) under laboratory and experimental field conditions. *Freshwat. Biol.* **10**: 297–315.

Hart, R.C., 1981. Population dynamics and production of the tropical freshwater shrimp *Caridina nilotica* (Decapoda: Atyidae) in the littoral of Lake Sibaya. *Freshwat. Biol.* **11**: 531–547.

Hartoto, D.I., 1986. Local and spatial distribution of *Puntius binotatus* and *Rasbora lateristriata* in Ci Taman Jaya and Ci Binua, Ujung Kulon, Indonesia. *Berita Biol.* **3**: 261–267 (in Indonesian).

Hasan, R., 1988. Annual production and productivity of the macrophytes of the River Champanala, a side spill channel of the Ganges at Bhagalpur. *Acta Hydrochim. Hydrobiol.* **16**: 573–578.

Hauer, F.R. & Benke, A.C., 1991. Rapid growth of snag-dwelling chironomids in a blackwater river: the influence of temperature and discharge. *J. N. Am. Benthol. Soc.* **10**: 154–164.

Hawkes, H.A., 1975. River zonation and classification. In: Whitton, B.A. (Ed.), *River Ecology*. Blackwell Scientific publications, Oxford: 312–374.

Hayashi, F., 1988a. Life history variation in a dobsonfly, *Protohermes grandis* (Megaloptera: Corydalidae): effects of prey availability and temperature. *Freshwat. Biol.* **19**: 205–216.

Hayashi, F., 1988b. Prey selection by the dobsonfly larva, *Protohermes grandis* (Megaloptera: Corydalidae). *Freshwat. Biol.* **20**: 19–29.

Hayashi, F., 1996. Life cycle of *Protohermes immaculatus* (Megaloptera: Corydalidae) accelerated by warm water overflowing a dam. *Aquatic Insects* **18**: 101–110.

Hayashi, F. & Nakane, M., 1989. Radio tracking and activity monitoring of the dobsonfly larva, *Protohermes grandis* (Megaloptera: Corydalidae). *Oecologia* **78**: 468–472.

Hechtel, F.O.P. & Sawyer, R.T., 1991. Freshwater and terrestrial leeches (Euhirudinea) of Nam Cat Tien. In: Spitzer, K., Leps, J. & Zacharda, M. (Eds), *Nam Cat Tien. Czechoslovak Vietmanese Expedition, November, 1989*. Research Report, Czechoslovak Academy of Sciences, Ceske Budejovice: 30–31.

Hefti, D. & Tomka, I., 1989. Comparative morphological and electrophoretic studies on *Afronurus zebratus* (Hagen, 1864) comb. nov. and other European Heptageniidae (Ephemeroptera), including a key to the European genera of Heptageniidae. *Aquatic Insects* **11**: 115–124.

Hellawell, J.M., 1986. *Biological Indicators of Freshwater Pollution and Environmental Management*. Elsevier Applied Science Publishers, London: 546 pp.

Heller, J., 1993. Hermaphroditism in molluscs. *Biol. J. Linnean Soc.* **48**: 19–42.

Heller, J. & Farstey, V., 1990. Sexual and parthenogenetic populations of the freshwater snail *Melanoides tuberculata* in Israel. *Israel J. Zool.* **37**: 75–87.

Henderson, G.S. & Rouysungnern, S., 1985. Erosion and sedimentation in

Thailand. In: O'Loughlin, C.L. & Pearce, A.J. (Eds), *Symposium on Effects of Forest Land Use on Erosion and Slope Stability, 1984, Honolulu*. East-West Center, Environment & Policy Institute, Honolulu: 31–39.

Hendrich, L. & Balke, M., 1997. Taxonomische Revision der südostasiatischen Arten der Gattung *Neptosternus* SHARP, 1882 (Coleoptera: Dytiscidae: Laccophilinae). *Koleopterologische Rundschau* **67**: 53–97.

Henry, T.J. & Froeschner, R.C., 1988. *Catalog of the Heteroptera, or True Bugs, of Canada and the Continental United States*. E.J. Brill, Leiden: 958 pp.

Henshaw, J., 1994. The barasingha, or swamp deer, in Suklaphanta Wildlife Reserve, Nepal. *Oryx* **28**: 199–206.

Herrmann, R., 1983. Environmental implications of water transfer. In: Biswas, A.K., Zuo, D., Nickum, J.E. & Liu, C. (Eds), *Long-distance Water Transfer: a Chinese Case Study and International Experiences*. Tycooly International Publishers Limited, Dublin: 151–158.

Hewavisenthi, A.C.D., 1992. Mahaweli water resources project. *Water International* **17**: 33–43.

Hickling, C.L., 1961. *Tropical Inland Fisheries*. Longmans, London: 287 pp.

Hideux, P., Elouard, J.-M. & Troubat, J.-J., 1991. Elavage et éléments de biologie des larves de *Cheumatopsyche digitata* (Trichoptère: Hydropsychidae). *Arch. Hydrobiol.* **122**: 493–512.

Higler, L.W.G., 1981. Caddisfly systematics up to 1960 and a review of the genera (Insecta: Trichoptera). In: Morretti, G.P. (Ed.), *Proceedings of the 3rd International Symposium on Trichoptera*. Dr W. Junk Publishers, The Hague: 117–126.

Higler, L.W.G., 1992. A check-list of the Trichoptera recorded from India and a larval key to the families. *Oriental Insects* **26**: 67–102.

Hilborn, R. & Walters, C.J., 1981. Pitfalls of environmental baseline and process studies. *Environmental Impact Assessment Review* **2**: 265-278.

Hill, M.T., 1995. Fisheries ecology of the lower Mekong River: Myanmar to Tonle Sap River. *Nat. Hist. Bull. Siam Soc.* **43**: 263–288.

Hilsenhoff, W.L., 1991. Diversity and classification of insects and Collembola. In: Thorp, J.H. & Covich, A.P. (Eds), *Ecology and Classification of North American Freshwater Invertebrates*. Academic Press, San Diego: 593–663.

Hinton, H.E., 1955. On the respiratory adaptations, biology and taxonomy of the Psephenidae, with notes on some related families (Coleoptera). *Proc. Zool. Soc. Lond.* **123**: 543–568.

Hiremath, K.G. & Shetty, S.M., 1988. Studies on the phytoplankton in some of the streams at the iron ore mining areas of Goa, India. In: Trivedy, R.K. (Ed.), *Ecology and Pollution of Indian Rivers*. Ashish Publishing House, New Delhi: 303–320.

Hirose, Y. & Itoh, S., 1993. *A Guide to the Dragonflies of Hokkaido*. Selfpublishing, Abashiti/Shizunai, Hokkaido: 187 pp. (in Japanese).

Hirth, F., 1911. *The Ancient History of China to the End of the Chou*

Dynasty. The Columbia University Press, New York: 383 pp.

Hjorth, P. & Nguyen, T.D., 1993. Environmentally sound urban water management in developing countries: a case study of Hanoi. *Water Resources Development* **9**: 453–465.

Ho, S.C., 1975. Some aspects of the bacteriological conditions of Sungai Renggam, Shah Alam, Selangor. *Malay. Nat. J.* **29**: 70–82.

Ho, S.C., 1976. Periphyton production in a tropical lowland stream polluted by inorganic sediments and organic wastes. *Arch. Hydrobiol.* **77**: 458–474.

Ho, S.C., 1994. Status of limnological research and training in Malaysia. *Mitt. Internat. Verein. Limnol.* **24**: 129–145.

Ho, S.C., 1996. Vision 2020: towards and environmentally sound and sustainable development of freshwater resources in Malaysia. *GeoJournal* **40**: 73–84.

Jacobs, J.W., 1994. Toward sustainability in Lower Mekong River Basin development. *Water International* **19**: 43–51.

Ho, K.K. & Hsu, T.T., 1977. Studies of the aquatic insects of Hsin Tein Stream in Taipei area. *Ann. Taiwan Mus.* **20**: 1–42 (in Chinese).

Ho, Y.C., 1987. Control and management of pollution of inland waters in Malaysia. *Arch. Hydrobiol. Beih., Ergebn. Limnol.* **28**: 547–556.

Hoagland, K.E. & Davis, G.M., 1979. The stenothyrid radiation of the Mekong River. 1.: The *Stenothyra mcmulleni* complex. *Proc. Acad. Natur. Sci. Philad.* **131**: 191–230.

Hobbs, H.H., 1991. Decapoda. In: Thorp, J.H. & Covich, A.P. (Eds), *Ecology and Classification of North American Freshwater Invertebrates*. Academic Press, San Diego: 823–858.

Hobbs, H.H., Jass, J.P. & Huner, J.V., 1989. A review of global crayfish introductions with particular emphasis on two North American species (Decapoda, Cambridae). *Crustaceana* **56**: 299–316.

Hodgson, A.N. & Heller, J., 1990. Spermatogenesis and sperm structure of the normally parthenogenetic freshwater snail *Melanoides tuberculata*. *Israel J. Zool.* **37**: 31–50.

Hofer, T., 1993. Deforestation, changing river discharge, and increasing floods: myth or reality? *Mountain Research & Development* **13**: 213–233.

Hoffman, W.E., 1931. Studies on the bionomics of the water-bug, *Lethocerus indicus* (Hemiptera, Belostomatidae), in China. *Verh. Intenat. Verein. Limnol.* **5**: 661–667.

Hoffman, W.E., 1933. A preliminary list of the aquatic and semi-aquatic Hemiptera of China, Chosen (Korea), and Indo-China. *Lingnan Sci. J.* **12** (**Suppl.**): 243–258.

Hoffman, W.E., 1936a. The life history *Limnogonus fossarum* (Fabr.) in Canton (Hemiptera: Gerridae). *Lingnan Sci. J.* **15**: 477–482.

Hoffman, W.E., 1936b. Life history notes on *Rhagadotarsus kraepelini* Breddin (Hemiptera: Gerridae) in Canton. *Lingnan Sci. J.* **15**: 289–299.

Hoffman, W.E., 1936b. The occurrence of *Limnometra gigas* China (Hemiptera: Gerridae) in Hainan Island with description of the apterous form. *Lingnan Sci. J.* **15**: 489–492.

Hoggarth, D.D. & Utomo, A.D., 1994. The fisheries ecology of the Lubuk Lampan River floodplain in south Sumatra, Indonesia. *Fisheries Research* **20**: 191–213.

Holeman, J.N., 1968. Sediment yield of major rivers of the world. *Water Resources Res.* **4**: 737–747.

Holmen, M. & Vazirani, T.G., 1990. Notes on the genera *Neptosternus* SHARP and *Copelatus* ERICHSON from Sri Lanka and India with the description of new sepcies (Coleoptera: Dytiscidae). *Koleopterologische Rundschau* **60**: 19–31.

Holopainen, A-L. & Huttunen, P., 1992. Effects of forest clear-cutting and soil disturbance on the biology of small forest brooks. *Hydrobiologia* **243/244**: 457–464.

Holt, P.C., 1986. Newly established families of the order Branchiobdellida (Annelida: Clitellata) with a synopsis of the genera. *Proc. Biol. Soc. Wash.* **99**: 676–702.

Holt, P.C., 1989. Comments on the classification of the Clitellata. *Hydrobiologia* **180**: 1–5.

Holthuis, L.B., 1950. The Decapoda of the *Siboga* Expedition, Part X: the Palaemonidae collected by the *Siboga* and *Snellius* expeditions, with remarks on other species, Part I: subfamily Palaemoninae. *Siboga-Expeditie* **39a⁹**: 1–268.

Holthuis, L.B., 1978a. Zoological results of the British Speleological Expedition to Papua New Guinea 1975. 7. Cavernicolous shrimps (Crustacea Decapoda, Natantia) from New Ireland and the Philippines. *Zool. Meded. (Leiden)* **53**: 209–224.

Holthuis, L.B., 1978b. A collection of decapod Crustacea from Sumba, Lesser Sunda Islands, Indonesia. *Zool. Verh.* **162**: 3–55.

Holthuis, L.B., 1982. Freshwater Crustacea Decapoda of New Guinea. In: Gressitt, J.L. (Ed.), *Biogeography and Ecology of New Guinea. Volume 2.* Dr W. Junk Publishers, The Hague: 603–619.

Holthuis, L.B., 1984. Freshwater prawns (Crustacea: Decapoda: Natantia) from subterranean waters of the Gunung Sewu area, Central Java, Indonesia. *Zool. Med., Leiden* **58**: 141–148.

Hora, S.L., 1923. Observations of the fauna of certain torrential streams in the Khasi Hills. *Rec. Indian Mus.* **25**: 579–600.

Hora, S.L., 1927. Animal life in torrential streams. *J. Bombay Nat. Hist. Soc.* **32**: 111–126.

Hora, S.L., 1930. Ecology, bionomics and evolution of the torrential fauna, with special reference to the organs of attachment. *Phil. Trans. R. Soc. Lond.* **B218**: 171–282.

Hora, S.L., 1932. Classification, bionomics and evolution of homalopterid fishes. *Mem. Indian Mus.* **12**: 262–330.

Hora, S.L., 1936. Nature of substratum as an important factor in the ecology of torrential fauna. *Proc. Natn. Inst. Sci. India* **2**: 45–47.

Hora, S.L., 1952. Major problems of the fisheries of India with suggestions for their solution. *J. Asiatic Soc. (Sci.)* **18**: 83–101.

Hori, H., 1993. Development of the Mekong River Basin, its problems and future prospects. *Water International* **18**: 110–115.

Hortle, K., 1995. A survey of the Barito River fishery near Mount Muro, Central Kalimantan, Indonesia. In: Timotius, K.H. & Göltenboth, F. (Eds), *Tropical Limnology Volume III. Tropical Rivers, Wetlands and Special Topics.* Satya Wacana University Press, Salatiga, Indonesia: 15–27.

Horvat, B., 1994. *Dolichocephala sinica* sp. n. (Diptera: Empididae: Clinocerinae) from Sichuan (China). *Aquatic Insects* **16**: 201–203.

Houbrick, R.S., 1986. Transfer of *Quadrasia* from the Planaxidae to the Buccinidae (Mollusca: Gastropoda: Prosobranchia). *Proc. Biol. Soc. Wash.* **99**: 359–362.

Houbrick, R.S., 1991. Anatomy and systematic placement of *Faunus* Montfort 1810 (Prosobranchia: Melanopsinae). *Malacol. Rev.* **24**: 35–54.

Howarth, F.G., 1985. Biosystematics of the *Culicoides* of Laos (Diptera: Ceratopogonidae). *Int. J. Entomol.* **27**: 1–96.

Hsia, Y.J., 1987. Changes in strom hydrographs after clearcutting at a small hardwood forest watershed in central Taiwan. *Forest Ecology & Management* **20**: 117–133.

Hsu, Y.C., 1931. Two new species of mayflies from China (Order Ephemeroptera). *Peking Nat. Hist. Bull.* **6**: 30–33.

Hsu, Y.C., 1935. New Chinese mayflies from Kiangsi Province (Ephemeroptera). *Peking Nat. Hist. Bull.* **10**: 319–326.

Hsu, Y.C., 1936a. The mayflies of China (Order Ephemeroptera). *Peking Nat. Hist. Bull.* **11**: 129–148.

Hsu, Y.C., 1936b. Mayflies of Hong Kong with descriptions of two new species (Ephemeroptera). *Hong Kong Naturalist* **7**: 233–238.

Hsu, Y.C., 1936c. The mayflies of China (Order Ephemeroptera). *Peking Nat. Hist. Bull.* **11**: 287–296.

Hsu, Y.C., 1936d. The mayflies of China (Order Ephemeroptera). *Peking Nat. Hist. Bull.* **11**: 433–440.

Hsu, Y.C., 1937a. The mayflies of China. *Peking Nat. Hist. Bull.* **12**: 53–56

Hsu, Y.C., 1937b. The mayflies of China. *Peking Nat. Hist. Bull.* **12**: 123–126

Hsu, Y.C., 1937c. The mayflies of China. *Peking Nat. Hist. Bull.* **12**: 221–224.

Hu, M., Stallard, R.F. & Edmond, J.M., 1982. Major ion chemistry of some large Chinese rivers. *Nature* **298**: 550–553.

Huang, R., 1983. On the problem of water supply on the Hai-Luan Plain. In: Biswas, A.K., Zuo, D., Nickum, J.E. & Liu, C. (Eds), *Long-distance Water Transfer: a Chinese Case Study and International Experiences.* Tycooly International Publishers Limited, Dublin: 309–319.

Huang, S. & Ferng, J., 1990a. Applied land classification for surface water quality management: I. Watershed classification. *J. Environ. Manag.* **31**: 107–126.

Huang, S. & Ferng, J., 1990b. Applied land classification for surface water quality management: II. Land process classification. *J. Environ. Manag.* **31**: 127–141.

Huang, W., 1985. The demarcation line between the Palaearctic and Oriental Regions in eastern China. In: Kawamichi, T. (Ed.), *Contemporary Mammalogy in China and Japan*. The Mammal Society of Japan, Osaka: 133–141.

Huang, Z., Lin, H. & Zhang, S., 1986. Analysis of the Landsat sensing images of the types of habitats of Yangtze alligators. *Chin. J. Oceanol. Limnol.* **4**: 19–26 (in Chinese).

Hubbard, M.D., 1983. Ephemeroptera of Sri Lanka: Ephemeridae. *Syst. Entomol.* **8**: 383–392.

Hubbard, M.D., 1984. A revision of the genus *Povilla* (Ephemeroptera: Polymitarcyidae). *Aquatic Insects* **6**: 17–35.

Hubbard, M.D., 1989. *Matsumuracloeon*, a replacement name for *Pseudocloeon* Matsumura, 1931 (Ephmeroptera: Baetidae). *Florida Entomol.* **72**: 388.

Hubbard, M.D., 1990. *Mayflies of the World. A Catalog of the Family and Genus Group Taxa (Insects: Ephemeroptera)*. Sandhill Crane Press, Gainesville: 119 pp.

Hubbard, M.D., 1994. The mayfly family Behningiidae (Ephemeroptera: Ephemeroidea): keys to the recent species with a catalog of the family. *Great Lakes Entomol.* **27**: 161–168.

Hubbard, M.D. & Pescador, M.L., 1978. A catalog of the Ephemeroptera of the Philippines. *Pacific Insects* **19**: 91–99.

Hubbard, M.D. & Peters, W.L., 1976. The number of genera and species of mayflies (Ephemeroptera). *Ent. News* **87**: 245.

Hubbard, M.D. & Peters, W.L., 1978. A catalogue of the Ephemeroptera of the Indian subregion. *Oriental Insects, Suppl.* **9**: 1–43.

Hubbard, W.D. & Peters, W.L., 1984. Ephemeroptera of Sri Lanka: an introduction to their ecology and biogeography. In: Fernando, C.H. (Ed.), *Ecology and Biogeography in Sri Lanaka*. Dr. W. Junk Publishers, The Hague: 257–274.

Hulme, M., Wigley, T., Jiang, T., Zhao, Z., Wang, F., Ding, Y., 1992. *Climate Change due to the Greenhouse Effect and the Implications for China*. World Wide Fund for Nature, Switzerland: 57 pp.

Humphrey, C.L. & Dostine, P.L., 1994. Development of biological monitoring programs to detect mining-waste impacts upon aquatic ecosystems upon aquatic ecosystems of the Alligator Rivers Region, Northern Territory, Australia. *Mitt. Internat. Verein. Limnol.* **24**: 293–314.

Humphrey, C.L., Faith, D.P. & Dostine, P.L., 1995. Baseline requirements for assessment of mining impact using biological monitoring. *Aust. J. Ecol.* **20**: 150–166.

Hung, T., 1987. Study on heavy metals in rivers and estuaries of western Taiwan. *Arch. Hydrobiol. Beih., Ergebn. Limnol.* **28**: 181–192.

Hunt, P., 1992. Sweet smell of death on Thailand's rivers. *New Scientist* **9 May 1992**: 7.

Hurlbert, S.H., 1984. Pseudoreplication and the design of ecological field experiments. *Ecol. Monogr.* **54**: 187–211.

Hutchinson, G.E., 1981. Thoughts on aquatic insects. *BioScience* **31**: 495–500.

Hutchinson, G.E., 1993. *A Treatise on Limnology: Volume IV. The Zoobenthos.* John Wiley & Sons, New York: 944.

Hutchinson, R., 1991. Observations d'araignees (Araneae) qui capturent des demoiselles et des libellules (Odonata: Zygoptera, Anisoptera) au Quebec et importance du phenomene. *Pirata* **1**: 130–133.

Huang, B. & Zhang, L., 1990. The distribution and environment of unionids from Dianchi and Erhai Lakes in Yunnan Province. *Trans. Chinese. Soc. Malacol.* **3**: 69–75 (in Chinese).

Huang, M., Luo, X. & Liu, J. 1986. On the freshwater crabs from Sichuan, *Sichuan Journal of Zoology* **5**: 4–7 (in Chinese).

Hubendick, B. 1951. Recent Lymnaeidae. Their variation, morphology, taxonomy, nomenclature, and distribution. *K. Svenska Vetenskapsakad. Handl. (Ser. 4)* **3**: 1–223.

Hubendick, B., 1955. Phylogeny in the Planorbidae. *Trans. Zool. Soc. Lond.* **28**: 453–542.

Hubendick, B., 1957. *Patelloplanorbis*, a new genus of Planorbidae (Mollusca: Pulmonata). *Koninkl. Nederl. Akad. Wetens. Amsterdam (3)* **60**: 90–95.

Hubendick, B., 1964. Studies on Ancylidae. The subgroups. *Göteburgs K. Vetensk.-o. vitterh Samh., Handl. Ser. B* **9**: 1–72.

Hudson, P.L., 1987. Unusual larval habitats and life history of chironomid (Diptera) genera. *Ent. Scand., Suppl.* **29**: 369–373.

Hufschmidt, M.M., 1993. Water resource management. In: Bonell, M., Hufschmidt, M.M. & Gladwell, J.S. (Eds), *Hydrology and Water Management in the Humid Tropics.* UNESCO/Cambridge University Press, New York: 471–495.

Huisman, J., 1991. A study of Trichoptera in Sabah and Sarawak. In: Tomaszewski, C. (Ed.), *Proceedings of the 6th International Symposium on Trichoptera.* Adam Mickiewicz University Press, Lodz-Zakopane, Poland: 275–278.

Huisman, J., 1992. New species of *Apsilochorema* (Trichoptera: Hydrobiosidae) from Sabah, East Malaysia. *Zool. Med., Leiden* **66**: 127–137.

Huisman, J., 1993. New species of *Gunungiella* (Trichoptera: Philopotamidae) from Sabah, East Malaysia. *Zoologische Mededelingen* **67**: 75–89.

Hung, M., Chan, T. & Yu, H., 1993. Atyid shrimps (Decapoda: Caridea) of Taiwan, with descriptions of three new species. *J. Crustacean Biol.* **13**: 481–503.

Hunter, F.F. & Maier, A.K., 1994. Feeding behaviour of predatory larvae of *Atherix lantha* (Diptera: Athericidae). *Can. J. Zool.* **72**: 1695–1699.

Hwang, C.L., 1957. Descriptions of Chinese caddis flies (Trichoptera). *Acta Entomologica Sinica* 7: 373–404 (in Chinese).

Hwang, C.L., 1958. Descriptions of Chinese caddsiflies (Trichoptera). *Acta Zoologica Sinica* 10: 279–285 (in Chinese).

Hwang, C.L., 1963. A review of the Chinese Stenopsychidae (Trichoptera). *Acta Ent. Sinica* 12: 476–489 (in Chinese).

Hwang, C.L. & Tian, L., 1982. Trichoptera. *Insects of Xizang* 2: 1–10 (in Chinese).

Hwang, J.J. & Mizue, K., 1985. Fresh-water crabs of Taiwan. *Bull. Fac. Fish., Nagasaki Univ.* 57: 1–21.

Hwang, J.J. & Takeda, M., 1986. A new freshwater crab of the family Grapsidae from Taiwan. *Proceeding of the Japanese Society of Systematic Zoology* 33: 11–18.

Hynes, J.D., 1975. Annual cycles of macroinvertebrates of a river in southern Ghana. *Freshwat. Biol.* 5: 71–83.

Hynes, H.B.N., 1970. *The Ecology of Running Waters*. Liverpool University Press, Liverpool: 555 pp.

Hynes, H.B.N., 1975. The stream and its valley. *Verh. Internat. Verein. Limnol.* 19: 1–15.

Hynes, H.B.N., 1976. Biology of Plecoptera. *Ann. Rev. Entomol.* 21: 135–153.

Hynes, H.B.N., 1984. Taxonomy and ecology. In: Resh, V.H. & Rosenberg, D.M. (Eds), *The Ecology of Aquatic Insects*. Praeger Publishers, New York: 2–23.

Hynes, H.B.N., 1989. Keynote address. *Can. Spec. Publ. Fish. Aquat. Sci.* 106: 5–10.

Ichikawa, A. & Kawakatsu, M., 1963. *Polycelis akkeshi*, a new freshwater planarian, from Hokkaido. *Publications from the Akkeshi Marine Biological Station* 12: 1–18.

Ichikawa, N., 1988. Male brooding behaviour of the giant water bug *Lethocerus deyrollei* Vuillefroy (Hemiptera: Belostomatidae). *J. Ethol.* 6: 121–127.

Ichikawa, N., 1995. Male counterstrategy against infanticide of the female giant water bug *Lethocerus deyrollei* (Hemiptera: Belostomatidae). *J. Insect Behav.* 8: 181–188.

Ida, M., 1994. The life cycle of *Scopura montana* (Plecoptera: Scopuridae). *Jap. J. Limnol.* 55: 23–25 (in Japanese).

Idris, B.A.G., 1983. *Freshwater Zooplankton of Malaysia (Crustacea: Cladocera)*. U.P.M. Monograph, Penerbit Universiti Pertanian, Malaysia, 153 pp.

Ilango, K., 1994. *Neotelmatoscopus ctenophorus*, a new species from India (Diptera; Psychodidae). *Aquatic Insects* 16: 141–145.

Illies, J., 1965. Phylogeny and zoogeography of the Plecoptera. *Ann. Rev. Entomol.* 10: 117–140.

Illies, J., 1966. Katalog der rezenten Plecoptera. *Das Tierreich* 82: 1–631.

Imamura, T., 1983. A new subfamily and two new species of water mite (Acari: Hydrachnellae) from Papua New Guinea. *Proceedings of the Japanese Society of Systematic Zoology* **26**: 11–18.

Imamura, T., 1986. More hyporheic water mites from Kyushu, Japan. *Zoological Science (Tokyo)* **3**: 1110.

Imamura, T. & Kikuchi, Y., 1986. Studies on water mites and harpacticoid copepods as the indicative organisms for environmental factor in inland water habitats. *Research Projects in Review 1985, Nissan Science Foundation* **8**: 317–331 (in Japanese).

Imanishi, K., 1934. Mayflies from Japanese torrents IV. Notes on the generus *Epeorus*. *Annot. Zool. Japon.* **14**: 380–394.

Imanishi, K., 1936. Mayflies from Japanese torrents VI. Notes on the genera *Ecdyonurus* and *Rhithrogena*. *Annot. Zool. Japon.* **17**: 23–36.

Imanishi, K., 1938. Mayflies from Japanese torrents IX. Life forms and life zones of mayfly nymphs. I. Introduction. *Annot. Zool. Japon.* **17**: 23–36.

Imanishi, K., 1941. Mayflies from Japanese torrents X. Life forms and life zones of mayfly nymphs. II. Ecological structure indicated by life zone arrangement. *Mem. Coll. Sci. Kyoto Imp. Univ. (B)* **16**: 1–35.

Indrasena, H.H.A., 1970. Limnological and freshwater fisheries work in Ceylon. In: *Regional Meeting of Inland Water Biologists in Southeast Asia. Proceedings.* Unesco Field Science Office for Southeast Asia, Djakarta: 45–47.

Inger, R.F. & Chin, P.K., 1962. The freshwater fishes of North Borneo. *Fieldlandia (Zoology)* **45**: 1–268.

Inger, R.F., Shaffer, H.B., Koshy, M. & Bakde, R., 1987. Ecological structure of a herpetological assemblage in South India. *Amphib-Reptilia* **8**: 189–202.

Ingle, R.W., 1986. The Chinese mitten crab *Eriocheir sinensis* H. Milne Edwards — a contentious immigrant. *London Naturalist* **65**: 101–105.

Inoue, K. & Eda, S., 1984. To Dr Syoziro Asahina on his 70th birthday. *Odonatologica* **13**: 187–213.

Iqbal, S.H., Bhatty, S.F. & Malik, K.S., 1979. Freshwater hyphomycetes on decaying plant debris submerged in some streams of Pakistan. *Trans. Mycol. Soc. Jap.* **20**: 51–61.

Iqbal, S.H., Bhatty, S.F. & Malik, K.S., 1979. Freshwater hyphomycetes on decaying pine needles in Pakistan. *Trans. Mycol. Soc. Jap.* **21**: 321–327.

Irons, J.G., Oswood, M.W., Stout, R.J. & Pringle, C.M., 1994. Latitudinal patterns in leaf litter breakdown: is temperature really important? *Freshwat. Biol.* **32**: 401–411.

Ishida, S. & Ishida, K., 1985. Odonata. In: Kawai, T. (Ed,), *An Illustrated Book of Aquatic Insects of Japan.* Tokai University Press, Tokyo: 33–124 (in Japanese).

Ishwaran, N., 1993. Ecology of the Asian Elephant in lowland dry zone habitats of the Mahaweli River Basin, Sri Lanka. *J. Trop. Ecol.* **9**: 169–182.

Ito, T., 1978. Morphological and ecological studies on the caddisfly genus Dinanthrodes in Hokkaido, Japan (Trichoptera, Lepidostomatidae) I. The larval development and the cases of four species of *Dinanthrodes*. *Kontyû* 46: 574–584.

Ito, T., 1984. On the genus *Goerodes* (Trichoptera: Lepidostomatidae) in Japan. *Kontyû* 52: 506–515.

Ito, T., 1985a. Two new species of the *naraensis* group of *Goerodes* (Trichoptera, Lepidostomatide). *Kontyû* 53: 507–515.

Ito, T., 1985b. Description, geographical variation and ecology of *Goerodes naraensis* (Tani) (Trichoptera, Lepidostomatidae). *Jpn. J. Limnol.* 46: 199–211.

Ito, T., 1985c. Morphology and ecology of three species of *orientalis* group of *Goerodes* (Trichoptera, Lepidostomatide). *Kontyû* 53: 12–24.

Ito, T., 1985d. Females, pupae and larvae of the *japonicus* group of *Goerodes* (Trichoptera, Lepidostomatide). *Kontyû* 53: 261–269.

Ito, T., 1986. Three lepidostomatid caddisflies from Nepal, with descriptions of two new species. *Kontyû* 54: 485–494.

Ito, T., 1987. A review of bionomics of the Japanese Lepidostomatidae (Trichoptera) with particular reference to the change of case materials and its ecological significance. In: Bournaud, M. & Tachet, H. (Eds), *Proceedings of the 5th International Symposium on Trichoptera*. Dr W. Junk Publishers, The Hague: 269–273.

Ito, T., 1989. Lepidostomatid caddsiflies (Trichoptera) from the Tsushisma Islands of Japan, with description of a new species. *Jpn. J. Ent.* 57: 46–60.

Ito, T., 1990a. Taxonomic notes on Japanese lepidostomatid caddisflies (Trichoptera). *Jpn. J. Ent.* 58: 781–793.

Ito, T., 1990b. Lepidostomatid caddsiflies (Trichoptera) from the Yakushima and Amami-oshshima Islands of Japan, with descripotions of three new species. *Jpn J. Ent.* 58: 361–373.

Ito, T., 1991a. Morphology and bionomics of *Palaeagapetus flexus* n. sp. from northern Japan (Trichoptera, Hydroptilidae). In: Tomaszewski, C. (Ed.), *Proceedings of the 6th International Symposium on Trichoptera*. Adam Mickiewicz University Press, Lodz-Zakopane, Poland: 419–426.

Ito, T., 1991b. Description of a new species of *Palaeagapetus* from Central Japan, with notes on bionomics (Trichoptera, Hydroptilidae). *Jpn, J. Ent.* 59: 357–366.

Ito, T., 1992a. Taxonomic notes on some Asian Lepidostomatidae (Trichoptera), with descriptions of two new species. *Aquatic Insects* 14: 97–106.

Ito, T., 1992b. Lepidostomatid caddisflies (Trichoptera) from the Ryukyu Islands of southern Japan, with descriptions of two new species. *Jpn J. Ent.* 60 333–342.

Ito, t., 1995. Description of a boreal caddisfly *Micrasema gelidum* McLachlan (Trichoptera, Brachycentridae), from Japan and Mongolia, with notes on bionomics. *Jap. J. Entomol.* **3**: 493–502.

Ito, T. & Kawamula, H., 1980. Morphology and biology of the immature stages of *Hydroptila itoi* KOBAYASHI (Trichoptera, Hydroptilidae). *Aquatic Insects* **2**: 113–122.

Ito, T. & Kawamula, H., 1984. Morphology and ecology of immature stages of *Oxyethira acuta* KOBAYASHI (Trichoptera, Hydroptilidae). *Jap. J. Limnol.* **43**: 313–317.

Ito, T. & Hattori, T., 1986. Description of a new species of *Palaeagapetus* (Trichoptera, Hydroptilidae) from northern Japan, with notes on bionomics. *Kontyû, Tokyo* **54**: 131–151.

Ito, T., Tanida, K. & Nozaki, T., 1993. Checklists of Trichoptera in Japan 1. Hydroptilidae and Lepidostomatidae. *Jpn. J. Limnol.* **54**: 141–150.

Ittekkot, V. & Laane, R.W.P.M., 1991. Fate of riverine particulate organic matter. In: Degens, E.T., Kempe, S. & Richey, J.E. (Eds), *Biogeochemistry of Major World Rivers*. SCOPE/John Wiley & Sons, Chichester: 233–243.

Ittekkot, V. & Zhang, S., 1989. Pattern of particulate nitrogen transport in world rivers. *Global Biogeochemical Cycles* **3**: 383–391.

Ittekkot, V., Safiullah, S. Mycke, B. & Seifert, R., 1985. Seasonal variability and geochemical significance of organic matter in the River Ganges, Bangladesh. *Nature* **317**: 800–802.

Iverson, T.M., Thorup, J. & Skriver, J., 1982. Inputs and transformation of allochthonous particulate organic matter in a headwater stream. *Holarct. Ecol.* **5**: 10-19.

Iwata, M., 1927. Trichopterous larvae from Japan. *Annot. Zool. Japon.* **11**: 203–233.

Iwata, M., 1928. Five new species of trichopterous larvae from Formosa. *Annot. Zool. Japon.* **11**: 341–343.

Iyer, R.R., 1994. Indian federalism and water resources. *Water Resources Development* **10**: 191–202.

Jäch, M.A., 1982. Neue Dryopoidea und Hydraenidae aus Ceylon, Nepal, Neu Guinea und der Turkei (Col.). *Koleopterologische Rundschau* **56**: 89–114.

Jäch, M.A., 1983. *Podelmis xanthogramma* n. sp. aus Malaya (Elmidae, Col.). *Zeitschrift der Arbeitsgemeinschaft Österreichischer Entomologen* **34**: 111–112.

Jäch, M.A., 1984a. Die Koleopterenfauna der Bergbäche von Südwest-Ceylon. *Arch. Hydrobiol., Suppl.* **69**: 228–332.

Jäch, M.A., 1984b. Die Gattung *Grouvellinus* im Himalaya und in Südostasien (Elmidae, Col.). *Koleopterologische Rundschau* **57**: 107–127.

Jäch, M.A., 1985. Beitrag zur Kenntnis der Elmidae und Dryopodae Neu Guineas (Coleoptera). *Rev. Suisse Zool.* **92**: 229–254.

Jäch, M.A., 1993. *Ancyronyx* (Coleoptera: Elmidae) — the Spider Riffle Beetle of the Malaysia forest rivers. *Nature Malaysiana* **18**: 86–89.

Jäch, M.A., 1994a. Description of *Gondraena* gen. n. from South India and Madagascar and *Davidraena* gen. n. from South India (Coleoptera:

Hydraenidae). *Koleopterologische Rundschau* **64**: 85–102.

Jäch, M.A., 1994b. A taxonomic review if the Oriental species of the genus *Ancyronyx* Erichson, 1847 (Coleoptera, Elmidae). *Rev. Zool. Suisse* **101**: 601–622.

Jäch, M.A., 1995a. Hydraenidae (Coleoptera). In: Jäch, M.A. & Ji, L. (Eds), *Water Beetles of China*. Zoologisch-Botanische Gesellschaft & Wiener Coleopterologenverein, Vienna: 173–180.

Jäch, M.A., 1995b. Hydrochidae (Coleoptera). In: Jäch, M.A. & Ji, L. (Eds), *Water Beetles of China*. Zoologisch-Botanische Gesellschaft & Wiener Coleopterologenverein, Vienna: 181–183.

Jäch, M.A., 1995c. Hygrobiidae (Coleoptera). In: Jäch, M.A. & Ji, L. (Eds), *Water Beetles of China*. Zoologisch-Botanische Gesellschaft & Wiener Coleopterologenverein, Vienna: 109–110.

Jäch, M.A., 1995d. Hydroscaphidae (Coleoptera). In: Jäch, M.A. & Ji, L. (Eds), *Water Beetles of China*. Zoologisch-Botanische Gesellschaft & Wiener Coleopterologenverein, Vienna: 33–34.

Jäch, M.A., 1995e. Eulichadidae: synopsis of the genus Eulichas JACOBSON from China, Laos and Vietnam (Coleoptera). In: Jäch, M.A. & Ji, L. (Eds), *Water Beetles of China*. Zoologisch-Botanische Gesellschaft & Wiener Coleopterologenverein, Vienna: 359–388.

Jäch, M.A., 1997. Daytime swarming of rheophilic water beeltes in Austria (Coleoptera: Elmidae, Hydraenidae, Haliplidae). *Latissimus* **September 1997**: 10–11.

Jäch, M. A. & Boukal, D.S., 1995a. Elmidae: 2. Notes on Macronychini, with descriptions of four new genera from China (Coleoptera). In: Jäch, M.A. & Ji, L. (Eds), *Water Beetles of China*. Zoologisch-Botanische Gesellschaft & Wiener Coleopterologenverein, Vienna: 299–323.

Jäch, M. A. & Boukal, D.S., 1995b. *Rudielmis* gen. n. from South India (Coleoptera: Elmidae). *Koleopterologische Rundschau* **65**: 149–157.

Jäch, M. A. & Boukal, D.S., 1995b. Description of two new riffle beetle genera from Peninsular Malaysia (Coleoptera: Elmidae). *Koleopterologische Rundschau* **66**: 179–189.

Jäch, M. A. & Boukal, D.S., 1997a. Description of two new genera of Macronychini: *Aulacosolus* and *Nesonychus* (Coleoptera: Elmidae). *Koleopterologische Rundschau* **67**: 207–224.

Jäch, M. A. & Boukal, D.S., 1997b. *Prionosolus* and *Podonychus*, two new genera of Macronychini (Coleoptera: Elmidae). *Entomological Problems* **28**: 9–23.

Jäch, M.A. & Jeng, L., 1995. *Nematopsephus* gen. n., a new genus of Psephenoidinae from Asia (coleoptera: Psephenidae). *Koleopterologische Rundschau* **65**: 159–167.

Jäch, M. A. & Kodada, J., 1995. Elmidae: 1. Checklist and bibliography of the Elmidae of China (Coleoptera). In: Jäch, M.A. & Ji, L. (Eds), *Water Beetles of China*. Zoologisch-Botanische Gesellschaft & Wiener Coleopterologenverein, Vienna: 289–298.

Jäch, M.A. & Kodada, J., 1996a. Three new riffle beetle genera from Borneo: *Homalosolus*, *Loxostirus* and *Rhopalonychus* (Coleoptera: Elmidae). *Annalen des Naturhistorischen Museums Wein* **98b**: 399–419.

Jäch, M.A. & Kodada, J., 1996b. *Graphosolus* gen. nov. from Southeast Asia (Coleoptera: Elmidae). *Entomological Problems* **27**: 93–98.

Jäch, M.A. & Lee, C.-F., 1994. Description of *Granuleubria*, a new genus of Eubriinae from West and South Asia (Coleoptera: Psephenidae). *Koleopterologische Rundschau* **64**: 223–232.

Jacob, J., 1957. Cytological studies of Melaniidae (Mollusca) with special reference to parthenogenesis and polyploidy, I. Oogenesis of the parthenogenetic species of *Melanoides* (Prosobranchia: Gastropoda). *Trans Roy. Soc. Edinburgh* **53**: 341–352.

Jacob, J., 1958. Cytological studies of Melaniidae (Mollusca) with special reference to parthenogenesis and polyploidy, I. A study of meiosis in the rare males of the polyploid race of *Melanoides tuberculatus* and *Melanoides lineatus*. *Trans Roy. Soc. Edinburgh* **63**: 433–444.

Jacob, U. & Glazaczow, A., 1986. *Pseudocentroptiloides*, a new baetid genus of Palaearctic and Oriental distribution. *Aquatic Insects* **8**: 197–206.

Jaensch, R., 1994. The middle Fly wetlands, Papua New Guinea. *Asian Wetland News* **7** (**2**): 14–15.

Jahi, J. & Sani, S., 1987. Development process, soil erosion and flash floods in the Kelang Valley Region, Peninsular Malaysia: a general consideration. *Arch. Hydrobiol. Beih., Ergebn. Limnol.* **28**: 399–405.

Jalal, K.F., 1987. Regional water resources situation: quantitative and qualitative aspects. In: Ali, M., Radosevich, G.E. & Ali Khan, A. (Eds), *Water Resources Policy for Asia*. A.A. Balkema Publishers, Boston: 13–34.

Jalihal, D.R. & Sankolli, K.N., 1975. On the abbreviated metamorphosis of the freshwater prawn *Macrobrachium hendersodayanum* (Tiwari) in the laboratory. *Karnatak Univ. J. Sci.* **20**: 283–291.

Jalihal, D.R., Sankoli, K.N. & Shenoy, S., 1993. Evolution of larval development patterns and the process of freshwaterization in the prawn genus *Macrobrachium* Bate, 1868 (Decapoda, Palaemonidae). *Crustaceana* **65**: 363–376.

Janaki Ram, K. & Radhakrishna, Y., 1984. The distribution of freshwater molluscs in Guntur District (India) with a description of *Scaphula nagarjunai* sp. n. (Arcidae). *Hydrobiologia* **119**: 49–55.

Jansson, A, 1982. Notes on some Corixidae (Heteroptera) from New Guinea and New Caledonia. *Pacific Insects* **24**: 95–103.

Jassby, A.D. & Powell, T.H., 1990. Detecting changes in ecological time series. *Ecology* **71**: 2044–2052.

Jayawardena, A.W. & Peart, M.R., 1989. Spatial and temporal variation of rainfall and runoff in Hong Kong. In: Roald, L., Nordseth, K. & Hassel, K.A. (Eds), *FRIENDS in Hydrology*. IAHS Press, Wallingford, U.K.: 409–418.

Jebensan. A. & Selvanayagam, M., 1994. Population dynamics of aquatic hemipterans in the River Cooum, Madras. *J. Environ. Biol.* **15**: 213–220.

Jeng, M.L. & Yang, P.S., 1991. Elmidae of Taiwan part I: two new species of the genus *Stenelmis* (Coleoptera: Dryopoidea) with notes on the group of *Stenelmis hisamatsui. Entomological News* **102**: 236–252.

Jeng, M.L. & Yang, P.S., 1993. Elmidae of Taiwan part II: redescription of *Leptelmis formosana* (Coleoptera: Dryopoidea). *Entomological News* **104**: 53–59.

Jewett, S.G., 1958. Stoneflies from the Philippines (Plecoptera), *Fieldiana (Zoology)* **42**: 77–87.

Jewett, S.G., 1970. Stonefly records from the Northwest (Punjab) Himalaya. *Oriental Insects* **4**: 481–482.

Jha, P.K., Vaithiyanathan, P. & Subramanian, V., 1993. Mineralogical characteristics of the sediments of a Himalayan River — Yamuna River — a tributary of the Ganges. *Environmental Geology* **22**: 13–20.

Jhingran, A.G., 1980. Riverine fishery resources of India and their socio-cultural impact. In: Furtado, J.I. (Ed.), *Tropical Ecology and Development. Proceedings of the Vth International Symposium of Tropical Ecology.* International Society of Tropical Ecology, Kuala Lumpur: 747–756.

Jhingran, A.G. & Ghosh, K.K., 1978. The fisheries of the Ganga River system in the context of Indian aquaculture. *Aquaculture* **14**: 141–162.

Jhingran, V.G., 1975. *Fish and Fisheries of India.* Hindustan Publishing Corporation, New Delhi: 954 pp.

Jhingran, V.G., Ahmad, S.H. & Singh, A.K., 1989. Application of Shannon-Wiener index as a measure of pollution of River Ganga at Patna, Bihar, India. *Curr. Sci. (Bangalore)* **58**: 717–720.

Ji, L. & Jäch, M. A., 1996. Amphizoidae (Coleoptera). In: Jäch, M.A. & Ji, L. (Eds), *Water Beetles of China.* Zoologisch-Botanische Gesellschaft & Wiener Coleopterologenverein, Vienna: 103–108.

Jin, X, Liu, H., Tu, Q., Zhang, Z. & Zhu, X., 1990. *Eutrophication of Lakes in China.* Chinese Research Academy of Environmental Sciences, Beijing: 652 pp.

Johanson, K.A., 1994. *Helicopsyche siama* sp. n. from Thailand (Trichoptera: Heliocpsychidae). *Aquatic Insects* **16**: 17–20.

Johanson, K.A., 1997. Zoogeography and diversity of the snail case caddisflies (Trichoptera: Helicopsychidae). In: Holzenthal, R.W. & Flint, O.S. (Eds), *Proceedings of the 8th International Symposium on Trichoptera.* Ohio Biological Survey, Columbus: 205–212.

John, L. & Fernandez, T.V., 1989. Influence of environmental factors on the burrowing behaviour of an estuarine bivalve, *Villorita cyprinoides*, from Veli Lake, S.W. coast of India. *Journal of Ecobiology* **1**: 137–148.

Johnsingh, A.J.T. & Joshua, J., 1989. The threatened gallery forest of the River Tambiraparani, Mundanthurai Wildlife Sanctuary, South India. *Biological Conservation* **47**: 273–280.

Johnson, D.S., 1957. A survey of Malayan freshwater life. *Malay. Nat. J.* **12**: 57–65.

Johnson, D.S., 1961a. A synopis of the Decapoda Caridea and Stenopodidea of Singapore, with notes on their distribution and a key to the genera of Caridea occurring in Malayan waters. *Bull. National Mus., Singapore* **30**: 44–79.

Johnson, D.S., 1961b. Notes on the freshwater Crustacea of Malaya. I. The Atyidae. *Bull. Raffles Mus., Singapore* **26**: 120–153.

Johnson, D.S., 1964. Distributional and other notes on some freshwater prawns (Atyidae and Palaemonidae) mainly from the Indo-West Pacific Region. *Bull. National Mus., Singapore* **32**: 5–30.

Johnson, D.S., 1967a. On the chemistry of freshwaters in southern Malaya and Singapore. *Arch. Hydrobiol.* 63: 447–496.

Johnson, D.S., 1967b. Distributional patterns of Malayan freshwater fish. *Ecology* **48**: 722–730.

Johnson, D.S., 1968. Water pollution in Malaysia and Singapore: some comments. *Malay. Nat. J.* **21**: 221–222.

Jones, T.C. & Lester, R.J.G., 1992. The life history and biology of *Diceratocephala boschmai* (Platyhelminthes; Temnocephalida), an ectosymbiont on the redclaw crayfish *Cherax quadricarinatus*. *Hydrobiologia* **248**: 193–199.

Joseph, G.L. & Job, S.V., 1983. Studies on spawning migration and spawning of hillstream fish *Discognathus mullya* (Sykes). *J. Bombay Nat. Hist. Soc.* 80: 80–85.

Joshi, P.C., 1991. The degrading fish habitats of the Kumaun Himalaya. In: Bhatt, S.D. & Pande, R.K. (Eds), *Ecology of the Mountain Waters*. Ashish Publishing House, New Delhi: 311–320.

Joshi, P.C. & Kanna, S.S., 1982b Seasonal changes in the ovary of a freshwater crab, *Potamon koolooense* (Rathbun). *Proc. Indian Acad. Sci., Anim. Sci.* **91**: 451–462.

Joshi, P.C. & Kanna, S.S., 1982a. Seasonal changes in the testes of a freshwater crab, *Potamon koolooense* (Rathbun). *Proc. Indian Acad. Sci., Anim. Sci.* **91**: 439–450.

Jowett, I.G., 1997. Instream flow methods: a comparison of approaches. *Regulated Rivers: Research & Managment* **13**: 115–127.

Joy, C.M., 1990. Toxicity testing with freshwater algae in River Periya (India). *Bull environ. Contam. Toxicol.* **45**: 915–922.

Julka, J.M., Vasisht, H.S. & Bala, B., 1988. Ecology of invertebrates of two hill streams of Solan (H.P.). *Res. Bull. (Sci.) Panjab Univ.* **39**: 139–143.

Junk, W.J., Bayley, P.B. & Sparks, R.E., 1989. The flood pulse concept in river-floodplain systems. *Can. Spec. Publ. Fish. Aquat. Sci.* **106**: 110–127.

Kakati, V.A., 1988. Larval development of the Indian spider crab *Elamenopsis demeloi* (Kemp) (Brachyura, Hymenosomatidae) in the laboratory. *Mahasagar* **21**: 219–227.

Kale, V.S., Ely, L.L., Enzel, Y. & Baker, V.R., 1994. Geomorphic and hydrologic aspects of monsoon floods on the Narmada and Tapi Rivers in central India. *Geomorphology* **10**: 157–168.

Kaltenbach, A., 1973. Results of the Austrian-Ceylonese hydrological expedition, 1970. Part XIII. Some remarkable ripicol insects of the Ceylonese Fauna. *Bull. Fish. Res. Stn, Sri Lanka (Ceylon)* **24**: 125–128.

Kamita, T., 1956. Ecological notes on the freshwater shrimps and prawns of Japan. I. On the shrimp *Caridina japonica* De Man. *Zool. Mag., Zool. Soc. Japan* **65**: 440–444 (in Japanese).

Kamita, T., 1957. Ecological notes on the freshwater shrimps and prawns of Japan. III. On the shrimps *Caridina leucosticta* Stimpson and *C. serratirostris celebensis* De Man. *Zool. Mag., Zool. Soc. Japan* **66**: 327–331 (in Japanese).

Kamita, T., 1959. Ecological notes on the freshwater shrimps and prawns of Japan. IV. On the shrimp *Caridina typus* H. Milne-Edwards. *Zool. Mag., Zool. Soc. Japan* **68**: 21–24 (in Japanese).

Kang, S.C. & Yang, C.T., 1994a. Three new species of the genus Ameletus from Taiwan (Ephemeroptera: Siphlonuridae). *Chinese J. Entomol.* **14**: 261–269.

Kang, S.C. & Yang, C.T., 1994b. The nymph of *Isonychia formosana* (Ulmer, 1912) (Ephemeroptera: Oligoneuridae). *J. Taiwan Mus.* **47**: 1–3.

Kang, S.C. & Yang, C.T., 1994c. Heptageniidae of Taiwan (Ephemeroptera). *J. Taiwan. Mus.* **47**: 5–36.

Kang, S.C. & Yang, C.T., 1994d. Leptophlebiidae of Taiwan. *J. Taiwan Mus.* **47**: 57–82.

Kang, S.C. & Yang, C.T., 1994e. Caenidae of Taiwan (Ephemeroptera). *Chinese J. Entomol.* **14**: 93–113.

Kang, S.C. & Yang, C.T., 1996a. Two new species of *Baetis* Leach (Ephemeroptera: Baetidae) from Taiwan. *Chinese J. Entomol.* **16**: 61–66.

Kang, S.C. & Yang, C.T., 1996b. A new species of *Caenis* Stephens (Ephemeroptera: Caenidae) from Taiwan. *Zhonghua Kunchong* **16**: 55–59.

Kang, S.C., Chang, H.C. & Yang, C.T., 1994. A revision of the genus *Baetis* in Taiwan (Ephemeroptera, Baetidae). *J. Taiwan Mus.* **47**: 9–44.

Kang, Z., 1983. A new genus and three new species of the family Hydrobiidae (Gastropoda: Prosobranchia) from Hubei Province, China. *Oceanol. Limnol. Sinica* **14**: 499–505 (in Chinese).

Kang, Z., 1986a. On two new species of the genus *Akiyoshia* (Gastropoda: Hydrobiidae) from China. *Oceanol. Limnol. Sinica* **17**: 276–282 (in Chinese).

Kang, Z., 1986b. Descriptions of eight new minute freshwater snails and a new and rare species of land snail from China (Prosobranchia: Pomatiopsidae, Hydrobiidae, Hydrocenidae). *Arch. Mollusk.* **117**: 73–91.

Kannan, K., Sinha, R.K., Tanabe, S., Ichihashi, H. & Tatsukawa, R., 1993. Heavy metals and organochlorine residues in Ganges River dolphins. *Mar. Pollut. Bull.* **26**: 159–162.

Kapur, A.P & Kripalani, M.B., 1961. The mayflies (Ephemeroptera) from the north-western Himalaya. *Rec. Indian Mus.* **59**: 183–221.

Karaman, G., 1984. Remarks to the freshwater Gammarus species (Fam. Gammaridae) from Korea, China, Japan and some adjacent regions (contribution to the knowledge of the Amphipoda 134). *The Montenegrin Academy of Sciences and Arts, Glasnik of the Section of Natural Sciences* **4**: 139–162.

Karlson, R.H., 1991. Recruitment and local persistence of a freshwater bryozoan in stream riffles. *Hydrobiologia* **226**: 119–128.

Kaul, B.K., Sharma, S.C. & Bali, R.K., 1987. Some observations on the ecology of streams of Kangra Valley. *Indian J. Ecol.* **14**: 136–139.

Kaushik, N.K. & Hynes, H.B.N., 1968. Experimental study on the role of autumn shed leaves in aquatic environments. *J. Ecol.* **52**: 229-243.

Kaushik, N.K. & Hynes, H.B.N., 1971. The fate of the dead leaves that fall into streams. *Arch. Hydrobiol.* **68**: 465-515.

Kawai, T., 1961. A plecopteran nymph from Thailand. In: Kira, T. & Umesao, T. (Eds), *Nature and Life in Southeast Asia, Volume 1.* Japanese Society for the Promotion of Science, Tokyo: 199–201.

Kawai, T., 1963. Stoneflies (Plecoptera) from Afghanistan, Karakoram, and Punjab Himalaya. *Results of the Kyoto University Scientific Expedition to the Karakoram and Hindukush, 1955* **3**: 53–86.

Kawai, T., 1966. Plecoptera from the Hindukush. *Results of the Kyoto University Scientific Expedition to the Karakoram and Hindukush, 1955* **8**: 203–216.

Kawai, T., 1968a. Stoneflies (Plecoptera) from Thailand and India with descriptions of a new genus and two new species. *Oriental Insects* **2**: 107–139.

Kawai, T., 1968b. A new species of *Capnia* from Mt. Everest. *Kumbu Himal* **3**: 29–36.

Kawai, T., 1968c. Stoneflies (Plecoptera) from Taiwan in the Bishop Museum, Honolulu. *Pacific Insects* **10**: 241–248.

Kawai, T., 1969. Stoneflies (Plecoptera) from Southeast Asia. *Pacific Insects* **11**: 613–625.

Kawai, T, 1973. Plecoptera from Ceylon. *Ent. Scand., Suppl.* **4**: 5–78.

Kawai, T., 1974a. The second species of the genus *Scopura* (Plecoptera, Scopuridae). *Bull. Natn. Sci. Mus. Tokyo* **17**: 275–281.

Kawai, T., 1974b. A note of the nymph of *Sopkalia yamadae* (OKAMOTO) (Plecoptera, Perlodidae). *Kontyû, Tokyo* **42**: 1–6.

Kawai, T., 1976. A catalogue of Japanese Plecoptera. *Nara Hydrobiol.* **5**: 5–46.

Kawai, T., 1985a. *An Illustrated Book of Aquatic Insects of Japan.* Tokai University Press, Tokai: 408 pp. (in Japanese).

Kawai, T., 1985b. Hymenoptera. In: Kawai, T. (Ed.) *An Illustrated Book of Aquatic Insects of Japan.* Tokai University Press, Tokai: 261–262 (in Japanese).

Kawai, T. & Isobe, Y., 1985. Plecoptera. In: Kawai, T. (Ed.), *An Illustrated Book of Aquatic Insects of Japan.* Tokai University Press, Tokai: 125–158 (in Japanese).

Kawakatsu, M., 1964. On the ecology and distribution of freshwater planarians from the Japanese Islands, with special reference to their vertical distribution. *Hydrobiologia* **26**: 349–410.

Kawakatsu, M., 1973. Report on freshwater planarians from Pakistan. *Bull. Fuji Women's Coll. (Series II)* **11**: 79–95.

Kawakatsu, M. & Basil, J.A., 1971. Records of freshwater and land planarians from India. *Bull. Fuji Women's Coll. (Series II)* **9**: 41–50.

Kawakatsu, M. & Iwaki, S., 1968. Report on freshwater planaria from Taiwan (Formosa). *Bull. Fuji Women's Coll. (Series II)* **6**: 129–137.

Kawakatsu, M. & Mitchell, R.W., 1989a. *Dugesia deharvengi* sp. nov., a new trogloditic freshwater planarian from Tam Kubio Cave, Thailand (Turbellaria; Tricladida; Paluicola). *Bull. Biogeog. Soc. Japan* **44**: 175–182.

Kawakatsu, M. & Mitchell, R.W., 1989b. Record of a trogloditic planarian from Tanette Cave located in the Maros Karst, Sulawesi (Celebes), Indonesia (Turbellaria, Tricladida, Paludicola). *Bull. Fuji Women's Coll. (Series II)* **27**: 35–40.

Kawakatsu, M. & Ogawara, G., 1974. Additional report on freshwater planarians from north Borneo, Malaya, Sri Lanka, India, and South Africa. *Bull. Fuji Women's Coll. (Series II)* **12**: 69–86.

Kawakatsu, M. & Mitchell, R.W., Oki, I., Tamura, S. & Yussov, S., 1989. Taxonomic and karyological studies of *Dugesia batuensis* Ball, 1970 (Turbellaria, Tricladida, Paludicola), from the Batu Caves, Malaysia. *J. Speleol. Soc. Japan* **14**: 1–14.

Kawakatsu, M., Oki, I., Tamura, S., Yamayoshi, T., Lue, K.Y. & Hagiya, M., 1979. Additional report on freshwater planarians from Taiwan. *Bull. Fuji Women's Coll. (Series II)* **17**: 59–91.

Kato, H., 1997. Habitat selection by *Micrasema uenoi* in a Japanese mountain stream (Trichoptera: Brachycentridae). In: Holzenthal, R.W. & Flint, O.S. (Eds), *Proceedings of the 8th International Symposium on Trichoptera*. Ohio Biological Survey, Columbus: 213–219.

Keffer, S.L., Polhemus, J.T. & McPherson, J.E., 1989. Notes on critical character states in *Telmatotrephes* (Heteroptera: Nepidae). *Florida Entomologist* **72**: 626–629.

Keffer, S.L., Polhemus, J.T. & McPherson, J.E., 1990. What is *Nepa hoffmanni* (Heteroptera: Nepidae)? Male genitalia hold the answer, and delimit species groups. *J. N.Y. Entomol. Soc.* **98**: 154–162.

Kelley, R.W., 1984. Phylogeny, morphology and classification of the micro-caddisfly genus *Oxyethira* Eaton (Trichoptera: Hydroptilidae). *Trans. Am. Entomol. Inst.* **110**: 435–463.

Kernas, B.L., Peckarsky, B.L. & Anderson, C., 1995. Estimates of mayfly mortality: is stonefly predation a significant source? *Oikos* **74**: 315–323.

Kettle, D.S., 1977. Biology and bionomics of bloodsucking ceratopogonids. *Ann. Rev. Entomol.* **22**: 33–51.

Khalil, G.M., 1990. Floods in Bangladesh: a question of disciplining the rivers. *Natural Hazards* **3**: 379–401.

Khan, A.A. & Kulshrestha, S.K., 1993. Benthic fauna in relation to pollution: a case study at River Chambal near Kota in Central India. *Environment International* **19**: 597–610.

Khan, I.S.A.N., 1990a. Diatom distribution and inter-site relationship in the Linggi River Basin, peninsular Malaysia. *Malay. Nat. J.* **44**: 85–95.

Khan, I.S.A.N., 1990b. Assessment of water pollution using diatom community structure and species distribution — a case study in a tropical river basin. *Int. Rev. ges. Hydrobiol.* **75**: 317–338.

Khan, I.S.A.N., 1991. Efect of urban and industrial wastes on species diversity of the diatom community in a tropical river, Malaysia. *Hydrobiologia* **224**: 175–184.

Khan, I.S.A.N., 1992. A study of the impact of urban and industrial development in the Linggi River Basin, Malaysia. 2. Chemical environment. *Int. Rev. ges. Hydrobiol.* **77**: 391–419.

Khan, I.S.A.N. & Lim, R.P., 1991. Distribution of metals in the Linggi River Basin, Malaysia, with reference to pollution. *Aust. J. Mar. Freshwat. Res.* **42**: 435–449.

Khan, I.A.S.N., Furtado, J.I. & Lim, R.P., 1987. Periphyton on artificial and natural substrates in a tropical river. *Arch. Hydrobiol. Beih., Ergebn. Limnol.* **28**: 473–484.

Khan, L.R., 1991. Impacts of recent floods on the rural environment of Bangladesh: a case study. *Water Resources Development* **7**: 45–52.

Khan, M.H., 1924. Observations on the breeding habits of some freshwater fishes in the Punjab. *J. Bombay Nat. Hist. Soc.* **29**: 958–962.

Khan, S.M., 1985. Management of river and reservoir sedimentation in Pakistan. *Water International* **10**: 18–21.

Khattak, G.M., 1983. The watershed management program in Mansehra, Pakistan. In: Hamilton, L.S. (Ed.), *Forest and Watershed Development and Conservation in Asia and the Pacific*. Westview Press, Boulder, Colorado: 359–410.

Khondker, M., 1994. The status of limnological research in Bangladesh. *Mitt. Internat. Verein. Limnol.* **24**: 147–154.

Khoo, K.H., Leong, T.S., Soon, F.L., Tan, S.P. & Wong, S.Y., 1987. Riverine fisheries in Malaysia. *Arch. Hydrobiol. Beih., Ergebn. Limnol.* **28**: 261–268.

Khulbe, R.D., 1991. Watermoulds of the uplands: an ecological approach. In: Bhatt, S.D. & Pande, R.K. (Eds), *Ecology of the Mountain Waters*. Ashish Publishing House, New Delhi: 138–148.

Kiffney, P.M., 1996. Main and interactive effects of invertebrate density, predation, and metals on a Rocky Mountain stream macroinvertebrate community. *Can. J. Fish. Aquat. Sci.* **53**: 1595–1601.

Kikuchi, M & Sasa, M., 1990. Studies on the chironomid midges (Diptera, Chironomidae) of the Lake Toba area, Sumatra, Indonesia. *Jap. J. Sanit. Zool.* **41**: 291–329.

Kim, J.W., 1974. On the larvae of Trichoptera from Korea. *Korean J. Limnol.* **7**: 1–42 (in Korean).

Kimmins, D.E., 1947. New species of Himalayan Plecoptera. *Ann. Mag. Nat. Hist.* Ser. **11** (13): 721–740.

Kimmins, D.E., 1950. Some Assamese Plecoptera, with descriptions of new species of Nemouridae. *Ann. Mag. Nat. Hist.* Ser. **13** (3): 194–209.

Kimmins, D.E., 1952. Entomological results from the Swedish Expedition 1934 to Burma and British India. Trichoptera, Part 1. Families Phryganeidae, Limnephilidae and Sericostomatidae. *Arkiv för Zoologi* **3**: 173–178.

Kimmins, D.E., 1953a. Entomological results from the Swedish Expedition 1934 to Burma and British India. Trichoptera. The genus *Rhyacophila* PICTET (Fam. Rhyacophilidae). *Arkiv för Zoologi* **4**: 505–555.

Kimmins, D.E., 1953b. Entomological results from the Swedish Expedition 1934 to Burma and British India. Trichoptera (Rhyacophilidae, subfamilies Hydrobiosinae, Glossosomatinae and Agapetinae). *Arkiv för Zoologi* **6**: 167–183.

Kimmins, D.E., 1955a. Entomological results from the Swedish Expedition 1934 to Burma and British India. Trichoptera (Philopotamidae, genera *Wormalia* McLACHLAN, *Doloclanes* BANKS and *Dolophilodes* ULMER). *Arkiv för Zoologi* **9**: 67–92.

Kimmins, D.E., 1955b. Results of the Oxford University Expedition to Sarawak, 1932. Order Trichoptera. *The Sarawak Museum Journal* **6**: 374–442.

Kimmins, D.E., 1957. Entomological results from the Swedish Expedition 1934 to Burma and British India. Trichoptera. The genus *Chimarra* STEPHENS (Fam. Philopotamidae). *Arkiv för Zoologi* **11**: 53–75.

Kimmins, D.E., 1962. Miss L.E. Cheesman's expeditions to New Guinea. *Bull. British Mus. (Nat. Hist.)* **11**: 97–187.

Kimmins, D.E., 1963. On the Leptocerinae of the Indian Sub-continent and North-east Burma. *Bull. Br. Mus. nat. Hist. (Ent.)* **14**: 263–316.

Kirmani, S.S., 1990. Water, peace and conflict management: the experience of the Indus and Mekong river basins. *Water International* **15**: 200–206.

Kitikoon, V., Sornmani, S. & Schneider, C.R., 1981. Studies on *Tricula aperta* and related taxa, the snail intermediate hosts of *Schistosoma mekongi* 1. Geographical distribution and habits. *Malacol. Rev.* **14**: 1–10.

Kjaerandsen, J. & Netland, K., 1997. *Pahamunaya occidentalis*, new species, from West Africa, with a redescription of *Cyrnodes scotti* Ulmer (Trichoptera: Polycentropodidae). In: Holzenthal, R.W. & Flint, O.S. (Eds), *Proceedings of the 8th International Symposium on Trichoptera*. Ohio Biological Survey, Columbus: 249–257.

Klausnitzer, B., 1976. Ergenbisse der Bhutan-Expedition 1972 des Naturhistorischen Museums in Basel (Coleoptera: Fam. Helodidae). *Deutsche Entomologishe Zeitschrift* **23**: 213–220.

Klausnitzer, B., 1979. Neue Arten der Gattung *Cyphon* Paykull aus Neuguinea

(Coleoptera, Helodidae) (48. Beitrag zur Kenntnis der Helodidae). *Reichenbachia* **17**: 1–8.

Klausnitzer, B., 1980a. Neue Arten der Gattung *Cyphon* Paykull aus Sumatra und Neuguinea (Coleoptera, Helodidae) (64. Beitrag zur Kenntnis der Helodidae). *Entomologische Berichten* (Amsterdam) **40**: 169–175.

Klausnitzer, B., 1980b. Zur Kenntnis der Helodidae des Himalaja-Gebietes (Col.) (61. Beitrag zur Kenntnis der Helodidae). *Entomologica Basiliensia* **5**: 195–214.

Klausnitzer, B., 1980c. Zur Kenntnis der Helodidae von Vietnam (Coleoptera). *Folia Entomologica Hungarica* **33**: 87–94.

Klausnitzer, B., 1980d. Sudostasiatische neue Arten aus der Gattung *Cyphon* Paykull, 1799 (Coleoptera, Helodidae) (60. Beitrag zur Kenntnis der Helodidae). *Reichenbachia* **18**: 219–226.

Klausnitzer, B., 1980e. Eine neue Artengruppe der Gattung *Flavohelodes* Klausnitzer, 1980 (Coleoptera, Helodidae) (55. Beitrag zur Kenntnis der Helodidae). *Reichenbachia* **18**: 85–87.

Klausnitzer, B., 1995. Scirtidae (Coleoptera). In: Jäch, M.A. & Ji, L. (Eds), *Water Beetles of China*. Zoologisch-Botanische Gesellschaft & Wiener Coleopterologenverein, Vienna: 109–110.

Klausnitzer, B. & Pospisil, P., 1991. Larvae of *Cyphon* sp. (Coleoptera, Helodidae) in ground water. *Aquatic Insects* **13**: 161–165.

Klepper, O., 1992. Model study of the Negara River Basin to assess the regulating role of its wetlands. *Regulated Rivers: Research & Management* **7**: 311–325.

Klepper, O., Chairuddin, G.T., Iriansyah & Rijksen, H.D., 1992. Water quality and the distribution of some fishes in an area of acid sulphate soils, Kalimantan, Indonesia. *Hydrobiol. Bull.* **25**: 217–224.

Kluge, N.Yu., 1982. The mayflies (Ephemeroptera) of the Taymyr National Territory. *Entomol. Rev.* **59**: 41–59.

Kluge, N.Yu., 1985. Mayflies of the subgenus *Euthraulus* Barn. (Ephemeroptera, Leptophlebiidae, genus *Choroterpes*) of the USSR. *Entomol. Rev.* **63**: 56–62.

Kluge, N.Yu., 1988. Generic revison of the family Heptageniidae (Ephememeroptera). I. Diagnosis of tribes, genera and subgenera of the subfamily Heptageniinae. *Entomol. Obozr.* **67**: 291–313 (in Russian).

Kluge, N.Yu., 1993. Generic revison of the family Heptageniidae (Ephememeroptera). II. Phylogeny. *Entomol. Obozr.* **72**: 39–54 (in Russian).

Kluge, N.Yu. & Novikova, 1992. Revision of the Palaearctic genera and subgenera of mayflies of the subfamily Cloeoninae (Ephemeroptera, Baetidae) with description of new species from the USSR. *Entomol. Obozr.* **71**: 60–83 (in Russian).

Kluge, N.Y., Studemann, D., Landolt, P. & Gonser, T., 1995. A reclassification of Siphlonuridae (Ephemeroptera). *Mitt. Schweiz. Entomol. Gesell.* **68**: 103–132.

Knott, B., 1986. In: Botosaneanu, L. (Ed.), *Stygofauna Mundi. A Faunistic, Distributional and Ecological Synthesis of the World Fauna Inhabiting Subterranean Waters (Including the Marine Interstitial)*. E.J. Brill, Leiden: 486–492.

Kobayashi, J., 1959. Chemical investigation on river waters of southeastern Asiatic countries (Report I). The quality of waters in Thailand. *Berichte d. Ohara Instituts* **11**: 167–233.

Kobayashi, M., 1980. A revision of the family Philopotamidae from Japan (Trichoptera: Insecta). *Bull. Kanagawa Pref. Mus.* **12**: 85–104.

Kobayashi, M., 1982. A classification for Japanese species of *Glossosoma* (Trichoptera, Insecta). *Bull. Kanagawa Pref. Mus.* **13**: 1–30.

Kobayashi, M., 1987. Systematic study of the caddisflies from Taiwan with descriptions of 11 new species. *Bull. Kanagawa Pref. Mus.* **17**: 37–48.

Kobayashi, T., 1995. *Eurycnemus* sp. (Diptera: Chironomidae) larvae ectoparasitic on pupae of *Goera japonica* (Trichoptera: Limnephilidae). In: Cranston, P., (Ed.), *Chironomids: From Genes to Ecosystems*. CSIRO Publications, Melbourne: 317–322.

Kodada, J., 1992. *Pseudamophilus davidi* sp.n. from Thailand (Coleoptera: Elmidae). *Linzer biologische Beiträge* **24**: 359–365.

Kodada, J., 1993. *Dryopomorphus siamensis* sp. nov., a new riffle beetle from Thailand (Coleoptera: Elmidae) and remarks on the morphology of the mouthparts and hind wing venation of *D. bishopi* Hinton. *Entomological Problems* **24**: 51–58.

Kodada, J. & Jäch, M.A., 1995a. Dryopidae: 1. Checklist and bibliography of the Dryopidae of China (Coleoptera). In: Jäch, M.A. & Ji, L. (Eds), *Water Beetles of China*. Zoologisch-Botanische Gesellschaft & Wiener Coleopterologenverein, Vienna: 325–328.

Kodada, J. & Jäch, M.A., 1995b. Dryopidae: 2. Taxonomic review of the Chinese species of the genus *Helichus* ERICHSON (Coleoptera). In: Jäch, M.A. & Ji, L. (Eds), *Water Beetles of China*. Zoologisch-Botanische Gesellschaft & Wiener Coleopterologenverein, Vienna: 329–339.

Kolasa, J., 1991. Flatworms: Turbellaria and Nemertea. In: Thorp, J.H. & Covich, A.P. (Eds), *Ecology and Classification of North American Freshwater Invertebrates*. Academic Press, San Diego: 145–163.

Konar, S.K., 1977. Pesticides and aquatic ecosystems. *Indian J. Fish.* **22**: 80–95.

Kondo, T., Matsamura, N., Hashimoto, M. & Nagata, Y., 1987. Emergence of the rose bitterling larvae from the host mussel. *Memoirs of Osaka Kyoiku University III, Natural Science & Applied Science* **367**: 23–26 (in Japanese).

Kondo, T., Yamashita, J. & Kano, M., 1984. Breeding ecology of five species of bitterling (Pisces: Cyprinidae) in a small creek. *Physiology & Ecology Japan* **21**: 53–62.

Kondratieff, B.C. & Voshell, J.R., 1983. Subgeneric and species-group classification of the mayfly genus *Isonychia* in North America

(Ephemeroptera: Oligoneuriidae). *Proc. Entomol. Soc. Wash.* **85**: 128–138.

Kortmulder, K., 1987. Ecology and behaviour in tropical freshwater fish communities. *Arch. Hydrobiol. Beih., Ergebn. Limnol.* **28**: 503–513.

Kortmulder, K., Feldbrügge, E.J. & de Silva, S.S., 1978. A combined field study of *Barbus* (= *Puntius*) *nigrofasciatus* Günther (Pisces, Cyprinidae) and water chemisty of its habitat in Sri Lanka. *Neth. J. Zool.* **28**: 111–131.

Kortmulder, K., Padmanabhan, C. & de Silva, S.S., 1990. Patterns of distribution and endemism in some cyprinid fishes as determined by the geomorphology of South-west Sri Lanka and South Kerala (India). *Ichthyol. Explor. Freshwaters* **1**: 97–112.

Koteswaram, P., 1974. Climate and meteorology of humid tropical Asia. In: *Natural Resources of Humid Tropical Asia*. Natural Resources Research, XII, Unesco, Paris: 27–85.

Kottelat, M., 1984. Revision on the Indonesian and Malaysian loaches of the subfamily Noemacheilinae. *Jap. J. Ichthyol.* **31**: 225–260.

Kottelat, M., 1985. Freshwater fishes of Kampuchea. *Hydrobiologia* **121**: 249–279.

Kottelat, M., 1988. Indian and Indochinese species of *Balitora* (Osteichthyes: Cypriniformes) with descriptions of two new species and comments on the family-group names Balitoridae and Homalopteridae. *Rev. Suisse Zool.* **95**: 487–504.

Kottelat, M., 1989. Zoogeography of the fishes from Indochinese inland waters with an annotated checklist. *Bull. zool. Mus., Univ. Amsterdam* **12**: 1–56.

Kottelat, M. & Whitten, A.J., 1993. *Freshwater Fishes of Western Indonesia and Sulawesi*. Periplus Editions (HK) Ltd, Hong Kong: 291 pp.

Kottelat, M. & Whitten, T., 1996. Freshwater biodiversity in Asia with special reference to fish. *World Bank Technical Paper* **343**: 1–59.

Kovac, D. & Yang, C.M., 1990. A preliminary checklist of the semiaquatic and aquatic Hemiptera (Heteroptera: Gerromorpha and Nepomorpha) of Ulu Kinchin, Pahang, Malaysia. *Malay. Nat. J.* **43**: 282–288.

Kreis, H.A., 1936. Suswasser-nematoden aus der umbegung von Madras (India). *Rev. Suisse Zool.* **43**: 6741–645.

Krishna, J.H., 1983. Water resources development in Thailand. *Water International* **8**: 154–157.

Krishna Murthy, D. & Rajagopal, K.V., 1990. Food and feeding habits of the freshwater prawn, *Macrobrachium equidens* (Dana). *Indian J. Anim. Sci.* **69**: 118–122.

Kruuk, H., Kanchanasaka, B., O'Sullivan, S. & Wanghongsa, S., 1994. Niche separation in three sympatric otters *Lutra perspicillata*, *L. lutra* and *Aonyx cinerea* in Huai Kha Khaeng, Thailand. *Biological Conservation* **69**: 115–120.

Kubo, I., 1938. On the Japanese atyid shrimps. *Journal of the Imperial Fisheries Institute* **33**: 73–83.

Kuiper, J.G.J., 1982. Mollusca, Lamellibranchiata, Sphaeriidae: *Pisidium* from Sri Lanka (Ceylon). *Entomologica Scandanavia Supplementum* **11**: 28–29.

Kulshrestha, S.K., Adholia, U.N., Khan, A.A., Bhatnagar, A., Saxena, M. & Baghail, M., 1989. Community structure of planktons and macrozoobenthos, with special reference to pollution in the River Khan (India). *Int. J. Environ. Stud.* **35**: 83–96.

Kulshrestha, S.K., Adholia, U.N., Bhatnagar, A., Khan, A.A. & Baghail, M., 1991. Community structure of zooplankton at River Chambal near Nagda with reference to industrial pollution. *Acta Hydrochim. Hydrobiol.* **19**: 181–191.

Kumanski, K., 1979. Trichoptera (Insecta) from New Guinea. *Aquatic Insects* **1**: 193–219.

Kumanski, K., 1980. A contribution to the knowledge of Trichoptera (Insecta) of the Causcasus. *Acta Zoologica Bulgarica* **14**: 32–48.

Kumanski, K., 1992. Studies on Trichoptera of Korea (North) 3. Superfamily Hydropsychoidea. *Insecta Koreana* **9**: 52–77.

Kumar, A., 1970. Bionomics of *Orthetrum pruinosum neglectum* Rambur (Odonata: Libellulidae). *Bull. Ent.* **11**: 85–93.

Kumar, A., 1973a. Descriptions of the last instar larvae of Odonata from the Dehra Dun Valley (India), with notes on biology I. Suborder Zygoptera. *Oriental Insects* **7**: 83–118.

Kumar, A., 1973b. Descriptions of the last instar larvae of Odonata from the Dehra Dun Valley (India), with notes on biology II. Suborder Anisoptera. *Oriental Insects* **7**: 291–331.

Kumar, A., 1976. Biology of Indian dragonflies with special reference to seasonal regulation and larval development. *Bull. Entomol.* **17**: 37–47.

Kumar, A., 1977. Last instar larvae of two Odonata species from Western Himalayas. *Entomon* **2**: 225–230.

Kumar, A., 1984. Studies on the life history of Indian dragonflies, *Diplacodes trivialis* (Rambur, 1842) (Libellulidae: Odonata). *Rec. zool. Surv. India* **81**: 13–22.

Kumar, A. & Kanna, V., 1983. A review of the taxonomy and ecology of Odonata larvae from India. *Oriental Insects* **17**: 127–157.

Kumar, A. & Prasad, M., 1977. On the larvae of *Rhinocypha* (Odonata: Chlorocyphidae) from Garhwal Hills. *Oriental Insects* **11**: 547–554.

Kumar, A. & Prasad, M., 1981. Field ecology, zoogeography and taxonomy of the Odonata of Western Himalaya, India. *Records of the Zoological Survey of India, Misc. Publ. Occ. Pap.* **29**: 1–118.

Kumar, K., 1987. Observations on seasonal variations of benthic organisms in two trout streams of Kashmir. *Proc. Indian natn. Sci. Acad.* **B53**: 227–234.

Kumar, K.P. & John, P.A., 1984. Seasonal variations of the allochthonus fauna and flora and their influence on the diet of *Rasbora daniconius* (Hamilton). *Arch. Hydrobiol.* **102**: 537–542.

Kumar, N. & Dobriyal, A., 1992. Some observations on the water mites of a hillstream Khandagad in Garhwal Himalaya. *J. Freshwat. Biol.* **4**: 193–197.

Kumar, S. & Verma, K.R., 1991. The impact of vegetal cover on streamflow with special reference to Kumaun Himalaya. In: Bhatt, S.D. & Pande, R.K. (Eds), *Ecology of the Mountain Waters*. Ashish Publishing House, New Delhi: 348–358.

Kurian, C.V. & Sebastian, V.O., 1982. *Prawns and Prawn Fisheries of India*. Hindustan Publishing Corporation, New Delhi: 286 pp.

Kuroda, T., Fujimoto, T. & Watanabe, N.C., 1984. Longitudinal distribution and life cycle of three species of *Ephemera* in the Kazuradani River, Kagawa Prefecture. *Kagawa Seibutsu* **12**: 15–21 (in Japanese).

Kushari, D.P & Sinhababu, A., 1987. Seasonal effect of two trees leaf leaching on growth of *Azolla pinnata*. *Pol. Arch. Hydrobiol.* **34**: 163–170.

Kuwayama, S., 1930. The Stenopsychidae of Nippon. *Insecta Matsumurana* **4**: 109–119.

Kwak, T.J., Wiley, M.J., Osborne, L.L. & Larimore, R.W. 1992. Application of diel feeding chronology to habitat suitability analysis of warmwater stream fishes. *Can. J. Fish. Aquat. Sci.* **49**: 1417–1430.

La Rivers, I., 1956. Aquatic Orthoptera. In: Usinger, R.L. (Ed.) *Aquatic Insects of California*. University of California Press, Berkeley: 154.

La Rivers, I., 1970a. Two new species of *Laccocoris* from British North Borneo (Hemiptera: Naucoridae: Laccocorinae). *Proc. Ent. Soc. Wash.* **72**: 496–499.

La Rivers, I., 1970b. A new Philippine *Sagocoris* (Hemiptera: Naucoridae). *Pan-Pacific Entomologist* **46**: 167–169.

LaBounty, J.F., 1984. Assessment of the environmental effects of constructing the Three Gorges Project on the Yangtze River. *Water International* **9**: 10–17.

Ladle, M., 1981. Organic detritus and its role as a food source in chalk streams. *Freshwat. Biol. Assoc. (UK) Ann. Rep.* **50**: 30-37.

Ladle, M. & Radke, R., 1990. Burrowing and feeding behaviour of the larva of *Ephemera danica* Müller (Ephemeroptera: Ephemeridae). *Entomologist's Gazette* **41**: 113–118.

Lai, H.T., Shy, J.Y. & Yu, H.P., 1986. Morphological observation on the development of larval *Eriocheir japonica* de Haan (Crustacea, Decapoda, Grapsidae) reared in the laboratory. *J. Fisheries Soc. Taiwan* **13**: 12–21 (in Chinese).

Lake, P.S., 1982. Ecology of the macroinvertebrates of Australian upland streams — a review of current knowledge. *Bull. Aust. Soc. Limnol.* **8**: 1–15.

Lake, P.S. & Barmuta, L.A., 1986. Stream benthic communities: persistent presumptions and current speculations. In: De Deckker, P. & Williams, W.D. (Eds), *Limnology in Australia*. CSIRO, Melbourne & Dr. W. Junk Publishers, Dordrecht: 263–276.

Lakshminarayana, J.S., 1965a. Studies on the phytoplankton of the River Ganges, Varanasi, India Part I. The physico-chemical characteristics of the River Ganges. *Hydrobiologia* **25**: 119–137.

Lakshminarayana, J.S., 1965b. Studies on the phytoplankton of the River Ganges, Varanasi, India Part II. The seasonal growth and succession of the planktonic algae in the River Ganges. *Hydrobiologia* **25**: 138–165.

Lang, C.R., 1996. Problems in the making. A critique of Vietnam's tropical forestry action plan. In: Parnwell, M.J.G. & Bryant, R.L. (Eds), *Environmental Change in South-east Asia. People, Politics and Sustainable Development.* Routledge, London: 225–234.

Lange, W.H., 1996. Aquatic and semiaquatic Lepidoptera. In: Merritt, R.W. & Cummins, K.W. (Eds), *An Introduction to the Aquatic Insects of North America. Third Edition).* Kendall/Hunt Publishing Company, Iowa: 387–398.

Langer, R.K., 1986. Food and feeding habits of *Schizothorax logipinnis* Heckel from River Jhelum, Kashmir. *Indian J. Fish. Assoc.* **14/15**: 41–48.

Lanly, J.P., 1982. *Tropical Forest Resources.* Forestry Paper No. 30, FAO, Rome: 106 pp.

Lansbury, I., 1968. The *Enithares* (Hemiptera-Heteroptera: Notonectidae) of the Oriental Region. *Pacific Insects* **10**: 353–442.

Lansbury, I., 1972a. A review of the Oriental species of *Ranatra* Fabricius (Hemiptera-Heteroptera: Nepidae). *Trans. R. Ent. Soc. Lond.* **124**: 287–341.

Lansbury, I., 1972b. A revision of the genus *Telmatotrephes* Stål (Hemiptera-Heteroptera: Nepidae). *Zoologica Scripta* **1**: 271–286.

Lansbury, I., 1973a. A review of the genus *Cercotmetus* Amyot & Serville, 1843 (Hemiptera-Heteroptera: Nepidae). *Tijdschrift voor Entomologie* **116**: 83–106.

Lansbury, I., 1973b. *Montonepa* gen. n. from India with notes on the genus *Borborophyes* Stål (Hemiptera-Heteroptera: Nepidae). *Zoologica Scripta* **2**: 111–118.

Lansbury, I., 1992. Notes on the marine-freshwater gerrid genus *Rheumatometroides* (Hemiptera, Gerridae) of Papua New Guinea. *Tijdschrift voor Entomologie* **135**: 1–10.

Larson, J.S. 1991. Downstream environmental impacts. In: Ryder, G. (Ed.), *Damming the Three Gorges — What the Dam-builders Don't Want You to Know.* Probe International, Toronto: 67–78.

Le, C.K., 1994. Native freshwater vegetation communities in the Mekong Delta. *Internat. J. Ecol. Environ. Sci.* **20**: 55–71.

Le Roy, A., 1978. The habits of *Thalassius spenceri. British Arachnological Society Newsletter* **23**: 7–8.

Leach, G.J. & Osborne, P.L., 1985. *Freshwater Plants of Papua New Guinea.* The University of Papua New Guinea Press, Port Moresby: 254 pp.

Lean, G., Hinrichsen, D. & Markham, A., 1990. *Atlas of the Environment.* Arrow Books Ltd and WWF-World Wide Fund for Nature, London: 194 pp.

Lee, C.F. & Jäch, M.A., 1996. Review of the genera *Eubria* and *Microeubria* Lee & Yang (Coeloptera: Psephenidae: Eubriinae). *Coleopterists' Bulletin* 50: 39–51.

Lee, C.F. & Satô, M., 1996. *Nipponeubria yoshitomii* Lee & Satô, a new species in a new genus of Eubriinae from Japan, with notes on the immature stages and description of the larva of *Ectopria opaca* (Kiesenwetter). *Coleopterists' Bulletin* 50: 122–134.

Lee, C.F. & Yang, P.S., 1990. Five new species of the genus *Eubrianax* from Taiwan (Coleoptera: Psephenidae). *Journal of the Taiwan Museum* 43: 79–88.

Lee, C.F. & Yang, P.S., 1993. A revision on the genus *Homoeogenus* Waterhouse with notes on the immature stages of *H. laurae* sp.n. (Coleoptera: Psephenidae: Eubriinae). *Systematic Entomology* 18: 351–361.

Lee, C.F. & Yang, P.S., 1996. Taxonomic revision of the Oriental species of *Dicranopselaphus* Guerin-Meneville (Coleoptera: Psephenidae: Eubriinae). *Ent. Scand.* 27: 169–196.

Lee, C.F. & Yang, P.S., 1994a. *Microeubria*, a new genus of Oriental Eubriinae (Coleoptera: Psephenidae). *Coleopterists' Bulletin* 48: 325–329.

Lee, C.F. & Yang, P.S., 1994b. A review of the genus *Ectopria* LeConte from East Asia, with the description of a new species from Taiwan (Coleoptera: Psephenidae: Eubriinae). *Coleopterists' Bulletin* 48: 381–389.

Lee, C.F., Yang, P.S. & Brown, H.P., 1990. Notes on the genus *Mataeopsephus* (Coleoptera: Psephenidae) in Taiwan with description of a new species. *Journal of the Taiwan Museum* 43: 73–78.

Lee, C.F., Yang, P.S. & Brown, H.P., 1993. Revision of the genus *Schinostethus* Waterhouse with notes on the immature stages and ecology of *S. satoi*, n.sp. (Coleoptera: Psephenidae). *Ann. Entomol. Soc. Am.* 86: 683–693.

Lee, S.J., Yoon, I.B. & Bae, Y.L., 1995. Altitudinal distribution of *Ephemera strigata* Eaton and *E. orientalis* McLachlan (Ephemeroptera: Ephemeridae). *Korean J. Entomol.* 25: 201–208 (in Korean).

Legris, P., 1974. Vegetation and and floristic composition of humid tropical continental Asia. In: *Natural Resources of Humid Tropical Asia*. Natural Resources Research, XII, Unesco, Paris: 217–238.

Leichtfried, M.P.S. & Kristyanto, A.I.A., 1995. Organic matter in sediments of a Javanese mountain stream (project JAVADAT). In: Timotius, K.H. & Göltenboth, F. (Eds), *Tropical Limnology Volume III. Tropical Rivers, Wetlands and Special Topics*. Satya Wacana University Press, Salatiga, Indonesia: 51–60.

Lelek, A., 1985. About a source of nutrients in the tropical River Rajang (N. Borneo) in relation to future impoundment. *Verh. Internat. Verein. Limnol.* 22: 2115–2118.

Lelek, A., 1987. Freshwater and swamp fisheries on the Rajang River, Sarawak, Malaysia. *Arch. Hydrobiol. Beih., Ergebn. Limnol.* 28: 247–260.

Leong, C.Y., 1961. The small water boatman, *Micronecta*. *Malay. Nat. J.* 15: 168–172.

Leong, C.Y. & Fernando, C.H., 1962. The genus *Anisops* (Hemiptera: Notonectidae) in Ceylon. *J. Bombay Nat. Hist. Soc.* **59**: 513–519.

Li, B., 1992. Building a new life in wake of Three Gorge. *China Environment News* **33** (**April 1992**): 4–5.

Li, C., 1994. *Three Gorges Project: Strategic Patterns of Environment in China*, Unpublished manuscript, 6th International Planning History Conference, The University of Hong Kong, June 1994.

Li, G., Shen, Q., & Xu, Z., 1993. Morphometric and biochemical variation of the mitten crab, *Eriocheir*, in southern China. *Aquaculture* **111**: 103–115.

Li, S., 1981. *Studies on Zoogeographical Divisions for Freshwater Fishes of China*. Science Press, Beijing: 292 pp. (in Chinese).

Li, Y., 1989. Chinese forestry: crisis and options. *Liaowong (Outlook)* **12**: 9–10.

Li, Y. & Dudgeon, D., 1988. Four new species of the genus *Cheumatopsyche* from China (Trichoptera: Hydropsychidae). *Journal of Nanjing Agricultural University* **11**: 1–46 (in Chinese).

Li, Y. & Morse, J.C., 1997a. Phylogeny and classification of Psychomyiidae (Trichoptera) genera. In: Holzenthal, R.W. & Flint, O.S. (Eds), *Proceedings of the 8th International Symposium on Trichoptera*. Ohio Biological Survey, Columbus: 271–276.

Li, Y. & Morse, J.C., 1997b. Species of the genus *Ecnomus* (Trichoptera: Ecnomidae) from the People's Republic of China. *Trans. Am. Entomol. Soc.* **123**: 85–134.

Li, Y. & Tian, L., 1990. Note on the subgenus *Ceratopsyche* (Trichoptera: Hydropsychidae: *Hydropsyche*) from China with descriptions of nine new species. *Entomotaxonomia* **12**: 127–138 (in Chinese).

Li, Y., Tian, L. & Dudgeon, D., 1990. Notes on six new specoes of Hydropsychae (Trichoptera). *Journal of Nanjing Agricultural University* **13**: 37–42 (in Chinese).

Li, Y.H., 1976. Denudation of Taiwan Island since the Pliocene epoch. *Geology* **4**: 105–107.

Liang, X., 1984. Successful introduction of a new aquatic mollusc, *Ampullaria gigas* Spix., from the Amazon. *Ecologic Science* **1984** (**2**): 122 (in Chinese).

Liang, X. & Yan, S., 1983. A new species of *Caridina* (Crustacea, Decapoda) from Guangxi, China. *Acta Zootaxonomica Sinica* **8**: 252–254 (in Chinese).

Liang, X. & Yan, S.L., 1991. A new genus and two new species of freshwater prawns (Crustacea, Decapoda) from Guangxi. *Acta Zootaxonomica Sinica* **6**: 31–35 (in Chinese).

Liang, X., Hong , F. & Yang, Z., 1990. A new species of *Caridina* from Sichuan Province (Decapoda, Atyidae). *Acta Zootaxonomica Sinica* **15**: 161–164 (in Chinese).

Liang, X. & Zhou, J., 1993. Study on a new atyid shrimp (Decapoda, Caridea) from Guangxi, China. *Acta Hydrobiologica Sinica* **17**: 231–239.

Liang, X., Yan, S. & Wang, Z., 1987. Description of a new species of

Caridina (*Caridina*) from Yunnan, China (Decapoda, Atyidae). *Acta Zootaxonomica Sinica* **12**: 133–135 (in Chinese).

Liang, Y., 1963. Studies on the aquatic Oligochaeta of China. 1. Descriptions of new naids and branchiobdellids. *Acta Zoologica Sinica* **15**: 560–570 (in Chinese).

Liang, Y., 1987. Preliminary study of the aquatic Oligochaeta of the Changjiang (Yangtze) River. *Hydrobiologia* **155**: 195–198.

Liang, Y., Wang, J. & Hu, C., 1988. Hydrobiology of a flooding ecosystem, Lake Chenhu in Hanyang, Hubei, with preliminary estimation of its potential fishery production capacity. *Chin. J. Oceanol. Limnol.* **6**: 1–14 (in Chinese).

Liang, Z., Chang, J. & Chen, H., 1986. The spawning habit and the embryonic development of Chinese gudgeon, *Gnathopogon argentatus* (Sauvage & Darby), in Pearl River. *Transactions of the Chinese Ichthyological Society* **5**: 35–45 (in Chinese).

Liao, G.Z., 1980. Preliminary approach to the water conservancy project and the rescue of fish. *Freshwater Fisheries* **1**: 1–4 (in Chinese).

Liao, G.Z., Lu, K.X. & Xiao, X.Z., 1989. Fisheries resources of the Pearl River and their exploitation. *Can. Spec. Publ. Fish. Aquat. Sci.* **106**: 561–568.

Lieftinck, M.A., 1933. Notes on the larvae of two interesting Gomphidae (Odon.) from the Malay Peninsula. *Bull. Raffles Mus., Singapore* **7**: 102–115.

Lieftinck, M.A., 1934. Notes on the genus *Drepanosticta* (Laid.) with description of the larvae and of new Malaysian species (Odon., Zygoptera). *Treubia* **14**: 436–476.

Lieftinck, M.A., 1939. On the true position of the genus *Orolestes* MacLachlan, with notes on *O. wallacei* (Kirby), its habits and life history (Odonata: Lestidae). *Treubia* **17**: 45–61.

Lieftinck, M.A., 1940. On some Odonata collected in Ceylon, with descriptions of new species and larvae. *Ceylon. J. Sci. (B)* **22**: 79–117.

Lieftinck, M.A., 1941. Studies on Oriental Gomphidae (Odon.) with descriptions of new or interesting larvae. *Treubia* **18**: 233–251.

Lieftinck, M.A., 1950. Further studies on Southeast Asiatic species of *Macromia* Rambur, with notes on their ecology, habits and life history, and with descriptions of larvae and two new species (Odon., Epopthalmiinae). *Treubia* **20**: 657–716.

Lieftinck, M.A., 1965. The species group of *Vestalis amoena* Selys in Sundaland (Odonata, Calopterygidae). *Tijdschrift voor Entomologie* **108**: 325–364.

Lieftinck, M.A., Lien, J.C. & Maa, T.C., 1984. *Catalogue of Taiwanese Dragonflies (Insecta: Odonata)*. Asian Ecological Society, Taichung, Taiwan: 81 pp.

Lien, J.C. & Matsuki, K., 1983. Description of the larvae of *Orolestes selysi* McLachlan from Taiwan (Lestidae: Odonata). *Tombo* **25**: 13–15.

Lien, J.C. & Matsuki, K., 1988. Description of the larvae of *Cratilla lineata assidua* Lieftinck from Taiwan (Libellulidae: Odonata). *Tombo* **31**: 35–36.

Lim, K.K.P. & Ng, P.K.L., 1990. *A Guide to the Freshwater Fishes of Singapore*. Singapore Science Centre, Singapore: 160 pp.

Lim, R.P., 1980. Limnological research and education with reference to natural ecosystems in Malaysia. In: Mori, S. & Ikusima, I. (Editors), *Proceedings of the First Workshop for the Promotion of Limnology in Developing Countries (1979, Kyoto)*. Societas Internationalis Limnologiae, Kyoto, Japan: 57–65.

Lim, R.P., 1987. Water quality and faunal composition in the streams and rivers of the Ulu Endau area, Johore, Malaysia. *Malay. Nat. J.* **41**: 337–347.

Lin, C., 1982. Preliminary notes on May-stream aquatic insects. *Quart. J. Taiwan Mus.* **35**: 167–169.

Lindberg, D.R., 1990. Systematics of *Potamacmaea fluviatilis* (Blanford): a brackish water patellogastropod (Patteloidinae: Lottidae). *J. Moll. Stud.* **56**: 309–316.

Liu, B., 1993. Effect of acidic water on survival growth and reproduction of freshwater snails. *Chinese J. Appl. Ecol.* **4**: 313–318 (in Chinese).

Liu, C. & Liu, E., 1983. Water balance in the water transfer region. In: Biswas, A.K., Zuo, D., Nickum, J.E. & Liu, C. (Eds), *Long-distance Water Transfer: a Chinese Case Study and International Experiences*. Tycooly International Publishers Limited, Dublin: 215–232.

Liu, C. & Zuo, D., 1983. Impact of south-to-north water transfer upon the natural environment. In: Biswas, A.K., Zuo, D., Nickum, J.E. & Liu, C. (Eds), *Long-distance Water Transfer: a Chinese Case Study and International Experiences*. Tycooly International Publishers Limited, Dublin: 169–179.

Liu, C.H. & Zheng, Y.J., 1988. Notes on the Chinese paddlefish, *Psephurus gladius* (Martens). *Copeia* **1988**, 482–484.

Liu, H. & Wang, Y., 1987. Water environment and its management in China. *Mitt. Geol.-Palaont. Inst. Univ. Hamburg, SCOPE/UNEP Sonderbd.* **64**: 131–137.

Liu, J.K., 1980. *China's Largest River*. Foreign Languages Press, Beijing: 165 pp.

Liu, J.K., 1984. Lakes of the middle and lower basins of the Chang Jiang (China). In: F.B. Taub (Editor), *Ecosystems of the World 23: Lakes and Reservoirs*. Elsevier, Amsterdam: 331–355.

Liu, J.K. & Yu, Z.T., 1992. Water quality changes and effects on fish populations in the Hanjiang River, China, following hydroelectric dam construction. *Regulated Rivers: Research & Management* 7: 359–368.

Liu, L., Wu, G. & Wang, Z., 1990. Reproduction ecology of *Coreius heterodon* (Bleeker) and *Coreius guichenoti* (Sauvage et Dabry) in the mainstream of the Changjiang River after the construction of Gezhouba Dam. *Acta*

Hydrobiol. Sinica **14**: 205–215 (in Chinese).

Liu, S.C., 1984. Descriptions of two new species of the genus *Stephanodrilus* from north-east China and notes on *S. truncatus* Liang from Guangdong Province (Oligochaeta: Branchiobdellidae) *Acta Zootaxonomica Sinica* **9**: 351–355 (in Chinese).

Liu, S.C. & Zhang, D.C., 1983. Three new species of the genus *Branchiobdella* (Oligochaeta: Branchiobdellidae) from China. *Acta Zootaxonomica Sinica* **8**: 246–251 (in Chinese).

Liu, S.C., 1964. A second report of Brachiobdellidae in Liaoning Province with descriptions of three new species. *Acta Zoologica Sinica* **8**: 246–251 (in Chinese).

Liu, Y.Y. & Zhang, W.Z., 1979. On new flukes and species of freshwater snails harbouring cercaria of lung flukes from China. *Acta Zootaxonomica Sinica* **4**: 132–136 (in Chinese).

Liu, Y.Y., Wang, Y.X. & Zhang, W.Z., 1980a. On new species and new records of freshwater snails of the family Hydrobiidae from Yunnan, China. *Acta Zootaxonomica Sinica* **5**: 358–368 (in Chinese).

Liu, Y.Y., Wang, Y.X.& Zhang, W.Z., 1980b. A new record of Chinese buccinoid snail. *Acta Zootaxonomica Sinica* **5**: 252 (in Chinese).

Liu, Y.Y., Zhang, W.Z., Wang, Y.X., Chen, C.E. & Chen, S.Z., 1982a. Discovery of *Akiyoshia* Kuroda & Habe (Hydrobiidae: Mollusca) from China with descriptions of two new species. *Acta Zootaxonomica Sinica* **7**: 364–367 (in Chinese).

Liu, Y.Y., Huang, Y.S.& Zhuang, W.Q., 1982b. The discovery of *Biomphalaria straminea* (Dunker), an intermediate host of *Schistosoma mansoni*, from China. *Acta Zootaxonomica Sinica* **7**: 256 (in Chinese).

Liu, Y.Y., Zhang, W.Z. & Wang, Y.X., 1983. Three new species of Hydrobiidae (Gastropoda, Prosobranchia) from China. *Acta Zootaxonomica Sinica* **8**: 366–369 (in Chinese).

Liu, Y.Y., Zhang, W.Z. & Wang, Y.X., 1993. *Medical Malacology*. China Ocean Press, Beijing: 157 pp. (in Chinese).

Livingstone, D.A., 1963. Chemical composition of rivers and lakes. *U.S. Geol. Surv. Prof. Pap.* **440–G**: 1–41.

Livshits, G. & Fishelson, L., 1983. Biology and reproduction of the freshwater snail *Melanoides tuberculata* (Gastropoda: Prosobranchia) in Israel. *Israel J. Zool.* **32**: 571–577.

Lo, K.S.L. & Chen, H.H., 1991. Water quality management of Keelung River, northern Taiwan. *Water Science & Technology* **24**: 109–116.

Lodge, D.M., Brown, K.M., Klowiewski, S.P., Stein, R.A., Covich, A.P., Leathers, B.K. & Bronmark, C., 1987. Distribution of freshwater snails: spatial scale and the relative importance of physicochemical and biotic factors. *Amer. Malacol. Bull.* **5**: 73–84.

Loffler, E., 1977. *Geomorphology of Papua New Guinea*. CSIRO & Australian National University Press, Canberra: 195 pp.

Logan, P. & Brooker, M.P., 1983. The macroinvertebrate faunas of riffles and pools. *Water Res.* **17**: 263–270.

Lohmann, L., 1990. Remaking the Mekong. *The Ecologist* **20**: 61–66.

Lohmann, L., 1991. Engineers move in on the Mekong. *New Scientist* **131** (**1777**): 44–47.

Lohmann, L., 1993. Freedom to plant. Indonesia and Thailand in a globalizing pulp and paper industry. In: Parnwell, M.J.G. & Bryant, R.L. (Eds), *Environmental Change in South-east Asia. People, Politics and Sustainable Development*. Routledge, London: 23–48.

Low, K.S., 1993. Urban water resources in the humid tropics: an overview of the ASEAN region. In: Bonell, M., Hufschmidt, M.M. & Gladwell, J.S. (Eds), *Hydrology and Water Management in the Humid Tropics*. UNESCO/Cambridge University Press, New York: 526–534.

Low, K.S. & Peh, C.H., 1987. Hydro-geomorphological considerations in water resources management in Malaysia. *Arch. Hydrobiol. Beih., Ergebn. Limnol.* **28**: 539–546.

Lowe-McConnell, R.H., 1970. Ecological studies on tropical freshwater food fishes. *Regional Meeting of Inland Water Biologists in Southeast Asia. Proceedings*. Unesco Field Science Office for Southeast Asia, Djakarta: 91–103.

Lowe-McConnell, R.H., 1987. *Ecological Studies of Tropical Fish Communities*. Cambridge University Press, Cambridge: 382 pp.

Lowen, R.G. & Flannagan, J.F., 1992. Nymphs and imagoes of four North American species of *Procloeon* Bengtsson with description of a new species (Ephemeroptera, Baetidae). *Can. Ent.* **124**: 97–108.

Lucas, J.S., 1980. Spider crabs of the family Hymenosomatidae (Crustacea; Brachyura) with particular reference to Australian species: systematics and biology. *Rec. Aust. Mus.* **33**: 148–247.

Lue, K.Y., 1989. The freshwater planarian of Taiwan. *Chinese Bioscience* **32**: 29–42 (in Chinese).

Luk, S. & Whitney, J. 1993a. Introduction. In: Luk, S. & Whitney, J. (Eds), *Megaproject. A Case Study of China's Three Gorges Project*. M.E. Sharpe Inc., New York: 3–39.

Luk, S. & Whitney, J., 1993b. *Megaproject. A Case Study of China's Three Gorges Project*. M.E. Sharpe Inc., New York: 236 pp.

Lundblad, O., 1933. Zur Kenntnis der aquatilen und semi-aquatilen Hemiptera von Sumatra, Java und Bali. *Arch Hydrobiol., Suppl.* **12**: 1–195 & 263–489.

Mahajan, K.K., 1988 Deteriorating nation's rivers. In: Trivedy, R.K. (Ed.), *Ecology and Pollution of Indian Rivers*. Ashish Publishing House, New Delhi: 1–38

Maheshwari, G., 1989. Bioecological study of *Tanytarsus ipei* (Diptera: Chironomidae). *Journal of Bioecology* **1**: 109–114.

Maitland, P.S., 1995. The conservation of freshwater fish: past and present experience. *Biological Conservation* **72**: 259–270.

Majhi, A. & Dasgupta, M., 1989. A study on the food and feeding habits of

Pillia indica Yazdani. *Arq. Mus. Bocage (Nov. Ser.)* **1**: 1–7.

Maketon, M. & Stewart, K.W., 1988. Patterns and evolution of drumming behavior in the stonefly families Perlidae and Peltoperlidae. *Aquatic Insects* **10**: 77–98.

Makhan, D., 1995. Descriptions of ten new species of *Hydrochus* from different parts of the world (Coleoptera: Hydrochidae). *Phegea* **23**: 187–193.

Malas, D. & Wallace, J.B., 1977. Strategies for coexistence in three species of net-spinning caddisflies in a second-order southern Appalachian stream. *Can. J. Zool.* **55**: 1829–40.

Malicky, H., 1973. Results of the Austrian-Ceylonese hydrological mission, 1970. Part XVI: the Ceylonese Trichoptera. *Bull. Fish. Res. Stn., Sri Lanka (Ceylon)* **24**: 153–177.

Malicky, H., 1983a. Caddisflies (Trichoptera) from Parakrama Samudra, an ancient man-made lake in Sri Lanka. In: Schiemer, F. (Ed.), *Limnology of Parakrama Samudra — Sri Lanka*. Dr W. Junk Publishers, The Hague: 227–228.

Malicky, H., 1983b. *Atlas of European Trichoptera*. Dr W. Junk Publishers, The Hague: 298 pp.

Malicky, H., 1989a. Köcherfliegen (Trichoptera) von Sumatra und Nias: Die Gattungen *Chimarra* (Philopotamidae) und *Marilia* (Odontoceridae), mit Nachträgen zu *Rhyacophila* (Rhyacophilidae). *Mitt. Schweiz. Entomol. Gesell.* **62**: 131-143.

Malicky, H., 1989b. Odontoceridae aus Thailand (Trichoptera). *Opusc. zool. flumin.* **36**: 1–16.

Malicky, H., 1991. Some unusual caddisflies (Trichoptera) from Southeastern Asia. In: Tomaszewski, C. (Ed.), *Proceedings of the 6th International Symposium on Trichoptera*. Adam Mickiewicz University Press, Lodz-Zakopane, Poland: 381–384.

Malicky, H., 1993a. Neu asiatische Köcherfliegen (Trichoptera: Rhyacophilidae, Philopotamidae, Economidae und Polycentropodidae). *Entomologische Berichte Luzern* **29**: 77–88.

Malicky, H., 1993b. Neu asiatische Köcherfliegen (Trichoptera: Philopotamidae, Polycentropodidae, Psychomyiidae, Economidae Hydropsychidae, Leptoceridae). *Linzer biol. Beitr.* **25**: 1099–1136.

Malicky, H., 1994a. Neue Trichopteren aus Nepal, Vietnam, China von den Philippinen und vom Bismark-Archipel (Trichoptera). *Entomologische Berichte Luzern* **31**: 163–172.

Malicky, H., 1994b. Ein Beitrag zur Kenntnis asiatischer Calamoceratidae (Trichoptera). *Zeitschrift der Arbeitsgemeinschaft Österreichischer Entomologen* **46**: 62–79.

Malicky, H., 1995a. Neue Köcherfliegen (Trichoptera, Insecta) aus Vietnam. *Linzer biol. Beitr.* **27**: 851–885.

Malicky, H., 1995b. Weitre neue Köcherfliegen (Trichoptera) aus Asien. *Braueria* **22**: 11–26.

Malicky, H., 1997. Weitre neue Köcherfliegen-Arten (Trichoptera) aus Asien. *Linzer biol. Beitr.* **29**: 217–238.

Malicky, H. & Chantaramongkol, P., 1989a. Einge Rhyacophilidae aus Thailand (Trichoptera). *Ent. Z.* **99**: 17–24.

Malicky, H. & Chantaramongkol, P., 1989b. Beschreibung von neuen Köcherfliegen (Trichoptera) aus Thailand und Burma. *Entomologische Berichte Luzern* **22**: 117–126.

Malicky, H. & Chantaramongkol, P., 1991. Beschreibung von *Trichomacronema paniae* n sp. (Trichoptera, Hydropsichidae) aus Nord-Thailand und Beobachtungen uber ihre Lebensweise. *Entomologische Berichte Luzern* **25**: 113–122.

Malicky, H. & Chantaramongkol, P., 1991. Einige Goera (Trichoptera, Goeridae) aus Südasien. *Entomologische Berichte Luzern* **27**: 141–150.

Malicky, H. & Chantaramngkol, P., 1992. Neue Köcherfliegen (Trichoptera) aus Thailand und angrenzenden Länderen. *Braueria* **19**: 13–23.

Malicky, H. & Chantaramongkol, P., 1993a. The altitudinal distribution of Trichoptera species in Mae Klang catchment on Doi Inthanon, northern Thailand: stream zonation and cool- and warm-adapted groups. *Rev. Hydrobiol. trop.* **26**: 279–291.

Malicky, H. & Chantaramongkol, P., 1993b. Neue Trichopteren aus Thailand. Teil 1: Rhyacophilidae, Hydrobiosidae, Philopotamidae, Polycentropodidae, Economidae, Psychomyidae, Arctopsychidae, Hydropsychidae. *Linzer biol. Beitr.* **25**: 433–487.

Malicky, H. & Chantaramongkol, P., 1993c. Neue Trichopteren aus Thailand. Teil 2: Rhyacophilidae, Philopotamidae, Polycentropodidae, Economidae, Psychomyidae, Xiphocentronidae, Helicopsychidae, Odontoceridae. *Linzer biol. Beitr.* **25**: 1137–1187.

Malicky, H. & Chantaramongkol, P., 1994. Neue Lepidostomatidae aus Asien (Insecta: Trichoptera: Lepidostomatidae). *Ann. Naturhist. Mus. Wien* **96B**: 349–368.

Malicky, H. & Chantaramongkol, P., 1996. Neue Köcherfliegen aus Thailand. *Entomologische Berichte Luzern* **36**: 119–128.

Malicky, H. & Chantaramongkol, P., 1997. Weitere neue Köcherfliegen (Trichoptera) aus Thailand. *Linzer biol. Beitr.* **29**: 203–215.

Malzacher, P., 1984. Die europaischen Arten der Gattung *Caenis* Stephens (Insecta: Ephemeroptera). *Stuttgarter Beitrage zur Naturkunde Serie A (Biologie)* **373**: 1–48.

Malzacher, P., 1991. Genital-morphological features in the Caenidae. In: Alba-Tercedor & Sanchez-Ortega, A. (Eds), *Overview and Strategies of Ephemeroptera and Plecoptera*. The Sandhill Crane Press, Gainesville: 73–85.

Malzacher, P., 1992. Mayflies from Israel (Insecta, Ephemeroptera) II. — Caenidae. *Mitt. Schweiz. Entomol. Gesell.* **65**: 385–394.

Malzacher, P., 1993. Caenidae der athiopischen Region (Insecta: Ephemeroptera). Teil 2. Systematische Zusammenstellung aller bisher bekannten Arten. *Mitt. Schweiz. Entomol. Gesell.* **66**: 379–416.

Marchant, R., 1982. Life spans of two species of tropical mayfly nymph (Ephemeroptera) from Magela Creek, Northern Territory. *Aust. J. Mar. Freshwat. Res.* **33**: 173–179.

Marchant, R., Metzeling, L., Graesser, A. & Suter, P., 1985. The organization of macroinvertebrate communities in the major tributaries of the LaTrobe River. *Freshwat. Biol.* **15**: 315–331.

Marlier, G., 1962. Genera des Trichoptères de l'Afrique. *Musee Royal de L'Afrique centrale, Tervuren, Belgique Annales, Science Zoologiques* **109**: 1–262.

Marshall, J.E., 1979. A review of the genera of the Hydroptilidae (Trichoptera). *Bull. Br. Mus. Nat. Hist. (Ent.)* **39**: 135–239.

Martynov, A.B., 1914. Contributions à la faune des Trichoptères de la Chine. *Annuaire du Musée Zoologique de l'"Academie Imperiale des Science, St Petersburg* **19**: 323–339 (in Russian).

Martynov, A.B., 1926. On the family Stenopsychidae Mart., with a revision of the genus *Stenopsyche* McLachl. (Trichoptera). *Eos* **2**: 281–308.

Martynov, A.B., 1931. Report on a collection of insects of the order Trichoptera from Siam and China. *Proc. U.S. Natn Mus.* **79**: 1–20.

Martynov, A.B., 1935. On a collection of Trichoptera from the Indian Museum. Part I. Annulipalpia. *Rec. Indian Mus.* **37**: 93–209.

Martynov, A.B., 1936. On a collection of Trichoptera from the Indian Museum. Part II. Integripalpia. *Rec. Indian Mus.* **37**: 239–306.

Marwoto, R.M., 1987. Some freshwater mussels family Unionidae and gastropods from Rengas and Sungkai Rivers in Jambi Province. *Berita Biologi* **3**: 306–309.

Mascagni, A., 1990. Two new species of Heteroceridae from Nepal (Coleoptera). *Bollettino della Societa Entomologica Italiana* **122**: 107–110.

Mascagni, A., 1991. Contributo alla conoscenza degli Heteroceridae del Burma e della Cambogia con descrizione di *Heterocerus anulatus* n. sp. (Insecta: Coleoptera: Heteroceridae). *Redia* **74**: 15–28.

Mascagni, A., 1995a. Heteroceridae: checklist of the Heteroceridae of China and neighbouring countries and descriptions of two new species. In: Jäch, M.A. & Ji, L. (Eds), *Water Beetles of China*. Zoologisch-Botanische Gesellschaft & Wiener Coleopterologenverein, Vienna: 341–348.

Mascagni, A., 1995b. Contributi all conoscenza degli Heteroceridae del Nepal, con descrizione di sie nuove specie (Coleoptera). *Opusc. Zool. Flumin.* **135**: 1–12.

Mashiko, K., 1981. Sexual dimorphism of the cheliped in the prawn *Macrobrachium nipponense* (De Haan) and its significance in reproductive behaviour. *Zool. Mag. (Tokyo), Zool. Soc. Japan* **90**: 333–337.

Mashiko, K., 1983a. Differences in the egg and clutch sizes of the prawn *Macrobrachium nipponense* (De Haan) between brackish and fresh waters of a river. *Zool. Mag. (Tokyo), Zool. Soc. Japan* **92**: 1–9.

Mashiko, K., 1983b. Comparison of growth pattern until sexual maturity between the estuarine and upper freshwater populations of the prawn *Macrobrachium nipponense* (De Haan) within a river. *Jap. J. Ecol.* **33**: 207–212.

Mashiko, K., 1990. Diversified egg and clutch sizes among local populations of the freshwater prawn *Macrobrachium nipponense* (de Haan). *J. Crustacean Biol.* **10**: 306–314.

Mashiko, K. & Numachi, K.I., 1993. Genetic evidence for the presence of distinct fresh-water prawn (*Macrobrachium nipponense*) populations in a single river system. *Zoological Science* **10**: 161–167.

Matëna, J. & Soldán, T., 1986. New findings of the genus *Epoicocladius* (Diptera, Chironomidae). *Dipterologica Bohemoslovaca (Ceské Budëjovice)* **4**: 31–41 (in Czech).

Matsuki, K., 1978. Taxonomic studies if the larval stage of Gomphidae (Odonata) in Taiwan. *Science Yb. Taiwan Mus.* **21**: 133–180 (in Chinese).

Matsuki, K., 1985. Larvae of *Ceriagrion f. fallax* of Taiwan (Odonata, Coenagrionidae). *Tombo* **28**: 33–34 (in Japanese).

Matsuki, K., 1986a. The larvae of *Gynacantha hyalina* Selys from Taiwan (Odonata, Aeshnidae). *Gekkan-Mushi* **189**: 31 (in Japanese).

Matsuki, K., 1986b. Description of the possible larvae of *Gynacantha hyalina* Selys from Taiwan (Odonata, Aeshnidae). *Gekkan-Mushi* **182**: 26–27 (in Japanese).

Matsuki, K., 1987. Descriptions of the larvae of four species of the genus *Anax* in Taiwan (Odonata, Aeshnidae). *Tombo* **20**: 25–28 (in Japanese).

Matsuki, K., 1988a. Description of *Zygonyx iris insignis* (Kirby) from Hong Kong (Libellulidae, Odonata). *Gekkan-Mushi* **210**: 24–25 (in Japanese).

Matsuki, K., 1988b. Description of the larva of *Tetracanthagyna waterhousei* (McLachlan) from Hong Kong and Thailand (Aeshnidae: Odonata). *Tombo* **31**: 37–40 (in Japanese).

Matsuki, K., 1989a. Description of the possible larva of *Planaeschna ishigakiana ishigakiana* Asahina (Aeshnidae: Odonata). *Tombo* **22**: 29–32 (in Japanese).

Matsuki, K., 1989b. Descriptions of the larvae of two species of the genus *Onychogomphus* in Hong Kong (Gomphidae, Odonata). *Gekkan-Mushi* **215**: 30–32 (in Japanese).

Matsuki, K., 1989c. Description of the larva of *Macromidia genialis shanensis* Fraser from Thailand (Corduliidae, Odonata). *Gekkan-Mushi* **219**: 28–29 (in Japanese).

Matsuki, K., 1990a. Description of the larva of *Heliogomphus scorpio* (Ris, 1912) in Hong Kong (Gomphidae: Odonata). *Nature & Insects* **25**: 9–12 (in Japanese).

Matsuki, K., 1990b. Description of the larva of *Gomphidia perakensis* Laidlaw from Thailand (Gomphidae, Odonata). *Gekkan-Mushi* **228**: 32–33 (in Japanese).

Matsuki, K., 1991. Description of the larval stage of *Prodasineura croconota*

(Ris) from Taiwan (Protoneuridae, Odonata). *Tombo* **34**: 27–28 (in Japanese).

Matsuki, K. & Kitagawa, K., 1986. Description of the possible larva of *Zygonyx iris malayanus* (Laidlaw) from Thailand (Libellulidae: Odonata). *Gekkan-Mushi* **183**: 24–25 (in Japanese).

Matsuki, K. & Kuwahara, H., 1989. Taxonomic notes on Taiwanese *Polycanthagyna melanictera* (Selys) (Aeshnidae: Odonata). *Gekkan-Mushi* **222**: 20–22 (in Japanese).

Matsuki, K. & Lien, J.C., 1978. Descriptions of the larvae of three families of Zygoptera breeding in the streams of Taiwan (Synlestidae, Euphaeidae and Calopterygidae). *Tombo* **21**: 15–26.

Matsuki, K. & Lien, J.C., 1984. Descriptions of the larvae of two species of the genus *Coeliccia* in Taiwan (Odonata, Platycnemididae). *Tombo* **27**: 21–22.

Matsuki, K. & Lien, J.C., 1984. Description of the larvae of two species of the genus *Planaeschna* in Taiwan (Aeshnidae: Odonata). *Chô Chô* **8**: 2–7 (in Japanese).

Matsuki, K. & Yamamoto, T., 1990. Additional notes on the larva of *Planaeschna ishigakiana ishigakiana* Asahina (Aeshnidae: Odonata). *Tombo* **23**: 27–32 (in Japanese).

Mathis, M.L. & Bowles, D.E., 1994. A description of the immature stages of *Paduniella nearctica* (Trichoptera: Psychomyiidae) with notes on its biology. *J. N.Y. Entomol. Soc.* **102**: 361–366.

Mathur, D., Bason, W.H., Purdy, E.J. & Silver, C.D., 1985. A critique of the instream flow incremental methodology. *Can. J. Fish. Aquat. Sci.* **42**: 825-831.

Mayer, K., 1934. Ceratopogoniden-Metamorphosen (C. Intermediae and C. Vermiformes) der Deutschen Limnologischen Sunda-Expedition. *Arch. Hydrobiol., Suppl.* **13**: 166–202.

Mazzoldi, P., 1995. Gyrinidae: Catalogue of Chinese Gyrinidae (Coleoptera). In: Jäch, M.A. & Ji, L. (Eds), *Water Beetles of China*. Zoologisch-Botanische Gesellschaft & Wiener Coleopterologenverein, Vienna: 155–172.

Mazzoldi, P. & Van Vondel, B.J., 1997. *Haliplus samosirensis* VONDEL recorded from Kalimantan. *Koleopterologische Rundschau* **67**: 119–120.

McCafferty, W.P., 1973. Systematic and zoogeographic aspects of Asiatic Ephemeridae (Ephemeroptera). *Oriental Insects* **7**: 49–67.

McCafferty, W.P., 1981. *Aquatic Entomology. The Fishermem's and Ecologists' Illustrated Guide to Insects and their Relatives*. Science Books International, Boston: 448 pp.

McCafferty, W.P., 1989. Characterization and relationships of the subgenera of *Isonychia* (Ephemeroptera: Oligoneuriidae). *Entomol. News* **100**: 72–78.

McCafferty, W.P., 1991a. Towards a phylogenetic classification of the Ephemeroptera (Insecta): a commentary on systematics. *Ann. Entomol. Soc. Amer.* **84**: 343–360.

McCafferty, W.P., 1991b. Synopsis of the Oriental mayfly genus *Eatonigenia* (Ephemeroptera: Ephemeridae). *Oriental Insects* **25**: 179–181.

McCafferty, W.P. & Bae, Y.J., 1992. Filter-feeding habits of the larvae of *Anthopotamus* (Ephemeroptera: Potamanthidae). *Annls Limnol.* **28**: 27–34.

McCafferty, W.P. & Bae, Y.J., 1994. Life history aspects of *Anthopotamus verticis* (Ephemeroptera: Potamanthidae). *Great Lakes Entomol.* **27**: 57–67.

McCafferty, W.P. & Edmunds, G.F., 1973. Subgeneric classification of *Ephemera* (Ephemeroptera: Ephemeridae). *Pan-Pacific Entomol.* **49**: 300–307.

McCafferty, W.P. & Edmunds, G.F., 1979. The higher classification of the Ephemeroptera and its evolutionary basis. *Ann. Entomol. Soc. Am.* **72**: 5–12.

McCafferty, W.P. & Waltz, R.D., 1990. Revisionary synopsis of the Baetidae (Ephemeroptera) of North and Middle America. *Trans. Amer. Entomol. Soc.* **116**: 769–799.

McCafferty, W.P. & Waltz, R.D., 1995. *Labiobaetis* (Ephemeroptera: Baetidae): new status, new North American species, and related new genus. *Ent. News* **106**: 19–28.

McCafferty, W.P. & Wang, T.Q., 1994. Phylogenetics and classification of the *Timpanoga* complex (Ephemeroptera: Ephemerellidae). *J. N. Am. Benthol. Soc.* **13**: 569–579.

McCafferty, W.P. & Wang, T.Q., 1995a. A new genus and species of Tricorythidae (Ephemeroptera: Pannota) from Madagascar. *Annales de Limnologie* **31**: 179–183.

McCafferty, W.P. & Wang, T.Q., 1995b. Relationships of the genera *Acanthametropus*, *Analetris* and *Siphluriscus*, and re-evaluation of their higher classification. *Great Lakes Entomol.* **27**: 209–215.

McDowell, R.M., 1981. The relationships of Australian freshwater fishes. In: Keast, A. (Ed.), *Ecological Biogeography of Australia. Volume 2*. Dr W. Junk Publishers, The Hague: 1251–1273.

McKay, G.M., 1984. Ecology and biogeography of mammals. In: Fernando, C.H. (Ed.), *Ecology and Biogeography in Sri Lanka*. Dr. W. Junk Publishers, The Hague: 413–429.

McMahon, R.F., 1991. Mollusca: Bivalvia. In: Thorp, J.H. & Covich, A.P. (Eds), *Ecology and Classification of North American Freshwater Invertebrates*. Academic Press, San Diego: 315–399.

McMichael, D.F., 1956. Notes on the freshwater mussels of New Guinea. *The Nautilus* **70**: 38–48.

McNeely, J.A., 1987. How dams and wildlife can coexist: natural habitats, agriculture, and major water resource development projects in tropical Asia. *Conservation Biology* **1**: 228–238.

Medway, Lord, 1978. *The Wild Mammals of Malaya (Peninsula Malaysia) and Singapore, Second Edition*. Oxford University Press, Kuala Lumpur: 128 pp.

Meier, P.G., 1980. Diel periodicity in the feeding activity of *Potamanthus myops* (Ephemeroptera). *Arch. Hydrobiol.* **88**: 1–8.

Meier-Brook, C., 1974. A snail intermediate host of *Schistosoma mansoni* introduced into Hong Kong. *Bull. W.H.O.* **51**: 661.

Meier-Brook, C., 1983. Taxonomic studies on *Gyraulus* (Gastropoda: Planorbidae). *Malacologia* **24**: 1–113.

Meier-Brook, C., 1979. The planorbid genus *Gyraulus* in Eurasia. *Malacologia* **18**: 67–72.

Meisner, J.D. & Shuter, B.J., 1992. Assessing potential effects of global climate change on tropical freshwater fishes. *GeoJournal* **28**: 21–27.

Melkania, N.P., 1991. Mountain water ecosystem: ecological status and future perspectives. In: Bhatt, S.D. & Pande, R.K. (Eds), *Ecology of the Mountain Waters*. Ashish Publishing House, New Delhi: 1–32.

Melville, D.S., 1994. Management of Jangxi Poyang Lake National Nature Reserve, China. *Mitt. Int. Verein. Limnol.* **24**: 237–242.

Melville, D.S., MacKinnon, J.R., Wang, B., Wang, H. & Song, X., 1992. *Management Plan for Jiangxi Poyang Lake National Nature Reserve.* Ministry of Forestry, People's Republic of China & WWF-World Wide Fund for Nature, Hong Kong: 130 pp.

Mendis, A.S. & Fernando, C.H., 1962. A guide to the freshwater fauna of Ceylon. *Bull. Fish. Res. Stn, Ceylon* **12**: 1–160.

Mer, G.S. & Sati, S.C., 1989. Seasonal fluctuation in species composition of aquatic hyphomycetous flora in a temperate freshwater stream of central Himalaya, India. *Int. Rev. ges. Hydrobiol.* **74**: 433–437.

Merritt, R.W. & Cummins, K.W., 1978. Introduction. In: Merritt, R.W. & Cummins, K.W. (Eds), *An Introduction to the Aquatic Insects of North America*. Kendall/Hunt Publishing Co., Iowa: 1–3.

Metcalfe, J.R., 1989. Biologcal water quality assessment of running waters based on macroinvertebrate communities: history and present state in Europe. *Environ. Pollut.* **60**: 101–139.

Mey, W., 1986. Drie neue Köcherfliegen aus Mittelasien (Trichoptera). *Dtsch. ent. Z.* **33**: 65–70.

Mey, W., 1989. *Setodes fluviovivens* sp. n. — a new potamobiontic species from the Mekon River of Kambodia (Trichoptera: Leptoceridae). *Aquatic Insects* **11**: 125–127.

Mey, W., 1990a. Neue Köcherfliegen von den Philippen (Trichoptera). *Opusc. zool. flumin.* **57**: 1–19.

Mey, W., 1990b. Neue und wenig bekannte Arten der Gattung *Hydropsyche* PICTET von den Philippinen (Trichoptera, Hydropsychidae). *Dtsch. ent. Z.* **37**: 413–424.

Mey, W., 1993a. Beschreibung von vier neuen Köcherfliegen aus Nord-China (Trichoptera, Annulipalpia). *Dtsch. ent. Z.* **40**: 333–340.

Mey, W., 1993b. *Macrostemum thomasi* n. sp., eine neue Köcherfliegen aus Sikkim, Nordindien (Trichoptera, Hydropsychidae). *Nachr. entomol. Ver. Apollo, Frankfurt/Main N.F.* **13**: 393–400.

Mey, W., 1995a. Beitrag zur Kenntnis der Köcherfliegenfauna der Philippinen, 1 (Trichoptera). *Dtsch. ent. Z.* **42**: 191–209.

Mey, W., 1995b. Bearbeitung einer kleinen Kollektion von Köcherfliegen aus Vietnam (Trichoptera). *Entomologische Zeitschrift* **105**: 209–218.

Mey, W., 1996a. Die Köcherfliegenfauna des Fan Si Pan-Massivs in Nord-Vietnam. 1. Beschreibung neuer und endemischer Arten aus den Unterordnungen Spicipalpia und Annulipalpia (Trichoptera). *Beitr. Entomol.* **46**: 39–65.

Mey, W., 1996b. Zur Kenntnis der *Hydropsyche pluvialis*-Gruppe in Sudostasien (Trichoptera: Hydropsychidae). *Entomologische Zeitschrift* **106**: 144–152.

Mey, W., 1996c. Die Arten der Gattung Nothopsyche in Vietnam und ihre biogeographische und phylogenetische Stellung (Trichoptera: Limnephilidae). *Entomological Problems* **27**: 99–109.

Mey, W., 1997a. Revision of the type-species of Hydropsychinae and Diplectroninae described by N. Banks from the Philippines (Trichoptera: Hydropsychidae). In: Holzenthal, R.W. & Flint, O.S. (Eds), *Proceedings of the 8th International Symposium on Trichoptera*. Ohio Biological Survey, Columbus: 303–308.

Mey, W., 1997b. A second species of *Apatidelia* Mosely from China (Trichoptera, Apataniidae). *Aquatic Insects* **19**: 14.

Mey, W., 1997c. Phylogeny of the *Arctopsyche composita*-group (Insecta, Trichoptera: Arctopsychidae) with the description of three new species from Vietnam. *Aquatic Insects* **19**: 155–164.

Mey, W., 1997d. Die Köcherfliegen des Fan Si Pan-Massivs in Nord-Vietnam. 2. Beschreibung neuer und endemischer Arten aus der Unterordnung Integripalpia (Insecta: Trichoptera). *Zeitschrift für Entomologie* **18**: 197–211.

Mey, W. & Dulmaa, A., 1986. Die Larve von *Asynarchus iteratus* McLachlan (Trichoptera, Limnephilidae). *Reichenbachia* **23**: 145–149.

Mey, W. & Levanidova, I.M., 1989. Revision der Gattung *Apataniana* Mosely 1936 (Trichoptera, Limnephilidae). *Dtsch. ent. Z.* **36**: 65–98.

Meybeck, M., 1984. Variabilite geographique de la composition chimique naturelle des eaux courantes. *Verh. Internat. Verein. Limnol.* **22**: 1766–1774.

Meybeck, M., 1985. Variabilite dens le temps de la composition chimique des rivieres et de leurs transports en solution et en suspension. *Revue Francaise des Sciences de l'Eau* **4**: 93–121.

Meybeck, M., 1987. The water quality of world rivers through the GEMS program. *Mitt. Geol.-Palaont. Inst. Univ. Hamburg, SCOPE/UNEP Sonderbd.* **64**: 1–17.

Meybeck, M., Chapman, D. & Helmer, R., 1989. *Global Freshwater Quality. A First Assessment*. Basil Blackwell Ltd, Oxford: 306 pp.

Meyer, J.L. & Edwards, R.T., 1990. Ecosystem metabolism and turnover of organic carbon along a blackwater river continuum. *Ecology* **71**: 668–677.

Miller, R.C. & McClure, F.A., 1931. The fresh-water clam industry of the Pearl River. *Lingnan Sci. J.* **10**: 307–322.

Milliman, J.D. & Meade, R.H., 1983. World-wide delivery of river sediment to the oceans. *J. Geol.* **91**: 1–21.

Milliman, J.D., Xie, Q. & Yang, Z., 1984. Transfer of particulate organic carbon and nitrogen from the Yangtze River to the ocean. *Amer. J. Sci.* **284**: 824–834.

Ministry of Water Resources & Power, 1987. *Irrigation and Drainage in China.* China Water Resources & Electric Power Press, Beijing: 113 pp.

Minshall, G.W., Petersen, R.C., Cummins, K.W., Bott, T.L., Sedell, J.R., Cushing. C.E. & Vannote, R.L., 1983. Interbiome comparison of stream ecosystem dynamics. *Ecol. Monogr.* **53**: 1–25.

Minshall, G.W., Cummins, K.W., Petersen, R.C., Cushing, C.E., Bruns, D.A., Sedell, J.R. & Vannote, R.L., 1985. Developments in stream ecosystem theory. *Can. J. Fish. Aquat. Sci.* **42**: 1045–1055.

Mishra, G.P. & Yadav, A.N., 1978. A comparative study of physiochemical characteristics of river and lake water in Central India. *Hydrobiologia* **59**: 275–278.

Mizue, K. & Iwamoto, Y., 1961. On the development and growth of *Neocaridina denticulata* De Haan. *Bull. Fac. Fish. Nagasaki Univ.* **10**: 15–24.

Mizuno, T. & Mori, S., 1970. Preliminary hydrobiological survey of some Southeast Asian inland waters. *Biol. J. Linn. Soc.* **2**: 77–117.

Mohan, M. & Bisht, R.S., 1991a. Taxo-ecology of aquatic entomofauna in freshwater ecosystem with special reference to River Bhagirathi and Bhilangana in Garhwal Himalaya. In: Bhatt, S.D. & Pande, R.K. (Eds), *Ecology of the Mountain Waters.* Ashish Publishing House, New Delhi: 251–265.

Mohan, M. & Bisht, R.S., 1991b. A report on predation of *Ephemera (E.) supposita* eaten by terrestrial spiders in riverine ecosystem of Bhilangana Tehri-Garhwal, Himalaya. *Geobios New Reports* **10**: 64–65.

Mohan, M., Bisht, R.S. & Das, S.M., 1989. Physico-chemical parameters in relation to aquatic entomofauna of River Bhagirathi and Bhilangana, Tehri-Garhwal, Himalaya. In: Khulbe, R.D. (Ed.), *Perspectives in Aquatic Biology.* Papyrus Publishing House, New Delhi: 197–203.

Mol, A.W.M., 1987. *Afronurus sibuyanensis* spec. nov., a new mayfly from the Philippines (Ephemeroptera: Heptageniidae). *Opusc. zool. flumin.* **15**: 1–9.

Mol, A.W.M., 1989. *Echinobaetis phagas* gen.nov., spec. nov., a new mayfly from Sulawesi (Ephemeroptera: Baetidae). *Zoologische Mededelingen (Leiden)* **63**: 61–72.

Monkolprasit, S. & Roberts, T.R., 1990. *Himantura chaophraya*, a new freshwater stingray from Thailand. *Jap. J. Ichthyol.* **37**: 203–208.

Montague, J.J., 1983. Influence of water level, hunting pressure and habitat type on crocodile abundance in the Fly River, Papua New Guinea.

Biological Conservation **26**: 309–340.

Mooney, H.A. & Drake, J.A., 1989. Biological invasions: a SCOPE program overview. In: Drake, J.A., Mooney, H.A., di Castri, F., Groves, R.H., Kruger, F.J., Rejmanek, M. & Williamson, M. (Eds) *Biological Invasions: a Global Perspective.* John Wiley & Sons, Chichester: 491–506.

Moore, J. & Gibson, R., 1985. The evolution and comparative physiology of terrestrial and freshwater nemerteans. *Biol. Rev.* **60**: 257–312.

Moore, J. & Gibson, R., 1988. Further studies on the evolution of land and freshwater nermerteans: generic relationships among the paramonostiliferous taxa. *J. Zool, Lond.* **216**: 1–20.

Moore, J.P., 1924. Notes on some Asiatic leeches (Hirudinea) principally from China, Kashmir, and British India. *Proc. Acad. Natur. Sci. Philad.* **76**: 343–388.

Moore, J.P., 1927. Arhynchobdellae. In: Shipley, A.E. (Ed.), *The Fauna of British India, including Ceylon and Burma. Hirudinea.* Taylor & Francis, London: 97–295.

Moore, J.P., 1930a. Leeches (Hirudinea) from China, with description of new species. *Proc. Acad. Natur. Sci. Philad.* **82**: 169–192.

Moore, J.P., 1930b. The leeches (Hirudinea) of China. *Peking Nat. Hist. Bull.* **4**: 39–43.

Moore, J.P., 1935. Leeches from Borneo and the Malay Peninsula. *Bull Raffles Mus.* **10**: 67–79.

Moore, J.P., 1937. Leeches (Hirudinea) especially from the Malay Peninsula, with descriptions of new species. *Bull Raffles Mus.* **14**: 64–80.

Morihara, D.K. & McCafferty, W.P., 1979. The *Baetis* larvae of North America (Ephemeroptera: Baetidae). *Trans. Amer. Entomol. Soc.* **105**: 139–221.

Morrisey, D.J., Howitt, L., Underwood, A.J. & Stark, J.S., 1992a. Spatial variation in soft-sediment benthos. *Marine Ecology Progress Series* **81**: 197–204.

Morrisey, D.J., Howitt, L., Underwood, A.J. & Stark, J.S., 1992b. Temporal variation in soft-sediment benthos. *J. Exp. Mar. Biol. Ecol.* **164**: 233–245.

Morrison, J.P.E., 1954. The relationship of New World and Old World melanians. *Proc. U.S. Nat. Mus.* **103**: 357–394.

Morse, J.C., 1997. Phylogeny of Trichoptera. *Ann. Rev. Entomol.* **42**: 427–450.

Morse, J.C. & Yang, L., 1992. Higher classification of the Chinese Glossosomatidae (Trichoptera). In: Otto, C. (Ed.), *Proceedings of the 7th International Symposium on Trichoptera.* Umea, Sweden: 139–148.

Morton, B., 1975. The colonization of Hong Kong's raw water supply system by *Limnoperna fortunei* (Dunker) (Bivalvia: Mytilaceae) from China. *Malacol. Rev.* **8**: 91–105.

Morton, B., 1977a. An estuarine bivalve (*Modiolus striatulus*) fouling raw water supply systems in West Bengal, India. *J. Inst. Water Engin. Sci.* **31**: 441–453.

Morton, B., 1977b. The population dynamics of *Limnoperna fortunei* (Dunker 1857) (Bivalvia: Mytilaceae) in Plover Cove Reservoir, Hong Kong. *Malacologia* **16**: 165–182.

Morton, B., 1977c. The population dynamics of *Corbicula fluminea* (Müller 1774) (Bivalvia: Corbiculaceae) in Plover Cove Reservoir, Hong Kong. *J Zool., Lond.* **181**: 21–42.

Morton, B., 1979. *Corbicula* in Asia. In: Britton, J.C. (Ed.), *Proceedings, First International Corbicula Symposium*. Texas Christian University Research Foundation, Fort Worth: 15–38.

Morton, B., 1982. Some aspects of the population structure and sexual strategy of *Corbicula cf. fluminalis* (Bivalvia: Corbiculaceae) from the Pearl River, People's Republic of China. *J. Moll. Stud.* **48**: 1–23.

Morton, B., 1983. The sexuality of *Corbicula fluminea* (Bivalvia: Corbiculaceae) in lentic and lotic waters in Hong Kong. *J. Moll. Stud.* **49**: 81–83.

Morton, B., 1984. A review of *Polymesoda* (*Geloina*) Gray 1842 (Bivalvia: Corbiculaceae) from Indo-Pacific mangroves. *Asian Marine Biology* **1**: 77–86.

Morton, B., 1985. The population dynamics, reproductive strategy and life history of *Musculium lacustre* (Pisidiidae) in Hong Kong. *J. Zool., Lond.* **207**: 581–603.

Morton, B., 1986. The population dynamics and life history tactics of *Pisidium clarkeanum* and *P. annandalei* (Bivalvia: Pisidiidae) sympatric in Hong Kong. *J. Zool., Lond.* **210**: 427–449.

Morton, B. 1992. The salinity tolerance of *Cyrenobatissa subsulcata* (Bivalvia: Corbiculoidea) from China. *Malacol. Rev.* **25**: 103–108.

Mosely, M.E., 1942. Chinese Trichoptera: a collection made by Mr. M.S. Yang in Foochow. *Trans. R. ent. Soc., Lond.* **92**: 343–362.

Moubayed, Z., 1983. Sur le dévelopment larvaire de *Grovellinus coyei* Allard 1868 (Coleoptera-Elmidae) récolté dans la Bekaa-Liban. *Annls Limnol.* **19**: 115–119.

Moubayed, Z., 1989. Description of five new species of Chironominae (Dipt., Chironomidae) from Near East and the Oriental Region. *Acta Biologica Debrecina Supplementum Oecologica Hungarica* **2**: 275–283.

Moubayed, Z., 1990. Chironomids from running waters of Thailand: description of *Rheotanytarsus thailandensis* sp. n. and *Tanytarsus thaicus* sp. n. (Dipt., Chironomidae). *Hydrobiologia* **203**: 29–33.

Moulton, S.R. & Stewart, K.W., 1997. A new species and first record of the caddisfly genus *Cnodocentron* Schmid (Trichoptera: Xiphocentronidae) north of Mexico. In: Holzenthal, R.W. & Flint, O.S. (Eds), *Proceedings of the 8th International Symposium on Trichoptera*. Ohio Biological Survey, Columbus: 343–347.

Moyle, P.B. & Leidy, R.A., 1992. Loss of biodiversity in aquatic ecosystems: evidence from fish faunas. In: Fielder, P.L. & Jain, S.A. (Ed.), *Conservation Biology: the Theory and Practice of Nature Conservation, Preservation and Management*. Chapman & Hall, New York: 128–169.

Moyle, P.B. & Senanayake, F.R., 1984. Resource partitioning among the fishes of rainforest streams in Sri Lanka. *J. Zool., Lond.* **202**: 195–224.

Mukai, H., Backus, B.T. & Wood, T.S., 1990. Comparative studies of American, European and Japanese forms of *Plumatella emarginata*, a freshwater bryozoan. *Proc. Jap. Soc. Syst. Zool.* **42**: 51–59.

Müller, H.G., 1992. Anthuridae of the genera *Apanthura* and *Cyathura* from Malaysian coral reefs, with descriptions of two new species (Crustacea: Isopoda: Anthuridae). *Zoologisher Anzeiger* **228**: 156–166.

Müller-Liebenau, I., 1978. *Raptobaetopus*, eine neue carnivore Ephemeropteren-Gattung aus Malaysia (Insecta, Ephemeroptera: Baetidae). *Arch. Hydrobiol.* **82**: 465–481.

Müller-Liebenau, I., 1980a. *Jubabaetis* gen. n. and *Platybaetis* gen. n.: two new genera of the family Baetidae from the Oriental region. In: Flannagan, J.E. & Marshall, K.E. (Eds), *Advances in Ephemeroptera Biology*. Plenum Press, New York: 103–114.

Müller-Liebenau, I., 1980b. A new species of the genus *Platybaetis* Müller-Liebenau 1980, *P. bishopi* sp. n., from Malaysia (Insecta, Ephemeroptera). *Gewässer und Abwässer* **66/67**: 95–101.

Müller-Liebenau, I., 1981. Review of the original material of the baetid genera *Baetis* and *Pseudocloeon* from the Sunda Islands and the Philippines described by G. ULMER, with some general remarks (Insecta: Ephemeroptera). *Mitt. Hamburg Zool. Mus. Inst.* **78**: 197–208.

Müller-Liebenau, I., 1982a. New species of the family Baetidae from the Philippines (Insecta, Ephemeroptera). *Arch. Hydrobiol.* **94**: 70–82.

Müller-Liebenau, I., 1982b. Five new species of *Pseudocloeon* Klapálek, 1905 (Fam. Baetidae), from the Oriental region (Insecta: Ephmeroptera) with some general remarks on *Pseudocloeon*. *Arch. Hydrobiol.* **95**: 282–298.

Müller-Liebenau, I., 1982c. A new genus and species of Baetidae from Sri Lanka (Ceylon): *Indocloeon primium* gen. nov., sp. nov. (Insecta, Ephmemeroptera). *Aquatic Insects* **4**: 125–129.

Müller-Liebenau, I., 1983. Three new species of the genus *Centroptella* Braasch & Soldán, 1980, from Sri Lanka (Insecta: Ephemeroptera). *Arch. Hydrobiol.* **97**: 486–500.

Müller-Liebenau, I., 1984a. New genera and species of the family Baetidae from West-Malaysia (River Gombak) (Insecta: Ephemeroptera). *Spixiana* **7**: 253–284.

Müller-Liebenau, I., 1984b. Baetidae from Sabah (East Malaysia) (Ephemeroptera). In: Landa, V. (Ed.), *Proceedings of the Fourth International Conference on Ephemeroptera*. CSAV, Czechoslovakia: 85–99.

Müller-Liebenau, I., 1985. Baetidae from Taiwan with remarks on *Baetiella* UENO, 1931 (Insecta, Ephemeroptera). *Arch. Hydrobiol.* **104**: 93–110.

Müller-Liebenau, I. & Heard, W.H., 1979. *Symbiocloeon*: a new genus of Baetidae from Thailand. In: Sowa, R. (Ed.), *Proceedings of the 2nd*

International Conference on Ephemeroptera, 1975. Warszawa-Krakow: 57–65.

Müller-Liebenau, I. & Hubbard, M.D., 1986. Baetidae from Sri Lanka with some general remarks on the Baetidae of the Oriental region (Insecta: Ephemeroptera). *Florida Entomol.* **68**: 537–561.

Müller-Liebenau, I. & Morihara, D.K., 1982. *Indobaetis*: a new genus of Baetidae from Sri Lanka (Insecta: Ephemeroptera) with two new species. *Gewässer und Abwässer* **68/69**: 26–34.

Munroe, E.G., 1995. *Usingeriessa onyxalis* (Dyar) (Lepidoptera: Crambidae: Nymphulinae), a moth with presumably aquatic larvae, newly recorded from Hawaii, with a synopsis of Hawaiian Nymphulinae. *Bishop Museum Occasional Papers* **42**: 39–42.

Muraoka, K., 1977. The larval stages of *Halicarcinus orientalis* Sakai and *Rhynchoplax messor* reared in the laboratory (Crustacea, Brachyura, Hymenosomatidae). *Zoological Magazine (Tokyo)* **86**: 94–99.

Murray, D.A., 1995. *Conchapelopia insolens*, n. sp., a new species of Tanypodinae (Diptera: Chironomidae) from Sulawesi. In: Cranston, P., (Ed.), *Chironomids: From Genes to Ecosystems*. CSIRO Publications, Melbourne: 417–423.

Mustow, S.E., Wilson, R.S., Sannarm, G., 1997. Chironomid assemblages in two Thai water courses in relation to water quality. *Nat. Hist. Bull. Siam Soc.* **45**: 53–64.

Muttamara, S. & Sales, C.L., 1994. Water quality management of the Chao Phraya River (a case study). *Environmental Technology* **15**: 501–516.

Mwango, J., Williams, T. & Wiles, R., 1995. A preliminary study of the predator-prey relationships of watermites (Acari: Hydrachnida) and blackfly larvae (Diptera: Simuliidae). *Entomologist* **114**: 107–117.

Nagasaki, O., 1992. Life history traits and resource partitioning between two coexisting aquatic pyralid moths, *Elophila interruptalis* (Pryer) and *Neoshoenobia decoloralis* Hampson (Lepidoptera). *Jap. J. Ecol.* **42**: 263–274 (in Japanese).

Nagatomi, A., 1962. Studies in the aquatic snipe flies of Japan. Part V. Biological notes (Diptera, Rhagionidae). *Mushi* **36**: 103–149.

Nagatomi, A., 1984. Taxonomic notes on *Atrichops* (Athericidae: Diptera). *Memoirs of the Kagoshima University Research Center for the South Pacific* **5**: 10–24.

Nagatomi, A., 1985. Notes on Athericidae (Diptera). *Memoirs of the Kagoshima University Research Center for the South Pacific* **5**: 87–106.

Nagayasu, Y. & Ito, T., 1997. Life history of *Dicosmoecus jozankeanus* in northern Japan, with particular reference to the difference between spring brook and mountain stream populations (Trichoptera: Limnephilidae: Dicosmoecinae). In: Holzenthal, R.W. & Flint, O.S. (Eds), *Proceedings of the 8th International Symposium on Trichoptera*. Ohio Biological Survey, Columbus: 365–372.

Naidu, K.V., 1963. Studies on the freshwater Oligochaeta of South India I.

Aeolosomatidae and Naididae, Part 5. *J. Bombay Nat. Hist. Soc.* **60**: 201–227.

Naidu, K.V. & Naidu, K.A., 1981a. Some fresh-water Oligochaeta from Bombay city and environs. *J. Bombay Nat. Hist. Soc.* **78**: 524–538.

Naidu, K.V. & Naidu K.A., 1981b. Some aquatic oligoichaetes of the Niligris, south India. *Hydrobiologia* **75**: 113–118.

Naidu, K.V. & Srivastava, H.N., 1980. Some fresh-water Oligochaeta of Nagpur, India. *Hydrobiologia* **72**: 261–271.

Nair, N.B., Arunachalam, M., Madhusoodanan Nair, K.C. & Suryanarayanan, H., 1989. A spatial study of the Neyyar River in the light of the River-Continuum-Concept. *Trop. Ecol.* **30**: 101–110.

Naiyanetr, P., 1989. *Phricotelphusa sirindhorn* n. sp., a new freshwater crab from Thailand (Decapoda, Brachyura, Gecarcinucidae). *Crustaceana* **56**: 225–229.

Naiyanetr, P., 1992. *Demanietta sirikit* n. sp., a new freshwater crab from Thailand (Decapoda, Brachyura, Potamidae). *Crustaceana* **62**: 113–120.

Naiyanetr, P. & Ng, P.K.L., 1995. The river crabs of the genus *Mekhongthelphusa* Naiyanetr, 1985 (Crustacea: Decapoda: Brachyura: Parathelphusidae), with description of a new species from Thailand. *Zool. Meded., Leiden* **29**: 365–374.

Nakahara, W., 1958. The Neurorthinae, a new subfamily of the Sisyridae (Neiroptera). *Nushi* **32**: 19–32.

Nakane, T., 1985. New or little-known Coleoptera from Japan and its adjacent regions, 38. *Fragmenta Coleopterologica* **38**: 153–164.

Nakane, T., 1996. The beetles collected by Drs. Keizo Kojima and Shingo Nakamura in Taiwan (Insecta, Coleoptera) 2. *Miscellaneous Reports of the Hiwa Museum for Natural History* **34**: 129–140.

Nandeesha, M.C., 1994. Fishes of the Mekong River — conservation and need for aquaculture. *Naga, the ICLARM Quarterly* **17**: 17–18.

Nanadeesha, M.C., Keshavanath, P., Basavaraja, N., Varghese, T.J. & Srikanth, G.K., 1989. Biological control and utilization of aquatic weeds in India: a review. *Indian J. Anim. Sci.* **59**: 1191–1198.

Natarajan, A.V., 1989. Environmental impact of Ganga Basin development on gene-pool and fisheries of the Ganga River system. *Can. Spec. Publ. Fish. Aquat. Sci.* **106**: 545–560.

Nautiyal, P., 1984. Studies on riverine ecology of torrential waters in the uplands of the Garhwal region. II. Seasonal fluctuations in diatom density. *Proc. Indian Acad. Sci., Anim. Sci.* **93**: 671–674.

Nautiyal, P., 1985. Studies on the riverine ecology of torrential waters in the Indian uplands of the Garhwal region. I. Seasonal variations in percentage occurrence of planktonic algae. *Uttar Pradesh J. Zool.* **5**: 14–19.

Nautiyal, P., 1986. Studies on the riverine ecology of torrential waters in the Indian uplands of the Garhwal region. III. Floristic and faunistic survey. *Trop. Ecol.* **27**: 157–165.

Nautiyal, P., Nautiyal, R. & Singh, H.R., 1991. Proposed Tehri Dam and reservoir fisheries: an ecological perspective. In: Bhatt, S.D. & Pande, R.K. (Eds), *Ecology of the Mountain Waters*. Ashish Publishing House, New Delhi: 365–374.

Nawawi, A., 1975. *Triscelophorus acuminatus* sp. nov. *Trans. Br. mycol. Soc.* **64**: 345–348.

Nawawi, A., 1976. A new genus of hyphomycetes. *Trans. Br. mycol. Soc.* **66**: 344–347.

Nawawi, A., 1985. Another aquatic hyphomycete genus from foam. *Trans. Br. mycool. Soc.* **85**: 174–177.

Nawawi, A. & Kuthubutheen, A.J., 1988. *Camposporidium* new genus, a phragmoconidial genus of hyphomycetes. *Mycotaxonomia* **32**: 161–168.

Nawawi, A. & Kuthubutheen, A.J., 1989a. A new taxon in *Colispora* hyphomycetes from Malaysia. *Mycotaxonomia* **34**: 497–502.

Nawawi, A. & Kuthubutheen, A.J., 1989b. *Quadricladium aquaticum* new genus new species, an aquatic hyphomycete with tetraradiate conidia. *Mycotaxonomia* **34**: 498–496.

Nazneen, S. & Begum, F., 1988. Hydrological studies of Lyari River. *Pakistan J. Sci. Ind. Res.* **31**: 26–29.

Nazneen, S. & Begum, F., 1994. Distribution of molluscs in Layari River (Sindh), Pakistan. *Hydrobiologia* **273**: 95–100.

Neame, P.A., 1988. Hydrochemistry of the lower Solo River, Indonesia. *Verh. Internat. Verein. Limnol.* **23**: 1372–1379.

Neboiss, A., 1980. First record of the subfamily Hyalopsychinae from Australia (Trichoptera: Polycentropodidae). *Arch. Hydrobiol.* **90**: 357–361.

Neboiss, A., 1984a. Notes on New Guinea Hydrobiosidae (Trichoptera). *Aquatic Insects* **6**: 177–184.

Neboiss, A., 1984b. Review of taxonomic positin of Australian and New Guinean species previously ascribed to *Macronema* (Trichoptera: Hydropsychidae). *Proc. R. Soc. Victoria* **96**: 127–139.

Neboiss, A., 1986. *Atlas of Trichoptera of the SW Pacific — Australian Region*. Dr W. Junk Publishers, Dordrecht: 286 pp.

Neboiss, A., 1987. Preliminary comparison of New Guinea Trichoptera with the faunas of Sulawesi and Cape York Peninsula. In: Bournaud, M. & Tachet, H. (Eds), *Proceedings of the 5th International Symposium on Trichoptera*. Dr W. Junk Publishers, The Hague: 103–105.

Neboiss, A., 1989a. Caddis-flies (Trichopera) of the families Polycentropodidae and Hyalopsychidae from Bumoga-Bone National Park, Sulawesi, Indonesia, with comments on identity of *Polycentropus orientalis* McLachlan. *Bull. Zool. Mus. Univer. Amsterdam* **12**: 101–109.

Neboiss, A., 1989b. The *Oecetis reticulata* species-group from the South-West Pacific area (Trichoptera: Leptoceridae). *Bijdragen tot de Dierkunde* **59**: 191–202.

Neboiss, A., 1990. Trichoptera of the families Goeridae and Lepidostomatidae from Sulawesi, Indonesia. *Mem. Mus. Victoria* **51**: 87–92.

Neboiss, A. 1992. Revised definitions of the genera *Nyctiophylax* Brauer and

Paranyctiophylax Tsuda (Trichoptera: Polycentropodidae). In: Otto, C. (Ed.), *Proceedings of the 7th International Symposium on Trichoptera*. Umea, Sweden: 107–111.

Neboiss, A.A., 1993. New species of the genus *Molanna* Curtis from Sulawesi (Trichoptera: Molannidae). *Tijdschrift voor Entomologie* **136**: 257–258.

Neboiss, A.A. & Botosaneanu, L., 1988. Caddisflies (Trichoptera) of the families Rhyacophilidae, Hydrobiosidae and Glossosomatidae from Sulawesi. *Bull. Zool. Mus. Univ. Amsterdam* **11**: 157–167.

Needham, J.G., 1930. *A Manual of the Dragonflies of China. A Monographic Study of the Chinese Odonata*. The Fan Memorial Institute of Biology, Bejing: 344 pp.

Needham, J.G., 1941. Observations on Chinese gomphine dragonflies. *Peking Nat. Hist. Bull.* **16**: 143–156.

Negi, M. & ingh, H.R., 1990. Substratum as determining factor for bottom fauna in the River Alaknanda. *Proc. Indian J. Natl. Sci. Acad. (B Biol. Sci.)* **56**: 417–423.

Nel, A. & Paicheler, C., 1994. Les Libelluloidea autres que Libellulidae fossiles un inventaire critique (Odonata, Corduliidae, Macromiidae, Synthemistidae, Chlorogomphidae et Mesophlebiidae). *Nouv. Rev. Entomol.* **11**: 321–334.

Nelson, C.H., 1988. Note on the phylogenetic systematics of the family Pteronarcyidae (Plecoptera), with a description of the eggs and nymphs of the Asian species. *Ann. Entomol. Soc. Am.* **81**: 560–576.

Nelson, D.J. & Scott, D.C., 1962. Role of detrius in the productivity of a rock-outcrop community in a Piedmont stream. *Limnol. Oceanogr.* **7**: 396-413.

Nelson, H.G., 1989. *Postelichus*, a new genus of Nearctic Dryopidae (Coleoptera). *Coleopterists' Bulletin* **43**: 19–24.

Nelson, H.G., 1990. *Pomatinus*, a name reclaimed for Helichus substriatus (PH. MÜLLER) (Coleoptera: Dryopidae). *Coleopterists' Bulletin* **44**: 223–224.

Nepszy, S.J. & Leach, J.H., 1973. First records of the Chinese mitten crab, *Eriocheir sinensis* (Crustacea: Brachyura) from North America. *J. Fish. Res. Bd Can.* **30**: 1909–1910.

Nesemann, H. 1995. On the morphology and taxonomy of the Asian leeches (Hirudinea: Erpobdellidae, Salifidae). *Acta Zoologica Academiae Scientarium Hungaricae* **41**: 165–182.

Newbold, J.D., Elwood, J.W., O'Neill, R.V. & Sheldon, A.L., 1983a. Phosphorus dynamics in a woodland stream ecosystem: a study of nutrient spiralling. *Ecology* **64**: 1249–1265.

Newbold, J.D., Elwood, J.W., O'Neill, R.V. & Van Winkle, W. 1983b. Resource spiraling: an operational paradigm for analysing lotic ecosystems. In: Fontaine, T.D. & Bartell, S.M. (Eds), *Dynamics of Lotic Ecosystems*. Ann Arbour Science Publishers, Michigan: 3–27.

Newbold, J.D., Erman, D.C. & Roby, K.B., 1980. Effects of logging on macroinvertebrates in streams with and without buffer strips. *Can. J. Fish. Aquat. Sci.* **37**: 1076–1085.

Newby, E., 1989. *Slowly Down the Ganges*. Hodder & Stoughton (Textplus Edition), London: 284 pp.

Ng, P.K.L., 1986a. *Balssiathelphusa cursor* sp. nov., a new species of freshwater crab from East Kalimantan (Decapoda: Parathelphusidae). *Treubia* **29**: 207–213.

Ng, P.K.L., 1986b. New freshwater crabs of the genus *Isolapotamon* Bott, 1968 from Kalimantan (Decapoda: Potamidae). *Treubia* **29**: 215–223.

Ng, P.K.L., 1986c. *Perithelphusa lehi* sp. nov., a new species of highland freshwater crab from Sarawak, Borneo (Crustacea, Decapoda, Brachyura, Parathelphusidae). *Sarawak Museum Journal* **36**: 291–295.

Ng, P.K.L., 1986d. *Terrapotamon* gen. nov., a new genus of freshwater crabs from Malaysia and Thailand, with description of a new species, *Terrapotamon aipooae* sp. nov. (Crustacea: Decapoda: Brachyura: Potamidae). *Journal of Natural History* **20**: 445–451.

Ng, P.K.L., 1987. Freshwater crabs of the genus *Isolapotamon* Bott, 1968 from Sarawak, Borneo (Crustacea, Decapoda, Brachyura, Potamidae). *Sarawak Museum Journal* **37**: 139–153.

Ng, P.K.L., 1988a. *The Freshwater Crabs of Peninsular Malaysia and Singapore*. Department of Zoology, National University of Singapore: 156 pp.

Ng, P.K.L., 1988b. *Allopotamon*, a new genus for the freshwater crab *Potamon (Potamonautes) tambelanensis* Rathbun, 1905 (Crustacea: Decapoda: Potamidae) from the Tambelan Islands, Proc. Biol. Soc. Washington **101**: 861–865.

Ng, P.K.L., 1989a. *Terrathelphusa*, a new genus of semiterrestrial freshwater crabs from Borneo and Java (Crustacea: Decapoda: Brachyura: Sundathelphusidae). *Raffles Bull. Zool.* **37**: 116–131.

Ng, P.K.L., 1989b. The identity of the cavernicolous freshwater crab *Potamon (Thelphusa) bidiense* Lanchester, 1900 (Crustacea: Decapoda: Gecarcinucidae) from Sarawak, Borneo, with description of a new genus. *Raffles Bull. Zool.* **37**: 63–72.

Ng, P.K.L., 1989c. A new cavernicolous freshwater cab, *Thelphusula styx* sp. nov. (Crustacea: Decapoda: Brachyura: Gecarcinucidae), from Gunung Mulu, Sarawak, Borneo. *Zool. Meded. (Leiden)* **63**: 53–59.

Ng, P.K.L., 1990a. Freshwater crabs and prawnbs of Singapore. In: Chou, L.M. & Ng, P.K.L. (Eds), *Essays in Zoology*. Department of Zoology, National University of Singapore, Singapore: 25–37.

Ng, P.K.L., 1990b. *Currothelphusa asserpes*, new genus new species (Crustacea: Decapoda: Brachyura: Sundathelphusidae) from a cave in Hamlmahera, Moluccas (Indonesia). *Bull. Mus. Natn Hist. Nat. Sec A., Zool., Biol. & Ecol. Anim.* **12**: 177–186.

Ng, P.K.L., 1990c. *Parathelphusa reticulata sp. nov.*, a new species of freshwater crab from blackwater swamps in Singapore (Crustacea:

Decapoda: Brachyura: Gecarcinucidae). *Zool. Meded. (Leiden)* **63**: 241–254.

Ng, P.K.L., 1991a. *Cancrocaeca xenomorpha*, new genus and species, a blind troglobitic hymenosomatid (Crustacea: Decapoda: Brachyura) from Sulawesi, Indonesia. *Raffles Bull. Zool.* **39**: 59–63.

Ng, P.K.L., 1991b. Bornean freshwater crabs of the genus *Arachnothelphusa* gen. nov. (Crustacea: Decapoda: Brachyura: Gecarcinucidae). *Zool. Meded. (Leiden)* **65**: 1–12.

Ng, P.K.L., 1991c. On two new species of *Archipelothelphusa* Bott, 1969 (Crustacea: Decapoda: Brachyura: Sundathelphusidae) from Luzon, Philippines. *Zool. Meded. (Leiden)* **65**: 13–24.

Ng, P.K.L., 1991d. A note on the taxonomy of two Malayan freshwater crabs, *Stoliczia rafflesi* (Roux, 1936), and *Stoliczia changamanae* Ng, 1988 (Crustacea: Decapoda: Brachyura: Potamidae). *Verh. Naturfordschenden Gesellschaft Basel* **100**: 91–97.

Ng, P.K.L., 1992a. On a collection of freshwater cabs (Crustacea: Brachyura) from Terengganu, Peninsular Malaysia, with description of a second species of *Geithusa* (Parathelphusidae). *Raffles Bull. Zool.* **40**: 95–101.

Ng, P.K.L., 1992b. The freshwater crabs and palaemonid prawns (Crustacea: Decapoda) of Batam Island, Riau Archipelago, Indonesia, Including descriptions of two new species. *Proc. Biol. Soc. Washington* **105**: 788–794.

Ng, P.K.L., 1993. Freshwater crabs allied to *Stoliczia tweediei* (Roux, 1934) (Crustacea: Decapoda: Brachyura: Potamidae) with descriptions of two nes species from Keda and Perak, Peninsular Malaysia. *Verh. Naturfordschenden Gesellschaft Basel* **103**: 81–95.

Ng, P.K.l., 1995a. The freshwater crabs and prawns (Crustacea: Decapoda) of Bako National Park, Sarawak, Malaysia, with descriptions of one new genus and three new species. *Raffles Bull. Zool.* **43**: 181–205.

Ng, P.K.L., 1995b. On one new genus and three new species of freshwater crabs (Crustacea: Deapoda: Brachyura: Potamidae and Grapsidae) from Lanjak-Entimau, Sarawak, East Malaysia, Borneo. *Zool. Meded. (Leiden)* **69**: 57–72.

Ng, P.K.L., 1996. *Nemoron nomas*, a new genus and new species of terrestrial crab (Crustacea: Decapoda: Brachyura: Potamidae) from central Vietnam. *Raffles Bull. Zool.* **44**: 29–36.

Ng, P.K.L. & Dudgeon, D., 1992. The Potamidae and Parathelphusidae (Crustacea: Decapoda: Brachyura) of Hong Kong. *Invertebr. Taxon.* **6**: 741–768.

Ng, P.K.L. & Goh, R., 1987. Cavernicolous freshwater crabs (Crustacea, Decapoda, Brachyura) from Sabah, Borneo. *Stygologia* **3**: 313–330.

Ng, P.K.L. & Lim, K.K.P., 1992. The conservation status of the Nee Soon freshwater swamp forest of Singapore. *Aquatic Conservation* **2**: 255–266.

Ng, P.K.L. & Lim, R.P., 1986. Description of a new genus and species of lowland freshwater crab from Peninsular Malaysia (Crustacea, Decapoda, Brachyura, Parathelphusidae). *Indo-Malayan Zoology* 3: 97–103.

Ng, P.K.L. & Lim, R.P., 1987. The taxonomy and biology of the nepenthiphilous sesarmine freshwater crab, *Geosesarma malayanum* Ng & Lim, 1986 (Crustacea, Decapoda, Brachyura, Grapsidae), from Peninsular Malaysia. *Malay. Nat. J.* 41: 393–402.

Ng, P.K.L. & Naiyanetr, P. 1993. New and recently described freshwater crabs (Crustacea: Decapoda: Brachyura: Potamidae, Gecarcinucidae and Parathelphusidae) from Thailand. *Zool. Meded. (Leiden)* 284: 3–117.

Ng, P.K.L. & Naiyanetr, P. 1995. *Pudaengon*, a new genus of terrestrial crabs (Crustacea: Decapoda: Brachyura: Potamidae) from Thailand and Laos, with descriptions of seven new species. *Raffles Bull. Zool.* 43: 355–376.

Ng, P.K.L. & Stuebing, R., 1989. Description of a new species of montane freshwater crab of the genus *Sundathelphusa* Bott, 1969 (Crustacea: Decapoda: Brachyura: Gecarcinucoidea) from Borneo. *Malay. Nat. J.* 43: 13–19.

Ng, P.K.L. & Stuebing, R., 1990. *Thelphusula dicerophilus* sp. nov., a new species of freshwater crab (Crustacea: Decapoda: Brachyura: Gecarcinucidae) found in mud wallows of the Sumatran rhinoceros from Sabah, Borneo. *Indo-Malayan Zoology* 6: 45–51.

Ng, P.K.L. & Takeda, M., 1992a. On some freshwater crabs (Crustacea: Brachyura: Potamidae, Parathelphusidae and Grapsidae) from Peninsular Malaysia. *Bull. Natn Sci. Mus., Ser. A (Zool.)* 18: 103–116.

Ng, P.K.L. & Takeda, M., 1992b. A new freshwater crab of the genus *Geosesarma* de Man, 1892 from the Philippines (Crustacea, Brachyura, Grapsidae). *Proc. Jap. Soc. Syst. Zool.* 47: 29–32.

Ng, P.K.L. & Takeda, M., 1992c. The freshwater crab fauna (Crustacea, Brachyura) of the Philippines. 1. The family Potamidae Ortmann, 1896. *Bull. Natn Sci. Mus., Ser. A (Zool.)* 18: 149–166.

Ng, P.K.L. & Takeda, M., 1993a. The freshwater crab fauna (Crustacea, Brachyura) of the Philippines. 2. The genus *Parathelphusa* H. Milne Edwards, 1853 (family Parathelphusidae). *Bull. Natn Sci. Mus., Ser. A (Zool.)* 19: 1–19.

Ng, P.K.L. & Takeda, M., 1993b. The freshwater crab fauna (Crustacea, Brachyura) of the Philippines. 3. The identity of *Telphusa cummingii* Miers, 1884, and its placement in the genus *Ovitamon* Ng & Takeda, 1992 (family Potamidae). *Bull. Natn Sci. Mus., Ser. A (Zool.)* 19: 111–116.

Ng, P.K.L. & Tan, C.G.S., 1995. *Geosesarma notophorum* sp. nov. (Decapoda, Brachyura, Grapsidae, Sesarminae), a terrestrial crab from Sumatra, with novel brooding behaviour. *Crustaceana* 68: 390–395.

Ng, P.K.L. & Tan, L.W.H., 1991. *Irmengardia didacta*, a new freshwater crab (Crustacea: Decapoda: Brachyura: Parathelphusidae) from Johor, Peninsular Malaysia. *Raffles Bull. Zool.* 39: 135–140.

Ng, P.K.L. & Wowor, D., 1990. *Terrathelphusa adipis*, new species (Crustacea:

Decapoda: Brachyura: Sundathelphusidae) from Kalimantan, Borneo. *Raffles Bull. Zool.* **38**: 263–268.

Ng, P.K.L. & Yang, C.M., 1986. A new freshwater crab from the genus *Isolapotamon* Bott, 1968 (Decapoda, Brachyura, Potamidae) from Sarawak, Borneo. *Indo-Malayan Zoology* **3**: 15–18.

Ng, P.K.L., Chou, L.M. & Lam, T.J., 1993. The status and impact of introduced freshwater animals in Singapore. *Biological Conservation* **64**: 19–24.

Ng, P.K.L., Tay, J.B. & Lim, K.K.P., 1994. Diversity and conservation of blackwater fishes in Peninsular Malaysia, particularly in the north Selangor peat swamp forest. *Hydrobiologia* **285**: 203–218.

Nickum, J.E., 1982. *Irrigation Management in the People's Republic of China.* World Bank Staff Working Paper no. 545, The World Bank, Washington D.C.: 106 pp.

Nieser, N. & Chen, P., 1991. Naucoridae, Nepidae and Notonectidae, mainly from Sulawesi and Pulau Buton. *Tijdschrift voor Entomologie* **134**: 47–67.

Nieser, N. & Chen, P., 1992a. Notes on Indonesian waterbugs (Nepomorpha and Gerromorpha). *Storkia* **1**: 30–40.

Nieser, N. & Chen, P., 1992b. Revision of *Limnometra* Mayr (Gerridae) in the Malay Archipelago. *Tijdschrift voor Entomologie* **135**: 11–26.

Nilsson, A.N., 1995. Noteridae and Dytiscidae: annotated check list of the Noteridae and Dytiscidae of China (Coleoptera). In: Jäch, M.A. & Ji, L. (Eds), *Water Beetles of China.* Zoologisch-Botanische Gesellschaft & Wiener Coleopterologenverein, Vienna: 35–96.

Nilsson, A.N. & Wewalka, 1994. Two new species of the genera *Allopachria* and *Agabus* from Taiwan (Coleoptera, Dytiscidae). *Linzer biologische Beiträge* **26**: 991–998.

Nilsson, A.N., Roughley, R.E. & Brancucci, M., 1989. A review of the genus- and family-group names of the family Dytiscidae Leach (Coleoptera). *Ent. Scand.* **20**: 287–316.

Nirmalakumari, K.R. & Balakrishnan-Nair, N., 1984. Relative abundance of predatory aquatic insects in the Chackai Canal, Trivandrum, India. *Comparative Physiology & Ecology* **9**: 105–113.

Nishimoto, H., 1989. A new species of *Moropsyche* (Trichoptera, Limnephilidae) from Japan, with some notes on the genus. *Jpn. J. Ent.* **57**: 695–702.

Nishimura, N., 1966. Ecological studies on the net-spinning caddisfly, *Stenopsyche griseipennis* McLachlan (Trichoptera, Stenopsychidae) 1. Life history and habit. *Mushi* **39**: 103–114.

Nishimura, N., 1981. Ecological studies on the net-spinning caddisfly, *Stenopsyche marmorata* Navas (Trichoptera, Stenopsychidae) 5. On the upstream migration of adult. *Kontyû, Tokyo* **49**: 192–204 (In Japanese).

Nishimura, N., 1984. Ecological studies on the net-spinning caddisfly, *Stenopsyche marmorata* Navas (Trichoptera, Stenopsychidae) 6. Larval and pupal density in the Maruyama River, Central Japan, with special reference to floods and after-flood recovery processes. *Physiol. Ecol. Japan* **21**: 1–34.

Nishino, M., 1981. Brood habits of two subspecies of a freshwater shrimp *Paratya compressa* (Decapoda, Atyidae), and their geographical variations. *Japanese Journal of Limnology* **42**: 201–219.

Nishiwaki, S., Hirata, T., Ueda, H., Tsuchiya, Y. & Sato, T., 1991a. Egg-laying season and monthly change in egg capsule production of *Clithon retropictus* (Prosobranchia: Neritidae) in the Naka River of Izu Peninsula. *Venus — Jap. J. Malacol.* **50**: 197–201 (in Japanese).

Nishiwaki, S., Hirata, T., Ueda, H., Tsuchiya, Y. & Sato, T., 1991b. Studies on the migratory direction of *Clithon retropictus* (Prosobranchia: Neritidae) by marking-and-recapture method. *Venus — Jap. J. Malacol.* **50**: 202–210 (in Japanese).

Nolen, J.A. & Pearson, R.G., 1992. Life history studies of *Anisocentropus kirramus* Neboiss (Trichoptera: Calmoceratidae) in a tropical Australian rainforest stream. *Aquatic Insects* **14**: 213–221.

Nolen, J.A. & Pearson, R.G., 1993. Factors influencing litter processing by *Anisocentropus kirramus* (Trichoptera: Calamoceratidae) from an Australian tropical rainforest stream. *Freshwat. Biol.* **29**: 469–479.

Nontji, A., 1994. The status of limnology in Indonesia. *Mitt. Internat. Verein. Limnol.* **24**: 95–113.

Noonan, R., 1980. The swift spider that is nature's smallest 'angler'. *Smithsonian* **11** (4): 78–83.

Norling, U., 1982. Structure and ontogeny of the lateral abdominal gills and the caudal gills in Euphaeidae (Odonata: Zygoptera) larvae. *Zool. Jb. Anat.* **107**: 343–389.

Novikova, E.A. & Kluge, N. Yu., 1995. Mayflies of the subgenus *Nigrobaetis* (Ephemeroptera, Baetidae, *Baetis Leach*, 1815). *Entomological Review* **74**: 16–39.

Nowell, W.R., 1980. A new species of *Dixa* from Peninsular Thailand (Diptera: Dixidae). *Pacific Insects* **22**: 174–177.

Nozaki, T. & Kagaya, T., 1994. A new *Ernodes* (Trichoptera, Beraeidae) from Japan. *Jap. J. Entomol.* **62**: 193–200.

Nozaki, T. & Shimada, T., 1997. Nectar feeding by adults of *Nothopsyche ruficollis* (Ulmer) (Trichoptera: Limnephilidae) and its effect on their reproduction. In: Holzenthal, R.W. & Flint, O.S. (Eds), *Proceedings of the 8th International Symposium on Trichoptera*. Ohio Biological Survey, Columbus: 379–386.

Nozaki, T., Ito, T. & Tanida, K., 1994. Checklists of Trichoptera in Japan. 2. Glossosomatidae, Beraeidae, Odontoceridae and Molannidae. *Jap. J. Limnol.* **55**: 297–305.

Nyholm, T., 1981. Helodiden aus Birma, gesammelt von René Malaise. 1. Die Arten der Gattung *Hydrocyphon* Redtenbacher (Coleoptera). *Ent. Scand., Suppl.* **15**: 253–267.

Nykvist, N., Grip, H., Sim, B.L., Malmer, A. & Wong, F.K., 1994. Nutrient losses in forest plantations in Sabah, Malaysia. *Ambio* **23**: 210–215.

O'Connor, J.P. & Ashe, P., 1992. *Jabitrichia wellsae* sp. n. (Trichoptera, Hydrotilidae) from Tasek Bera, Malaysia. *Aquatic Insects* **14**: 255–257.

Ochs, G., 1937. Nachtrag zur Gyriniden-Fauna Javas und der benachbarten Sunda-Inseln nebst Beschreibung einiger Verwandten des *Orectochilus spiniger* Rég aus Hinter-Indien und Ceylon. *Arch. Hydrobiol., Suppl.* **15**: 109–118.

Oda, S., 1990. Life cycle of *Pectinatella magnifica*, a freshwater Bryozoan. *Adv. Invert. Reprod.* **5**: 43–48.

Office of the National Economic and Social Development Board, 1994. *Study of Potential Development of Water Resources in the Mae Khong River Basin (Executive Summary)*. Water Resources Engineering Program, Asian Institute of Technology, Bangkok, Thailand: 29 pp.

Ogawa, H. & Male, J.W., 1990. Evaluation framework for wetland regulation. *J. Environ. Manag.* **30**: 95–109.

Ogawa, Y. & Kakuda, S., 1986. On the population growth and life span of the Oriental river prawn *Macrobrachium nipponense* inhabiting the Lake Kojima (Japan). *Bull. Jpn. Soc. Sci. Fish.* **52**: 777–786 (in Japanese).

Okada, H. & Hotta, M., 1987. Species diversity at wet tropical environments II. Speciation of *Schismatoglottis okadae* (Araceae): an adaptation to the rheophytic habitat of mountain stream in Sumatra, Indonesia. *Contrib. Biol. Lab. Kyoto Univ.* **27**: 153–170.

Okada, H. & Nakasuji, F., 1993a. Patterns of local distribution and coexistence of two giant water bugs, *Diplonychus japonicus* and *D. major* (Hemiptera, Belostomatidae) in Okayama, western Japan. *Jap. J. Ent.* **61**: 79–84.

Okada, H. & Nakasuji, F., 1993b. Comparative studies on the seasonal occurrence, nymphal development and food menu in two giant water bugs, *Diplonychus japonicus* Vuillefroy and *Diplonychus major* Esaki (Hemiptera, Belostomatidae). *Researches on Population Ecology* **35**: 15–22.

Oláh, J., 1987. Two new Trichoptera from Bhutan. *Folia Entomologica Hungarica* **66**: 139–141.

Oláh, J., 1987. Seven new *Rhyacophila* species from Vietnam (Trichoptera: Rhyacophilidae). *Folia Entomologica Hungarica* **68**: 141–149.

Oláh, J., 1988. Eight new *Agapetus* species from Vietnam (Trichoptera: Glossosomatidae). *Folia Entomologica Hungarica* **69**: 157–166.

Oláh, J, 1989. Thirty-five new hydroptilid species from Vietnam (Trichoptera, Hydroptilidae). *Acta zool. Hung.* **35**: 255–293.

Oliver, D.R., 1971. Life history of the Chironomidae. *Ann. Rev. Entomol.* **16**: 211–230.

Oliver, D.R. & Roussel, M.E., 1983. *The Insects and Arachnids of Canada Part 11. The Genera of Larval Midges of Canada*. Biosystematics Research Institute, Ottawa: 263 pp.

Olmi, m., 1986. New species of *Dryops* from Asia (Coleoptera, Dryopidae).

Frustula Entomologica **7/8**: 589–593.

Orchymont, A.d', 1934. Aquatic insects of China. Article XX. Catalogue of Chinese Palpicornia (Order Coleoptera). *Peking Nat. Hist. Bull.* **9**: 185–225.

Ormerod, S.J., Rundle, S.D., Wilkinson, S.M., Daly, G.P., Dale, K.M. & Juttner, I., 1994. Altitudinal trends in the diatoms, bryophytes, macroinvertebrates and fish of a Nepalese river system. *Freshwat. Biol.* **32**: 309–322.

Osborne, P.L., Polunin, N.V.C. & Nicholson, K., 1988. Geochemical traces of riverine influence on a tropical lateral lake. *Verh. Internat. Verein. Limnol.* **23**: 207–211.

Oswood, M.W., 1976. Comparative life histories of the Hydropsychidae (Trichoptera) in a Montana lake outlet. *Amer. Midl. Nat.* **96**: 493–497.

Pace, G.L., 1973. Freshwater snails of Taiwan (Formosa). *Malac. Rev., Suppl.* **1**: 1–118.

Padgett, D.E., 1976. Leaf decomposition by fungi in a tropical rainforest stream. *Biotropica* **8**: 166–178.

Paepke, H.J., 1984. Zur aktuellen Verbreitung von *Eriocheir sinensis* (Crustacea, Decapoda, Grapsidae) in der DDR. *Mitt. Zool. Mus. Berlin* **60**: 103–113.

Pahwa, D.V., 1979. Studies on the distribution of the benthic macrofauna in the stretch of the River Ganaga. *Indian J. Anim. Sci.* **49**: 212–219.

Pahwa, D.V. & Mehrotra, S.N., 1966. Observations on fluctuations in the abundance of plankton in relation to certain hydrological conditions of River Ganga. *Proc. natn Acad. Sci. India* **36B**: 157–189.

Paclt, 1994. *Ephacerella*, a replacement name for *Acerella* Allen, 1971 (Ephemeroptera) nec Berlese, 1909 (Protura). *Ent. News* **105**: 283–284.

Palharya, J.P. & Malviya, S., 1988. Pollution in the Narmada River at Hoshangabad in Madhya Pradesh and suggested measures for control. In: Trivedy, R.K. (Ed.), *Ecology and Pollution of Indian Rivers*. Ashish Publishing House, New Delhi: 55–86.

Palupi, K., Sumengen, S., Inswiasri, S., Augustina, L., Nunik, S.A., Sunarya, W. & Quraisyn, A., 1995. River water quality study in the vicinity of Jakarta. *Water Science & Technology* **31**: 17–25.

Pancho, J.V. & Soerjani, M., 1978. *Aquatic Weeds of Southeast Asia*. National Publishing Corporation Incorporated, Quezon City, Philippines: 130 pp.

Pande, R.K., 1991. Impact of vegetation on basin geomorphology: the experience of Central Himalaya, India. In: Bhatt, S.D. & Pande, R.K. (Eds), *Ecology of the Mountain Waters*. Ashish Publishing House, New Delhi: 357–364.

Pandey, B.N., Lal, R.N., Mishra, P.K. & Jha, A.K., 1992. Seasonal rhythm in the physico-chemical properties of Mahananda River, Katihar, Bihar. *Environ. Ecol.* **10**: 354–357.

Pandit, A.K. & Qadri, S.S., 1990. Floods threatening Kashmir wetlands. *J. Environ. Manag.* **31**: 299–311.

Pant, M.M., 1983. Harvesting water for ravaged siwaliks in India. In: L.S. Hamilton, L.S. (Ed.), *Forest and Watershed Development and Conservation in Asia and the Pacific*. Westview Press, Boulder, Colorado: 411–483.

Pantulu, V.R., 1970. Some biological considerations related to the lower Mekong development. *Regional Meeting of Inland Water Biologists in Southeast Asia. Proceedings*. Unesco Field Science Office for Southeast Asia, Djakarta: 113–119.

Pantulu, V.R., 1986a. The Mekong River system. In: Davies, B.R. & Walker, K.F. (Eds), *The Ecology of River Systems*. Dr W. Junk Publishers, The Hague: 695–719.

Pantulu, V.R., 1986b. Fish of the lower Mekong basin. In: Davies, B.R. & Walker, K.F. (Eds), *The Ecology of River Systems*. Dr W. Junk Publishers, The Hague: 721–741.

Papácek, M., 1994. *Idiotrephes maior* sp. n., a new species of water bug from Vietnam with morphological notes on *I. chinai* (Heteroptera: Helotrephidae). *Eur. J. Entomol.* **91**: 419–428.

Papácek, M., 1995. *Idiotrephes meszarosi* sp. n., a new helotrephid (Heteroptera: Helotrephidae) from Vietnam. *Aquatic Insects* **17**: 105–111.

Papácek, M., Stys, P. & Tonner, M., 1988. A new subfamily of Helotrephidae (Heteroptera, Nepomorpha) from Southeast Asia. *Acta Entomol. Bohemoslov.* **85**: 120–152.

Papácek, M., Stys, P. & Tonner, M., 1989. A new genus and species of Helotrephidae from Afghanistan and Iran (Heteroptera: Nepomorpha). *Vest. Cs. Spol. Zool.* **53**: 107–122.

Parameswaran, S. & Murugesan, V.K., 1976. Breeding season and seed resources of murrels in swamps of Karnataka State. *J. Inland Fish. Soc. India* **8**: 60–67.

Pareek, A. & Sharma, S., 1988. Phytogeographical affinities of the aquatic flora of Rajasthan. *Acat Bot. Indica* **16**: 19–22.

Parenti, L.R. & Allen, G.R., 1991. Fishes of the Gogol River and other coastal habitats, Madang Province, Papua New Guinea. *Ichthyol. Explor. Freshwat.* **1**: 307–320.

Parker, C.R. & Voshell, J.R., 1979. *Cardiocladius* (Diptera: Chironomidae) larvae ectoparasitic on pupae of Hydropsychidae (Trichoptera). *Environ. Entomol.* **8**: 808–809.

Parker, C.R. & Wiggins, G.B., 1987. Revision of the caddisfly genus *Psilotreta* (Trichoptera: Odontoceridae). *Royal Ontario Museum Life Sciences Contribution* **144**: 1–55.

Parnwell, M.J.G. & Taylor, D.M., 1996. Environmental degradation, non-timber forest products and Iban communities in Sarawak. Impact, response and future prospects. In: Parnwell, M.J.G. & Bryant, R.L. (Eds), *Environmental Change in South-east Asia. People, Politics and Sustainable Development*. Routledge, London: 269–300.

Pastorino, G., Darrigran, G. & Martin, S.M., 1993. *Limnoperna fortunei*

(Dunker, 1857) (Mytilidae), nuevo bivalvo invasor en aguas del Rio de la Plata. Neotropica (La Plata) **39**: 34.

Patil, S.G., Singh, D.F. & Harshey, D.K., 1986. Impact of gelatine factory effluent on the water quality and biota of a stream near Jabalpur, M.P. *J. Environ. Biol.* **7**: 61–65.

Patra, A.K., Nayak, L. & Patnaik, E., 1984. Seasonal primary production of River Mahanadi at Sambalpur in Orissa, India. *Trop. Ecol.* **25**: 153–157.

Patralekh, L.N., 1991a. Bacterial population of three freshwater ecosystems. *Environ. Ecol.* **9**: 218–220.

Patralekh, L.N., 1991b Periodicity of phytoplankton in the River Ganga at Bhagalpur, Bihar, India. *Environ. Ecol.* **9**: 84–86.

Payne, J.B., Francis, C.M. & Phillips, K., 1985. *A Field Guide to the Mammals of Borneo*. Sabah Society & WWF-World Wide Fund for Nature Malaysia, Kuala Lumpur: 332 pp.

Peckarsky, B.L., 1984. Sampling the stream benthos. In: Downing, J.A. and Rigler, F.H. (Eds), *A Manual on Methods for Assessment of Secondary Productivity in Freshwaters*. Blackwell Scientific Publications, Oxford: 131–160.

Peckarsky, B.L., Cooper, S.D. & McIntosh, A.R., 1997. Extrapolating from individual behavior to populations and communities in streams. *J. N. Am. Benthol. Soc.* **16**: 375–390.

Pederzani, F., 1995. Keys to the identification of the genera and subgenera of adult Dytiscidae (*sensu lato*) of the world (Coleoptera: Dytiscidae). *Atti della Accademia Rovertana Degli Agaiti Serie 7 B Classe di Scienze Matematiche Fisiche e Naturali* **4B**: 5–83.

Pelletier, C. & Pelletier, F.X., 1980. Rapport sur l'expedition Delphinasia (septembre 1977 — septembre 1978). *Ann. Soc. Sci. Nat. Charente-Marit.* **6**: 647–679.

Pemberton, R.W., 1988. The use of the Thai giant waterbug, *Lethocerus indicus* (Hemiptera, Belostomatidae), as human food in California. *Pan-Pacific Entomologist* **64**: 81–82.

Pennak, R.W., 1989. *Freshwater Invertebrates of the United States. Protozoa to Mollusca (3rd Edition)*. John Wiley & Sons, New York: 628 pp.

Penney, J.T. & Racek, A.A., 1968. Comprehensive revision of a world-wide collection of sponges. *Bull. U.S. Nat. Hist. Mus.* **272**: 1–184.

Pereira, H.C., 1981. Rehabilitating eroded hill lands in the People's Republic of China. *World Crops* **33**: 96.

Pereira, H.C., 1989. *Policy and Practice in the Management of Tropical Watersheds*. Westview Press, Boulder, Colorado & Belhaven Press, London: 237 pp.

Pernetta, J.C. & Burgin, S., 1983. The status and ecology of crocodiles in the Purari. In: Petr, T. (Ed.), *The Purari — Tropical Environment of a High Rainfall River Basin*. Dr W. Junk Publishers, The Hague: 409–428.

Perrin, W.F., Brownell, R.L. Jr, Zhou, K. & Liu, J., 1989. *Biology and Conservation of the River Dolphin*. Occasional Papers of the IUCN (World

Conservation Union) Species Survival Commission (SSC) No. 3: 180 pp.

Peterman, R.M., 1990. Statistical power analysis can improve fisheries research management. *Can. J. Fish. Aquat. Sci.* **47**: 2–15.

Peters, R.H. 1976. Tautology in evolution and ecology. *Amer. Nat.* **110**: 1–12.

Peters, T.M. & Savary, R.W., 1994. A new genus of *Dixidae* (Diptera) from the Philippines. *Proc. Entomol. Soc. Wash.* **96**: 22–26.

Peters, W.L., 1967. New species of Prosopistoma from the Oriental Region (Prosopistomatoidea: Ephemeroptera). *Tijdschrift voor Entomologie* **110**: 207–222.

Peters, W.L. & Edmunds, G.F., 1970. Revision of the generic classification of the Eastern Hemisphere Leptophlebiidae (Ephemeroptera). *Pacific Insects* **12**: 157–240.

Peters, W.L. & Edmunds, G.F., 1990. A new genus and species of Leptophlebiidae: Atalophlebiinae from the Celebes (Sulawesi). In: Campbell, I.C. (Ed.), *Mayflies and Stoneflies: Life Histories and Biology.* Kluwer Academic Publishers, Dordrecht: 327–335.

Peters, W.L. & Gillies, M.T., 1991. The male imago of *Protobehningia* Tshernova from Thailand (Ephemeroptera: Behningiidae). In: Alba-Tercedor, J. & Sanchez-Ortega, A. (Eds), *Overview and Strategies of Ephemeroptera and Plecoptera.* Sandhill Crane Press, Gainesville: 207–216.

Peters, W.L. & Peters, J.G., 1993. Some changes in Leptohyphidae and Tricorythidae (Ephemeroptera). *Aquatic Insects* **15**: 45–48.

Peters, W.L. & Tsui, T.P.T., 1972. New species of *Thraulus* from Asia (Leptophlebiidae: Ephemeroptera). *Oriental Insects* **6**: 1–17.

Petersen, R.C. & Sköglund, E., 1990. *Wetlands Management Programme. Study to Formulate Plans for the Management of the Wetlands in the Lower Mekong Basin (1.3.13/88/SWE).* The Mekong Secretariat, Bangkok, Thailand: 245 pp.

Peterson, C.H., 1993. Improvement of environmental impact analysis by application of principles derived from manipulative ecology: lessons from coastal marine case histories. *Aust. J. Ecol.* **18**: 21–52.

Pethiyagoda, R., 1994. Threats to the indigenous freshwater fishes of Sri Lanka and remarks on their conservation. *Hydrobiologia* **285**: 189–201.

Petr, T., 1976. Some chemical features of two Papuan fresh waters (Papua New Guinea). *Aust. J. Mar. Freshwat. Res.* **27**: 467–474.

Petr, T., 1983a. Limnology of the Purari basin Part 1. The catchment above the delta. In: Petr, T. (Ed.), *The Purari — Tropical Environment of a High Rainfall River Basin.* Dr W. Junk Publishers, The Hague: 141–177.

Petr, T., 1983b. Dissolved chemical transport in major rivers of Papua New Guinea. *Mitt. Geol.-Palaont. Inst. Univ. Hamburg, SCOPE/UNEP Sonderbd.* **55**: 477–481.

Petr, T., 1983c. Limnology of the Purari basin Part 2. The delta. In: Petr, T. (Ed.), *The Purari — Tropical Environment of a High Rainfall River Basin.* Dr W. Junk Publishers, The Hague: 179–203.

Petr, T., 1983d. Aquatic pollution in the Purari basin. In: Petr, T. (Ed.), *The Purari — Tropical Environment of a High Rainfall River Basin.* Dr W. Junk Publishers, The Hague: 325–339.

Phang, S.M., 1987. Agro-industrial wastewater reclamation in Peninsular Malaysia. *Arch. Hydrobiol. Beih., Ergebn. Limnol.* **28**: 77–94.

Phang, S.M. & Leong, P., 1987. Freshwater algae from the Ulu Endau Area, Johore, Malaysia. *Malay. Nat. J.* **41**: 145–157.

Pholprasith, S., 1983. Induced breeding of *pla buk* (*Pangasianodon gigas*). *Thai Fisheries Gazette* **36**: 347–360.

Pholprasith, S., 1993. Development techniques for induced spawning of the Mekong giant catfish. *Thai Fisheries Gazette* **46**: 399–415 (in Thai).

Pickup, G., 1983. Sedimentation processes in the Purari River upstream of the delta. In: Petr, T. (Ed.), *The Purari — Tropical Environment of a High Rainfall River Basin.* Dr W. Junk Publishers, The Hague: 205–225.

Pickup, G. & Chewings, V.H., 1983. The hydrology of the Purari and its environmental implications. In: Petr, T. (Ed.), *The Purari — Tropical Environment of a High Rainfall River Basin.* Dr W. Junk Publishers, The Hague: 123–139

Pillai, C.K. & Subramonian, T. 1984. Monsoon-dependent breeding in the field crab *Paratelphusa hydrodromous* (Herbst). *Hydrobiologia* **119**: 7–14.

Pilleri, G. & Pilleri, O., 1987. Indus and Ganges river dolphins *Platanista indi* (Blyth, 1859) and *Platanista gangetica* (Roxburgh, 1801). *Invest. Cetacea* **20**: 17–35.

Pinder, L.C.V., 1986. Biology of freshwater Chironomidae. *Ann. Rev. Entomol.* **31**: 1–23.

Pinratana, Bro. A., Kiasuta, B. & Hämäläinen, M., 1988. *List of the Odonata of Thailand and Annotated Bibliography.* The Viratham Press, Bangkok (unpaginated).

Pippet, J.R. & Fernando, C.H., 1961. Hairworms — a zoological curiosity. *Malay. Nat. J.* **15**: 148–151.

Poinar, G.O., 1991. Nematoda and Nematomorpha. In: Thorp, J.H. & Covich, A.P. (Eds), *Ecology and Classification of North American Freshwater Invertebrates.* Academic Press, San Diego: 249–283.

Polhemus, D.A., 1986. A review of the genus *Coptocatus* Montandon (Hemiptera: Naucoridae). *Pan-Pacific Entomologist* **62**: 248–256.

Polhemus, D.A., 1990. A revision of the genus *Metrocoris* Mayr (Heteroptera: Gerridae) in the Malay Archipelago and the Philippines. *Ent. Scand.* **21**: 1–28.

Polhemus, D.A., 1994a. A new species of *Aphelocheirus* from Sumatra, and addenda to the world checklist (Heteroptera: Naucoridae). *J. N.Y. Entomol. Soc.* **102**: 74–78.

Polhemus, D.A., 1994b. Taxonomy, phylogeny and zoogeography of the genus *Cylindrostethus* in the Paleotropical region (Hemiptera: Gerridae). *Bishop Museum Occasional Papers* **38**: 1–34.

Polhemus, D.A., 1996. Two new species of *Rhagovelia* from the Philippines, with a discussion of zoogeographic relationships between the Philippines and New Guinea (Heteroptera: Veliidae). *J. N.Y. Entomol. Soc.* **103**: 55–68.

Polhemus, D.A. & Polhemus, J.T., 1985. Naucoridae (Hemiptera) of New Guinea I. A review of the genus *Nesocricos*, with descriptions of two new species. *Int. J. Entomol.* **27**: 197–203.

Polhemus, D.A. & Polhemus, J.T., 1986a. Naucoridae (Hemiptera) of New Guinea. 2. A review of the genus *Idiocarus* Montandon, with descriptions of three new species. *J. N.Y. Entomol. Soc.* **94**: 39–50.

Polhemus, D.A. & Polhemus, J.T., 1986b. Naucoridae (Hemiptera) of New Guinea. III. A review of the genus *Tanycricos* Montandon, with descriptions of three new species. *J. N.Y. Entomol. Soc.* **94**: 163–173.

Polhemus, D.A. & Polhemus, J.T., 1987. A new genus of Naucoridae (Hemiptera) from the Philippines, with comments on zoogeography. *Pan-Pacific Entomologist* **63**: 256–269.

Polhemus, D.A. & Polhemus, J.T., 1988. The Aphelocheirinae of tropical Asia (Heteroptera: Naucoridae). *Raffles Bulletin of Zoology* **36**: 167–300.

Polhemus, D.A. & Polhemus, J.T., 1989. Naucoridae (Hemiptera) of New Guinea. IV. A review of the genus *Cavocoris* Montandon, with descriptions of four new species. *J. N.Y. Entomol. Soc.* **94**: 73–86.

Polhemus, D.A. & Polhemus, J.T., 1997. A review of the genus *Limnometra* Mayr in New Guinea, with the description of a very large new species (Heteroptera: Gerridae). *J. New York Entomol. Soc.* **105**: 24–39.

Polhemus, J.T., 1979. Results of the Austrian-Ceylonese hydrological expedition, 1970. Part XXIX. Aquatic and semiaquatic Hemiptera of Sri Lanka from the Austrian Indo-Pacific Expedition, 1970–71. *Bull. Fish. Res. Stn, Sri Lanka (Ceylon)* **29**: 89–113.

Polhemus, J.T., 1990a. Surface wave communication in water striders; field observations of unreported taxa (Heteroptera: Gerridae, Veliidae). *J. N.Y. Entomol. Soc.* **98**: 383–384.

Polhemus, J.T., 1990b. A new tribe, a new genus and three new species of Helotrephidae (Heteroptera) from Southeast Asia, and a world checklist. *Acta Entomol. Bohemoslov.* **87**: 45–63.

Polhemus, J.T., 1994. Stridulatory mechanisms in aquatic and semiaquatic Heteroptera. *J. N.Y. Entomol. Soc.* **102**: 270–274.

Polhemus, J.T., 1997. Seven new species of *Hydrotrephes* China (Helotrephidae: Heteroptera) from Sulawesi. *Tijdschrift voor Entomologie* **140**: 43–54.

Polhemus, J.T. & Andersen, N.M., 1984. A revision of *Amemboa* Esaki with notes on the phylogeny and ecological evolution of eotrechine water striders (Insecta, Hemiptera, Gerridae). *Steenstrupia* **10**: 65–111.

Polhemus, J.T. & Karunaratne, P.B., 1993. A review of the genus *Rhagadotarsus*, with descriptions of three new species (Heteroptera:

Gerridae). *Raffles Bulletin of Zoology* **41**: 95–112.

Polhemus, J.T. & Lansbury, I., 1997. Revision of the genus *Hydrometra* Latreille in Australia, Malanesia, and the Southwest Pacific (Heteroptera: Hydrometridae). *Bishop Museum Occasional Papers* **47**: 1–67.

Polhemus, J.T. & Polhemus, D.A., 1987. The genus *Valleriola* Distant (Hemiptera: Leptopodidae) in Australia, New Caledonia and Papua New Guinea with notes on zoogeography. *J. Aust. Ent. Soc.* **26**: 209–214.

Polhemus, J.T. & Polhemus, D.A., 1988. Zoogeography, ecology, and systematics of the genus *Rhagovelia* Mayr (Heteroptera: Veliidae) in Borneo, Celebes, and the Moluccas. *Insect Mundi* **2**: 161–230.

Polhemus, J.T. & Polhemus, D.A., 1989. A new mesoveliid genus and two new species of *Hebrus* (Heteroptera: Mesoveliidae, Hebridae) from intertidal habitats in Southeast Asian mangrove swamps. *Raffles Bulletin of Zoology* **37**: 73–82.

Polhemus, J.T. & Polhemus, D.A., 1990. Zoogeography of the aquatic Heteroptera of Celebes: regional relationships versus insular endemism. In: Knight, W.J. & Holloway, J.D. (Eds), *Insects and the Rain Forests of South East Asia (Wallacea)*. The Royal Entomological Society of London, London: 49–56.

Polhemus, J.T. & Polhemus, D.A., 1993. The Trepobatinae (Heteroptera: Gerridae) of New Guinea and surrounding regions, with a review of the world fauna. 1. Tribe Merobatini. *Ent. Scand.* **24**: 241–284.

Polhemus, J.T. & Polhemus, D.A., 1994a. The Trepobatinae (Heteroptera: Gerridae) of New Guinea and surrounding regions, with a review of the world fauna. Part 2. Tribe Naboandelini. *Ent. Scand.* **25**: 333–359.

Polhemus, J.T. & Polhemus, D.A., 1994b. Four new genera of Microveliinae (Heteroptera) from New Guinea. *Tijdschrift voor Entomologie* **137**: 57–74.

Polhemus, J.T. & Polhemus, D.A., 1995a. The Trepobatinae (Heteroptera: Gerridae) of New Guinea and surrounding regions, with a review of the world fauna: part 3. Tribe Trepobatini. *Ent. Scand.* **26**: 97–117.

Polhemus, J.T. & Polhemus, D.A., 1995. Revision of the genus *Hydrometra* Latreille in Indochina and the western Malay Archipelago (Heteroptera: Hydrometridae). *Bishop Museum Occasional Papers* **43**: 9–72.

Polhemus, J.T. & Polhemus, D.A., 1996. The Trepobatinae (Heteroptera: Gerridae) of New Guinea and surrounding regions, with a review of the world fauna. Part 4. The marine tribe Stenobatini. *Ent. Scand.* **27**: 279–346.

Polhemus, J.T., & Zettel, H., 1997. Five new *Potamometropsis* species (Insecta: Heteroptera: Gerridae) from Borneo. *Ann. Naturhist. Mus. wien* **99B**: 21–40.

Powell, C.B., 1979. Three alpheid shrimps of a new genus from West African fresh and brackish waters: taxonomy and ecological zonation (Crustacea Decapoda Natantia). *Rev. Zool. afr.* **93**: 116–150.

Prakash, S. & Agarwal, G.P., 1985. A note on the fishery of the North Indian freshwater prawn, *Macrobrachium birmanicum choprai*, in the middle stretch of the River Ganga. *Indian J. Fish.* **32**: 139–144.

Prashad, B., 1929. Revision of the Asiatic species of the genus Corbicula. III. The species of the genus *Corbicula* from China, South-eastern Russia, Tibet, Formosa, and the Philippine Islands. *Memoirs of the Indian Museum* **9**: 49–68.

Pringle, C.M. & Blake, G.A., 1994. Quantitative effects of atyid shrimp (Decapoda: Atyidae) on the depositional environment in a tropical stream: use of electricity for experimental exclusion. *Can. J. Fish. Aquat. Sci.* **51**: 1443–1450.

Pringle, C.M., Blake, G.A., Covich, A.P., Buzby, K.M. & Finley, 1993. Effects of omnivorous shrimp in a montane tropical stream: sediment removal, disturbance of sessile invertebrates and enhancement of understorey algal biomass. *Oecologia* **93**: 1–11.

Pritchard, G., 1983. Biology of Tipulidae. *Ann. Rev. Entomol.* **28**: 1–22.

Provonsha, a.V. & McCafferty, W.P., 1995. New brushlegged caenid mayflies from South Africa (Ephemeroptera: Caenidae). *Aquatic Insects* **17**: 241–251.

Prowse, G.A., 1962. Diatoms of Malayan freshwaters. *Gardens' Bull. Singapore* **19**: 105–145.

Prowse, G.A., 1968. Pollution in Malayan waters. *Malay. Nat. J.* **21**: 149–158.

Pu, C.L., 1942. Three new species of Palpicornia from Yunnan. *Lingnan Science Journal* **20**: 167–176.

Pu, C.L., 1951. The subgenus *Bilimneus* Rey of South and Southwestern China (Coleoptera, Hydraenidae, *Limnebius*). *Lingnan Science Journal* **23**: 159–163.

Pu, C.L., 1956. The genus *Hydraena* Kugel. of China (Coleoptera, Palpicornia). *Acta Entomologica Sinica* **6**: 299–310 (in Chinese).

Pu, C.L., 1958. The genus *Ochthebius* Leach of China (Coleoptera, Palpicornia). *Acta Entomologica Sinica* **8**: 247–265 (in Chinese).

Pu, C.L., 1964. The genus *Anacaena* Thomson of China (Coleoptera, Hydrophilidae). *Acta Entomologica Sinica* **13**: 396–400 (in Chinese).

Pu, C.L., 1981. Coleoptera: Hydrophilidae. *Insects of Xizang* **1**: 337–338 (in Chinese).

Qasim, S.Z. & Qayyum, A., 1961. Spawning frequencies and breeding seasons of some freshwater fishes with special reference to those occurring in the plains of northern India. *Indian J. Fish.* **8**: 24–43.

Qi, S., 1987. Some ecological aspects of aquatic oligochaetes in the Lower Pearl River (People's Republic of China). *Hydrobiologia* **155**: 199–208.

Qi, S. & Erseus, C., 1985. Ecological survey of the aquatic oligochaetes in the Lower Pearl River (People's Republic of China). *Hydrobiologia* **128**: 39–44.

Qi, S. & Lin, M., 1985. Further assessment of pollution status of Pearl River, Guangzhou, by using benthic macroinvertebrates. *Acta Scientiae Circumstantiae* **5**: 354–359.

Qi, S. & Lin, M., 1987. Studies on some ecological Aspects of the population of *Corbicula fluminea* (Müller) (Mollusca) in the Pearl River, Guangzhou. *Acta Ecologica Sinica* **7**: 161–169 (in Chinese).

Qi, S. & Zhao, H., 1984. Ecological survey of the zooplankton (mainly Cladocera and Copepoda) in the Guangzhou reach of the Pearl River. *J. Sci. Med. Jinan Univ.* **7**: 59–72 (in Chinese).

Qi, S., Lin, M. & Li, K., 1982. Water pollution assessment of Pearl River, Guangzhou, by using benthic invertebrates. *Acta Scientiae Circumstantiae* **2**: 181–189 (in Chinese).

Qian, N., Zhang, R. & Chen, Z., 1993. Some aspects of sedimentation at the Three Gorges Project. In: Luk, S. & Whitney, J. (Eds), *Megaproject. A Case Study of China's Three Gorges Project*. M.E. Sharpe Inc., New York: 161–175.

Quinn, J.T. & Harrington, J.J., 1992. Generating alternative designs for interjurisdictional natural resource devlopment schemes in the Greater Ganges River Basin. *Papers in Regional Science* **71**: 373–391.

Qureshi, S.A., Saksena, A.B. & Singh, V.P., 1980. Acute toxicity of four heavy metals to benthic fish food organisms from the River Khan, Ujjain. *Int. J. Environ. Stud.* **15**: 59–61.

Rabeni, C.F., Davies, S.P. & Gibbs, K.E., 1985. Benthic invertebrate response to pollution abatement: structural changes and functional implications. *Water Resour. Bull.* **21**: 489–497.

Rao, A.R. & Prasad, T., 1994. Water resources development of the Indo-Nepal region. *Water Resources Development* **10**: 157–173.

Rachmatika, I. & Soetikno, W., 1988. Feeding ecology of *Glyptothorax major* (Bagaridae: Siluriformes) in the Atlas River, Southeast Aceh, Sumatra, Indonesia. *Berita Biol.* **3**: 396–399.

Radhakrishna, Y., 1984. Introduction. *Hydrobiologia*, **119**: 5.

Rahim Ismail, A. & Edington, J.M., 1990. Redescription of *Hydromanicus malayanus* Banks, 1931 (Trichoptera, Hydropsychidae). *Aquatic Insects* **12**: 193–198.

Rahim Ismail, A., Edington, J.M. & Green, P.C., 1996. Descriptions of the pupae and larvae of *Stenopsyche siamensis* Martynov, 1931 (Trichoptera: Stenopsychidae) with notes on larval biology. *Aquatic Insects* **18**: 241–252.

Rahman Khan, H., 1987. Water resource development in Bangladesh: problems and prospects. In: Ali, M., Radosevich, G.E. & Ali Khan, A. (Eds), *Water Resources Policy for Asia*. A.A. Balkema Publishers, Boston: 165–181.

Rai, L.C., 1978. Ecological studies of algal communities of the Ganges River at Varanasi. *Indian J. Ecol.* **5**: 1–6.

Rainboth, W.J., 1991a. Cyprinids of South East Asia. In: Winfield, I.J. & Nelson, J.S. (Eds), *Cyprinid Fishes: Systematics, Biology and Exploitation*. Chapman & Hall, London: 156–210.

Rainboth, W.J., 1991b. *Aaptosyax grypus*, a new genus and species of large piscivorous cyprinds from the middle Mekong River. *Jap. J. Ichthyol.* **38**: 231–237.

Rainboth, W.J. & Kottelat, M., 1987. *Rasbora spiolocerca*, new species of cyprinid from the Mekong River, Southeast Asia. *Coepia* **1987**: 417–423.

Raj, U. & Fergusson, J.E., 1980. Osmotic and ionic cmposition of a tropical freshwater mussel *Batissa violacea* Lamarck (Lamellibranchiata: Spahaeriidae). *New Zealand Journal of Science* **23**: 199–204.

Rajashekhar, M. & Kaveriappa, K.M., 1993. Ecological observations on water-borne hyphomycetes of Cauvery River and its tributaries in India. *Arch. Hydrobiol.* **126**: 487–497.

Rajyalakshmi, T., 1980. Comparative study of the biology of the freshwater prawn *Macrobrachium malcolmsonii* of Godvari and Hoogly river systems. *Proc. Indian natn. Sci. Acad.* **B45**: 77–89.

Ram Vaidya, S., 1995. Studies on present status of limnology in Nepal and drinking water problem in Kathmandu City. In: Timotius, K.H. & Göltenboth, F. (Eds), *Tropical Limnology Volume III. Tropical Rivers, Wetlands and Special Topics*. Satya Wacana University Press, Salatiga, Indonesia: 169–177.

Rama Rao, S.V., Singh, V. P. & Mall, L.P., 1978. Pollution studies of River Khan (Indore), India — I. Biological assessment of pollution. *Water Res.* **12**: 555–559.

Rama Rao, S.V., Singh, V.P. & Mall, L.P., 1979a. The effect of sewage and industrial waste discharges on the primary production of a shallow turbulent river. *Water Res.* **13**: 1017–1021.

Rama Rao, S.V., Sharma, S.K., Singh, V.P. & Mall, L.P., 1979b. The importance of ecosystem structure and function in the management of water quality in India. *Environmental Conservation* **6**: 293–296.

Ramakrishnan, P.S., 1991. *Ecology of Biological Invasions in the Tropics*. International Scientific Publications, New Delhi: 195 pp.

Raman, K., 1967. Observations on the fishery and biology of the giant freshwater prawn *Macrobrachium rosenbergii* de Man. *Proc. Symp. Crustacea, Ernakulam (1965), Mar. Biol. Assoc. India, Symp. Ser.* **2**: 649–669.

Ramesh, R. & Subramanian, V., 1993. Geochemical characteristics of the major tropical rivers of India. In: Gladwell, J.S. (Ed.), *Hydrology of Warm Humid Regions*. IAHS Press, Wallingford, U.K.: 157–164.

Rana, B.C. & Palria, S., 1988. Assessment, evaluation and abatement studies of a polluted River — the Bandi (Rajasthan). In: Trivedy, R.K. (Ed.), *Ecology and Pollution of Indian Rivers*. Ashish Publishing House, New Delhi: 345–360.

Ranjit, R., 1995. Present status of water pollution in Nepal. In: Timotius, K.H. & Göltenboth, F. (Eds), *Tropical Limnology Volume III. Tropical Rivers, Wetlands and Special Topics*. Satya Wacana University Press, Salatiga, Indonesia: 161–167.

Rao, K.S., 1973. Studies of freshwater Bryozoa — III. The Bryozoa of the Narmada River system. In: Larwood, G.P. (Ed.), *Living and Fossil Bryozoa. Recent Advances in Research*. Academic Press, London: 529–537.

Rao, K.S., Agrawal, V., Diwan, A.P. & Shrivastava, P., 1985. Studies on freshwater Bryozoa. V. Observations on central Indian materials. In: Nielsen, C. & Larwood, G.P. (Eds), *Bryozoa: Ordivician to Recent*. Olsen & Olsen, Denmark: 257–264.

Rao, K.S., Diwan, A.P. & Shrivastava, P., 1978. Structure and environmental relations of sclerotized structures in freshwater Bryozoa. III. Observations on *Plumatella casmiana* (Ectoprocta: Phylactolaemata). *J. Anim. Morph. Physiol.* **25**: 8–15.

Rao, K.S., Srivastava, S., Choubey, U. & Rao, B.S., 1991. Observations on biological monitoring concepts of some riverine ecosystems of central India. In: *National Symposium on New Horizons in Freshwater Aquaculture, 23–25 January, 1991. Proceedings*. ICAR/CIFA, Bhubaneswar (India): 146–148.

Rapp, A., Li, J. & Nyberg, R., 1991. Mudflow disasters in mountainous areas. *Ambio* **20**: 210–218.

Rasmussen, E., 1987. A review of the distribution and occurrence of the Chinese mitten crab (*Eriocheir sinensis* M.-Edw.) in Denmark. *Flora og Fauna* **93**: 51–58 (in Danish).

Rathore, H.S. & Rama, S.V., 1979. A note on the feeding ecology of *Chironomus* (Diptera) larvae in the River Kshipra, India. *Bangladesh J. Zool.* **7**: 73–74.

Ratnasabapathy, M., 1975. Preliminary observations on Gombak River algae at the Field Studies Centre, University of Malaya. *Phytos* **14**: 15–23.

Raut, S.K. & Saha, T.C., 1986a. Life-history pattern of the freshwater leech *Glossophonia weberi* (Blanchard) (Hirudinea: Glossiphonidae). *J. Bombay Nat. Hist. Soc.* **83**: 260–263.

Raut, S.K. & Saha, T.C., 1986b. Growth rate in a freshwater leech *Glossiphonia weberi. Science & Culture* **52**: 119–120.

Ravindranath, K., 1980. Shrimps of the genus *Acetes* H. Milne Edwards (Crustacea, Decapoda, Sergestidae) from the estuarine system of River Krishna. *Proc. Indian Acad. Sci., Animal Sci.* **86**: 253–273.

Ravindranath, K., 1981. Larval stages of a freshwater shrimp, *Caridina rajadhari* Bouvier (Crustacea, Decapoda, Atyidae). *Proc. Indian Acad. Sci., Anim. Sci.* **90**: 683–702.

Ray, P., Singh, S.B. & Sehgal, K.L., 1966. A study of some aspects of the Rivers Ganga and Yamuna at Allahabad (U.P.) in 1958–59. *Proc. Indian natn. Acad. Sci.* **B36**: 235–272.

Realon, C.B.R., 1980. An ecological study of mayfly nymphs in Molawin Creek, Mt. Makiling, Laguna. *Philipp. Ent.* **4**: 233–291.

Reckhow, K., 1990. Bayesian inference in non-replicated ecological studies. *Ecology* **71**: 2053–2059.

Reddy, P.M. & Venkateswarlu, V., 1986. Ecology of algae in the paper mill effluents and their impact on the River Tungabhadra. *Indian J. Environ. Biol.* **7**: 215–224.

Reeves, R.R. & Leatherwood, S., 1994. Dams and river dolphins: can they coexist? *Ambio* **23**: 172–175.

Reeves, R.R., Chaudhry, A.A. & Khalid, U., 1991. Competing for water on the Indus Plain: is there a future for Pakistan's river dolphins? *Environmental Conservation* **18**: 341–350.

Regier, H.A., Welcomme, R.L., Steedman, R.J. & Henderson, H.F., 1989. Rehabilitation of degraded river ecosystems. *Can. Spec. Publ. Fish. Aquat. Sci.* **106**: 86–97.

Reichholf, J., 1973. Results of the Austrian-Ceylonese hydrobiological mission, 1970. Part VIII. Larval stages of some water moths (Lepid., Pyralidae, Nymphulinae) from torrents of Ceylon and some South-Pacific islands. *Bull. Fish. Res. Stn, Sri Lanka (Ceylon)* **24**: 75–81.

Reiss, F., 1988. Die Gattung *Kloosia* Kruseman, 1933 mit der Neubeschreibung zweiter Arten (Diptera, Chironomidae). *Spixania, Suppl.* **14**: 35–44.

Resh, V.H., Wood, J.R., Bergey, E.A., Feminella, J.W., Jackson. J.K. & McElravy, E.P., 1997. Biology of *Gumaga niricula* (McL.) in a northern California stream. In: Holzenthal, R.W. & Flint, O.S. (Eds), *Proceedings of the 8th International Symposium on Trichoptera*. Ohio Biological Survey, Columbus: 401–410.

Revelle, R., 1980. Energy dilemmas in Asia. *Science* **209**: 164–173.

Reyes, M.R. & Mendoza, V.B., 1983. The Pantabangan watershed management and erosion control project. In: Hamilton, L.S. (Ed.), *Forest and Watershed Development and Conservation in Asia and the Pacific*. Westview Press, Boulder, Colorado: 485–556.

Reynolds, J.D., Blackith, R.E. & Blackith, R.M., 1988. Dietary observations on some tetrigids (Orthoptera: Caelifera) from Sulawesi (Indonesia). *J. Trop. Ecol.* **4**: 403–406.

Richards, O.W. & Davies, R.G., 1977a. *Imm's General Textbook of Entomology, Tenth Edition. Volume 1: Structure, Physiogy and Development*. Chapman & Hall, London: 418 pp.

Richards, O.W. & Davies, R.G., 1977b. *Imm's General Textbook of Entomology, Tenth Edition. Volume 2: Classification and Biology*. Chapman & Hall, London: 1354 pp.

Roback, S.S. & Coffman, W.P., 1989. Tanypodinae pupae from southern India (Diptera: Chironomidae). *Proc. Acad. Nat. Sci. Philad.* **141**: 85–113.

Roberts, T.R., 1978. An ichthyological survey of the Fly River in Papua New Guinea with descriptions of new species. *Smithson. Contrib. Zool.* **281**: 1–70.

Roberts, T.R., 1986. Systematic review of the Mastacembelidae or spiny eels of Burma and Thailand, with a description of two new species of *Macrognathus*. *Jap. J. Ichthyol.* **33**: 95–109.

Roberts, T.R., 1988. *Danioella translucida*, the world's smallest cyprinid fish. *Flora Fauna (Stockh.)* **83**: 212–215 (in Swedish).

Roberts, T.R., 1989. The freshwater fishes of Western Borneo (Kalimantan Barat, Indonesia). *Mem. Calif. Acad. Sci.* **14**: 1–210.

Roberts, T.R., 1992. Revision of the Southeast Asian cyprinid fish genus *Probarbus*, with two new species threatened by proposed construction of dams on the Mekong River. *Ichthyological Exploration of Freshwaters* **3**: 37–48.

Roberts, T.R., 1993a. Artisanal fisheries and fish ecology below the great waterfalls in the Mekong River in southern Laos. *Nat. Hist. Bull. Siam Soc.* **41**: 31–62.

Roberts, T.R., 1993b. Just another dammed river? Negative impacts of Pak Mun Dam on the fishes of the Mekong basin. *Nat. Hist. Bull. Siam Soc.* **41**: 105–133.

Roberts, T.R., 1994. *Osphronemus exodon*, a new species of giant gouramy with extraordinary dentition from the Mekong. *Nat. Hist. Bull. Siam Soc.* **42**: 67–77.

Roberts, T.R., 1995. Mekong mainstream hydropower dams: run-of-the-river or ruin-of-the-river? *Nat. Hist. Bull. Siam Soc.* **43**: 9–19.

Roberts, T.R. & Baird, I.G., 1995. Traditional fisheries and fish ecology on the Mekong River at Khone Waterfalls in southern Laos. *Nat. Hist. Bull. Siam Soc.* **43**: 219–262.

Roberts, T.R. & Karnasuta, J., 1987. *Dasyatis laoensis*, a new whiptailed stingray (family Dasyatidae) from the Mekong River of Laos and Thailand. *Environ. Biol. Fish.* **20**: 161–167.

Roberts, T.R. & Vidthayanon, C., 1991. Systematic revision of the Asian catfish family Pangasiidae, with biological observations and descriptions of three new species. *Proc. Acad. Nat. Sci. Philad.* **143**: 97–144.

Roberts, T.R. & Warren, T.J., 1994. Observations of fish and fisheries in southern Laos and northeastern Cambodia, October 1993 — February 1994. *Nat. Hist. Bull. Siam Soc.* **42**: 87–115.

Rocchi, S., 1976. Ditiscidi del Bengla-Desh con descrizione de *Copelatus brivioi* n. sp. (Coleoptera, Dytiscidae). *Boll. Soc. ent. ital., Genova* **108**: 177–180.

Rocchi, S., 1982. Ditiscidi raccolti nel Nepal dal Dr. Enrico Migliaccio et dal Dr. Guido Sabatinelli (Coleoptera, Dytiscidae). *Boll. Ass. Romana Entomol.* **35**: 57–60.

Rocchi, S., 1986. Ditiscidi di Birmania, Thailandia e Sri Lanka, con descrizione di due nuove specie (Coleoptera). *Boll. Soc. ent. ital., Genova* **118**: 31–34.

Roland, C. & Rovner, J.S., 1983. Chemical and vibratory communication in the aquatic pisaurid spider *Dolomedes triton*. *J. Arachnol.* **11**: 77–85.

Rolfe, W.D.I., 1985. Early terrestrial arthropods: a fragmentary record. *Phil. Trans. R. Soc. Lond.* **B309**: 207–218.

Room, P.M., 1990. Ecology of a simple plant-herbivore system: biological control of *Salvinia*. *Trends in Ecology & Evolution (TREE)* **5**: 74–79.

Roonwal, M.C., 1979. Field bioecology and morphometry of some Central Indian grasshoppers (Acridoidea), with notes on a swimming species (Tetrigoidea). *Proc. Zool. Soc. (Calcutta)* **32**: 97–106.

Rose, H.S. & Pajni, H.R., 1987. Studies on the external genitalia of some species of subfamily Nymphulinae from North India (Lepidoptera: Pyraustidae). *Research Bulletin of the Punjab University (Science)* **37**: 1–10.

Rosenberg, D.M. & Resh, V.H., 1993. *Freshwater Biomonitoring and Benthic Macroinvertebrates.* Chapman & Hall, New York: 488 pp.

Ross, H.H., 1956. *Evolution and Classification of the Mountain Caddisflies.* University of Illinois Press, Urbana: 213 pp.

Ross, H.H., 1967. The evolution and past dispersal of the Trichoptera. *Ann. Rev. Entomol.* **12**: 169–206.

Rothe, S.P., Bhivare, V.N., Patil, T.S. & Jain, B.K., 1986. The vegetation of Gomai River, India. *Indian Bot. Rep.* **5**: 123–125.

Rounick, J.S., Winterbourn, M.J. & Lyon, G.L. 1982. Differential utilization of allochthonous and autochthonous inputs by aquatic invertebrates in some New Zealand streams: a stable carbon isotope study. *Oikos* **39**: 191–198.

Rout, J. & Gaur, J.P., 1990. Identification of nutrient-limited algal growth in two streams at Shillong (India). *Acta Oecol.* **11**: 631–642.

Rout, J. & Gaur, J.P., 1994. Composition and dynamics of epilithic algae in a forest stream at Shillong (India). *Hydrobiologia* **291**: 61–74.

Rowe, R.J., 1992. Larval development in *Diplacodes bipunctata* (Brauer) (Odonata: Libellulidae). *J. Aust. ent. Soc.* **31**: 351–355.

Roy, H.K., 1955. Plankton ecology of the River Hooghly at Palta, West Bengal. *Ecology* **36**: 169–175.

Rundle, S.D., Jenkins, A. & Ormerod, S.J., 1993. Macroinvertebrate communities in streams in the Himalaya, Nepal. *Freshwat. Biol.* **30**: 169–180.

Ruttner, F., 1931. Hydrographische und hydrochemische Beobachtungen auf Java, Sumatra und Bali. *Arch. Hydrobiol., Suppl.* **8**: 197–454.

Ryder, G., 1991. *Damming the Three Gorges — What the Dam-builders Don't Want You to Know.* Probe International, Toronto: 135 pp.

Sabhasri, S., 1968. Preliminary watershed management research in Northern Thailand. In: Talbot, L.M. & Talbot, M.H. (Eds), *Conservation in Tropical South East Asia.* IUCN Publications New Series No. 10, IUCN, Morges, Switzerland: 111–114.

Saeni, M.S., Sutamihardja, R.T.M. & Sukra, J., 1980. Water quality of the Musi River in the city area of Palembang. In: J.I. Furtado (Ed.), *Tropical Ecology and Development. Proceedings of the Vth International Symposium of Tropical Ecology.* International Society of Tropical Ecology, Kuala Lumpur: 717–724.

Saether, O.A. & Wang, X., 1992. *Euryhapis fuscipropes* sp. n. from China and *Tokyobrilla andersoni* sp. n. from Tanzania, with a review of genera near *Irisobrilla* Oliver (Diptera: Chironomidae). *Annls Limnol.* **28**: 209–223.

Saether, O.A. & Wang, X., 1993. *Xiaomyia, Shangomyia* and *Zhouomyia,* three new and unusual genera of Chironomini from Oriental China (Diptera: Chironomidae). *Ent. Scand.* **24**: 185–195.

Safiullah, S., Chowdhury, M.I., Mafizuddin, M., Ali, I. & Karim, M., 1985. Monitoring of the Pamda (Ganges), the Jamuna (Brahmaputra) and the Baral in Bangladesh. *Mitt. Geol.-Palaont. Inst. Univ. Hamburg, SCOPE/UNEP Sonderbd.* **58**: 519–524.

Safiullah, S., Mofizuddin, M., Ali, I. 7 Enamul Kabir, S., 1987. Carbon transport in the Ganges and Brahmaputra River systems. *Mitt. Geol.-Palaont. Inst. Univ. Hamburg, SCOPE/UNEP Sonderbd.* **64**: 435–422.

Saha, T.C. & Raut, S.K., 1992. Bioecology of the waterbug *Sphaerodema annulatum* Fabricius (Heteroptera: Belostomatidae). *Arch. Hydrobiol.* **124**: 239–253.

Salagar, M.S. & Hosetti, B.B., 1990. Nutrient budgets of Kasari-Panchaganga River, Kolhapur. *J. Nat. Conserv.* **2**: 155–159.

Sameshima, O. & Sato, H., 1994. Life cycles of *Glossosoma inops* and *Agapetus yasensis* (Trichoptera: Glossosomatidae) at Kii Peninsula, southern Honshu, Japan. *Aquatic Insects* **16**: 65–74.

Sankaralingam, A. & Venkatesan, P., 1989. Larvicidal properties of water bug *Diplonychus indicus* Venkatesan & Rao and its use in mosquito control. *Indian J. Exp. Biol.* **27**: 174–176.

Sarin, M.M. & Krishnaswami, S., 1984. Major ion chemistry of the Ganga-Brahmputra river systems, India. *Nature* **312**: 538–541.

Sartori, M. & Sowa, R., 1992. New data on some *Rhithrogena* species from the Near- and Middle East (Ephemeroptera: Heptageniidae). *Aquatic Insects* **14**: 31–40.

Sasa, M. & Kikuchi, M., 1995. *Chironomidae [Diptera] of Japan*. University of Tokyo Press, Tokyo: 333 pp.

Satô, M., 1981. Dryopoidea (Coleoptera) of Nepal. 1. Family Dryopidae. *Bulletin of the National Science Museum Tokyo, Series A (Zoology)* **7**: 51–56.

Satô, M., 1983. A new *Eubrianax* sp. (Col., Psephenidae) from the Philippines. *Aquatic Insects* **5**: 65–69.

Satô, M., 1985. Taxonomic notes on the aquatic Coleoptera of Japan. III. *Coleopterists' News* **69**: 1–5 (in Japanese).

Satô, M., 1992. Notes on the coleopteran fauna of Malaysia. 1. The Malaysian species of the genus *Elmomorphus* (Dryopidae). *Elytra* **20**: 67–71.

Satô, M. & Chûgô, M., 1961. Coleoptera from Southeast Asia/Family Hydrophilidae. In: Kira, T. & Umesao, T. (Eds), *Nature and Life in Southeast Asia, Volume 1*. Japanese Society for the Promotion of Science, Tokyo: 314–320.

Sawyer, R.T., 1986. *Leech Biology and Behaviour*. Clarendon, Oxford: 1065 pp.

Schefter, P.W., Wiggins, G.B & Unzicker, J.D., 1986. A proposal for assignment of *Ceratopsyche* as a subgenus of *Hydropsyche*, with a new synonym and a new species (Trichoptera: Hydropsychidae). *J. N. Am. Benthol. Soc.* **5**: 67–84.

Schefter, P.W. & Wiggins, G.B., 1986. *A systematic study of the Nearctic larvae of the* Hydropsyche morosa *group (Trichoptera: Hydropsychidae)*. Life Sciences Miscellaneous Publications, Royal Ontario Museum: 94 pp.

Schmid, F., 1958a. Trichoptères de Ceylan. *Arch. Hydrobiol.* **54**: 1–173.

Schmid, F., 1958b. Trichoptères du Pakistan. *Tijdschrift voor Entomologie* **101**: 181–221.

Schmid, F., 1958c. Quelques Diptères Nématocères nouveaux ou intéressants (Thaumaléides et Limnobiides). *Bull. Inst. r. Sci. nat. Belg.* **34**: 1–23.

Schmid, F., 1959a. Quelques Trichoptères de Chine. *Mitt. Zool. Mus. Berlin* **35**: 317–345.

Schmid, F., 1959b. Trichoptères du Pakistan. II. *Tijdschrift voor Entomologie* **102**: 231–253.

Schmid, F., 1960. Trichoptères du Pakistan. III. *Tijdschrift voor Entomologie* **103**: 89–109.

Schmid, F., 1961. Trichoptères du Pakistan. IV. *Tijdschrift voor Entomologie* **104**: 187–230.

Schmid, F., 1963. Quelques *Himalopsyche* indiennes (Trichoptères: Rhyacophilidae). *Bonner zool. Beitr.* **14**: 206–223.

Schmid, F., 1964. Quelques Trichoptères asiatiques. *Can. Ent.* **96**: 825–840.

Schmid, F., 1965a. Quelques Trichoptères asiatiques II. *Tijdschrift voor Entomologie* **86**: 28–35.

Schmid, F., 1965b. Quelques Trichoptères de Chine II. *Bonner zool. Beitr.* **16**: 127–154.

Schmid, F., 1965c. D'étranges Goerodes, les *Larcasia* Navas (Trichoptera). *Ent. Tidskr.* **86**: 260–265.

Schmid, F., 1965d. Ergebnisse der zoologischen Forschungen von Dr Z. Kaszab in Mongolei. 63. Trichoptera. *Reichenbachia* **7**: 201–203.

Schmid, F., 1966. A propos des limites de la zone Paléarctique dans L'Himalaya on les Limnephilines en Inde (Trichoptera). *Acta Zoologica Academiae Scientarium Hungaricae* **12**: 363–369.

Schmid, F., 1968a. Le genre Poecilopsyche n. gen. (Trichoptera, Leptoceridae). *Ann. Soc. Ent. Québec* **13**: 3–31.

Schmid, F., 1968b. Les genres *Neurocyta* Navas et *Phryganopsyche* Wiggins en Inde. *Nat. Can.* **95**: 723–726.

Schmid, F., 1968c. La sous-famille des Apataniinea en Inde (Trichoptera, Limnephlidae). *Can. Ent.* **100**: 1233–1269.

Schmid, F., 1968d. La famille des Arctopsychides (Trichoptera). *Mém. Soc. Ent. Québec* **1**: 1–84.

Schmid, F., 1968e. Le genre *Gunungiella* Ulmer (Trichoptères: Philopotamidae). *Can. Ent.* **100**: 897–957.

Schmid, F., 1969. La famille des Sténopsychides (Trichoptera). *Can. Ent.* **101**: 187–224.

Schmid, F., 1970a. Sur quelques *Apsilochorema* orientaux (Trichoptera, Hydrobiosidae). *Tijdschrift voor Entomologie* **113**: 261–271.

Schmid, F., 1970b. Le genre *Rhyacophila* et la famille des Rhyacophilidae (Trichoptera). *Mém. Soc. Ent. Can.* **66**: 1–230.

Schmid, F., 1971. Quelques nouveaux *Glossosoma* orientaux (Trichoptera: Glossosomatidae). *Nat. Can.* **98**: 607–631.

Schmid, F., 1972. Sur quelques nouvelles Psychomyiines tropicales (Trichoptera: Psychomyiidae). *Nat. Can.* **99**: 143–172.

Schmid, F., 1982. La famille des Xiphocentronides (Trichoptera: Annulipalpia). *Mém. Soc. Ent. Can.* **121**: 1–127.

Schmid, F., 1983. Encore quelques *Stactobia* McLachlan (Trichoptera, Hydroptilidae). *Nat. Can.* **110**: 239–283.

Schmid, F., 1984. Essai d'évaluation de la faune mondiale des Trichoptères. In: Morse, J.C. (Ed.), *Proceedings of the 4th International Symposium on Trichoptera.* Dr W. Junk Publishers, The Hague: 337.

Schmid, F., 1987. Considerations diverses sur quelques genres Leptocerins (Trichoptera, Leptoceridae). *Bull. Inst. R. Sci. Nat. Belg., Ent. Suppl.* **57**: 1–147.

Schmid, F., 1989. Les Hydrobiosides (Trichoptera, Annulipalpia). *Bull. Inst. R. Sci. Nat. Belg., Ent. Suppl.* **59**: 1–154.

Schmid, F., 1991a. Quelques Philopatamides orientaux nouveaux ou peu commus (Trichoptera, Annulipalpia). *Beaufortia* **42**: 89–107.

Schmid, F., 1991b. Quelques nouveaux Trichoptères indienes (Trichoptera). *Can. Nat.* **117**: 239–251.

Schmid, F., 1991c. Les Goerides en Inde (Trichoptera, Integripalpia). *Rev. Hydrobiol. Trop.* **24**: 305–326.

Schmid, F., 1992. Les brachycentrides en Inde (Trichoptera, Integripalpia). *Bijdragen Tot de Dierkunde* **62**: 99–109.

Schmid, F., 1993a. Quatre genres des Trichoptères forlignants. *Fabreries* **18**: 37–48.

Schmid, F., 1993b. Considerations sur les helicopsychides (Trichoptera: Integripalia). *Beaufortia* **43**: 65–100.

Schmid, F., 1994a. Quelques *Adicella* indiennes (Trichoptera: Leptoceridae). *Fabreries* **19**: 85–127.

Schmid, F., 1994b. Les *Adicella* du groupe de pulcherrima (Trichoptera, Intergripalpia, Leptoceridae). *Fabreries* **19**: 37–44.

Schmid, F., 1994c. Le genre *Triaenodes* McLachlan en Inde (Trichoptera: Leptoceridae). *Fabreries* **19**: 1–11.

Schmid, F., 1995a. Le genre *Oecetis* en Inde (Trichoptera: Leptoceridae). *Fabreries* **20**: 113–151.

Schmid, F., 1995b. Les *Oecetis* du groupe d'*eburnea* en Inde (Trichoptera: Leptoceridae). *Fabreries* **20**: 41–56.

Schmid, F., 1995c. Les *Oecetis* du groupe de *testacea* en Inde (Trichoptera: Leptoceridae). *Fabreries* **20**: 57–78.

Schmid, F. & Botosaneanu, L., 1966. Le genre *Himalopsyche* Banks

(Trichoptera, Rhyacophilidae). *Mém. Soc. Ent. Québec* **11**: 123–174.

Schmid, F. & Denning, D.G., 1979. Descriptions of new Annulipalpia (Trichoptera) from southeastern Asia. *Can. Ent.* **111**: 243–249.

Schneider, W., 1937. Freilebende nematoden der Deutschen Limnologischen Sunda Expedition nach Sumatra, Java und Bali. *Arch Hydrobiol., Suppl.* **15**: 30–108.

Schödl, S., 1992. Revision der Gattung *Berosus* LEACH 2. Teil: Die orentalischen Arten der Untergattung *Enoplurus* (Coleoptera: Hydrophilidae). *Koleopterologische Rundschau* **62**: 137–164.

Schödl, S., 1993. Revision der Gattung *Berosus* LEACH 3. Teil: Die paläarktischen und orentalischen Arten der Untergattung *Berosus* s. str. (Coleoptera: Hydrophilidae). *Koleopterologische Rundschau* **63**: 189–223.

Schödl, S., 1995. *Tylomicrus* gen. n. *costatus* n. sp. aus Malaysia. *Koleopterologische Rundschau* **65**: 145–148.

Schödl, S., 1997. Description of two new *Berosus* SHARP from Southeast Asia, with faunistic notes on *Berosus nigropictus* RÉGIMBART (Coleoptera: Hydrophilidae). *Koleopterologische Rundschau* **67**: 195–200.

Schödl, S. & Ji, L., 1995. Hydrophilidae: 2. Synopsis of *Hydrocassis* DEYROLLE & FAIRMAIRE and *Ametor* SEMENOV, with descriptions of three new species (Coleoptera). In: Jäch, M.A. & Ji, L. (Eds), *Water Beetles of China*. Zoologisch-Botanische Gesellschaft & Wiener Coleopterologenverein, Vienna: 221–243.

Schönmann, S., 1994. Revision der Gattung *Pelthydrus* ORCHYMONT 1. Teil: *Globipelthydrus* subgen. n. (Coleoptera: Hydrophilidae). *Koleopterologische Rundschau* **64**: 189–222.

Schut, J., de Silva, S.S. & Kortmulder, K., 1984. Habitat, associations and competition of eight *Barbus* (= *Puntius*) species (Cyprinidae) indigenous to Sri Lanka. *Neth. J. Zool.,* **34**: 159–181.

SCMP, 1993. Power-hungry Thais eye mighty Mekong. *South China Morning Post* **6 November 1993**: 10.

SCMP, 1994. Flood control funds spent on economy. *South China Morning Post* **11 August 1994**: 6.

SCMP, 1995a. Rare river dolphin found. *South China Morning Post* **23 December 1995**: 8.

SCMP, 1995b. Drought plan in comeback. *South China Morning Post* **26 November 1995**: 5.

SCMP, 1995c. Pact signed to protect Mekong. *South China Morning Post* **6 April 1995**: 10.

SCMP, 1995d. Death sentence for 'serious pollution'. *South China Morning Post* **19 November 1995**: 5.

SCMP, 1996a. Swollen lake bursts banks. *South China Morning Post* **21 July 1996**: 6.

SCMP, 1996b. China's flood of uncetainty. *South China Morning* **28** *Post* **28 July 1996**: 9.

SCMP, 1996c. US bank refuses loans for dam. *South China Morning Post* **1 June 1996**: 1.

SCMP, 1996d. Funds rethink after dolphin dies. *South China Morning Post* **11 August 1996**: 2.

SCMP, 1996e. PM agrees to review dam-building project. *South China Morning Post* **3 August 1996**: 11.

SCMP, 1996f. Electors 'don't give a dam'. *South China Morning Post* **24 August 1996**: 13.

SCMP, 1996g. Yangtze erosion cash plea. *South China Morning Post* **3 February 1996**: 7.

SCMP, 1996h. Island faces disaster as mine waste destroys river. *South China Morning Post* **30 March 1996**: 13.

Scott, D.A. & Poole, C.M., 1989. *A Status Overview of Asian Wetlands.* Asian Wetland Bureau, Kuala Lumpur: 140 pp.

Scott, K.M.F., 1983. On the Hydropsychidae (Trichoptera) of southern Africa with keys to African genera of imagos, larvae and pupae and species lists. *Ann. Cape Prov. Mus. (Nat. Hist.)* **14**: 299–422.

Sebastian, A.C., 1992. One too many Muggers at croc bank. *Asian Wetland News* **5 (1)**: 8.

Sedell, J.R., Richey, J.E. & Swanson, F.J., 1989. The River Continuum Concept: a basis for the expected ecosystem behavior of very large rivers? *Can. Spec. Publ. Fish. Aquat. Sci.* **106**: 49–55.

Sehgal, K.L., 1983. Fishery resources and their management. In: Singh, T.V. & Kaur, J. (Eds), *Studies in Ecodevelopment: Himalayas Mountains and Men.* Print House, Lucknow: 225–272.

Sehgal, K.L., 1991. Distributional pattern, structural modifications and diversity of benthic biota of North Western Himalayas. In: Bhatt, S.D. & Pande, R.K. (Eds), *Ecology of the Mountain Waters.* Ashish Publishing House, New Delhi: 198–250.

Sehgal, K.L., Kumar, K. & Sunder, S., 1984. Food preferences of Brown trout *Salmo trutta* in Kashmir, India. *Indian J. Anim. Sci.* **54**: 675–682.

Selvanayagam, M & Rao, T.K.R., 1988. Biological aspects of two species of gerrids, *Limnogonus fossarum fossarum* and *L. nitidus* Mayr. *J. Bombay Nat. Hist. Soc.* **85**: 474–484.

Seshadri, A.R., 1955. An extraordinary outbreak of 'caddice flies' (Trichoptera) in the Mettur-Dam Township area in Salem District, South India. *Indian J. Ent.* **17**: 337–340.

Setamanit, S., 1987. Thailand. In: Chia, L.S. (Ed.), *Environmental Management in Southeast Asia. Directions and Current Status.* Faculty of Science, National University of Singapore: 169–211.

Shah, R.B., 1994. Inter-state river water disputes: a historical review. *Water Resources Development* **10**: 175–189.

Shaji, C. & Patel, R.J., 1991. Phytoplankton species diversity of Sabarmati River near Ahmedabad, Gujarat as an index of environmental changes. *Ann. Biol.* **7**: 15–20.

Sharma, B.K. & Naik, L.P., 1995. Species composition, distribution and abundance of rotifers in River Narmada, Madhya Pradesh (Central India). In: Timotius, K.H. & Göltenboth, F. (Eds), *Tropical Limnology Volume III. Tropical Rivers, Wetlands and Special Topics.* Satya Wacana University Press, Salatiga, Indonesia: 77–87.

Sharma, K.D. & Vangani, N.S., 1982. Flash flood of July 1979 in the Luni Basin, *Hydrol. Sci. J. (IAHS)* **27**: 285–298.

Sharma, R.C., 1984a. Potamological studies on lotic environment of the upland River Bhagirathi of Garhwal Himalaya. *Environ. Ecol.* **2**: 239–242.

Sharma, R.C., 1984b. Fish ecology and dynamics of snowfed River Bhagirathi, India 1. Seasonal variations in feeding of *Pseudecheneis sulcatus. Geobios (Jodhpur)* **11**: 49–52.

Sharma, R.C., 1984c. Dynamics of food and feeding habits of *Crossocheilus latius* (Hamilton) in fluvial ecosystem of Garhwal Himalaya. *Comp. Physiol. Ecol.* **9**: 305–308.

Sharma, R.C., 1984d. Studies on spawning ecology of a hillstream fish, *Glyptothorax pectinopterus* (McClelland) of Garhwal Himalaya. *J. Adv. Zool.* **5**: 28–33.

Sharma, R.C., 1985. Seasonal abundance of phytoplankton in the Bhagirathi River, Garhwal Himalaya. *Indian J. Ecol.* **12**: 157–160.

Sharma, R.C., 1986. Effect of physico-chemical factors on benthic fauna of Bhagirathi River, Garhwal Himalaya. *Indian J. Ecol.* **13**: 133–137.

Sharma, R.C., 1990. Trophic ecology of *Garra gotyla gotyla* (Gray) from upland streams of India. *Pol. Arch. Hydrobiol.* **38**: 127–134.

Sharma, R.C., 1991. Rithronology of Bhagirathi, Garhwal Himalaya (India). In: Bhatt, S.D. & Pande, R.K. (Eds), *Ecology of the Mountain Waters.* Ashish Publishing House, New Delhi: 125–137.

Sharma, S., Saxena, M.N., Mathur, R. & Mathur, A., 1990. Seasonal and domestic sewage induced changes in entomofauna of mesosprobic water in Morar (Kalpi) River, Gwalior, India. In: Agrawal, V.P. & Das, P. (Eds), *Recent Trends in Limnology.* Society of Biosciences, Muzaffarnagar (India): 185–196.

She, S. & You, D. (1988) A new species of *Isonychia* from China (Ephmeroptera: Oligoneuriidae). *Pan-Pacific Entomol.* **64**: 29–31.

Shelford, R., 1909. Notes on some amphibious cockroaches. *Rec. Indian Mus.* **3**: 125–127.

Shen, G., 1992. Three Gorges needs to power ahead. *China Environment News* **32** (**March 1992**): 4–5.

Shen, H., Mao, Z., Gu, G. & Xu, P., 1983. The effect of south-to-north transfer on saltwater intrusion in the Chang Jiang estuary. In: Biswas, A.K., Zuo, D., Nickum, J.E. & Liu, C. (Eds.), *Long-distance Water Transfer: a Chinese Case Study and International Experiences.* Tycooly International Publishers Limited, Dublin: 351–359.

Shen, S. & Qi, S., 1982. Fresh-water polychaetes from the Pearl River of

Guangzhou. *J. Sci. Med. Jinan Univ.* **5**: 66–77 (in Chinese).

Shen, Z., Liu, X. & Lu, J., 1987. Distribution of inorganic nitrogen and phosphate in the water downstream of the Changjiang River and their removal processes in the estuary. *Stud. Mar. Sinica* **28**: 69–77 (in Chinese).

Shepard, W.D., 1992. A redescription of *Ordobrevia nubifera* (Fall) (Coleoptera: Elmidae). *Pan-Pacific Entomologist* **68**: 140–143.

Shibusawa, A.H., 1987. Co-operation in water resources development in the Ganges Brahmaputra basins. *Mountain Research & Development* **7**: 319–322.

Shimizu,T., 1994. *Indonemoura nohirae* (OKAMOTO, 1922), comb. nov. (Plecoptera, Nemouridae) newly recorded from Japan, with a redescription of *Amphinemura longispina* (OKAMOTO, 1922). *Jap. J. Entomol.* **62**: 619–627.

Shiraishi, Y., 1970. The migration of fishes in the Mekong River. *Regional Meeting of Inland Water Biologists in Southeast Asia. Proceedings.* Unesco Field Science Office for Southeast Asia, Djakarta: 135–140.

Shokita, S., 1976. Early life-history of the land-locked atyid shrimp, *Caridina denticulata ishigakiensis* Fujino & Shokita, from the Ryukyu Islands. *Carcinol. Soc. Japan (Odawara Carcinol. Mus.), Res. Crustacea* **7**: 1–10.

Shrestha, B.D., Van Ginnekan, P. & Sthapit, K.M., 1983. *Watershed Condition of the Districts of Nepal.* Ministry of Forests and Soil Conservation, Kathmandu, Nepal. (Not seen.)

Shrestha, D.L. & Paudyal, G.N., 1992. Water resources planning in the Karnali River Basin, Nepal. *Water Resources Development* **8**: 195–203.

Shrestha, R.R., 1995. A review of limnological studies and research in Nepal 1994. In: Timotius, K.H. & Göltenboth, F. (Eds), *Tropical Limnology Volume III. Tropical Rivers, Wetlands and Special Topics.* Satya Wacana University Press, Salatiga, Indonesia: 221–234.

Shrestha, T.K., 1991. Ecological, physico-chemical and biological monitoring of Mahseer habitat with reference to conservation and management in Nepal. In: Bhatt, S.D. & Pande, R.K. (Eds), *Ecology of the Mountain Waters.* Ashish Publishing House, New Delhi: 321–333.

Shrestha, T.K., 1994. Habitat ecology of the Mai Pokhari wetlands in Nepal and management for survival of the Himalayan newt, *Tylototriton verrucosus.* In: Mitsch, W.J. (Ed.), *Global Wetlands Old World and New.* Elsevier, Amsterdam: 857–862.

Shrivastav, A.K., Ambasht, R.S. & Kumar, R., 1993. Net biomass production, energy and nutrients of *Potamogeton crispus* L. in unpolluted and polluted waters ofthe Ganga River at Varanasi, India. *Geo-Eco-Trop* **17**: 137–147.

Shrivastava, P. & Rao, K.S., 1985. Ecology of *Plumatella emarginata* (Ectoprocta: Phylactolaemata) in the surface waters of Madhya Pradesh with a note on its occurrence in the protected waterworks of Bhopal (India). *Environmental Pollution (Series A)* **39**: 123–130.

Shrivastava, S., Rao, K.S. & Dubey, R., 1992. Diel feeding patterns of some freshwater catfish from Kshipra River, India. *Pak. J. Zool.* **24**: 27–29.

Shukla, S.C., Tripathi, B.D., Mishra, B.P. & Chaturvedi, S.S., 1992. Physico-chemical and bacteriological properties of the water of the River Ganga at Ghazipur. *Comparative Physiology & Ecology* **17**: 92–96.

Shulka, D.C., 1994. Habitat characteristics of wetlands of the Betwa Basin, India, and wintering populations of endangered waterfowl species. In: Mitsch, W.J. (Ed.), *Global Wetlands Old World and New*. Elsevier, Amsterdam: 863–868.

Shultz, J.W., 1987. Walking and surface film locomotion in terrestrial and semi-aquatic spiders. *J. Exp. Biol.* **128**: 427–444.

Shunmugavela, M. & Palanichamy, S., 1992. Note on the aquatic spider *Argyroneta aquatica* (Araneae: Agelenidae). *Journal of Ecobiology* **4**: 79–80.

Shy, J.Y. & Yu, H.P., 1992. Complete larval development of the mitten crab *Eriocheir rectus* Stimpson, 1858 (Decapoda, Brachyura, Grapsidae) reared in the laboratory. *Crustaceana* **63**: 277–290.

Shy, J.Y., Ng, P.K.L. & Yu, H.P., 1994. Crabs of the genus *Geothelphusa* Stimpson, 1858 (Crustacea: decapoda: Brachyura: Potamidae) from Taiwan, with descriptions of 25 new species. *Raffles Bull. Zool.* **42**: 781–846.

Siebert, D.J., 1991. Revision of *Acanthopsoides* Fowler, 1934 (Cypriniformes: Cobitidae), with the description of new species. *Jap. J. Ichthyol.* **38**: 97–114.

Sierwald, P., 1988. Notes on the behaviour of *Thalassius spinosissimus* (Arachnida: Araneae: Pisauridae. *Psyche (Cambridge)* **95**: 243–252.

Sigurdsson, J.B. & Yang, C.M, 1990. Marine mammals of Singapore. In: Chou, L.M. & Ng, P.K.L. (Eds), *Essays in Zoology*. Department of Zoology, National University of Singapore, Singapore: 25–37.

Sih, A. & Wooster, D.E., 1994. Prey behavior, prey dispersal, and predator impacts on stream prey. *Ecology* **75**: 1199–1207.

Silva, E.I.L. & Davies, R.W., 1986. Movements of some indigenous riverine fish in Sri Lanka. *Hydrobiologia* **137**: 265–270.

Singh, A.K., Sinha, R.K. & Sharan, R.K., 1988. Occurrence of freshwater polychaete *Nephtys oligobranchia* in the River Ganga at Patna, India. *Environment & Ecology (Kalyani)* **6**: 1032–1034.

Singh, H.R., 1990. Altitudinal changes and the impact of municipal sewage on the community structure of macrobenthic insects in the torrential reaches of the River Ganges in the Garhwal Himalaya (India). *Acta Hydrobiol.* **32**: 407–421.

Singh, H.R., Dobriyal, A.K. & Kumar, N., 1991. Hillstream fishery potential and development issues. *J. Inland Fish. Soc. India* **23**: 60–68.

Singh, K., Sandhu, H.S. & Singh, N. & Kumar, B., 1991. Kandi Watershed and Area Development Project: cost-benefit analysis of investments in two watersheds. *Indian J. Agricultural Economics* **46**: 132–141.

Singh, S.R. & Srivastava, V.K., 1989. Observations on the bottom fauna of the Ganga River (between Buxar and Ballia) with special reference to its role in the seasonal abundance of fresh water prawn *Macrobrachium birmanicum choprai* (Tiwari). *Acta Hydrochim. Hydrobiol.* **17**: 159–166.

Singh, S.R. & Srivastava, V.K., 1991a. Observations on the gonosomatic index and fecundity of the Ganaga River prawn, *Macrobrachium birmanicum choprai* (Tiwari). *Journal of Advanced Zoolology* **12**: 50–55.

Singh, S.R. & Srivastava, V.K., 1991b. On the length-weight relationship of the Ganga River prawn *Macrobrachium birmanicum choprai* (Tiwari). *Acta Hydrochim. Hydrobiol.* **19**: 419–423.

Sinha, B., 1984. Role of watershed management in water resources development planning -need for intergrated approach to development of catchment and command of irrigation projects. *Water International* **9**: 158–160.

Sioli, H., 1965. Demerkung zur Typologie amazonischer Flüsse. *Amazoniana* **1**: 74–83.

Sipahimalani, A.T., Chada, M.S., Joshi, K.N. & Mamdapur, V.R., 1970. Steroids in the defensive secretion of the water beetle *Cybister limbatus*. *Naturwissenschaften* **57**: 40.

Sites, R.W., Nichols, B.J. & Permkam, S., 1997. The Naucoridae (Heteroptera) of southern Thailand. *Pan-Pacific Entomologist* **73**: 127–134.

Sivankutty-Nair, G., 1975. Studies on the rate of growth of *Villorita cyprinoides* var. *cochinensis* (Hanley) from the Cochin backwaters. *Bulletin of the Department of Marine Sciences, University of Cochin* **7**: 919–929.

Sivankutty-Nair, G. & Shynamma, C.S., 1975. Studies on the salinity tolerance of *Villorita cyprinoides* var. *cochinensis* (Hanley). *Bulletin of the Department of Marine Sciences, University of Cochin* **7**: 537–542.

Sivaramakrishnan, K.G., 1984. A new genus and species of Atalophlebiinae: Leptophlebiidae from southern India (Ephemeroptera). *Int. J. Entomol.* **26**: 194–203.

Sivaramakrishnan, K.G., 1985. New genus and species of Atalophlebiinae (Ephemeroptera: Leptophlebiidae) from southern India. *Ann. Entomol. Soc. Am.* **78**: 235–239.

Sivaramakrishnan, K.G. & Peters, W.L., 1984. Description of a new species of *Notophlebia* from India and reassignment of the ascribed nymph of *Nathanella* (Ephemeroptera: Leptophlebiidae). *Aquatic Insects* **6**: 115–121.

Sivaramakrishnan, K.G. & Venkataraman, K., 1987. Biosystematic studies of south Indian Leptophlebiidae and Heptageniidae in relation to egg ultrastructure and phylogenetic interpretations. *Proc. Indian Acad. Sci. (Anim. Sci.)* **96**: 637–646.

Sivaramakrishnan, K.G. & Venkataraman, K., 1990. Abundance, altitudinal distribution and swarming of Ephemeroptera in Palni Hills, South India. In: Campbell, I.C. (Ed.), *Mayflies and Stoneflies: Life Histories and Biology*. Kluwer Academic Publishers, Dordrecht: 209–213.

Sivaramakrishnan, K.G., Venkataraman, K. & Balasubramanian, C., 1996. Biosystematics of the genus *Nathanella* Demoulin (Ephemeroptera:

Leptophlebiidae: Atalophlebiinae) from South India. *Aquatic Insects* **18**: 19–28.

Sivasothi, N. & Nor, B.H.J., 1994. A review of otters (Carnivora: Mustelidae: Lutrinae) in Malaysia and Singapore. *Hydrobiologia* **285**: 151–170.

Sivec, I., 1981a. A new species of *Nemoura* (Plecoptera: Nemouridae) from China. *Aquatic Insects* **3**: 79–80.

Sivec, I., 1981b. Contribution to the knowledge of Nepal stoneflies (Plecoptera). *Aquatic Insects* **3**: 245–257.

Sivec, I., 1981c. Some notes about Nemouridae larvae (Plecoptera) from Nepal. *Entomologica Basiliensa* **6**: 108–119.

Sivec, I., 1982. Notes on Himalayan stoneflies from the collection of Zoologische Staatssammlung Muenchen. *Spixiana* **5**: 181–186.

Sivec, I., 1984. Study of genus *Neoperla* (Plecoptera: Perlidae) from the Philippines. *Scopolia* **7**: 1–44.

Sivec, I., 1995. *Cryptoperla fujianica* spec. nov., and *Cryptoperla stilifera* spec. nov., two new Peltoperlidae species from Fujian, China. *Opusc. zool. flumin.* **132**: 1–5.

Sivec, I. & Zwick, P., 1987a. Some *Neoperla* from Taiwan (Plecoptera: Perlidae). *Beitr. Ent., Berlin* **37**: 391–405.

Sivec, I. & Zwick, P., 1987b. Addition to the knowledge of genus *Chinoperla* (Plecoptera: Perlidae). *Aquatic Insects* **11**: 11–16.

Sivec, I., Stark, B.P. & Uchida, S., 1988. Synopsis of the World genera of Perlinae (Plecoptera: Perlidae). *Scopolia* **16**: 1–66.

Sket, B., 1982. New Protojaniridae (Isopoda, Asellota) from Sri Lanka and some corrections of the taxonomy of the family. *Bioloski Vestnik (Ljubljana)* **30**: 127–142.

Slater, J.A. & Polhemus, D.A., 1987. Two new species of *Botocudo* from vertical rock faces in Indonesia (Hemiptera: Lygaeidae). *Proc. Entomol. Soc. Wash.* **89**: 483–488.

Slobodkin, L.B., 1988. Intellectual problems of applied ecology. *BioScience* **38**: 337–342.

Slobodkin, L.B. & Bossert, P.E., 1991. The freshwater Cnidaria — or coelenterates. In: Thorp, J.H. & Covich, A.P. (Eds), *Ecology and Classification of North American Freshwater Invertebrates*. Academic Press, San Diego: 125–143.

Smart, J. & Clifford, E.A., 1969. Simuliidae (Diptera) of Sabah (British North Borneo). *Zool. J. Linn. Soc.* **48**: 9–47.

Smil, V., 1993. *China's Environmental Crisis. An Inquiry into the Limits of National Development.* M.E. Sharpe Inc., New York: 257 pp.

Smith, B.P., 1988. Host-parasite interaction and impact of larval water mites on insects. *Ann. Rev. Entomol.* **33**: 487–507.

Smith, H.M., 1945. The freshwater fishes of Siam, or Thailand. *Bull. U.S. Natn Mus.* **188**: 1–622.

Smith, I.M. & Cook, D.R., 1991. Water mites. In: Thorp, J.H. & Covich, A.P. (Eds), *Ecology and Classification of North American Freshwater*

Invertebrates. Academic Press, San Diego: 523–592.

Smith, I.M. & Oliver, D.R., 1986. Review of parasitic associations of larval water mites (Acari: Parasitengona: Hydrachnida) with insect hosts. *Can. Ent.* **118**: 407–472.

Smith, M.A.K., 1991. Models of seasonal growth of the equatorial carp *Labeo dussumieri* in response to the river flood cycle. *Environ. Biol. Fish.* **31**: 157–170.

Smith, M.J. & Williams, W.D., 1981. The occurrence of *Antecaridina lauensis* (Edmondson) (Crustacea, Decapoda, Atyidae) in the Solomon Islands. An intriguing biogeographical problem. *Hydrobiologia* **85**: 49–58.

Smith, R.E.W. & Bakowa, K., 1994. Utilization of floodplain water bodies by the fishes of the Fly River, Papua New Guinea. *Mitt. Internat. Verein. Limnol.* **24**: 187–196.

Smith, R.E.W. & Hortle, K.G., 1991. Assessment and prediction of the impacts of the Ok Tedi copper mine on fish catches in the Fly River system, Papua New Guinea. *Environmental Monitoring & Assessment* **18**: 41–68.

Smith, R.E.W. & Morris, T.F., 1992. The impacts of changing geochemistry on the fish assemblages of the lower Ok Tedi and middle Fly River, Papua New Guinea. *Science of the Total Environment* **125**: 321–344..

Smith, R.E.W., Ahsanullah, M. & Batley, G.E., 1990. Investigations of the impact of effluent from the Ok Tedi copper mine on the fisheries resource in the Fly River, Papua New Guinea. *Environmental Monitoring & Assessment* **14**: 315–331.

Soerjani, M., 1980. Aquatic plant management in Indonesia. In: Furtado, J.I. (Ed.), *Tropical Ecology and Development. Proceedings of the Vth International Symposium of Tropical Ecology*. International Society of Tropical Ecology, Kuala Lumpur: 725–737.

Soetikno, W., 1987. The river ecosystem in the forest area at Ketambe, Gunung Leuser National Park, Indonesia. *Arch. Hydrobiol. Beih., Ergebn. Limnol.* **28**: 239–246.

Soetikno, W. & Atmowidjojo, A.H., 1985. Insect community of stream ecosystem of Ketambe forest, Gunung Leuser National Park, Indonesia. *Berita Biol.* **3**: 111–115.

Soh, C.L., 1969. Abbreviated development of a non-marine crab *Sesarma (Geosesarma) perracae* (Brachyura: Grapsidae) from Singapore. *J. Zool., Lond.* **158**: 357–370.

Soldán, T., 1978a. New genera and species of Caenidae (Ephemeroptera) from Iran, India and Australia. *Acta ent. bohemoslov.* **75**: 119–129.

Soldán, T., 1978b. Revision of genus *Palingenia* in Europe (Ephemeroptera, Palingeniidae). *Acata ent. bohemoslov.* **75**: 272–284.

Soldán, T., 1983. Two new species of *Clypeocaenis* (Ephemeroptera, Caenidae) with a description of adult stage and biology of the genus. *Acata ent. bohemoslov.* **80**: 196–205.

Soldán, T., 1986. A revision of the Caenidae with occular tubercles in the nymphal stage. *Acta Universitatis Carolinae — Biologica* **1982–1984**: 289–362.

Soldán, T. & Braasch, D., 1984. Two new species of the genus *Prosopistoma* (Ephemeroptera, Prosopistomatidae) from Vietnam. *Acta ent. bohemoslov.* **81**: 370–376.

Soldán, T. & Braasch, D., 1986. *Rithrogeniella tonkinensis* sp. n. (Ephemeroptera, Heptageniidae) from Vietnam, with descriptions of the nymphal stages and biology of the genus. *Acta ent. bohemoslov.* **83**: 202–212.

Soldán, T. & Landa, V., 1977. Three new species of the genus Oligoneuriella (Ephemeroptera, Oligoneuriidae). *Acta ent. bohemoslov.* **74**: 10–15.

Soldán, T. & Landa, V., 1991. Two new species of Caenidae (Ephemeroptera) from Sri Lanka. In: Alba-Tercedor & Sanchez-Ortega, A. (Eds), *Overview and Strategies of Ephemeroptera and Plecoptera*. The Sandhill Crane Press, Gainesville: 235–243.

Soldán, T., Braasch, D. & Luu, T.M., 1987. Two new species of *Centroptella* (Ephemeroptera, Baetidae) from Vietnam, with a description of adult stages of the genus. *Acta ent. bohemoslov.* **84**: 342–349.

Somashekar, R.K., 1988a. Ecological studies on the two major rivers of Karnataka. In: Trivedy, R.K. (Ed.), *Ecology and Pollution of Indian Rivers*. Ashish Publishing House, New Delhi: 39–54.

Somashekar, R.K., 1988b. On the possible utilization of diatoms as indicators of water quality — a study of River Cauvery. In: Trivedy, R.K. (Ed.), *Ecology and Pollution of Indian Rivers*. Ashish Publishing House, New Delhi: 375–382.

Somasundaram, M.V. & Sappanimuthu, T., 1991. The hydrochemistry of a tropical pernnial river and its a adjoining wells as influenced by seasonal rains. *Water International* **16**: 75–82.

Song, D.X. & Chen, Z.F., 1991. A new species of the genus *Dolomedes* from Zhejiang, China (Araneae: Dolomedidae). *Acta Zootaxonomica Sinica* **16**: 15–17 (in Chinese).

Song, T.H. & Yang, T., 1978. *Leeches of China*. Science Press, Beijing. 176 pp. (in Chinese).

Soos, A., 1965. Identification key to the leech (Hirudinoidea) genera of the world, with a catalogue of the species. I. Family: Piscicolidae. *Acta Zoologica Academiae Scientarium Hungaricae* **11**: 417–464.

Soos, A., 1966. Identification key to the leech (Hirudinoidea) genera of the world, with a catalogue of the species. III. Family: Erpobdellidae. *Acta Zoologica Academiae Scientarium Hungaricae* **12**: 371–407.

Soos, A., 1967. Identification key to the leech (Hirudinoidea) genera of the world, with a catalogue of the species. IV. Family: Haemodipsidae. *Acta Zoologica Academiae Scientarium Hungaricae* **13**: 417–432.

Soos, A., 1969a. Identification key to the leech (Hirudinoidea) genera of the world, with a catalogue of the species. V. Family: Hirudinidae. *Acta Zoologica Academiae Scientarium Hungaricae* **15**: 151–201.

Soos, A., 1969b. Identification key to the leech (Hirudinoidea) genera of the world, with a catalogue of the species. VI. Family: Glossiphoniidae. *Acta Zoologica Academiae Scientarium Hungaricae* **15**: 397–454.

Spackman, S.C. & Hughes, J.W., 1995. Assessment of minimum stream corridor width for biological conservation: species richness and distribution along mid-order streams in Vermont, USA. *Biological Conservation* **71**:325–332.

Spangler, P.J., 1985. A new species of the aquatic beetle genus *Dryopomorphus* from Borneo (Coleoptera: Elmidae: Larinae). *Proc. Biol. Soc. Wash.* **98**: 416–421.

Spence, J.R., 1989. The habitat template and life history strategies of pondskaters (Heteroptera: Gerridae): reproductive potential, phenology and wing dimorphism. *Can. J. Zool.* **67**: 2432–2447.

Spence, J.R. & Andersen, N.M., 1994. Biology of water striders: interactions between systematics and ecology. *Ann. Rev. Entomol.* **39**: 101–128.

Spencer, J.E., 1954. *Asia East by South.* John Wiley & Sons, New York: 453 pp.

Spencer, J.E., 1973. *Oriental Asia: Themes Toward a Geography.* Prentice-Hall Inc., Englewood Cliffs, New Jersey: 146 pp.

Spiridonov, S.E., 1993. *Gordius annamensis* sp. n. (Nematomorpha: Parachordodidae), a new species of gordian worm from central Vietnam. *Parazitologiya (St Petersburg)* **27**: 429–432 (in Russian).

Sridhar, K.R. & Kaveriappa, K.M., 1984. Seasonal occurrence of water-borne fungi in Konaje stream (Mangalore), India. *Hydrobiologia* **119**: 101–105.

Sridhar, K.R. & Kaveriappa, K.M., 1989. Colonization of leaves by water-borne hyphomycetes in a tropical stream. *Mycological Research* **92**: 392–396.

Srivardhana, R., 1987. The Nam Pong case study. Some lessons to be learned. *Water Resources Development* **3**: 238–246.

Srivastava, K.N., Srivastava, P. & Sinha, A.K., 1990. Zooplankton studies of Ganga River between Kalakanker (Pratapgarh) and Phaphamau (Allahbad) (Uttar Pradesh). In: Agrawal, V.P. & Das, P. (Eds), *Recent Trends in Limnology.* Society of Biosciences, Muzaffarnagar (India): 129–134.

Srivastava, U.S. & Kulshrestha, A.K., 1990. Seasonal variations in certain physico-chemical parameters in Ganga, Yamuna and Tons in Allahabad Region (U.P., India). In: Agrawal, V.P. & Das, P. (Eds), *Recent Trends in Limnology.* Society of Biosciences, Muzaffarnagar (India): 351–364.

St Clair, R.M., 1993. Life histories of six species of Leptoceridae (Insecta: Trichoptera) in Victoria. *Aust. J. Mar. Freshwat. Res.* **44**: 363–379.

St Clair, R.M., 1994a. Some larval Leptoceridae (Trichoptera) from south-eastern Australia. *Records of the Australian Museum* **46**: 171–226.

St Clair, R.M., 1994b. Diets of some larval Leptoceridae (Trichopera) in south-eastern Australia. *Aust. J. Mar. Freshwat. Sci.* **45**: 1023–1032.

St Quentin, D., 1973. Results of the Austrian-Ceylonese hydrobiological mission, 1970. Part XII. Contributions to the ecology of the larvae of some Odonata from Ceylon. *Bull. Fish. Res. Stn, Sri Lanka (Ceylon)* **24**: 113–124.

Stamp, L.D., 1962. *Asia: a Regional and Economic Geography*. Methuen, London: 730 pp.

Stanger, J.A. & Baumann, R.W., 1993. A revision of the stonefly genus *Taenionema* (Plecoptera, Taeniopterygidae). *Trans. Am. Entomol. Soc.* **119**: 171–229.

Stark, B.P., 1983. Descriptions of Neoperlini from Thailand and Malaysia (Plecoptera: Perlidae). *Aquatic Insects* **5**: 99–114.

Stark, B.P., 1987. Records and descriptions of Oriental Neoperlini (Plecoptera: Perlidae). *Aquatic Insects* **9**: 45–50.

Stark, B.P., 1989a. Oriental Peltoperlidae (Plecoptera): a generic review and descriptions of a new genus and seven new species. *Ent. Scand.* **19**: 503–525.

Stark, B.P., 1989b. *Perlesta placida* (Hagen), an eastern Nearctic species complex (Plecoptera: Perlidae). *Ent. Scand.* **20**: 263–286.

Stark, B.P. & Nelson, C.R., 1994. Systematics, phylogeny and zoogeography of genus *Yoraperla* (Plecoptera: Peltoperlidae). *Ent. Scand.* **25**: 241–273.

Stark, B.P. & Sivec, I., 1991. Descriptions of Oriental Perlini (Plecoptera: Perlidae). *Aquatic Insects* **13**: 151–160.

Stark, B.P. & Szczytko, W., 1979. A new species of *Neoperla* (Plecoptera: Perlidae) from Burma. *Aquatic Insects* **1**: 221–224.

Stark, B.P. & Szczytko, W., 1981a. Contributions to the systematics of *Paragnetina* (Plecoptera: Perlidae). *J. Kansas Ent. Soc.* **54**: 625–648.

Stark, B.P. & Szczytko, W., 1981b. *Skwala brevis* (Koponen) from Japan (Plecoptera: Perlodidae). *Aquatic Insects* **3**: 61–63.

Starmühlner, F., 1974. Results of the Austrian-Ceylonese hydrobiological mission, 1970. Pt XVII. The freshwater gastropods of Ceylon. *Bull. Fish. Res. Stn, Sri Lanka (Ceylon)* **25**: 97–181.

Starmühlner, F., 1977. The genus *Paludomus* in Ceylon. *Malacologia* **16**: 261–264.

Starmühlner, F., 1984a. Mountain stream fauna, with special reference to Mollusca. In: Fernando, C.H. (Editor), *Ecology and Biogeography in Sri Lanaka*. Dr. W. Junk Publishers, The Hague: 215–255.

Starmühlner, F., 1984b. Checklist and longitudinal distribution of the meso- and macrofauna of mountain streams of Sri Lanka (Ceylon). *Arch. Hydrobiol.* **101**: 303–325.

Statzner, B., 1984. Keys to adult and immature Hydropsychinae in the Ivory Coast (West-Africa) with notes on their taxonomy and distribution. *Spixiana* **7**: 23–50.

Statzner, B., 1976. Die Köcherfliegen-Emergenz (Trichoptera, Insecta) aus dem zentralafrikanischen Bergbach Kalengo. *Arch. Hydrobiol.* **78**: 102–137.

Statzner, B. & Gibon, F.-M., 1984. Keys to adult and immature Macronematinae (Insecta: Trichoptera) from the Ivory Coast (West Africa) with notes on their taxonomy and distribution. *Rev. Hydrobiol. trop.* **17**: 129–151.

Statzner, B. & Higler, B., 1985. Questions and comments on the River Continuum Concept. *Can. J. Fish. Aquat. Sci.* **42**: 1038-1044.

Statzner, B. & Higler, B., 1986. Stream hydraulics as a major determinant of benthic invertebrate zonation patterns. *Freshwat. Biol.* **16**: 127–139.

Statzner, B. & Resh, V.H., 1993. Multiple-site and -year analyses of stream insect emergence: a test of ecological theory. *Oecologia* **96**: 65–79.

Stewart, K.W. & Maketon, M., 1991. Structures used by Nearctic stoneflies (Plecoptera) for drumming, and their relationship to behavioural pattern diversity. *Aquatic Insects* **13**: 33–53.

Stewart, K.W., Abbott, J.C., Kirchner, R.F. & Moulton, S.R., 1995. New descriptions of North American euholognathan stonefly drumming (Plecoptera) and first Nemouridae ancestral call discovered in *Soyedina carolinensis* (Plecoptera: Nemouridae). *Ann. Entomol. Soc. Am.* **88**: 234–239.

Stewart-Oaten, A., Murdoch, W. & Parker, K., 1986. Environmental impact assessment: 'Pseudoreplication' in time? *Ecology* **67**: 929–940.

Stites, D.L. & Benke, A.C., 1989. Rapid growth rates of chironomids in three habitats of a subtropical blackwater river and their implications for *P:B* ratios. *Limnol. Oceanogr.* **34**: 1278–1289.

Stock, J.H., 1991. A new species of *Psammogammarus* (Amphipoda, Melitidae) from river alluvia in Luzon, Philippines. *Stygologia* **6**: 227–233.

Stone, B., 1983. The Chang Jiang diversion project: an overview of economic and environmental issues. In: Biswas, A.K., Zuo, D., Nickum, J.E. & Liu, C. (Eds), *Long-distance Water Transfer: a Chinese Case Study and International Experiences*. Tycooly International Publishers Limited, Dublin: 193–214.

Stout, J., 1980. Leaf decomposition rates in Costa Rican lowland tropical rainforest streams. *Biotropica* **12**: 264–272.

Stout, R.J., 1989. Effects of condensed tannins on leaf processing in mid-latitude streams: a theoretical approach. *Can. J. Fish. Aquat. Sci.* **46**: 1097–1106.

Studemann, D., Landolt, P. & Tomka, I., 1994. Biochemical investigations of Siphlonuridae and Ameletidae (Ephemeroptera). *Arch Hydrobiol.* **130**: 77–92.

Stuebing, R.B., Ismail, G. & Ching, L.H., 1994. The distribution and abundance of the Indo-Pacific crocodile *Crocodylus porosus* Schneider in the Klias River, Sabah, East Malaysia. *Biological Conservation* **69**: 1–7.

Su, B. & Li, O., 1988. Water quality assessment of the Beijiang River by using benthic macroinvertebrates. *Proc. 3rd Chinese Oceanol. Limnol. Sci. Conf.*: 168–174 (in Chinese).

Su, C.R. & You, D.S., 1989. A new species of *Ephemerella* subgenus *Ephemerella* from Jilin Province, China (Ephemeroptera: Ephemerellidae). *Acta Zootaxonomica Sinica* **14**: 181–185 (in Chinese).

Su, S. & Yang, P., 1992. Morphology and life history of the giant water bug (*Sphaerodema rustica* Fabricius). *Chinese J. Entomol.* **12**: 49–61 (in Chinese).

Subramanian, V. & Ittekkot, V., 1991. Carbon transport by the Himalayan rivers. In: Degens, E.T., Kempe, S. & Richey, J.E. (Eds), *Biogeochemistry of Major World Rivers*. SCOPE/John Wiley & Sons, Chichester: 157–168.

Sultana, P. & Thompson, P.M., 1997. Effects of flood control and drainage of fisheries in Bangladesh and the design of mitigating measures. *Regulated Rivers: Research & Management* **13**: 43–55.

Sun, C. & Yang, L., 1994. A new species of the genus *Himalopsyche* Banks, 1940 (Trichoptera: rhyacophilidae) from China. *Braueria* **21**: 8.

Sun, C. & Yang, L., 1995. Studies on the genus *Rhyacophila* (Trichoptera) in China (I). *Braueria* **22**: 27–32.

Sun, F.C., 1985. Sediment problems and their management in Peninsular Malaysia. *Water International* **10**: 3–6.

Sun, H. & Wang, G., 1992. NPC passes Three Gorges project. *China Environment News* **33** (**April 1992**): 1.

Sunder, S. & Subla, B.A., 1986. Macrobenthic fauna of a Himalaya river. *Indian J. Ecol.* **13**: 127–132.

Suren, A.M., 1994. Macroinvertebrate communities of streams in western Nepal: effects of altitude and land use. *Freshwat. Biol.* **32**: 323–336.

Suryadiputra, I.N.N. & Suyasa, G., 1995. Some limnological aspects of the River Ayung in Bali, Indonesia. In: Timotius, K.H. & Göltenboth, F. (Eds), *Tropical Limnology Volume III. Tropical Rivers, Wetlands and Special Topics*. Satya Wacana University Press, Salatiga, Indonesia: 29–38.

Sutadipradja, E. & Hardjowitjitro, H., 1984. Watershed rehabilitation program related to the management of river and reservoir sedimentation in Indonesia. *Water International* **9**: 146–149.

Suvatti, C. & Menasveta, D., 1968. Threatened species of Thailand's aquatic fauna and preservation problems. In: Talbot, L.M. & Talbot, M.H. (Eds), *Conservation in Tropical South East Asia*. IUCN Publications New Series No. 10, IUCN, Morges, Switzerland: 332–336.

Suzuki, H., 1980. An atyid shrimp living in anchialine pool on Kuro-shima, the Yaeyama Group, Okinawa Prefecture. *Proceedings of the Japanese Society of Systematic Zoology* **18**: 47–53 (in Japanese).

Suzuki, K., 1984. Character displacement and evolution of the Japanese *Mnias* damselflies (Zygoptera: Calopterygidae). *Odonatologica* **13**: 287–300.

Sweeney, B.W. & Vannote, R.L., 1982. Population synchrony in mayflies: a predator satiation hypothesis. *Evolution* **36**: 810–821.

Sweeney, B.W. & Vannote, R.L., 1984. Influence of food quality and temperature on life history characteristics of the parthenogenetic mayfly, *Cloeon triangulifer*. *Freshwat. Biol.* **14**: 621–630.

Swegman, B.G. & Coffman, W.P., 1980. *Stenopsyche kodaikanalensis*: a new species of *Stenopsyche* from Souh India (Trichoptera: Stenopsychidae). *Aquatic Insects* 2: 73–79.

Tai, A.Y. & Manning, R.B.M., 1984. A new species of *Potamocypoda* (Crustacea: Brachyura: Ocypodidae) from Malaysia. *Proc. Biol. Soc. Washington* 97: 615–617.

Takami. A., 1991. The birth frequency, number and size of newborns in three species of the genus *Semisulcospira* (Prosobranchia: Pleuroceridae). Venus — *Jap. J. Malacol.* 50: 218–232.

Takaoka, H. & Davies, D.M., 1995. *The Black Flies (Diptera: Simuliidae) of West Malaysia*. Kyushu University Press, Fukuoka-shi, Japan: 175 pp.

Takaoka, H. & Davies, D.M., 1996. The black flies (Diptera: Simuliidae) of Java, Indonesia. *Bishop Museum Bulletins in Entomology* 6: 1–81.

Takaoka, H & Hadi, U.K., 1991. Two new blackfly species of *Simulium (Simulium)* from Java, Indonesia. *Jpn. J. Trop. Med. Hyg.* 19: 357–370.

Takaoka, H. & Sigit, S.H., 1992. A new blackfly species of *Simulium (Gomphostilbia)* from Java, Indonesia (Diptera: Simuliidae). *Jpn. J. Trop. Med. Hyg.* 20: 135–142.

Takaoka, h. & Suzuki, H., 1984. The blackflies (Diptera: Simuliidae) from Thailand. *Jap. J. Sanit. Zool.* 35: 7–45.

Takaoka, H., Davies, D.M. & Dudgeon, D., 1995. Black flies (Diptera: Simuliidae) from Hong Kong: taxonomic notes with descriptions of two new species. *Jpn. J. Trop. Med. Hyg.* 23: 189–196.

Takemon, Y. & Tanida, K., 1994. New data on *Nymphomyia alba* (Diptera: Nymphomyiidae) from Japan with notes on the larvae and micro-habitat. *Aquatic Insects* 16: 119–124.

Takemoto, K., 1983. On the life history of a caddisfly, *Stenopsyche japonica* Martynov (Trichoptera: Stenopsychidae) in River Hikigawa, Wakayama. 1. Ages, generations and growth. *Nankai-Seibutsu* 25: 210–216 (in Japanese).

Takenouchi, T., 1966. A hydrologic study of the Mekong basin. In: Fujioka, Y. (Ed.), *Water Resource Utilization in Southeast Asia*. The Center for Southeast Asian Studies, Kyoto University, Kyoto: 55–66.

Takeuchi, K., 1993. Analyses of the flow regime of the Chao Phraya River. In: Gladwell, J.S. (Ed.), *Hydrology of Warm Humid Regions*. IAHS Press, Wallingford, U.K.: 181–193.

Taki, Y., 1978. An analytical study of the fish fauna of the Mekong Basin as a biological production system in nature. *Research Institute of Evolutionary Biology, Tokyo, Special Publication* 1: 1–74.

Tan, E.S.P., 1980. Ecological aspects of some Malaysian riverine cyprinids in relation to their aquaculture potential. In: Furtado, J.I. (Ed.), *Tropical Ecology and Development. Proceedings of the Vth International Symposium of Tropical Ecology*. International Society of Tropical Ecology, Kuala Lumpur: 757–762.

Tan, T.H., Pai, J.Y. & Hsha, K.C., 1987. The recovery of fouling clam,

Limnoperna fortunei, from Taiwan. *Bulletin of Malacology, Republic of China* **13**: 97–100 (in Chinese).

Tanida, K., 1980. Life history and sitribution of three species of *Hydropsyche* (Trichoptera: Hydropsychidae) in the River Kibune (Kyoto, central Japan), with perticular references to the variations in their life cycles and the relation of larval growth to their density. *Jap. J. Limnol.* **41**: 95–111 (in Japanese).

Tanida, K., 1984. Larval microlocation on stone faces of three *Hydropsyche* species (Insecta: Trichoptera), with a general consideration on the relation of systematic groupings to the ecological and geographical distribution among the Japanese *Hydropsyche* species. *Physiol. Ecol. Japan* **21**: 115–130.

Tanida, K., 1985. Trichoptera. In: Kawai, T. (Ed,), *An Illustrated Book of Aquatic Insects of Japan*. Tokai University Press, Tokyo: 167–215 (in Japanese).

Tanida, K., 1986a. A revision of Japanese species of the genus *Hydropsyche* (Trichopera, Hydropsychidae) I. *Kontyû* **54**: 467–484.

Tanida, K., 1986b. A revision of Japanese species of the genus *Hydropsyche* (Trichopera, Hydropsychidae) II. *Kontyû* **54**: 624–633.

Tanida, K., 1987a. A revision of Japanese species of the genus *Hydropsyche* (Trichopera, Hydropsychidae) III. *Kontyû* **55**: 59–70.

Tanida, K., 1987b. An introduction to the Trichoptera fauna of Japan, with a tentative checklist of genera. In: Bournaud, M. & Tachet, H. (Eds), *Proceedings of the 5th International Symposium on Trichoptera*. Dr W. Junk Publishers, The Hague: 119–124.

Tanida, K., 1989a. Reconsideration of the *sumiwake* concept. In *Recent Progress of Aquatic Entomology in Japan, with Special Reference to Speciation and* Sumiwake: 1–16. Tanida, K. & Sibatani, A. (Eds). Tokyo, Tokai University Press (in Japanese).

Tanida, K., 1989b. The ecology of *Hydropsyche* larvae (Trichoptera: Hydropsychidae) — *sumiwake* and microlocatioñ on stone faces. In: Tanida, K. & Sibatani, A. (Eds), *Recent Progress of Aquatic Entomology in Japan, with Special Reference to Speciation and* Sumiwake. Tokai University Press, Tokyo: 117–129 (in Japanese).

Tandon, R.S., 1988. River pollution and fish — a report on pollution calamity of Uttar Pradesh (India). In: Trivedy, R.K. (Ed.), *Ecology and Pollution of Indian Rivers*. Ashish Publishing House, New Delhi: 383–391.

Tani, K. & Miyatake, Y., 1979. The discovery of *Epiophlebia laidlawi* Tillyard, 1921 in the Kathmandu Valley, Nepal (Anisozygoptera: Epioplebiidae). *Odonatologica* **8**: 329–332.

Tariq, J., Ashraf, M. & Jaffar, M., 1993. Metal pollution status of the River Chennab, Pakistan, through fish, water and sediment analysis. *Toxicological & Environmental Chemistry* **38**: 175–181.

Tariq, J., Jaffar, M., Ashraf, M., 1994. Trace metal concentration, distribution and correlation in water, sediment and fish from the Ravi River, Pakistan.

Fisheries Research **19**: 131–139.

Tarjan, A.C., Esser, R.P. & Chang, S.L., 1977. An illustrated key to nematodes found in freshwater. *J. Water Pollut. Control Fed.* **49**: 2318–2337.

Tejwani, K.G., 1993. Water management issues: population, agriculture and forests — a focus on watershed management. In: Bonell, M., Hufschmidt, M.M. & Gladwell, J.S. (Eds), *Hydrology and Management in the Humid Tropics*. Cambridge University Press, Cambridge: 496–525.

Teng, K., 1990. Biological studies on the horseflies in Beijing (China) (Diptera: Tabanidae). *Sinozoologia* **0(7)**: 319–324.

Terada, M., 1977. On the zoea larvae of four crabs of the family Hymenosomatidae. *Zoological Magazine (Tokyo)* **86**: 174–184.

Terada, M., 1981. Zoea larvae in five crabs in the subfamily Varuninae. *Researches on Crustacea* **11**: 66–76 (in Japanese).

Thapa, G.B. & Weber, K.E., 1994. Managing mountain watersheds in Nepal: issues and policies. *International Journal of Water Resources Development* **10**: 474–495.

Thapa, G.B. & Weber, K.E., 1995. Status and management of watersheds in Upper Pokhara Valley, Nepal. *Environmental Management* **19**: 497–513.

Thew, T.B., 1960. Revision of the genera of the family Caenidae. *Ann. Entomol. Soc. Am.* **86**: 187–205.

Thiemmedh, J., 1970. Hydrobiology and inland fisheries problems in Thailand. *Proceedings of the Regional Meeting of Inland Water Biologists in Southeast Asia*. Unesco Field Science Office for Southeast Asia, Djakarta: 141–157.

Thirakhupt, K. & Van Dijk, P.P., 1994. Species diversity and conservation of turtles in western Thailand. *Nat. Hist. Bull. Siam Soc.* **42**: 207–259.

Thomas, A., 1992. *Gratia sororculaenadinae* n. gen., n. sp., Ephéméroptère nouveau de Thaïlande (Ephemeroptera, Baetidae). *Bull. Soc. Hist. Nat., Toulouse* **128**: 47–51.

Thomas, A., 1993. Athericidae oest palearctiques: le genre *Atherix* Meigen, 1803 1. Description d' *A. nicolae* n. sp. du Nepal (Diptera, Brachycera Orthorrhpha). *Bulletin de la Societe d'Histoire Naturelle de Toulouse* **129**: 51–54.

Thorpe, W.H., 1950. Plastron respiration in aquatic insects. *Biol. Rev.* **25**: 334–390.

Tian, F. & Lin, F., 1993. Population resettlement and economic development in the Three Gorges reservoir. In: Luk, S. & Whitney, J. (Eds), *Megaproject. A Case Study of China's Three Gorges Project*. M.E. Sharpe Inc., New York: 3–39.

Tian, L., 1985. Two new species of genus *Stenopsyche* McLachlan from Xizang Plateau (Trichoptera: Stenopsychidae). *Journal of Nanjing Agricultural University* **1**: 23–25 (in Chinese).

Tian, L., 1988. A review of the Chinese genus *Stenopsyche* McLachlan (Trichoptera: Stenopsychidae). *Acta Entomlogica Sinica* **31**: 194–202 (in Chinese).

Tian, L. & Li, Y., 1985. A preliminary survey of caddisflies from Fujian Province, with descriptions of two new species (order Trichoptera). *Wuyi Science Journal* 5: 51–58 (in Chinese).

Tian, L. & Li, Y., 1986. Four new species of the caddisflies on Hengduan Mountain, China. *Journal of Nanjing Agricultural University* 2: 50–54 (in Chinese).

Tian, L. & Li, Y., 1987a. Trichoptera. *Agricultural Insects, Spiders, Plant Diseases and Weeds of Xizang* 1: 241–250.

Tian, L. & Li, Y., 1987b. A preliminary study of the subfamily Hydropsychinae (Trichoptera: Hydropsychidae) in China. In: Bournaud, M. & Tachet, H. (Eds), *Proceedings of the 5th International Symposium on Trichoptera*. Dr W. Junk Publishers, The Hague: 125–129.

Tian, L. & Li, Y., 1988. Trichoptera: Rhyacophilidae, Philopotamidae, Stenopsychidae, Hydropsychidae, Phryganeidae, Limnehilidae, Polycentropodidae, Sericostomatidae. In: Hwang, F., Wang, P., Yin, W., Yu, P., Lee, T., Yang, C. & Wang, X. (Eds), *Insects of Mt. Namjagbarwa region of Xizang*. Science Press, Beijing: 377–383.

Tian, L. & Zheng, J., 1989. Description of two new species of the genus *Stenopsyche* McLachlan from China (Trichoptera). *Journal of Nanjing Agricultural University* 12: 50–52 (in Chinese).

Tian, L., Li, Y. & Sun, C., 1992. Notes on two new species of Trichoptera from China. *Journal of Nanjing Agricultural University* 15: 28–29 (in Chinese).

Tian, L., Li, Y., Yang, L. & Sun, C., 1993. Trichoptera. In: Chen, S. (Ed.), *Insects of the Hengduan Mountains region. Volume 2*. Science Press, Beijing: 867–892 (in Chinese).

Tian, R.C., Hu, F.X. & Martin, J.M., 1993. Summer nutrient fronts in the Changjiang (Yangtze River) estuary. *Estuarine, Coastal and Shelf Science* 37: 27–41.

Tikkanen, P., Muotka, T., Huhta, A. & Juntunen, A., 1997. The roles of active predator choice and prey vulnerability in determining the diet of predatory stonefly (Plecoptera) nymphs. *J. Anim. Ecol.* 66: 36–48.

Tillyard, R.J., 1921. On an anisozygoterous larva from the Himalayas (order Odonata). *Rec. Indian Mus.* 23: 93–107.

Timm, T., 1980. Distribution of aquatic oligochaetes. In: Brinkhurst, R.O. & Cook, D.G. (Eds), *Aquatic Oligochaete Biology*. Plenum Press, New York: 55–77.

Tiunova, T.M., 1991. Description of the imago of *Potamanthellus rarus* (Ephemeroptera, Neoephemeridae) from the Primorski Krai. *Zoologicheskii Zhurnal* 70: 136–138 (in Russian).

Tiunova, T.M., 1992. Descriptions of a new species of mayfly of the genus *Cinygmula* McD. (Ephemeroptera, Heptageniidae) and the larva of *Rhoenanthus rohdendorfi* Tshern. (Potamanthidae) from Maritime Territory. *Entomological Review* 70: 104–109.

Tiunova, T.M. & Levanidova, I.M., 1989. Description of a new mayfly species (Ephemeroptera) from the Soviet Far East. *Aquatic Insects* 4: 241–245.

Tiwari, N.C., 1990. Periphytonic succession in perennial snow fed river Amrit Ganga of Garhwal Himalaya, India. *Environ. Ecol.* 8: 180–183.

Tokeshi, M., 1993. On the evolution of commensalism in the Chironomidae. *Freshwat. Biol.* 29: 481–489.

Tokeshi, M., 1995a. Life cycles and population dynamics. In: Armitage, P., Cranston, P.S. & Pinder, L.C.V., (Eds), *The Chironomidae: Biology and Ecology of Non-biting Midges.* Chapman & Hall, London: 225–268.

Tokeshi, M., 1995b. Life cycles and population dynamics. In: Armitage, P., Cranston, P.S. & Pinder, L.C.V., (Eds), *The Chironomidae: Biology and Ecology of Non-biting Midges.* Chapman & Hall, London: 226–296.

Trivedy, R.K., 1988. *Ecology and Pollution of Indian Rivers.* Ashish Publishing House, New Delhi: 463 pp.

Tomka, I. & Zurwerra, A., 1985. Key to the genera of Heptageniidae (Ephemeroptera) of the Holarctic, Oriental and Ethiopean region. *Entomologische Berichte Luzern* 14: 113–126.

Tonapi, G.T., 1959. Studies on the aquatic insect fauna of Poona (Aquatic Heteroptera). *Proc. Natl. Inst. Sci. India* 25B: 321–332.

Tong, X., Hu, H. & Chen, S., 1995. Use of aquatic insects to evaluate water quality in the streams of Mt. Nankun. *J. South China Agric. Univ.* 16: 6–10 (in Chinese).

Traver, J.R., 1939. Himalayan mayflies (Ephemeroptera). *Ann. Mag. Nat. Hist., Ser. II* 4: 32–56.

Trihadiningrum, Y., De Pauw, N., Tjondronegoro, I. & Verheyen, R.F., 1995. Use of benthic macroinvertebrates for water quality assessment of the Blawi River (East Java, Indonesia). In: Timotius, K.H. & Göltenboth, F. (Eds), *Tropical Limnology Volume III. Tropical Rivers, Wetlands and Special Topics.* Satya Wacana University Press, Salatiga, Indonesia: 89–110.

Truesdale, F.M. & Mermilliod, W.J., 1979. The river shrimp *Macrobrachium ohione* (Smith) (Decapoda, Palaemonidae): its abundance, reproduction, and growth in the Atchafalya River Basin of Louisiana, U.S.A. *Crustaceana* 36: 61–73.

Tsai, C., Islam, M.N., Karim, R. & Rahman, K.U.M.S., 1981. Spawning of major carps in the lower Halda River, Bangladesh. *Estuaries* 4: 127–138.

Tshernova, O.A., 1972. Some new Asiatic species of mayflies (Ephemeroptera, Heptageniidae, Ephemerellidae). *Entomological Review* 51: 364–369.

Tshernova, O.A., 1976. A nymphal key to the genera of the Heptageniidae (Ephemeroptera) of the Holarctic and the Oriental Region. *Entomol. Rev.* 55: 47–56.

Tsuda, M., 1961. Important role of net-spinning caddis-fly larvae in Japanese running water. *Verh. Internat. Verein. Limnol.* 14: 376–377.

Tsuda, S., 1991. *A Distributional List of World Odonata.* Published by the author, Osaka: 362 pp.

Tsui, P.T.P. & Peters, W.L., 1970. The nymph of *Habrophlebiodes gilliesi* Peters (Ephemeroptera: Leptophlebiidae). *Proc. R. ent. Soc. Lond. (A)* **45**: 89–90.

Tundisi, J.G., 1984. Tropical limnology. *Verh. Internat. Verein. Limnol.* **22**: 60–64.

Tyler, P.A., 1984. Water chemistry of the River Kwai, Thailand. *Hydrobiologia* **111**: 65–73.

Uchida, S., 1983. A new species of *Calineuria* (Plecoptera, Perlidae) from Japan, with notes on the Japanese species of the genus. *Kontyû, Tokyo* **51**: 622–627.

Uchida, S., 1990. Distribution of Plecoptera in the Tama-Gawa River system, Central Japan. In: Campbell, I.C. (Ed.), *Stoneflies and Mayflies: Life Histories and Biology*. Kluwer Academic Publishers, Dordrecht: 181–188.

Uchida, S., 1991. Relationship between the water temperature and the distribution of Plecoptera in the Tama Rivers. *Insectarium* **28**: 20–25 (in Japanese).

Uchida, S. & Isobe, Y., 1988. *Cryptoperla* and *Yoraperla* from Japan and Taiwan (Plecoptera: Pletoperlidae). *Aquatic Insects* **10**: 17–31.

Uchida, S. & Isobe, Y., 1989. Styloperlidae, stat. nov. and Microperlinae, subfam. nov. with a revised system of the family group Systellognatha. *Spixiana* **12**: 145–182.

Uchida, S. & Isobe, Y., 1991. Designation of a neotype for *Kamimuria tibialis* (Pictet, 1841), and *K. uenoi* Kohno, 1947, spec. propr., stat. n. (Plecoptera, Perlidae). *Aquatic Insects* **13**: 65–77.

Uchida, S. & Maruyama, H., 1987. What is *Scopura longa* Uéno, 1929 (Insecta, Plecoptera)? A revision of the genus. *Zoological Science* **4**: 699–709.

Uchida, S. & Yamasaki, T., 1989. Some Perlinae (Plecoptera: Perlidae) from the Malay Peninsula and Thailand with the redescription of *Neoperla hamata* from Assam. *Bull. Biogeog. Soc. Japan* **44**: 135–143.

Uchida, T. & Imamura, T., 1951. Some water-mites from China. *Journal of the Faculty of Science, Hokkaido University, Series VI, Zoology* **10**: 324–358.

Uéno, M., 1966. Mayflies collected by the Kyoto University Pamir-Hindukush Expedition, 1960. *Results of the Kyoto University Scientific Expedition to the Karokorum and Hindukush, Additional Reports* **8**: 299–326.

Uéno, M., 1969. Mayflies (Ephmeroptera) from various regions of Southeast Asia. *Oriental Insects* **3**: 221–238.

Ullah, S.M., Nuruzzaman, M. & Gerzabek, M.H., 1995. Heavy metal and microbiological pollution of water and sediments by industrial wastes, effluents and slums aound Dhaka City. In: Timotius, K.H. & Göltenboth, F. (Eds), *Tropical Limnology Volume III. Tropical Rivers, Wetlands and Special Topics*. Satya Wacana University Press, Salatiga, Indonesia: 179–186.

Ulmer, G., 1912. H. Sauter's Formosa-Ausbeute. Ephemeriden. *Entomologische*

Mitteilungen **1**: 369–375.

Ulmer, G., 1911. Die von Herrn Hans Sauter auf Formosa gesammelten Trichopteren (Neur.). *Deutsche Entomologische Zeitschrift* **1911**: 369–401.

Ulmer, G., 1915. Trichopteren des Ostens, besonders von Ceylon und Neu-Guinea. *Deutsche Entomologische Zeitschrift* **1915**: 41–75.

Ulmer, G., 1919. Neue Ephemeropteren. *Archiv für Naturgeschichte* **85** (**A 11**): 1–80.

Ulmer, G., 1920. Übersicht über die Gattungen der Ephemeropteren, nebst Bemerkungen über einzelne Arten. *Stettinger Entomologiche Zeitung* **81**: 97–144.

Ulmer, G., 1925. Beiträge zur Fauna sinica Bewirkt von Dr R. Mell. III. Trichopteren.

und Ephemeropteren. *Archiv für Naturgeschichte* **91** (**A 5**): 19–110.

Ulmer, G., 1927. Einige neue Trichopteran aus Asien. *Entomologische Mitteilungen* **16**: 172–182.

Ulmer, G., 1930. Trichoptern von den Philippen und von den Sunda-Inseln. *Treubia* **11**: 373–498.

Ulmer, G., 1932a. Aquatic insects of China. Article VI. Revised key to the genera of Ephemeroptera. *Peking Nat. Hist. Bull.* **7**: 195–218.

Ulmer, G., 1932b. Aquatic insects of China. Article III. Neue chinesische Trichopteren, Nebst Übersicht über die Bisher aus China Bekannten Arten. Teil I. Beschreibung der neuen Arten und Aufzählung neuer Funfort aus den Sammlungen C.F. Wu, Museum Stockholm, Museum Hamburg, British Museum, R. Mell, Ebsen-Petersen und G. Ulmer. *Peking Nat. Hist. Bull.* **7**: 39–70.

Ulmer, G., 1932c. Aquatic insects of China. Article IV. Neue chinesische Trichopteren, Nebst Übersicht über die Bisher aus China Bekannten Arten. Teil II. Liste der chinesischen Trichopteren, mit Angaben Über synonymie, Litteratur und Geographische Verbreitung. *Peking Nat. Hist. Bull.* **7**: 135–157.

Ulmer, G., 1935. Neue chinesische Ephemeropteren, Nebst Übersict uber die Bisher aus China Bekannten Arten. *Peking Nat. Hist. Bull.* **10**: 201–215.

Ulmer, G., 1940. Eintagsfliegen (Ephemeropteren) von den Sunda-Inseln. *Arch. Hydrobiol., Suppl.* **16**: 444–578 & 581–692.

Ulmer, G., 1951. Köcherfliegen (Trichopteren) von den Sunda Inseln. Teil I. *Arch. Hydrobiol., Suppl.* **19**: 1–528.

Ulmer, G., 1955. Köcherfliegen (Trichopteren) von den Sunda Inseln. Teil II. *Arch. Hydrobiol., Suppl.* **21**: 408–608.

Ulmer, G., 1957. Köcherfliegen (Trichopteren) von den Sunda Inseln. Teil III. *Arch.*

Hydrobiol., Suppl. **23**: 109–170.

Underwood, A.J., 1981. Techniques of analysis of variance in experimental marine biology and ecology. *Ann. Rev. Oceanogr. Mar. Biol.* **19**: 513–605.

Underwood, A.J., 1991. Beyond BACI: experimental designs for detecting human environmental impacts on temporal variations in natural populations. *Aust. J. Mar. Freshwat. Res.* **42**: 569–587.

Underwood, A.J., 1993. The mechanics of spatially replicated sampling programmes to detect environmental impacts in a variable world. *Aust. J. Ecol.* **18**: 99–116.

Underwood, A.J., 1995. Ecological research and (and research into) environmental management. *Ecological Applications* **5**: 232–247.

United Nations, 1957. *Development of Water Resources in the Lower Mekong Basin.* Flood Control Series No. 12, Bangkok. (Not seen.)

Unni, S.K., 1995. Physico-chemical limnology of a stretch of Narmada River with a mainstream impoundment. In: Timotius, K.H. & Göltenboth, F. (Eds), *Tropical Limnology Volume III. Tropical Rivers, Wetlands and Special Topics.* Satya Wacana University Press, Salatiga, Indonesia: 39–49.

Usher, A.D., 1996. The race for power in Laos. The Nordic connections. In: Parnwell, M.J.G. & Bryant, R.L. (Eds), *Environmental Change in Southeast Asia. People, Politics and Sustainable Development.* Routledge, London: 123–144.

Vaas, K.F., Sachlan, M. & Wiraatmadja, G., 1953. On the ecology and fisheries of some inland waters along the rivers Ogan and Komering in South-east Sumatra. *Contr. Inland Fish. Res. Stn. Jakarta-Bogor* **3**: 1–32.

Vadhanaphuti, B., Klaikayai, T., Thanopanuwat, S. & Hungspreug, N., 1992. Water resources planning and management of Thailand's Chao Phraya River basin. *World Bank Technical Paper* **175**: 197–202.

van Benthem Jütting, W.S.S., 1953. Systematic studies on the non-marine Mollusca of the Indo-Australian Archipelago, IV. Critical Revision of the freshwater bivalves of Java. *Treubia* **22**: 19–73.

van Benthem Jütting, W.S.S., 1956. Systematic studies on the non-marine Mollusca of the Indo-Australian Archipelago, V. Critical Revision of the Javanese freshwater gastropods. *Treubia* **23**: 259–477.

van Bronckhorst, B., 1985. Working a new balance in the deteriorating upper Citarum basin. In: Lundqvist, J., Lohm, U. & Falkenmark, M. (Eds), *Strategies for River Basin Management.* D. Reidel Publishing Co., Dordrecht: 71–79.

van der Leeden, F., 1975. *Water Resources of the World.* Water Information Centre Inc., New York: 568 pp.

Van Slyke, L.P., 1988. *Yangtze: Nature, History and the River.* Addison-Wesley Publishing Company, Reading, Massachusetts: 211 pp.

van Tol, J., 1992. An annotated checklist to names of Odonata used in publications by M.A. Lieftinck. *Zoologische Verhandlungen* **279**: 1–263.

Van Vondel, B.J., 1992. Revision of the Palaearctic and Oriental species of *Peltodytes* Regimbart (Coleoptera: Haliplidae). *Tijdschrift voor Entomologie* **135**: 275–297.

Van Vondel, B.J., 1993. Revision of the *Liaphlus* species of the Oriental region excluding China (Coleoptera: Haliplidae). *Tijdschrift voor*

Entomologie **136**: 289–316.

Van Vondel, B.J., 1995. Revision of the Haliplidae (Coleoptera) of the Australian region and the Moluccas. *Rec. S. Aust. Mus. (Adelaide)* **28**: 61–101.

Van Zwieten, P.A.M., 1994. Biology of the cardinalfish *Glossamia gjellerupi* (Perciformes: Apogonidae) from the Sepik-Ramu River Basin, Papua New Guinea. *Environ. Biol. Fish.* **42**: 161–179.

Vannote, R.L., Minshall, G.W., Cummins, K.W., Sedell, J.R. & Cushing, C.E., 1980. The River Continuum Concept. *Can J. Fish. Aquat. Sci.* **37**: 130–137.

Vazirani, T.G., 1967. Contribution to the study of aquatic beetles (Coleoptera). 1. On a collection of Dytiscidae from the Western Ghats with descriptions of two new species. *Oriental Insects* **1**: 99–112.

Vazirani, T.G., 1977. Catalogue of Oriental Dytiscidae. *Rec. Zool. Surv. India, Occ. Pap.* **6**: 1–111.

Vazirani, T.G. 1984. *The Fauna of India. Coleoptera. Family Gyrinidae and Family Haliplidae.* Zoological Survey of India, Calcutta: 136 pp.

Venkataraman, K. & Sivaramakrishnan, K.G., 1987. A new species of *Thalerosphyrus* from South India (Ephemeroptera: Heptageniidae). *Current Science* **56**: 1126–1129.

Venkatesan, P. & Rhaghunatha-Rao, T.K., 1981. Description of a new species and a key to Indian species of Belostomatidae. *J. Bombay Nat. Hist. Soc.* **77**: 299–303.

Venkateswarlu, V. & Jayanti, T.V., 1968. Hydrobiological studies of the River Sabarmati to evaluate water quality. *Hydrobiologia* **31**: 442–448.

Venkateswarlu, V., Kumar, P.T.S. & Kumari, J.N., 1990. Ecology of algae in the River Moosi, Hyderbad, India: a comprehensive study. *J. Environ. Biol.* **11**: 79–92.

Venkateswarlu, V., Reddy, G.B. & Reddy, P.M., 1987. Nutrient status and algae in the rivers of Andhra Pradesh, India. *Indian J. Bot.* **10**: 37–48.

Venu, P., Kumar, V. & Bhasin, M.K., 1985. Water chemistry and production studies on Testa River and its two tributaries in Sikkim Himalayas, India. *Acta Bot. Indica* **13**: 158–164.

Verghese, S. & Furtado, J.I.,1987. Decomposition of leaf litter in a tropical freshwater swamp, the Tasek Bera, Malaysia. *Arch. Hydrobiol. Beih., Ergebn. Limnol.* **28**: 425–434.

Verma, S.R. & Dalela, R.C., 1975a. Studies on the pollution of the Kalinadi by industrial wastes near Mansurpur: Part 1. Hydrometric and physico-chemical characteristics of the wastes and river water. *Acta Hydrochim. Hydrobiol.* **3**: 239–257.

Verma, S.R. & Dalela, R.C., 1975b. Studies on the pollution of the Kalinadi by industrial wastes near Mansurpur: Part 2. Biological index of pollution and biological characteristics of the water. *Acta Hydrochim. Hydrobiol.* **3**: 259–274.

Verma, S.R., Sharma, A.K. & Goel, D.P., 1987. Diversity as a measure of water pollution and an aid for biological water analysis. *Acta Hydrochim. Hydrobiol.* **15**: 559–576.

Vidal, J., 1998. Woman power halts work on Indian dam. *Guardian Weekly* **January 18, 1998**: 4.

Vidrine, M.F., 1984. *Fulleratax*, new subgenus (Acari: Unionicolidae: Unionicolinae: *Unionicola*), in southeast Asia. *Int. J. Acarology* **10**: 229–233.

Viets, K., 1987. *Die Milbem des Süsswassers. Hydrachnellae und Halacaridae (Part), Teil 2: Katalog.* Paul Parey, Hamburg: 1012 pp.

Villavicencio, V., 1987. Philippines. In: Chia, L.S. (Ed.), *Environmental Management in Southeast Asia. Directions and Current Status.* Faculty of Science, National University of Singapore: 77–107.

Viner, A.B., 1982. A quantitative assessment of nutrient phosphate transport transported by particles in a tropical river. *Rev. Hydrobiol. trop.* **15**: 3–8.

Viner, A.B., 1987. Nutrients transported on silt in rivers. *Arch. Hydrobiol. Beih., Ergebn. Limnol.* **28**: 63–71.

Vines, G., 1993. Classic Chinese firsts (an interview with Joseph Needham). *Times Higher Education Supplement* **26 November 1993**: 18–19.

Vineyard, R.N. & Wiggins, G.B., 1988. Further revision of the caddisfly family Uenoidae (Trichoptera): evidence for inclusion of Neophylacinae and Thremmatidae. *Syst. Entomol.* **13**: 361–372.

Vyas, N., Rathore, S.S. & Nama, H.S., 1989. First record of the occurrence of freshwater planaria in Rajasthan. *Geobios New Reports* **8**: 20–22.

Wagner, R., 1982. Eine neue *Dixa* — Art aus der orientalischen Region Nepals (Insecta: Diptera: Nematocera: Dixidae). *Senckenbergiana Biologica* **63**: 181–183.

Wah, T.T., Wee, Y.C. & Phang, S.M., 1987. Freshwater diatoms of Ulu Endau, Johore, Malaysia. *Malay. Nat. J.* **41**: 159–172.

Wallace, J.B. & O'Hop, J., 1979. Fine particle suspension-feeding capabilities of *Isonychia* spp. (Ephemeroptera: Siphlonuridae). *Ann. Entomol. Soc. Amer.* **72**: 353–357.

Wallace, J.B. & Merritt, R.W., 1980. Filter-feeding ecology of aquatic insects. *Ann. Rev. Entomol.* **25**: 103–132.

Wallace, J.B., Sherberger, S.R. & Sherberger, F.F., 1976. The larval dwelling-tube, capture net and food of *Phylocentropus placidus* (Trichoptera: Polycentropodidae). *Ann. Entomol. Soc. Am.* **69**: 149–154.

Wallace, J.B., Webster, J.R. & Woodall, W.R., 1977. The role of filter-feeders in flowing waters. *Arch. Hydrobiol.* **79**: 506–532.

Walsh, C.J., 1993. Larval development of *Paratya australiensis* Kemp, 1917 (Decapoda, Caridea, Atyidae), reared in the laboratory, with comparisons of fecundity and egg and larval size between estuarine and riverine environments. *J. Crustacean Biol.* **13**: 456–480.

Walter, C., 1930. Hydracarinen der Inseln Luzon, Philippen. *Philipp. J. Sci.* **41**: 159–167.

Walters, C.J. & Holling, C.S., 1990. Large-scale management experiments and learning by doing. *Ecology* **71**: 2060-2068.

Waltz, R.d., 1996. *Acentrella feminalis*, new combination for an Oreintal Baetis (Ephemeroptera: Baetidae). *Aquatic Insects* **18**: 111–112.

Waltz, R.D. & McCafferty, W.P., 1984. Illustration and lectotype designation for *Baetis feminalis* Eaton (Ephemeroptera: Baetidae). *Oriental Insects* **18**: 335–337.

Waltz, R.D. & McCafferty, W.P., 1985a. Redescription and new lectotype designation for the type species of *Pseudocloeon, P. kraepelini* Klapálek (Ephemeroptera: Baetidae). *Proc. Entomol. Soc. Wash.* **87**: 800–804.

Waltz, R.D. & McCafferty, W.P., 1985b. A new species of *Procloeon* from Taiwan. *Oriental Insects* **19**: 121–123.

Waltz, R.D. & McCafferty, W.P., 1987a. Systematics of *Pseudocloeon, Acentrella, Baetiella*, and *Liebebiella*, new genus (Ephemeroptera: Baetidae). *J. New York Entomol. Soc.* **95**: 553–568.

Waltz, R.D. & McCafferty, W.P., 1987b. New genera of Baetidae for some Nearctic species previously included in *Baetis* Leach (Ephemeroptera) *Ann. Entomol. Soc. Amer.* **80**: 667–670.

Waltz, R.D. & McCafferty, W.P., 1987c. Revision of the genus *Cloeodes* Traver (Ephemeroptera: Baetidae). *Ann. Entomol. Soc. Amer.* **80**: 191–207.

Waltz, R.D. & McCafferty, W.P., 1987d. A generic revision of *Cloeodes* Traver and a description of two new genera (Ephemeroptera: Baetidae). *Proc. Entomol. Soc. Washington* **89**: 177–184.

Waltz, R.D. & McCafferty, W.P., 1987e. New genera of Baetidae (Ephemeroptera) from Africa. *Proc. Entomol. Soc. Washington* **89**: 95–99.

Waltz, R.D. & McCafferty, W.P., 1989. New species, redescriptions, and cladistics of the genus *Pseudocentroptiloides* (Ephemeroptera: Baetidae). *J. New York Entomol. Soc.* **98**: 151–158.

Waltz, R.D. & McCafferty, W.P., 1994. *Cloeodes* (Ephemeroptera: Baetidae) in Africa. *Aquatic Insects* **16**: 165–169.

Waltz, R.D., McCafferty, W.P. & Thomas, A., 1994. Systematics of *Alainites* n. gen., *Diphetor, Indobaetis, Nigrobaetis* n. stat., and *Takobia* n. stat. (Ephemeroptera, Baetidae). *Bull. Soc. Hist. Nat., Toulouse* **130**: 33–36.

Wang, C., 1993. Comprehensive assessment of the ecological and environmental impact of the Three Gorges Project. In: Luk, S. & Whitney, J. (Eds), *Megaproject. A Case Study of China's Three Gorges Project.* M.E. Sharpe Inc., New York: 71–109.

Wang, H., 1987. The water resources of lakes in China. *Chinese Journal of Oceanology & Limnology* **5**, 263–280 (in Chinese).

Wang, H., 1996. Status and conservation of Reeves' shad resources in China. *Naga, the ICLARM Quarterly* **19**: 20–22.

Wang, J., 1989. Water pollution and water shortage problems in China. *J. Appl. Ecol.* **26**: 851–857.

Wang, J., 1993. How much investment is required by the Three Gorges Project?. In: Luk, S. & Whitney, J. (Eds), *Megaproject. A Case Study of China's Three Gorges Project*. M.E. Sharpe Inc., New York: 196–206.

Wang, J.T., Liu, M.C. & Fang, L.S., 1995. The reproductive biology of an endemic cyprinid, *Zacco pachycephalus*, in Taiwan. *Environmental Biology of Fishes* **43**: 135–143.

Wang, L. & Zhou, W., 1989a. The growth characteristics of juvenile of woolly handed crab (*Eriocheir sinensis*) cultivated in ponds. *Journal of Fisheries of China* **13**: 17–23 (in Chinese).

Wang, L. & Zhou, W., 1989b. Effect of water temperature on the growth features of young crabs. *Journal of Ecology (China)* **8**: 22–25 (in Chinese).

Wang, S., 1987. Chironomidae research in China. *Ent. Scand., Suppl.* **29**: 35–38.

Wang, S., 1990. *Pseudosmitta aizaiensis*, new species Orthocladiinae (Diptera, Simuliidae) from Hunan Province of China. *Chin. J. Oceanol. Limnol.* **8**: 272–279.

Wang, S.T., 1963. On some Trichoptera from Lake Tunghu, Wuchang. *Acta Hydrobiologica Sinica* **12** :55–66 (in Chinese).

Wang, T.Q. & McCafferty, W.P., 1995. First larval descriptions, new species, and evaluation of the Southeast Asian genus *Atopopus* (Ephemeroptera, Heptageniidae). *Bull. Soc. Hist. Nat., Toulouse* **131**: 19–25.

Wang, T.Q., McCafferty, W.P. & Edmunds, G.F., 1995a. Larva and adult of *Teloganella* (Ephemeroptera: Pannota) and assessment of familial classification. *Ann. Entomol. Soc. Amer.* **88**: 324–327.

Wang, T.Q. & McCafferty, W.P., 1995b. Specific assignments in *Ephemerellina* and *Vietnamella* (Ephemeroptera: Ephemerellidae). *Entomological News* **106**: 1193–1194.

Wang, T.Q. & McCafferty, W.P., 1996. New genus of Teloganodinae (Ephemeroptera: Pannota: Ephemerellidae) from Sri Lanka. *Bull. Soc. Hist. Nat. Toulouse* **132**: 15–18.

Wang, X., 1995. *Rheocricotopus (R.) orientalis*, a new species from China (Diptera: Chironomidae). *Aquatic Insects* **17**: 37–40.

Wang, X. & Saether, O.A., 1993a. A new species of the 'marine' genus *Thalassosmittia* Strenzke & Remmert from Xizang (Tibet), China (Diptera: Chironomidae). *Ent. Scand.* **24**: 211–214.

Wang, X. & Saether, O.A., 1993b. First Palaearctic and Oriental Records of the orthoclad genus *Antillocladius* Saether (Diptera: Chironomidae). *Ent. Scand.* **24**: 227–230.

Wang, X. & Saether, O.A., 1993c. *Limnophyes* Eaton from China, with description of five new species (Diptera: Chironomidae). *Ent. Scand.* **24**: 215–226.

Wang, X. & Zheng, L., 1990a. Two new species of the genus *Rheocricotopus* from China (Diptera: Chironomidae). *Entomotaxonomia* **11**: 311–313 (in Chinese).

Wang, X. & Zheng, L., 1990b. A new species of the genus *Cladotanytarsus* from China (Diptera: Chironomidae). *Acta Zootaxonomica Sinica* **15**: 480–482 (in Chinese).

Wang, X. & Zheng, L., 1990c. Two new species of *Mesosmittia* from China (Diptera: Chironomidae). *Acta Entomologica Sinica* **33**: 486–489 (in Chinese).

Wang, X. & Zheng, L., 1990d. A new species of the genus *Neozavrelia* from China (Diptera: Chironomidae). *Acta Zootaxonomica Sinica* **15**: 355–357 (in Chinese).

Wang, X. & Zheng, L., 1990e. Notes on genus *Paratrichocladius* from China (Diptera: Chironomidae). *Acta Entomologica Sinica* **33**: 243–246 (in Chinese).

Wang, X. & Zheng, L., 1991. Notes on the genus *Rheocricotopus* from China (Diptera: Chironomidae). *Acta Zootaxonomica Sinica* **16**: 99–105 (in Chinese).

Wang, X. & Zheng, L., 1993. Checklist of Chironomidae records from China. *Neth. J. Aquat. Ecol.* **26**: 247–255.

Wang, W. & Chen, J., 1987. Carbon transport and the relationship between carbon and other elements in the Changjiang River. *Mitt. Geol.-Palaont. Inst. Univ. Hamburg, SCOPE/UNEP Sonderbd.* **64**: 217–231.

Ward, J.V. & Stanford, J.A., 1979. The serial discontinuity concept of lotic ecosystems. In: Fontaine, T.D. & Bartell, S.M. (Eds). *Dynamics of Lotic Ecosystems.* Ann Arbour Science Publishers, Michigan: 29–42.

Warwick, R.M., 1993. Environmental impact studies on marine communities: pragmatical considerations. *Aust. J. Ecol.* **18**: 63–80.

Watanabe, N.C., 1985. Distribution of *Ephemera* nymphs in Kagawa Prefecture, Japan, in relation to altitude and gradient. *Kagawa Seibutsu* 13: 1–7 (in Japanese).

Watanabe, N.C., 1988. Life history of *Potamanthodes kamonis* in a stream of central Japan (Ephemeroptera: Potamanthidae). *Verh. Internat. Verein. Limnol.* **23**: 2118–2125.

Watanabe, n.C., 1989. Seasonal and diurnal changes in emergence of *Potamanthodes kamonis* in a stream of central Japan (Ephemeroptera: Potamanthidae). *Jap. J. Limnol.* **50**: 157–161.

Watanabe, N.C. & Kuroda, T., 1985. Change in growth of a mayfly nymph, *Ephemera japonica*, along the stream length and thermal effect on it. *Mem. Fac. Educ. Kagawa Univ., Part II* **35**: 47–54.

Watson, D.J., 1982. Subsistence fish exploitation and implications for management in the Baram River system, Sarawak, Malaysia. *Fish. Res.* **1**: 299–310.

Watson, D.J. & Balon, E.K., 1984a. Structure and production of fish communities in tropical rain forest streams of northern Borneo. *Can. J. Zool.* **62**: 927–940.

Watson, D.J. & Balon, E.K., 1984b. Ecomorphological analysis of fish taxocenes in rain-forest streams of northern Borneo. *J. Fish. Biol.* **25**: 371–384.

Watson, J.A.L. & Dyce, A.L., 1978. The larval habit of *Podopteryx selysi* (Odonata: Megapodagrionidae). *J. Aust. Entomol. Soc.* **17**: 361–3 62.

Watson, J.A.L., Theischinger, G. & Abbey, H.M., 1991. *The Australian Dragonflies*. CSIRO, Canberra and Melbourne: 278 pp.

Weaver, J.S. III, 1985. The Oriental Lepidostomatidae (Trichoptera) described by Banks and Hagen. *Psyche* **92**: 237–254.

Weaver, J.S. III, 1987. New species of *Stenopsyche* from the northeastern Orient (Trichoptera: Stenopsychidae). *Aquatic Insects* **9**: 161–168.

Weaver, J.S. III, 1989. Indonesian Lepidostomatidae (Trichoptera) collected by Dr E.W. Diehl. *Aquatic Insects* **11**: 47–63.

Weaver, J.S. III, 1992. Theliopsychinae, a new subfamily, and *Zephropsyche*, a new genus of Lepidostomatidae (Trichoptera). In: Otto, C. (Ed.), *Proceedings of the 7th International Symposium on Trichoptera*. Umea, Sweden: 133–138.

Weaver, J.S. III & Huisman, J., 1992a. A review of the Lepidostomatidae (Trichoptera) of Borneo. *Zool. Med. Leiden* **66**: 529–560.

Weaver, J.S. III & Huisman, J., 1992b. New species and descriptions of Lepidostomatidae (Trichoptera) from Sulawesi. *Zool. Med. Leiden* **66**: 429–439.

Weaver, J.S. III & Malicky, H., 1994. The genis *Dipseudopsis* Walker from Asia (Trichoptera: Dipseudopsidae). *Tijdschrift voor Entomologie* **137**: 95–142.

Weaver, J.S. III & Morse, J.C., 1986. Evolution of feeding and case-making behaviour in Trichoptera. *J. N. Amer. Benthol. Soc.* **5**: 150–158.

Webb, D.W., 1994. The immature stages of *Suragina concinna* (Williston) (Diptera: Atheridicae). *J. Kansas Entomol. Soc.* **67**: 421–425.

Webster, J.R. & Benfield, E.F., 1986. Vascular plant breakdown in freshwater ecosystems. *Ann. Rev. Ecol. Syst.* **17**: 567–594.

Welch, D.N. & Lim, T.K., 1987. Water resources development and management in Malaysia. In: Ali, M., Radosevich, G.E. & Ali Khan, A. (Eds), *Water Resources Policy for Asia*. A.A. Balkema Publishers, Boston: 71–81.

Welcomme, R.L., 1979. *Fisheries Ecology of Floodplain Rivers*. Longman, London: 317 pp.

Welcomme, R.L., 1988. *International Introductions of Inland Aquatic Species*. FAO Fisheries Technical Paper 294, FAO, Rome. 318 pp.

Welcomme, R.L., 1987. Riverine and swamp fisheries. *Arch. Hydrobiol. Beih., Ergebn. Limnol.* **28**: 237–238.

Wells, A., 1985. Larvae and pupae of Australian Hydrotilidae (Trichoptera), with observations on general biology and relationships. *Aust. J. Zool., Suppl. Ser.* **113**: 1–69.

Wells, A., 1990a. The hydrotilid tribe Stactobiini in New Guinea (Trichoptera: Hydroptilidae: Hydrotilinae). *Invertebr. Taxon.* **3**: 817–849.

Wells, A., 1990b. The micro-caddisflies (Trichoptera: Hydrotilidae) of North

Sulawesi. *Invertebr. Taxon.* **3**: 363–400.

Wells, A., 1991. The hydroptilid tribes Hydroptilini and Orthotrichini in New Guinea (Trichoptera: Hydroptilidae: Hydrotilinae). *Invertebr. Taxon.* **5**: 487–526.

Wells, A., 1992. The first parasitic Trichoptera. *Ecological Entomology* **17**: 299–302.

Wells, A., 1993. Micro-caddisflies (Trichoptera: Hydrotilidae) from Bali, Indonesia. *Zool. Med., Leiden* **67**: 351–359.

Wells, A. & Cartwright, D., 1993. Females and immatures of the Australian caddisfly *Hyalopsyche disjuncta* Neboiss (Trichoptera), and a new family placement. *Trans. R. Soc. South Australia* **117**: 97–104.

Wells, A. & Dudgeon, D., 1990. Hydroptilidae (Insecta) from Hong Kong. *Aquatic Insects* **12**: 161–175.

Wells, A. & Malicky, H., 1997. the micro-caddiflies of Sumatra and Java (Trichoptera: Hydroptilidae). *Linzer biol. Beitr.* **29**: 173–202.

Wells, A. & Huisman, J., 1992. Micro-caddisflies in the tribe Hydrotilini (Trichoptera: Hydroptilidae: Hydroptilinae) from Malaysia and Brunei. *Zool. Med., Leiden* **66**: 91–126.

Wells, A. & Huisman, J., 1992. Malaysian and Bruneian micro-caddisflies in the tribes Stactobiini and Orthotrichiini (Trichoptera: Hydroptilidae: Hydroptilinae). *Zool. Med., Leiden* **67**: 91–125.

Weninger, G., 1972. Results of the Austrian-Ceylonese hydrobiological mission, 1970. Pt II. Hydrochemical studies on mountain rivers in Ceylon. *Bull. Fish. Res. Stn, Sri Lanka (Ceylon)* **23**: 77–100.

Wesenburg-Lund, E., 1958. Lesser Antillean polychaetes, chiefly from brackish water, with a survey and a bibliography of fresh and brackish-water polychaetes. *Studies on the Fauna of Curaçao and other Caribbean Islands* **8**: 1–41.

West, G.J. & Glucksman, J., 1976. Introductions and distribution of exotic fish in Papua New Guinea. *Papua New Guinea Agric. J.* **27**: 19–48.

Wezeman, B., 1978. *Rhodeus ocellatus* from Taiwan. *Aquarium (Den Haag)* **48**: 290–296 (in Dutch).

Wewalka, G., 1992. Die *Canthydrus flavus* (Motschulsky)-Gruppe aus Sudostasien (Coleoptera: Noteridae). *Linzer Biologische Beitrage* **24**: 803–811.

Wewalka, G., 1997. Taxonomic revision of *Microdytes* BALFOUR-BROWN (Coleoptera: Dytiscidae). *Koleopterologische Rundschau* **67**: 13–51.

Wewalka, G. & Biström, O., 1994. *Hyphovatus* gen. n. from Southeast Asia, with description of two new species (Coleoptera: Dytiscidae). *Koleopterologische Rundschau* **64**: 37–43.

Wewalka, G. & Brancucci, M., 1995. Dytiscidae: notes on Chinese *Platambus* THOMSON, with description of two new species (Coleoptera). In: Jäch, M.A. & Ji, L. (Eds), *Water Beetles of China*. Zoologisch-Botanische Gesellschaft & Wiener Coleopterologenverein, Vienna: 97–102.

Whitaker, R. & Whitaker, Z., 1978. A preliminary survey of the Salt water crocodile (*Crocodylus porosus*) in the Andaman Islands, India. *J. Bombay Nat. Hist. Soc.* **75**: 43–49.

Whitaker, R. & Whitaker, Z., 1984. Reproductive biology of the Mugger crocodile, *Crocodylus palustris. J. Bombay Nat. Hist. Soc.* **81**: 297–317.

White, D.S., 1989. Defense mechanisms in riffle beetles (Coleoptera: Dryopoidea). *Ann. Entomol. Soc. Am.* **82**: 237–241.

White, D.S. & Brigham, W.U., 1996. Aquatic Coleoptera. In: Merritt, R.W. & Cummins, K.W. (Eds), *An Introduction to the Aquatic Insects of North America. Third Edition.* Kendall/Hunt Publishing Co., Iowa: 399–473.

Whitley, G.P., 1938. Descriptions of some New Guinea fishes. *Rec. Aust. Mus.* **20**: 221–233.

Whitten, A.J., Damanik, S.J., Anwar, J. & Hisyam, N., 1984. *The Ecology of Sumatra.* Gadjah Mada University Press, Yogyakarta: 583 pp.

Whitten, A.J., Mustafa, M. & Henderson, G.S., 1987. *The Ecology of Sulawesi.* Gadjah Mada University Press, Yogyakarta: 777 pp.

Whitton, B.A. & Catling, H.D., 1986. Algal ecology of deepwater rice-fields in Thailand. *Arch. Hydrobiol.* **105**: 289–297.

Whitton, B.A. & Rother, J.A., 1988. Diel changes in the environment of a deepwater rice-field in Bangaldesh. *Verh. Internat. Verein. Limnol.* **23**: 1074–1079.

Whitton, B.A., Rother, J.A. & Paul, A.R., 1988a. Ecology of deepwater rice-fields in Bangladesh 2. Chemistry of sites at Manikganj and Sonargaon. *Hydrobiologia* **169**: 23–30.

Whitton, B.A., Aziz, A., Kawecka, B. & Rother, J.A., 1988b. Ecology of deepwater rice-fields in Bangladesh 3. Associated algae and macrophytes. *Hydrobiologia* **169**: 31–42.

Whitton, B.A., Aziz, A., Francis, P., Rother, J.A., Simon, J.W. & Tahmida, Z.N., 1988c. Ecology of deepwater rice-fields in Bangladesh 1. Physical and chemical environment. *Hydrobiologia* **169**: 3–22.

Wiberg-Larsen, P., 1993. Notes on the feeding biology of *Ecnomus tenellus* (Rambur, 1842) (Trichoptera, Ecnomidae). *Entomologiske Meddelelser* **61**: 29–38.

Wiberg-Larsen, P., Iversen, T.M. & Thorup, J., 1991. First Danish *Ptilocolepus granulatus*, new record (Pictet) (Trichoptera, Hydroptilidae). *Entomologiske Meddelelser* **59**: 45–50.

Wiebach, F., 1974. Indische Susswässer-Bryozoen. *Gewässer und Abwässer* **53/54**: 69–84.

Wiederholm, T. (Ed.), 1983. Chironomidae of the Holarctic Region. Keys and diagnoses. Part 1. Larvae. *Ent. Scand., Suppl.* **19**: 1–457.

Wiederholm, T. (Ed.), 1986. Chironomidae of the Holarctic Region. Keys and diagnoses. Part 1. Pupae. *Ent. Scand., Suppl.* **28**: 1–482.

Wiggins, G.B., 1959. A new family of Trichoptera from Asia. *Can. Ent.* **91**: 745–757.

Wiggins, G.B., 1968. Contributions to the systematics of the caddisfly family

Molannidae in Asia (Trichoptera). *Royal Ontario Museum Life Sciences Contribution* **72**: 1–26.

Wiggins, G.B., 1969. Contributions to the biology of the Asian caddisfly famly Limnocentropodidae. *Royal Ontario Museum Life Sciences Contribution* **74**: 1–29.

Wiggins, G.B., 1996. *Larvae of the North American Caddisfly Genera (Trichoptera). 2nd Edition.* University of Toronto Press, Toronto: 457 pp.

Wiggins, G.B., Tani, K. & Tanida, K., 1985. *Eobrachycentrus*, a genus new to Japan, with a review of the Japanese Brachycentridae (Trichoptera). *Kontyû* **53**: 59–74.

Wiggins, G.B., Weaver, J.S. III & Unicker, J.D., 1985. Revision of the caddsifly family Uenoidae. *Can. Ent.* **117**: 763–800.

Wikramanayake, E.D., 1990. Ecomorphology and biogeography of a tropical stream fish assemblage: evolution of assemblage structure. *Ecology* **71**: 1756–1764.

Wikramanayake, E.D. & Moyle, P.B., 1989. Ecological structure of tropical fish assemblages in wet-zone streams of Sri Lanka. *J. Zool., Lond.* **218**: 503–526.

Wilcox, R.S., 1995. Ripple communication in aquatic and semiaquatic insects. *Ecoscience* **2**: 109–115.

Wiles, R., 1991. Rheophilic watermites (Acari: Hydrachnidia) from mainland Malaysia. *Acarologia (Paris)* **32**: 41–56.

Wilkinson, S.M., Rundle, S.D., Brewin, P.A. & Ormerod, S.J., 1995. A study of the whirligig beetle *Dineutus indicus* (Aube) (Gyrinidae) in a Nepalese hillstream. *Entomologist* **114**: 131–137.

Will, P.E. 1985. State intervention in the administration of a hydraulic infrastructure: the example of Hubei Province in Late Imperial times. In: Schram, S.R. (Ed.), *The Scope of State Power in China.* School of Oriental and African Studies, London: 295–347.

Williams, D.S., 1979. The feeding behaviour of New Zealand *Dolomedes* species (Araneae: Pisauridae). *N.Z. J. Zool.* **6**: 95–105.

Williams, W.D, 1980. *Australian Freshwater Life (2nd Edition).* MacMillan Company, Melbourne: 321 pp.

Williams, W.D., 1987. Salinization of rivers and streams: an important environmental hazard. *Ambio* **16**: 180–185.

Williams, W.D., 1988. Limnological imbalances: an antipodean viewpoint. *Freshwat. Biol.* **20**: 407–420.

Williams, W.D., 1994. Constraints to the conservation and management of tropical inland waters. *Mitt. Internat. Verein. Limnol.* **24**: 357–363.

Wilson, K.D.P., 1995a. *Hong Kong Dragonflies.* The Urban Council of Hong Kong, Hong Kong: 211 pp.

Wilson, K.D.P., 1995b. The gomphid dragonflies of Hong Kong, with descriptions of two new species. *Odonatologica* **24**: 319–340.

Winterbourn, M.J., Rounick, J.S. & Cowie, B., 1981. Are New Zealand stream ecosystems really different? *N.Z. J. Mar. Freshwat. Res.* **15**: 321–328.

Wirawan, N., 1986. Protecting the *Pesut* (freshwater dolphin) in the Mahakam River of Kalimantan, Borneo. *Wallaceana* **44**: 3–6.

Wirth, W.W. & Grogan, W.L., 1988. *The Predaceous Midges of the World (Diptera: Ceratopogonidae; Tribe Ceratopogonini).* Flora & Fauna Handbook No. 4, E.J. Brill, Leiden: 160 pp.

Wirth, W.W. & Hubert, A.A., 1989. The *Culicoides* of Southeast Asia (Diptera: Ceratopogonidae). *Mem. Am. Entomol. Inst.* **44**: 1–510.

Wirth, W.W. & Ratanaworabhan, N.C., 1993. The Oriental biting midges of the genus *Atrichopogon* related to *Dolichohelea polita* Edwards (Diptera: Ceratopogonidae). *Oriental Insects* **27**: 317–334.

Wirth, W.W., Ratanaworabhan, N.C. & Blanton, F.S. 1974. Synopsis of the genera of Ceratopogonidae (Diptera). *Annls Parasit. hum. comp.* **49**: 595–613.

Wong, J.T.Y., 1987. Salinity tolerance of *Macrobrachium nipponense* (De Haan) from Hong Kong. *Aquacult. Fish. Manag.* **18**: 203–207.

Wong, J.T.Y., 1989. Abbreviated larval development of *Macrobrachium hainanense* (Parisi, 1919) reared in the laboratory (Decapoda, Caridea, Palaemonidae). *Crustaceana* **56**: 18–30.

Wong, K.M., 1987. The bamboos of the Ulu Endau Area, Johore, Malaysia. *Malay. Nat. J.* **41**: 249–256.

Wongratana, T., 1988. A case of total albinism in *Synaptura panoides* Bleeker (Pisces, Teleostei, Soleidae) from the Chao Phraya River, Thailand. *Nat. Hist. Bull. Siam Soc.* **36**: 55–60.

Wood, J.R., 1993. India's Narmada River dams: Sardar Sarovar under siege. *Asian Survey* **33**: 968–984.

Wood, T., 1991. Bryozoans. In: Thorp, J.H. & Covich, A.P. (Eds), *Ecology and Classification of North American Freshwater Invertebrates.* Academic Press, San Diego: 481–499.

World Resources Institute, 1994. *World Resources 1994–95.* Oxford University Press, New York: 400 pp.

Wu, C.F., 1931. Aquatic insects of China. Article I. Catalogue of Chinese Gyrinidae. *Peking Nat. Hist. Bull.* **6**: 63–73.

Wu, C.F., 1932a. Aquatic insects of China. Article XI. Catalogue of Chinese Haliplidae. *Peking Nat. Hist. Bull.* **7**: 339–343.

Wu, C.F., 1932b. Aquatic insects of China. Article XIII. Catalogue of Chinese Heteroceridae. *Peking Nat. Hist. Bull.* **7**: 339–343.

Wu, C.F., 1932c. Aquatic insects of China. Article XIV. Catalogue of Chinese Heteroceridae. *Peking Nat. Hist. Bull.* **7**: 351–352.

Wu, C.F., 1935. *Catalogus Insectorum Sinensium. Volume I.* The Fan Memorial Institute of Biology, Beijing: 378 pp.

Wu, C.F., 1936a. The stoneflies of China (order Plecoptera). *Peking Nat. Hist. Bull.* **11**: 49–82.

Wu, C.F., 1936b. The stoneflies of China (order Plecoptera). *Peking Nat.*

Hist. Bull. **11**: 163–179.

Wu, C.F., 1936c. The stoneflies of China (order Plecoptera). *Peking Nat. Hist. Bull.* **11**: 297–307.

Wu, C.F., 1936d. The stoneflies of China (order Plecoptera). *Peking Nat. Hist. Bull.* **11**: 441–443.

Wu, C.F., 1937a. The stoneflies of China (order Plecoptera). *Peking Nat. Hist. Bull.* **12**: 57–70.

Wu, C.F., 1937b. The stoneflies of China (order Plecoptera). *Peking Nat. Hist. Bull.* **12**: 127–166.

Wu, C.F., 1937c. The stoneflies of China (order Plecoptera). *Peking Nat. Hist. Bull.* **12**: 225–252.

Wu, C.F., 1937d. The stoneflies of China (order Plecoptera). *Peking Nat. Hist. Bull.* **12**: 319–351.

Wu, C.F., 1938a. The stoneflies of China (order Plecoptera). Appendix A. Additions. Appendix B. Corrections. Appendix C. Systematic index. *Peking Nat. Hist. Bull.* **13**: 53–87.

Wu, C.F., 1938b. *Catalogus Insectorum Sinensium. Volume IV*. The Fan Memorial Institute of Biology, Beijing: 1007 pp.

Wu, C.F., 1939. First supplement to the stoneflies of China (order Plecoptera). *Peking Nat. Hist. Bull.* **14**: 153–157.

Wu, C.F., 1940. Second supplement to the stoneflies of China (order Plecoptera). *Peking Nat. Hist. Bull.* **14**: 331–333.

Wu, C.F., 1947. Third supplement to the stoneflies of China (order Plecoptera). *Peking Nat. Hist. Bull.* **16**: 265–273.

Wu, C.F., 1948a. Fourth supplement to the stoneflies of China (order Plecoptera). *Peking Nat. Hist. Bull.* **17**: 75–83.

Wu, C.F., 1948b. Fifth supplement to the stoneflies of China (order Plecoptera). *Peking Nat. Hist. Bull.* **17**: 145–151.

Wu, C.F., 1949. Sixth supplement to the stoneflies of China (order Plecoptera). *Peking Nat. Hist. Bull.* **17**: 251–258.

Wu, C.F., 1962. Results of the Zoologico-botanical Expedition to Southwest China, 1955–1957 (Plecoptera). *Acta Entomological Sinica, Suppl.* **11**: 139–160.

Wu, C.F., 1973. New species of Chinese stoneflies (order Plecoptera). *Acta Entomologica Sinica* **16**: 97–118.

Wu, C.F. & Claassen, P.W, 1934. Aquatic insects of China. Article XVIII. New species of Chinese stoneflies (order Plecoptera). *Peking Nat. Hist. Bull.* **9**: 111–129.

Wu, S.K., 1981. The leeches (Annelida: Hirudinea) of Taiwan. Part 2. *Hirudinaria manillensis* (Lesson). *Quart. J. Taiwan Mus.* **34**: 207–211.

Wu, S.K., 1979. The leeches (Annelida: Hirudinea) of Taiwan. Part 1. Introduction and description of two hirudinid species. *Quart. J. Taiwan Mus.* **32**: 193–207.

Wu, T., Chen, C.F., Cong, N. & You, D.S., 1985. Three species of nymphs of the genus *Cinygmina* from Yi Xing. *Journal of the Nanjing Teachers' College (Natural Science)* **8**: 65–70 (in Chinese).

Wu, T. & Chen, Q., 1986. Densities and distribution of zoobenthos in the lower reaches (Nanjing — Jiangyin section) of the Chanjiang (the Yangtze) River. *Acta Hydrobiol. Sinica* **10**: 73–85 (in Chinese).

Wu, T. & You D.S., 1986. A new species of the genus *Cinygmina* from China (Ephemeroptera: Ecdyoneuridae). *Acta Zootaxonomica Sinica* **11**: 280–283 (in Chinese).

Wu, T. & You, D.S., 1989. Two new species of the genus *Choroterpes* from China (Ephemeroptera: Leptophlebiidae). *Acta Zootaxonomica Sinica* **14**: 91–95 (in Chinese).

Wu, T. & You, D.S., 1992. A new species of the genus *Choroterpes* from Anhui Province, China (Ephemeroptera: Leptophlebiidae). *Acta Zootaxonomica Sinica* **17**: 64–66 (in Chinese).

Wu, T., Ying, X.D. & Chen, H.D., 1987. Two species of nymphs of *Choroterpes* from Yi Xing. *Journal of Nanjing Normal University (Natural Science)* **4**: 81–84 (in Chinese).

Wu, W.L., 1982. Phylogenetic studies of Taiwan freshwater mussels (Bivalvia: Unionidae). *Bulletin of the Institute of Zoology Academia Sinica (Taipei)* **21**: 145–153.

Wu, X.Y., 1987. A new species of *Potamanthus* (Ephemeroptera: Potamanthidae). *Acta Zootaxonomica Sinica* **12**: 421–423 (in Chinese).

Wu, X.Y. & You, D.S., 1986b. A new genus and species of Potamanthidae from China (Ephemeroptera). *Acta Zootaxonomica Sinica* **11**: 401–405 (in Chinese).

Xie, Z., 1993. The impacts of the Three-Gorge project on the endangered and endemic plant *Adiantum reniforme* var. *sinense* and the conservation strategies. *Chinese Biodiversity* **1**: 16–18.

Xiong, Y., Shang, J.C., Cheng, T.W., Zhang, J.Z., Zhao, C.N. & Li, Y.Y., 1989. *The Rivers of China*. Geography of China Series, Ren Min Jiao Yu Publishers, Beijing: 306 pp. (in Chinese).

Xu, J.Z., You, D.S., Su, C.R. & Hsu, Y.C., 1980. Two new species of *Ephemerella* (Ephemeroptera: Ephemerellidae). *Journal of the Nanjing Teachers' College (Natural Science)* **3**: 60–63 (in Chinese).

Xu, J.Z., You, D.S. & Hsu, Y.C., 1984. A new species of *Ephemerella* (Ephemeroptera: Ephemerellidae). *Acta Zootaxonomica Sinica* **9**: 413–415 (in Chinese).

Xu, Y. & Hong, J., 1983. Impact of water transfer on the natural environment. In: Biswas, A.K., Zuo, D., Nickum, J.E. & Liu, C. (Eds.), *Long-distance Water Transfer: a Chinese Case Study and International Experiences*. Tycooly International Publishers Limited, Dublin: 159–167.

Yadav, U.R. & Mishra, P.N., 1982. Studies on the freshwater leeches of Kathmandu Valley, Nepal. *Journal of the Natural History Museum of Tribhuvan University* **6**: 119–123.

Yan, J. & Ye, C., 1977. Notes on the larvae of some chironomid midges

(Diptera: Tendipedidae) and two nes species from Bai-Yang-Dian Lake in Hopei Province. *Acta Entomologica Sinica* **20**: 183–198.

Yang, C. & Yang, D., 1986. New fishflies from Guangxi, China. *Entomotaxonomia* **8**: 85–95 (in Chinese).

Yang, C. & Yang, D., 1991. New species and a new record of the fishflies from China (Megaloptera: Corydalidae). *Acta Entomologica Sinica* **34**: 74–75 (in Chinese).

Yang, C.M., Ng., P.K.L. & Lua, H.K., 1990. Record of a hairworm (phylum Nematomorpha) from a tettigoniid grasshopper in Ulu Kinchin, Pahang, Malaysia. *Malay. Nat. J.* **43**: 268–270.

Yang, D., 1995. The Chinese *Odontomyia* (Diptera: Stratiomyidae). *Entomotaxonomia, Suppl.* **17**: 58–72.

Yang, D., 1996a. New species of *Hercostomus* and *Ludovicius* from North China (Diptera: Dolichopodidae). *Dtsch. ent. Z.* **43**: 235–244.

Yang, D., 1996b. The genus *Dolichopus* from Southwest China (Diptera, Dolichopododae). *Bulletin de l'Institut Royal des Sciences Naturelles de Belgique, Entomologie* **66**: 79–83.

Yang, D., 1996c. Six new species of Dolichopodinae from China (Diptera, Dolichopodidae). *Bulletin de l'Institut Royal des Sciences Naturelles de Belgique, Entomologie* **66**: 85–89.

Yang, D. & Nagatomi, A., 1991. A study on the Chinese *Suragina* (Diptera, Athericidae). *Jap. J. Entomol.* **59**: 755–762.

Yang, D. & Nagatomi, A., 1992. Asuragina, a new genus of Athericidae (Insecta: Diptera). *Proc. Jap. Soc. Syst. Zool.* **48**: 54–62.

Yang, D. & Yang, C., 1990. Eight new species of the genus *Chelipoda* from China (Diptera: Empididae). *Acta Zootaxonomica Sinica* **15**: 483–488 (in Chinese).

Yang, D. & Yang, C., 1991. Five new species of the genus *Hemerodromia* from China (Diptera: Empididae). *Acta Entomologica Sinica* **34**: 234–237 (in Chinese).

Yang, D. & Yang, C., 1992. Megaloptera: Corydalidae. In: Peng, J. & Liu, Y. (Eds), *Iconography of Forest Insects in Hunan China*. Academia Sinica & Hunan Forestry Institute, Hunan: 640–643 (in Chinese).

Yang, D. & Yang, C., 1993. The fishflies (Megaloptera: Corydalidae) from Maolan, Guizhou. *Entomotaxonomia* **15**: 246–248 (in Chinese).

Yang, D. & Yang, J., 1992. Plecoptera: Perlidae. In: Huang, F. (Ed.), *Insects of Wuling Mountains Area, Southwestern China*. Science Press, Beijing: 62–65 (in Chinese).

Yang, D. & Yang, J., 1993. New and little known species of Plecoptera from Guizhou Province. *Entomotaxonomia* **15**: 235–238 (in Chinese).

Yang, D. & Yang, J., 1995a. Three new species of Plecoptera from Hainan Province. *Acta Agriculturae Universitatis Pekinensis* **21**: 223–226 (in Chinese).

Yang, D. & Yang, J., 1995b. Plecoptera: Leuctridae. *Insects and Macrofungi of Gutianshan, Zhejiang*: 20–24 (in Chinese).

Yang, J.[2] & Yang, D., 1990. One new species of the genus *Styloperla* from Jiangxi (Plecoptera: Styloperlidae). *Acta Agriculturae Universitatis Jiangxiensis* **12**: 45–46 (in Chinese).

Yang, L. & Armitage, BJ., 1996. The genus *Goera* (Trichoptera: Goeridae) in China. *Proc. Ent. Soc. Washington* **98**: 551–569.

Yang, C.M. & Polhemus, D.A., 1994. Notes on *Rhagovelia* Mayr (Hemiptera: Veliidae) from Singapore, with description of a new species. *Raffles Bulletin of Zoology* **42**: 987–993.

Yang, L. & Morse, J.C., 1988. *Ceraclea* of the People's Republic of China (Trichoptera: Leptoceridae). *Contr. Amer. Ent. Inst.* **23**: 1–69.

Yang, L. & Morse, J.C., 1989. Setodini of the People's Republic of China (Trichoptera: Leptoceridae: Leptocerinae). *Contr. Amer. Ent. Inst.* **25**: 1–77.

Yang, L. & Morse, J.C., 1991. Leptoceridae (Trichoptera) of the People's Republic of China. In: Tomaszewski, C. (Ed.), *Proceedings of the 6th International Symposium on Trichoptera*. Adam Mickiewicz University Press, Lodz-Zakopane, Poland: 255–258.

Yang, L. & Morse, J.C., 1992. Phylogenetic outline of Triaenodini (Trichoptera: Leptoceridae). In: Otto, C. (Ed.), *Proceedings of the 7th International Symposium on Trichoptera*. Umea, Sweden: 161–167.

Yang, L. & Morse, J.C., 1997. Six new species of Integriplpia (Trichoptera) from southern China. *Insecta Mundi* **11**: 45–51.

Yang, L. & Tian, L., 1987. Descriptions of three new species of *Ceraclea* (Trichoptera: Leptoceridae). *Entomotaxonomia* **9**: 213–216 (in Chinese).

Yang, L. & Tian, L., 1989. Four new species and two new records of genus *Ceraclea* Stephens (Trichoptera: Leptoceridae). *Entomotaxonomia* **11**: 293–297. (in Chinese).

Yang, L. & Weaver, J.S., 1997. An annotated checklist of the Lepidostomatidae (Trichoptera) of China, with new collection records. In: Holzenthal, R.W. & Flint, O.S. (Eds), *Proceedings of the 8th International Symposium on Trichoptera*. Ohio Biological Survey, Columbus: 481–487.

Yang, L. & Xue, Y., 1992. Six new species of Hydroptilidae (Insecta: Trichoptera) from China. *Entomotaxonomia* **14**: 26–34 (in Chinese).

Yang, L. & Xue, Y., 1992. Six new species of *Hydroptila* (Insecta: Trichoptera) from China. *Braueria* **21**: 9–11.

Yang, L., Kelley, R.W. & Morse, J.C., 1997. Six new species of *Oxyethira* from southern China. *Aquatic Insects* **19**: 91–105.

Yang, P.S., Wong, K.C. & Hsieh, S.H., 1990. Survey on the resource and ecology of aquatic insects in Pei-Shih Stream I. Aquatic insect fauna and related ecological study. *Chinese J. Entomol.* **10**: 209–224 (in Chinese).

Yano, K., Miyamoto, S. & Gabriel, B.P., 1981. Faunal and biological studies on the insects of paddy fields in Asia. IV. Aquatic and semiaquatic Heteroptera from the Philippines. *Esakia* **16**: 5–32.

---

[2] Yang, J. (= Jikun) and Yang, C. (= Chi-kun) are the same individual.

Yao, B. & Chen, C., 1983. Some aspects of the necessity and feasibility of China's south-to-north water transfer. In: Biswas, A.K., Zuo, D., Nickum, J.E. & Liu, C. (Eds.), *Long-distance Water Transfer: a Chinese Case Study and International Experiences*. Tycooly International Publishers Limited, Dublin: 321–331.

Yen, M.D., 1985. Species composition and distribution of the freshwater fish fauna of the north of Vietnam. *Hydrobiologia* **121**: 281–286.

Yen, M.D. & van Trong, 1988. Species composition and distribution of the freshwater fish fauna of southern Vietnam. *Hydrobiologia* **160**: 45–51.

Yeo, D.C.J. & Ng, P.K.L., 1996. A new species of freshwater snapping shrimp, *Alpheus cyanoteles* (Decapoda: Caridea: Alpheidae) from Peninsular Malaysia and a redescription of *Alpheus paludicola* Kemp, 1915. *Raffles Bull. Zool.* **44**: 37–63.

Yie, S., 1933.Observations on Japanese *Deuterophlebia* (Diptera). *Transactions of the Natural History Society of Formosa* **23**: 271–296 (in Japanese).

Yipp, M.W., 1990. Distribution of the schistosome vector snail *Biomphalaria straminea* (Pulmonata: Planorbidae) in Hong Kong. *J. Moll. Stud.* **56**: 47–55.

Yipp, M.W., 1991. The relationship between hydrological factors and distribution of freshwater gastropods in Hong Kong. *Verh. Internat. Verein. Limnol.* **24**: 2954–2959.

Yipp, M.W., Cha, M.W. & Liang, X.Y., 1992. A preliminary impact assessment of the introduction of two species of *Ampullaria* (Gastropoda: Ampullariidae) into Hong Kong. In: Meier-Brook, C. (Ed.), *Proceedings of the Tenth International Malacological Congress, Tübingen 1989*. UNITAS Malacologia, Baja, Hungary: 393–397.

Yoon, I.B. & Bae, Y.J, 1985. The classification of Ephemeroidea (Ephemeroptera) in Korea. *Ent. Res. Bulletin* **11**: 93–109.

Yoon, I.B. & Bae, Y.J., 1988. The classification of the Ephemerellidae (Ephemeroptera) in Korea. *Ent. Res. Bulletin* **14**: 23–44.

Yoon, I.B. & Dong, S.K., 1990. Systematic study of the dragonfly (Odonata) larva from Korea (I) — superfamily Aeshnoidea. *Korean J. Entomol.* **20**: 55–81.

Yoon, I.B. & Ki, H.K, 1989a. A taxonomic study of the caddisfly larvae in Korea (II). *Korean J. Entomol.* **19**: 299–318.

Yoon, I.B. & Ki, H.K., 1989b. A systematic study of the caddsifly larvae in Korea (I). *Korean J. Entomol.* **19**: 25–40.

Yoon, I.B. & Kim, J.I., 1992. Systematics on the larvae of crane flies (Tipulidae: Diptera) in Korea. *Entomological Research Bulletin* **18**: 39–53.

Yoon, I.B. & Yeon, J.B., 1984. The classification of Heptageniidae (Ephemeroptera) in Korea. *Ent. Res. Bulletin* **10**: 1–34.

Yoshida, T., Sugimoto, K. & Hayashi, F., 1985. Notes on the life history of the dobsonfly, *Protohermes grandis* THUNBERG (Megaloptera, Corydalidae). *Kontyû, Tokyo* **53**: 734–742 (in Japanese).

Yoshitomi, H. & Satô, M., 1996a. A new species of the genus *Elodes*

(Coleoptera, Scirtidae) from Guangxi Province, South China. *Jap. J. Syst. Ent.* **2**: 245–249.

Yoshitomi, H. & Satô, M., 1996b. Two new species of the genus *Flavohelodes* (Coleoptera, Scirtidae) from Taiwan. *Elytra* **24**: 303–309.

Yoshiyasu, Y., 1985. Lepidoptera. In: Kawai,T. (Ed.), *An Illustrated Book of Aquatic Insects of Japan*. Tokai University Press, Tokai: 217–226 (in Japanese).

Yoshiyasu, Y., 1987. The Nymphulinae (Lepidoptera: Pyralidae) from Thailand, with descriptions of a new genus and six new species. *Microlepidoptera of Thailand* **1**: 133–184.

You, D.S., 1984. A revision of genus *Potamanthodes* with a description of two new species (Ephemeroptera, Potamanthidae). In: Landa, V. (Ed.), *Proceedings of the 4th International Conference on Ephemeroptera*. CSAV, Czechoslovakia: 101–107.

You, D.S., 1987. A preliminary study of Xizang Ephemeroptera. *Agricultural Insects, Spiders, Plant Diseases and Weeds of Xizang* **1**: 29–36 (in Chinese.)

You, D.S., 1990. Two new species of genus *Rhithrogena* (Ephemeroptera, Heptageniidae) of Fujian Province. *Journal of Nanjing Normal University (Natural Scence)* **13**: 60–63.

You, D.S. & Gui, H., 1995. *Economic Insect Fauna of China. Fasc. 48. Ephemeroptera*. Science Press, Beijing: 152 pp. (in Chinese).

You, D.S. & Su, C.R., 1987a. Descriptions of the nymphs of *Choroterpes nanjingensis*, *Potamanthodes fujianensis* and *Isonychia kiangsinensis*. *Acta Zootaxonomica Sinica* **12**: 332–336 (in Chinese).

You, D.S. & Su, C.R., 1987b. A new species of *Vietnamella* from China (Ephemeroptera: Ephemerellidae). *Acta Zootaxomonica Sinica* **12**: 176–180 (in Chinese).

You, D.S., Wu, T., Gui, H. & Hsu, Y.C., 1980a. A new species of the genus *Choroterpes* from Nanjing (Ephemeroptera: Leptophlebiidae). *Acta Zootaxonomica Sinica* **5**: 388–391 (in Chinese).

You, D.S., Wu, T., Gui, H., Xu, Y.Q. & Hsu, Y.C., 1980b. A new species of genus *Potamanthodes* (Ephemeroptera: Potamanthidae). *Journal of the Nanjing Teachers' College (Natural Science)* **2**: 56–59 (in Chinese).

You, D.S., Wu, T., Gui, H. & Hsu, Y.C., 1981. Two new species and diagnostic haracters of genus *Cinygmina* (Ephemeroptera: Ecdyoneuridae). *Journal of the Nanjing Teachers' College (Natural Science)* **4**: 26–32 (in Chinese).

You, D.S., Su, C.R. & Hsu, Y.C., 1982a. A new species of the genus *Afronurus* from Fujian Province (Ephemeroptera: Heptageniidae). *Journal of the Nanjing Teachers' College (Natural Science)* **5**: 61–64 (in Chinese).

You, D.S., Wu, T., Gui, H. & Hsu, Y.C. , 1982b. Genus *Potamanthodes* and two new species (Ephemeroptera: Potamanthidae). *Acta Zootaxonomica Sinica* **7**: 410–415 (in Chinese).

Young, J.O. & Gibson, R., 1975. Some ecological studies on two populations of the freshwater hoplonemertean *Prostoma eilhardi* (Montgomery 1894) from Kenya. *Verh. Internat. Verein. Limnol.* **19**: 2803–2810.

Yu, K.C., Ho, S.T., Chang, J.K. & Lai, S.D., 1995. Multivariate correlation of water quality, sediment and benthic bio-community components in Ell-Ren river system, Taiwan. *Water, Air & Soil Pollution* **84**: 31–49.

Yu, P. & Stork, N.E., 1991. New evidence on the phylogeny and biogeography of the Amphizoidae: discovery of a new species from China (Coleoptera). *Systematic Entomology* **16**: 253–256.

Yu, X., 1983. Possible effects of the proposed eastern transfer route on the fish stock of the principal water bodies along the course. In: Biswas, A.K., Zuo, D., Nickum, J.E. & Liu, C. (Eds.), *Long-distance Water Transfer: a Chinese Case Study and International Experiences.* Tycooly International Publishers Limited, Dublin: 373–388.

Yu, Z., Deng, Z., Xu, Y., Wei, X., Zhou, C. Liang, Z. & Huang, H., 1981. A study on fish resources in Hanshi River after construction of the Danjiangkou Dam. *Trans. Chinese Ichthyol. Soc.* **1**: 77–96 (in Chinese).

Yu, Z., Zhou, C. Deng, Z. & Xu, Y., 1985a. On spawning grounds of four Chinese farm fishes in River Changjiang after damming at Gezhouba area. *Trans. Chinese Ichthyol. Soc.* **4**: 1–12 (in Chinese).

Yu, Z., Deng, Z., Zhou, C., Xu, Y. & Zhao, Y., 1985b. Prognosis of the effects of the Gezhouba Hydroelectric Project on fish resources of the River Chang Jiang. *Trans. Chinese Ichthyol. Soc.* **4**: 193–208 (in Chinese).

Yu, Z., Xu, Y., Deng, Z., Zhou, C. and Xiang, Y., 1986. Study on reproductive ecology of Zhonghua sturgeon (*Acipenser sinensis* Gray) in the downstreams of Gezhouba Hydroelectric Project. *Trans. Chin. Ichthyol. Soc.* **5**: 1–16 (in Chinese).

Yu, Z., Deng, Z., Cai, M., Deng, X., Jiang, H., Yi, J. & Tian, J., 1988. Preliminary report on reproductive biology and artificial propagation of the Chinese Sucker (*Myxocyprinus asiaticus*) in the downstreams of Gezhouba Hydroelectric Dam. *Acta Hydrobiol. Sinica* **12**: 87–89 (in Chinese).

Yuan, H.W., 1992a. Contribution of bacterioplankton, phytoplankton, zooplankton and detritus to organic seston carbon load in a Changjiang floodplain lake (China). *Arch. Hydrobiol.* **126**: 213–238.

Yuan, H.W., 1992b. Seston in a seasonally flooded lake of the central Changjiang River (China). *Hydrobiologia* **242**: 95–104.

Yule, C., 1985. Comparative study of the life cycles of six species of *Dinotoperla* (Plecoptera: Gripopterygidae) in Victoria. *Aust. J. Mar. Freshwat. Res.* **36**: 717–735.

Yule, C.M., 1995a. The impact of sediment pollution on the benthic invertebrate fauna of the Kelian River, East Kalimantan, Indonesia. In: Timotius, K.H. & Göltenboth, F. (Eds), *Tropical Limnology Volume III. Tropical Rivers, Wetlands and Special Topics.* Satya Wacana University Press, Salatiga, Indonesia: 61–75

Yule, C.M., 1995b. The ecology of an aseasonal tropical river on Bougainville Island, Papua New Guinea. In: Timotius, K.H. & Göltenboth, F. (Eds),

Tropical Limnology Volume III. Tropical Rivers, Wetlands and Special Topics. Satya Wacana University Press, Salatiga, Indonesia: 1–14.

Yule, C.M., 1996. Trophic relationships and food webs of the benthic invertebrate fauna of two aseasonal streams on Bougainville Island, Papua New Guinea. *J. Trop. Ecol.* **12**: 517–534.

Yusop, Z. & Suki, A., 1994. Effects of selective logging methods on suspended solids concentrations and turbidity level in streamwater. *Journal of Tropical Forest Science* **7**: 199–219.

Zafer, M. & Mahmood, N., 1989. Studies on the distribution of zooplankton communities in the Satkhira estuarine system (Bangladesh). *Chittagong University Studies Part II, Science* **13**: 115–122.

Zakaria-Ismail, M., 1987. The fish fauna of the Ulu Endau River system, Johore, Malaysia. *Malay. Nat. J.* **41**: 403–411.

Zakaria-Ismail, M., 1994. Zoogeography and biodiversity of the freshwater fishes of Southeast Asia. *Hydrobiologia* **285**: 41–48.

Zettel, H., 1993. Zur Kenntnis der *Aphelocheirus*-Arten in Borneo (Sarawak) und Sulawesi (Heteroptera: Naucoridae). *Zeitschrift der Arbeitsgemeinschaft Österreichischer Entomologen* **45**: 81–86.

Zhang, J., 1988. On the discovery of the genus *Eatonigenia* (Ephemeroptera: Ephemeridae) in China. *J. Nanjing Normal University (Nat. Sci.)* **3**: 68–72 (in Chinese).

Zhang, J. & Cai, W., 1991. Notes on the genus *Cinygmina* (Ephemeroptera: Heptageniidae) from Hunan Province, China. *Entomotaxonomia* **13**: 237–239 (in Chinese).

Zhang, J. & Lin, Z. 1992. *Climate of China*. John Wiley & Sons, New York: 376 pp.

Zhang, J. & Sun, X., 1979. Studies on the larval development of six freshwater prawn species in the middle and lower Chang Jiang (Yangtze) valley. *Acta Zoologica Sinica* **25**: 143–153 (in Chinese).

Zhang, J., Huang, W.W., Liu, M.G. & Cui, J.Z., 1994. Eco-social impact and chemical regimes of large Chinese rivers — a short discussion. *Water Research* **28**: 609–617.

Zhang, S., Dong, W., Zhang, L. & Chen, X., 1989. Geochemical characteristics of heavy metals in the Xiangjiang River, China. *Hydrobiologia* **176/177**: 253–262.

Zhang, Q., Lin, F. Li, X. & Hu, M., 1987. Major ion chemistry and fluxes of dissolved solids with rivers in southern coastal China. *Mitt. Geol.-Palaont. Inst. Univ. Hamburg, SCOPE/UNEP Sonderbd.* **64**: 243–249.

Zhang, T. & Wang, D., 1991a. A new species of *Simulium* from Fujian, China (Diptera: Simuliidae). *Acta Zootaxonomica Sinica* **16**: 109–113 (in Chinese).

Zhang, T. & Wang, D., 1991b. A preliminary survey of blackflies (Diptera: Simuliidae) from Fujian Province. *Acta Entomologica Sinica* **34**: 483–491 (in Chinese).

Zhao, J., Zheng, G., Wang, H. & Xu, J., 1990. *The Natural History of*

China. McGraw-Hill Publishing Co., New York: 224 pp.

Zhao, X., 1990. *The Gomphid Dragonflies of China (Odonata: Gomphidae)*. Contributions of the Biological Control Research Institute, Fujian Agricultural College, Special Publication No. 1, Science & Technology Publishing House, Fujian: 486 pp. (in Chinese).

Zheng, H., Su, L. & Xie, S., 1983. On the freshwater molluscs from Chingqing region. *Transactions of the Chinese Society of Malacology* **1**: 69–72 (in Chinese).

Zhiltzova, L.A. & Zwick, P., 1993. On the genus *Kyphopteryx* Kimmins (Plecoptera: Taeniopterygidae). *Aquatic Insects* **15**: 193–198.

Zhou, F., 1987. Transport and deposition of suspended sediments in the Chiangjiang estuary. *Mitt. Geol.-Palaont. Inst. Univ. Hamburg, SCOPE/ UNEP Sonderbd.* **64**: 175–184.

Zhou, K., Pilleri, G. & Li, Y., 1979. Observations on the *Baiji* (*Lipotes vexillifer*) and the Finless porpoise (*Neophocaena asiaeorientalis*) in the Chanjiang (Yangtze) River between Nanjing and Taiyangzhou, with remarks on some physiological adaptations of the Baiji to its environment. *Invest. Cetacea* **10**: 109–120.

Zhu, H, 1991. A new species of the genus *Davidius* from southern Shaanxi (Odonata: Gomphidae). *Entomotaxonomia* **13**: 175–177 (in Chinese).

Zhu, H. & Han, F., 1992. A new species of the genus *Cordulegaster* Leach (Odonata: Cordulegastridae) from Shaanxi. *Entomotaxonomia* **14**: 18–21.

Zhu, H., Shen, G., Yu, Y., Zhang, L. & Wang, Y., 1987. Prediction of impacts of the Three Gorges Project on the ecological function and environment of Poyang Lake. In: *Essays on the Impacts and Strategies of the Three Gorges Project to the Environment*. Science Press, Beijing: 319–332 (in Chinese).

Zhu, S., Wang, Z. & Hseung, Y., 1983. Effect of diverting water from south-to-north on the ecosystem of the Huang-Huai-Hai Plain. In: Biswas, A.K., Zuo, D., Nickum, J.E. & Liu, C. (Eds.), *Long-distance Water Transfer: a Chinese Case Study and International Experiences*. Tycooly International Publishers Limited, Dublin: 389–394.

Zimmermann, M. & Spence, J.R., 1989. Prey use of the fishing spider *Dolomedes triton* (Pisauridae, Araneae): an important predator of the neuston community. *Oecologia* **80**: 187–194.

Zimmermann, M. & Spence, J.R., 1992. Adult population dynamics and reproductive effort of the fishing spider *Dolomedes triton* (Araneae, Pisauridae) in central Alberta. *Can. J. Zool.* **70**: 2224–2233.

Zuckerman, Lord, 1992. Between Stockholm and Rio. *Nature* **358**: 273–276.

Zuo, D., 1983. China's south-to-north water transfer proposals. In: Biswas, A.K., Zuo, D., Nickum, J.E. & Liu, C. (Eds.), *Long-distance Water Transfer: a Chinese Case Study and International Experiences*. Tycooly International Publishers Limited, Dublin: 91–96.

Zurwerra, A. & Tomka, I., 1985. *Electrogena* gen. nov., eine neue Gattung der Heptageniidae (Ephemeroptera). *Entomologische Berichte Luzern* **13**: 99–104.

Zurwerra, A., Metzler, M. & Tomka, I., 1987. Biochemical systematics and evolution of the European Heptageniidae. *Arch. Hydrobiol.* **109**: 481–510.

Zurwerra, A., Tomka, I. & Lamper, G., 1986. Morphological and enzyme electrophoretic studies on the relationships of the European *Epeorus* species (Ephemeroptera, Heptageniidae). *Systematic Entomology* **11**: 255–266.

Zwick, P., 1976. *Neoperla* (Plecoptera, Perlidae) emerging from a mountain stream in Cental Africa. *Int. Revue ges. Hydrobiol.* **61**: 683–697.

Zwick, P., 1980. The genus *Neoperla* (Plecoptera: Perlidae) from Sri Lanka. *Oriental Insects* **14**: 263–269.

Zwick, P., 1981. The South Indian species of *Neoperla* (Plecoptera: Perlidae). *Oriental Insects* **15**: 113–126.

Zwick, P., 1982a. Contribution to the knowledge of *Chinoperla* (Plecoptera: Perlidae: Neoperlini). *Aquatic Insects* **4**: 167–170.

Zwick, P., 1982b. A revision of he Oriental stonefly genus *Phanoperla* (Plecoptera: Perlidae). *Systematic Entomology* **7**: 87–126.

Zwick, P., 1982c. Notes on Plecoptera (4) some overlooked Philippine species, and an amendment to my revision of *Phanoperla*. *Aquatic Insects* **4**: 20.

Zwick, P., 1982d. Notes on Plecoptera (6) *Etrocorema nigrogeniculatum* (Enderlein). *Aquatic Insects* **4**: 104.

Zwick, P., 1983a. The *Neoperla* of Sumatra and Java (Indonesia) (Plecoptera: Perlidae). *Spixiana* **6**: 167–204.

Zwick, P., 1983b. Notes on Plecoptera (8) *Nemoura sumatrensis* Navás. *Aquatic Insects* **5**: 16.

Zwick, P., 1984a. The genera *Tetropina* and *Neoperlops* (Plecoptera: Perlidae). *Aquatic Insects* **6**: 169–176.

Zwick, P., 1984b. Notes on the genus *Agnetina* (= *Phasganophora*) (Plecoptera: Perlidae). *Aquatic Insects* **6**: 71–79.

Zwick, P., 1986a. The Bornean species of the stonefly genus *Neoperla* (Plecoptera: Perlidae). *Aquatic Insects* **8**: 1–53.

Zwick, P., 1986b. Contribution to the knowledge of *Phanoperla* BANKS, 1938 (Plecoptera: Perlidae). *Mitt. Schweiz. Entomol. Gessell.* **59**: 151–158.

Zwick, P., 1988a. Species of *Neoperla* from the South-East Asian mainland. *Ent. Scand.* **18**: 393–407.

Zwick, P., 1988b. Notes on Plecoptera (16) *Tylopyge* Klapálek, a synonym of *Paragnetina* Klapálek. *Aquatic Insects* **10**: 201–203.

Zwick, P. 1990a. Emergence, maturation and upstream oviposition flights of Plecoptera from the Breitenbach, with notes on the adult phase as a possible control of stream insect populations. *Hydrobiologia* **194**: 207–223.

Zwick, P., 1990b. Systematic notes on Holarctic Blephariceridae (Diptera). *Bonn. zool. Beitr.* **41**: 231–257.

Zwick, P., 1991. Notes on some types of Indian Blephariceridae (Diptera) named by B.K. Kaul. *Aquatic Insects* **13**: 129–132.

Zwick, P., 1992. Stream habitat fragmentation — a threat to biodiversity. *Biodiversity and Conservation* **1**: 80–97.

Zwick, P., 1996. Variable egg development of *Dinocras* spp. (Plecoptera, Perlidae) and the stonefly seed bank theory. *Freshwat. Biol.* **35**: 81–100.

Zwick, P. & Sivec, I., 1980. Beiträge zur Kenntnis der Plecoptera des Himalaja. *Entomologica Basiliensia* **5**: 59–138.

Zwick, P. & Sivec, I., 1985. Supplements to the Perlidae (Plecoptera) of Sumatra. *Spixiana* **8**: 123–133.

Zwick, P. & Hortle, K.G., 1989. First records of net-winged midges (Diptera: Blephariceridae) from Papua New Guinea, with descriptions of a new species. *Aust. J. Mar. Freshwat. Res.* **40**: 361–367.

Subject Index

To facilitate use of this book, the index includes brief definitions of terms which may be unfamiliar to the general reader. Entries relating to invertebrate anatomy indicate places in the text where body parts are figured or structures are described.

Organism Index

QM LIBRARY
(MILE END)

WITHDRAWN
FROM STOCK
QMUL LIBRARY

WITHDRAWN
FROM STOCK
QMUL LIBRARY